COMPUTATIONAL METHODS IN VISCOUS FLOWS

Ce que nous connaissons est peu de chose
Ce que nous ignorons est immense

Pierre Simon Laplace

COMPUTATIONAL METHODS IN VISCOUS FLOWS

VOLUME 3 IN THE SERIES
Recent Advances in Numerical Methods in Fluids

Edited by:

W. G. Habashi
Concordia University, Montreal, Canada
&
Consultant, Pratt and Whitney Canada

PINERIDGE PRESS

Swansea, U.K.

First Published in 1984 by
Pineridge Press Limited
54, Newton Road, Mumbles, Swansea, U.K.

British Library Cataloguing in Publication Data

Computational methods in viscous flows.
(Recent advances in numerical methods; Volume 3)
1. Fluid dynamics 2. Numerical calculations
I. Habashi, W. G. II. Service
532'.051'01511 QA911

ISBN 0–906674–27–1

Printed and bound in Great Britain by
Dotesios (Printers) Ltd., Bradford-on-Avon, Wiltshire

CONTENTS OF VOLUME 3

COMPUTATIONAL METHODS IN VISCOUS FLOW

Section 2: Page

NAVIER-STOKES EQUATIONS, COMPRESSIBLE. 223

Section 3:

VISCOUS-INVISCID INTERACTION, INCOMPRESSIBLE 341

Section 4:

VISCOUS-INVISCID INTERACTION, COMPRESSIBLE 417

Section 4: (Continued) Page

Section 5:

PREFACE TO VOLUMES 3 AND 4

When I received an invitation from Dr. Cedric Taylor to edit an upcoming volume of the series "Recent Advances in Numerical Methods in Fluids" I enthusiastically accepted, not fully realizing the magnitude of the effort involved. It was evident to me that there was a dearth of books in the field of Numerical Computations in viscous flows and transonics. Because of the rapid advances in these two areas it is a daunting task to write a textbook that includes both theory and modern applications while remaining useful for a number of years. Edited books on the other hand sometimes wind up being a collection of unrelated articles. The challenge therefore was to solicit some of the best known people in these two fields and coordinate their contributions into a volume useful both to the student and the practitioner. I was blessed with two things from the outset: the willingness of such top individuals to participate and the invaluable help of my friend and colleague Mohamed Hafez. Almost everyone invited accepted to contribute and although the books have no pretense of including all the major names in their area, I have managed to involve several names from the recognized academic centers of excellence in the U.S.A. and Europe and names from the major research centers such as NASA, FDL, O.N.E.R.A., NLR, NAL, DFVLR, FFA and others as well as names from industry such as Boeing, Grumman, Lockheed, Dassault/Bréguet, Dornier and Pratt & Whitney. It soon became clear that we had enough contributors to produce two books: Volume 3: "Computational Methods in Viscous Flow" and Volume 4: "Advances in Computational Transonics". A new round of invitations were sent out to complement the existing articles and to complete some gaps in the books.

All participants were provided with a complete list of names and addresses of other contributors and were encouraged to contact each other to coordinate and streamline their efforts. The continuity of the manuscript bears witness to their collaboration.

During the course of assembling and editing the contributions to the book, I have come to the sober realization that although we have learned so much in numerical methods in the last decade or two, there remains that much more to understand and explore. Each breakthrough in computer technology opens countless doors of opportunity in solving numerical problems. New ideas are quickly replaced by better ones. Lamartine, writing in a different context perhaps best summarizes what scientific effort and thirst are about and hence how fast knowledge is superseded:

Marchez. L'humanité ne vit pas d'une idée
Elle éteint chaque soir celle qui l'a guidée
Elle en allume une autre à l'immortel flambeau
Comme ces morts vêtus de leur parure immonde
Les générations emportent de ce monde
Leurs vêtements dans leur tombeau.

Lamartine: "Les Révolutions"

Finally I would like to thank all the participants for their invaluable contributions and for their patience during the many delays in assembling the manuscript. I would also like to thank Ms. L. Mathieu and Ms. E. Takla for their technical help. The support of Concordia University and Pratt & Whitney Canada are kindly acknowledged.

In the following, the fifty articles in the two volumes are briefly summarized.

VOLUME 3: COMPUTATIONAL METHODS IN VISCOUS FLOW

Volume 3 starts with a review article by Lomax and Mehta on the effects of viscosity as it relates to drag, flow separation, vortical flows and unsteady flows. From the physical point of view they discuss the important scales used and describe the relatively large scale organized structures that have recently been identified in transitional and fully developed turbulent flows. From the analytical point of view they emphasize some simplifications necessary for the practical computation of the compressible Navier-Stokes equations. They conclude with a simplified description of some turbulence models. The results discussed in this important article testify to the increased capability provided by modern computers in actually allowing sophisticated calculation of the details of complex flow fields. These authors conclude that the rapid progress achieved in computational aerodynamics will make it become the primary design tool, with experiments used to verify the results or provide answers when the computational techniques cannot.

In the section entitled Incompressible Flows, Rubin gives a detailed review of finite difference methods for the Incompressible Navier-Stokes equations (INSE) and Parabolized Navier-Stokes equations (PNSE) using both the primitive variables and the vorticity-stream function formulations. He emphasizes his personal belief that any discretization procedure, including finite element, can be included in the broad category of finite differences. His chapter discusses the difficulties encountered in early Navier-Stokes calculations and identifies methods of solution, coupling, boundary conditions and grid generation as culprits and then presents logical approaches to eliminate these difficulties. He uses the results of the driven cavity problem to compare different methods. A new composite velocity procedure is developed in the spirit of matched asymptotic expansions and discretization follows boundary layer and potential flow procedures; some results for the boattail problem are presented. For convergence acceleration a coupled strongly implicit procedure is proposed as a preconditionning for a conjugate gradient method. For accuracy improvement a fourth order spline collocation procedure is recommended. For the sake of efficiency the PNSE are preferred. The multiple sweep-global marching procedure with a forward difference approximation of the pressure seems to be attractive judging by the results presented for separation bubbles, flow around finite flat plates and troughs.

The paper by Ghia and Ghia addresses the Multigrid technique for accelerating the solution of INSE. High Reynolds number results for the driven cavity problem as well as for asymptotic flows in curved ducts of square cross-section are obtained using a stream function-vorticity formulation. Formulation of the Neumann problem for the pressure requires careful construction of the restriction and prolongation operators to guarantee the satisfaction of an integral constraint, at each level, essential to the success of multigrid in such problems. Comparison of the multigrid solution convergence behaviour with the convergence history of simple grid solutions confirms the computational advantage of the technique for elliptic problems. Application of the multigrid techniques to 3-D flows is the subject of current investigation by these authors and the key item seems to be the construction of a good smoothing operator to achieve the utmost effectiveness of multigrid. The strongly implicit procedure seems to be a candidate and its success remains to be seen for this application.

The chapter by Elsaesser and Peyret focuses on the accuracy problem. Hermitian methods for the solution of the INSE (in primitive variables) provide higher order approximations while preserving a compact differencing leading to the solution of simple or tridiagonal systems of equations. Results of numerical experiments for flows in a square cavity and a channel are obtained through a combination of artificial compressibility and ADI method.

In the second part of this section, Finite Element Methods for INSE are discussed by Gunzburger and Nicolaides. They address the question of choosing finite element velocity/pressure elements to achieve stable discretizations leading to convergent pressure approximations. They also study the role of artificial viscosity in viscous flow calculations. Results are presented for the driven cavity problem at a Reynolds number of 1000 and compared to the results of Ghia and Ghia. They make use of the capability of the finite element method in efficiently using non-uniform grids (hence fewer unknowns) to resolve the viscous layer without any use of explicit artificial viscosity or upwinding the convective terms.

Quartapelle, Donea and Morgan note the scarcity of studies aimed at investigating the discretization properties of numerical schemes for the fully multi-dimensional INSE. They attribute this, in the case of finite differences, to a certain difficulty in dealing with the discrete representation of the condition of incompressibility viewed either as a solenoidal constraint for the velocity field or a Poisson equation for the pressure. They analyse a linearized model of the unsteady INSE using a fractional step finite difference approach in time and the finite element method in space and evaluate the properties of various discrete approximations by comparing the response of the fully-discretized equations against the exact response of the continuum model equations.

The section on Compressible Navier-Stokes equations (CNSE) starts with a rather challenging chapter by MacCormack who predicts that the computer power required to calculate the viscous flow field about a complete aircraft at flight Reynolds numbers will become available within the next year or two. MacCormack believes that the complex topological problems of fitting a mesh about an aircraft can be solved and the numerical procedures required for such a flow calculation can also be developed within the same time frame. His paper outlines some interesting approaches to these issues.

In his chapter, Chaussée discusses the process of developing approximate-factorization algorithms in delta form for both unsteady and steady viscous flows. These algorithms, cast in conservation law form, are simplified by using a thin-layer approximation to the Navier-Stokes equations (TLNSE). The implementation of implicit surface viscous boundary conditions is discussed in detail. The complete development of inviscid implicit boundary conditions is also referenced. An example is included showing the advantage of using these implicit boundary conditions. Three-dimensional results from the steady form of the algorithm are presented and compared with experiment. A number of references to other results for both the unsteady and the steady computer codes show that the delta form of the approximate-factorization algorithm is robust and applicable to a wide variety of configurations.

Deiwert discusses some of the considerations and problems associated with numerically simulating unsteady interactive flows of aerodynamic interest. Attention is focused on solutions to the time-dependent compressible Reynolds-averaged Navier-Stokes equations using empirical eddy viscosity models to account for the effects of turbulence. The importance of writing the equations in strong conservation law form for a generalized body-oriented coordinate system is pointed out. Some considerable discussion of time and length scales inherent in the class of flows considered is given. He points out that, to date, simulations have been performed for unsteady flows with narrow frequency bands. The treatment of many flows with broad-band unsteadiness has not been attempted yet and poses a serious challenge. The numerical schemes used to solve the governing equations are of explicit, implicit or hydrid type. Several examples of simulated unsteady interacting flows are given covering such aerodynamic phenomena as buffet, stall and buzz.

Baker presents an implicit finite element numerical algorithm for solution of CNSE. Orthogonality of the semi-discrete approximation error, to the space of functions employed for the approximation, forms the basic theoretical approach. A highly phase selective artificial dissipation mechanism is introduced, to control non-linear dispersion error, by requiring the convective semi-discrete approximation error to be orthogonal to a modified basis set. His algorithm is cast in a generalized coordinates framework, and is expressed in a FORTRAN-like notation using tensor summation index convention for addressing multi-dimensional problem descriptions. Numerical results demonstrate solution accuracy, convergence with respect to grid size, grid sensitivity, and boundary conditions as well as applications to compressible flows with shocks.

The section entitled Viscous-Inviscid Interaction is divided into two subsections: incompressible and compressible. Veldman, in his chapter, discusses the Goldstein singularity from a numerical point of view. He contends that the singularity is definitely a mathematical artifice which can be avoided by a careful viscous-inviscid coupling algorithm. The convergence of the overall iteration is greatly enhanced if the viscous and inviscid problems are solved in a quasi-simultaneous fashion.

Pletcher demonstrates the capability of incompressible viscous-inviscid interaction procedures in analyzing the leading edge separation bubble and separated flow in a channel. He solves the boundary layer equations for the viscous flow and the Laplace equation for the outer inviscid flow. In specified non-interaction regions standard boundary layer (parabolic) methods are used, while in interaction regions an inverse calculation with specified δ^* is used. The interaction procedure is also discussed in good detail.

In the work of Halim and Hafez, the boundary layer equation (a fourth order equation in terms of stream function) is integrated using an efficient Gaussian elimination for a pentadiagonal scalar system with various boundary conditions such as direct, inverse and hybrid. For viscous-inviscid interaction a patching procedure is used where a partially-parabolized Navier-Stokes equation is solved iteratively, coupled with a Poisson equation for inviscid rotational flow, using a line relaxation method. They also discuss a new implicit coupling procedure if it is desired to solve the viscous and inviscid problems separately in order to take advantage of the different length scales involved.

For the compressible part, Le Balleur discusses his interacting defect integral (IDI) approach. He utilizes two explicit relaxation techniques for the numerical viscous-inviscid coupling. These direct and semi-inverse relaxations ensure stability control with respect to the local flow conditions and mesh size, and allow to switch the viscous solution towards a direct or inverse formulation according to a criterion of local shape parameter in the viscous layer. Examples are given for complex flows involving problems of airfoils, wings, multiple separation, trailing edges and shock waves. The coupling is achieved using either a potential or an Euler inviscid flow solver.

Whitfield and Thomas, in their viscous-inviscid interaction calculation, have used Jameson's Euler code and an inverse integral compressible boundary layer method and couple them by Carter's displacement thickness iteration procedure. They present results for separated flows over airfoils.

Inger describes the features of an approximate non-asymptotic triple deck theory of shock-turbulent boundary layer interaction that accurately describes non-separating two dimensional flows over a wide range of practical Reynolds numbers and its application as an element in the overall viscous flow analysis of a body. Two main aspects of the problem are examined: (1) the local interactive thickening and skin friction drop in the shock foot region, including the effects of the incoming boundary layer shape factor, wall curvature , an improved "viscous ramp" model of the interaction and an approximate prediction of incipient separation behaviour, (2) the significant influence of such interaction of the subsequent downstream turbulent boundary layer thickening, profile shape and skin friction behaviour. Comparisons with experimental data are given and applications are presented for both supercritical airfoils and transonic bodies of revolution.

Finally, a multi-domain approach for the computation of viscous transonic flows by unsteady type methods is presented by Cambier, Ghazzi, Veuillot and Viviand. The complete Navier-Stokes equations are used in the viscous region while the Euler equations are used in the inviscid core. Careful treatment at boundaries and artificial interfaces are discussed and numerical results of shock wave/turbulent boundary layer interaction in a channel are given.

In the section on special methods Hankey and Shang study self-excited oscillations in fluid flows. Some interesting numerical simulation of these phenomena are shown and the configurations investigated include a cylinder, airfoil, open cavity, spike-tipped body, inlet and cone. They conclude that linear theory is useful in predicting the frequency of the instability and providing a qualitative description of the phenomenon and that quantitative results are possible through the numerical solution of the time-dependent Navier-Stokes equations.

Taylor and Morgan present turbulent flow and heat transfer in coupled solid/fluid systems. A two equation model of turbulence (k,ε) is used to evaluate an effective kinematic viscosity. The spatial domain, either solid or fluid, is discretized utilizing conventional eight nodes isoparametric finite elements and a simple relaxation technique is used to solve the discretized equations. Examples are given on flow in a sudden expansion and on cooling of rod bundles.

Vittal and Tabakoff study the effect of solid particle size on viscous two-phase flow around a circular cylinder as an example for particle ingestion in aircraft helicopter engines and similar situations of vehicles operating in dusty atmospheres. They particularly account for the rebounding character of the particles when they impact on solid boundaries. A stream function/vorticity formulation is used and the system is solved using a factored ADI scheme. Interesting results are obtained due to the coupling and the accounting for viscosity.

Hamed and Abdallah introduce a new method for 3-D viscous calculations in terms of three stream-like functions and the vorticity transport equation. Results are presented for uniform entrance flow in a square duct with a cascade entrance and they represent an extension of their results for inviscid rotational flows.

CONTRIBUTING AUTHORS TO VOLUME 3

COMPUTATIONAL METHODS IN VISCOUS FLOW

S. ABDALLAH , Pennsylvania State University
State College, Pennsylvania, U.S.A.

A.J. BAKER , University of Tennessee,
Knoxville, Tennessee, U.S.A.

L. CAMBIER , O.N.E.R.A.,
Châtillon-sous-Bagneux, France.

D.S. CHAUSSEE , NASA-Ames Research Center,
Moffett Field, California, U.S.A.

G.S. DEIWERT , NASA-Ames Research Center,
Moffett Field, California, U.S.A.

J. DONEA , Joint Research Center,
Ispra, Italy.

E. ELSAESSER , O.N.E.R.A.,
Châtillon-sous-Bagneux, France.

W. GHAZZI , O.N.E.R.A.,
Châtillon-sous-Bagneux, France.

K.N. GHIA , University of Cincinnati,
Cincinnati, Ohio, U.S.A.

U. GHIA , University of Cincinnati,
Cincinnati, Ohio, U.S.A.

M.D. GUNZBURGER, Carnegie-Mellon University,
Pittsburgh, Pennsylvania, U.S.A.

M.M. HAFEZ , Computer Dynamics Inc.,
Virginia Beach, Virginia, U.S.A.

A. HALIM , George Washington University,
Hampton, Virginia, U.S.A.

A. HAMED , University of Cincinnati,
Cincinnati, Ohio, U.S.A.

W.L. HANKEY , Flight Dynamics Laboratory,
Wright-Patterson Air Force Base,
Ohio, U.S.A.

G.R. INGER , West Virginia University,
Morgantown, West Virginia, U.S.A.

J.C. LE BALLEUR, O.N.E.R.A.,
Châtillon-sous-Bagneux, France.

H. LOMAX , NASA-Ames Research Center,
Moffett Field, California, U.S.A.

R.W. MACCORMACK, University of Washington,
Seattle, Washington, U.S.A.

U.B. MEHTA , NASA-Ames Research Center,
Moffett Field, California, U.S.A.

K. MORGAN , University College of Swansea,
Swansea, Wales, U.K.

R.A. NICOLAIDES, Carnegie-Mellon University,
Pittsburgh, Pennsylvania, U.S.A.

R. PEYRET , Université de Nice,
Nice, France.

R.H. PLETCHER , Iowa State University,
Ames, Iowa, U.S.A.

L. QUARTAPELLE , Politecnico di Milano,
Milano, Italy.

S.G. RUBIN , University of Cincinnati,
Cincinnati, Ohio, U.S.A.

J.S. SHANG , Flight Dynamics Laboratory,
Wright-Patterson Air Force Base,
Ohio, U.S.A.

W. TABAKOFF , University of Cincinnati,
Cincinnati, Ohio, U.S.A.

C. TAYLOR , University College of Swansea,
Swansea, Wales, U.K.

J.L. THOMAS , NASA-Langley Research Center,
Hampton, Virginia, U.S.A.

A.E.P. VELDMAN , National Aerospace Laboratory, NLR,
Amsterdam, The Netherlands.

J.P. VEUILLOT , O.N.E.R.A.,
Châtillon-sous-Bagneux, France.

B.V.R. VITTAL , University of Cincinnati,
Cincinnati, Ohio, U.S.A.

H. VIVIAND , O.N.E.R.A.,
Châtillon-sous-Bagneux, France.

D.L. WHITFIELD , Mississippi State University,
Mississippi, Mississippi, U.S.A.

SOME PHYSICAL AND NUMERICAL ASPECTS OF COMPUTING THE EFFECTS OF VISCOSITY ON FLUID FLOW

HARVARD LOMAX* AND UNMEEL B. MEHTA†

1. INTRODUCTORY REMARKS

We have just entered a period in the evolution of aerodynamics in which computational fluid dynamics will be playing one of the dominant roles. The primary emphasis will be on determining the effect of viscosity as it relates to drag, flow separation, vortical flows, and unsteady flows. The purpose of this article is to discuss some physical, analytical, and computational aspects of viscous flows, as well as to give some examples of computed flows which represent the state of the art in advanced areas.

The discussion of physical aspects starts with the development of the important scales used to reference the flow phenomena in laminar and turbulent shear layers. This is followed by a brief development of the usefulness of the concepts of circulation and vorticity. Then comes a description of some of the relatively large-scale organized structures that have recently been identified in transitional and fully developed turbulent flows.

The discussion of analytical aspects starts with a compact presentation of the compressible Navier-Stokes equations, and the Reynolds-averaged form of these equations, continues with a brief description of the simplifications necessitated for their practical application to strongly interacting, high-Reynolds-number flows, and concludes with a simplified description of some forms of turbulence models.

The discussion of numerical aspects is quite short for reasons given therein, and the chosen examples speak for themselves.

2. PHYSICAL ASPECTS OF VISCOUS FLOWS

This section is composed of three parts: the first is devoted to a brief discussion of the various scales that are used in classifying and evaluating both laminar and turbulent viscous flows; the second is concerned with the role of the concepts of circulation and vorticity; and the third discusses some of the organized structures that have recently been observed in turbulent flow fields.

---

* Chief, Computational Fluid Dynamics Branch, NASA Ames Research Center, Moffett Field, Calif.

† Research Scientist, NASA Ames Research Center, Moffett Field, Calif.

2.1 Spatial and Temporal Scales

In this section we present some of the spatial and temporal scales that are found in various flow regimes. These scales help determine the experimental and computational space-time resolutions required to explore and analyze the various domains. Viscous laminar flows are usually scaled according to the thickness and extent of a shear layer. On the other hand, fully developed, high-Reynolds-number turbulent flows have three sets of scales that are commonly used for classification. In a statistical sense these are identified with three ranges: (1) the driving, high-energy range associated with the geometry, free-stream speed, etc., of the aerodynamic event; (2) the inertial range, supposedly of a universal nature and independent of the geometry; and (3) the dissipation range which is controlled by the local molecular viscosity. Quantifications that identify these ranges are determined by evaluating [1] the local (1) integral scale, (2) Taylor microscale, and (3) Kolmogorov microscale.

2.1(a) Spatial Scales

When they can be clearly determined, there are two lengths that are used to define the outer scale. One is δ, the thickness of an established viscous boundary layer, and the other is L, the distance of the boundary layer from its origin. Under these conditions, the boundary layer has an edge and the velocity used to scale these flows is the mean velocity at this edge. We designate this velocity by the symbol U_e in this paper. A Reynolds number, Re, can be established using the length L, the velocity U_e, and other local conditions at the edge.

Using x for the streamwise coordinate and y for its normal, we can consider a boundary layer, or shear layer, to be thin if $\partial\delta/\partial x$ is of the order δ/L and $<< 1$. In a thin shear layer the Navier-Stokes equations can be simplified by neglecting all but the normal viscous stresses and Reynolds stresses. The resulting equations are referred to as the thin-shear-layer equations. When these equations are applicable, the ratio of neglected to retained stress gradients is $O([\partial\delta/\partial x]^2)$ for laminar flows and $O(\partial\delta/\partial x)$ for turbulent flows (see, for example, [2]). Since $O(\partial\delta/\partial x)$ is generally larger in turbulent flows than in laminar ones, the approximation is generally less accurate in turbulent flows.

A shear layer is considered to be slender, if it is thin in two directions normal to a streamline. An example is the region near an interior corner. In slender shear layers, only the streamwise gradients of viscous and Reynolds stresses can be neglected in the governing equations. This results in the slender-shear-layer equations; see [3].

Finally there is a large body of theory, based on asymptotic expansions, that is bounded in all three dimensions. In laminar flows, this is usually referred to as triple-deck theory [4]. This theory is used to study phenomena in the vicinity of a separation point or a trailing edge. The streamwise lengths for which this analysis is valid is $O(Re^{-3/8}L)$. The lateral extent is $O(Re^{-\alpha}L)$ where $3/8 < \alpha < 5/8$ depending on the position of the deck. For flows with moderate or high Reynolds numbers having small regions of separation, it would be prudent to resolve these length scales in numerical computations. Burggraf et al. [5] have demonstrated this requirement on grid-spacing.

Triple-deck theory has a counterpart that applies to turbulent boundary layers; it has been reviewed by Melnik [6]. The overall asymptotic structure for shock wave and turbulent boundary-layer interactions, and for turbulent trailing-edge flows is similar. In the turbulent case, the length scales are $O([lnRe]^{-\alpha}L)$, where $-1/2 < \alpha < 5/2$. Nonasymptotic length scales may be found in a paper by Inger in this volume.

Flows in which shear layers can be identified may contain interactions of these layers with other types of phenomena such as inviscid outer layers, or shock waves. If the various phenomena can be studied separately with good approximation, the flow is said to have a weak interaction. When the phenomena cannot be treated separately because of strong coupling, the flow is classified as one with a strong interaction Papers dealing with strong interactions appear in this book. The nature of viscous-inviscid interaction is identified by the magnitude of the shear layer or the displacement thickness. If these layers are either $5.3Re^{-1/2}L$ for laminar flow or $f([lnRe]^{-1})L$ [3] for turbulent flow, a viscous-inviscid interaction is considered to be weak. When the layers rapidly thicken and become larger than this magnitude, an interaction is classified as a strong viscous-inviscid interaction.

For a laminar velocity profile we can use the concept of Reynolds similarity, but such a concept does not hold for a turbulent mean-velocity profile. In fact, these two velocity profiles greatly differ in both external and internal flows. Let us look more closely at the the structure of a turbulent mean-velocity boundary layer, because in certain regions the scales of importance in this boundary layer are much different than those discussed above.

First of all, it has been observed that a time-averaged, turbulent velocity profile is made of two layers which are referred to as the inner layer and the outer layer. The inner layer is adjacent to the wall and can, in turn, be divided into three layers. Proceeding away from the wall these are the linear sublayer, the buffer layer, and the inertial sublayer or log-wall region. The outer layer is often also subdivided, this time into two regions. Proceeding away from the wall these are referred to as the wake-law layer and the viscous superlayer. The velocity and length scales in all of these layers are based on measurements and dimensional analysis, and are identified below for a constant-density, two-dimensional mean flow over an impermeable smooth wall. They are valid, for both internal and external flows, provided the total shear stress varies slowly in the normal direction, and provided the surface-pressure gradient at the wall is small compared with the normal shear-stress gradient.

Consider first the inner layer. It includes about 10% — 20% of the turbulent boundary-layer profile. The characteristic velocity scale and length scale usually used in this layer are, respectively, u_τ and (ν/u_τ), where ν is the kinematic viscosity, u_τ is given by $\sqrt{\tau_w/\rho}$, and τ_w is the shear (i.e., the slope of the mean-velocity profile) at the wall. The term u_τ is referred to as the friction velocity. The scales can be combined to form $y^+ = u_\tau y/\nu$, which is a Reynolds number related to the energy-containing turbulence. Any variable that is given the $+$ superscript has been made dimensionless in this way, and is said to be expressed in "wall units." The scales just defined are referred to as the inner boundary-layer scales or, where appropriate, simply as the inner scales. In terms of these scales the thickness of the linear sublayer, the buffer layer, and the log-law region are, respectively, $(0 < y^+ < 5)$, $(5 < y^+ < 40)$, and $(40 < y^+ < 500$ to $1000)$, where the bounds are, of course, approximate. Comparing the definitions of δ and y^+, we see that 750 wall units $=$ 0.15δ gives an approximate relation between the inner scales and the outer scales.

We briefly note what has been observed about the nature of the flow within the inner layer. In the linear sublayer the Reynolds stresses are negligible, the pressure gradient has little effect, and the averaged velocity varies linearly. In the buffer layer, both viscous and Reynolds shear stresses are of the same order of magnitude, and there is no simple variation of velocity with the normal distance. In the log-law region the viscous stresses are negligible relative to the Reynolds stresses. Both the buffer and inertial layer are affected by the pressure gradient. Finally, at the outer edge of the inner layer the mean velocity is about 70% of U_e.

The outer layer extends over the remaining 80% — 90% of the turbulent velocity profile. In this layer we return to the outer boundary-layer scales, U_e and δ, for the referencing system. It has been observed for external flows that there are two subdivisions, the wake-law layer and the viscous superlayer. The wake-law layer is fully turbulent in the first half and then becomes intermittently so; the viscous superlayer is laminar-like. In fully developed internal flows there is no superlayer and, with decreasing Reynolds number, both the viscous sublayer and superlayer become thick and the outer-law layer disappears [7].

2.1(b) Temporal Scales

Temporal scales are determined by the frequency of an unsteadiness. The unsteadiness may be externally induced, such as the flow generated by the movement of a flap, or self-induced, such as the flow generated by vortex shedding. The two forms may be coupled. For example, the unsteady forces owing to vortex shedding may cause a body to move, which in turn may change the shedding frequency. Clearly, unsteadiness can be completely stochastic, highly organized, or a combination of random and periodic components interacting with one another. From a computational point of view the critical time-scale relates to the highest frequency that must be resolved, since all other frequencies are then accommodated. Experiments have provided some quantitative information which permits the identification of time-scales in certain cases.

When a flow field is oscillating with a dominant frequency f, the Strouhal number, $S_t = fL/U_\infty$, is used to identify the time-scale of importance, where U_∞ is the free-stream velocity. This number characterizes the relative importance of the local velocity with respect to the convective velocity, that is, the time-scale of physical motion to the basic fluid dynamic time-scale. If S_t is referenced to the speed of sound, instead to U_∞, a necessary condition for the flow to be incompressible is that $S_t^2 << 1$. For an oscillating body in an incompressible flow, the boundary-layer thickness is $O(\sqrt{\nu/f})$. The typical value of the Strouhal number in the wake of a circular cylinder is about 0.2 .

If a body undergoes a sudden change in position that is rapid relative to the diffusion process, there is an unsteadiness that is confined to a layer, called the Stokes layer; the Stokes layer is thinner than the boundary layer. In such a case the diffusive time-scale is $O(\delta^2/\nu)$, and it pertains to the diffusion of vorticity and momentum outward from the body surface. For a body moving from rest, the layer thickness is $O(\sqrt{\nu t})$ at small times.

When the flow is decomposed into a mean field and a fluctuating field, and the fluctuating components are made to satisfy the Reynolds conditions [8], time-averaged Navier-Stokes equations can be appropriate for the computation of unsteady, as well as steady, compressible and incompressible fluid flows. This occurs when the averaging time-interval is large relative to the periods characteristic of time scales that can be neglected, but small relative to the period of the mean unsteady motion. This double decomposition of the velocity field is satisfactory for many important physical problems involving relatively small amplitude and low-frequency oscillations.

Furthermore, when the turbulence structure is unaffected by externally induced unsteadiness, there is a critical frequency below which steady-flow turbulence models can also be used to predict unsteady turbulent flows. In zero pressure-gradient flows, the value of this critical frequency ranges from 20% to 100% of $F_b = U_\infty/(5\delta)$ which is roughly the burst frequency based on the outer scales ([9]-[12]). In adverse pressure gradients, this critical frequency appears to be well below this burst frequency. Values of 6% to 28% of F_b have been reported ([13]-[15]).

Additional discussion concerning temporal scales is presented by Carr [16], Chapman [17], and in the article written in this volume by Deiwert.

2.2 Generation of Circulation and Vorticity

Vorticity and circulation, which is the flux of vorticity, play important roles in the analysis of flow fields that are governed by the compressible or incompressible Navier-Stokes equations. It is worthwhile to examine them in context with the role that viscosity plays in similar flow fields. Let us define the symbols ρ, p, μ, Ω, and U to be the density, pressure, molecular viscosity, vorticity vector, and velocity vector, respectively. The vorticity vector is defined to be the curl of the velocity vector and its magnitude and direction comprise the vorticity field. At high Reynolds numbers, the vorticity field is confined to thin layers or fine tubes, although the velocity vector is, of course, present in the whole flow field. In general, the effect of viscosity in a laminar or turbulent flow is transmitted through the vorticity field.

Although one cannot see the vorticity field and its measurement is difficult, its presence in aerodynamic applications is easily detected by the determination of circulation, Γ, which is given by

$$\Gamma = \oint \underline{U} \cdot d\underline{\ell} \tag{1}$$

The time rate of change of circulation is given by Kelvin's theorem (see [18])

$$\frac{D\Gamma}{Dt} = \oint \left(-\frac{\nabla p}{\rho}\right) \cdot d\underline{\ell} + \oint \frac{1}{\rho}[-\nabla \times [\mu\underline{\Omega}] + \nabla[(4/3)\mu\nabla \cdot \underline{U}]] \cdot d\underline{\ell} \tag{2}$$

in which we have omitted the body force term.

If the pressure is a function of the density alone, the pressure forces cannot produce circulation. This would be the case if the flow were incompressible or isentropic. However, this would not be the case if there were some entropy producing mechanism in the flow, such as a curved shock or heat addition. It is also not the case for stratified flows.

In the absence of sources of vorticity resulting from pressure and body forces, circulation cannot be created in the interior of the fluid (see [19] and [20]). In such cases circulation can only be generated by viscous forces acting at a solid boundary or at a free surface.

Vorticity is related to circulation by the equation

$$\Gamma = \oint \oint \underline{\Omega} \cdot \underline{n}\, ds \tag{3}$$

where the surface is any area bounded by the closed curve in Eq. (1). In a Newtonian fluid the vorticity field is governed by the vorticity equation which is found by taking the curl of both sides of the equation of motion; it can be written

$$\frac{D\underline{\Omega}}{Dt} = (\underline{\Omega} \cdot \nabla)\underline{U} + \nabla \times \left(-\frac{\nabla p}{\rho} \right) + \nabla \times \left[\frac{1}{\rho} [-\nabla \times (\mu \underline{\Omega}) + \nabla[(4/3)\mu \nabla \cdot \underline{U}]] \right] \tag{4}$$

When a solid body is passed through a viscous fluid that is void of vorticity, vorticity is imparted to the fluid through the mechanism of the viscosity. It is then both diffused and convected into the interior of the fluid. In classical boundary-layer theory the convection mechanism is neglected and the flow is usually diffusion dominated. For high Reynolds numbers and for most flows of interest in aerodynamics, the molecular viscosity can be neglected everywhere except in the boundary layer where it is essential for the initiation of vorticity. If at some point the flow separates, the vorticity is streamed out into the interior of the fluid by the process of convection and the flow in that region is said to be convection dominated.

Let us now examine the difference between Eqs. (2) and (4). Both have a term related to the pressure field, and in both cases this term produces neither circulation nor vorticity if there are no entropy producing mechanisms. Both have terms that involve the molecular viscosity, and in both cases these terms produce neither circulation nor vorticity except along a solid body surface (or free surface). The principal difference between them appears in the term $(\underline{\Omega} \cdot \nabla)\underline{U}$, which has no counterpart in the equation for the time rate of change of circulation. It is interesting to notice that this term is zero in two dimensions, and the equation $D\underline{\Omega}/Dt = 0$ is often used to study two-dimensional flows at high Reynolds numbers. When used numerically, this form of analysis is Lagrangian and referred to as vortex tracing (see Sec. 5.4).

In three dimensions, the term $(\underline{\Omega} \cdot \nabla)\underline{U}$ is not zero and, in fact, represents a mechanism that is a fundamental cause of transition and ultimately of turbulence. The local effect of this mechanism is referred to as vortex stretching. What can happen as a result of vortex stretching is described in the following simple example. An elongated, essentially two-dimensional, vortex-filament is formed in a uniform laminar boundary layer at a solid surface by the no-slip boundary condition. This filament is diffused and convected away from the wall. As time passes, the stretching term induces an instability at sufficiently high Re, any small perturbation causes the filament to kink, the instability causes more intense kinking, and soon the filament becomes elaborately distorted and highly stretched. This is occurring away from the wall and, by Eq. (2), the circulation of the filament is not changed during the process. The stretching greatly decreases the cross-sectional area of the filament and the only way in which Eq. (3) can be satisfied is by a correspondingly large increase in the vorticity. This continuing process of vortex stretching and distortion causes an intense increase in the magnitude of the local vorticity field, causing a cascade of energy from large-scale vortex motions to small-scale motions. As a result, larger and larger velocity gradients are formed, being limited only by eventual viscous dissipation in the very small scales.

2.3 Organized Structures in Turbulent Flows

Organized structures are also known as "coherent" structures. Coherent means having the quality of cohering, or being logically consistent, or having a definite phase relationship. As organized flow structures are not necessarily coherent, they are not called coherent structures in this article.

2.3(a) Transitional Flows

Transitional flow refers to the flow regime between a region that is completely laminar and one that is completely turbulent. In its study we are primarily concerned with how the turbulence starts to form and how it enters the surrounding laminar flow.

The primary structure observed in a wall-bounded, **external** transitional flow is the turbulent "spot," the existence of which was first proposed by Emmons [21]. Each turbulent spot is surrounded by nonturbulent fluid. In plan view, it has a kidney shape, much more elongated in the streamwise direction than in spanwise direction, pointing downstream like a rounded-off elongated delta. The upstream and downstream edges travel at constant but not equal velocities. These velocities are larger than the velocity of the undisturbed laminar flow. Therefore, the fluid downstream of the spot is overtaken by the spot, becomes turbulent, then returned to laminar after the passage of the spot. The motion within the spot is just like that in a turbulent boundary layer with streamwise streaks next to the wall. The ensemble-averaged spot is essentially a single horseshoe vortex superimposed on small-scale motions ([22] and [23]).

The main structure in an **internal** transitional flow is somewhat different from that in an external flow. For example, in a pipe a turbulent "slug," which is not the same as a turbulent spot, can occur. Slugs have been observed in smooth or slightly disturbed inlets for $Re > 3200$

[24]. These structures occupy the entire crosssection of the pipe, and are elongated in the streamwise direction. The leading edge and the trailing edge of slugs propagate at velocities that are larger than the undisturbed laminar flow. Another type of intermittent turbulent flow, referred to as a "puff," is also found in pipes. A turbulent puff occurs when a large disturbance is introduced at the entrance of the pipe for $2000 \leq Re \leq 2700$. A puff has been described as an incomplete relaminarization process [1].

2.3(b) Turbulent Flows

Townsend [25] first identified organized structures in turbulent flows. He described a "double structure" of fully developed turbulence consisting of more or less organized large eddies, containing about one-fifth of the total turbulent energy, in an essentially unorganized, small-scale, turbulent background. Organized structures are defined only in a statistical sense, but in this sense they can be identified and even numerically predicted and experimentally reproduced. Such structures have been identified in several ways, for example, by special forms of flow visualization, conditional sampling, and the method of Eulerian space-time correlation [26]. Recently identification has been attempted by means of a technique based on statistical approaches using an orthogonal decomposition theorem of probability theory [27]. It appears that there is some form of organized structure in all types of turbulent shear flows, although, in general, these structures contain only a small part of the total energy and play a secondary role in controlling transport [27].

At present there is a lack of consensus concerning the definition, identification, physical dimensions, and origin of organized structures in turbulent boundary layers. However, certain characteristics of these structures are beginning to be established. It appears that there are five main parts of the turbulent boundary layer: streamwise vortices and streaks next to the wall, pockets of energetic fluid near the wall, large-scale motions in the main part of the boundary layer, mushroom-shaped intermediate scale motions of energetic fluid at its edge, and the viscous superlayer. These structures are briefly discussed below for a boundary layer in which there is no streamwise curvature, pressure gradient, or roughness (see also [28]-[32]).

Next to the wall, the flow structure consists of a fluctuating array of counterrotating vortices aligned in the streamwise direction [33] and separated by alternating high- and low-speed regions called streaks. High-speed fluid moving into the region next to the wall, called a sweep, and low-speed fluid moving out of this region, call an ejection, are primary sources of the continual production of turbulence in the region between the wall and a y^+ of about 100. Observations of the length of these structures vary from 100 to 2000 wall units ([34]-[36]), their vertical extent varies from 20 to 50 wall units, and the mean spanwise extent between vortex centers is about 50 wall units (with a possible dependence on Reynolds number [37]). The streamwise streaks interact with the outer flow during a "bursting" process in which the streaks break up following a lifting and an oscillation [38]. For the last 10 years, the mean dimensionless time between bursts was presented in terms of the outer scales as $U_e T/\delta \approx 6$, see [39]. Recently it has been shown that the inner scales are more appropriate ([28] and [40]).

The bursting process leaves pockets of intermediate-scale energetic motions in a region $y^+ \approx 5$ to 40, which lead to the formation of streamwise vortices and hairpin vortices. Their period of formation is $u_\tau{}^2 T/\nu \approx 25$. The dimensions of these pockets are not easy to identify. Cantwell [29] estimates the streamwise extent to be from 20 to 40, and the normal extent to be from 15 to 20 wall units. However, Falco [28] reports lengths between 50 and 90 wall units. There is no direct information about the spanwise extent. These energetic pockets persist over a streamwise distance between one-half and one boundary-layer thickness δ, and travel at about $0.65U_e$. Instantaneous maximum Reynolds shear stress is very large compared to the local mean value during bursting; for example, at $y^+ \approx 30$, it can exceed 60 times the local mean value [41].

Two phenomena have been observed in the outer layer. One occurs along both its inner and outer edges, and the other in its interior. Along the edges, mushroom-shaped, intermediate-scale, energetic regions appear. These are called "typical eddies" by Falco [42]. They have a streamwise and normal length of about 200 and $O(100)$ wall units, respectively [42], although the lengths are Reynolds-number dependent,and at very low Reynolds numbers they become of $O(\delta)$ and appear as large-scale motions. These energetic regions exist over a distance that is approximately 5 times their own streamwise extent, and they travel at about $0.85U_e$. They account for a major part of the Reynolds stresses [42].

In the interior of the outer layer, relatively large-scale energetic motions occur at a height of about 0.8δ. They have a streamwise and spanwise extent of between δ and 2δ and between 0.5δ and δ, respectively, and their centers are spaced from two to three δ in the spanwise direction. The distance over which they persist is about 1.8δ and they travel with a speed of about $0.85U_e$. Instantaneous maximum Reynolds shear stress may exceed 10 times the local mean value in these regions.

The outer layer is bounded by the laminar-like viscous superlayer of the order of the Kolmogorov length scale, $\eta \equiv (\nu^3/\epsilon)^{(1/4)}$, where ϵ is the dissipation rate [43]. Outside the superlayer, away from the wall, the vorticity fluctuations are zero (although the velocity fluctuations are not).

From the description of the three-dimensional, unsteady, turbulent, boundary layer, it is concluded that both the wall layer and the outer energetic regions are equally important. When the fluid is incompressible, the boundary layer is an elliptic region. The wall layer and the outer layer interact and the cause and effect cannot be separated.

3. ANALYTICAL ASPECTS OF VISCOUS FLOWS

This section includes a compact expression for the Navier-Stokes equations in two or three dimensions. It also contains some simplified approximate forms of the compressible Navier-Stokes equations for cases in which parts of the flow are inviscid, other parts are viscous, and both are interacting strongly with one another. It concludes with some simple turbulent models. Additional discussion of these and the other recent advances in analytical aspects of viscous flows may be found elsewhere in this volume. For discussion of both analytical and numerical aspects of the thin-shear-layer theory refer to [44]-[50].

3.1 The Navier-Stokes Equations

The continuum incompressible fluid mechanics is described by combining the classical Navier-Stokes equations, which express the conservation of momentum, with an equation expressing the conservation of mass. Compressible fluid mechanics is described by these equations, properly modified to take into account variations in density and temperature, together with an equation for the conservation of energy and a thermodynamic equation of state. In this article both of these systems are referred to as the Navier-Stokes equations. Such expressions are considered to be the proper fundamental equations for viscous flows, as long as scales of interest are many times the mean free paths of fluid molecules.

Problems for incompressible, laminar fluids can be formulated in terms of the vorticity vector, $\underline{\Omega}$, the velocity vector $\underline{U}$, and the stream function $\underline{\Psi}$. This has the advantage of eliminating the pressure. These equations are well known and can be written

$$\begin{aligned} \frac{D\underline{\Omega}}{Dt} &= (\underline{\Omega}\cdot\nabla)\underline{U} + \nu\nabla^2\underline{\Omega} \\ \nabla^2\underline{\Psi} &= -\underline{\Omega} \\ \underline{U} &= \nabla\times\underline{\Psi} \end{aligned} \tag{5}$$

Problems for turbulent flows, compressible or incompressible, are usually formulated in terms of quantities representing the mean and fluctuating parts of a variable. This is expressed by the relation

$$f = \overline{f} + f' \tag{6a}$$

where the over-bar designates a mean or averaged value, and the prime designates the remaining fluctuating value. Monin and Yaglom [8] present a general space-time averaging procedure represented for any function $f(\underline{x}, t)$ by the equation

$$\overline{f}(\underline{x}, t) = \langle f(\underline{x}, t) \rangle = \iint_{-\infty}^{\infty} f(\underline{x} - \underline{\varsigma}, t - \tau)\, g(\underline{\varsigma}, \tau) d\underline{\varsigma} d\tau \qquad (6b)$$

where the underscore indicates a vector field. The non-negative weighting function, g, satisfies the normalizing condition

$$\iint_{-\infty}^{\infty} g(\underline{\varsigma}, \tau) d\underline{\varsigma} d\tau = 1 \qquad (6c)$$

The choice of this weighting function determines the significance of the averaged quantities. The term $\overline{f}(\underline{x}, t)$ is referred to as a time-averaged quantity if $g(\underline{\varsigma}, \tau)$ is set equal to $g(\tau)\,\delta(\underline{\varsigma})$ and $g(\tau)$ is constant over some segment of time and zero elsewhere. (In this article $\delta(\ )$ represents the Dirac delta function of its argument.) Space-averaged equations follow from setting $g(\underline{\varsigma}, \tau) = g(\underline{\varsigma})\delta(\tau)$ and letting $g(\underline{\varsigma})$ be a piecewise, continuously-differentiable function which tends to zero at least as fast as $1/\underline{\varsigma}^4$.

For an incompressible fluid without external forces, both space-averaged and time-averaged Navier-Stokes equations can be written in dimensional form as

$$\frac{\partial \overline{u}_i}{\partial x_i} = 0$$

$$\frac{\partial \overline{u}_i}{\partial t} + \frac{\partial \langle \overline{u}_i \overline{u}_j \rangle}{\partial x_j} = -\frac{1}{\overline{\rho}}\frac{\partial \overline{p}}{\partial x_i} + \frac{\partial \sigma_{ij}}{\partial x_j} \qquad (7)$$

where $\langle \overline{u}_i \overline{u}_j \rangle = \langle\langle u_i \rangle \langle u_j \rangle\rangle$. The symbol σ_{ij} is referred to as the stress tensor and is usually broken down into two terms:

$$\sigma_{ij} = 2\nu S_{ij} - R_{ij} \qquad (8)$$

which is composed of the mean flow rate of strain tensor, S_{ij}, and the Reynolds stress tensor $-\overline{\rho}R_{ij}$. These are related to the mean and fluctuating terms by

$$S_{ij} = \frac{1}{2}\left(\frac{\partial \overline{u}_i}{\partial x_j} + \frac{\partial \overline{u}_j}{\partial x_i}\right)$$

$$R_{ij} = \langle u'_i u'_j \rangle + \langle \overline{u}_i u'_j \rangle + \langle u'_i \overline{u}_j \rangle \qquad (9)$$

Note: No approximations have been made if we express the Navier-Stokes equations in terms of the mean and fluctuating quantities using the stress and strain tensors defined above. The Reynolds-averaged Navier-Stokes equations are an approximating subset of this system found when it is assumed that

$$
\begin{aligned}
< \overline{u}_i \overline{u}_j > &\rightarrow \overline{u}_i \overline{u}_j \\
< \overline{u}_i u'_j > = < u'_i \overline{u}_j > &\rightarrow 0
\end{aligned}
\tag{10}
$$

For compressible fluids, space-averaged Navier-Stokes equations have yet to be derived. However, the time-averaged Navier-Stokes equations have been derived. These contain second-order moments, such as $< \rho' u' >$, and third-order moments, such as $< \rho' u' u' >$, owing to fluctuations in the fluid density [51]. In order to remove such moments, mass-weighted, time-averaged equations have been formulated. In this formulation a mass-weighted velocity $\overline{u}_i$ is defined such that

$$
\overline{u}_i = < \rho u_i > / < \rho > \tag{11}
$$

with similar definitions for the other quantities. This averaging procedure eliminates the moment terms such as $< \rho' u' >$ from the equations but it does not, of course, remove the averaged density fluctuations from turbulence. Mass-weighting was first used in the study of atmospheric turbulence by Hesselberg [52]. A comprehensive discussion of the procedure is presented in [53]-[55]. Henceforth the equations derived from this type of averaging are combined with the assumptions given in Eq. (10), and the result is referred to as the compressible, Reynolds-averaged Navier-Stokes equations. These equations exhibit a term-by-term correspondence with the equations for incompressible fluids. They are also identical to the equations used to determine laminar flows, except for the Reynolds stress tensor and turbulent heat-flux vector, which are given by

$$
\begin{aligned}
\sigma_{ij} &= 2\nu\left(S_{ij} - \frac{1}{3}\delta_{ij}\frac{\partial \overline{u}_k}{\partial x_k} \right) - \mathbf{R}_{ij} \\
q_j &= -\frac{\nu}{Pr_l}\frac{\partial \overline{h}}{\partial x_j} + \frac{< \rho u'_j h' >}{\overline{\rho}} \\
\mathbf{R}_{ij} &= \frac{< \rho u'_i u'_j >}{\overline{\rho}}
\end{aligned}
\tag{12}
$$

where δ_{ij}, h, and Pr_l are, the Kronecker delta, mass-weighted mean enthalpy, and the laminar Prandtl number, respectively.

The incompressible Reynolds-averaged Navier-Stokes equations are presented above in Cartesian coordinates. They can also be expressed in arbitrary curvilinear coordinates, $(\underline{\xi}, \tau)$, in conservation-law form. Below, the compressible Reynolds-averaged Navier-Stokes equations in conservative Cartesian variables are presented. They are made nondimensional by using reference values of length, velocity, density, etc., such that the dimensional and nondimensional equations are identical, except for the appearance of the Reynolds and Prandtl numbers in the nondimensional equations. If d represents the number of dimensions, equations for the variables Q_m, C_{mi}, and V_{mi} can be written

$$\frac{\partial Q_m}{\partial \tau} + \sum_{i=1}^{d} \frac{\partial C_{mi}}{\partial \xi_i} = \frac{1}{Re} \sum_{i=1}^{d} \frac{\partial V_{mi}}{\partial \xi_i} \tag{13}$$

where

$$Q_m \equiv \mathcal{D}[\overline{\rho}, \overline{\rho u}_1, ..., \overline{\rho u}_d, \overline{e}]^T, \qquad C_{mi} = Q_m \mathcal{U}_i + \overline{p}\Phi_{mi}$$

$$V_{mi} = \mathcal{D} \sum_{j=1}^{d} R_{mj} \frac{\partial \xi_i}{\partial x_j}, \qquad \mathcal{U}_i = \frac{\partial \xi_i}{\partial t} + \sum_{j=1}^{d} \overline{u}_j \frac{\partial \xi_i}{\partial x_j}$$

$$\Phi_{mi} \equiv \mathcal{D} \sum_{j=1}^{d} \frac{\partial \xi_i}{\partial x_j} [0, \delta_{j1}, ..., \delta_{jd}, \overline{u}_j]^T, \qquad \mathcal{D} \equiv J\left(\frac{x_1, ..., x_d}{\xi_1, ..., \xi_d}\right)$$

$$R_{mi} \equiv [0, \sigma_{i1}, ..., \sigma_{id}, \sum_{j=1}^{d} \overline{u}_j \sigma_{ij} - q_i]^T$$

$$e_I = \frac{\overline{e}}{\overline{\rho}} - \sum_{k=1}^{d} \frac{\overline{u}_k \overline{u}_k}{2}, \qquad \overline{p} = (\gamma - 1)\overline{\rho} e_I$$

$$\frac{\partial \xi_i}{\partial t} = -\sum_{j=1}^{d} \frac{\partial \xi_i}{\partial x_j} \frac{\partial x_j}{\partial \tau}, \qquad \frac{\partial \xi_i}{\partial x_j} \equiv \frac{1}{\mathcal{D}} J\left(\frac{x_{j+1}, x_{j+2}}{\xi_{i+1}, \xi_{i+2}}\right)$$

In the previous expression, subscripts (i, i+1, i+2), and (j, j+1, j+2) vary in a cyclic order, (1, 2, 3), (2, 3, 1), etc. The symbol J represents the Jacobian,and the Stokes hypothesis of local thermodynamic equilibrium, $(3\lambda + 2\mu = 0)$, has been used. The total energy-per-unit volume and the internal energy-per-unit mass are represented by e and e_I, respectively. Further, the stress tensor and heat flux vector are given by

$$\begin{aligned} \sigma_{ij} &= \mu \sum_{k=1}^{d} \left(\frac{\partial \xi_k}{\partial x_j} \frac{\partial \overline{u}_i}{\partial \xi_k} + \frac{\partial \xi_k}{\partial x_i} \frac{\partial \overline{u}_j}{\partial \xi_k} - \frac{2}{3} \sum_{n=1}^{d} \frac{\partial \xi_n}{\partial x_k} \frac{\partial \overline{u}_k}{\partial \xi_n} \delta_{ij} \right) - \mathbf{R}_{ij} \\ q_i &= -\frac{\gamma \mu}{Pr_l} \sum_{j=1}^{d} \frac{\partial \xi_j}{\partial x_i} \frac{\partial e_I}{\partial x_j} + \frac{< \rho u_i' h' >}{\overline{\rho}} \end{aligned} \tag{14}$$

Unfortunately, the terms $\mathbf{R}_{ij}$ and $< \rho u_i' h' > / \overline{\rho}$ cannot be expressed in terms of the other variables and this leads to the classical "closure" problem and the necessity to construct turbulence models.

3.2 The Approximating Equations for Strong Interactions (The Composite Equations)

As discussed in Sec. 2.1(a), many viscous flow problems that cannot be solved by the boundary-layer approximations via the thin-shear-layer equations, can be solved with an intermediate set of equations that fall between the simple shear-layer equations and the complete Navier-Stokes equations. This intermediate set of equations is applicable to both inviscid and viscous regions, and, therefore, they can be used to compute strong interactions between the two. Since they couple the viscous and inviscid approximations to a fluid flow, they have been referred to as composite equations [56]. The main feature of these equations is the presence of a nonzero normal pressure gradient, a necessary term for coupling the viscous and inviscid regions.

There are several forms of composite equations. They are referred to by different names such as thin-layer, slender-layer, parabolized, or conical Navier-Stokes equations. The adjectives, thin-layer, slender-layer, and conical are based on physics, and the adjective parabolized is based on analytical properties.

The composite thin-shear-layer equations are often referred to in the literature as the thin-layer Navier-Stokes equations. They are obtained from Eq. (13) by neglecting all streamwise and spanwise derivatives of the viscous and turbulence stress and heat flux terms. This is justified either on a physical order of magnitude analysis or on a computational accuracy argument. The latter argument amounts, in simple terms, to the observation that since the neglected terms cannot be computed correctly with the available grid resolution anyway, why keep them? On the other hand, the composite slender-layer Navier-Stokes equations are derived by neglecting only the streamwise derivatives of the viscous and turbulence stress and heat flux terms. For a compressible fluid, these sets of equations can be written for the three-dimensional case as

$$\frac{\partial Q_m}{\partial \tau} + \sum_{i=1}^{3} \frac{\partial C_{mi}}{\partial \xi_i} = \frac{1}{Re} \sum_{i=2}^{\ell} \frac{\partial V_{mi}}{\partial \xi_i} \tag{15}$$

in which the composite thin-shear-layer, and slender-shear-layer equations are given by the value of ℓ equal to 2 and 3, if the streamwise direction and the direction normal to the surface are represented by ξ_1 and ξ_2, respectively. By the same arguments, the upper and lower limits for the k and j summations in Eq. (14) are replaced by ℓ and 2, respectively.

The approximation referred to as the parabolized Navier-Stokes equations (PNS) omits the time-derivative and marches the solution in the streamwise direction. For this approximation to be valid, the inviscid outer flow must be supersonic and the upstream influence of the streamwise pressure gradient in the subsonic viscous flow must either be treated approximately or accounted for by introducing some iterative or relaxation procedure. A special subset of this approximation has been called the parabolic Navier-Stokes equations. In these equations, the streamwise pressure-gradient term is deliberately specified in the subsonic region. The form of the thin-layer parabolized Navier-Stokes equations should be obvious from the the above discussion.

When a flow field is surrounded by conical boundaries, there is no significant length scale in the conical direction for inviscid flows. This conical flow assumption has been applied locally to viscous flows. The conical Navier-Stokes equations are determined by using a conical transformation

$$\alpha = \xi_1, \quad \beta = \frac{\xi_2}{\xi_1}, \quad \gamma = \frac{\xi_3}{\xi_1}, \quad \tau^* = \tau \tag{16}$$

together with the assumption of local conical self-similarity in which derivatives with respect to α are neglected. This leads to a set of equations having a form similar to that for the slender shear layer, but with a source term. For these flows the local Reynolds number is determined by the position where the solution is computed.

Partially parabolized Navier-Stokes equations are usually formulated for an incompressible fluid. Terms in the momentum equations are neglected in order to make these equations parabolic, and the pressure field is determined from an elliptic equation. A discussion of composite equations for an incompressible fluid, and solution procedures for these equations, are given by Rubin in this volume.

3.3 Turbulence Modeling

As discussed in the section dealing with scales of motions, turbulent flows are characterized by a wide range of scales. Some way to cope with this range of velocity and length scales is required for the successful modeling of turbulence. Most turbulence modeling methods employ a single local velocity and a single local length to describe the various interactions. These methods are called single-scale methods. The magnitudes of the local velocity and length vary greatly in various regions of the flow according to some formula that usually relates them to the mean-flow quantities and, possibly, the distance from a solid surface. Methods that introduce two or more time or length scales are referred to as multiscale methods. They are not discussed here.

There are two approaches to turbulence modeling: the so-called first-order approach, in which the Reynolds stress tensor is modeled in terms of the mean flow quantities, and the second-order approach, in which the terms in this tensor are carried along in the computations as dependent variables having been expressed in terms of a higher order tensor which is modeled. Here we consider only first-order approaches. They form the basis for the conventional zero-equation (or algebraic), one-equation and two-equation models. In practice, the actual form of various basic turbulence models and the manner of applying them generally differ in detail from investigator to investigator. General definitions and characteristics of both first-order and second-order models are available from [55]-[65].

We consider below some simple examples of zero- and two-equation models. All of these models are expressed in terms of a turbulence velocity scale v and a turbulence length scale ℓ. On dimensional grounds, a combination of these scales determines the value of the kinetic eddy viscosity

$$\nu_t = c_t v \ell \tag{17}$$

where c_t is sometimes referred to as the Clauser constant. From this expression we can determine the dynamic eddy viscosity, $\mu_t = \rho\nu_t$.

Algebraic models relate ν_t directly to averaged field quantities without additional partial differential equations. Both one- and two-equation models contain additional partial differential equations for the turbulence scales. One-equation models use a prescribed, empirical formula for the local length-scale, together with a single partial differential equation for the velocity scale, v. Two-equation models use additional partial differential equations to find both v and ℓ. Methods using partial differential equations for the turbulence scales are also called transport-equation methods.

There are many zero-equation models. The simplest types are constructed using, for the turbulence velocity and length scales, equations that apply uniformly throughout the flow. Such models are used, for example, in the so-called large-eddy simulations (LES). An example based on space averaging and used by Moin and Kim [66] following Schumann [67] is given by

$$\begin{aligned} v &= \ell f(S_{ij}) \\ \ell &= f(\Delta_1\Delta_2\Delta_3)\big(1 - f(-y^+)\big) \end{aligned} \tag{18}$$

where Δ_i is twice the local grid-spacing in the i-direction, and the precise form of the functional dependences is given in the reference. These models are often referred to as sub-grid-scale models. Without the exponential damping function, the form of this eddy-viscosity model is the same as a model suggested by Smagorinsky [68] for atmospheric studies. The damping function, which accounts for unresolved turbulence near a wall, was first suggested by Van Driest [69]. An evaluation of sub-grid-scale models is given by Kaneda and Leslie [70].

A more elaborate zero-equation turbulence model used in the Reynolds-averaged formulation for two- and three-dimensional flows that are steady is due basically to Cebeci [71] and is briefly outlined next. The turbulent boundary layer is assumed to consist of an inner layer and an outer layer. In each layer the distributions of v and ℓ are prescribed by different empirical expressions. In the inner (log-law) layer, ℓ is proportional to y, the distance normal to the wall, and in the outer layer, ℓ is proportional to the total shear-layer thickness. The proportionality of ℓ to y in the inner layer is modified by a damping function [69] that has the same form as that used in Eq. (18). However, in this case the parameters are adjusted to take into account the effect of pressure gradient, as done, for example, in [72]. In the outer layer the vorticity is used to define the shear-layer thickness [73].

In the inner layer, $0 \leq y \leq y_c$, the expressions for v and l are

$$\begin{aligned} (v)_{inner} &= \ell|\overline{\Omega}| \\ (\ell)_{inner} &= \alpha_1 y\left[1 - e^{(-y^+/A^+)}\right] \\ A^+ &= 26\left[1.0 - 11.8p^+\right]^{(-1/2)} \\ p^+ &= -\left(\frac{\mu}{Re\rho^2 u_\tau^3}\frac{\partial \overline{p}}{\partial x}\right) \end{aligned} \tag{19}$$

with $\alpha_1 = 0.40$ and $c_t = 1.0$ in Eq. (17). In this case, the wall coordinate y^+ can be based on the local value of the density, Cebeci [71], or the value of the density at the wall [73].

In the outer region, $y > y_c$, the expressions for v and ℓ are

$$\begin{aligned} (v)_{outer} &= \text{minimum of}\left(L_{max},\ \frac{U^2_{diff}}{4L_{max}}\right) \\ (\ell)_{outer} &= y_{max} C_{BL}\,\alpha_2 \end{aligned} \tag{20}$$

If we define the function $L(y)$ to be

$$L(y) = y|\overline{\Omega}|\left[1 - e^{(-y^+/A^+)}\right] \tag{21}$$

the quantity L_{max} is the maximum value of $L(y)$ that occurs in this equation, and y_{max} is the value of y at which it occurs. The quantity U_{diff} is the difference between the maximum and minimum value of the absolute velocity, the value of C_{BL} is 1.6, and the Klebanoff intermittency factor, α_2, is given by

$$\alpha_2 = \left[1 + 5.5\left(\frac{C_K y}{y_{max}}\right)^6\right]^{-1} \tag{22}$$

where the Klebanoff constant $C_K = 0.3$. The Clauser constant, c_t in Eq. (17), is taken to be $= 0.0168$.

The region of validity of the inner and outer scales is determined by y_c, which is defined to be the smallest value of y at which the values of the inner and outer eddy viscosity are the same. The value of α_1 in the inner layer and of c_t in the outer layer are assumed to be universal constants for

$Re_\theta > 5000$, where Re is based on the momentum thickness. At lower Reynolds numbers, they are functions of Reynolds number (see [74]).

The model presented above is based on the absolute value of the vorticity, and it works about as well as a zero-equation model can be expected to for many cases. However, in the absence of strain and body forces, it is inappropriate for irrotational shear flow, because it gives an infinite, rather than finite, value of the time-scale; and it is inappropriate for a purely rotating flow, because it gives a finite, rather than an infinite, value of the time-scale [75]. It is also clear that the turbulence length scale given above needs to be modified for flows in corners and other flows bounded by two or more walls (see for example, [76]-[78]). On top of these limitations, the manner of determining y_{max} has been questioned recently in some flows ([79] and [80]), and the constants C_{BL} and C_K have been found to be a function of Mach number [80]. Thus, the range of validity of a turbulence model in terms of types of flow problems and of flow parameters needs to be established and questioned.

A two-equation model due to Wilcox and Rubesin [81] is presented next. In this model, partial differential equations are written for twice the turbulent kinetic energy ρv^2, which represents indirectly the role of the velocity scale, and the specific energy dissipation $\rho\omega^2$, which in combination with the former gives the length scale. Thus,

$$\frac{\partial(\rho v^2)}{\partial t} + \frac{\partial(\rho\overline{u}_j v^2)}{\partial x_j} = -2\rho\mathbf{R}_{ij}\frac{\partial\overline{u}_i}{\partial x_j} - \beta_1\rho\omega v^2 + \frac{\partial[(\mu+\beta_2\mu_t)\partial v^2/\partial x_j]}{\partial x_j} \tag{23}$$

$$\frac{\partial(\rho\omega^2)}{\partial t} + \frac{\partial(\rho\overline{u}_j\omega^2)}{\partial x_j} = -\beta_3\ell^{-2}\rho\mathbf{R}_{ij}\frac{\partial\overline{u}_i}{\partial x_j} - \left[\beta_4 + \beta_5\left(\frac{\partial\ell}{\partial x_k}\right)^2\right]\rho\omega^3 + \frac{\partial[(\mu+\beta_5\mu_t)\partial\omega^2/\partial x_j]}{\partial x_j} \tag{24}$$

where the length scale is defined by $\ell = v/\omega$. The eddy viscosity is again computed from Eq. (17). The above model requires that the coefficients β_1, β_2, β_3, β_4, β_5, and β_6, in addition to c_t, be specified. Wilcox and Rubesin [81] recommend the following values for these coefficients:

$$\beta_1 = 0.09, \quad \beta_2 = \beta_5 = 0.5, \quad \beta_4 = 0.15, \quad \beta_6 = 1/11,$$
$$c_t = \left[1-(1-\beta_6^2)e^{(-Re_t/2)}\right]/2 \quad with \quad Re_t = v\ell/\nu$$
$$\beta_3 = \beta_7\left[1-(1-\beta_6^2)e^{(-Re_t/4)}\right]/c_t \quad where \quad \beta_7 = 10/9$$

The most popular of the first-order models are the so-called "eddy-viscosity" models, based on the Newtonian assumption of linearity between the stress and rate-of-strain tensors. Similarly, the transport of heat owing to the time-averaged product of fluctuating enthalpy and fluctuating velocity is modeled by an eddy-conductivity model assuming the turbulent heat flux follows a law similar to Fourier's law. Further, it is generally assumed that the eddy viscosity and eddy conductivity have the same functional relationship with temperature. Although the turbulent Prandtl number varies across the boundary layer, it is commonly considered to be a constant. For example, these eddy models remove the Reynolds stress and turbulent heat flux terms from Eq. (14) and approximates their effect by adding some empirically based function, μ_t, to the molecular viscosity, μ. This leads to the alternative approximate form of Eq. (14),

$$\begin{aligned} \sigma_{ij} &= (\mu + \mu_t) \sum_{k=1}^{d} \left(\frac{\partial \xi_k}{\partial x_j} \frac{\partial \overline{u}_i}{\partial \xi_k} + \frac{\partial \xi_k}{\partial x_i} \frac{\partial \overline{u}_j}{\partial \xi_k} - \frac{2}{3} \sum_{l=1}^{d} \frac{\partial \xi_l}{\partial x_k} \frac{\partial \overline{u}_k}{\partial \xi_l} \delta_{ij} \right) \\ q_i &= -\frac{\gamma}{Pr_l} \left(\mu + \frac{Pr_l}{Pr_t} \mu_t \right) \sum_{j=1}^{d} \frac{\partial \xi_j}{\partial x_i} \frac{\partial e_I}{\partial x_j} \end{aligned} \tag{25}$$

where Pr_t and μ_t represent the turbulent Prandtl number and eddy viscosity, respectively.

Two-equation models like the above one are the simplest models, for which the length scale is not prescribed empirically, and they are, therefore, likely to be less restrictive than the zero- or one-equation models. Examples in which they are useful are flows with mutiple-shock waves, problems with more than a single surface, recirculating flows, and flows with multiple length scales. However, these models are still weak because the model constants are not universal, and they are found to be applicable to a limited class of problems. Perhaps, their range of applicability can be extended as explained below.

In these examples of eddy-viscosity type models, the eddy viscosity is assumed to be isotropic. Such models cannot account for any directional influences of turbulence. However, some generalizations to the two-equations models have been developed which do attempt to account for this influence (see [81] and [82]). These models are economical compared with the second-order, Reynolds stress models, and those based on algebraic stress relations [82] do maintain some of the universality of the latter.

Finally, it should be noted that there is literature on the possibility of predicting transition with eddy-viscosity methods in the Reynolds-averaging approach (see [3], [55], and [58]).

4. COMPUTATIONAL ASPECTS OF VISCOUS FLOWS

This subject is much too broad to cover here. At the present time it is an area of numerical research still very much in a state of development, having many aspects that are controversial, or at best, poorly understood. It covers physical ranges, such as compressible and incompressible, low and high Reynolds numbers, and steady and unsteady flows, and analytical ranges, such as the full Navier-Stokes, the composite, and the boundary-layer equations. For maximum efficiency all of these would be treated by different numerical techniques and different codes that would take full advantage of the individual properties involved. A wide variety of basic techniques have been used, for example, finite difference, finite volume, finite element, and spectral. Each has to be integrated into a mesh and combined with a turbulence model where necessary. In general the computation of viscous phenomena, with any useful degree of practical resolution, leads to systems of equations that for efficiency, have to be solved using some form of an implicit numerical method; this tends to complicate the structure of the codes.

The numerical methods for transition simulations, direct turbulence simulations, and large-eddy simulations must resolve the unsteady, fluctuating nature of the flow. At present these studies are generally based on the assumption that the flow is periodic in at least two of the space directions, and spectral methods are used accordingly. The execution times required to produce an event that has the appropriate statistical properties is prohibitive from an industry-use point of view. For example, one of the original LES simulations carried out on the ILLIAC IV computer required about 92 hr of total machine time with a $63 \times 64 \times 128$ mesh [66]. At present, because of computer limitations and expense, usefulness of the above simulations is limited to basic research into the structure of turbulence at the major research centers. Even in basic research, they must be judiciously used. The codes for these simulations are very carefully constructed, according to both numerical and physical issues. Whenever possible their results are compared extensively to those from experiments.

The computations that are probably of major interest to the aircraft industry at the present writing would make use of the Reynolds-averaged equations with the thin-layer approximations to compute flows with at least mild forms of separation caused by shocks or trailing edges. Many codes that will do this have been constructed, but they are mostly in the research stage, and to get meaningful information from them the user must be aware of their dependence on mesh system and turbulence-modeling assumptions. This is not to say that they should be avoided. As some of the examples show, when properly applied, they can give very useful results. At the present time, the algorithms being used for these computations are slow to converge. For example, the fastest known, compressible Navier-Stokes code is claimed to be running at about 3.5 min per case (160×50 mesh) on a CRAY-1S for two-dimensional simulations of flow past airfoils [83]. Another code takes about 65 min per case (40

$\times$ 30 $\times$ 30 mesh) for three-dimensional simulations with supersonic free streams [84].

5. SOME EXAMPLES OF FLOW SIMULATION ATTEMPTS

In 1933, Thom [85] made the first successful attempt at a numerical solution of the Navier-Stokes equations for an incompressible fluid. In 1953, Kawaguti [86] reported an effort of about 1.5 years working 20 hr every week to numerically integrate these equations. Both of them solved for flow past a circular cylinder. In 1963, Fromm and Harlow [87] presented a study of a vortex street development behind a plate in channel. In 1970, MacCormack [88] computed the interaction of a shock wave with a laminar boundary layer. There are numerous computational milestones like these regarding viscous flows, most of them occurring in the recent past. Some examples demonstrating the present computational capabilities for viscous flows are presented in this introductory article. These examples are primarily taken from computations done at NASA Ames Research Center. Other examples may be found in other articles of this volume and also, for example, in [57] and [89]-[92].

5.1 Simulation of the Onset of Transition

Laminar flows can be solved using numerical methods on the Navier-Stokes equations without approximations other than those resulting from discretization. Such flows extend into the initial stages of the laminar to turbulent transition process. Many of the conceptual discrepancies concerning this process can be resolved, and a proper conceptual picture can be formed by numerical simulations. The main provisions are that (1) the discretization errors do not contaminate physical phenomena, and (2) the artificial boundaries imposed by the limited size of the computational domain are adequate representations of the true physical boundaries. These demanding requirements limit numerical solutions to the early stages of the transition process. Several studies have been carried out (see for instance, [93]-[96]).

We present here the results of Wray and Hussaini [97] who studied the instability of a flat-plate boundary layer in a manner analogous to the vibrating-ribbon experiment of Kovasznay et al. [98]. The initial conditions are nearly the same for both the physical and numerical experiments. The Reynolds number, based on the displacement thickness, is about 1100, and the streamwise and spanwise wavelengths of the perturbation waves to the Blassius profile are about $25.2\delta^*$ and $30\delta^*$, respectively. The ratio of the amplitudes of the two- and three-dimensional perturbations and the total rms amplitude are within the experimental range. The computation differs from the experiment in two aspects: the flow is assumed to be periodic in the streamwise and spanwise directions, and the evolution of perturbed flow occurs in time rather than in space. Discrepancies brought about by these effects are considered not to be large over the time frame of interest [97].

Figure 1 compares computed and experimental contours of $\frac{\partial u}{\partial y}$ at the so-called one spike-stage. The computational contours are taken at $t = 340\delta^*/U_\infty$, the time at which these contours matched the best with those in the experiment. The contours are shown in the vertical

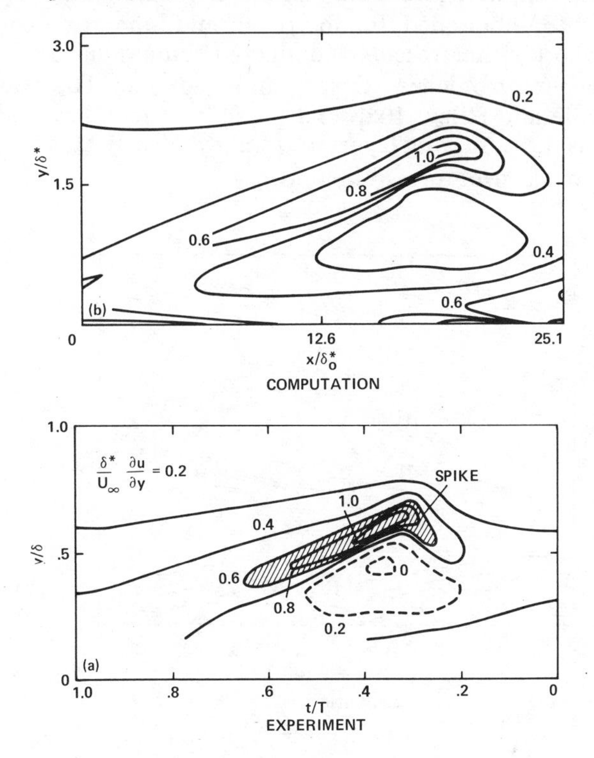

Figure 1. Contours of $\partial u/\partial y$ at the one-spike stage [97].

streamwise plane that contains the maximum initial perturbation. A single spike of high-intensity shear suggests the existence of a distinct shear layer in the outer part of the boundary layer. Figure 2 compares experimental and computed contours of maximum values of $\frac{\partial u}{\partial y}$ for $y \geq 0.35\delta^*$ at each x amd z location. Two streamwise and spanwise wavelengths are shown. Figure 3 shows the so-called two-spike stage in the vertical plane containing the maximum initial perturbation. Wray and Hussaini show contours that indicate the existence of three spikes at a later time. However, Fig. 3 also shows that in some parts of the computational domain the resolution of flow structure is already poor at this stage and the solution is becoming contaminated with numerical error.

5.2 Direct Turbulence Simulations

As in the case of the transition simulations, these simulations represent an attempt to resolve all scales of motion accurately both in time and space. This means that the only errors made are numerical ones, and there is no turbulence modeling. The principal problem is to encompass a broad enough range of scale to make the simulated event meaningful. Computer resources limit such simulations to low Reynolds numbers and to unbounded, homogeneous and inhomogeneous turbulent flows. Rogallo [99] has recently conducted some numerical experiments in homogeneous turbulence. Orszag and Pao [100], Riley and Metcalfe [101], Metcalfe and Riley [102], and possibly others, have dealt with inhomogeneous flows, mixing layers and wakes. Nearly all simulations have been done for an incompressible fluid.

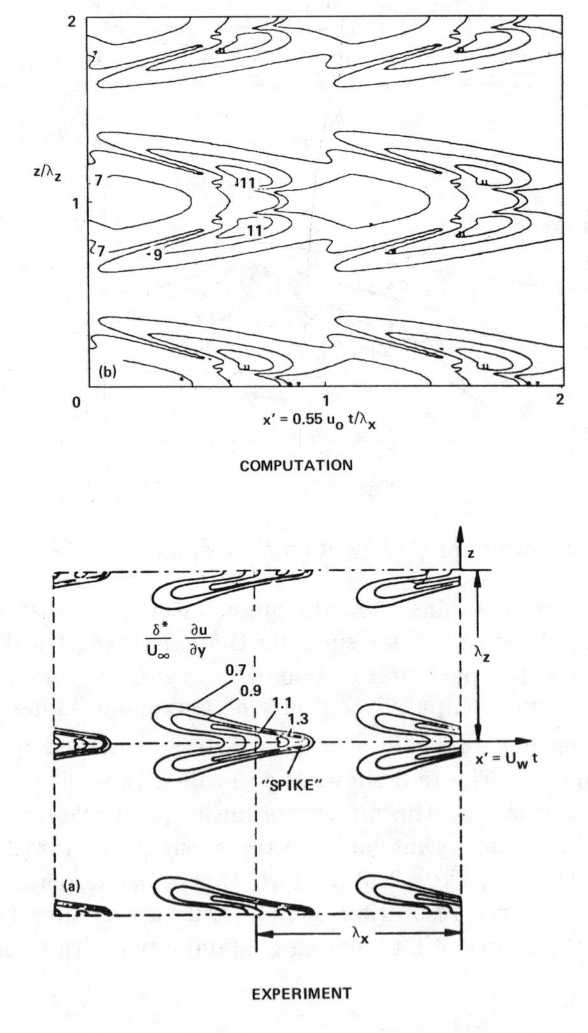

Figure 2. Contours of maximum $\partial u/\partial y$ at the one-spike stage [97].

We present here the results of some numerical experiments due to

Rogallo for a homogeneous flow. In a homogeneous turbulent flow statistical quantities are independent of position in space, the mean velocity is constant, and there is no net convection or diffusion transport. The computations were carried out on a 128 × 128 × 128 grid for periodic boundary conditions using spectral methods in all directions. This grid allows a resolution of $R_\lambda \leq 61.97$ in each direction, where R_λ is the Reynolds number based on the Taylor microscale. Computed results of the trace of the Reynolds stress tensor at $t = 10$ and $R_\lambda = 76$ are compared in Fig. 4 with the experimental results of Tavoularis and Corrsin for $t \approx 12.6$ and $R_\lambda = 266$ [103]. Although the computed results at $t = 10$ are marginally resolved, the agreement between some statistical quantities is good. For a more detailed evaluation, the reader is referred to the paper.

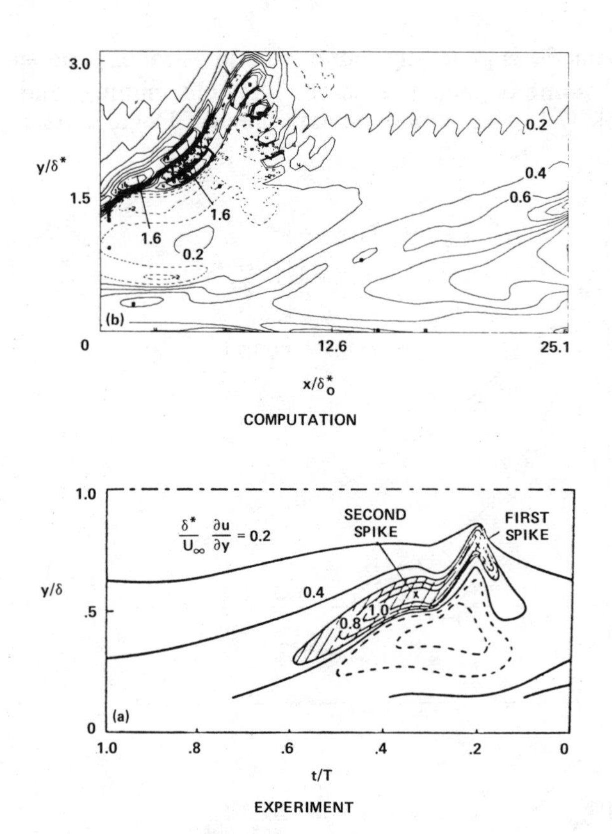

Figure 3. Contours of $\partial u / \partial y$ at the two-spike stage [97].

Direct simulations have the potential of providing new insight into the nature of turbulence, developing and verifying turbulence models, and determining the model constants, provided these simulations are validated by available physical experimental information. As an example, Rotta [104] suggested modeling the tensor sum of the pressure-strain correlation, which is responsible for the return of anisotropic turbulence to isotropy, and the deviator of the dissipation tensor in terms of the mean flow quantities by the equation

$$\phi_{ij} = \beta_R b_{ij} \tag{26}$$

where

$$b_{ij} = \frac{< u'_i u'_j >}{< u'_k u'_k >} - \frac{1}{3}\delta_{ij}$$

$$\epsilon_{ii}\phi_{ij} = -2 < p' S'_{ij} > +2\left(\epsilon_{ij} - \frac{1}{3}\epsilon_{ii}\delta_{ij}\right)$$

$$\epsilon_{ij} = \nu < \frac{\partial u'_i}{\partial x_k}\frac{\partial u'_j}{\partial x_k} >$$

The constant β_R is generally found from physical experiments to be about 3.25 with some dependence on the Reynolds number and level of anisotropy [105]. Using his numerical experiments, Rogallo attempted to

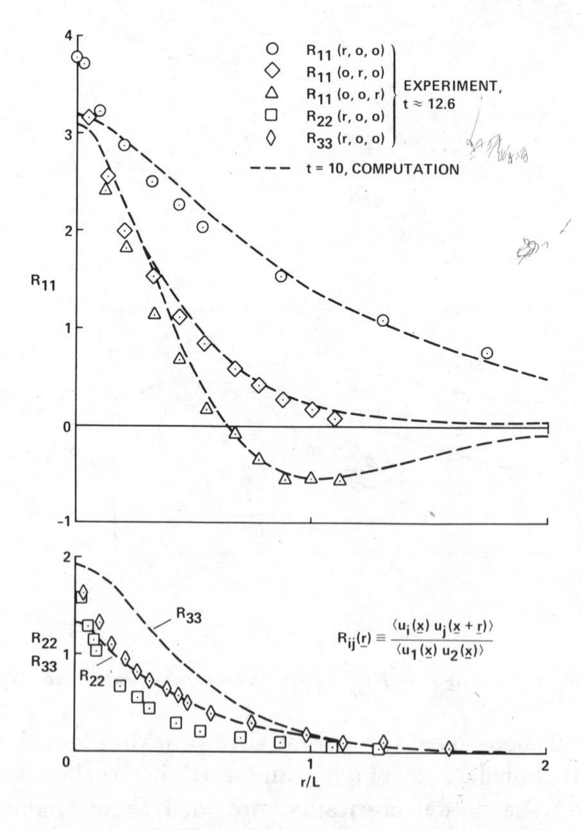

Figure 4. Self-similarity of the autocorrelations in homogeneous shear turbulence [99].

correlate ϕ_{ij} and b_{ij} for the relaxation toward isotropy of turbulence subjected to axisymmetric strain, plane-strain, uniform shear, and uniform rotation. In case of axisymmetric strain and plain shear, these tensors do correlate, but they do not in the other two cases. Figure 5 shows results for the shear flow. The data points consist of four flows with different values of uniform shear, each case having reached a different stage of evolution in time.

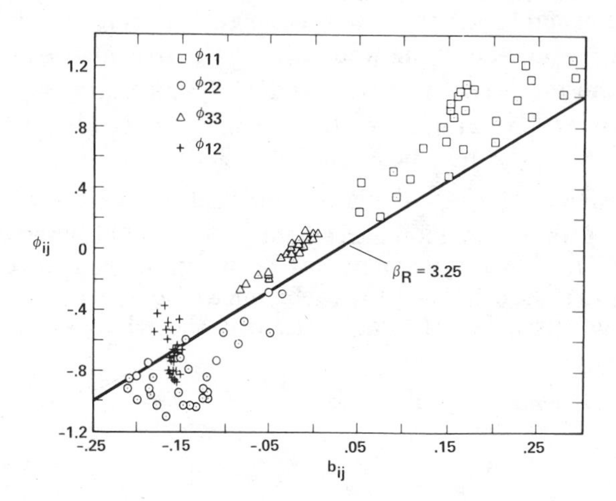

Figure 5. Correlation of ϕ_{ij} and b_{ij} for homogeneous shear.

5.3 Large-Eddy Simulations

Fully developed channel flow has been extensively studied numerically using the so-called large-eddy simulation approach see [66], [67], and [106]-[108]. Early investigations ([67], [106] and [107]), because of computer resources, could not compute the viscous sublayer. Deardorff [106] used the Smagorinsky's model [68] for the entire flow field. Schumann [67] applied this model to homogeneous subgrid stresses, but treated inhomogeneities in the near-wall region differently. The wall-function approximations were used in both of these studies to place boundary conditions at the beginning of the log-law region. The results compare well with experimental data in the region outside of the viscous sublayer. Considerable statistical information about the structure of turbulence was extracted from these initial investigations. Later studies ([66] and [108]) extended the computational domain all the way to the channel walls. At present, the LES approach has been limited to fully developed turbulent flows, to incompressible fluids, and to periodic boundary conditions in the streamwise and spanwise directions.

Present capabilities and limitations of the LES approach can be judged from a recent study presented in [66]. The study appertained to a channel flow at a Reynolds number, based on centerline velocity and on

h, the channel half-width, of about 13,800. In this study, two different space-averaging operators were applied to the convective derivatives in Eq. (7). One averaging operator was applied to the velocity product appearing in the streamwise and spanwise derivatives, and another one for this product appearing in the derivative normal to the wall. The computational domain was bounded by $2h \times 2\pi h \times \pi h$ in directions normal to the wall, along the stream, and along the span, respectively. The number of grid points in these directions was $63 \times 64 \times 128$, respectively. In the direction normal to the wall, a variable grid-spacing was used with minimum spacing being about one wall unit. In the streamwise and spanwise directions the mesh size was 62.8 and 15.7 wall units, repectively. Consequently, the grid-spacing is relatively coarse for resolving the flow structures observed in the wall layer (see Sec. 2.3(b)).

As seen in Fig. 6 the calculated horizontal-, space- and time-averaged velocity profile is in good agreement with the experimental time-averaged data [109]. This agreement has been reproduced for different computational grid systems, using the turbulence model given in Eq. (18), which uses characteristic length scales based on the grid system itself.

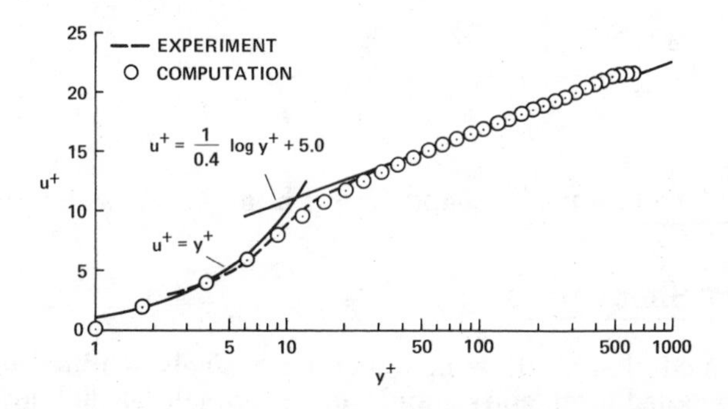

Figure 6. Horizontal-, space- and time-averaged velocity profile compared with time-averaged velocity profile [66].

One of the physical experiments that is numerically reproduced, is the flow-field visualization made possible by tracing passive markers introduced along vertical and horizontal wires. Figures 7 and 8 show computed and experimental results where the experimental particles were formed by a pulsed injection of hydrogen bubbles along the wires. The experimental flow visualizations were for Re of $O(10^5)$, and the horizontal wires were at $y^+ \approx 6$ in both cases. The computed visualization is so realistic that it is difficult to distinguish it from the experimental one. Inflectional velocity profiles are clearly visible, the formation and breakdown of a streamwise vortex is discernible, and the low- and high-speed, wall-layer streaks are clearly evident.

It is extremely difficult to measure the fluctuating pressure with any accuracy in a turbulent flow, and so far no measurements, except at the wall, have been made. This information is useful for turbulence modeling, since the average of the products of the fluctuating pressure and fluctuating velocities and their derivatives appear in the equations for the Reynolds stresses. The pressure-strain correlations, which govern the exchange of energy between components of resolvable kinetic energy, are shown in Fig. 9. A negative value of this correlation indicates transfer of energy to other components, whereas a positive value is a gain. Next to the wall, downward moving fluid collides with the wall, producing high local pressure points; and energy from the vertical component of turbulence intensity is transferred to the horizontal components.

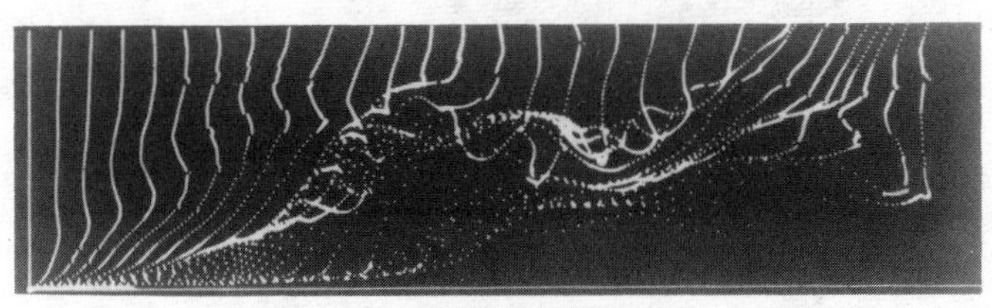

COMPUTATION

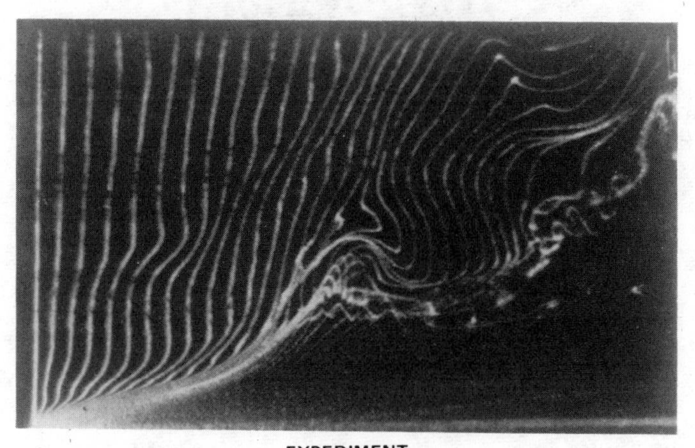

EXPERIMENT

Figure 7. Structure of turbulent boundary layer visualized by passive markers introduced along a vertical wire - computation [110] and experiment (Courtesy of S. J. Kline).

Typically, numerical experiments using LES techniques require considerable computer resources (see Sec. 4). However, the algorithms being used are the most efficient ones known and the flow problems are not, in the classical sense, stiff. The only known way to reduce the execution time is to provide more powerful computers.

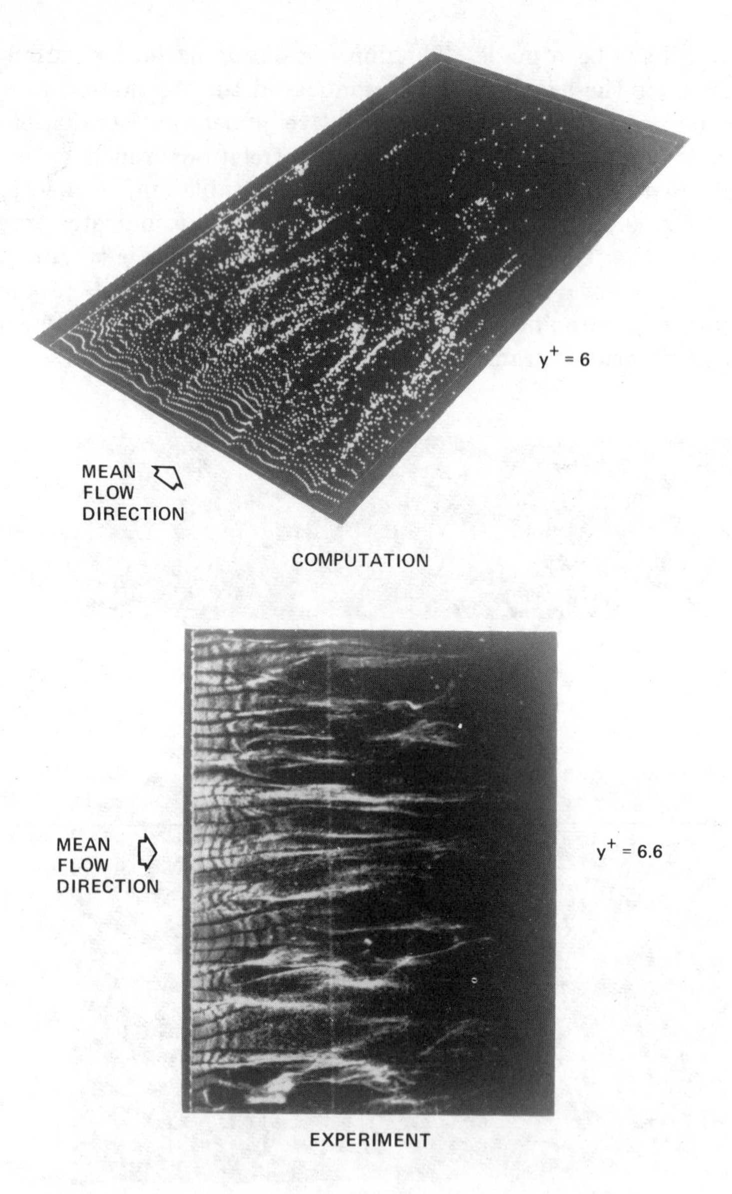

Figure 8. Structure of turbulent boundary layer visualized by passive markers introduced along a horizontal wire - computation [110] and experiment (Courtesy of S. J. Kline).

5.4 Simulations by Vortex Tracing

Some flows can be represented by computing the interaction between viscous and inviscid models. In vortex-tracing techniques, the vorticity is treated as a dependent variable identified by the location of discrete point vortices, vortex blobs (vortices with finite core), vortex filaments, or vortex sheets. These interact with one another and with the surface from which they were generated. The viscous model accounts for the initiation of the vorticity and its diffusion within the boundary layer. For moderate-to-high aerodynamic Reynolds numbers, its diffusion outside the boundary layer can be neglected, and in two dimensions no additional vorticity can be created. Discussions of some aspects of these modeling and interaction procedures, and computed examples may be found, for instance, in [111]-[117]. Simulations by vortex tracing have been limited to incompressible fluids.

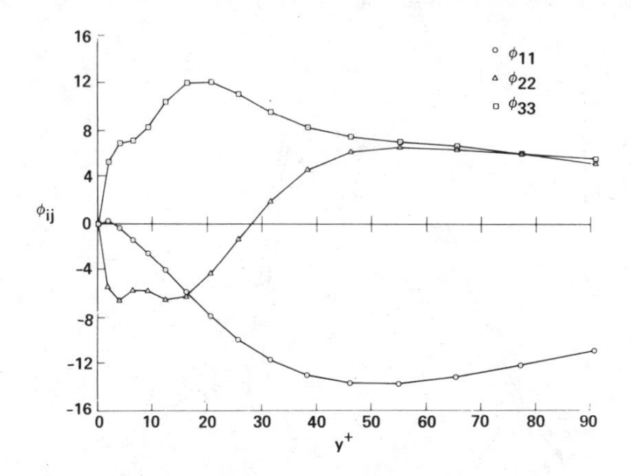

Figure 9. Profiles of the diagonal elements of the resolvable pressure-strain correlation tensor [66].

Here we present some recent results found by Spalart et al. [118] who coupled an inviscid outer flow computation having vortex blobs with a viscous boundary-layer model. The boundary-layer model was made to approximate the transition from laminar to turbulent flow. Both integral and differential boundary-layer methods were used. The inviscid and viscous regions are computed by two entirely different approaches, so their interaction requires the matching of velocity fields and a scheme for transferring vorticity across the boundary-layer edge, into the vortex blobs.

In the vortex-tracing approach, there are a number of parameters that must be tuned by trial comparisons with experiments, as is the case in the development of any empirical model. Spalart et al. were able to design a procedure that gives very good qualitative results. Some idea of its quantitative accuracy can be judged by the following examples.

Figure 10 shows normal force and moment coefficients acting on an NACA 0012 airfoil, oscillating about an axis at its quarter chord with its incidence varying from 5° to 25° at a reduced frequency of 0.25. The Reynolds number is 2.5×10^6. The computations made use of an integral method for the viscous region and were carried out before the experimental data were available. The experimental data [119] are phased-averaged over a large number of cycles. Figure 11 shows drag coefficient and Strouhal number for flow past a circular cylinder with the Reynolds number varying through the subcritical, critical, and postcritical regimes. In this case, the viscous region is solved with a differential approach, using the Baldwin-Lomax turbulence model [73] with a transition approximation. The computed drag coefficient is in a better agreement with the experimental values than that computed by Chorin [121] who also used a vortex-tracing method but without a boundary-layer code to initiate the vorticity.

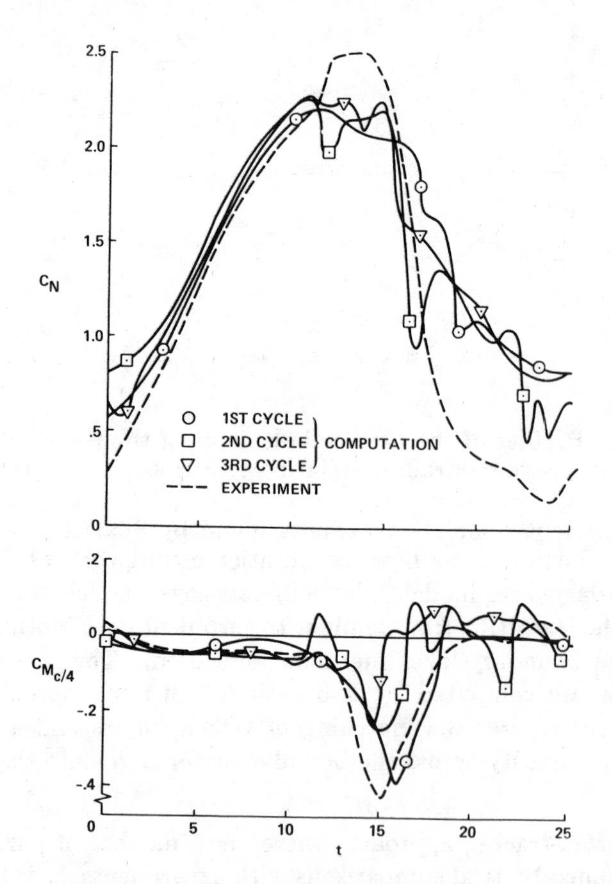

Figure 10. Normal force and moment coefficients during dynamic stall [118].

5.5 Reynolds-Averaged Simulations of Strong Interactions

In this section we present one example based on the parabolized Navier-Stokes equations and another found from the steady-state solution of some composite equations representing a complex three-dimensional flow. In both cases a zero-equation turbulence model was used and in both cases the flow had regions of separation.

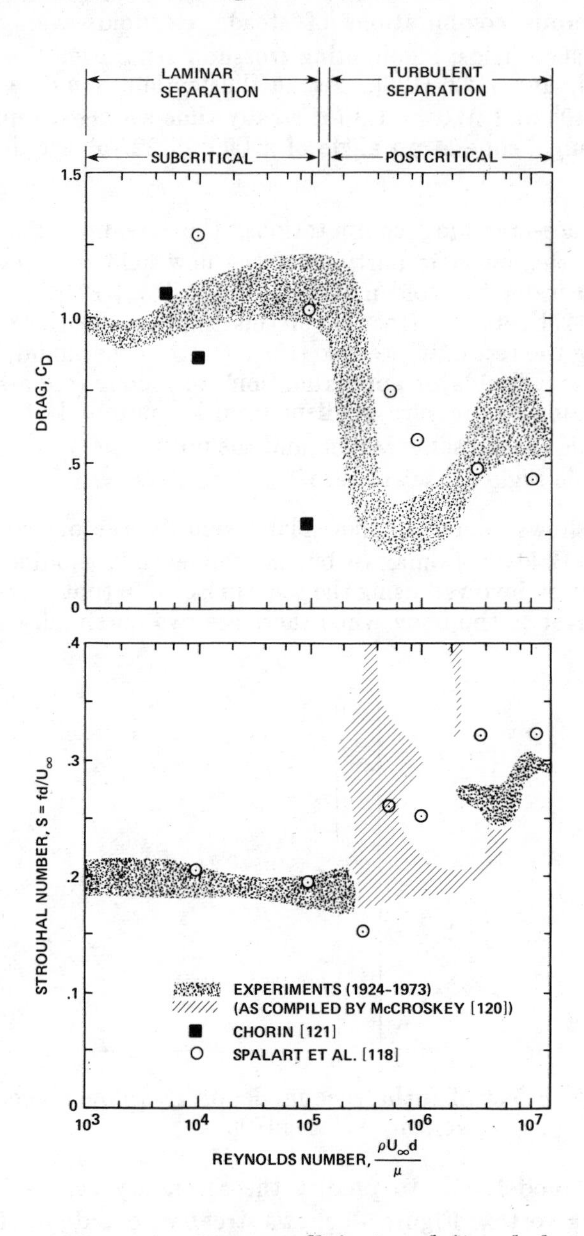

Figure 11. Drag coefficient and Strouhal number for rigid circular cylinder.

The parabolized Navier-Stokes approach is usually applied to computations of high-incidence, supersonic and hypersonic flows around bodies such as military aircraft, missiles, and the Space Shuttle. The flow fields under these conditions typically contain three-dimensional, unsteady, separated regions, accompanied by crossflow separation. There have been numerous computations of steady Reynolds-averaged approximations to such fields, highlighting crossflow separation, see for instance, [79], [92], and [122]-[126]. As an illustration, some results at $Re_x = 21.25 \times 10^6$ and $M_\infty = 1.8$ for steady time-averaged separation on a 12.5° half-angle cone at an angle of attack of 22.75° are discussed next.

In these space-marching computations, the streamwise pressure-gradient term in the subsonic portions of the flow field requires a special treatment in order to avoid numerical instability ([127] and [128]). Rakich et al. [125] retain a fraction of this term in the the subsonic region, neglecting the rest of it, see also [128]. On the other hand, Degani and Schiff [79] use a "sublayer approximation" technique (see also [92]), wherein the streamwise pressure gradient term is computed at a supersonic point outside the subsonic region, and the normal pressure gradient inside the subsonic region is set to zero.

Figure 12 shows some crossflow-plane velocity vectors computed using the original Baldwin-Lomax turbulence model and a modification of it. [The modification involved using the y_{max} in Eq. (20) that corresponds to the L_{max} nearest to the body when there are two maxima].

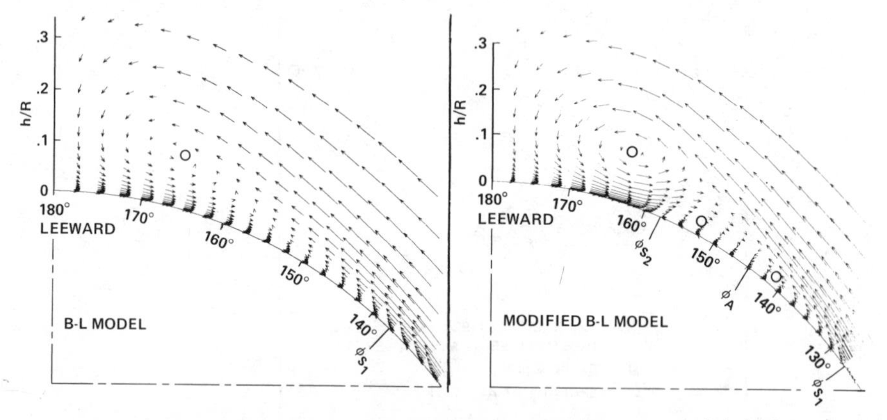

Figure 12. Effect of turbulence model on crossflow plane velocity vectors [79].

The unmodified model fails to predict the secondary separation and the corresponding vortex. Figure 13 shows streamwise and circumferential velocity components before and after primary separation. Solutions found by using the two models agree very well with the experimental data upstream of separation, but only the solutions produced using the modified model agree with the data on the downstream side. Figure 14 compares circumferential surface-pressure-coefficient distribution and surface flow direction from [79] and [125]. For this case, the Cebeci model and the modified Baldwin-Lomax model predict about the same results.

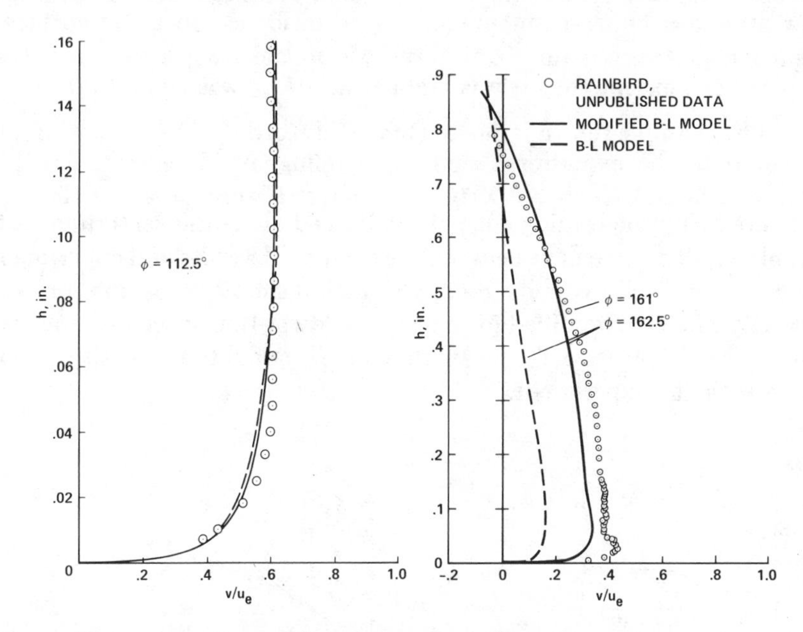

Figure 13. Comparison of computed and experimental circumferential velocity profile [79].

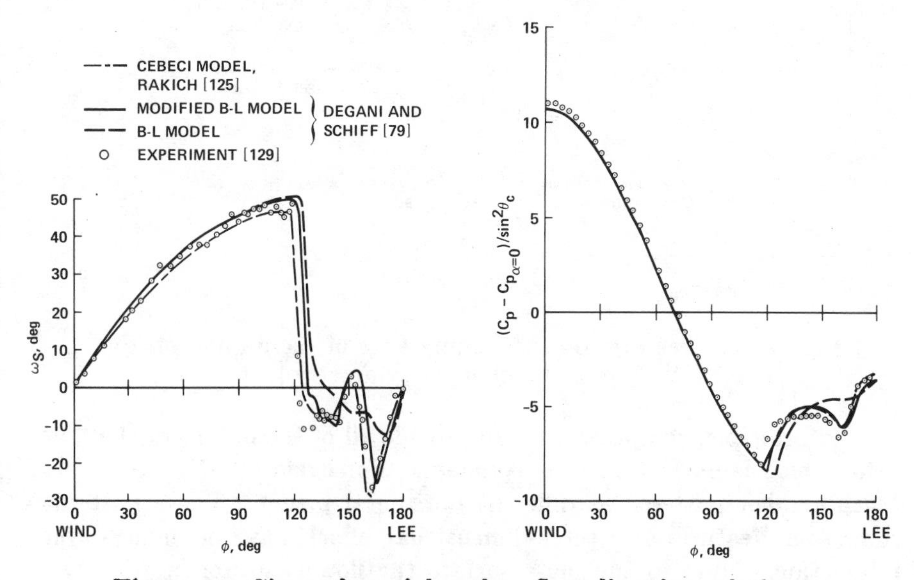

Figure 14. Circumferential surface flow direction relative to local cone generator and surface-pressure distribution.

An example of a composite approach based on the Reynolds-averaged Navier-Stokes equations excluding the viscous cross-derivative terms. is presented as a final example. It demonstrates the capability of computing a complex, three-dimensional horseshoe vortex (averaged) flow around a blunt fin mounted at zero incidence on a flat surface in a supersonic free stream. (Briley and McDonald [130] have computed a similar flow, except that it was laminar and M_∞ was equal to 0.2).

The example shown is due to Hung and Kordulla [84] who attempted to compute the experiment done by Dolling and Bogdonoff [131] for $M_\infty = 2.95$ and $Re = 6.3 \times 10^7$ per meter. Figure 15 shows the static pressure along the leading edge of the fin and along the centerline on the flat plate. The agreement between the computed results and experimental time-averaged data is quite good, although a simple zero-equation eddy-viscosity model, a modified form of the Baldwin-Lomax model, was used. This does not necessarily mean that other computed quantities would agree with the experiment.

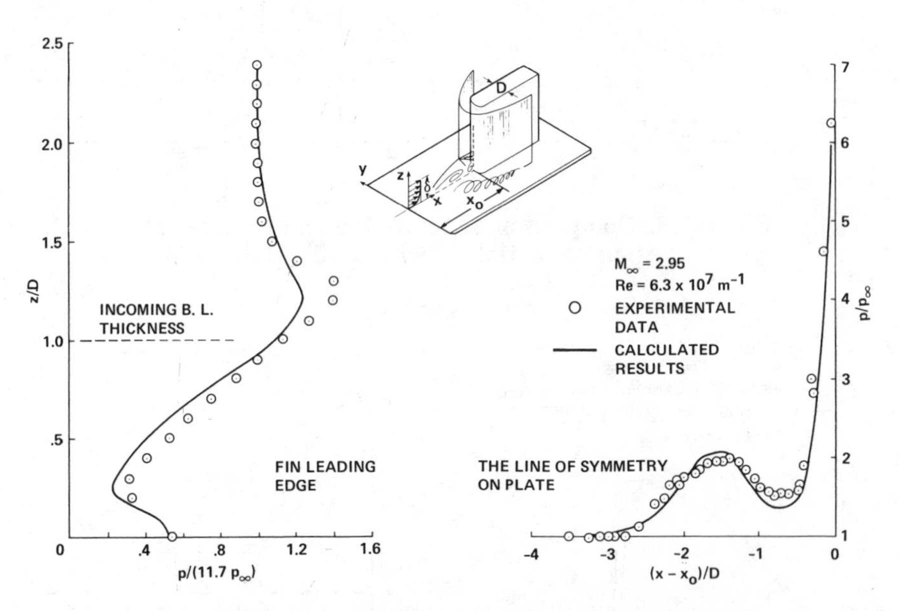

Figure 15. Pressures on the leading edge of the fin and on the flat plate in the plane of symmetry [84].

In principle, computations can provide all of the details of the flow field, which is never true in experiments. For example, Fig. 16 shows Mach-number patterns at different horizontal planes. These patterns show some features of a three-dimensional, shock wave boundary-layer interaction. Next to the plate surface the flow is subsonic, with two regions of nearly incompressible flow in front of the blunt fin. Between these regions, the flow field has high subsonic Mach numbers. As we move away from the surface, the larger of these two regions decreases in size and eventually disappears, while the high subsonic region develops

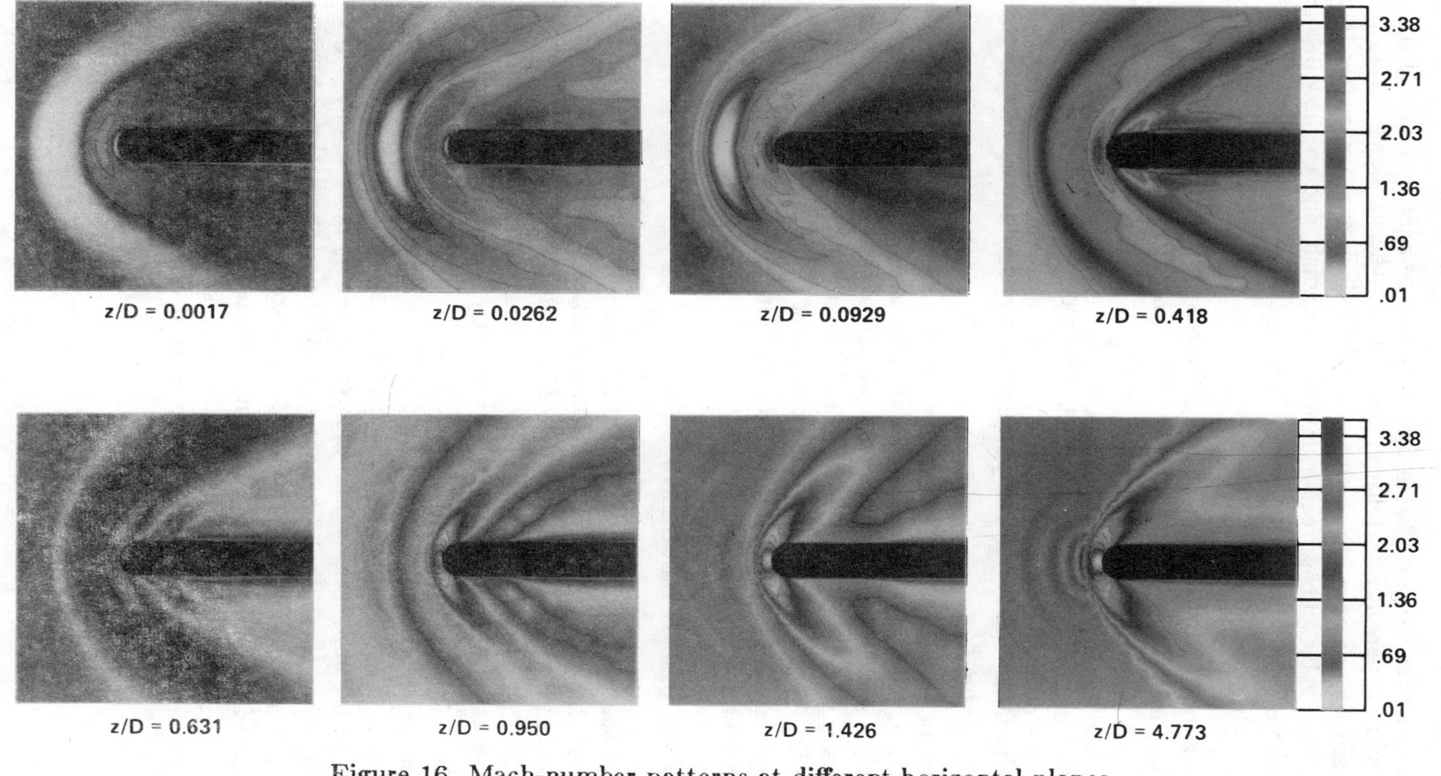

Figure 16. Mach-number patterns at different horizontal planes (Courtesy of C. M. Hung).

into a supersonic region. From $z/D = 0.418$ to $z/D = 1.426$, there are two highly supersonic regions on the sides of the fin downstream from the blunt leading edge. In these regions, the Mach number is close to the free-stream value. These regions eventually vanish, and the flow becomes two-dimensional with the upstream Mach number approaching the free-stream value.

Figure 17 shows Mach-number patterns and the separating particle path line in the plane of symmetry. The bow shock is discernible. The most striking feature in this figure is the existence of two separate supersonic zones that are embedded in a subsonic region, as experimentally observed in [132]. The particle path line [133] leaving the plate surface forms a spiral, showing the (statistical) existence of the horseshoe vortex. From the Mach-number pattern we see that a part of this vortex has supersonic flow.

In the experiment [131] this flow was observed to be unsteady, having frequencies that could be resolved by the computation. It is believed that this unsteadiness was suppressed from the computation because symmetry was imposed along the vertical plane through the leading edge of the fin. Nevertheless, it is remarkable that the averaged values of such a complicated flow can be computed with some accuracy in the relatively coarse grid of $40 \times 32 \times 32$.

6. CONCLUDING REMARKS

In the past, the effects of viscosity on nonboundary-layer types of fluid flow have been, for the most part, studied indirectly. For example, the separation bubble is often implied from the surface-pressure distribution. With the increased capability provided by modern computers, we can now begin to study these effects directly by actually computing the details of the flow field. Future computers will permit the studies to be even more direct, with primary emphasis on unsteady flows encountered in separated and vortical regions. Experiments will always be needed to verify both computational and theoretical aerodynamics, and to provide answers when these disciplines can not. However, the dominant theme in aerodynamics during the lifetime of a research worker born at the time of this writing will probably be computational aerodynamics. Although this is most likely to provide the most new knowledge in aerodynamics, the computational tasks required to carry it out will be very difficult.

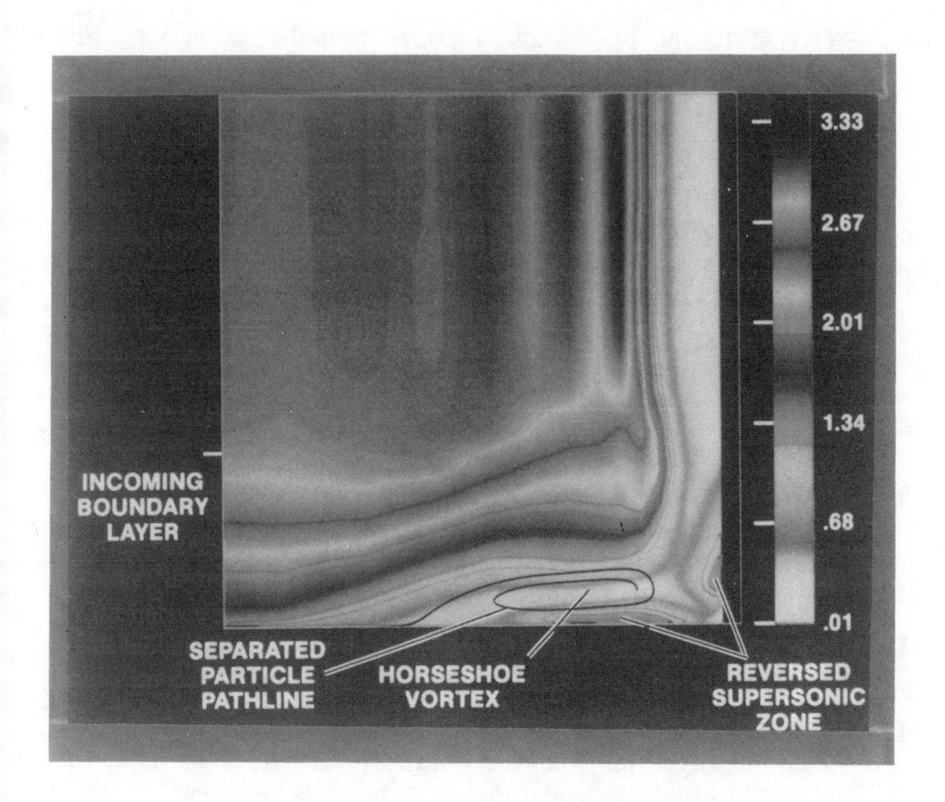

Figure 17. Mach-number patterns (Courtesy of C. M. Hung) and flow structure in the plane of symmetry [133].

REFERENCES

1. HINZE, J. O. - Turbulence, second edition, McGraw-Hill, Inc., 1975.

2. BRADSHAW, P. - Effects of Streamline Curvature on Turbulent Flow, AGARD-AG-169, Aug. 1973.

3. CEBECI, T. and BRADSHAW, P. - Momentum Transfer in, Boundary Layers, Hemisphere Publishing Corporation, 1977.

4. STEWARTSON, K. - Multistructured Boundary Layers on Flat Plates and Related Bodies, Adv. in Applied Mech., Vol. 14, 1974.

5. BURGGRAF, O. R., RIZZETTA, D., WERLE, M. J., and VASTA, V. N. - Effect of Reynolds Number on Laminar Separation of a Supersonic Stream, AIAA Journal, Vol. 17, pp. 336-343, 1979.

6. MELNIK, R. E. - Turbulent Interactions on Airfoils at Transonic Speeds - Recent Developments, Viscous-Inviscid Interactions, AGARD-CP-291, 1980.

7. COLES, D. E. - The Turbulent Boundary Layer in a Compressible Fluid, Rand Corp. R-403-PR, 1962. (Also AD 285-651.)

8. MONIN, A. S. and YAGLOM, A. M., Statistical Fluid Mechanics of Turbulence, The MIT Press, Vol. 1, pp. 205-209, 1971.

9. ACHARYA, M. and REYNOLDS, W. C. - Measurements and Prediction of a Fully Developed Turbulent Channel Flow with Imposed Controlled Oscillations, Rept. TF-8, Mechanical Engineering Dept., Stanford University, Stanford, Calif., 1975.

10. KARLSSON, S. K. F. - An Unsteady Turbulent Boundary Layer, Ph. D. Thesis, Johns Hopkins University, Baltimore, Md., 1958.

11. RAMAPRIAN, B. R. and TU, S.-W. - An Experimental Study of Oscillatory Pipe Flow at Transitional Reynolds Numbers, Journal of Fluid Mechanics, Vol. 100, pp. 513-544, 1980.

12. MIZUSHINA, T., MARUYAMA, T., and SHIOZAKI, Y. - Pulsating Turbulent Flow in a Tube, Journal of Chemical Engineering of Japan, Vol. 6, No. 6, 1973.

13. COUSTEIX, J., HOUDEVILLE, R. and RAYNAUD, M. - Oscillating Turbulent Boundary Layer with Strong Mean Pressure Gradient, Turbulent Shear Flows, Vol. 2, Eds. Durst, F., Launder, B. E., Schmidt, F. W., Bradbury, L. J. S., and Whitelaw, J. H., Springer-Verlag, 1980.

14. PARIKH, P. G., REYNOLDS, W. C., and JAYARAMAN, R. - On the Behavior of an Unsteady Turbulent Boundary Layer, Numerical and Physical Aspects of Aerodynamic Flows, Ed. Cebeci, T., 1982.

15. SIMPSON, R. L., CHEW, Y. T., AND SHIVAPRASAD, B. G. - Measurements of Unsteady Turbulent Boundary Layers with Pressure Gradients, SMU Rept. WT-6, Southern Methodist University, Dallas, Tex., Aug. 1980.

16. CARR, L. W. - A Review of Unsteady Turbulent Boundary-Layer Experiments, NASA TM-81297, June 1981.

17. CHAPMAN, D. R. - Computational Aerodynamics Development and Outlook, AIAA Journal, Vol. 17, pp. 1293-1313, 1979.

18. KELVIN, LORD - On Vortex Motion, Mathematics and Physics Papers, Vol. 4, p. 49, 1869.

19. LIGHTHILL, M. J. - Introduction - Boundary Layer Theory, Laminar Boundary Layers, Ed. Rosenhead, L., Oxford University Press, 1963.

20. BATCHELOR, G. K. - An Introduction to Fluid Dynamics, Cambridge University Press, 1967.

21. EMMONS, H. W. - The Laminar-Turbulent Transition in a Boundary Layer, Part I, Journal of the Aeronautical Sciences, Vol. 18, pp. 490-498, 1951.

22. COLES, D. E. and BARKER, S. J. - Some Remarks on a Synthetic Turbulent Boundary Layer, Turbulent Mixing in Nonreactive and Reactive Flows, Ed. Murthy, S. N. B., pp. 285-292, 1975.

23. WYGNANSKI I., SOKOLOV, N., and FREIDMAN, D. - On the Turbulent "Spot" in a Boundary Layer Undergoing Transition, Journal of Fluid Mechanics, Vol. 78, pp. 785-819, 1976.

24. WYGNANSKI, I. J. and CHAMPAGNE, F. H. - On Transition in a Pipe, Part I, The Origin of Puffs and Slugs and the Flow in a Turbulent Slug, J. Fluid Mech., Vol. 59, pp. 281-335, 1973.

25. TOWNSEND, A. A. - The Structure of Turbulent Shear Flow, first edition, Cambridge University Press, 1956.

26. FAVRE, A. J., GAVIGLIO, J. J., and DUMAS, R. - Space-Time Double Correlation and Spectra in a Turbulent Boundary Layer, Journal of Fluid Mechanics, Vol. 2, pp. 313-341, 1957.

27. LUMLEY, J. L. - Coherent Structures in Turbulence, Transition and Turbulence, Ed. Meyer, R. E., Academic Press, 1981.

28. FALCO, R. E. - New Results, a Review and Synthesis of the Mechanism of Turbulence Production in Boundary Layers and its Modification, AIAA Preprint No. 83-0377, Presented at 21st Aerospace Sciences Meeting, Reno, Nev., Jan. 1983.

29. CANTWELL, B. J. - Organized Motion in Turbulent Flow, Annual Review of Fluid Mechanics, Vol. 13, pp. 457-515, 1981.

30. Coherent Structure of Turbulent Boundary Layers, Eds. Smith, C. R., and Abbott, D. E., An AFOSR/Lehigh University Workshop, Nov. 1978.

31. WILLMARTH, W. W. and BOGAR, T. J. - Survey and New Measurements of Turbulent Structure Near the Wall, The Physics of Fluid, Vol. 20, No. 10, Pt. II, pp. S9-S21, 1977.

32. WILLMARTH, W. W. - Structure of Turbulence in Boundary Layers, Adv. in Applied Mech., Vol. 15, pp. 159-254, 1975.

33. BAKEWELL, H. P. and LUMLEY, J. L. - Viscous Sublayer and Adjacent Region in Turbulent Pipe Flow, Physics of Fluids, Vol. 10, pp. 1880-1889, 1967.

34. BLACKWELDER, R. F. - The Bursting Process in Turbulent Boundary Layers, Coherent Structure of Turbulent Boundary Layers, Eds. Smith, C. R., and Abbott, D. E., An AFOSR/Lehigh University Workshop, Nov. 1978.

35. PRATURI, A. K. and BRODKEY, R. E. - A Stereoscopic Visual Study of Coherent Structures in Turbulent Shear Flow, Journal of Fluid Mechanics, Vol. 89, pp. 251-272, 1978.

36. BLACKWELDER, R. F. and ECKELMANN, H. - Streamwise Vortices Associated with the Bursting Phenomenon, Journal of Fluid Mechanics, Vol. 94, pp. 577-594, 1979.

37. GUPTA, A. K., LAUFER, J., and KAPLAN, R. E. - Spatial Structure in the Viscous Sublayer, Journal of Fluid Mechanics, Vol. 50, pp. 493-512, 1971.

38. KLINE, S. J., REYNOLDS, W. C., SCHRAUB, F. A., and RUNSTADLER, P. W. - The Structure of Turbulent Boundary Layers, Journal of Fluid Mechanics, Vol.30, pp. 741-773, 1967.

39. RAO, K. N., NARASIMHA, R., and NARAYANAN, M. A. B. - Bursting in a Turbulent Boundary Layer, Journal of Fluid Mechanics, Vol. 48, pp. 339-352, 1971.

40. BLACKWELDER, R. F. and HARITONIDIS, J. H. - Scaling of the Bursting Frequency in Turbulent Boundary Layers, Journal of Fluid Mechanics, Vol. 132, pp. 87-103, 1983.

41. WILLMARTH, W. W. and LU, S. S. - Structure of the Reynolds Stress Near the Wall, J. Fluid Mech., Vol. 55, pp. 65-69, 1972.

42. FALCO, R. E. - Coherent Motions in the Outer Region of Turbulent Boundary Layers, Physics of Fluids, Vol. 20, No. 10, pp. 5124-5132, 1977.

43. CORRSIN, S. and KISTLER, A. - The Free Stream Boundaries of Turbulent Flows, NACA TN-3133, 1954. (Also NACA Technical Report 1244, 1955.)

44. DWYER, H. A. - Some Aspects of Three-Dimensional Laminar Boundary Layers, Ann. Rev. Fluid Mech., Vol. 13, pp. 217-230, 1981.

45. TELIONIS, D. P. - Unsteady Viscous Flows, Springer-Verlag, 1981.

46. BUSHNELL, D. M., CARY, A. M., Jr., and HARRIS, J. E. - Calculation Methods for Compressible Turbulent Boundary Layers - 1976, NASA SP-422, 1977.

47. BLOTTNER, F. G. - Computational Techniques for Boundary Layers, AGARD Lecture Series 73, 1975.

48. KELLER, H. B. - Numerical Methods in Boundary-Layer Theory, Annual Review of Fluid Mechanics, Vol. 10, pp. 417-433, 1978.

49. EICHELBRENNER, E. A. - Three-Dimensional Boundary Layers, Annual Review of Fluid Mechanics, Vol. 5, pp. 339-360, 1973.

50. NICKEL, K. - Prandtl's Boundary-Layer Theory from the Viewpoint of a Mathematician, Annual Review of Fluid Mechanics, Vol. 5, pp. 405-428, 1973.

51. VAN DREIST, E. R. - Turbulent Boundary Layers in Compressible Fluids, J. Aero. Sci., Vol. 18, pp. 145-160, 1951.

52. HESSELBERG, Th. - Die Gesetze der Ausgeglichene Atmosharischen Bewegungen, Beitrage Physik Freien Atmosphare Vol. 12, pp. 141-160, 1926.

53. FAVRE, A. - Equations des Gaz Turbulents Compressibles, Journal de Mecanique, Vol. 4, pp. 361-390, 1965.

54. RUBESIN, M. W. and ROSE, W. C. - The Turbulent Mean-Flow, Reynolds-Stress, and Heat-Flux Equations in Mass-Averaged Dependent Variables, NASA TMX-62,248, March 1973.

55. CEBECI, T. and SMITH, A. M. O. - Analysis of Turbulent Boundary Layers, Academic Press, Inc., 1974.

56. VAN DYKE, M. - Perturbation Methods in Fluid Mechanics, annotated edition, The Parabolic Press, 1975.

57. The 1980-81 AFOSR-HTTM-Stanford Conference on Complex Turbulent Flows : Comparison of Computation and Experiment, Vol. 2, Taxonomies, Reporters' Summaries, Evaluation and Conclusions, Eds. Kline, S. J., Cantwell, B. J., and Lilley, G. M., Thermoscience Division, Stanford University, Stanford, Calif., 1982.

58. BRADSHAW, P., CEBECI, T. and WHITELAW, J. H.

- Engineering Calculation Methods for Turbulent Flow, Academic Press, 1981.

59. RODI, W. - Turbulence Models and Their Application in Hydraulics, International Association for Hydraulic Research, Deft, The Netherlands, 1980.

60. LAUNDER, B. E. - Turbulence Transport Models for Numerical Computation of Complex Turbulent Flows, von Karman Institute for Fluid Dynamics, Lecture Series 1980-3, 1980.

61. LUMLEY, J. L. - Computational Modeling of Turbulent Flows, Advances in Applied Mechanics, Vol. 18, pp. 123-176, 1978.

62. RUBESIN M. W. - Numerical Turbulence Modelling, AGARD LS-86, 1977.

63. REYNOLDS, W. C. and CEBECI, T. - Calculation of Turbulent Flows, Topics in Applied Physics, Vol. 12, Turbulence, Ed. Bradshaw, P., pp 193-229, 1976.

64. REYNOLDS, W. C. - Computation of Turbulent Flows, Annual Review of Fluid Mechanics, Vol. 8, pp. 183-208, 1976.

65. LAUNDER, B. E. and SPALDING, D. B. - Mathematical Models of Turbulence, Academic Press, Inc., 1972.

66. MOIN, P. and KIM, J. - Numerical Investigation of Turbulent Channel Flow, J. Fluid Mech., Vol. 118, pp. 341-377, 1982.

67. SCHUMANN, U. - Subgrid Scale Model for Finite Difference Simulations of Turbulent Flows in Plane Channels and Annuli, Journal of Computational Physics, Vol. 18, pp. 376-404, 1975.

68. SMAGORINSKY, J. - General Circulation Experiments with the Primitive Equations, Monthly Weather Review, Vol. 91, No. 3, pp. 99-164, March 1963.

69. VAN DREIST, E. R. - On Turbulent Flow Near a Wall, Journal of Aeronautical Sciences, Vol. 23, pp. 1007-1011, 1956.

70. KANEDA, Y. and LESLIE, D. C. - Tests of Subgrid Models in Near-Wall Region Using Represented Velocity Fields, Journal of Fluid Mechanics, Vol. 132, pp. 349-373, 1983.

71. CEBECI, T. - Calculation of Compressible Turbulent Boundary Layers with Heat and Mass Transfer, AIAA Journal, Vol. 9, pp. 1091-1097, 1971.

72. CEBECI, T. and MEIER, M. U., Modelling Requirements for the Calculation of the Turbulent Flow around Airfoils, Wings and Bodies of Revolution, AGARD CP-271, Sept. 1979.

73. BALDWIN, B. S. and LOMAX, H. - Thin Layer Approximation and Algebraic Model for Separated Turbulent Flows, Paper No. 78-257

presented at AIAA 16th Aerospace Sciences Meeting, Hunstville, Ala., Jan. 1978.

74. CEBECI, T. - Kinematic Eddy Viscosity at Low Reynolds Numbers, AIAA Journal, Vol. 11, No. 1, p. 102, 1973.

75. HANJALIC, K. - Velocity and Length Scales in Turbulent Flows – A Review of Approaches, The 1980-81 AFOSR-HTTM-Stanford Conference on Complex Turbulent Flows : Comparison of Computation and Experiment, Vol. 2, Taxonomies, Reporters' Summaries, Evaluation and Conclusions, Eds. Kline, S. J., Cantwell, B. J., and Lilley, G. M., Thermoscience Division, Stanford University, Stanford, Calif., 1982.

76. HUNG, C. M. and MacCORMACK, R. W. - Numerical Solution of Three-Dimensional Shock Wave and Turbulent Boundary-Layer Interaction, AIAA Journal, Vol. 16, No. 10, pp. 1090-1096, 1978.

77. COUSTEIX, J. - Turbulence Modelling and Boundary Layer Calculation Methods, von Karman Institute for Fluid Dynamics, Lecture Series 1981-1, Vol. 2, Jan. 1981.

78. KWON, O. K. and PLETCHER, R. H. - Prediction of the Incompressible Flow Over a Rearward-Facing Step, Technical Report HTL-26, CFD-4, ISU-ERI-AMES-82019, Iowa State U. of Science and Technology, Ames, Iowa, 1981.

79. DEGANI, D. and SCHIFF, L. B. - Computation of Supersonic Viscous Flows around Pointed Bodies at Large Incidence, AIAA Paper 83-0034, presented at the AIAA 21st Aerospace Sciences Meeting, Reno, Nev., Jan. 1983.

80. VISBAL, M. and KNIGHT, D. - Evalution of the Baldwin-Lomax Turbulence Model for Two-Dimensional Shock-Wave Boundary Layer Interactions, AIAA Paper 83-1697, Presented at the AIAA 16th Fluid and Plasma Dynamics Conference, Danvers, Mass., July 1983.

81. WILCOX, D. C. and RUBESIN, M. W. - Progress in Turbulence Modeling for Complex Flow Fields Including Effects of Compressiblility, NASA TP-1517, 1980.

82. RODI, W. - A New Algebraic Relation for Calculating the Reynolds Stresses, Zeitschr. Angewandte Math. Mech., Vol. 56, pp. 219-221, 1976.

83. COAKLEY, T. J. - Turbulence Modeling Methods for the Compressible Navier-Stokes Equations, AIAA Paper 83-1693, presented at the AIAA 16th Fluid and Plasma Dynamics Conference, Danvers, Mass., July 1983.

84. HUNG, C. M. and KORDULLA, W. - A Time-Split Finite-Volume Algorithm for Three-Dimensional Flow-Field Simulation, AIAA Paper 83-1957, presented at the AIAA 61h Computational Fluid Dynamics Conference, Danvers, Mass., July 1983.

85. THOM, A. - The Flow Past Circular Cylinders at Low Speeds, Proceedings of the Royal Society of London, Series A, Vol. 141, pp. 651-669, 1933.

86. KAWAGUTI, M. - Numerical Solution of the Navier-Stokes Equations for the Flow around a Circular Cylinder at Reynolds Number 40, Journal of the Physical Society of Japan, Vol. 8, No. 6, pp. 747-757, 1953.

87. FROMM, J. E. and HARLOW, F. H. - Numerical Solution of the Problem of Vortex Street Development, The Physics of Fluids, Vol. 6, No. 7, pp. 975-982, 1963.

88. MacCORMACK, R. W. - Numerical Solution of the Interaction of a Shock Wave with a Laminar Boundary Layer, Proceedings of the Second International Conference on Numerical Methods in Fluid Dynamics, Lecture Notes in Physics, No. 8, Springer-Verlag, pp. 151-163, 1971.

89. MARVIN, J. G. - Modeling of Turbulent Separated Flows for Aerodynamic Applications, NASA TM-84392, Aug. 1983. (Also Proceedings of an International Symposium on Recent Advances in Aerodynamics and Aeronautics, Joint Institute for Aeronautics and Acoustics, Stanford University, Stanford, Calif., Aug. 1983.)

90. CEBECI, T., STEWARTSON, K., and WHITELAW, J. H. - Calculation of Two-Dimensional Flow Past Airfoils, the Second Proceedings Volume of Numerical and Physical Aspects of Aerodynamic Flows, Ed. Cebeci, T., Springer-Verlag, 1983.

91. MEHTA, U. and LOMAX, H. - Reynolds Averaged Navier-Stokes Computations of Transonic Flows–the State-of-the-Art, Transonic Aerodynamics, Ed. Nixon, D., Vol. 81 of Progress in Astronautics and Aeronautics, pp. 297-375, 1982.

92. DAVIS, R. T. and RUBIN, S. G. - Non-Navier-Stokes Viscous Flow Computations, Computers and Fluids, Vol. 8, pp. 101-131, 1980.

93. ORSZAG, S. A. and PATERA, A. T. - Subcritical Transition to Turbulence in Planar Shear Flows, Transition and Turbulence, Ed. Meyer, R. E., Academic Press, pp. 127-146, 1981.

94. PATERA, A. T. and ORSZAG, S. A. - Transition and Turbulence in Plane Channel Flows, Proceedings of Seventh International Conference on Numerical Methods in Fluid

Dynamics, Eds. Reynolds, W. C., and MacCormack, R. W., Lecture Notes in Physics, Vol. 141, pp. 329-335, 1981.

95. ORSZAG, S. A. and KELLS, L. C. - Transition to Turbulence in Plane Poiseulle and Plane Couette Flow, Journal of Fluid Mechanics, Vol. 96, pp. 159-205, 1980.

96. ORSZAG, S. A. - Turbulence and Transition: A Progress Report, Proceedings of Fifth International Conference on Numerical Methods in Fluid Dynamics, Eds. A. I. van de Vooren and P. J. Zandbergen, Lecture Notes in Physics, Vol. 59, pp. 32-51, 1981.

97. WRAY, A. and HUSSAINI, M. Y. - Numerical Experiments in Boundary-Layer Stability, AIAA Paper 80-0275 , presented at the AIAA 18th Aerospace Sciences Meeting, Pasadena, Calif., Jan. 1980.

98. KOVASZNAY, L. S. G., KOMODA, H., and VASUDEVA, B. R. - Detailed Flow Field in Transition, Proceedings of the 1962 Heat Transfer and Fluid Mechanics Institute, pp. 1-26, Stanford University Press, Stanford, Calif., 1962.

99. ROGALLO, R. S. - Numerical Experiments in Homogeneous Turbulence, NASA TM-81315, Sept. 1981.

100. ORSZAG, S. A. and PAO, Y. H. - Numerical Computation of Turbulent Shear Flow, Advances in Geophysics, Vol. 18A, pp. 225-236, 1974.

101. RILEY, J. L. and METCALFE, R. W. - Direct Numerical Simulations of the Turbulent Wake of an Axisymmetric Body, in Turbulent Shear Flows, Vol. 2, Eds. Durst, F., Launder, B. E., Schmidt, F. W., Bradbury, L. J. S., and Whitelaw, J. H., Springer-Verlag, 1980.

102. METCALFE, R. W. and RILEY, J. L. - Direct Numerical Simulations of Turbulent Shear Flows, Proceedings of Seventh International Conference on Numerical Methods in Fluid Dynamics, Eds. Reynolds, W. C., and MacCormack, R. W., Lecture Notes in Physics, Vol. 141, pp. 279-284, 1981.

103. TAVOULARIS, S. and CORRSIN, S. - Experiments in Nearly Homogeneous Turbulent Shear Flow with a Uniform Mean Temperature Gradient, Part I, Journal of Fluid Mechanics, Vol. 104, pp. 311-347, 1981.

104. ROTTA, J. C. - Statistische Theorie Nichthomogener Turbulenz, Part I, Zs. Phys. Vol. 129, pp. 547-572, 1951. Statistische Theorie Nichthomogener Turbulenz, Part II, Zs. Phys. Vol. 132, pp. 51-77, 1951. (Also NASA TT F-14,560, 1972.)

105. LUMLEY, J. L. - Second Order Modeling of Turbulent Flows, Prediction Methods for Turbulent Flows, Ed. Kollman, W.,

Hemisphere Publishing Corporation, 1980.

106. DEARDORFF, J. W. - A Numerical Study of Three-Dimensional Turbulent Channel Flow at Large Reynolds Number, Journal of Fluid Mechanics, Vol. 41, pp. 453-480, 1970.

107. SCHUMANN, U. - Ein Verfahren zur Direkten Numerischen Simulation Turbulenter Strömungen in Platten- und Ringspaltkanälen und über seine Anwendung zur Untersuchung von Turbulenzmodellen, Ph. D. Thesis, Universitat Karlsruhe, Karlsruhe, West Germany, 1973. (Also NASA TT F-15,391.)

108. KIM, J. and MOIN, P. - Large Eddy Simulation of Turbulent Channel Flow - ILLIAC IV Calculation, AGARD-CP-271, 1980.

109. HUSSAIN, A. K. M. F. and REYNOLDS, W. C. - Measurements in Fully Developed Channel Flow, Journal of Basic Engineering, Vol. 97, p. 568, 1975.

110. ROGALLO, R. S. and MOIN, P. - Numerical Simulation of Turbulent Flows, Ann. Rev. Fluid Mech., Vol. 16, 1984.

111. STANSBY, P. K. and DIXON, A. G. - The Importance of Secondary Shedding in Two-Dimensional Wake Formation at Very High High Reynolds Number, Aero. Quart., Vol. 33, Part 2, pp. 105-123, 1982.

112. KIYA, M., SASAKI, K., and ARIE, M. - Discrete-Vortex Simulation of a Turbulent Separation Bubble, Journal of Fluid Mechanics, Vol. 120, pp. 219-244, 1982.

113. LEONARD, A. - Vortex Methods for Flow Simulation, J. Comp. Physics, Vol. 37, No. 3, pp. 289-335, 1980.

114. SAFFMAN, P. G. and BAKER, G. R. - Vortex Interactions, Annual Review of Fluid Mechanics, Vol. 11, pp. 95-122, 1979.

115. MAULL, D. J. - An Introduction to the Discrete Vortex Method, IUTAM/IAHR, Karlsruhe, 1979.

116. DEFFENBAUGH, F. D. and MARSHALL F. J. - Time Development of the Flow about an Impulsively Started Cylinder, AIAA Journal, Vol. 14, No. 7, pp. 908-913, July 1976.

117. CLEMENTS, R. R. and MAULL, D. J. - The Representation of Sheets of Vorticity by Discrete Vortices, Progress in Aerospace Sciences Vol. 16, pp. 129-146, 1975.

118. SPALART, P. R., LEONARD, A., and BAGANOFF, D. - Numerical Simulation of Separated Flows, NASA TM-84328, Feb. 1983.

119. McCROSKEY, W. J., McALISTER, K. W., CARR, L. W., and PUCCI, S. L. - An Experimental Study of Dynamic Stall on

Advanced Airfoil Sections, Vol. 1, Summary of the Experiment, NASA TM-84245, 1982.

120. McCROSKEY, W. J. - Some Current Research in Unsteady Fluid Dynamics, ASME, Journal of Fluids Engineering, Vol. 99, No. 1, pp. 8-38, March 1977.

121. CHORIN, A. J. - Numerical Study of Slightly Viscous Flow, Journal of Fluid Mechanics, Vol. 57, pp. 785-796, 1973.

122. RAKICH, J. V., VIGNERON, Y. C., and AGARWAL, R. - Computation of Supersonic Flows over Ogive-Cylinders at Angle of Attack, AIAA Paper 79-0131, Jan. 1979.

123. SCHIFF, L. B. and STEGER, W. B. - Numerical Simulation of Steady Supersonic Viscous Flow, AIAA Journal, Vol. 18, No. 12, pp. 1421-1430, Dec. 1980.

124. STUREK, W. B. and SCHIFF, L. B. - Numerical Simulation of Steady Supersonic Flow over Spinning Bodies of Revolution, AIAA Journal, Vol. 20, No. 12, pp. 1724-1731, Dec. 1982.

125. RAKICH, J. V., DAVIS, R. T., and BARNETT, M. - Simulation of Large Turbulent Vortex Structures with the Parabolic Navier-Stokes Equations, in Proceedings of the Eighth International Conference on Numerical Methods in Fluid Dynamics, Ed. Krause, E., Lecture Notes in Physics, Vol. 170, pp. 420-426, Springer-Verlag, 1982.

126. TANNEHILL, J. C., VENKATAPATHY, E., and RAKICH, J. V. - Numerical Solution of Supersonic Viscous Flow over Blunt Delta Wings, AIAA Journal, Vol. 20, No. 2, pp. 203-210, Feb. 1982.

127. LIGHTHILL, M. J. - On Boundary Layers and Upstream Influence, II, Supersonic Flows without Separation, Proceedings of the Royal Society of London, Series A, Vol. 217, 1953.

128. VIGNERON, Y. C., RAKICH, J. V., and TANNEHILL, J. C. - Calculations of Supersonic Flow over an Ogive-Cylinder-Boattail Body, AIAA Paper 78-1137, Presented at the AIAA 11th Fluid and Plasma Dynamics Conference, Seattle, Washington, July 1978. (Also NASA TM-78500, 1978.)

129. RAINBIRD, J. R. - Turbulent Boundary-Layer Growth and Separation on a Yawed Cone, AIAA Journal, Vol. 6, No. 12, pp. 2410-2416, Dec. 1968.

130. BRILEY, W. R. and McDONALD, H. - Computation of Three-Dimensional Horseshoe Vortex Flow Using the Navier-Stokes Equations, Proceedings of the Seventh International Conference on Numerical Methods in Fluid Dynamics, Eds. Reynolds, W. C., and

MacCormack, R. W., Lecture Notes in Physics, Vol. 141, pp. 91-98, 1981.

131. DOLLING, D. S. and BODGONOFF, S. M. - Blunt Fin-Induced Shock Wave/Turbulent Boundary-Layer Interaction, AIAA Journal, Vol. 20, No. 12, pp. 1674-1680, Dec. 1982.

132. VOLTENKO, D. M., ZUBKOV, A. I., and PANOV, Y. A. - The Existence of Supersonic Zones in Three-Dimensional Separated Flows, Journal of Fluid Dynamics, Vol. 2, No. 1, pp. 13-16, 1967.

133. HUNG, C. M. and BUNING, P. G. - Private communication, 1983.

Section 1:

NAVIER-STOKES EQUATIONS, INCOMPRESSIBLE

INCOMPRESSIBLE NAVIER-STOKES AND PARABOLIZED NAVIER-STOKES FORMULATIONS AND COMPUTATIONAL TECHNIQUES†

S.G. RUBIN

DEPARTMENT OF AEROSPACE ENGINEERING AND APPLIED MECHANICS
UNIVERSITY OF CINCINNATI
CINCINNATI, OHIO 45221 U.S.A.

1. INTRODUCTION

Although there has been considerable progress in the development of computational technique, the accurate numerical simulation of high Reynolds number interacting flow remains a difficult and costly exercise. For laminar conditions, the evaluation of circulatory, separating, and three-dimensional flow problems, with Re in excess of 10^3, is still considered noteworthy. For fully turbulent conditions, the inadequacy of existing closure models for these problems has obscured our ability to focus on the adequacy of numerical technique and of numerical solutions that have been found for much larger values of Re.

Numerical solutions of the laminar incompressible Navier-Stokes equations have been obtained, to one degree or another, by finite-difference [1-5],* finite-element [5,6], spectral or pseudo-spectral [7], and vortex or integral [8,9] methods. At the present time, there is no one generally accepted approach nor is there a particular algorithm that reflects the state-of-the-art of low-speed high Reynolds number simulation. The focus of the present review is on recent developments with "finite-difference" techniques. The quotation marks reflect the fact that any finite discretization procedure can be included in this category. Many so-called finite-element

---

† This review is a shortened and updated version of lecture notes prepared for presentation at the Series on Computational Fluid Dynamics, von Karman Institute for Fluid Dynamics, Brussels, Belgium, March 29-April 2, 1982.

* Typical references are presented throughout this review. These should not be considered as a complete listing of the literature pertinent to this subject.

collocation and galerkin methods can be reproduced by appropriate forms of the differential equations and discretization formulas.

Recent investigations have demonstrated that many of the difficulties encountered in early Navier-Stokes calculations were inherent not only in the choice of the difference equations (accuracy), but also in the method of solution or choice of algorithm (convergence and stability), in the manner in which the dependent variables or discretized equations are related (coupling), in the manner that boundary conditions are applied, in the manner that the coordinate mesh is specified (grid generation), and finally, in recognizing that for many high Reynolds number flows not all contributions to the Navier-Stokes equations are necessarily of equal importance (parabolization, preferred direction, pressure interaction, asymptotic and mathematical character). It is these elements that are reviewed in the following sections. In addition, several new Navier-Stokes and parabolized Navier-Stokes formulations are also presented.

2. EQUATIONS

(i) (u,v,p) [10]

The incompressible Navier-Stokes equations (2.1-2.3) are written in a generalized orthogonal coordinate frame (ξ,η) for the primitive variables, pressure p and velocities (u,v). The cartesian coordinates $\xi = \xi(x,y)$ and $\eta = \eta(x,y)$ are related to the (x,y) physical coordinate system through the transformation $\sigma = f(z)$ or $z = F(\sigma)$ where $\sigma = \xi+i\eta$ and $z = x+iy$.

continuity

$$(h_2h_3u)_\xi + (h_1h_3v)_\eta = 0 \tag{2.1}$$

ξ-momentum

$$\begin{aligned}\rho(h_2h_3u^2)_\xi + \rho(h_1h_2uv)_\eta + \rho uvh_3h_{1\eta} - \rho v^2h_3h_{2\xi} \\ = -h_2h_3p_\xi + (h_2h_3\tau_{11})_\xi + (h_1h_3\tau_{12})_\eta + h_3\tau_{12}(h_1)_\eta \\ - h_3\tau_{22}(h_2)_\xi - h_2\tau_{33}(h_3)_\xi\end{aligned} \tag{2.2}$$

η-momentum

$$\begin{aligned}h_1h_3p_\eta = -\rho(h_2h_3uv)_\xi - \rho(h_1h_3v^2)_\eta - \rho uvh_3h_{2\xi} + \rho u^2h_3h_{1\eta} \\ + (h_2h_3\tau_{12})_\xi + (h_1h_3\tau_{22})_\eta + h_3\tau_{12}h_{2\xi} - h_3\tau_{11}h_{1\eta} - h_1\tau_{33}h_{3\eta}\end{aligned} \tag{2.3}$$

where

$$\tau_{11} = 2\left(\frac{1}{h_1} u_\xi + \frac{v}{h_1 h_2} h_{1_\eta}\right)/Re$$

$$\tau_{22} = 2\left(\frac{1}{h_2} v_\eta + \frac{u}{h_1 h_2} h_{2_\xi}\right)/Re$$

$$\tau_{33} = 2\left(\frac{u}{h_3 h_1} h_{3_\xi} + \frac{v}{h_2 h_3} h_{3_\eta}\right)/Re$$

$$\tau_{12} = \left[\frac{h_2}{h_1}\left(\frac{v}{h_2}\right)_\xi + \frac{h_1}{h_2}\left(\frac{u}{h_1}\right)_\eta\right]/Re$$

The equations are shown here in a quasi-divergence or conservation form. Alternate forms are possible and may be more desirable in certain cases. The metrics $h_1(x,y)$ and $h_2(x,y)$ are determined from the transformation function $f(z)$:

$$h_1^2 = (x_\xi)^2 + (y_\xi)^2 , \quad h_2^2 = (x_\eta)^2 + (y_\eta)^2 \tag{2.4}$$

The metric $h_3 = y^\varepsilon$, where $\varepsilon = 0$ for two-dimensions and $\varepsilon = 1$ for axisymmetry.

If the coordinate lines are generated from a conformal mapping, then $f(z)$ is analytic so that $h_1 = h_2 = h$ and

$$h = |f'(z)| , \quad h_\xi - ih_\eta = h(f''/f') \tag{2.5}$$

In the form given by equations (2.1-2.3), the analytic expressions for the second derivatives of the metric coefficients will not be required in the discretized form of the equations. It is possible to replace (2.1-2.3) with a more conservative representation, so that even the first derivative expressions (2.5) are unnecessary. This can be particularly important when considering geometries having large axial curvature.

In certain problems, in order to insure necessary mesh refinement in all large gradient regions, it may be preferable to use a non-orthogonal coordinate frame. Numerical coordinate generators obtained from solutions of elliptic boundary value problems, of hyperbolic initial value problems, or with algebraic controls have been considered by Thompson and coworkers [11], Steger [12], Eiseman [13], among others. However, these techniques can introduce an additional level of numerical operation and may also increase the complexity of the governing equations, which are shown here only for orthogonal coordinates. Moreover, for large Reynolds number flow, the system (2.1-2.3) will describe the velocity components directly related to a streamline or body fitted system of equations. This may be very desirable both in the surface boundary layer and the outer inviscid region. Numerical grid generation technique has been the subject of several recent symposia [14] and is reviewed in a number of publications.

(ii) $(\psi,\omega;p)$

An alternate form of the equations (2.1-2.3) that has been applied extensively in two-dimensions is obtained with the vorticity (ω)-stream function (ψ) formulation:

$$\omega = -\frac{1}{h_1 h_2}[(h_1 u)_\eta - (h_2 v)_\xi] \qquad (2.6)$$

and

$$\frac{1}{h_1}\psi_\xi = -h_3 v\,, \qquad \frac{1}{h_2}\psi_\eta = h_3 u \qquad (2.7)$$

The continuity equation (2.1) is automatically satisfied with (2.7). The momentum equations (2.2, 2.3) are cross-differentiated to provide the vorticity transport equation (2.9), and the definition (2.6), with (2.7), leads to the stream function equation (2.8).

stream function

$$\frac{1}{h_3}D^2\psi = \frac{1}{h_1 h_2}[(\frac{h_2}{h_3 h_1}\psi_\xi)_\xi + (\frac{h_1}{h_3 h_2}\psi_\eta)_\eta] = -\omega \qquad (2.8)$$

vorticity

$$h_1 h_2 h_3 \omega_t - (\psi_\xi \omega)_\eta + (\psi_\eta \omega)_\xi + \frac{\omega}{h_3}(\psi_\xi h_{3_\eta} - \psi_\eta h_{3_\xi}) = \frac{1}{Re} D^2 (h_3 \omega) \qquad (2.9)$$

Significantly, the pressure no longer appears explicitly (it is uncoupled) in the (ψ,ω) system (2.8,2.9). The pressure solution, if required, is usually obtained from the poisson equation (2.10) for p generated by differentiating the momentum equations (2.2) and (2.3) and adding appropriately:

$$D^2 p = f(u,v;\ \mathrm{div}\ \vec{v}) \qquad (2.10)$$

This equation requires that the normal (n) pressure gradient, along the boundary ℓ enclosing the region S, satisfies the condition

$$\oint p_n\, d\ell \equiv \int_S f\, dS \qquad (2.11)$$

The same equation is generally used for the coupled (u,v,p) system (2.1-2.3). This is in lieu of the continuity equation (2.1), which is then only satisfied indirectly [16, 17]. The specification of appropriate boundary conditions for p, and a numerical approximation to insure that (2.11) is satisfied are of major importance.

The application of (2.10) to generate the pressure, and the secondary role of the continuity equation (2.1), is somewhat contrary to solution procedures for high Re boundary layer theory. As will be described in a later section on solution technique, it may be possible to develop a large Re Navier-Stokes strategy that reflects the asymptotic boundary layer philosophy. This is an important point of departure from conventional solution methods.

3. METHOD OF SOLUTION

Since it is not generally possible to solve any of the governing systems described in section 2 with an efficient direct solver, i.e. one that couples a large arbitrary set of discrete equations and boundary conditions, several important decisions are required. These include the ordering and degree of coupling of the governing equations, the accuracy and form of the discretization formulas for both the interior equations and boundary conditions, the specific choice of coordinates and grid, and the selection of an efficient inversion algorithm for the discrete equations, e.g., explicit [18] or implicit [18-22], transient consistent [20-22] or steady-state [19], direct [23] or iterative [19], factored [20-22] or preconditioned [24, 25], multi-grid [26-28], etc. The present review will concentrate only on solution techniques designed primarily for evaluation of the steady-state behavior. This is not to say that the solution algorithms are inconsistent in the transient, but rather that they do not have a small time step limitation, e.g., a CFL condition for stability or accuracy. Many of the early solution techniques belonged to this category and therefore are not considered in the present discussion. In this context, for the incompressible Navier-Stokes equations, the implicit iterative formulations appear to be the most successful. However, the procedures for ordering, coupling, factoring, pre-conditioning, grid generation, and improving accuracy and rates of convergence play a significant role in the degree of success. For example, similarity coordinates have proven useful for certain geometries [29]. For compressible low Mach number, laminar Navier-Stokes calculations, time-dependent implicit methods have been moderately successful [20-22].

(i) (ψ-ω) Formulation

The (ψ-ω) system (2.6-2.11) has been the most popular choice of past and even present day investigators. The mathematical character of the poisson equations is more amenable to available numerical methods, and in two-dimensions, the coupled system is smaller than the (u,v,p) system (2.1-2.3). Recall, that for steady-state solvers, the pressure equation is uncoupled and therefore must be evaluated only once. For three-dimensional flows, it is necessary to define multiple streamfunctions and therefore the (ψ,ω) system becomes somewhat more complex. For this reason, there have been relatively few (ψ,ω) calculations in three-dimensions.

Another difficulty, which is present in all formulations that use the poisson equation for p, concerns the pressure boundary conditions. These are usually obtained, from the respective normal momentum equation (2.2) or (2.3) and lead to Neumann conditions for p. The momentum equations (2.2) or (2.3) are never satisfied explicitly in the interior region. Converged and accurate numerical solutions for p, for large Reynolds number, have generally been difficult to obtain even though (ψ,ω) have been calculated to a high degree of accuracy. Recent analysis and calculations have demonstrated that the convergence properties for p are sensitive to the form of differencing of both the equations and Neumann conditions along the boundaries [18, 30]. In particular, the compatibility condition (2.11) for the surface normal pressure gradient, as obtained from (2.10), must be accurately satisfied.

We can trace the development of (ψ,ω) calculations through progress with the flow in a driven cavity. This problem has served as the prototype for almost every new innovation associated with the incompressible Navier-Stokes equations. The geometry and boundary conditions are shown in figure (3.1).

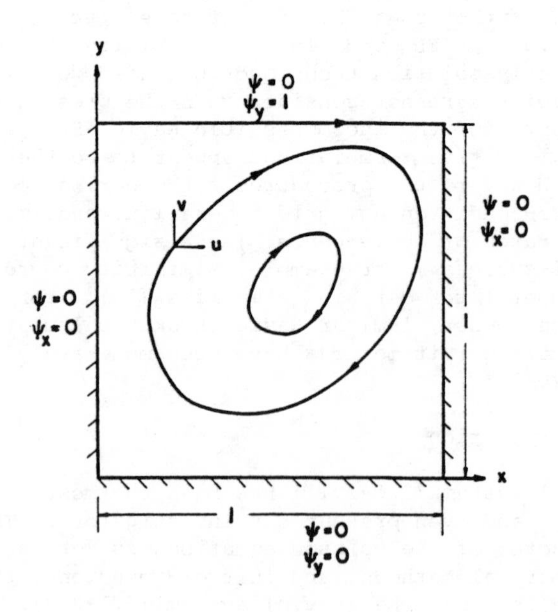

Fig. 3.1 Schematic of the Driven Cavity.

The existance of corner point singularities on the upper moving wall adds a significant degree of complexity for any numerical technique; however, the flow is confined. The difficulties with far field, inflow and outflow boundary conditions are removed, and the region of circulatory flow, even for very large Re, is limited by geometric constraints. For external problems, confined regions of laminar separated or circulatory flow, generally grow with increasing Re. This adds a significant degree of complexity to the numerical simulation.

Analysis and computation of the driven cavity flow started in the mid nineteen-sixties. In the subsequent years, the problem has been considered by upwind, central and upstream-weighted differencing [18], by various explicit methods, by SOR, ADI, hopscotch [18], direct solvers, spline, compact or Hermitian, CSIP, multi-grid and continuation methods, to name just a few. A number of observations result from the past 16 years of large Re cavity computation: (i) coarse grids can lead to spurious solutions associated with bifurcation phenomena; (ii) first-order upwinding on coarse grids provides rapid convergence but leads to the incorrect or spurious solution. The effect of numerical viscosity is dominant; (iii) iterative convergence rates for the alternating (ψ,ω) equations in central difference form decrease rapidly as Re increases; (iv) direct solvers for ψ significantly improve the convergence properties of the central-differenced equations and obviate the need for upwinding, even for large Re; (v) coupling the $(\psi-\omega)$ equations and boundary conditions, as with the CSIP [19] technique is more effective than any uncoupled (ψ,ω) method, including a direct ψ solver, for central and even higher-order differencing for $Re \leq 10^4$; (vi) combining CSIP with a multi-grid strategy [26] can significantly reduce computation times even for very fine grids and large Re; however, computer storage is increased; (vii) higher-order methods, such as compact [31], Hermitian [32], or spline [34-37] procedures, provide accurate solutions even with moderately coarse grids; (viii) continuation methods, although relatively new for application to problems in fluid mechanics, may be useful for reducing computational costs of high Re simulation [38]. Multi-grid techniques can be considered as continuation, with the mesh width as the relevant parameter.

The experience with $(\psi-\omega)$ computation leads to an important conclusion. For laminar flow with large Reynolds number (say $Re \leq 10^4$), stable, accurate and rapid finite-difference solutions can be obtained without requiring large numerical viscosity, e.g., first-order coarse grid upwinding. This is accomplished with stronger implicit algorithms, a greater degree of coupling of the discretized equations and boundary conditions, improved accuracy on coarser grids, and acceleration techniques for iterative algorithms, e.g., multi-grid, continuation, pre-conditioned conjugate gradient or improved incomplete LU decomposition.

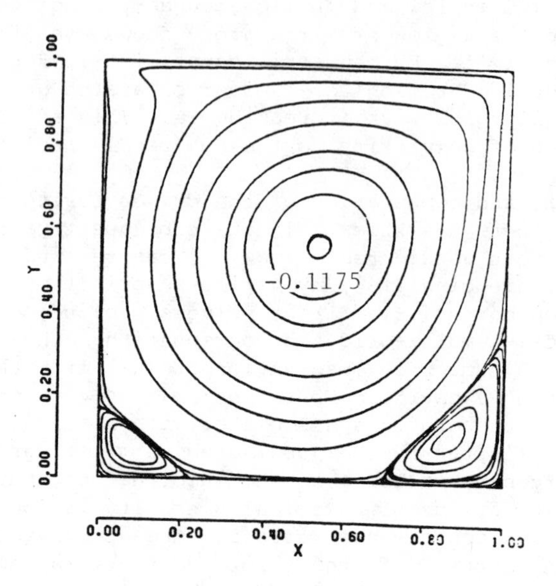

Fig. 3.2a Streamline Contours-Cavity CSIP-MG [26], Re = 1000 [129 x 129].

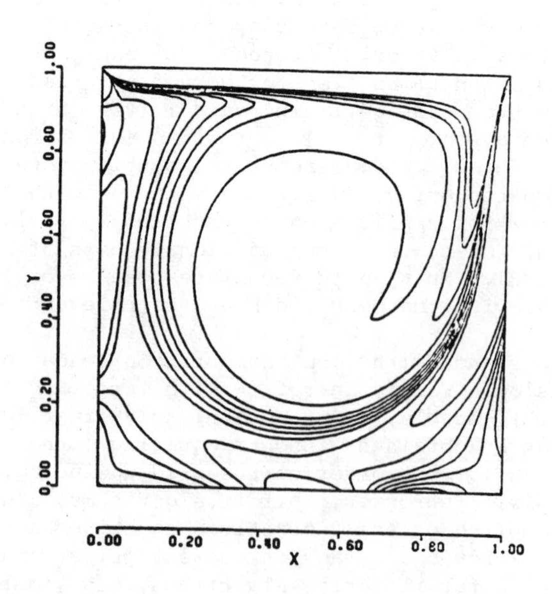

Fig. 3.2b Vorticity Contours-Cavity CSIP-MG [26], Re = 1000 [129 x 129].

The boundary conditions for the cavity geometry are $\psi_1 = 0$ on all boundaries; $(\psi_n) = \varepsilon$, where $\varepsilon = 0$ on fixed walls and $\varepsilon = 1$ on the moving wall. The coordinate n is normal to the surface. The no-slip condition on $(\psi_n)_1$ is satisfied indirectly through the vorticity condition, $\omega_1 = (\psi_{nn})_1$, where 1 denotes the wall, and 2 and 3 are interior points h and 2h from the wall, respectively. A second-order boundary condition for ω_1 is then given by

$$\omega_1^{k+1} = \frac{2}{h^2}(\psi_2-\psi_1)^{k+1} + \frac{1}{2h^2}(4\psi_2 - \psi_3 - 3\psi_1)^k - \frac{3\varepsilon}{h} \tag{3.1}$$

where k is the iteration index. For the coupled algorithms [19], (3.1) provides coupling of the boundary conditions as well as the interior points. Second-order accuracy is introduced iteratively as a deferred-corrector [19, 39]. This proved to be a more effective procedure for coupling the boundary condition with the algorithm. Some typical results are shown in figures (3.2-3.4). Although converged solutions can now be obtained even with relatively coarse meshes, the accuracy will depend on the level of discretization. For example, for Re = 1000 on a (17 x 17) grid, upwind solutions are very inaccurate; central differencing is somewhat better; spline results, [34-35] compare favorably with central differencing on a (129 x 129) grid. For large Re ($Re = 10^4$), the central difference results were obtained with the deferred corrector or KR procedure of [39]. This technique is now the basis of the spline methods as well [36, 37]. Multigrid strategy was required for large Re and very fine grids [26].

Due to the bifurcation phenomena [38] coarse meshes may lead to spurious or lower branch solutions, see figure (3.3).

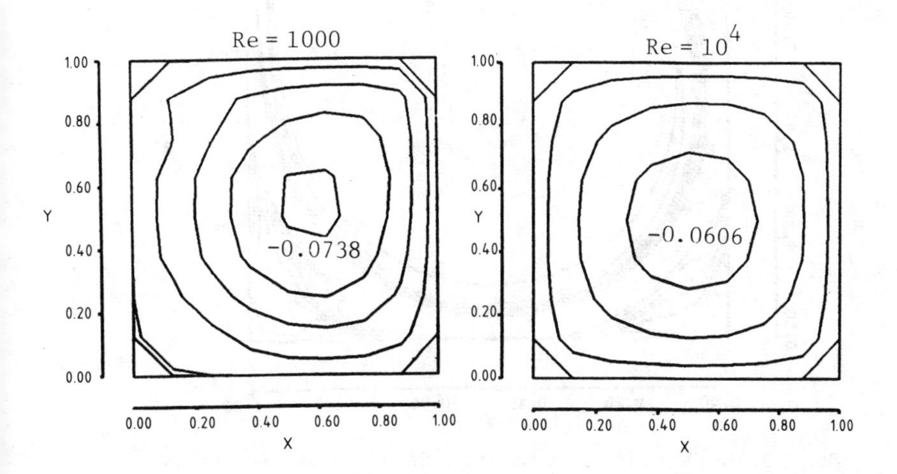

Fig. 3.3 Course Grid Isovels-Cavity, Re = 1000, 10000 [9x9] CSIP.

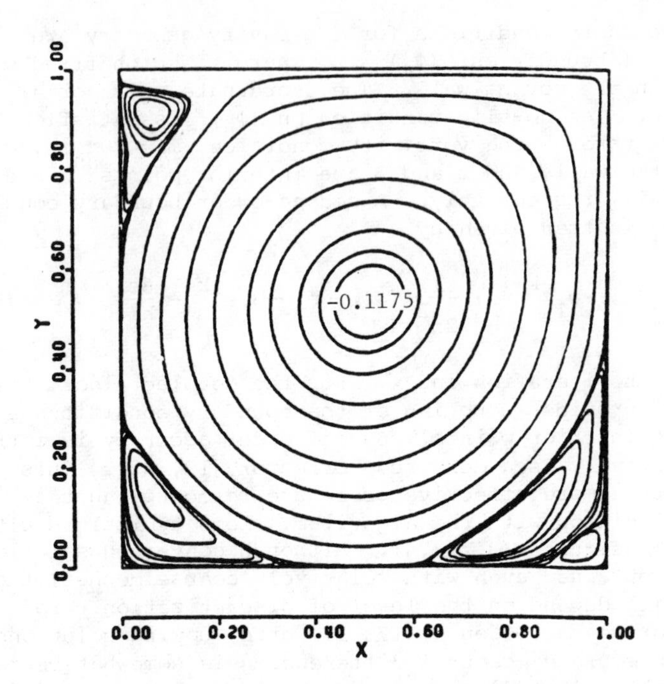

Fig. 3.4a Streamline Contours-Cavity CSIP-MG [26], Re = 7500 [257 x 257].

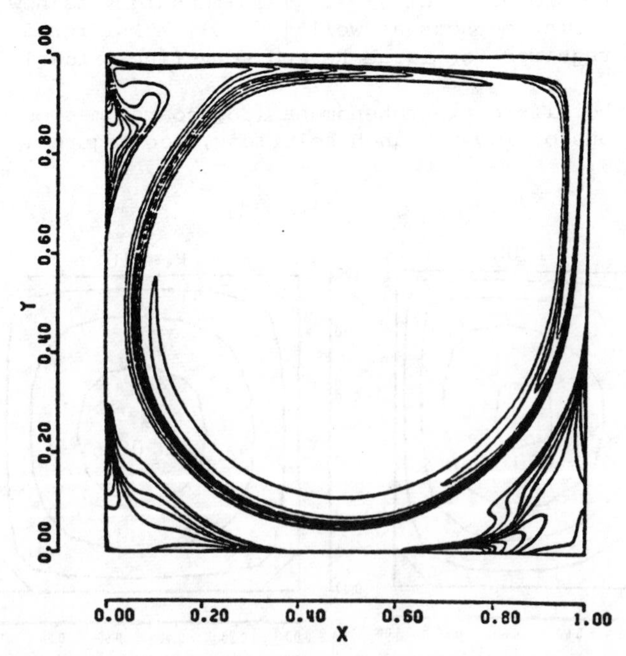

Fig. 3.4b Vorticity Contours-Cavity CSIP-MG [26], Re = 7500 [257 x 257].

As the degree of the discretization formula is increased, the same mesh may provide the correct or upper branch result. As the level of discretization decreases, upper branch solutions are obtained only with finer meshes and therefore increased computational costs. These can be reduced substantially with the acceleration techniques delineated previously.

(ii) (u,v,p) Formulation

The more natural (u,v,p) or primitive variables have experienced a much slower rate of development. Most of the early incompressible investigations involved explicit solution techniques and were relatively slowly convergent [18]. The difficulties with the poisson solver for the pressure, previously discussed for the (ψ,ω) formulation, are also more significant for the (u,v,p) development, as the pressure is required in (2.2) and (2.3) during each step of the iterative cycle. This is true even for implicit and steady-state computations. In addition, when the poisson equation (2.10) is applied, in lieu of the continuity equation (2.1), the latter can only be satisfied indirectly. The convergence rate and even the overall convergence of the procedure is, therefore, very much dependent on this indirect approach to local continuity and the related global compatibility condition (2.11) for the surface normal pressure gradient. The significance of staggered pressure-velocity grids to exactly satisfy the condition (2.11) has been discussed by several investigators [16, 19, 30].

Accurate implicit (u,v,p) solutions have recently been reported for internal separated flow with $Re = 10^3$ [12]. In order to satisfy the continuity equation (2.1) and the pressure condition (2.11) on a regular grid, it has been necessary to very carefully monitor the global compatibility condition for p, and enforce (2.1) at all boundaries where the Neumann conditions are imposed. This is accomplished through the discretized form of the velocity boundary conditions. On a regular (non-staggered) grid, the continuity (2.1) and pressure (2.11) conditions can at best be satisfied to grid accuracy (truncation error). Moreover, there is no analytic proof that the pressure-velocity iteration, between (2.10) and (2.2, 2.3), will be convergent, i.e., that the error in (2.1) will reduce to grid accuracy. However, numerical experience on a number of internal flow problems indicates that the method is convergent.

(iii) (u,v,p) Boundary Layer-Potential Flow Formulation:

As noted in the earlier discussion, one of the major problems with most of the current procedures for (u,v,p) is the lack of compatibility of the solution technique with the asymptotic boundary layer and/or inviscid (or potential) flow formulations. For large Reynolds number calculations, it would appear that such compatibility could be important and may possibly be necessary if accurate and rapidly convergent methods are to be developed.

In boundary layer calculations, the continuity and axial momentum equations are coupled to evaluate the two velocity components. The axial pressure variation is determined directly or interactively with the outer inviscid flow. Variations in the pressure across the layer can then be obtained from the normal momentum equation. In potential flow calculations, the continuity equation determines the potential function and thereby the velocity components. The pressure is obtained in an uncoupled fashion from the Bernoulli relation. For both boundary layer and potential flows, the continuity equation is an essential if not dominant element in the solution procedure. Conventional (u,v,p) methods satisfy this equation in an indirect and subsidiary manner.

A new composite velocity procedure proposed by P. Khosla and the present author attempts to incorporate the asymptotic large Reynolds number philosophy for the (u,v,p) Navier-Stokes equations. The method has been developed for both compressible [42] and incompressible [43] equations. The basic development is almost identical. This is also a feature unique to this solution technique. A velocity split procedure due to Dodge and co-workers [41] has several similiarities with the present composite approach; however, the formulations are in fact quite different. The composite representation is prescribed in the spirit of matched asymptotic expansions, and discretization follows boundary layer and potential flow procedures. No simplifying approximations are required for the pressure. The complete Navier-Stokes equations are considered.

This formulation has many desirable features for the calculation of high Reynolds number flow with a primary or dominant flow. In the present discussion, this is assumed to be in the ξ direction, i.e., along the transformed surface of the body; the gradients are then largest in the η or surface normal direction.

The flow outside the viscous region near the surface is essentially inviscid and can be represented by a "potential" function ϕ; therefore, following matched asymptotic expansion theory, a composite representation of the velocity field is prescribed.

$$u = \frac{U}{h_1}(1 + \phi_\xi) = U\, u_e \qquad (3.2)$$

$$v = \frac{1}{h_2}\phi_\eta \qquad (3.3)$$

With this composite approximation, the component U describes the behavior in all non-potential regions, e.g., the viscous boundary layer or irrotational inviscid flow. In the solution procedure, U is, in effect, evaluated from the primary or ξ-momentum equation (2.2), and, as in the case of boundary layer calculations, continuity (2.1) is used to determine the normal velocity component (ϕ_η). The normal momentum equation (2.3) is then applied as a (total) pressure corrector.

The composite-velocity representation allows the no-slip boundary condition $u = 0$ to be readily applied as $U = 0$, and assures that the solution outside the viscous regions ($U = 1$ for irrotational flow) reduces to that associated with the potential flow equation for the velocity components $(1 + \phi_\xi)$ and ϕ_η. The discretization formulas for the U and ϕ terms are chosen to reflect the appropriate differencing for boundary layer and potential flow calculations, respectively.

It should be emphasized that this procedure deviates significantly from classical time-dependent calculation methods, where the pressure is obtained indirectly, for incompressible flow, from the continuity equation, and the normal momentum equation provides the normal component of the velocity.

In discussing the development of the governing equations, only a conformal coordinate transformation f(z) is considered. This allows for $h_1 = h_2 = h$, and simplifies the equations. After substitution of the expressions (3.2, 3.3) into the Navier-Stokes system (2.1-2.3), and some rearranging of terms, the following equations for ϕ, U and G are obtained:

continuity

$$[h_3 U(1 + \phi_\xi)]_\xi + (h_3 \phi_\eta)_\eta = 0 \qquad (3.4)$$

ξ-momentum

$$u_t + \frac{1}{h^2 h_3} \{[h h_3 u_e^2 (U^2 - U)]_\xi + [h h_3 u_e v (U-1)]_\eta\}$$

$$+ \frac{h_\eta}{h^2} u_e v (U-1) + \frac{u_e}{h} (U-1) u_{e_\xi}$$

$$= - \frac{1}{h} \bar{G}'(\xi) - \frac{1}{h} G_\xi + \text{viscous terms} \qquad (3.5)$$

η-momentum

$$G_\eta = -(U-1) \{(\frac{u_e^2}{2})_\eta - \frac{h_\eta}{h} u_e^2 U\} + \text{viscous terms} \qquad (3.6)$$

where

$$G = \frac{p}{\rho} + \frac{u_e^2 + v^2}{2} - \bar{G}(\xi) \tag{3.7}$$

$\bar{G}(\xi)$ is evaluated at one of the boundaries. For external flow this is usually $\eta \to \infty$, where the flow is uniform and undisturbed, so that $\bar{G}$ = constant. For internal flow, $\bar{G}(\xi)$ must be determined from the mass-conservation condition (3.8)

$$\int_{A(\xi)} u \, dA = 1 \tag{3.8}$$

where $A(\xi)$ is the cross-sectional area of the channel. Therefore, $G = 0$ for $\eta \to \infty$ in external flow, or at one of the duct walls for internal flow; e.g., $\eta = 0$.

For purely inviscid flow $G \equiv$ constant. For the Navier-Stokes equations G is not constant, but is determined by the calculation procedure. It is significant that the η-momentum equation (3.6) for G contains only η derivatives of the convective terms. In the inviscid region, $U \to 1$ for irrotational motion and the continuity equation reduces to the well known potential flow equation; the momentum equations are then identically satisfied with G = constant. In the viscous region, the ξ-momentum equation determines U, while ϕ and therefore v is calculated from the continuity equation. The pressure, i.e., G, is obtained from the η-momentum equation. The two basic regions pertinent to large Reynolds number flows are appropriately described by the composite set of equations. This is in conformity with boundary-layer and inviscid flow theory. This method of defining v with a "potential" was first tested for the flat plate boundary-layer (U,ϕ) equations. The solution of the resulting two-point boundary value problem reproduced the results obtained with standard methods based on the velocities u and v.

Solution Procedure [43]

The governing equations have been discretized with second-order upwind or central-differencing for all the derivatives except the $(h_3U)_\xi$ term in the continuity equation; this is backward differenced using first or second-order accuracy. This term vanishes in the inviscid region. The resulting implicit algebraic system of equations has been solved iteratively using the CSIP algorithm [19]. The continuity and ξ-momentum equations for ϕ and U are solved in a coupled fashion, while the η-momentum equation for G is evaluated iteratively. In the ξ-momentum equation, G is fixed during each iterative cycle. The η-momentum equation is uncoupled and determines the updated value of G. Further coupling is possible, but proved to be unnecessary. Although G is treated explicitly, the pressure p is implicit in the coupled algorithm, as it is determined from the velocity components $1 + \phi_\xi$ and ϕ_η.

This pressure interaction removes the separation point singularity that appears when the pressure is prescribed. Furthermore, the elliptic character of the continuity equation (3.4), as well as other elliptic effects in (3.5), e.g., u_{e_ξ}, removes any departure solution typically appearing in marching calculations. This point will be discussed in greater detail in section 5 on the PNS equations.

The composite formulation follows almost an identical series of steps for compressible flow [42]. In fact the latter procedure will reproduce the incompressible results exactly for $M \to 0$. This property is unique to this compressible flow solver and is indicative of the boundary layer-potential flow character of the final equations.

Boundary Conditions

The composite formulation provides direct and natural boundary conditions. For external flow over finite geometries, far from the surface all disturbances decay so that $\phi \to 0$, $U \to 1$ and $G \to 0$. At the surface, the zero normal velocity condition is satisfied with $\phi_\eta = 0$, and the no-slip condition with $U = 0$. Inflow and outflow conditions depend on the particular geometry, i.e., external or internal flow, finite or infinite body. For internal flows, $\phi_\eta = 0$, $U = 0$ on both walls and $G = 0$ on one wall. The mass flow condition (3.8) determines $\bar{G}(\xi)$ and is completely compatible with the integrated continuity equation (3.4); i.e., if $v = 0$ on one wall, (3.8) insures that $v = 0$ on the other wall as well. The difficulties in satisfying the continuity and pressure compatibility condition of section 3(ii) no longer are present.

Solutions

Separated flow solutions for a typical boattail geometry are presented in the following figures. These solutions were obtained with the CSIP algorithm; a coarse (61 x 28) and more refined (91 x 31) grid were considered. The resolution in the recirculation region is significantly improved with mesh refinement. A more detailed paper on this subject is forthcoming. For the laminar flow conditions considered here, Reynolds numbers range from 1000 to 6000 based on the maximum body radius. For turbulent calculations, not presented here, Reynolds numbers of up to 2.2×10^6 have been considered.

The boundary conditions for the infinite boattail geometry were prescribed as follows: inflow ($\xi \to -\infty$), $U = 1$, $\phi = 0$; outflow ($\xi \to \infty$), $\phi_{\xi\xi} = 0$, $U_\xi = 0$; surface ($\eta = 0$), $U = 0$, $\phi_\eta = 0$; far field ($\eta \to \infty$), $U = 1$, $G = 0$, $\phi = 0$. The initial conditions were obtained from the potential flow calculation that results from the basic algorithm when U is set equal to unity on the surface of the body.

The grid, figure (3.5), was generated by the conformal Schwarz-Christoffel mapping procedure of Davis [44]. This technique was detailed in a previous VKI lecture series. This method is based on a unique extension of the Schwarz-Christoffel transformation applicable to smooth surfaces. This is given by:

$$\frac{dz}{d\sigma} = M \exp \left[\frac{1}{\pi} \int \log(\sigma - b) \, d\beta\right] \tag{3.9}$$

where the z-plane is the physical plane and σ the transformed plane. M is a parameter. A finite number of discontinuites in the body shape can easily be incorporated in the mapping routine. In the case of a boattail, one juncture point represents such a discontinuity. The nonlinear integro-differential equation is integrated numerically. We can represent the equation (3.9) by:

$$\frac{dz}{d\sigma} = f_1(\sigma) \; f_2(\sigma) \tag{3.10}$$

where $f_2(\sigma)$ is a smooth and well-behaved function and $f_1(\sigma)$ is written as $f_1(\sigma) = (\sigma - a_1)^{1/\pi}$ near the corner points a_1. Away from the juncture $f_1(\sigma)$ is well behaved along the boattail. The equation (3.10) is integrated numerically between any two grid points (σ_{i-1}, σ_i) by using the trapezoidal rule in the complex plane. Thus,

$$z_i - z_{i-1} = f_1 \left(\frac{\sigma_i + \sigma_{i-1}}{2}\right) f_2 \left(\frac{\sigma_i + \sigma_{i-1}}{2}\right) (\sigma_i - \sigma_{i-1}) \tag{3.11}$$

If $\sigma_{i-1} < a_1 < \sigma_i$,

$$z - z_{i-1} = f_2 \left(\frac{\sigma_i + \sigma_{i-1}}{2}\right) \frac{(\sigma_i - a_1)^{1-1/\pi} - (\sigma_{i-1} - a_1)^{1-1/\pi}}{1 + \frac{1}{\pi}} \tag{3.12}$$

The expression (3.12) is well behaved near the corner singularity at a_1. It should be noted that the accuracy of the mapping is determined solely by the integration formula that is applied (3.10). Typically the trapezoidal rule leads to second-order accurate mapping functions and derivatives.

Higher-order quadrature rules such as Simpson's rule or the one obtained from the application of spline interpolation can also be used when necessary. In spite of the fact that z is well behaved at $\sigma = a_1$, the mapping is nonanalytic at the boattail juncture. This introduces a considerable amount of complication in the flow field computations. The choice of the grid in the juncture region becomes highly critical. Mesh points very close to the juncture introduce large mapping derivatives thereby increasing the truncation error. A coarse grid may not properly resolve the recirculation region. This suggests that a proper balance between these two conflicting

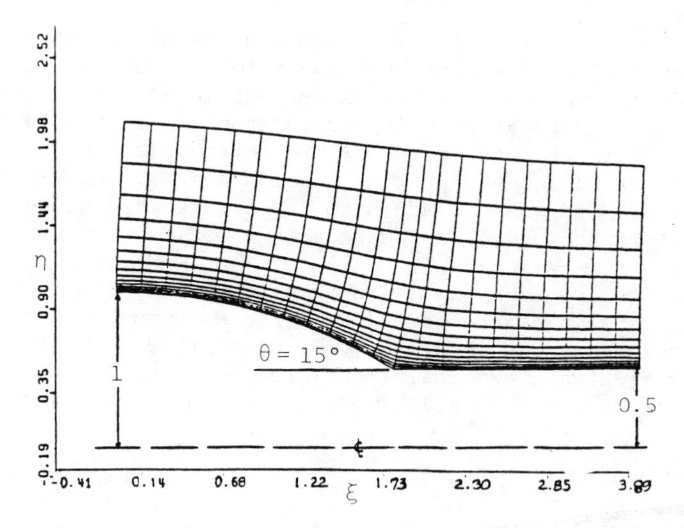

Fig. 3.5a Boattail Geometry and Grid Near Corner [42].

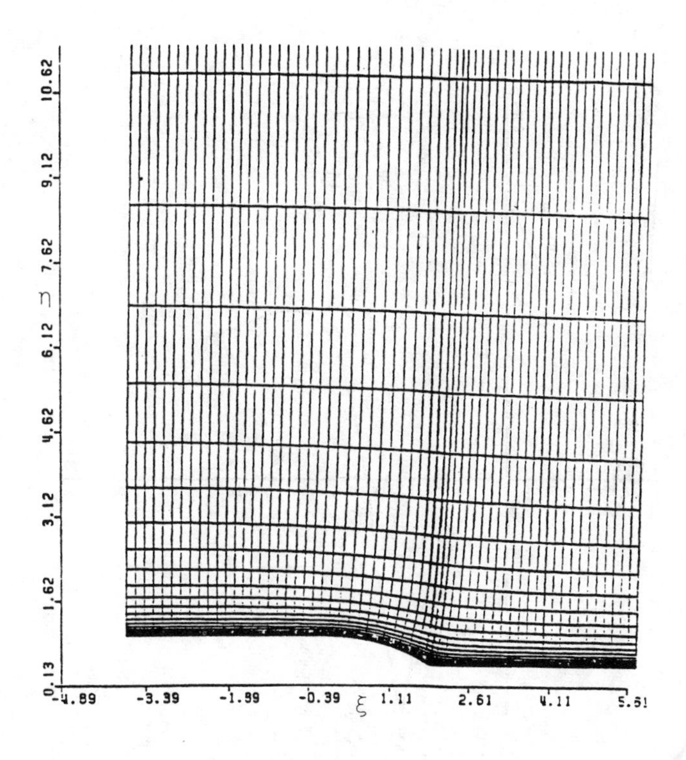

Fig. 3.5b Boattail Grid [42], $y_M = 15.5$.

effects must be considered for the accurate evaluation of the boattail flow field. Solutions are shown for the streamlines, skin-friction and pressure for a juncture angle of 30° in figure (3.6). The temporal step Δt varies from 0.1 initially to 10^6 after 50 iterations of the CSIP algorithm [19]. The changeover location depends on the Reynolds number. Explicit artificial viscosity was not added to the discretized equations for any of the cases considered ($Re \leq 10^4$).

The results are identical with those obtained from the compressible version of the code [42] for a Mach number M = 0.1. Additional solutions for internal ($\bar{G}(\xi)$ prescribed by mass-flow conservation) and external flows are given in [42, 43]. Typical solutions for a Joukowski airfoil are given in [43].

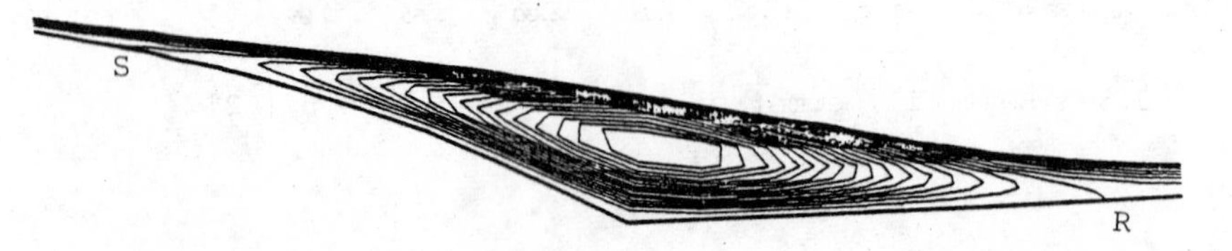

Fig. 3.6a Isovels for Boattail, $\theta = 15°$, Re = 7500 [43].

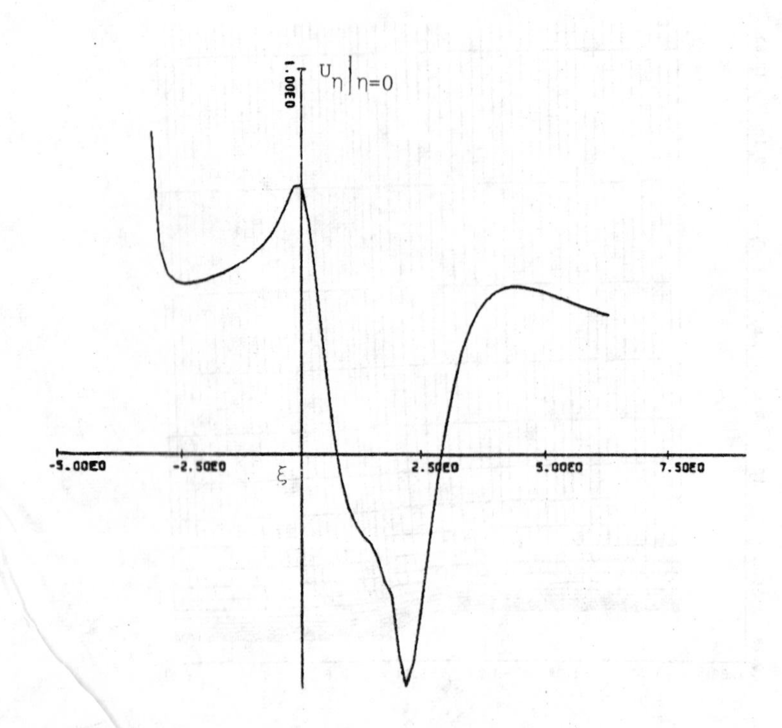

Fig. 3.6b Skin-Friction for Boattail, $\theta = 15°$, Re = 7500 [43].

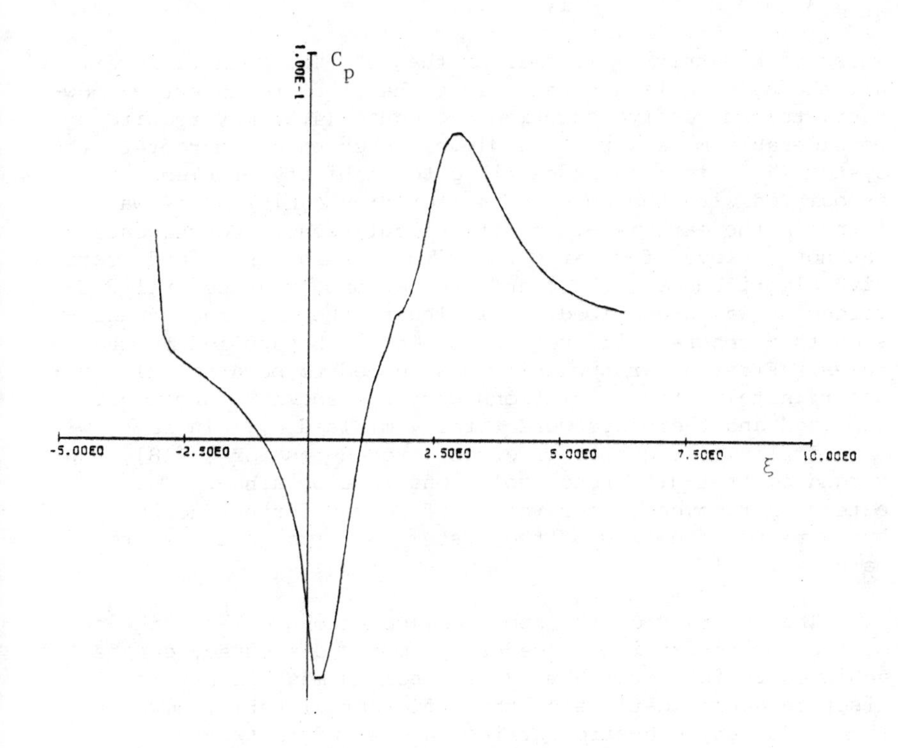

Fig. 3.6c Pressure for Boattail, $\theta = 15°$, $Re = 7500$ [43].

4. ACCURACY, ALGORITHM AND ACCELERATION

Once the equations (section 2) and methodology (section 3) have been prescribed, the questions of (1) the accuracy (numerical error) of the discretized equations and boundary conditions, (2) the "optimum" inversion algorithm for the linearized discrete system, and (3) the rate of convergence of the non-direct (iterative) solution procedure should be addressed. Since a direct inversion of the non-linear discrete equations is at the present time not practical, some form of iteration is required for both the non-linearity and the inversion of the coefficient matrix A of the linear discrete system (4.1)

$$Au = b \tag{4.1}$$

where u is the solution vector and A and b are functions of the discretization procedure, mesh prescription and solution values of u at the previous iteration.

If some of the dependent variables u are uncoupled, i.e., some equations are solved successively rather than simultaneously, then a third iteration is introduced and (4.1) becomes a system (4.2)

$$A_i u = b \qquad i = 1, 2, \ldots \qquad (4.2)$$

where each matrix A_i is smaller than A. The work to invert all the A_i once is generally less than that to invert A; however, the successive solution procedure (4.2) may require considerably more work to achieve convergence. Moreover, the system (4.2) is also susceptible to stability problems associated with the successive iteration cycle. This was found in the early (ψ-ω) cavity calculations. Convergence was not achieved for large Re, with certain meshes and iterative algorithms for the ψ and ω equations, when central differencing was prescribed. This led to the incorrect conclusion that central differencing could not be applied in these cases. First-order upwinding was deemed to be a useful cure; unfortunately, these solutions were filled with "numerical damping" and therefore were totally misleading. In fact, with more implicit or direct inversion procedures for ψ [18], converged central-difference solutions were obtained. The extension to even larger values of Re was further accomplished by increased coupling of the system (4.2) to better approximate (4.1) [19].

The accuracy of the numerical solution reflects the choice of the discretization procedure. Improved accuracy can be achieved by introducing many neighboring mesh points in the discrete approximations of the differential terms; however, this adds many non-zero entries into A and restricts the effectiveness of many iterative inversion algorithms. Discretization procedures that lead to block tridiagonal forms of A and, in particular, at most three central diagonals and two off-center diagonals are considered here; this form is associated with a five-point star in two-dimensions.

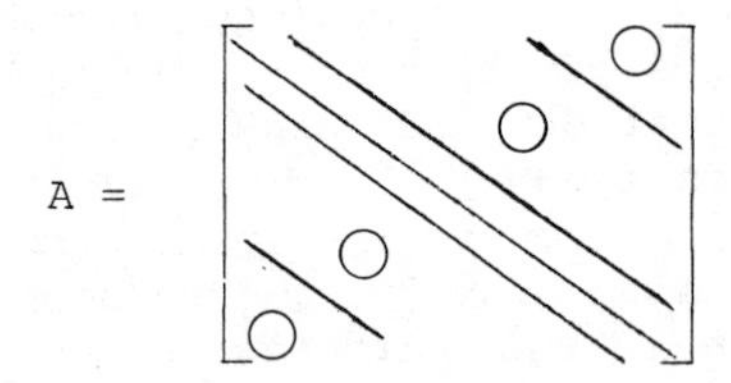

Second-order finite-difference and fourth-order spline [34-37] collocation procedures, which fall into this category, have been discussed in earlier studies.

A variety of iterative algorithms have been developed to invert A. These have been applied successively for the system (4.2). Coupled algorithms for (4.1) [19-22] generally involve some form of factoring or splitting of A. The results obtained for large Re are indicative of the strong positive influence that coupling provides on the convergence properties of the inversion iteration process. The factoring is typically of the form

$$A = \tilde{L}D\tilde{U} - E \qquad (4.3)$$

where $\tilde{L}$ is scalar tridiagonal or $\tilde{L} = L$ is lower triangular, D is diagonal and $\tilde{U}$ is upper triangular. E denotes the difference between the exact A and approximate $\tilde{L}D\tilde{U}$ forms. The system (4.1) becomes

$$\tilde{L}D\tilde{U}\, u^{k+1} = E\, u^{k} + b \tag{4.4}$$

or

$$\tilde{L}D\tilde{U}\, (u^{k+1} - u^{k}) = (E - \tilde{L}D\tilde{U})\, u^{k} + b \tag{4.5}$$

so that if $M = \tilde{L}D\tilde{U}$

$$M\, e^{k+1} = b - A\, u^{k} = r^{k} \tag{4.6}$$

where E operates on the solution vector u at the previous iteration level k; e^{k+1} is the error vector $u^{k+1} - u^{k}$; r^{k} is the residual $b - Au^{k}$, which reflects an error in (4.1) at any iteration level. The equation (4.6) can be designated as a preconditioned matrix form of (4.1).

The prescription (4.3) defines the method of iteration or the form of the preconditioning matrix M:

(i) $\tilde{L} = \tilde{U} = I$, $M = D$; point Jacobi

(ii) D, $\tilde{L} = I$, $\tilde{U} = U$; Gauss-Seidel

(iii) $D = I$, $\tilde{L}$, $\tilde{U}$ tridiagonal; ADI

(iv) $D = I$, $\tilde{L} = \hat{L}$, $\tilde{U} = \hat{U}$, $\hat{L}$, $\hat{U}$ are 3-diagonal lower and upper triangular, respectively; SIP

(v) $A = \bar{L} + \bar{U} + D$; $\tilde{L} = \bar{L} + \bar{D}$, $D = (\bar{D})^{-1}$, $\tilde{U} = \bar{D} + \bar{U}$, $E = \bar{L}(\bar{D})^{-1}\bar{U}$; SSOR

(vi) $M = LDU = A$; $E = 0$; direct solver. (4.7)

where I is the identity matrix. For the explicit methods (i), (ii) and (iv), $\tilde{L}$, $\tilde{U}$, $\tilde{D}$ are directly related to L, U and D of the original coefficient matrix A. For the ADI technique (iii), L and U are tridiagonal so that $L = L_x U_x$, and $U = L_y U_y$ where L_x, L_y, U_x, U_y are lower or upper triangular, and the entries are determined with an implicit one-dimensional recursive formula for L and U, respectively. Additional, one-dimensional storage is required for the recursion coefficients. For the SIP (v), L and U are 3-diagonal lower and upper triangular, respectively, and the entries are determined with a two-dimensional recursion formula. This adds to the implicitness of the algorithm but also increases the storage. The direct solver is totally implicit, but the LU decomposition requires considerable additional storage as well as work in the recursion procedure.

The character of E determines whether the factoring (4.3) is iteration consistent or inconsistent; i.e., if the truncation error at any level k tends to zero for Δt, $\Delta\xi$, $\Delta\eta \to 0$, the method is consistent. If the iteration procedure

(4.4) is consistent, then the non-linear differential or discrete system of equations can be quasi-linearized or expanded about a previous iteration (time) level to provide a specified degree of accuracy consistent with that of (4.4). These procedures [22-25] are then said to be "non-iterative"; i.e., for complete coupling, and with consistent factoring and linearization, the equation (4.4) or (4.6) is inverted only once at each level k. The appropriate (quasi) linearization results from a Taylor series expansion about the previous (k) iteration (time) level; i.e., for any operator F, second-order consistency is obtained with

$$F(u^{k+1}) = F(u^k) + \left(\frac{dF}{du}\right)^k (u^{k+1} - u^k) + O(\Delta t^2) \tag{4.8}$$

where Δt is the temporal increment between iterations. For steady-state solvers, the consistency of (4.4) and of the linearization process is not necessarily required. Of more importance, is a prescription (4.3) that results in $r^k \to 0$ most rapidly.

A comment at this juncture is appropriate. There have been numerous finite-difference "Navier-Stokes solvers" proposed in the open literature. The combinations and permutations are numerous, i.e., choice of equations, method of solution, discretization, linearization, degree of coupling and choice of algorithm. It is therefore important to distinguish between innovation and manipulation.

Finally, the rate of convergence of the factored system can be further improved by introducing an acceleration step into the inversion formula (4.4) or (4.6). Two effective methods, which have been applied to a limited degree for the incompressible Navier-Stokes equations, are the multi-grid technique [25-28] and the preconditioned conjugate gradient or Lanczos procedure [24, 25]; strongly-implicit or incomplete Choleski preconditioning is used to approximate M.

Rewriting (4.6) we have the following expression for the relative error e^{k+1}

$$e^{k+1} = M^{-1}b - M^{-1}A\,u^k = M^{-1}\,r^k \tag{4.9}$$

In this form, (4.9) can be viewed as the solution of

$$Cu = M^{-1}A\,u = M^{-1}\,b \tag{4.10}$$

so that M^{-1} is seen to "precondition" the coefficient matrix A. The factoring (4.3) is such that $M^{-1} = (\tilde{L}D\tilde{U})^{-1}$ can be easily evaluated. The conjugate gradient (CG) procedure can then be applied to the preconditioned equation (4.10) in lieu of (4.1). The preconditioning is used to restructure the coefficient matrix A, and with the factored form of M (4.3), the matrix $C\,e^k$ required in the CG procedure is easily evaluated from

$$C e^k \equiv M^{-1} (A e^k) \qquad (4.11)$$

In the multi-grid method, (4.4) or (4.6) is applied on a sequence of different grids in order to more effectively filter short and long wavelength errors.

Coupled Strongly Implicit Procedures

As discussed briefly in the previous sections, in addition to improved methodology, there is strong evidence to suggest that the efficiency of large Re Navier-Stokes solvers can be enhanced with a greater degree of implicitness for the inversion algorithm and with a greater degree of coupling between the flow variables. Coupled ADI (LBI, factored ADI, etc.) have been moderately successful for compressible turbulent flows, but less so for incompressible laminar flows.

In order to improve upon the one-dimensional ADI factoring, two-dimensional incomplete LU algorithms have been developed. These inversion techniques increase the implicitness of the solution procedure, but also require increased storage for the coefficients of the two-dimensional recursion formulas. Moreover, the storage requirements escalate significantly as the degree of coupling increases. When CADI procedures generally require only storage of the N coupled variables at two iteration or time levels, i.e., 2N two-dimensional storage functions plus one-dimensional storage for the recursion coefficients, CSIP procedures generally require 2N(N+1) storage functions for the coupled variables and two-dimensional recursion coefficients. Techniques for reducing the CSIP storage requirements are clearly subjects for future study [43]. It should be noted however, that CSIP storage is still less than for most incomplete LU inversion procedures and therefore finite element methods. Some improvement has been provided for three coupled variables [46] where one of the governing differential equations is first-order and integrated with the two-point trapezoidal rule, e.g., continuity equation or normal momentum equation when integrated for p_η or G_η in the (U,ϕ,G) formulation. Instead of a (3 x 3) coupled system, the CSIP algorithm can be written as (2 x 2)+1 so that storage is reduced from 2N(N+1) to N(2N-1), i.e., for N = 3 from 24 to 15 functions [46].

CSIP (2 x 2)

Finite-difference discretizations of the Navier-Stokes equations leads to an algebraic system of the following type

$$A_{i,j}W_{i,j-1} + B_{i,j}W_{i,j} + C_{i,j}W_{i,j+1} + D_{i,j}W_{i-1,j} + E_{i,j}W_{i+1,j} = G_{i,j} \qquad (4.12a)$$

where $W_{ij} = u_{ij}$ and $G_{ij} = b_{ij}$. In matrix form this can be written as

$$AW = b \quad , \tag{4.12b}$$

where W = u in (4.1). A is a N x N matrix of the following form

$$A = \begin{bmatrix} B_1 & C_1 & & E_1 & 0 & \\ A_2 & B_2 & C_2 & 0 & E_2 & \\ & & & & & E_3 \\ D_k & 0 & & & & \\ & D_{k+1} & & & & \\ 0 & & D_{k+2} & & & \end{bmatrix} \tag{4.12c}$$

The solution of these equations have been considered by all of the procedures discussed previously. Stone [47] developed the strongly implicit method, which is based on an incomplete LU decomposition of the matrix P.

Direct solvers, such as that due to Sweet and Schwarztrauber [48] are only applicable to a special class of differential equations and associated boundary conditions. These direct methods are not presently useful for the general finite-difference form of each of the Navier-Stokes equations and certainly not for a coupled system. Other variants of Gaussian elimination are extremely inefficient and time consuming and even susceptible to a large accumulation of round-off error. SOR and ADI techniques converge rather slowly unless optimum parameters or time steps are employed. These are difficult to determine for the general equation (4.12).

Stone's [47] strongly implicit iterative technique falls under the general category of "factorization methods". The underlying ideal of factorization is to replace the sparse matrix A by a modified form (A + E) (4.3-4.7) such that the resulting matrix can be decomposed into upper (U) and lower (L) triangular sparse matrices. This leads to the following general iterative procedure for the system (4.12)

$$(A + E)\, W^{n+L} = b + EW^n \tag{4.13a}$$

or if

$$A + E = M, \quad \text{then} \quad M\, W^{n+1} = b + EW^n \quad , \tag{4.13b}$$

where the superscript n denotes the iteration number. The rate of convergence of this iteration scheme depends upon the particular choice of the matrix E. The two essential requirements on the matrix E are as follows: (i) the elements of E should be small in magnitude so that the explicit perturbation is small and (ii) the resulting matrix M should be decomposable into sparse L, U factors. If the L and U factors are not sparse, which will be the case if E is a null matrix and A is formed from the finite-difference form of the Navier-Stokes equations, then the usual problems associated with the Gauss elimination procedure can reappear.

Although many other variants are possible, the primary contribution of Stone has been to devise the factorization in such a way that a certain degree of implicitness is associated with each coordinate direction and such that every element in EW^n is small; in particular, the elements are of order h^2, where h is the mesh width. However, Saylor [49] infers that a first-order factorization can be more useful than those that lead to second-order correction terms. In any event, EW^n should tend to zero as h vanishes. An undesirable feature of Stone's factorization is that the matrix A changes at each step so that two successive iterations are in a sense uncorrelated. Although, the ideal factorization may depend upon the particular problem being considered or upon previous experience, the algorithm presented herein is sufficiently general for all types of factorization. The final two-pass algorithm can be written as

$$(A+E)\,W^{n+1} = M\,W^{n+1} = b + E\,W^n \quad ,$$

$$Lv = b + E\,W^n, \qquad UW^{n+1} = v, \qquad M = LU \quad . \tag{4.14}$$

In Stone's factorization procedure E is prescribed such that L and U have only three non-zero diagonals. This leads to a solution of the following form

$$W^{n+1}_{i,j} = GM_{i,j} + \tilde{T}_{i,j}\,W^{n+1}_{i,j+1} + T_{i,j}\,W^{n+1}_{i+1,j} \quad . \tag{4.15}$$

This procedure has the distinct advantage of being implicit in both the i and j directions, as well as coupling all the boundary conditions. This technique generally converges more rapidly than do many of the more familiar, less implicit, iterative methods previously mentioned. Moreover, the lack of implicitness and coupling may explain some of the difficulties that have been encountered with calculations of high Reynolds number Navier-Stokes flows, e.g., the need for artificial viscosity and for under-relaxation. In view of these observations it would appear that a direct solver would be most suitable for the solution of the algebraic system (4.12) arising from the Navier-Stokes equations. Unfortunately, as noted previously, currently available efficient direct solvers are not applicable to coupled 2 x 2 nonlinear systems, while others are inefficient, time consuming and suffer from instability due to round-off accumulation. The strongly implicit iterative procedure can provide the necessary coupling of the dependent variables and boundary conditions, and add a significant degree of efficiency and speed of convergence. Computer storage requirements are however increased. This problem can be minimized by reducing the required number of grid points. With the higher-order methods to be presented in section (4.2) this becomes possible.

The factorization procedure has been developed for a coupled 2 x 2 system, e.g., the vorticity-stream function form or composite (U,ϕ) form of the two-dimensional Navier-Stokes equations. Experience with the coupled strongly implicit procedure (CSIP) [19] indicates that the use of iterative boundary conditions considerably reduces the applicability (convergence rate, stability) of the algorithm. Therefore, the 2 x 2 coupled strongly implicit algorithm includes coupling of the boundary conditions as well as the interior equations. CSIP solutions have previously been discussed for the cavity geometry with (ψ,ω), and with (U,ϕ) for the boattail configuration.

Further discussion of the numerical procedures can be found in the references cited or in the extended VKI version of this review.

5. PARABOLIZED NAVIER-STOKES (PNS) FORMULATION

Consider a body fitted coordinate system (ξ,η) as defined by the orthogonal mapping described in section 2. In this reference frame, a wide variety of large Re flow problems can be accurately described with a reduced (PNS) form (5.1) of the Navier-Stokes equations

$$u_t + uu_x + vu_y = - p_x + \frac{1}{Re} u_{yy} \tag{5.1a}$$

$$v_t + uv_x + vv_y = - p_y + \left(\cancelto{0}{\frac{1}{Re}} v_{yy}\right) \tag{5.1b}$$

$$u_x + v_y = 0 \tag{5.1c}$$

For simplicity, the PNS equations (5.1) are shown here only in two-dimensional cartesian coordinates. These equations differ from the full Navier-Stokes equations (2.1-2.3) by the neglect of the axial (x or ξ) diffusion terms and, therefore, a more general form can be obtained from (2.1-2.3) by setting $(\ )_{\xi\xi}$ terms to zero. Although the PNS equations are generally applied in this form, strictly speaking, the inclusion of the v_{yy} term in (5.1b) is inconsistent with the omission of u_{xx} in (5.1a); i.e., from (5.1b) $p \sim \frac{1}{Re} v_y + \ldots$, so that $p_x \sim \frac{1}{Re} v_{yx} + \ldots = -\frac{1}{Re} u_{xx} + \ldots$ from (5.1c). A more consistent description requires that viscous effects be neglected in the normal momentum equation. It is this PNS approximation that is adopted here. In this form, normal (η or y) pressure variations are essentially inviscid in origin. In this sense, PNS theory represents an extension of interacting boundary layer theory. The inviscid flow is now evaluated exactly. The neglected viscous terms would first appear in the third-order boundary layer equations. The class of flows for which these terms remain unimportant forms the basis of this PNS theory.

The PNS reduction as defined here, completely eliminates the elliptic diffusive character of the equations; however, the "elliptic" inviscid pressure interaction is retained in full. Therefore, pressure induced upstream influence and possible axial flow separation can still be described with this reduced system of equations. Recent publications [17, 29, 46, 50-52] have reviewed the early PNS development, as well as numerical computations. Both compressible and incompressible flows have been considered. For additional references, see [50-52].

In the PNS model, the normal momentum equation is convection or pressure driven and viscous effects are boundary layer-like. As discussed previously, these characteristics are compatible with both inviscid and boundary layer developments. The composite (U,ϕ,G) Navier-Stokes formulation [42, 43] of section 3, as well as a related compressible flow formulation [46], reflect this boundary layer or PNS character. Furthermore, other successful large Re Navier-Stokes solvers for $(\psi-\omega)$ [19] and (u,v,p) [17] also isolate the effects of axial diffusion in the solution procedure so that these terms can be easily removed. Therefore, for a spectrum of large Re flow problems, the PNS system (5.1), <u>when written in an appropriate body fitted coordinate frame</u>, can be expected to approximate the full Navier-Stokes equations with a minumum of error. The major advantage gained with the PNS model is a significant reduction in computer storage when these equations are solved with a global marching procedure.

Pressure Interaction

If the axial pressure gradient p_x in (5.1a) is prescribed, the pressure interaction is completely suppressed. This approximation is valid for weak pressure interactions. The system (5.1) is then truly parabolic and can be solved as an initial value problem. This was the case in the original PNS merged layer development [53] for cold wall ($p_x \approx 0$) hypersonic flows. It should be emphasized, however, that the axial pressure variation is still calculated from the momentum equation (5.1b). The primary assumption is that these pressure values are not important in modifying the assumed value of p_x in (5.1a). Again, this will be true for weakly interacting flows.

If p_x is prescribed everywhere, except at the outer boundary, where it is matched interactively, the elliptic effect can be restored; e.g., in interactive boundary layer theory $p_y \approx 0$, so that $p_x = p'(x)$ can be determined from the Cauchy integral of the inviscid flow [54]. This procedure has been applied to viscous flows with strong pressure interaction and even axial flow separation [54, 55].

If p_x is retained exactly in the PNS system (5.1), the equations are not strictly parabolic (p_x prescribed) and the full pressure interaction is retained. However, the similarity with the parabolic form has led to the expression parabolized Navier-Stokes equations.

When the pressure interaction is important, it has been shown from asymptotic ($Re \to \infty$) Navier-Stokes or so-called triple-deck theory, that for trailing edge, separated flows, etc. [56], a local triple layer description of the interaction applies. The extent of the triple-deck is $O(Re^{-3/8})$ in both the x and y directions. The lower deck, nearest the surface is $O(Re^{-5/8})$ and is described by the boundary layer equations; the middle deck, $O(Re^{-1/2})$, is an inviscid (vortical) displaced boundary layer that transmits the interaction between the upper (irrotational) inviscid deck and lower deck. Normal pressure gradients are important only for the inviscid outer region. The interactive disturbance vanishes outside the triple-deck structure.

Interacting boundary layer theory includes all of the inner and middle deck terms and prescribes an interactive matching with the outer deck. The PNS model includes all of the terms in the triple-deck structure, including a complete description of the inviscid pressure interaction. Triple deck theory reinforces the theoretical basis for the omission of both axial diffusion and normal momentum diffusion contributions for the large Re PNS formulation. When these effects are significant, but not dominant, the formulations of [29, 42, 43, 46] represent the natural extension to the full Navier-Stokes equations.

The elliptic pressure interaction was first examined for interacting boundary layers by Lighthill [57], who demonstrated the existence of exponential growing solutions in the interaction zone. Similar behavior was subsequently encountered with marching procedures for the PNS equations [58]. The primary difference between the interacting boundary layer and PNS equations is that for the latter the pressure interaction is manifested through both the outer pressure boundary condition and the normal momentum equation (5.1).

Single Sweep Marching

For problems where upstream influence and axial flow separation are not significant, it is natural to consider the system (5.1) by boundary layer marching techniques, i.e., backward differences are applied for all x-gradients. If

p_x is prescribed, this approach is quite acceptable as the equations are parabolic. For implicit numerical schemes, the marching calculation should be unconditionally stable for all Δx marching steps, see [50] for additional references. On the other hand, if p_x is assumed unknown, the pressure interaction is introduced and therefore the exponential behavior, representative of upstream influence, can be anticipated. Lubard and Helliwell [59] have examined the stability of the backward difference approximation for p_x in (5.1a) and they have shown that for $\Delta x < (\Delta x)_{min}$ exponential or departure solutions will occur. Similar results were found earlier in [58]. For $\Delta x > (\Delta x)_{min}$ the departure behavior is suppressed. Therefore $(\Delta x)_{min}$ would appear to represent a measure of the upstream elliptic interaction. For $\Delta x < (\Delta x)_{min}$ the marching scheme attempts to represent this interaction and therefore the exponential behavior should be recovered.

Since the backward difference formula for p_x does not provide any upstream contribution, it does not properly represent the differential form for $\Delta x < (\Delta x)_{min}$. When the backward difference approximation is less representative of p_x, the error introduced serves to reduce $(\Delta x)_{min}$. For example, at $x = x_i$, $(p_{i-1} - p_{i-2})/\Delta x$ is less severe than $(p_i - p_{i-1})/\Delta x$, and for p_x prescribed, $(\Delta x)_{min} = 0$. In view of this behavior, several investigators [60-62] have attempted to eliminate the pressure interaction, for problems where upstream influence is unimportant, by incorporating "small" inconsistencies into the difference approximation. Each of these techniques introduces some inconsistency into the difference equations; however, reasonable results have been obtained with these methods for certain problems. In order for these techniques to be effective, the inconsistency must be large enough to suppress the elliptic character, yet small enough to maintain an acceptable order of accuracy.

The PNS model has been considered in some detail in [46, 51, 52, 63]. For the system (5.1), with backward differences for $\partial/\partial x$ and central y differences, the linear von Neumann stability analysis leads to the following condition for the eigenvalues λ:

$$\frac{u(\lambda-1) + 4b\ \lambda\sin^2\frac{\beta}{2} + I\ c\ \lambda\sin\beta}{u(\lambda-1)\cos^2\frac{\beta}{2} + 4b\ \lambda\sin^2\frac{\beta}{2} + I\ c\ \lambda\sin\beta} = \frac{(\lambda-1)\ F(\lambda)}{4a^2\lambda^2\sin^2\frac{\beta}{2}} \tag{5.2}$$

where $a = \Delta x/\Delta y$, $b = \Delta x/\text{Re}\ \Delta y^2$, $c = av$, $I = \sqrt{-1}$, $F = (\lambda-1)$.

It can be shown that the value of λ_{max} is closely related to the highest frequency mode, so that when the number of grid points across the layer $N >> 1$, $\beta \approx \pi/(N-1)$. Equation (5.2) then takes the simplified form

$$\frac{(\lambda-1)\,F(\lambda)}{A^2\lambda^2} = 1 \quad , \qquad (5.3)$$

where $A = \pi\Delta x/y_M$ and y_M is the layer thickness; $y_M = (N)\Delta y$. The condition (5.2) indicates that $\Delta x/y_M$ is the relevant parameter. From (5.3), the marching procedure will not exhibit the departure behavior when

$$A = \pi\Delta x/y_M > 2 \quad .$$

Therefore

$$(\Delta x)_{min} \approx y_M \qquad (5.4)$$

The complete numerical solution of (5.2), for all β, has been obtained, and the analytic result $(\Delta x)_{min} \leq \frac{2}{\pi} y_M$ is confirmed. The extent of the elliptic interaction is of the order of the thickness of the total layer. If the system (5.1) is used to solve boundary layer problems, then $y_M = O(R^{-1/2})$ and therefore $(\Delta x)_{min} = O(R^{-1/2})$. For pressure interaction regions where triple deck structure is applicable, $(\Delta x)_{min} = \frac{2}{\pi} y_M = O(R^{-3/8})$. This would tend to establish the concept of a limited elliptic zone (triple-deck) contained in the PNS formulation. For $\Delta x > (\Delta x)_{min}$, this elliptic effect is suppressed. When the upstream influence is negligible, this inconsistency or those introduced in [60-62] should have little effect on the solution. For truly interactive flows the ellipticity must be retained and a global forward-difference concept is required.

Multiple Sweep Marching-Global Interaction

If global relaxation or multiple sweep marching is used, i.e., all ξ derivatives of velocities are backward differenced in non-separated regions, but some form of forward differencing is applied for p_ξ in (5.1a), the elliptic pressure interaction is recovered and the departure free limit $\Delta\xi > (\Delta\xi)_{min}$ is removed. Solutions can then be obtained for $\Delta\xi \to 0$, see [51, 52, 63]; the numerical procedure is consistent and any desired degree of accuracy can be specified. Finally, in order to circumvent the pressure singularity at separation, the p_ξ term must also allow for a local as well as a spatial interaction. For example, central differencing fails in this regard and, as discussed in [51, 63], is unstable globally. Forward differencing of p_ξ satisfies all constraints and moreover is consistent with the eigenvalue analysis of Vigneron et al. [60] which shows that for incompressible flow ($M \to 0$), there should not be any forward marched component of the p_ξ term; i.e., $\omega = 0$ in his analysis. Forward differencing and global relaxation was first applied successfully in [51, 63] for several model incompressible flow problems. The extension to compressible flows is discussed for a conical geometry in [46] and for flows with axial flow separation and strong pressure interaction in [64]. More detailed discussion and results are given in [52].

The difference scheme used in [51, 63] was developed from the following discrete grid:

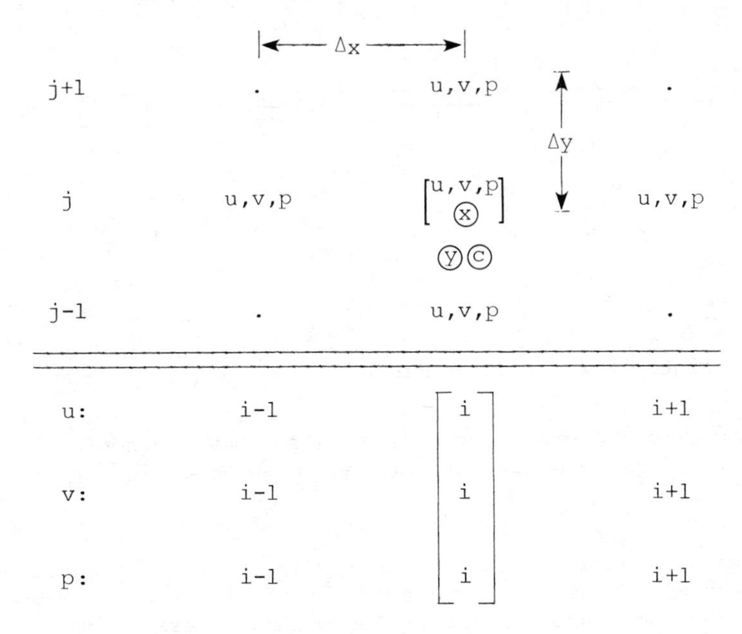

Figure 5.1. Difference Grid I.

The continuity, x and y momentum equations are centered at ⓒ, ⓧ and ⓨ, respectively.

An alternate and more accurate derivation of the equations and interpretation of "forward" differencing for p_ξ is given below for cartesian coordinates. This system was considered initially for inviscid flows [52] and resembles a slightly different development proposed by Israeli [65]. Consider the staggered difference grid as shown:

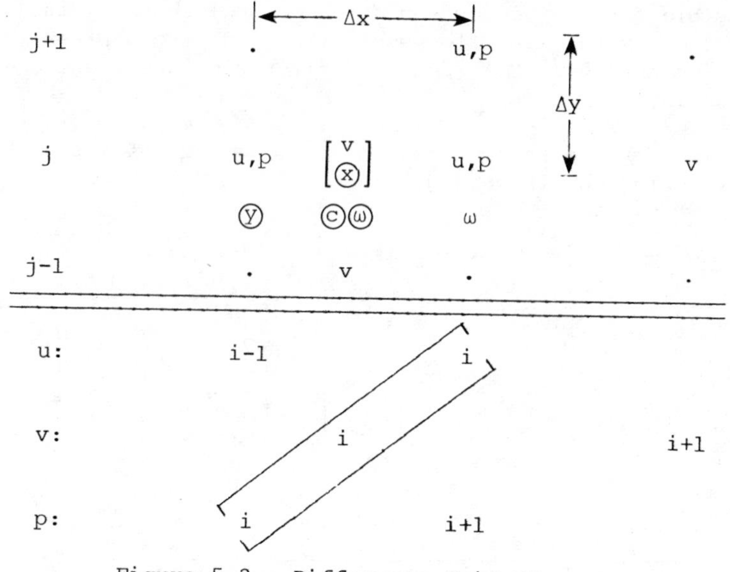

Figure 5.2. Difference Grid II.

The appropriate difference equations, also shown in non-conservative form, are now centered at somewhat different

ⓒ, ⓧ, ⓨ locations.

In this formulation full second-order accuracy is achieved. The primary modification is the averaging of the y-derivative terms in the momentum equations. The unknown pressure p_i is also shifted one point to the left of that given by the formulation of figure. This interpretation is more accurate and is consistent with the character of the interactive solutions.

In reference [52] more details are given on the difference equations. This includes the accuracy, convergence properties and difference forms of the vorticity and pressure solvers. In addition, several results for turbulent flow conditions are also presented.

Boundary Conditions

For the finite-difference grids in figures 5.1 or 5.2 the appropriate boundary conditions are specified as follows in the transformed body fitted coordinate system:

At a surface $y = 0$ $(j = 0)$, the velocities $u = v = 0$; at a symmetry line $y = 0$ $(j = 0)$, the velocities satisfy $u_y = v = 0$.

At the outer boundary $y = y_M$ $(j = M)$, where y_M lies outside of the extent of the interaction zone, e.g., triple deck, $p = p_\infty$, $u = u_\infty$ and a boundary condition on v is not required.

At the inflow boundary $x = 0$ ($i = 1$), $u = u(0,y)$, and $v_x(0,y) = 0$. For inviscid regions, where $u = u_\infty$, $v_x = 0$ is equivalent to a boundary layer approximation. The velocity v should not be specified at the inflow. This leads to inviscid vorticity production and has a destablizing effect on the global iteration procedure. The flow pressure is not prescribed and with the formulation of Figure 5.2 is unknown and a result of the calculation procedure.

Finally, the only boundary condition required at the outflow is the pressure or equivalent pressure gradient. This of course reflects the elliptic pressure interaction.

Solutions

In order to test the applicability of forward differencing for p_x several problems have been considered with the full PNS system (5.1).

Boundary Layer Examples

(i) Semi-Infinite Flat Plate: $u = p = 1$ at $y = y_M$;

$u = v = 0$ at $y = 0$

(ii) Separation Bubble: $u = p = 1$, $x < 0$

$u = 1-x$, $p_x = -uu_x$, $0 \leq x \leq 0.25$ at $y = y_M$

$u = 0.75$, $p_x = 0$, $x > 0.25$ (5.5)

Triple Deck Examples

(iii) Finite Flat Plate - same as (i) except $u_y = 0$ for $x > 1$, $y = 0$.

(iv) Carter-Wornom Trough - same as (i) except $u = v = 0$ at $y = y_b(x)$.

The pressure gradient p_x is forward differenced. All other x derivatives are backward differenced and are upwinded in regions of reverse flow. All y derivatives are central differenced, except for the continuity equation (5.1c) and momentum equation (5.1b) where the trapezoidal rule is used. It is significant that in this relaxation procedure it is necessary to store only the pressure field for each successive iteration level. The velocities are re-evaluated during each marching sweep. The results are in excellent agreement with published results for all problems. The free surface pressure interaction introduced by the y-momentum equation has eliminated the separation singularity for the separated flow examples. A value of Δx equal to one-sixth the boundary layer thickness $y_M = O(R^{-1/2})$ was adequate for problems [5.5(i) and (ii)]. For a smaller value of $(\Delta x)_{min} = y_M/60$, small variations in the solutions were obtained. From the stability results, we note that the value of Δx can be made arbitrarily small. Some typical results for the boundary layer examples are shown in figures (5.3, 5.4).

For the flat plate case, comparisons between the PNS and Blasius solution are shown, for $R = 10^3$ and 10^7, in figure (5.3a) for the velocity components, and in figure (5.3b) for the surface skin friction coefficient C_f. The agreement is quite reasonable. The maximum error occurs at the surface and this can be seen from the figures. Additional results for the pressure variation across the boundary layer are given in [63]. The pressure p_{i+1} of figure 5.1 (or p_i in figure 5.2) is updated during each sweep of the global iteration procedure. During the relaxation process small pressure variations are calculated across the boundary layer. Solutions for the separation bubble case are given in figure (5.4) for typical isovels. Solutions for the surface pressure variation are given in [63]. The predicted separation point value of $x_{sep} = 0.1180$ is close to the boundary layer value of 0.1198. Both separation and reattachment points exhibited smooth transitions and convergence. Since the outer boundary conditions were fixed and the second-order Cauchy integral displacement condition was not imposed, the free interaction was manisfested solely through the y-momentum equation (5.1b). Inclusion of the displacement boundary condition should have only a slight effect on the solutions. Convergence of the global relaxation procedure is quite rapid. Only five to ten iterations are required. Of course, for the problems considered here the pressure variation across the layer is small so that the initial guess is quite good. In view of the stability analysis previously discussed, and since the PNS system includes all of the elements of both boundary layer and triple deck equations, the present solutions with $y_M = O(R^{-3/8})$ should reproduce the results obtained with these approximations. This fact is examined very carefully for problems [5.5(iii) and (iv)].

If central differencing is used for p_x, the downstream point p_{i+1} is introduced once again; however, the value of p_i no longer appears and the y-momentum equation will be uncoupled from the velocities unless p_i is reintroduced. Several possibilities exist: (1) p_i appears directly through the outer boundary condition, as in interacting boundary layer theory [54, 55]; (2) a temporal relaxation term p_T is introduced in (5.1a) [55]; (3) a K-R approximation [39] where forward differencing is corrected during the relaxation sweeps, is applied.

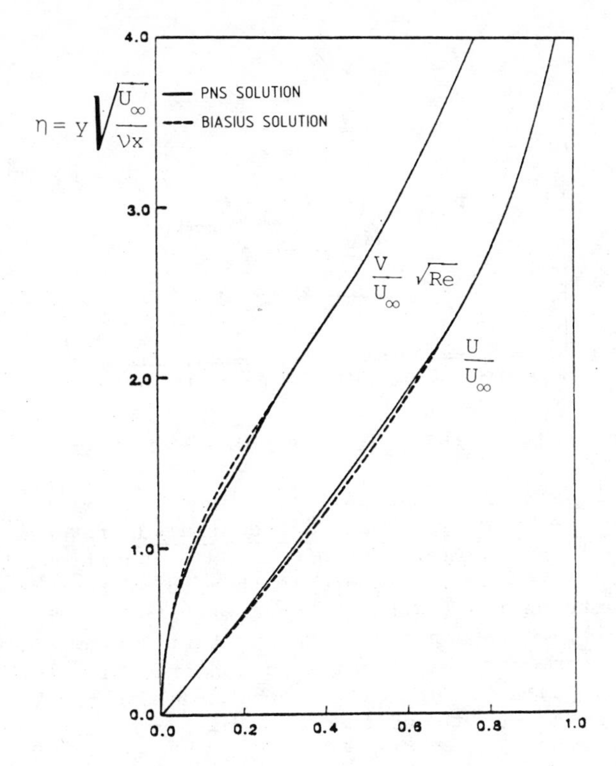

Fig. 5.3a Comparison of the PNS Velocity Profiles with the Blasius Solutions, Re = 10^3, 10^4 [51].

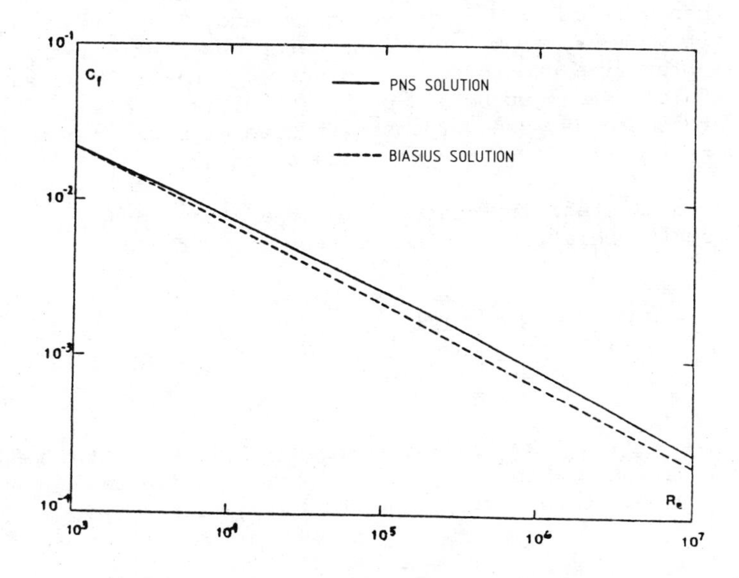

Fig. 5.3b Comparison of the Local Skin Friction Coefficient For a Flat Plate [51].

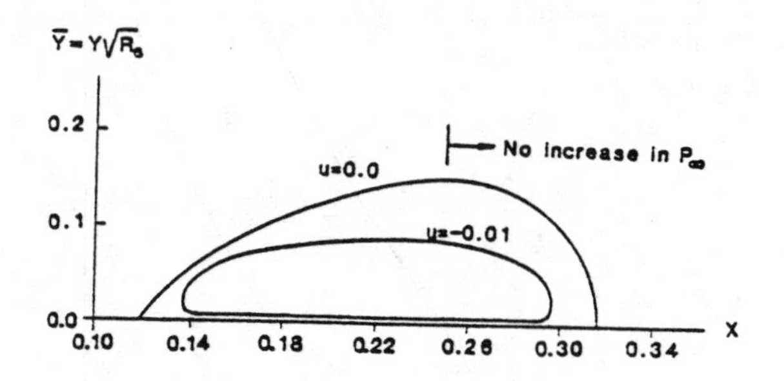

Fig. 5.4 Isovels for Separation Bubble: PNS Equations [51]

From the stability analysis for each sweep of the marching procedure, it is seen that with central differencing, the function $F(\lambda) = -1$ in (5.2). This is an unconditionally unstable condition and therefore further reinforces the need for a p_i contribution. With an appropriate p_i contribution, $F(\lambda) = (\sigma\lambda - 1)$, where σ reflects the p_i term. From (5.2), the marching procedure is stabilized conditionally for all $\sigma \neq 0$ (recall the earlier $(\Delta x)_{min}$ condition for $\sigma = 1$); however, for $\sigma \leq -1$, unconditional stability results.

Finally, a stability analysis for convergence of the global iteration procedure has also been considered. When p_x is treated implicitly, the global procedure is stable and convergence is assured. On the other hand, if p_x is treated explicitly, i.e., from a previous marching solution, the global iteration procedure will diverge. The pressure equation (5.1b) is uncoupled and p_x in (5.1a) is updated explicitly. This procedure has been proposed in several published analyses; however, the stability has not been discussed.

A local stability analysis for the global procedure leads to a complex relation for the amplification factor λ [52]. Assume

$$\begin{bmatrix} u \\ v \\ p \end{bmatrix}_{i,j} = \begin{bmatrix} A \\ B \\ C \end{bmatrix} e^{I(ik_x \Delta x + jk_y \Delta y)} \tag{5.6}$$

where $I = \sqrt{-1}$; k_x, k_y are wave numbers; i,j denote the mesh point location. With (5.6) in the forward differenced form of the PNS equations (5.1), the following matrix system results

$$G\begin{bmatrix} C \\ A \\ B \end{bmatrix}^{n+1} = \begin{bmatrix} G_{11} & G_{12} & 0 \\ 0 & G_{22} & G_{23} \\ G_{31} & G_{32} & G_{33} \end{bmatrix} \begin{bmatrix} C \\ A \\ B \end{bmatrix}^{n+1} = \begin{bmatrix} H_{11} & 0 & 0 \\ 0 & 0 & 0 \\ 0 & 0 & 0 \end{bmatrix} \begin{bmatrix} C \\ A \\ B \end{bmatrix}^{n} \tag{5.7}$$

The single eigenvalue λ is given by,

$$\lambda = \frac{H_{11}}{\det G} \det \begin{bmatrix} G_{22} & G_{23} \\ G_{32} & G_{33} \end{bmatrix} \tag{5.8}$$

The values of G_{22}, G_{23}, G_{32}, G_{33}, H_{11} are easily obtained from (5.6) and (5.1).

The condition $\lambda \leq 1$ is satisfied for all values of Δx. Therefore, with forward differencing of p_x, each marching sweep in the iteration process is departure free (5.3), and with (5.8) the multiple sweep or global procedure is stable. These conditions apply even for $\Delta x \to 0$, so that the relaxation technique is mathematically well-posed. Moreover, the pressure is treated implicitly through the coupling with the normal momentum equation.

The outer pressure boundary condition must be prescribed at a location beyond the extent of the interaction zone of influence, e.g., triple deck. The pressure boundary condition can be fixed during each sweep of the global procedure. This value will remain unchanged if the outer boundary is sufficiently far from, and unaffected by, the viscous interaction; alternatively, the pressure boundary condition can be updated prior to each sweep in order to account for viscous displacement effects. In conformal body fitted or streamline coordinates, this should be unnecessary. Unlike interactive boundary layer theory, where the outer pressure boundary value requires a local interactive treatment in order to circumvent the separation point singularity, a fixed outer pressure condition is acceptable with the PNS formulation. The normal momentum equation reflects the outer inviscid or interactive behavior and the separation singularity is automatically suppressed.

If p_x is treated explicitly or central differenced, the global procedure is unstable; if p_x is backward differenced, departure solutions appear for $\Delta x \to 0$, (5.3). With forward differencing all iterative procedures are stable and the global marching problem is well-posed.

In order to more critically examine the applicability of the global PNS procedure, with forward differencing of p_x, examples (iii) and (iv) have been considered [52]. Numerical solutions for the trailing edge and trough geometries have previously been obtained with triple deck and interacting boundary layer formulations, see for example [54, 55, 66]; these are used here to evaluate the results obtained with the PNS formulation. For the separated flow cases, solutions have been obtained with and without the inclusion of the convection terms in regions of reversed flow. For the former case, the convection derivatives are appropriately locally upwinded. Additional storage of the velocities u and v is required only for the reversed flow region. In fact, there is very little difference in the solutions with or without these terms.

For the trailing edge geometry the x and y grids were chosen so that adequate resolution was obtained in each of the layers of the triple deck structure, see references [54-56]. This requires very small values of Δx and Δy near the trailing edge and in the separation bubble for the trough geometry. The values of Δx were much smaller than $(\Delta x)_{min}$, see (5.4), for the single sweep marching procedure. This is to be expected in regions where upstream influence is large. Although the unconditional stability of the global procedure allows for $\Delta x \to 0$, in this limit, the asymptotic form of the amplification factor (5.8) is revealing. For $\lambda = \lambda_{max}$

$$\lambda \sim 1 - \frac{k_y^4}{k_x^2} \pi^2 \left(\frac{\Delta x}{y_M}\right)^4 N_x^2$$

or

$$\lambda \sim 1 - \frac{k_y^4}{k_x^2} \pi^2 \left(\frac{\Delta x}{y_M}\right)^2 \left(\frac{x_M}{y_M}\right) \tag{5.9}$$

where k_x, k_y are the wave numbers; x_M, y_M are the locations of the outer boundary in the x and y directions, respectively; i.e., $0 \leq x \leq x_m$, $0 \leq y \leq y_m$. $N_x = x_m/\Delta x$ is the number of mesh points in the x-direction. Although $\Delta x/y_M$ is not bounded below as for the single sweep method (5.3), the amplification factor λ_{max} approaches unity quadratically as $(\Delta x/y_M) \to 0$; i.e.,

$$\lambda_{max} - 1 = 0 \left(\frac{\Delta x}{y_M}\right)^{\alpha} \tag{5.10}$$

where $\alpha = 2$ for x_M/y_M fixed or $\alpha = 4$ for N_x fixed. Therefore, when increased resolution is required, i.e., $(\Delta x/y_M) << 1$, the rate of convergence of the relaxation procedure will deteriorate badly. The high frequency errors, described by (5.10), will decrease very slowly. This effect was evident in the numerical calculations, and in view of this behavior, the multi-grid strategy was applied. Since only the pressure is relaxed during the global iteration, the multigrid correction is only required for this variable. In addition, in order to simplify the programming, grid refinement was considered only for the axial (x) direction. The calculation is fully implicit in the y direction. This leads to somewhat greater storage requirements than a two-dimensional refinement; however, with only a single variable to be stored, this is not critical for the PNS application [52].

Solutions for the trailing edge geometry [example (iii)] are given in figures (5.5). The agreement with the interacting boundary layer results of [54, 55] are quite good. The finest grid includes 161 x 121 mesh points for (x,y), respectively. The coarsest grid was 41 x 121 and full convergence required only several global iterations. If the calculation was run on the finest grid alone, convergence was still not achieved after several hundred iterations. With the multigrid technique, full convergence to $0(10^{-4})$ for the maximum error in successive iterations was achieved in approximately ten to fifteen global iterations. The outer boundary y_M was chosen to lie outside the triple deck extent. If y_M violated this condition, the calculation diverged. The calculation was relatively insensitive to y_M, if y_M satisfied this condition.

The solutions for pressure and skin friction, both defined with triple deck normalization [54, 55] are shown for $Re = 10^5$. It is significant that the skin friction (velocity profile) is relatively insensitive to the grid and appears to be quite acceptable even on some of the coarser meshes. On the other hand, the pressure is extremely grid sensitive and requires the finest mesh in order to accurately represent the triple deck interaction.

The solutions for the trough geometry [66] $[y_b(x) = \varepsilon \text{ sech } 4(x-2.5)]$, $(0 \leq x \leq \infty)$ are shown in figures (5.6). Values of $\varepsilon = -0.015$ and $\varepsilon = -0.03$ were considered in [63]. Solutions were obtained for Reynolds numbers up to $Re = 3.6 \times 10^5$. Again, the agreement with the interacting boundary layer solutions is quite good. The insensitivity of C_f and the sensitivity of p to the grid is also repeated for this example. As the Reynolds number was increased, more smoothing was required on the coarser grids in order to achieve convergence to the prescribed tolerance. The outer undisturbed pressure boundary condition was held fixed at $y = y_M$ throughout the computation. There were no difficulties at separation or reattachment points. As with the trailing edge problem, full convergence was achieved in ten to fifteen global iterations. These four examples have established the validity of the forward differenced p_x, backward differenced u_x, v_x, PNS global procedure. This formulation has been examined for a more general variety of laminar and turbulent external flows in [52, 64] where results are shown for

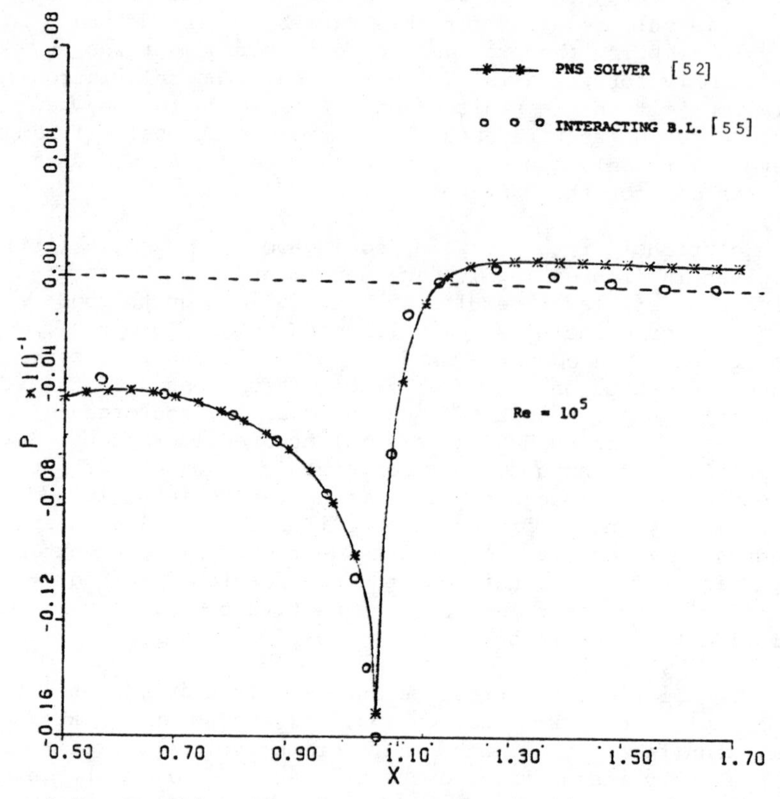

Fig. 5.5a Trailing Edge Pressure Distributions PNS Solver [52].

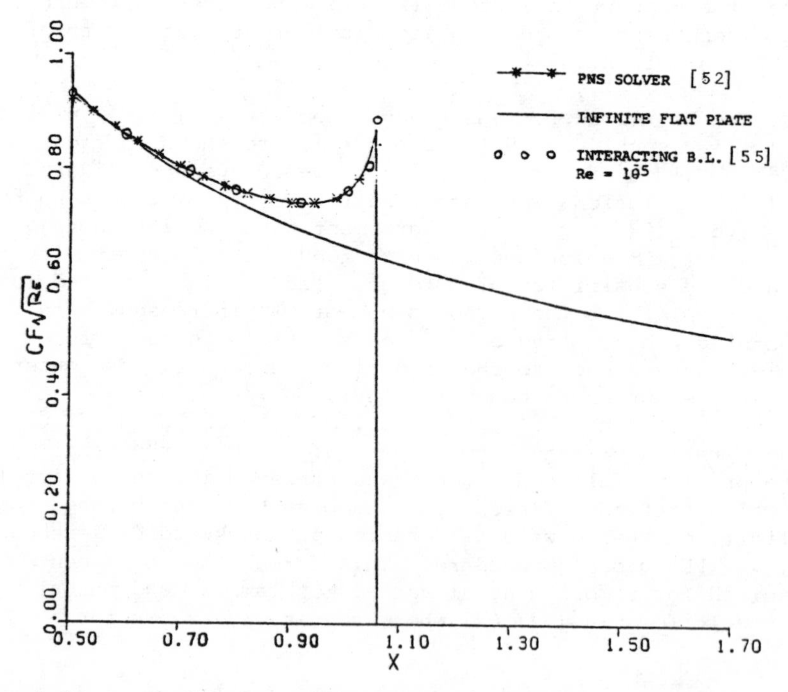

Fig. 5.5b Trailing Edge Skin Friction Solutions PNS Solver [52].

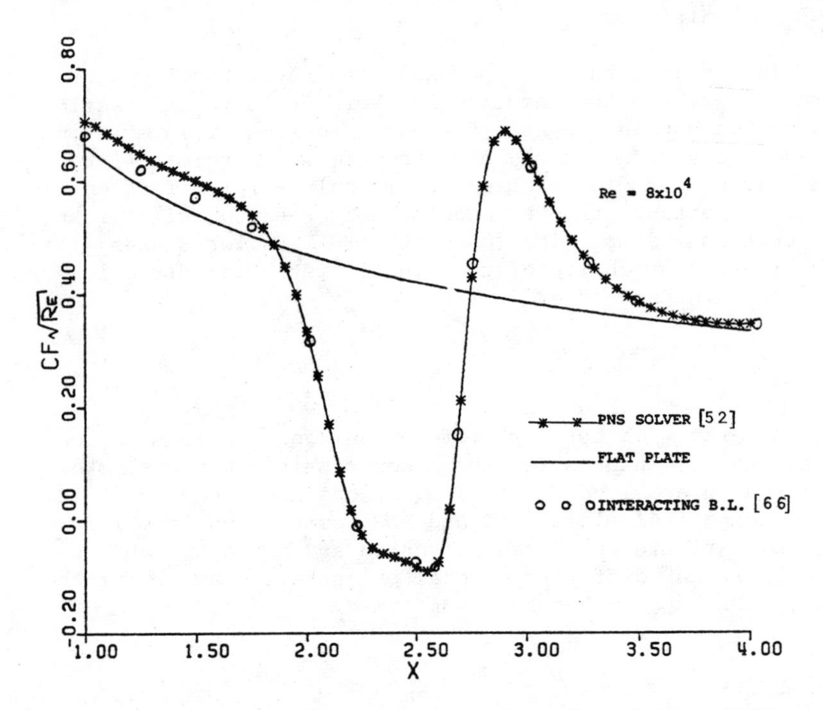

Fig. 5.6a Trough Skin-Friction Solutions PNS Solver [52]: $\varepsilon = -0.03$.

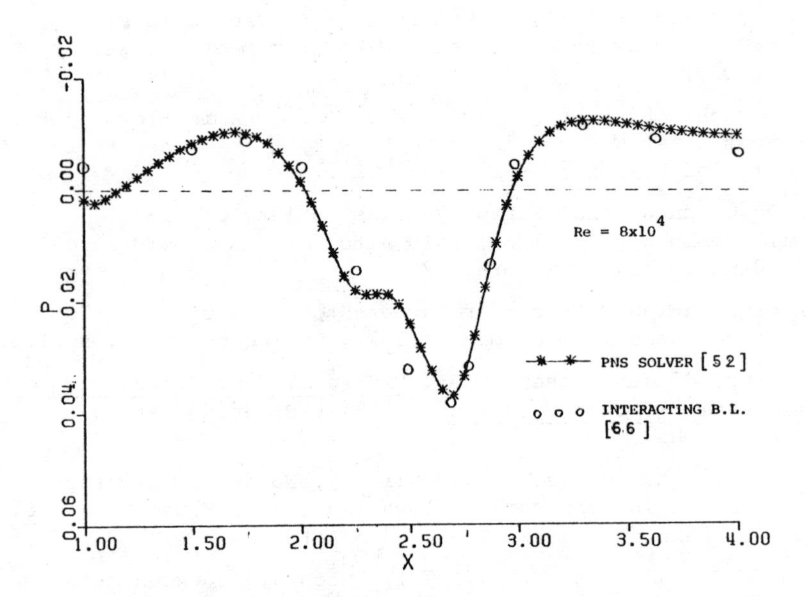

Fig. 5.6b Trough Pressure Solutions PNS Solver [52]: $\varepsilon = -0.03$.

NACA 0012 airfoils.

In summary, the PNS approximation retains all of the terms describing the complete inviscid pressure interaction, and can be considered as a further extension of interacting boundary layer and triple deck theories. Therefore, for a wide variety of flow problems where only axial diffusion is not important, the PNS model can represent a significant computational simplification of the full Navier-Stokes equations. The choice of coordinate system is extremely important in this regard.

6. SUMMARY

A variety of current differential formulations and computational techniques for the incompressible Navier-Stokes (NS) and parabolized NS (PNS) equations have been reviewed. A new composite velocity NS and PNS formulation in primitive (u,v,p) variables has been presented and the applicability of a "forward" difference global pressure iteration for the (u,v,p) PNS system has been confirmed.

REFERENCES

1. 5th ICNMFD (1976), 6th ICNMFD (1978), 7th ICNMFD (1980), Lecture Notes in Physics, 59, 90, 141, Springer-Verlag, Berlin.
2. Ghia, K. ed. (1981), Computers in Flow Predictions and Fluid Dynamics Experiments, ASME Winter Meeting, Washington, D.C.
3. Numerical and Physical Aspects of Aerodynamic Flows (1981), Symposium at California State Univ. Proceedings to be published by Springer-Verlag.
4. AIAA Computational Fluid Dynamics Conference (1979) Williamsburg, VA, (1981) Palo Alto, CA. Symposium Proceedings, AIAA, New York, NY.
5. Approximation Methods for Navier-Stokes Problems (1979). Lecture Notes in Mathematics, 771, Springer-Verlag, Berlin.
6. Taylor, C. and Hughes, T.G., Finite Element Programming of the Navier-Stokes Equations, (1981), Pineridge Press, Swansea, U.K.
7. Orszag, S.A. and Israeli, M. (1974), Numerical Simulation of Viscous Incompressible Flows, Ann. Rev. Fluid Mech., 6, p. 281, Annual Reviews Inc., Palo Alto, CA.
8. Wu, J.C. and Rizk, Y.M. (1979), Integral-Representation Approach for Time-Dependent Viscous Flows, 6th ICNMFD, Lecture Notes in Physics, 90, p. 558, Springer-Verlag, Berlin.

9. Leonard, A. (1976), Simulation of Three-Dimensional Separated Flows with Vortex Filaments, 5th ICNMFD, Lecture Notes in Physics, 59, Springer-Verlag, Berlin, p. 280.

10. Lagerstrom, P.A. (1964), Theory of Laminar Flows, Section B, p. 55, High Speed Aerodynamics and Jet Propulsion, IV, F. Moore, ed., Princeton University Press, Princeton, NJ.

11. Thompson, J. (1980), Numerical Solution of Flow Problems Using Body-fitted Coordinate Systems, p. 1, Computational Fluid Dynamics, Lecture at von Karman Institute, W. Kollmann, ed., Hemisphere Publication, New York, NY.

12. Steger, J.L. and Swanson, R.L. (1980), Use of Hyperbolic Partial Differential Equations to Generate Body Fitted Coordinates, Proc. NASA Workshop on Numerical Grid Generation Techniques, NASA, Hampton, VA, p. 463.

13. Eisemann, P. (1980), Geometric Methods in Computational Fluid Dynamics, Lecture Notes from von Karman Institute course on Shock-Boundary Layer Interaction in Turbomachines; also, ICASE Report No. 80-11, NASA Langley Research Center, Hampton, VA.

14. NASA Workshop on Numerical Grid Generation Technique (1980), NASA Langley Research Center, Hampton, VA.

15. Thompson, J. et al. (1982), (Review to appear in J. Comp. Phys.).

16. Hirt, C.W. and Harlow, F.H. (1967), A General Corrective Procedure for the Numerical Solution of Initial-Value Problems, J. Comp. Phys., 2, p. 114.

17. Ghia, U. et al. (1981), Study of Incompressible Flow Separation Using Primitive Variables, Computers and Fluids, 9, 2, p. 123.

18. Numerical Studies of Incompressible Flow in a Driven Cavity (1975), NASA SP-378, S. Rubin and J. Harris, eds.

19. Rubin, S.G. and Khosla, P.K. (1981), Navier-Stokes Calculations with a Coupled Strongly Implicit Method. I - Finite-Difference Solutions, Computers and Fluids, 9, 2, p. 163.

20. Briley, W.R. and McDonald, H. (1980), On the Structure and Use of Linearized Block Implicit Schemes, J. Comp. Phys., 34, 1, p. 54.

21. Briley, W.R. and McDonald, H. (1977), Solution of the Multidimensional Compressible Navier-Stokes Equations by a Generalized Implicit Method, J. Comp. Phys., 24, p. 372.

22. Beam, R.M. and Warming, R.F. (1978), An Implicit Factored Scheme for the Compressible Navier-Stokes Equations, AIAA Journal, 16, p. 393.

23. Osswald, G.A. (1981), Analysis and Numerical Solutions of Unsteady Two-Dimensional Navier-Stokes Equations in Primitive and Derived Variables, M.S. Thesis, Univ. of Cincinnati.

24. Khosla, P.K. and Rubin, S.G. (1981), A Conjugate Gradient Iterative Method. Computers and Fluids, 9, 2, p. 109; also, 7th ICNMFD, Lecture Notes in Physics, 141, Springer-Verlag, Berlin, p. 248.

25. Wesseling, P. and Sonneveld, P. (1979), Numerical Experiments with a Multiple Grid and a Preconditioned Lanczos Type Method. Lecture Notes in Mathematics, 771, Springer-Verlag, Berlin, p. 543.

26. Ghia, U., Ghia, K.N. and Shin, C.T. (1981), Solution of Incompressible Navier-Stokes Equations by Coupled Strongly Implicit-Multi Grid Method. Symposium on Multi Grid Methods, NASA Ames Research Center, Moffett Field, CA.

27. Brandt, A. and Dinar, N. (1979), Multi-Grid Solutions to Elliptic Flow Problems, Numerical Methods for Partial Differential Equations, S. Parker ed., Academic Press, New York, p. 333.

28. Mol, W.J.A. (1980), Numerical Solution of the Navier-Stokes Equations by Means of a Multi Grid Method and Newton-Iteration. 7th ICNMFD, Lecture Notes in Physics, 41, Springer-Verlag, Berlin, p. 285.

29. Ghia, U. and Davis, R.T. (1974), Navier-Stokes Solutions for Flow Past a Class of Two-Dimensional Semi-Infinite Bodies, AIAA Journal, 12, 2, p. 1659.

30. Alfrink, B.J. (1981), On the Neumann Problem for Pressure in a Navier-Stokes Model. Numerical Methods in Laminar and Turbulent Flow, C. Taylor and B.A. Schrefler, ed., Pineridge Press, Swansea, U.K.

31. Hirsh, R.S. (1975), Higher-Order Accurate Difference Solutions of Fluid Mechanics Problems by a Compact Differencing Technique, J. Comp. Phys., 19, p. 90.

32. Roux, B. et al. (1979), Optimization of Hermitian Methods for Navier-Stokes Equations in the Vorticity and Stream-Function Formulation,Lecture Notes in Mathematics, 771, Springer-Verlag, Berlin, p. 450.

33. Benjamin, A.S. and Denny, V.E. (1979), On the Convergence of Numerical Solutions for 2-D Flows in a Cavity at Large Re, J. Comp. Phys., 33, p. 340.

34. Rubin, S.G. and Khosla, P.K. (1976), Higher-Order Numerical Solutions Using Cubic Splines. AIAA J., 14, p. 851.

35. Rubin, S.G. and Khosla, P.K. (1977), Polynomial Interpolation Methods for Viscous Flow Calculations, J. Comp. Phys., 24, 3, p. 217.

36. Rubin, S.G. and Khosla, P.K. (1978), A Simplified Spline Solution Procedure. 6th ICNMFD, Lecture Notes in Physics, 90, Springer-Verlag, Berlin, p. 468.

37. Rubin, S.G. and Khosla, P.K. (1979), Navier-Stokes Calculations with a Coupled Strongly Implicit Procedure. II - Spline Deferred-Corrector Solutions. Lecture Notes in Mathematics, 771, Springer-Verlag, Berlin, p. 469.

38. Keller, H. (1981), Continuation Methods in Computational Fluid Dynamics. Symposium on Numerical and Physical Aspects of Aerodynamic Flows, Long Beach, CA (to be published by Srpinger-Verlag, T. Cebeci, ed.).

39. Khosla, P.K. and Rubin, S.G. (1974), A Diagonally Dominant Second-Order Accurate Implicit Scheme, Computers and Fluids, 2, 2, p. 207.

40. Pratap, V.S. and Spalding, D.B. (1976), Fluid Flow and Heat Transfer on Three-Dimensional Duct Flows. Int. J. Heat and Mass Transfer, 19, p. 1183.

41. Dodge, P.R. and Lieber, L.S. (1977), A Numerical Method for the Solution of Navier-Stokes Equations with Separated Flow. AIAA Paper No. 77-170.

42. Khosla, P.K. and Rubin, S.G. (1982), A Composite-Velocity/CSIP Procedure for the Compressible Navier-Stokes Equations. AIAA Paper No. 82-0099.

43. Rubin, S.G. and Khosla, P.K. (1982), A Composite-Velocity/CSIP Procedure for the Incompressible Navier-Stokes Equations, 8th ICNMFD, Aachen, W. Germany, July 1982, Springer-Verlag, Berlin.

44. Davis, R.T. (1979), Numerical Methods for Coordinate Generation Based on a Schwarz-Christoffel Mapping Technique," Lectures at von Karman Institute on Computational Fluid Dynamics, 1980, Brussels, Belgium, Hemisphere Press, Washington, D.C.

45. Dennis, S.C.R., Ingham, D.B. and Cook, R.N. (1979), Finite Difference Methods for Calculating Steady Incompressible Flows in Three-Dimensions. J. Comp. Phys., 33, p. 325.

46. Lin, A. and Rubin, S.G. (1982), Three-Dimensional Supersonic Viscous Flow Over a Cone at Incidence. AIAA J., 20, 11, pp. 1500-1507.

47. Stone, H.L. (1968), Iterative Solution of Implicit Approximations of Multidimensional Partial Differential Equations. SIAM J. Num. Anal., 5, p. 530.

48. Schwartztrauber, P.N. and Sweet, R.A. (1977), The Direct Solution of the Discrete Poisson Equation on a Disc. SIAM J. Num. Anal., 5, p. 900.

49. Saylor, P. (1974), Second-Order Strongly Implicit Symmetric Factorization Methods for the Solution of Elliptic Difference Equations. SIAM J. Num. Anal., 11, p. 894.

50. Davis, R.T. and Rubin, S.G. (1980), Non-Navier-Stokes Viscous Flow Computations. Computers and Fluids, 8, p. 101.

51. Rubin, S.G. (1981), A Review of Marching Procedures for Parabolized Navier-Stokes Equations. Proc. Numerical and Physical Aspects of Aerodynamics Flows, California State Univ., Long Beach, CA, Springer-Verlag, pp. 171-186.

52. Rubin, S.G. and Reddy, D.R. (1983), A Global PNS Solution Procedure for Laminar Interacting and Separated Flows. 2nd Symposium on Numerical and Physical Aerodynamic Flow, California State Univ., Long Beach, CA.

53. Rudman, S. and Rubin, S.G. (1968), Hypersonic Flow Over Slender Blades Having Sharp Leading Edges. AIAA Journal, 6, p. 1883.

54. Veldman, A.E.P. (1979), A Numerical Method for the Calculation of Laminar, Incompressible Boundary Layers with Strong Viscous-Inviscid Interaction. Nat. Aero. Lab. (Netherlands), Report No. NLR TR 79023 U.

55. Davis, R.T and Werle, M.J. (1981), Progress on Interacting Boundary Layer Calculations at High Reynolds Numbers. Proc. Numerical and Physical Aspects of Aerodynamic Flows, California State University, Long Beach, CA, Springer-Verlag, 187-210.

56. Stewartson, K. (1974), Multistructural Boundary Layers on Flat Plates and Related Bodies. Advances in Mechanics, Academic Press, New York, 14, p. 145.

57. Lighthill, M.J. (1953), On Boundary Layers and Upstream Influence. I. A Comparison Between Subsonic and Supersonic Flows. Proc. Roy. Soc., A, Vol. 217, 1953.

58. Lin, T.C. and Rubin, S.G. (1973), Viscous Flow Over a Cone at Moderate Incidence. Computers and Fluids, 1, 1, pp. 37-57, January 1973.

59. Lubard, S.C. and Helliwell, W.S. (1975), An Implicit Method for Three-Dimensional Viscous Flow with Application to Cones at Angle of Attack. Computers and Fluids, 3, 1, pp. 83-101, March 1975.

60. Vigneron, Y.C., Rakich, J. and Tannehill, T.C. (1978), Calculation of Supersonic Viscous Flow Over Delta Wings With Sharp Supersonic Leading Edges. AIAA Paper 78-1137, January 1978.

61. Yanenko, N.N., Kovenya, V.M., Tarnavsky, G.A. and Chernyi, S.G. (1980), Economical Methods for Solving the Problems of Gas Dynamics. Presented at 7th Intern. Conf. on Numerical Methods in Fluid Dynamics, Stanford Univ.

62. Lin, T.C. and Rubin, S.G. (1979), A Numerical Model for Supersonic 3-D Viscous Flow Over a Slender Vehicle. AIAA Paper No. 79-0205, January 1979.

63. Rubin, S.G. and Lin, A. (1980), Marching with the PNS Equations. Proceedings of 22nd Israel Annual Conference on Aviation and Astronautics, Tel Aviv, Israel, pp. 60-61, March 1980; also, Israel Journal of Technology, 18, 1980.

64. Khosla, P.K. and Lai, H. (1983), Global PNS Solutions for Subsonic Strong Interaction Flows, submitted for 6th AIAA CFD Conference, Danvers, MA.

65. Israeli, M. (1982), Private Communication.

66. Carter, J.E. and Wornom, S.F. (1975), Solutions for Incompressible Separated Boundary Layers Including Viscous-Inviscous Interaction. NASA SP-347, p. 125.

ACKNOWLEDGEMENT

The research of the present author was supported by the Air Force Office of Scientific Research, Grant No. AFOSR 80-0047, the Office of Naval Research, Contract No. N00014-79-C-0849, and the NASA Langley Research Center, Grant No. NAG-I-8.

MULTI-GRID TECHNIQUE FOR THE SOLUTION OF INCOMPRESSIBLE NAVIER-STOKES EQUATIONS

K.N. GHIA
DEPARTMENT OF AEROSPACE ENGINEERING AND APPLIED MECHANICS

U. GHIA
DEPARTMENT OF MECHANICAL ENGINEERING

UNIVERSITY OF CINCINNATI, CINCINNATI, OHIO 45221, U.S.A.

1. INTRODUCTION

The computation of numerical solutions of the Navier-Stokes equations experiences increasing difficulty as the flow Reynolds number increases, particularly for configurations encountering flow separation. The use of a coarse grid can frequently yield a converged numerical solution but such a solution may often be physically inaccurate. Refining the computational grid usually leads to improved representation of the physical flow phenomena, but is accompanied by deterioration in the convergence rate of conventional iterative methods. This has led several researchers to examine the multi-grid (MG) method as a means for obtaining rapidly converging solutions of elliptic boundary value problems.

The potential of the MG method has been adequately exposed and demonstrated ([1]-[6]) for the system of discretized equations arising from a single differential equation. However, application of the MG technique to the solution of a system of coupled nonlinear differential equations still poses several questions which are currently being studied by various investigators ([7]-[9]). The proceedings [10] of a recent conference on multi-grid methods provides a timely update of the current status of multi-grid technology and is suggested as further reading on this topic.

This chapter describes an approach for obtaining fine-mesh solutions of the Navier-Stokes equations for high-Re flow, including separation, in a computationally efficient manner. The approach consists of implementing a multi-grid solution procedure with suitably chosen elements, as described in the following section, such that the resulting overall solution technique is efficient as well as robust, i.e., useful for a wide range of problems and problem parameters. The robustness and the efficiency of the solution technique are demonstrated via application to three model problems. The first model problem considered is that of flow in a driven cavity. By virtue of its geometrical simplicity, this problem has been used by numerous researchers to test new numerical techniques. Consequently, a number of numerical solutions are available for this problem, the most significant amongst them being those due to Benjamin and Denny [11], Agarwal [12] and Schreiber and Keller [13], as these are for high Reynolds number. These solutions serve as a check on the accuracy of the present solutions and also enable evaluation of the efficiency of the solution technique developed. Recently, Peterson [14] and Gresho et al. [15] have obtained high-Re solutions for the shear-driven cavity problem using finite-element techniques including improvements to enhance computational efficiency. It is the present authors' understanding that the central idea behind the reduction-of-basis technique of Noor [16] employed by Peterson [14] may be interpreted to bear a striking resemblence to the principles on which the multigrid technique is based.

The second model problem considered is that of downstream-asymptotic flow in curved ducts of square and polar sections. Curved-duct flows have been investigated previously, e.g., Cheng, Lin and Ou [17], K. Ghia and Sokhey [18], and U. Ghia et al. [19]. However, computational limitations in the past have precluded resolving some questions related to the phenomenon of Dean's instability for these flows, as discussed in a later Section of this chapter. The recent fine-grid calculations performed by K. Ghia et al. [9] for this problem serve to conclusively determine, firstly, the critical value of the Dean number K at which Dean's instability occurs for square cross-section ducts and, secondly, whether Dean's instability is encountered for any combination of the parameters for polar cross-section ducts. These results are obtained using the velocity-vorticity formulation and are useful for verifying solutions obtained using different formulations and other numerical techniques.

In the third and last application considered, the MG technique is advanced for use with Neumann boundary-value problem in clustered curvilinear orthogonal coordinates. This comprises an important step in the analysis of viscous flows using the velocity-pressure formulation of the Navier-Stokes equations. In this formulation, pressure is frequently determined as the solution of a Neumann boundary-value problem. Such a problem possesses certain unique features which must be honored by its numerical solution procedure. Furthermore, clustered coordinates are generally used in viscous-flow calculations to resolve the high-gradient regions near solid boundaries. This model problem is also used to examine the influence of the basic iterative scheme used in conjunction with the MG procedure.

2. BASIC PRINCIPLE OF MULTI-GRID TECHNIQUE

The Multi-Grid (MG) technique is a numerical strategy for enhancing the convergence rate of an iterative procedure. The discretized equations corresponding to a boundary-value problem can be solved using a conventional iterative technique. Generally, the convergence of these schemes is fast during the first few iterations only, and is effective only on those Fourier components of the error whose wavelengths are smaller than or comparable to the grid increment. The more slowly varying components of the error are rather slow in being annihilated. The MG method is based primarily on this principle. It recognizes that a wave-length which is long relative to a fine mesh is short relative to a coarser mesh. It employs a hierarchy of grids and, as the convergence rate of the solution on the fine grid becomes slow, the MG procedure switches to a coarser grid, where the longer-wavelength errors are more effectively damped. The solution on the finer grid is then corrected appropriately to reflect the removal of these long-wavelength Fourier components of the error. Clearly, several possibilities present themselves in the development of the MG procedure. For example, the process may begin with the coarsest grid (Full MG procedure) or with the finest grid (cycling algorithms). The process may work with the full solution function (Full Approximation Storage) or only with the correction to the solution (Correction Storage); the latter is useful for linear problems while the former is needed in nonlinear problems. Choice has to be made for the basic iterative scheme to be used for high-frequency error damping, i.e., for smoothing the error. Operators need to be defined for transferring the error from the fine grid to the coarse grid (restriction operator) and the correction from the coarse to the fine grid (prolongation operator). Further, convergence and convergence rate need to be defined. Brandt [20] has provided guidelines for dealing with these and other issues pertinent to the MG process. U. Ghia et al. [21] and K. Ghia et al. [9] have applied Brandt's guidelines and investigated them further, specifically in relation to the solution of Navier-Stokes equations.

3. OUTLINE OF FULL MULTI-GRID (FMG) SOLUTION PROCEDURE

The general multi-grid solution technique as well as its possible variations have been described in a number of references (e.g. [2], [4], [6], [8]). The discussion given here describes the version employed by the present authors. The mathematical model describing the flow problem can be represented symbolically by the differential equation

$$LU = F \tag{1}$$

in a domain R, with the boundary conditions given as

$$\Lambda U = G \tag{2}$$

on the boundary ∂R of the solution domain R. In Eqs. (1) and (2), U is the solution vector, L and A are matrices whose elements contain at most second-order and first-order differential operators, respectively, which may well be nonlinear. The vectors F and G are prescribed functions of the independent variables only.

The MG procedure starts with selecting a suitable discretization operator whose order of accuracy is denoted as d. Next, a sequence of discrete grids D^k, $k = 1,2,\ldots,M$, with step size Δ_k, is defined such that $\Delta_{k+1}/\Delta_k = 1/2$ and, for the finest grid, $\Delta_M = \Delta$. The full multi-grid (FMG) procedure starts with the coarsest grid, i.e., $k = 1$, and is generally preferred when a good starting solution on the finest grid is not available. On the other hand, if a good initial guess for the fine-mesh solution is available, as in continuation problems involving increasing values of a flow parameter, it may be preferable to use the cycling MG procedure which starts with the finest grid. Either procedure becomes identical to the other after one pass through the complete sequence of grids employed, although their approach to convergence may sometimes differ. In the FMG procedure, a converged solution is first obtained on the coarsest grid, i.e.,

$$u^k = (L^k)^{-1} F^k, \qquad k = 1 \tag{3}$$

where u denotes the finite-difference approximation to U and superscript k denotes discretization on grid level k. The solution indicated in Eq. (3) can always be achieved; if simple iterative techniques fail, more powerful implicit techniques may be resorted to; even a direct solver may sometimes be useful on the coarsest grid. The solution u^k is then prolongated, using a prolongation operator P_k^{k+1}, to provide an estimate u_{est}^{k+1} for the solution on subsequent finer grid level (k+1) as

$$u_{est}^{k+1} = P_k^{k+1} u^k \quad . \tag{4}$$

With u_{est}^{k+1} as the initial guess, the solution u^{k+1} on grid level (k+1) is obtained by solving the governing differential equation (1), so that

$$u^{k+1} = (L^{k+1})^{-1} F^{k+1} \quad . \tag{5}$$

The steps indicated by Eqs. (4) and (5) are repeated until the solution on the finest grid level M is achieved. The description given so far is a simple-minded use of a sequence of grids; the idea has been traditionally employed in obtaining converged fine-mesh solutions for flow problems. However, the total procedure is not generally very efficient because, as k increases, the solution required by Eq. (5) converges rather slowly, in spite of the good initial guess provided by Eq. (4). It is in these circumstances that the MG procedure plays a significant role.

When Eq. (5) exhibits unsatisfactory convergence rate, a multi-grid cycle is interjected in the procedure. This consists of returning the evolving solution, denoted as u^{k+1}_{old}, to the coarser grid k using a restriction operator R^{k}_{k+1} such that

$$u^{k} = R^{k}_{k+1}\ u^{k+1}_{old} \ , \tag{6}$$

and correcting the evolving fine-grid solution u^{k+1}_{old} to provide an improved fine-grid solution u^{k+1}_{new} as

$$u^{k+1}_{new} = u^{k+1}_{old} + P^{k+1}_{k}\ (u^{k}_{k+1} - R^{k}_{k+1}\ u^{k+1}_{old}) \ . \tag{7}$$

Here, u^{k}_{k+1} is a correction, determined on the coarse grid k, and governed by the equation

$$L^{k}\ u^{k}_{k+1} = f^{k} \tag{8}$$

where

$$f^{k} \triangleq L^{k}(R^{k}_{k+1}\ u^{k+1}_{old}) + R^{k}_{k+1}\ (f^{k+1} - L^{k+1}\ u^{k+1}_{old}) \ . \tag{9}$$

It is important to note that, for the finest grid executed up to a given stage in the MG procedure, the governing equation corresponds to Eq. (5) and not to Eqs. (8) and (9).

If the solution of Eq. (8) itself exhibits slow convergence rate, yet another MG cycle may be interjected within the previous one, in order to obtain improved convergence rate for Eq. (8). Thus, a series of MG cycles may be nested, one within another, so as to obtain rapid convergence for the overall solution. In practice, it is not essential to achieve a fully converged solution for Eq. (8). It is only necessary that u^{k}_{k+1} be sufficiently smooth on grid k so that it can be meaningfully prolongated to grid (k+1), for use in Eq. (7). The required smoothing can be achieved by use of an appropriate smoothing operator.

The foregoing discussion provides an outline of the FMG procedure as used by the present authors. Since the first two model problems are governed by nonlinear equations, these problems are solved using the full-approximation scheme (FAS) whereas the correction scheme is employed for the third model problem which is linear. In the course of describing the procedure, several operators have been referred to, namely, those for discretization, prolongation, restriction and smoothing, as well as the operator in the coarse-grid correction equation (8). These, as well as the parameters used to define convergence and convergence rate in order to effect transfer between the coarse and fine grids, are discussed next.

4. OPERATORS AND PARAMETERS USED IN MULTI-GRID METHOD

4.1 Discretization Operator

A second-order accurate finite-difference scheme is generally employed. Hence, in terms of the nomenclature introduced in the preceding section for the order of accuracy of discretization, $d = 2$. Higher-order accuracy may be attained either by employing a discretization operator with $d > 2$ or by using 'τ-extrapolation' in the multi-grid process as discussed by Brandt [20].

4.2 Prolongation Operator P_k^{k+1}

Two types of prolongation operators are employed. Whenever the solution on a given coarse-grid has converged completely, it is desirable to retain as much of the information it contains as possible while prolongating it to the subsequent finer grid. For example, this is true of the solution on the coarsest grid when $k = 1$. In this case, the prolongation operator employed consists of cubic interpolation. For simplicity in multi-dimensional problems, only 1-D cubic-polynomial interpolation is used, first along one coordinate direction, to provide an interpolated value at the mid-point of a three-step interval, as indicated in Fig. 1.

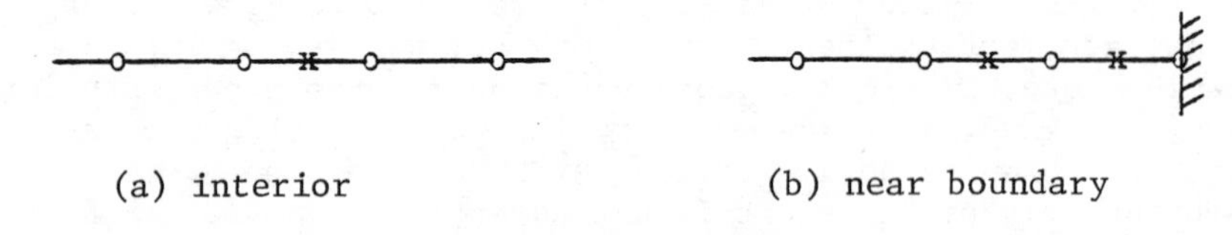

(a) interior (b) near boundary

o coarse-grid point

x additional fine-grid point

Fig. 1. 1-D Cubic Interpolation.

Adjacent to the boundary, the interpolating polynomial employed for the center-point is also used to provide an interpolated value at the fine-grid point next to the boundary. After all coordinate lines of one family have been considered, the procedure is repeated for all coordinate lines of the second family, including those formed by connecting the points indicated by the symbol x along the second coordinate direction.

For prolongating the coarse-grid correction u_{k+1}^{k} which, as mentioned earlier, is not required to be a fully converged solution of Eq. (8), it is sufficient to use linear interpolation. The procedure is analogous to that described for cubic interpolation, except for the fact that no special considerations are required adjacent to the boundaries.

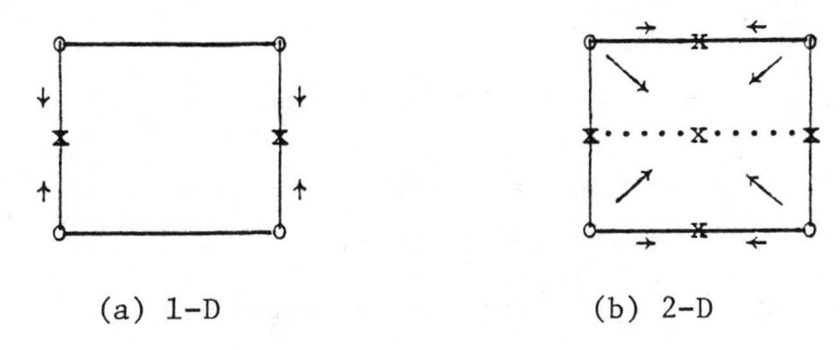

(a) 1-D (b) 2-D

Fig. 2. Linear Interpolation.

For 2-D problems, the linear interpolation operator has also been termed the 9-point prolongation operator in the MG-literature [6]. For the Neumann boundary-value problem considered in Section 7, linear interpolation is employed even for the coarsest grid, for reasons to be explained in that section.

4.3 Restriction Operator R_{k+1}^{k}

Injection is the simplest of the restriction operators. As the name suggests, injection consists of simply assigning to a coarse-grid point the value of the function at the corresponding fine-grid point. Its simplicity makes it attractive to use. But it is satisfactory mainly for problems with constant or slowly-varying coefficients. For nonlinear problems as well as for linear problems with rapidly varying coefficients or source terms, restriction to a coarse-grid point must include the influence of the neighboring fine-grid points. This is analogous to the concept of weighted averaging and leads to the 5-point and the 9-point restriction operators depicted in Fig. 3. Here a, b and c denote weighting factors associated with the fine-grid values at the points indicated in the figure.

Five-point restriction is superior to injection. However, in extreme cases such as some very high Reynolds number flows as well as for the Neumann problem, the use of 9-point restriction was vital to the success of the overall method.

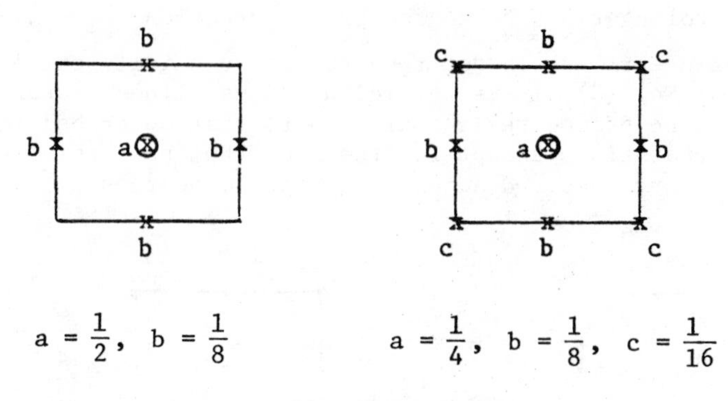

x fine-grid point

o coarse-grid point

(a) 5-Point Restriction (b) 9-Point Restriction

Fig. 3. Restriction Operator.

It is important to recognize that the 9-point prolongation operator and the 9-point restriction operator form a pair of adjoint operators. This is useful in defining the coarse-grid operator L^k.

4.4 Operator L^k in Coarse-Grid Correction Equation

This operator makes direct use of the discretization operator defined at the beginning of this section. For the constant-coefficient case, no additional information is needed for defining L^k. For non-constant-coefficient cases, as are of interest in fluid flows, it is necessary to provide the coefficients appropriate for use in L^k. It is found that the coefficients in L^k should be obtained from those in L^{k+1} using the same restriction operator as that used in Eqs. (6), (7) or (9) and defined in the preceding sub-section.

4.5 Smoothing Operator

During the first few sweeps of their application, conventional relaxation and iterative numerical methods are quite effective as smoothers. These include the Gauss-Seidel (GS) relaxation procedure (and its several variants, e.g., successive over-relaxation (SOR)) as well as the alternating-direction implicit (ADI) method which may be viewed as a locally 1-D implicit procedure. During each fractional time-step of the ADI method, a tridiagonal matrix needs to be inverted. This is efficiently accomplished by use of the Thomas algorithm.

However, for extreme values of the problem parameters and for highly refined meshes, the overall effectiveness of the ADI scheme becomes very sensitive to the time-step used. Generally, considerable reduction in the value of the time step becomes necessary, leading to slow convergence rate. In these cases, a strongly implicit (SI) solution procedure is employed. This has its origin in the work of Stone [22] and may be viewed as a generalization of the Thomas algorithm to two dimensions. The recent extension implemented by Rubin and Khosla [23] for simultaneous solution of two coupled differential equations using the coupled strongly implicit (CSI) procedure is particularly useful in conjunction with the MG method. The details of the SI and CSI methods are given in these references [22], [23], and, hence, are not included here.

4.6 Parameters in Convergence Definitions

The MG method requires that convergence as well as convergence rate of the solution be monitored at various stages of the procedure. In general, the criterion for convergence on grid level k may be stated as

$$e_k \leq \varepsilon_k \quad \text{where} \quad e_k = \|\tilde{R}_k\| = [\sum_{i,j} (\tilde{R}_k)^2_{i,j} \Delta_k^2]^{1/2} \tag{10}$$

with $\tilde{R}_k$ representing the dynamic residual in the solution (i.e., $\tilde{R}_k = Lu-f$) and ε_k the required tolerance which is defined next.

For the coarsest grid, convergence must be achieved in an absolute sense. Typically,

$$\varepsilon_k = 10^{-4} \quad \text{for} \quad k = 1 \ . \tag{11}$$

For convergence of the coarse-grid correction equation (8), ε_k is taken to be a fraction of the residual norm on the next finer grid, i.e.,

$$\varepsilon_k = \alpha \, e_{k+1} \quad \text{where} \quad \alpha < 1 \ . \tag{12}$$

When this condition is satisfied, the calculation returns to the next finer grid.

Convergence on the currently finest grid is defined to occur when the residual norm becomes smaller than the truncation error on the next coarser grid relative to the current finest grid, i.e.,

$$\varepsilon_{k+1} = \left(\frac{\Delta_{k+1}}{\Delta_k}\right)^{2q} \|f^k - R^k_{k+1} F^{k+1}\| \ . \tag{13}$$

Generally, $q = 1$ has been employed in the MG-literature. The present authors found it necessary to use $q > 1$ in order to obtain good comparison with reliable non-MG data available for the driven-cavity problem.

The criterion used for interjecting an MG-cycle in the procedure was based on the convergence rate of the solution on a given grid. Defining convergence rate as the ratio of the residual norms for two successive iterations N and (N+1), the convergence rate on grid k is considered to be satisfactory if

$$e_k^{N+1} < \lambda \, e_k^N \tag{14}$$

where $\lambda < 1$ denotes the theoretical convergence rate (or smoothing factor) of the iterative solution scheme employed. For increasing stiffness of the problem considered, λ is observed to increase towards unity.

The full-MG procedure was employed for the two model problems discussed in Sections 5 and 6, whereas a cycling-MG procedure was used for the third model problem described in Section 7 which, therefore, also includes some remarks pertinent to the cycling algorithm.

In applying the MG technique to a system of equations, convergence rate was tested separately for each of the governing equations of the system, but convergence was tested for the overall procedure as a whole. Hence, the symbol e in Eq. (10) must represent the arithmetic average of the root-mean-square (RMS) residual in all the governing equations, while that in Eq. (14) corresponds to the RMS residual in each individual equation.

5. APPLICATION TO FLOW IN DRIVEN CAVITY

5.1 Mathematical Representation

The Navier-Stokes equations for 2-D laminar incompressible flow, in terms of the vorticity ζ and the stream function ψ as dependent variables, can be written as

$$\zeta_t + u\,\zeta_x + v\,\zeta_y = \frac{1}{Re}\,(\zeta_{xx} + \zeta_{yy}) \tag{15}$$

and

$$\psi_{xx} + \psi_{yy} = -\,\zeta \tag{16}$$

where

$$u = \psi_y, \quad v = -\,\psi_x \quad \text{and} \quad \zeta = u_y - v_x\,. \tag{17}$$

The coordinates and the boundary conditions for the flow in a driven cavity are shown in Fig. 4.

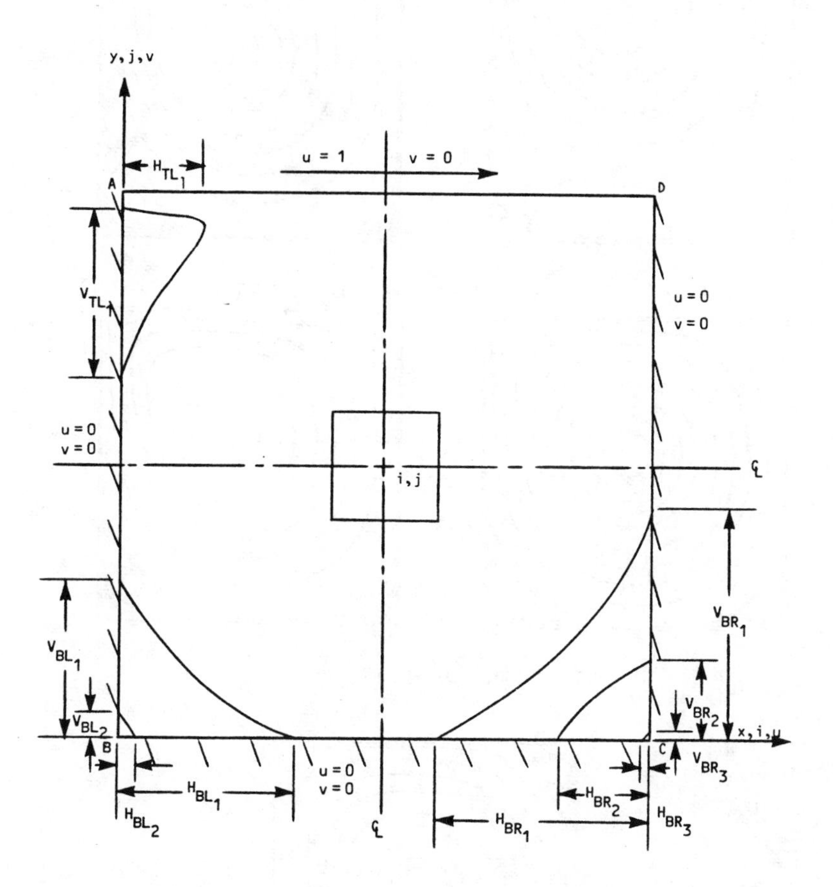

Fig. 4. Cavity-Flow Configuration, Coordinates, Nomenclature and Boundary Conditions.

5.2 Fine-Grid Results

Figure 5 shows the streamline contours for this flow at Re = 100, 400 and 1000 as well as the high value of Re = 10000. For Re = 400, the results for two mesh sizes are included. Comparison of these results for Re = 400 with (129 x 129) and (257 x 257) grids indicates that the (129 x 129) grid is adequate for Re of this order of magnitude. The cut-outs shown in this figure are enlargements of the secondary vortices. Letters T, B, L and R denote top, bottom, left and right, respectively; the numeral subscript denotes the hierarchy of secondary vortices. For example, BL_1 refers to the first of the secondary vortices that occur in the bottom left corner.

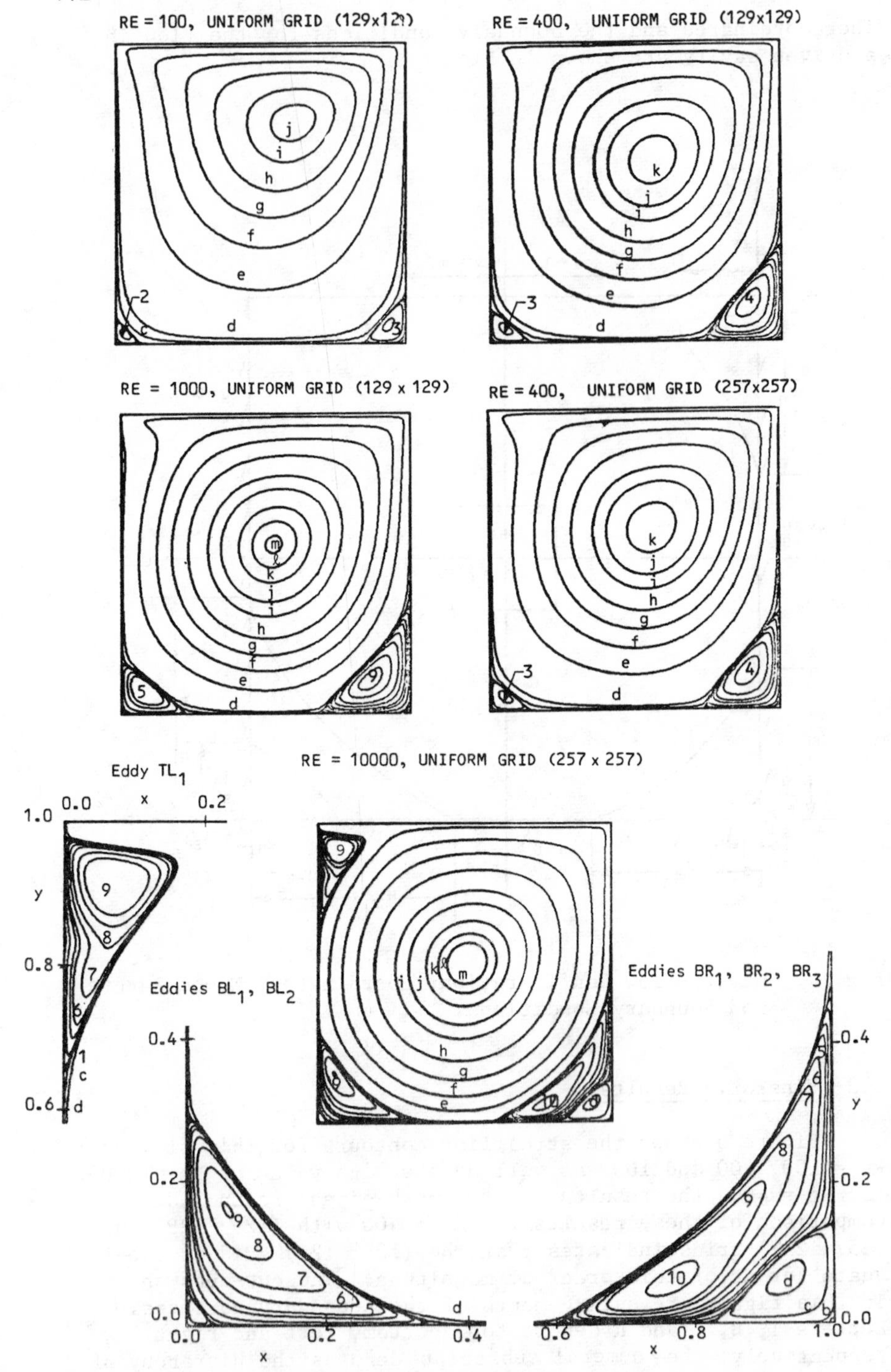

Fig. 5. Streamline Pattern for Primary, Secondary and Additional Corner Vortices.

As Re increases, the center of the primary vortex approaches the cavity center. All secondary vortices appear very near the corners (or near the wall, in case of TL_1) and move, very slowly, toward the cavity center as Re increases. Figure 6 shows the vorticity contours obtained for the lid-driven cavity flow. As Re increases to 5000 and beyond, thin regions of very high shear appear in the cavity; these regions are not aligned to the cavity boundaries. This necessitated the use of a uniform fine grid throughout the cavity. A flow-adaptive grid is highly desirable.

RE = 100, UNIFORM GRID (129 x 129)

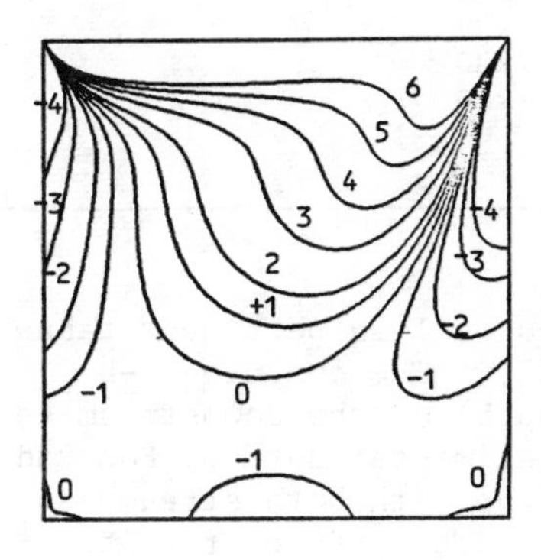

RE = 400, UNIFORM GRID (129 x 129)

RE = 1000, UNIFORM GRID (129 x 129)

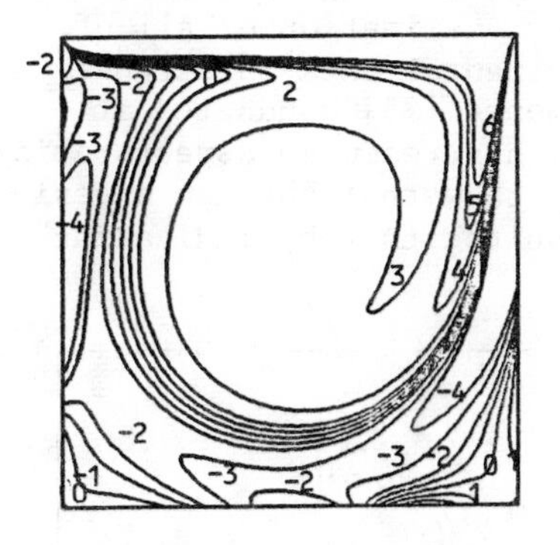

RE = 10000, UNIFORM GRID (257 x 257)

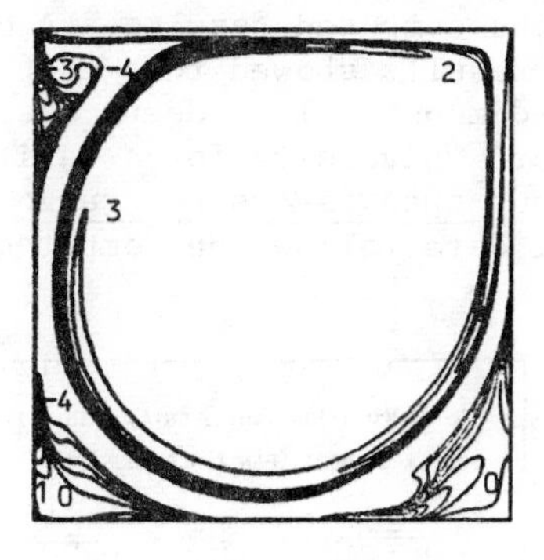

Fig. 6. Vorticity Contours for Flow in Driven Cavity.

TABLE
Values for Streamline and Vorticity Contours in Figs. 5 and 6

Stream function				Vorticity	
Contour letter	Value of ψ	Contour number	Value of ψ	Contour number	Value of ζ
a	-1.0×10^{-10}	0	1.0×10^{-8}	0	0.0
b	-1.0×10^{-7}	1	1.0×10^{-7}	±1	±0.5
c	-1.0×10^{-5}	2	1.0×10^{-6}	±2	±1.0
d	-1.0×10^{-4}	3	1.0×10^{-5}	±3	±2.0
e	−0.0100	4	5.0×10^{-5}	±4	±3.0
f	−0.0300	5	1.0×10^{-4}	5	4.0
g	−0.0500	6	2.5×10^{-4}	6	5.0
h	−0.0700	7	5.0×10^{-4}		
i	−0.0900	8	1.0×10^{-3}		
j	−0.1000	9	1.5×10^{-3}		
k	−0.1100	10	3.0×10^{-3}		
l	−0.1150				
m	−0.1175				

Additional graphical results as well as pertinent tabulated data are available in Ref. [21]. The computed 2-D results for the vertical extent (depth) of the downstream eddy are compared in Fig. 7 with the experimental data of Pan and Acrivos [24]. Their cavity was a cube with both streamwise and spanwise aspect ratios of unity. The present results agree well with the data up to Re = 400 but, for Re > 400, the data shows a trend completely opposite of that of the 2-D computed results. Although not shown here, the results of Benjamin and Denny [11] as well as those of Agarwal [12] agree with our computed results. A closer examination of all of these results showed that the experimental data is strongly three-dimensional. Recent measurements [25] conducted at Stanford University for a cavity with streamwise aspect ratio of 1 but spanwise aspect ratio of 3 show that the new experimental data follow the computed results, thereby indicating

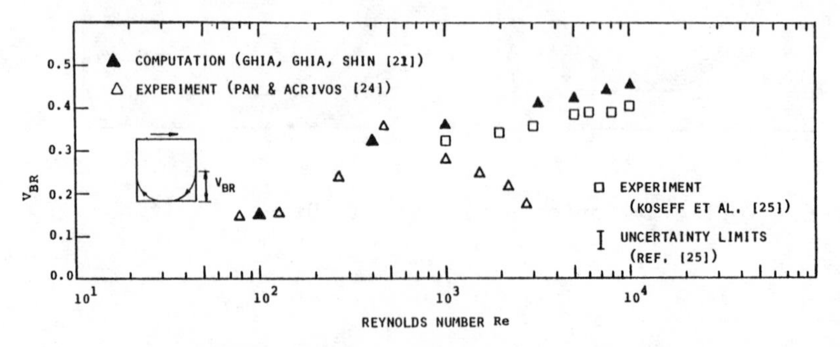

Fig. 7. Comparison of Vertical Extent of Secondary Downstream Eddy.

that spanwise aspect ratio is an important parameter contributing to three-dimensionality in the flow. It is important to note that inaccuracies in numerical computations can sometimes lead to erroneous results and, consequently, to wrong conclusions. For example, Nallasamy and Prasad [26] employed an upwind-differencing numerical method and a rather coarse computational grid and obtained results which followed the experimental data of Ref. [24], a coincidental agreement of inaccurate 2-D numerical results with a strongly 3-D flow configuration.

The computational advantage gained by use of the MG procedure is best illustrated in terms of the behavior of the root-mean square (RMS) value of the dynamic residuals of the discretized governing equations in the finest grid. Figure 8 shows the finest-grid RMS residuals for ψ and ζ obtained during a single-grid computation with $\Delta = 1/128$ (solid curve) as well as a multigrid calculation with $\Delta_M = 1/128$ and $M = 6$ (solid and dashed lines) where M is the total number of grids in the MG procedure. Flow configurations with Re = 100 and Re = 1000 are examined. In both cases, even the single-grid calculations exhibit a rapid initial decay of the RMS residuals for ψ as well as ζ during the first 4-6 iterations (work units). Thereafter, the solid curves show a marked decrease in their slope. Employing the multigrid process after these first 4-6 work units tends to retain the initial decay rate for the errors during the overall computation.

It is important to mention two points with respect to the MG curves in Fig. 8. First, the solid portions of the MG-curves correspond to the relaxation step (smoothing) on the finest grid while the dashed portions correspond to the coarse-grid correction due to the MG cycle. Second, although convergence was defined on the basis of the arithmetic average of the RMS residual in ζ and ψ, the convergence rate was examined in terms of the RMS residual in ζ alone. It is perhaps for this reason that ζ exhibits a much more desirable convergence behavior than ψ because the convergence rate is indeed the parameter that comprises the basis for interjecting an MG cycle in the solution procedure. Some further improvement in the overall convergence process may be possible by also including the convergence rate of ψ in the criterion controlling switching to the coarse-grid correction step. In the second model problem, the convergence rate of each of the governing equations is considered in the grid-switching process.

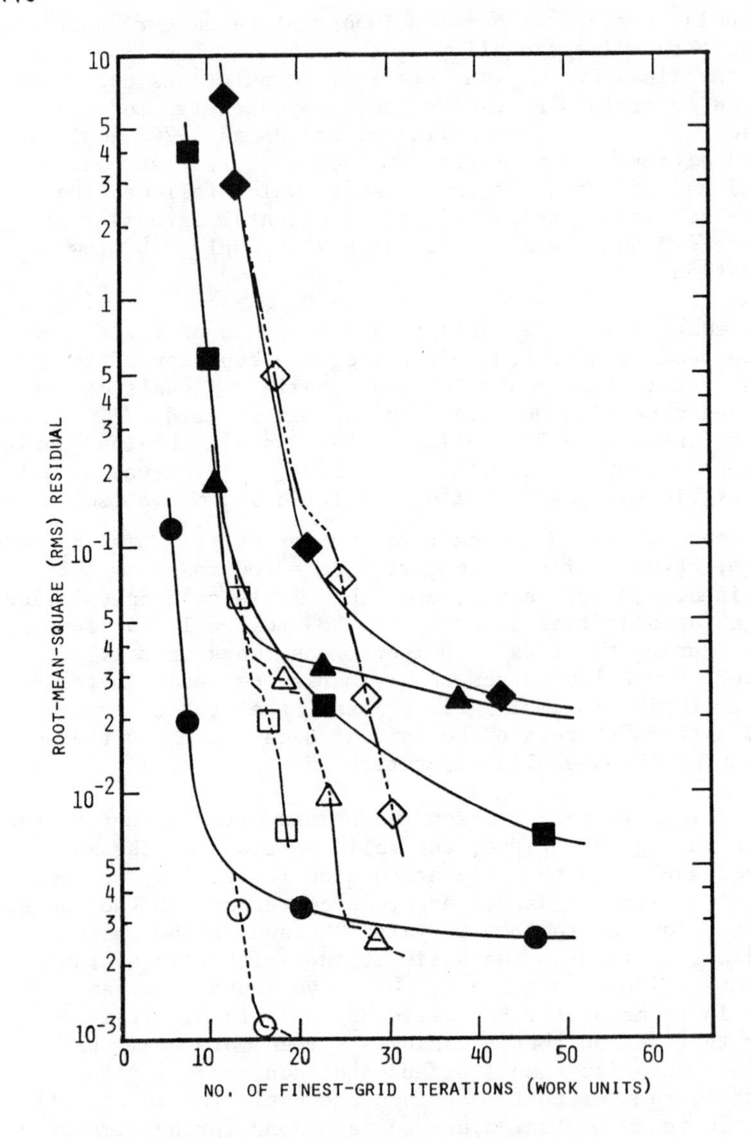

Fig. 8. Convergence of Single-Grid and Multi-Grid Computational Procedures.
Single Grid ($\Delta = 1/128$):
Re = 100: (●) e_ψ, (■) e_ζ; Re = 1000: (▲) e_ψ, (◆) e_ζ.
Multi-Grid ($\Delta_M = 1/128$, M = 6):
Re = 100: (○) e_ψ, (□) e_ζ; Re = 1000: (△) e_ψ, (◇) e_ζ.

6. APPLICATION TO ASYMPTOTIC FLOW IN CURVED DUCTS

In the asymptotically far downstream region of curved ducts, all three orthogonal components of the velocity vector continue to remain non-zero. However, they are functions of only the two cross-plane coordinates. This is significant because the flow problem remains numerically tractable and can serve as an excellent model problem. Moreover, this problem is free from the singularities associated with the driven-cavity problem considered in the previous section. Rectangular as well as polar cross-section ducts are considered here. Figure 9 shows the toroidal coordinates and the nomenclature used. The primary similarity parameter for these flows is the Dean number K. The cross flow exhibits markedly different characteristics for various ducts at different values of K.

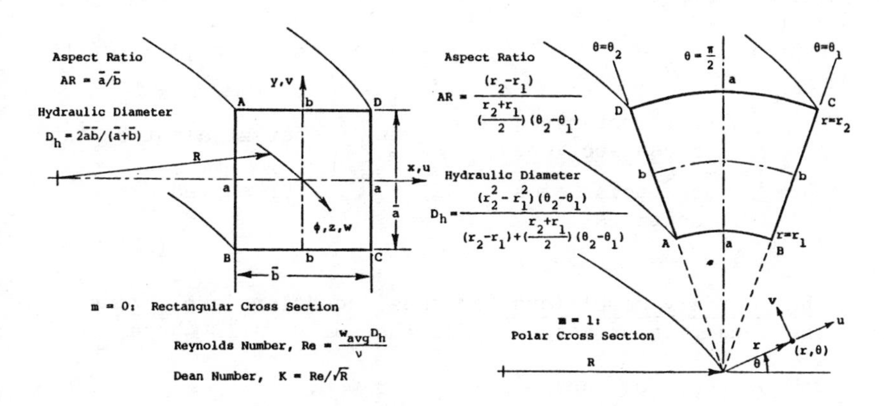

Fig. 9. Coordinates and Nomenclature for Asymptotic Curved-Duct Flows.

6.1 Governing Equations and Boundary Conditions

The flow problem is formulated using the streamwise velocity, w, the streamwise vorticity ζ, and the cross-flow stream function ψ, as the dependent variables. In terms of these nondimensional variables, the Navier-Stokes equations have the following form for fully developed curved-duct flows:

Streamwise Momentum Equation

$$\frac{1}{h_2h_3}[\psi_\theta w_r - \psi_r w_\theta] + [\frac{c^m}{h_2}\psi_\theta + ms\,\psi_r]\frac{nw}{h_3^2}$$
$$= -\frac{1}{h_3}p_\Phi + \frac{1}{Re}[w_{rr} + (\frac{m}{h_2} + \frac{nc^m}{h_3})w_r + \frac{1}{h_2^2}w_{\theta\theta} - \frac{mns}{h_2h_3}w_\theta - \frac{nw}{h_3^2}] \tag{18}$$

Streamwise Vorticity Equation

$$\frac{1}{h_2h_3}[\psi_\theta\zeta_r - \psi_r\zeta_\theta] - [\frac{c^m}{h_2}\psi_\theta + ms\,\psi_r]\frac{n\zeta}{h_3^2} + [\frac{c^m}{h_2}w_\theta + msw_r]\frac{2nw}{h_3^2}$$

(continued)

$$= \frac{1}{Re} [\zeta_{rr} + (\frac{m}{h_2} + \frac{nc^m}{h_3})\zeta_r + \frac{1}{h_2^2} \zeta_{\theta\theta} - \frac{mns}{h_2 h_3} \zeta_\theta - \frac{n\zeta}{h_3^2}] \quad (19)$$

Cross-Flow Stream-Function Equation

$$\psi_{rr} + (\frac{m}{h_2} - \frac{nc^m}{h_3}) \psi_r + \frac{1}{h_2^2} \psi_{\theta\theta} + \frac{mns}{h_2 h_3} \psi_\theta = - h_3 \zeta \quad (20)$$

In these equations,

$$h_2 = r^m, \quad h_3 = (R + rc^m)^n, \quad c = \cos\theta, \quad s = \sin\theta, \quad (21)$$

$$m = \begin{cases} 0 & \text{for rectangular cross sections,} \\ 1 & \text{for polar cross sections,} \end{cases} \qquad n = \begin{cases} 0 & \text{for straight ducts,} \\ 1 & \text{for curved ducts.} \end{cases}$$

(22),(23)

The boundary conditions for these equations are derived from the condition of zero slip at the solid walls where

$$w = 0, \quad \psi = 0 \quad \text{and} \quad \zeta \sim \psi_{rr} \text{ or } \psi_{\theta\theta}. \quad (24)$$

6.2 Normalization and Specification/Determination of p_Φ, Re

The formulation given by Eqs. (18)-(20) employs the hydraulic diameter D_h of the duct cross section as the reference length and the average velocity w^*_{avg} as the reference velocity. In this form, the problem contains the two parameters p_Φ and Re, of which only one is truly independent. Nevertheless, it is not necessary to use a trial-and-error approach to arrive at consistent values of p_Φ and Re. The following alternative forms of normalization for the dimensional streamwise velocity w* allow for one of these to be prescribed and the corresponding value of the other to be determined.

i. If the quantity ν/D_h, where ν is the fluid viscosity, is used to normalize w*, then Eqs. (18)-(20) contain only p_Φ. Hence, Eqs. (18)-(20) can be solved using a prescribed value of p_Φ. Thereafter, the corresponding value of Re is determined from the final solution as follows.

$$\frac{1}{A} \iint_A (\frac{w^*}{\nu/D_h}) dA = \frac{1}{A} \iint_A (\frac{w^*}{w^*_{avg}}) \, Re \, dA = Re \quad (25)$$

where A represents the duct cross-sectional area and $Re = w^*_{avg} D_h/\nu$.

In this approach, it is possible to solve all three coupled equations (18)-(20) simultaneously. However, the resulting value of Re at which solutions are obtained may not be round numbers as often desired in practice.

ii. In an alternate approach, the quantity $w^*_{avg}(\partial p/\partial \Phi)$ is used to normalize w^* in the streamwise momentum equation (18) whereas w^*_{avg} is used to normalize w^* in Eqs. (19) and (20) for ζ and ψ. This approach permits desired values of Re to be prescribed but requires that the streamwise momentum equation be solved uncoupled from the equations for ζ and ψ. As given by K. Ghia et al. [9], the MG-SI and the MG-CSI procedures were used to solve Eq. (18) and Eqs. (19)-(20), respectively.

6.3 Typical Results

Detailed results for this model problem are given in Ref. [9]. Only some typical results are included here in terms of the cross-flow streamlines as these are most illustrative of the main features of the cross flows.

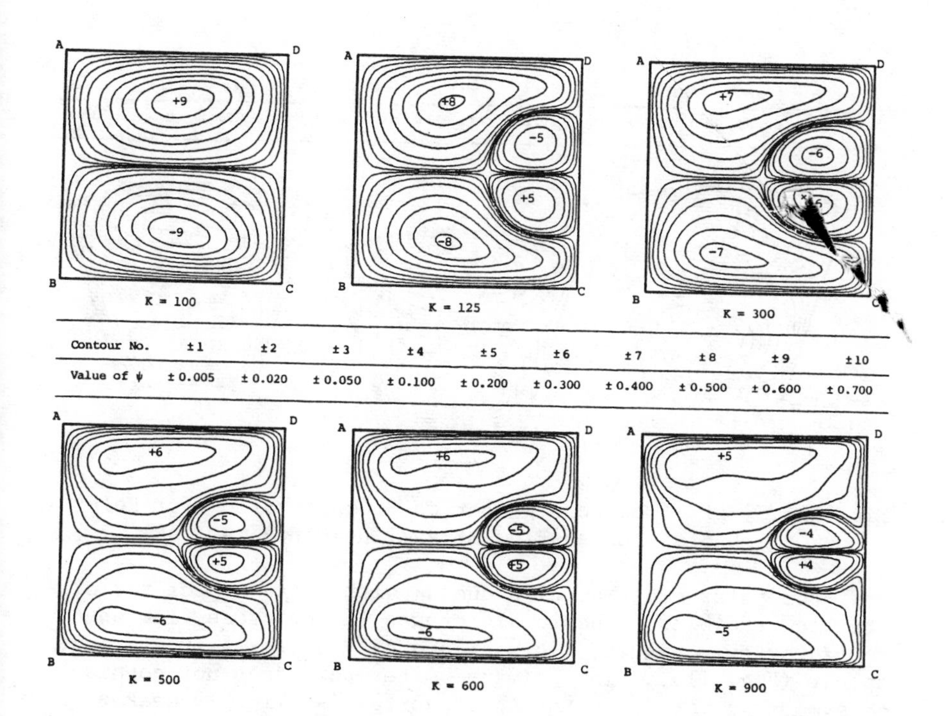

Fig. 10. Effect of Dean Number on Secondary Flow for Square Ducts - Cross-Flow Streamline Contours. R = 100.

In general, the cross flow consists of a pair of counter-rotating vortices in the cross plane of the duct. The flow near the axis is directed towards the outer curved wall and returns along the horizontal boundaries towards the inner curved wall. For square cross-section ducts, an additional pair of counter-rotating vortices suddenly makes its appearance in the cross flow when the Dean number K = 125 (Fig. 10). This phenomenon is known as Dean's instability and results in additional streamwise pressure drop. The strength $Re|\psi|_{max}/R$ of the primary vortex-pair increases as K increases except when K = 125 where it experiences a sudden drop due to the appearance of the secondary vortex-pair. The additional pair of vortices persists up to K = 900. Cheng et al. [17] had shown these to disappear at K = 520 with R = 4. It should be noted that R = 100 for the present results.

For a typical solution with an (81 x 81) grid, the AMDAHL 470/V6 computer CPU time was less than one minute for K < 125, while it was about 3 minutes for 125 < K < 400 and 6 minutes for K > 400. The CPU time of 3 minutes is to be approximately compared with the 40 CPU seconds required, for this range of values of K, by a coupled ADI scheme using a (21 x 21) grid, i.e., 16 times fewer grid points.

Contour No.	±1	±2	±3	±4	±5	±6	±7	±8	±9	±10
Value of ψ	±0.005	±0.020	±0.050	±0.100	±0.200	±0.300	±0.400	±0.500	±0.600	±0.700

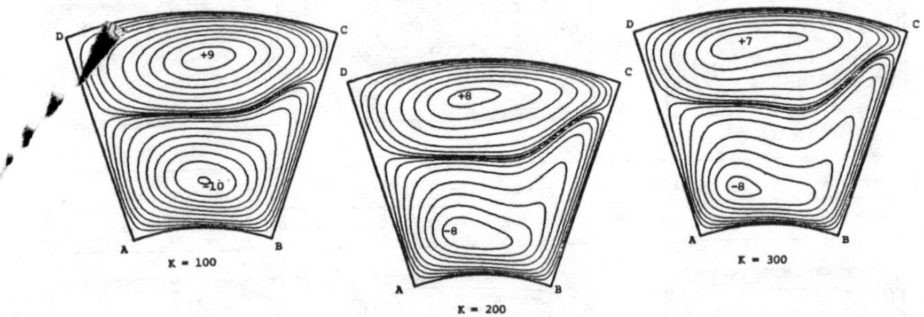

Fig. 11a. Effect of Dean Number on Secondary Flow for Polar Ducts - Cross-Flow Streamline Contours. AR = 1.

For polar cross-section ducts with unity aspect ratio, the flow exhibits no basic difference in the cross flow when K = 125 or even higher, i.e., a secondary vortex-pair does not appear (Fig. 11a). The primary vortex-pair does not consist of symmetric vortices; the upper vortex is slightly weaker than the lower vortex. As K increases from 100 to 300, the lower vortex is pulled upward towards the top right corner where the radii of longitudinal as well as transverse curvature attain their maximum values. The strengths $Re|\psi|_{max}/R$ for both vortices increase monotonically as K is increased. For aspect ratio of unity, the polar cross section employed a (121 x 81) grid in order to maintain $\Delta r = \Delta\theta$.

Contour No.	±1	±2	±3	±4	±5	±6	±7	±8	±9	±10
Value of ψ	± 0.005	± 0.020	± 0.050	± 0.100	± 0.200	± 0.300	± 0.400	± 0.500	± 0.600	± 0.700

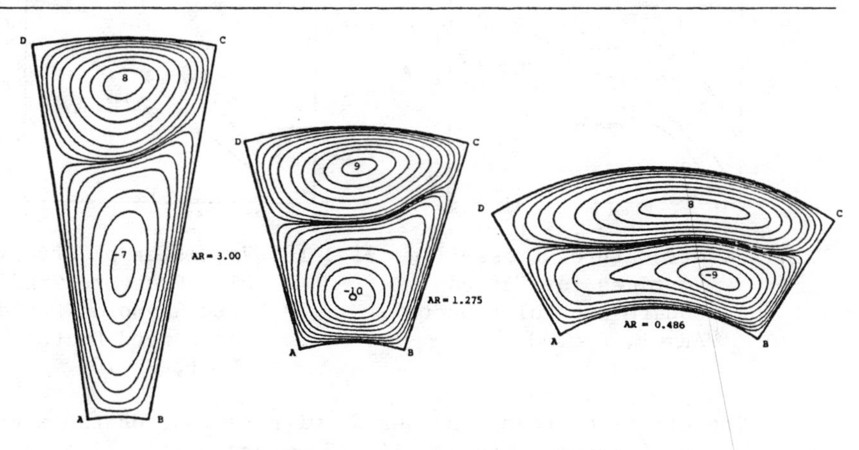

Fig. 11b. Effect of Aspect Ratio on Secondary Flow for Polar Ducts; K = 100, R = 100.

To investigate the possibility of Dean's instability for different aspect ratios of the duct cross section, calculations were made for polar ducts with cross-sectional aspect ratios ranging from 0.1926 to 6.7871. Again, no Dean's instability was observed. Figure 11b shows the streamline contours for three values of the aspect ratio. It is felt that, for certain combinations of the problem parameters, the flow for polar configurations should perhaps qualitatively develop some of the square-duct flow features. However, such a combination has not yet been arrived at.

The effect of Dean number on the strengths of various primary and secondary vortices for square and polar curved ducts is depicted in Fig. 12. The quantity plotted for the stream function is $|\psi|$ Re/R which is denoted here as ψ_C for convenience in writing. In terms of ψ_C, for the square duct, the strength of the primary vortex increases as K increases from 1 to 125 where it experiences an abrupt decrease, corresponding to the occurrence of Dean's instability. For $K > 125$, ψ_C again increases with K up to $K = 900$. The strength of the secondary vortex also increases with increase in K but the rate of increase is slower compared to that of the primary vortex. It is significant to observe that the curves of ψ_C for the primary as well as the secondary vortices exhibit a certain low-order discontinuity near $K \simeq 500$. This value is coincidentally very close to the value of K for which the results of Cheng et al. [17] had shown that the secondary vortex pair disappears. Also shown in Fig. 12 is the variation of the strengths of the upper and lower vortices occurring in the polar duct configurations for K up to 300.

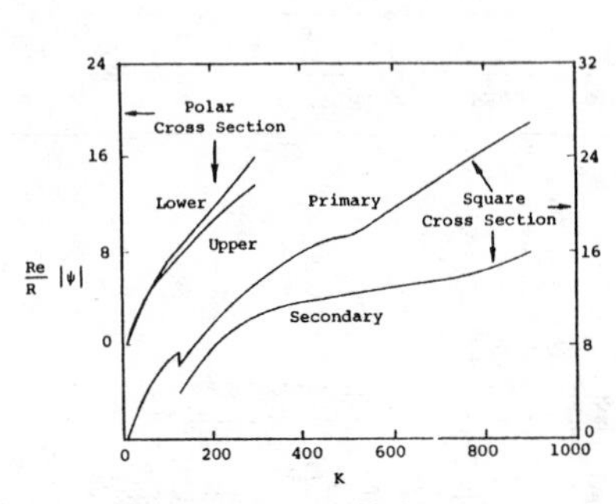

Fig. 12. Strength of Cross-Flow Vortices for Curved Square and Polar Ducts; AR = 1, R = 100.

Fig. 13. Variation of Streamwise Pressure Gradient with Reynolds Number for Curved Ducts; AR = 1, R = 100.

The effect of Dean's number K, with R = 100, on the magnitude of $\partial\bar{p}/\partial\Phi$ is depicted in Fig. 13, where $\bar{p}$ is defined as Re^2p/R. For the square duct, $|\partial\bar{p}/\partial\Phi|$ monotonically increases as Re increases from 10 to 9,000. In the vicinity of K = 125, i.e., Re = 1250, the variation of $|\partial\bar{p}/\partial\Phi|$ has also been plotted with an enlarged scale. This helps to show an abrupt increase in $|\partial\bar{p}/\partial\Phi|$ when Dean's instability occurs. Accounting for the shift in the origin of the curve for the duct of polar cross section with AR = 1, $|\partial\bar{p}/\partial\Phi|$ is almost identical to the corresponding square-duct values. The movement of the vortex center with increase in Dean number for both square and polar ducts is presented in Fig. 14.

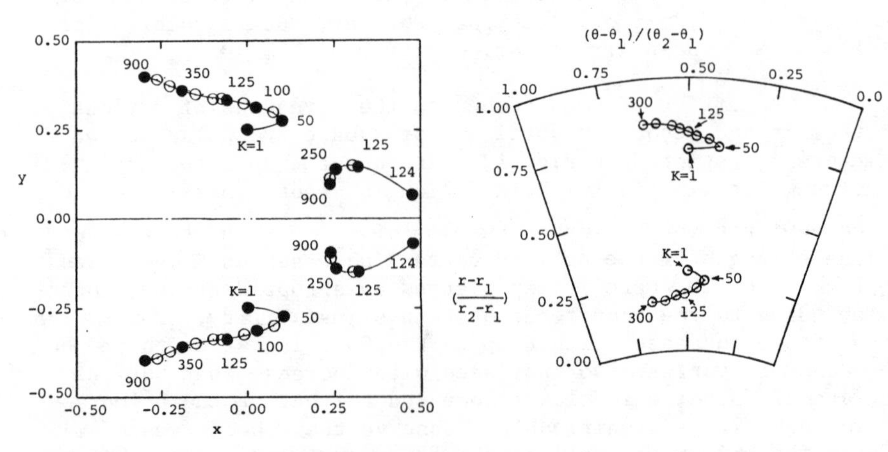

Fig. 14. Effect of Dean Number on Location of Vortex Centers For Curved Ducts; AR = 1, R = 100.

The feasibility of the occurrence of Dean's instability for polar ducts remains unsettled. The analogous situation for curved circular pipes has been resolved only recently. All investigations of curved pipes prior to 1978 revealed only a two-vortex solution. The analysis of Van Dyke [27] in 1978 cast doubt on these solutions for high Dean number. It was only thereafter that Dennis and Ng [28] and Nandakumar and Masliyah [29] obtained four-vortex solutions for this problem. Berger, Talbot and Yao [30] have compiled an excellent review of curved-pipe flows. Effort is presently underway by the present authors to resolve the question of dual solutions for curved polar-duct flow.

7. APPLICATION TO NEUMANN BOUNDARY-VALUE PROBLEM

7.1 Formulation of Neumann Problem for Pressure

In a velocity-pressure representation of the Navier-Stokes equations, the pressure is frequently determined as the solution of an elliptic problem formulated by forming the divergence of the momentum equations. With interest in duct flows, the configurations considered here correspond to parabolic flow through straight as well as curved ducts of simple rectangular or polar cross sections. These configurations have been studied earlier by the authors and their co-workers. In particular, the formulation used here follows that of U. Ghia et al. [19] who also employed grid-clustering coordinate transformations to resolve the high-gradient regions normal to the duct walls. With the nomenclature used in Ref. [19] for parabolic flow through ducts, the (ξ,η) cross-plane momentum equations may be represented as

$$\xi' \frac{\partial p}{\partial \xi} = F_1 \quad \text{and} \quad \frac{1}{r^m} \eta' \frac{\partial p}{\partial \eta} = F_2 \qquad (26a,b)$$

where p is the cross-plane pressure distribution,

ξ,η denote computational cross-plane coordinates corresponding to (x,y) or (r,θ), with primes denoting differentiation,

$m=0$ for rectangular cross-section ducts,

$m=1$ for polar or circular cross-section ducts,

and F_1,F_2 represent all the velocity terms in the momentum equations.

The details of the expressions for F_1, F_2 are not relevant for the present purpose and are, therefore, not given here; they are available in Ref. [19]. For the parabolic duct-flow configurations considered, the cross-plane Laplacian operator is defined as

$$\nabla^2 = \xi'^2 \frac{\partial^2}{\partial \xi^2} + \xi'' \frac{\partial}{\partial \xi} + \frac{m}{r} \xi' \frac{\partial}{\partial \xi} + \frac{1}{r^{2m}} [\eta'^2 \frac{\partial^2}{\partial \eta^2} + \eta'' \frac{\partial}{\partial \eta}]$$

$$+ \frac{n}{(R+rc^m)} [c^m \xi' \frac{\partial}{\partial \xi} - \frac{ms}{r} \eta' \frac{\partial}{\partial \eta}] \qquad (27)$$

where $n = \begin{cases} 0 & \text{for straight ducts,} \\ 1 & \text{for curved ducts,} \end{cases}$

$R =$ radius of longitudinal curvature,

$c = \cos\theta$, and $s = \sin\theta$.

Therefore, the pressure equation is obtained as

$$\nabla^2 p = F_p \qquad (28)$$

where

$$F_p = \xi' \frac{\partial F_1}{\partial \xi} + \frac{\xi' F_1}{r^m (R^n + nrc^m)} \frac{\partial}{\partial \xi} [r^m (R^n + nrc^m)]$$

$$+ \frac{\eta'}{r^m} \frac{\partial F_2}{\partial \eta} + \frac{n\eta' F_2}{(R^n + nrc^m)} \frac{\partial}{\partial \eta} [c^m] \quad . \qquad (29)$$

Equation (28) is a Poisson equation for pressure. The boundary conditions required for its solution are obtained by evaluating the normal gradients of pressure at the problem boundaries, using Eq. (26). Hence, Eq. (28) and Eq. (26) constitute a Neumann boundary-value problem for pressure.

The parabolic formulation of duct flows also involves a similar Neumann boundary-value problem of a correction-velocity potential function [19] in an effort to maintain local mass conservation in a non-iterative manner. The details of the solution of that Neumann problem parallel those for the pressure Neumann problem discussed here.

7.2 Remarks on Single-Grid Solution of Neumann Problems

A unique property of the Neumann problem is that the Laplacian operator in the governing differential equation and the normal derivative operator in the boundary values are related via the Gauss divergence theorem. Consequently, the source function F and the boundary-value function G are also similarly related by the integral theorem

$$\iint_A F \, dA = \oint_C G \, dn \qquad (30)$$

where C is the closed curve bounding the solution domain A. Failure to satisfy Eq. (30) causes eventual divergence of the resulting solution. Generally, Eq. (30) may not be satisfied exactly due to differences in truncation errors of discretization of the various terms in F and G. In order to ensure convergence, it is necessary to define a modified source function F' as

$$F' = F - \frac{\varepsilon}{A} \tag{31}$$

where

$$\varepsilon = \iint_A F \, dA - \oint_C G \, dn \tag{32}$$

i.e., ε represents the discrepancy in Eq. (30). Then, with F replaced by F', Eq. (30) is satisfied exactly.

The integral constraint (30) associated with the Neumann problem also implies some requirements to be satisfied by the discretization process. First of all, the ∇^2 operator must be in the conservation form as far as possible. Secondly, the form of the equation solved at each computational point must be guided by the form in which F appears in Eq. (30). Accordingly, the pressure equation for example, must be written at each computational point as

$$\sum_{i=1}^{I} a_i \nabla_i^2 p = \sum_{i=1}^{I} a_i F_{i_p} \tag{33}$$

where $a = r^m[R^n + nrc^m]/(\xi'\eta')$,

$$I = \begin{cases} 4 & \text{at general interior points,} \\ 2 & \text{at general boundary points,} \\ 1 & \text{at corner points.} \end{cases}$$

and i refers to center of cells adjoining the computational point under consideration. Also, the subscript i on ∇^2 and F denotes that the coefficients in these quantities are to be evaluated at the four possible positions indicated as $(r \pm \frac{\Delta r}{2}, \theta \pm \frac{\Delta\theta}{2})$. The conservation form of $\nabla^2 p$ requires that, in the defining Eq. (27), ξ'', η'' and s (= $\sin\theta$) have the following representation:

$$\xi'' = \frac{d(\xi')}{d\xi}\,\xi' \;; \qquad \eta'' = \frac{d(\eta')}{d\eta}\,\eta' \;,$$

and

$$s = \sin\theta = -\frac{d}{d\eta}(\cos\theta)\,\eta' \;. \tag{34}$$

Osswald et al. [31] have shown that, with the use of a staggered grid, the discretized formation of the divergence of the momentum equations leads to the pressure appearing in exactly the same form as would be obtained by directly discretizing the Laplacian of p since $\nabla\cdot(\nabla p) = \nabla^2 p$. This serves to minimize the discrepancy ε in Eq. (32) and, hence, minimizes the modification of the source term F. These considerations should be important when the pressure equation has the prime responsibility of satisfying local mass conservation. In the parabolic formulation of duct flows discussed here, local mass conservation is achieved via the use of a correction-velocity potential [19] and not via the pressure equation.

7.3 Remarks on Multi-Grid Solution of Neumann Problem

The application of the MG technique to the solution of Neumann boundary-value problems has been considered in detail by U. Ghia et al. [32]. In the MG solution of a Neumann problem, a convergent solution on the finest grid is ensured when the measures described in Section 7.2 are observed. No further modification of the problem may be made on the subsequent coarse grids. On these grids, convergence should be obtained by using an appropriate restriction operator and an appropriate coarse-grid-correction operator formulated so as to automatically satisfy the integral constraint [Eq. (30)] on all coarse grids. The derivation of the proper form for these operators is guided by the discretized representation of the integral constraint [Eq. (30)]. This integral constraint can be satisfied automatically on the coarser grid if the area integral and the line integral on a fine grid are made individually equal to the corresponding area integral and line integral on a subsequent coarse grid. To achieve this, it is useful to refer to the sketch in Fig. 15a for a typical computational region; here, numerals designate fine-grid points while alphabets designate coarse-grid points. With a mesh width ratio $\Delta_{k+1}/\Delta_k = 1/2$, the area integral AI^f on the fine grid can be made equal to the area integral AI^c on the coarse grid if the restriction operator R_k^{k-1} is defined such that the coarse-grid source term F^c is related to the fine-grid source term as follows.

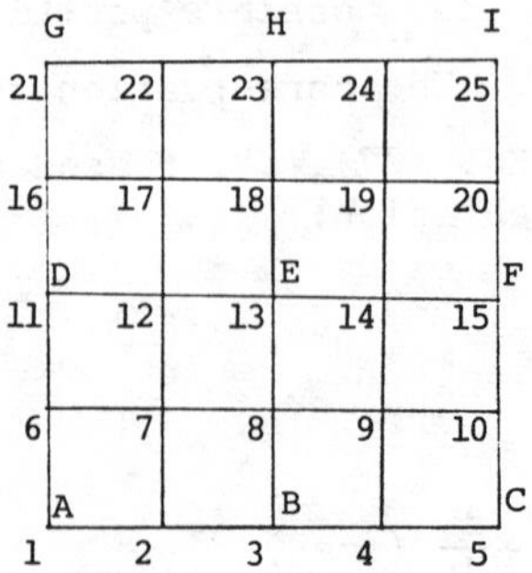

Fig. 15a. Grid Nomenclature.

<u>At a general interior point</u>,

F^c is a 9-point average of F^f.

<u>At a general boundary point</u>,

F^c is a 6-point average of F^f.

<u>At a corner</u>,

F^c is a 4-point average of F^f.

The appropriate weighting factors to be used in these averages are as shown in Fig. 15b.

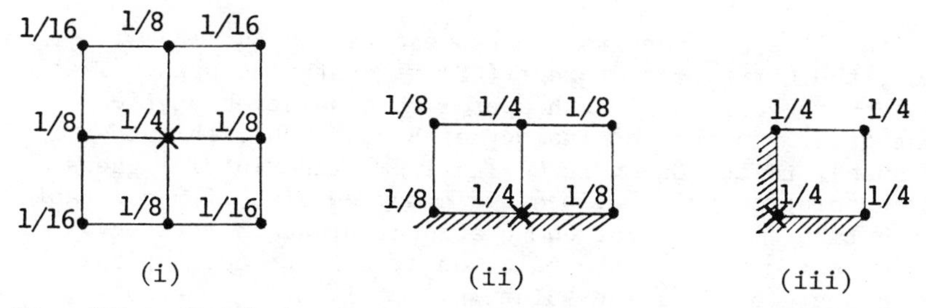

Fig. 15b. Weighting Factors for Restriction Operator for Source Function - (i) Interior Point, (ii) General Boundary Point, and (iii) Corner Point.

Similarly, the line integral I^c on the coarse grid can be made equal to the line integral I^f on the fine-grid if the coarse-grid boundary values are related to the fine-grid boundary values as follows.

<u>At a general boundary point</u>,

G^c is a 3-point average of G^f.

<u>At a corner</u>,

G^c is a 2-point average of G^f.

Again, the weighting factors to be employed in these averages are shown in Fig. 15c.

The restriction operators thus defined must also be used to define the coefficients of the various terms in the differential operator in the coarse-grid-correction equation. Accordingly, the discretized coarse-grid operators are of the same form as the corresponding fine-grid operators, with the coefficients in the former derived from the coefficients in

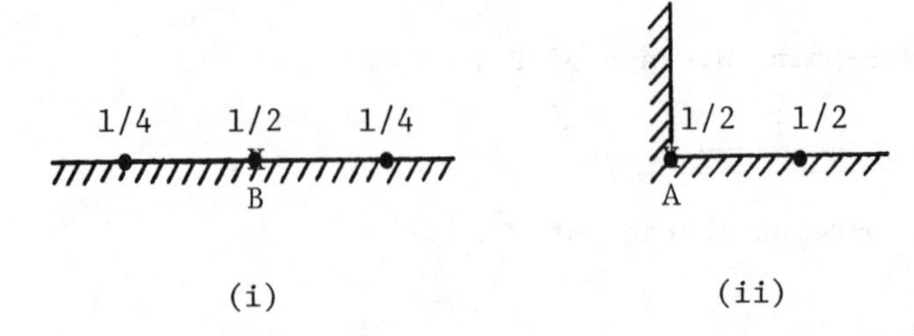

Fig. 15c. Weighting Factors for Restriction Operator for Boundary Values - (i) General Boundary Point, (ii) Corner Point.

the latter using the restriction operator defined in this section.

Since the Neumann problem represented by Eq. (33) is linear, the Correction-Storage (CS) algorithm is used. The procedure is started with an approximate solution on the finest grid. In the terminology of Brandt [20], this then corresponds to the CS-Cycle C algorithm. The initial guess for the solution on the finest grid may consist of a constant distribution at the first computational plane of the duct. For successive planes, the solution at the preceding plane comprises a good starting solution. It is also necessary to note that, in a Neumann problem, a non-zero residual exists in the boundary conditions at those boundaries which are treated explicitly in the relaxation process. Accordingly, non-zero boundary-value residuals are involved at all boundaries with an explicit SOR iteration procedure. On the other hand, the SI procedure that treats all boundary conditions implicitly, involves zero boundary-value residuals at all boundaries.

7.4 Convergence History of Solution

The primary purpose of this application was not so much to examine the flow features of this problem, but to study the convergence properties of the numerical solution. Hence, for discussion of the results here, a cross plane located very near the entrance section of the duct has been selected because, in earlier single-grid studies of this problem, the slowest convergence rates were experienced in this region with large cross-flow gradients near the duct walls. Also, the influence of the relaxation or smoothing operator on the overall efficiency of the solution is examined by employing GS, ADI as well as SI procedures as smoothing operators.

In the figures presented, the duct configurations are denoted by a sequence of three upper-case letters. The first letter indicates whether the duct is straight (S) or curved (C).

The second letter indicates whether the duct cross section is square (S) or polar (P). The last letter corresponds to uniform grids (U) or stretched grids (S). Thus, CPS corresponds to curved polar ducts with grid stretching. Also, MG denotes multi-grid solutions, while SG denotes single-grid solutions and GS denotes the Gauss-Seidel iterative procedure.

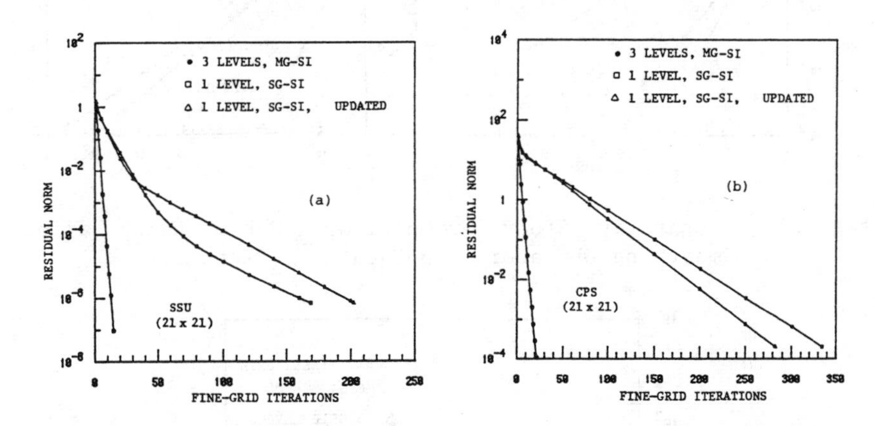

Fig. 16. Convergence History of p-Equation on Plane 3.

The convergence history for the solution for p on plane 3 is shown in Fig. 16 for the SSU and the CPS configurations for a (21 x 21) grid as the finest grid. Included in Fig. 16 are the corresponding curves for the SG-SI solutions as well as those for the SG-successive SI scheme [33] in which one of the recursion coefficients is updated during the back-substitution step of the procedure. The advantages of using the MG procedure are clearly seen from Fig. 16. It is not clear why the successive SI scheme is less efficient than the SI scheme for the SSU configuration; this point is presently being examined further. Although calculations were also made using the MG-GS as well as the MG-ADI schemes, use of the SI scheme as the smoothing operator was preferred, because of its relative insensitivity to the value of the iteration parameter Δt. The SI scheme has been described adequately in Refs. [22] and [23]. However, its adaptation for use with the Neumann problem required considerable reformulation of the various recursion relations involved at the boundaries; the details are provided in Ref. [34]. All boundary conditions were treated implicitly and with second-order accuracy. The scheme was then used, with $\Delta t \to \infty$.

A comparison of three smoothing operators employed in the MG procedure is presented next. Figure 17 shows the convergence history for the p solution on plane 3 for the MG scheme with the three smoothing procedures used, namely, GS, ADI and SI. For the SSU as well as the CPS configurations, the SI smoothing provides the fastest convergence while GS smoothing

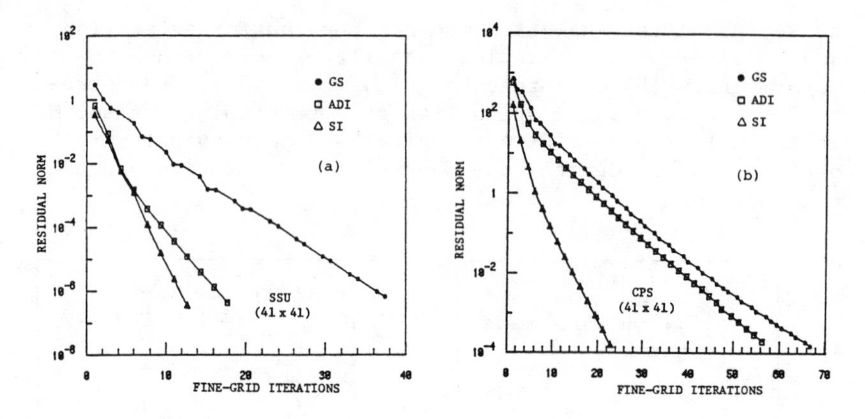

Fig. 17. Comparative Study for Multi-Grid Scheme Using Various Smoothing Operators - p-Equation, Plane 3.

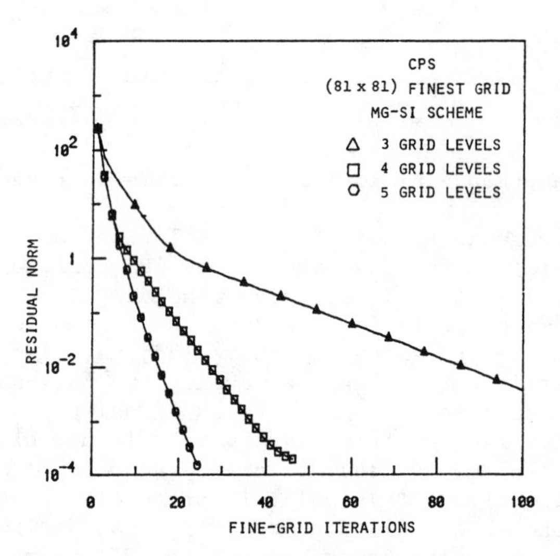

Fig. 18. Convergence History of p-Equation on Plane 3 with (81 x 81) Finest Grid - Effect of Number of Grid Levels.

contributes to the slowest convergence of the MG scheme. For the SSU case, the MG-ADI curve is nearer to the MG-SI curve, whereas for the CPS case, it shifts closer to the MG-GS curve. This is typical of the sensitivity of the ADI scheme to the parameters of the problem.

In order to examine the computational advantage of the MG procedure with increasing refinement of the finest grid, solutions were also obtained with an (81 x 81) finest grid. The resulting convergence history for the pressure solution on plane 3 is shown in Fig. 18. The three different curves included in this figure are useful for displaying the effect of the number of grid levels and the coarsest grid employed in the MG procedure. For the 3-grid scheme with the coarsest grid of (21 x 21) points, convergence requires in excess of 150 fine-grid iterations. This is to be compared with approximately 45 iterations required by the 4-grid scheme which includes the (11 x 11) grid as its coarsest grid. Inclusion of yet another grid results in the 5-grid scheme with the coarsest mesh of (6 x 6) points and convergence is then achieved in about 25 fine-grid iterations. Referring to Figs. 16 and 17, it is to be noted that approximately the same number, namely, 25, of fine-grid iterations were also required for convergence of the (21 x 21) finest-grid solutions as well as the (41 x 41) finest-grid solutions for the CPS duct configuration.

For the results presented in Figs. 16, 17 and 18, convergence was defined to occur when the L_2-norm of the residuals in the Poisson equation was below 7×10^{-7} for the SSU configuration and below 2×10^{-4} for the general CPS configuration. The corresponding values of the local relative error in the solution were approximately 10^{-10}.

The MG procedure outlined in this chapter is adaptive, in the sense that internal checks dictate the switching between the coarse and the fine grids, based on pre-assigned values of λ and α. A fixed MG procedure is one in which the internal checks are replaced by a pre-assigned number N^k of iterations on grids D^k where the value of N^k is also guided by the smoothing rate of the relaxation scheme. The main advantage of fixed algorithms is the associated saving of computational effort, since residuals are then needed only on the finest grid. The results shown in Figs. 16, 17 and 18 correspond to the fixed MG algorithm consisting of 1 iteration on each grid except the coarsest, on which 3 iterations were performed.

8. SUMMARY

Fine-mesh solutions have been obtained very efficiently for high-Re flow using the multi-grid method. The various operators and parameters in the multi-grid procedure were examined. It is observed that 9-point restriction is superior to 5-point restriction. The size of the finest mesh used continues to be the most significant parameter. The test value of the convergence rate λ increases with Re. Also, the test for convergence on the finest grid must be made stricter, i.e., $q > 1$.

For driven flow in a square cavity, detailed accurate results are obtained for Re as high as 10000 using as many as (257 x 257), i.e., 66049 mesh points. A uniformly fine grid was employed in order to satisfactorily resolve the physical scales of the problem without having to estimate the structure of this high-Re flow. However, a non-uniform grid is recommended when reliable information is available about the flow structure. The present detailed fine-grid results constitute benchmark solutions against which other numerical results can be assessed. Comparison of the present results with available experimental data provides a measure of the 3-D effects present in the experimental data. Other recent numerical solutions obtained using reasonably fine grids compare well with the present results but, accounting for the differences in the number of grid points and the computers employed, the present results are about 14 times as efficient as, for example, those in Ref. [11].

For asymptotic flow in curved ducts of square cross-section, numerical solutions are obtained for K up to 900, i.e., Re = 9000, using up to (81 x 81) mesh points. Dean's instability occurs at K = 125 and the additional vortex pair persists up to K = 900. For asymptotic flow in curved ducts of polar cross-sections with unity aspect ratio, solutions are obtained for K up to 600 with (121 x 81) mesh points. Results are also obtained for aspect ratio varying between 0.2 and 6.8 using similar fine grids. Dean's instability is not observed even for this wide range of the parameter values.

The multi-grid technique has also been studied for use with the generalized Neumann boundary-value problem in curvilinear coordinates with grid-clustering transformations. The restriction operator R_k^{k-1} at interior points, boundary points and corner points has been formulated appropriately so as to satisfy the discretized representation of the integral constraint associated with Neumann problems. The coefficients in the differential operators for the coarse-grid-correction equation are formulated using this restriction operator so as to satisfy the divergence theorem. Results have been obtained for the Neumann problems encountered in parabolic-flow analysis of various duct configurations.

Comparison of the MG solution convergence behavior with the convergence history of single-grid solutions confirms the computational advantage of the MG technique. The relaxation scheme used in the MG procedure is observed to influence the efficiency of the overall scheme. The GS smoother performs quite well but the SI scheme contributes to greater efficiency, while the ADI scheme continues to remain sensitive to the iteration parameters. Finally, the fixed MG algorithm proved to be more efficient than an adaptive one.

The successful implementation of the MG technique constitutes an important development in the efficient solution of the Navier-Stokes equations. Three-dimensional flow solutions, requiring a large number of grid points, may be facilitated, in the foreseeable future, by the MG technique with practical CPU-time requirements.

ACKNOWLEDGEMENT

The authors would like to acknowledge the contributions of their graduate students, Mr. C.T. Shin and Mr. R. Ramamurti, in the research reviewed herein. The research of the present authors was supported by AFOSR Grant 80-0160, NSRDC Contract No. N00167-79-M-4413, NASA Grant NSG 3267, and NASA Grant NAG-3-194.

REFERENCES

1. BAKHVALOV, N.S. - On the Convergence of a Relaxation Method with Natural Constraints on the Elliptic Operator. U.S.S.R. Comp. Math. Math. Phys., Vol. 6, No. 5, pp. 101-135, 1966.

2. BRANDT, A. - Multi-Level Adaptive Solutions to Boundary-Value Problems. Mathematics of Computation, Vol. 31, pp. 333-390, 1977.

3. FEDORENKO, R.P. - A Relaxation Method for Solving Elliptic Difference Equations. U.S.S.R. Comp. Math. Math. Phys., Vol. 1, pp. 1092-1096, 1962.

4. HACKBUSCH, W. - On the Multi-Grid Method Applied to Difference Equations. Computing, Vol. 20, pp. 291-306, 1978.

5. MERRIAM, M.L. - Formal Analysis of Multi-Grid Techniques Applied to Poisson's Equations in Three Dimensions. AIAA Paper, No. 81-1028, 1981.

6. WESSELING, P. - Theoretical and Practical Aspects of a Multi-Grid Method. Report NA-37, Delft University of Technology, Delft, The Netherlands, 1978.

7. BRANDT, A. and DINAR, N. - Multi-Grid Solutions to Elliptic Flow Problems. Numerical Methods for Partial Differential Equations, Editor: S. Parter, Academic Press, New York, 1979.

8. THUNNEL, T. and FUCHS, L. - Numerical Simulation of the Navier-Stokes Equations by Multi-Grid Techniques. Proceedings of Symposium on Numerical Methods in Laminar and Turbulent Flows, Editors: C. Taylor, K. Morgan and B.A. Schrefler, Venice, Italy, 1981.

9. GHIA, K.N., GHIA, U. and SHIN, C.T. - Study of Asymptotic Incompressible Flow in Curved Ducts Using a Multi-Grid Technique. To appear in ASME Journal of Fluids Engineering, 1983.

10. Proceedings of "International Multi-Grid Conference," Organized by S. McCormick, Institute for Computational Studies, Fort Collins, Colorado, April 1983.

11. BENJAMIN, A.S. and DENNY, V.E. - On the Convergence of Numerical Solutions for 2-D Flows in a Cavity at Large Re. Journal of Computational Physics, Vol. 33, pp. 340-358, 1979.

12. AGARWAL, R.K. - A Third-Order Accurate Upwind Scheme for Navier-Stokes Solutions at High-Reynolds Numbers. AIAA Paper No. 81-0112, 1981.

13. SCHREIBER, R. and KELLER, H.B. - Driven Cavity Flows by Efficient Numerical Techniques. Journal of Computational Physics, Vol. 49, No. 2, Feb. 1983.

14. PETERSON, J. - High-Reynolds Number Solutions for Incompressible Viscous Flow Using the Reduced-Basis Technique. Journal of Computational Physics, to appear.

15. GRESHO, P., CHAN, S.T., UPSON, C.D. and LEE, R.L. - A Modified Finite Element Method for Solving the Time-Dependent Incompressible Navier-Stokes Equations. UCRL-88937 Preprint, Lawrence Livermore National Laboratory, Livermore, California, 1983.

16. NOOR, A.K. - Recent Advances in Reduction Methods for Non-linear Problems. Computers and Structures, Vol. 13, pp. 31-44, 1981.

17. CHENG, K.C., LIN, R.C. and OU, J.W. - Fully Developed Laminar Flow in Curved Rectangular Channels. Journal of Fluids Engineering, pp. 41-48, March 1976.

18. GHIA, K.N. and SOKHEY, J.S. - Laminar Incompressible Viscous Flow in Curved Ducts of Regular Cross-Sections. Journal of Fluids Engineering, Vol. 99, No. 4, pp. 640-648, 1977.

19. GHIA, U., GHIA, K.N. and GOYAL, R.K. - Three-Dimensional Viscous Incompressible Flow in Curved Polar Ducts. AIAA Paper No. 79-1536, 1979.

20. BRANDT, A. - Multi-Level Adaptive Computations in Fluids Dynamics. AIAA Paper 79-1455, 1979.

21. GHIA, U., GHIA, K.N. and SHIN, C.T. - High-Re Solutions for Incompressible Flow Using the Navier-Stokes Equations and a Multi-Grid Method. Journal of Computational Physics, Vol. 48, No. 2, 1982.

22. STONE, H.L. - Iterative Solution of Implicit Approximations of Multi-Dimensional Partial Differential Equations. SIAM Journal of Numerical Analysis, Vol. 5, No. 3, pp. 530-559, 1968.

23. RUBIN, S.G. and KHOSLA, P.K. - Navier-Stokes Calculations with a Coupled Strongly Implicit Method - I: Finite Difference Solutions. Computers and Fluids, Vol. 9, pp. 163-180, 1981.

24. PAN, F. and ACRIVOS, A. - Steady Flows in Rectangular Cavities. Journal of Fluid Mechanics, Vol. 28, pp. 643-655, 1967.

25. KOSEFF, J.R., STREET, R.L. and RHEE, H. - Velocity Measurements in a Lid-Driven Cavity Flow. ASME Paper 83-FE-12, presented at ASME Applied Mechanics, Bioengineering and Fluids Engineering Conference, Houston, Texas, June 1983.

26. NALLASAMY, M. and PRASAD, K.K. - On Cavity Flow at High Reynolds Numbers. Journal of Fluid Mechanics, Vol. 79, Part 2, pp. 391-414, 1977.

27. VAN DYKE, M. - Extended Stokes Series: Laminar Flow Through a Loosely Coiled Pipe. Journal of Fluid Mechanics, Vol. 80, Part 1, pp. 129-146, 1978.

28. DENNIS, S.C.R. and NG, M. - Dual Solutions for Steady Laminar Flow Through a Curved Tube. Quart. J. Mech. Appl. Math., Vol. 35, pp. 305-324, 1982.

29. NANDAKUMAR, K. and MASLIYAH, J.H. - Bifurcation in Steady Laminar Flow Through Curved Tubes. J. Fluid Mech., Vol. 119, pp. 475-490, 1982.

30. BERGER, S.A., TALBOT, L. and YAO, L.S. - Flow in Curved Pipes. Annual Review of Fluid Mechanics, Vol. 15, pp. 461-512, 1983.

31. OSSWALD, G.A. and GHIA, K.N. - Study of Unsteady Two-Dimensional Incompressible Navier-Stokes Equations Using Semi-Direct Methods. Aerospace Engineering and Applied Mechanics Report AFL 81-9-58, University of Cincinnati, 1981.

32. GHIA, U., GHIA, K.N. and RAMAMURTI, R. - Multi-Grid Solution of Neumann Problem for Viscous Flows Using Primitive Variables. AIAA Paper 83-0557, presented at the AIAA 21st Aerospace Sciences Meeting, Reno, NV, 1983.

33. RUBIN, S.G. - Incompressible Navier-Stokes and Parabolized Navier-Stokes Solution Procedure and Computational Techniques. VKI Lecture Notes, Brussels, Belgium, 1982.

34. GHIA, U., GHIA, K.N. and RAMAMURTI, R. - Multi-Grid Solution of Neumann Pressure Problem for 3-D Viscous Flow in Curved Ducts. Aerospace Engineering & Applied Mechanics Report AFL 83-4-66, University of Cincinnati, 1983.

HERMITIAN METHODS FOR THE SOLUTION OF THE NAVIER-STOKES EQUATIONS IN PRIMITIVE VARIABLES.

E. ELSAESSER
ONERA, 92320 - Chatillon, France

R. PEYRET
C.N.R.S., Département de Mathématiques, Université de Nice,
06034 NICE Cedex, and ONERA, 95320 - Chatillon.

1. INTRODUCTION

The characteristic of Hermitian finite-difference methods [1] for the numerical solution of differential equations is to provide higher-order approximations while preserving a compact differencing leading to the solution of simple or block-tridiagonal systems.

The principle is to consider as unknowns not only the values of the function but also those of its derivatives. The system is closed by using relationships, called "Hermitian formulas" connecting the values of the function and those of its derivatives at neighbouring discretization points : with three points it is possible to obtain approximations for first and second derivatives up to sixth-order.

The introduction of relationships of Hermitian type is rather old (see [1]), but their use to the solution of flow problems is relatively recent. During the last ten years, a number of higher-order methods (generally of fourth-order accuracy) leading, in several cases, to identical approximations were proposed : Compact or Hermitian methods [2] - [6] , Mehrstellen and OCI methods [7] - [15] , Spline interpolation methods [16] - [19] . Applications to boundary layer equations were presented in [4],[7],[9],[16]-[18],[20]. Methods of solution of the Navier-Stokes equations have been proposed in [4],[16]-[19],[21]-[27] for the stream function - vorticity formulation and in [28]-[33] for the primitive variable formulation.

It must be noticed that applications concerning the unsteady Navier-Stokes equations make use of classical second-order approximation with respect to time (Crank-Nicolson scheme, 3-level scheme, A.D.I. methods...). Higher-order temporal approximations have been investigated in [34] , [35] in the case of simple parabolic equations (Burger's and heat equations), but problems related to stability, storage and computational time seem presently limit their use.

The present paper is devoted to a presentation and application of fourth-order accurate Hermitian methods, by recapitulating some of the results obtained in [28] - [31]. In the first part, the general Hermitian technique is described and applied to the solution of a linear ordinary differential equation. Questions related to boundary conditions are discussed and the accuracy of the resulting numerical solution is analyzed. The second part is devoted to application of the method to the solution of the steady Navier-Stokes equations. This solution is obtained through a combination of artificial compressibility and ADI methods. Numerical experiments are carried out for the following problems : an exact solution of the Navier-Stokes equations, flows in a square cavity and in a channel.

2. GENERAL HERMITIAN FORMULA.

We consider only the case of approximations for first and second order derivatives by using 3-point Hermitian formulas. Such formulas are relationships between the values of a function f and those of its derivatives f', f" at three neighbouring points $x_{i-1} = (i-1)h$, $x_i = ih$ and $x_{i+1} = (i+1)h$ (i = integer, h = const. > 0). The more general relation of this type is written as

$$H_i \equiv \sum_{j=-1}^{1} (a_j f_{i+j} + b_j f'_{i+j} + c_j f''_{i+j}) = 0, \qquad (2.1)$$

where the a_j, b_j and c_j are constants which are determined by requiring the relation (2.1) to represent, to some order of accuracy related to the mesh size h, the correct relationship existing between any regular fonction f(x) and its derivatives f' (x), f"(x) at the three points x_{i-1}, x_i and x_{i+1}. This is obtained by doing Taylor expansions around x_i in (2.1), so that this equation leads formally to :

$$\sum_{k=0}^{\infty} A_k f^{(k)} = 0. \qquad (2.2)$$

where $f^{(k)}$ denotes the k-th derivative. If formula (2.1) was exact, all the A_k's in (2.2) would be identically zero. Actually, only a finite number of A_k can be made equal to zero :

$$A_k = 0 \qquad k=0,\ldots,K \qquad (2.3)$$

The maximal value of K is limited by the number of arbitrary parameters in (2.1). If K=3, the resulting 4 homogeneous equations (2.3) determine 4 coefficients among a_j, b_j, c_j (j=-1,0,1) in terms of 5 others which remain arbitrary ; and after a change of notation for the parameters we have

$$a_{-1} = (15\beta-8\gamma)/h^2+(3\rho-2\Theta)/h, \quad a_o = 16\gamma/h^2+4\Theta/h,$$
$$a_1 =-(15\beta+8\gamma)/h^2-(3\rho+2\Theta)/h, \quad b_{-1} = (3\alpha+7\beta-5\gamma)/h+\rho-\Theta,$$
$$b_o = 16\beta/h+4\rho, \quad b_1=-(3\alpha-7\beta-5\gamma)/h+\rho+\Theta,$$
$$c_{-1} = \alpha+\beta-\gamma, \quad c_o=4\alpha, \quad c_1 = \alpha-\beta-\gamma.$$

The arbitrary parameters α,β,γ,ρ and Θ are constants independent of the mesh size h. The above expressions for the coefficients are substituted into formula (2.1) which will be written henceforth in the symbolic form :

$$H_i\left[\alpha,\beta,\gamma,\rho,\Theta\right] = 0 \qquad (2.4)$$

This Hermitian formula depends homogeneously on 5 parameters ; hence it is possible to express 4 of these parameters in terms of the remaining one which could be taken equal to 1. The error $\mathscr{E}_i$ associated with the formula (2.4) and defined by

$$\mathscr{E}_i = \sum_{k=K+1}^{\infty} A_k \, f^{(k)}$$

is equal to :

$$\mathscr{E}_i = \frac{h^3}{6}\Theta f^{iv} + \frac{h^4}{90}\left[3\rho\, f^{v} + (3\alpha-2\gamma)\, f^{vi}\right]$$
$$+ \frac{h^5}{630}(7\Theta\, f^{vi} - 2\beta\, f^{vii}) + O(h^6).$$

where the superscripts denote derivatives.

The general Hermitian formula (2.4) concerns a uniform mesh, but it is possible to derive an analogous formula for non uniform mesh. This is done in [5], [6], [17] for particular Hermitian formulas.

3. HERMITIAN METHODS FOR DIFFERENTIAL EQUATIONS.

Let us consider the solution of the differential problem

$$Lf \equiv -f''+af'+bf+c = 0 \qquad 0 < x < 1 \qquad (3.1.a)$$
$$f(0) = \varphi_o \quad f(1) = \varphi_1 \qquad (3.1.b)$$

where a, b and c are given functions of x. Here, we present the general formulation of Hermitian methods to solve the problem (3.1). The equation (3.1.a) is written at discretization points :

$$L f_i \equiv -f''_i + a_i f'_i + b_i f_i + c_i = 0, \quad i=1,\dots,I-1 \qquad (3.2)$$

where f_i, f'_i and f''_i are respectively approximation of the solution $f(x_i)$, $f'(x_i)$ and $f''(x_i)$ at the discretization point $x_i = ih$, $i=0,\dots,I$, $h=1/I$. To the above equation (3.2) we must add supplementary relations in order to connect the derivatives to the function itself. In usual finite difference methods, each of f'_i or f''_i is explicitely expressed in terms of the values of the function at x_i and neighbouring points ("finite-difference formulas"). Here, we consider instead two Hermitian formulas :

$$\left.\begin{array}{ll} H_i\,[\alpha_1,\beta_1,\gamma_1,\rho_1,\theta_1] = 0 & (3.3.a) \\ H_i\,[\alpha_2,\beta_2,\gamma_2,\rho_2,\theta_2] = 0 & (3.3.b) \end{array}\right\} i=1,\dots,I-1$$

which must be independent. The use of boundary conditions

$$f_0 = \varphi_0, \qquad f_I = \varphi_1 \qquad (3.4)$$

is not sufficient to close the system because of the presence of f'_i and f''_i as supplementary dependent variables. The manner to close the system will be considered below. So, the resulting system is generally a 3x3 block tridiagonal system for the unknowns (f_i, f'_i, f''_i) which can be easily solved by the block-factorization method [36].

The dimension of the blocks can be diminished as it will be shown in the next sections devoted to the description and study of special Hermitian approximations.

3.1. Method I : Explicit formula for second derivatives.

In this method [6] , the formula $H_i\,[0,0,0,1,0\,] = 0$, that is to say

$$f'_{i+1} + 4\,f'_i + f'_{i-1} - \frac{3}{h}\,(f_{i+1} - f_{i-1}) = 0 \qquad (3.5)$$

is used for the first derivatives, and the second derivatives are explicitely eliminated thanks to the formula $H_i\,[1,0,1,0,0\,] = 0$, let

$$f''_i = -\frac{1}{2h}(f'_{i+1} - f'_{i-1}) + \frac{2}{h^2}(f_{i+1} - 2\,f_i + f_{i-1}) \qquad (3.6)$$

The use of these formulas leads to approximations of the derivatives to $O(h^4)$. The main part of the error in the approximation of the derivative given by (3.5) is obtained by writing (3.5) in the form

$$f'_i = \delta_x f_i - \frac{h^2}{6} \delta_{xx} f'_i \tag{3.7.a}$$

with

$$\delta_x f_i = \frac{1}{2h}(f_{i+1} - f_{i-1}), \; \delta_{xx} f'_i = \frac{1}{h^2}(f'_{i+1} - 2f'_i + f'_{i-1}) \tag{3.7.b}$$

By replacing, in the R.H.S. of (3.7.a), f'_i by its own expression we get

$$f'_i = \delta_x f_i - \frac{h^2}{6} \delta_{xx}(\delta_x f_i - \frac{h^2}{6} \delta_{xx} f'_i) ;$$

then, by repeating once more the operation and by doing Taylor expansions in the result we obtain

$$f'_i = f'(x_i) - \frac{h^4}{180} f^V(x_i) + O(h^6) \tag{3.8}$$

where $f'(x_i)$ is the exact value of the derivative. For equation (3.6), we assume that first derivatives are approximated by means of (3.5) and we get :

$$f''_i = f''(x_i) + \frac{h^4}{360} f^{VI}(x_i) + O(h^6). \tag{3.9}$$

The system of equations made of (3.2), (3.5), (3.6) and (3.4), is closed by considering the noncentered, third order accurate, Hermitian formulas :

$$H_1[0,0,0,1,-1] = 0, \quad H_{I-1}[0,0,0,1,1] = 0 \tag{3.10.a,b}$$

that is to say :

$$2f'_1 + f'_o - (f_2 + 4f_1 - 5f_o)/(2h) = 0$$
$$f'_I + 2f'_{I-1} - (5f_I - 4f_{I-1} - f_{I-2})/(2h) = 0.$$

Let us make a remark concerning the uniqueness of relations (3.10). We consider for inner points the more general formula (2.1) which does not involve second derivatives

$$H_i[0,0,0,\rho,\Theta] = 0 \tag{3.11a}$$

and for the boundary i=1, the formula

$$H_i[0,0,0,\rho_o,\Theta_o] = 0 \tag{3.11b}$$

Now, if f'_{i+1} is eliminated from (3.11a) and (3.11b), the formula

(3.10a) is found whatever ρ,Θ,ρ_o and Θ_o provided $\rho\,\Theta_o - \rho_o\,\Theta \neq 0$, condition necessary to insure that (3.11a) and (3.11b) are independent.

To summarize, the equations (3.2), (3.5), (3.6) and the boundary conditions (3.4), (3.10) lead to a 2x2 block tridiagonal system for the unknowns (f_i, f'_i).

Now, it seems interesting to specify the error of the numerical solution given by the above method. The analysis is carried out in the special case

$$a = \text{const.}, \quad b=c=0, \ \varphi_o \neq 0, \ \varphi_1 = 0. \tag{3.12a}$$

for which the exact solution is

$$f(x) = \varphi_o \frac{e^{ax} - e^{a}}{1 - e^{a}} \tag{3.12b}$$

By doing suitable linear combinations of the equation (3.2) and the formulas (3.5), (3.6), (3.10), we can eliminate also the first order derivatives and we obtain the finite-difference problem :

$$f_{i+2} + 2(4-3R)f_{i+1} - 18f_i + 2(4+3R)f_{i-1} + f_{i-2} = 0, \ i=2,\dots,I-2 \tag{3.13a}$$

$$(12-9R+R^2)f_1 - (6-6R+5R^2)f_2 + R\ f_3 - (6-2R-R^2)f_o = 0 \tag{3.13b}$$

$$f_o = \varphi_o \tag{3.13c}$$

$$(12+9R+4R^2)f_{I-1} - (6+6R+5R^2)f_{I-2} - Rf_{I-3} - (6+2R+R^2)f_I = 0 \tag{3.13d}$$

$$f_I = \varphi_1 \tag{3.13e}$$

where R=ah. Obviously these equations, which are useful for a theoretical analysis, must not be used for practical computations since the compact character of the tridiagonal system is lost.

The truncation error of (3.13.a) is $-h^4 af^V/120 + O(h^6)$; on the other hand, the error of (3.13.b,d) is only of second order. This increase of the truncation error is due to the use of third-order approximations to first derivatives into the relations (3.6) giving f''_1 and f''_{I-1}. However, that does not destroy the accuracy of the numerical solution [37], [38]. The general solution of (3.13a) is

$$f_i = \sum_{j=1}^{4} C_j\, q_j^i$$

where the q_j' s are the roots of the characteristic equation associated with (3.13a). Two of these roots are $q_1 = 1$ and

$q_2 > 0$, corresponding respectively to the basic solutions 1 and e^{ax} of the differential equation. The two others roots q_3 and q_4 (which are spurious) are negative and, consequently, could produce an oscillatory solution. The constants C_j are determined by the boundary conditions (3.13b)-(3.13c). Finally, the error of the numerical solution where $h \longrightarrow 0$, is found to be :

$$E_i = f_i - f(x_i) = h^4 \varphi_o (\Phi+\Lambda) + O(h^5), i \in [1, I-1] \quad (3.14)$$

with

$$\Phi = \frac{a^5}{120(1-e^a)^2} [e^a(1-e^{ax_i}) - (1-e^a)x_i e^{ax_i}] \quad (3.15)$$

$$\Lambda = \frac{s\, a^4}{144(1-e^a)} \{e^{ax_i} - (-1)^i [\frac{e^{-ax_i/2}}{s^i} + (-1)^I \frac{e^{a(3-x_i)/2}}{s^{I-1}}]\}$$

where $s = 5+2\sqrt{6}$. The error term $E_I = \Phi+\Lambda$ is shown in the figure 1 for the case $a=-5$, $h=1/100$. The oscillatory character of Λ is not perceptible in the figure. As a matter of fact, the oscillatory part of Λ is exponentially small, when $h \longrightarrow 0$, except in a neighbourhood of the boundaries of the order h. Note that if the present problem was solved by using the standard second-order accurate finite-difference formulas $f'_i = \delta_x f_i$, $f''_i = \delta_{xx} f_i$ with δ_x and δ_{xx} defined in (3.7b), the error would be $E_i = -10h^2 \varphi_o \Phi/a^2 + O(h^4)$.

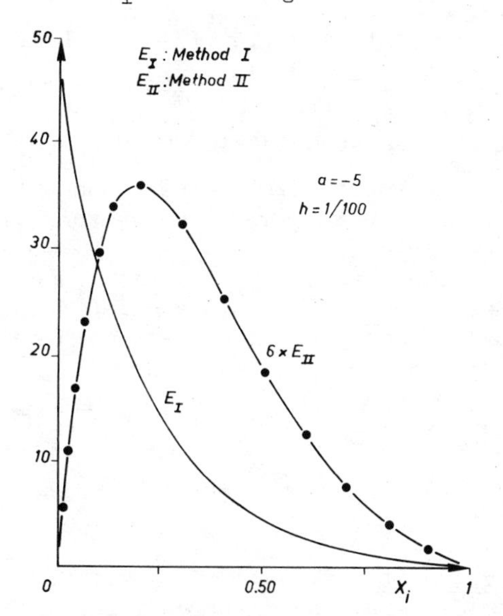

Fig.1. - Variations of the error terms E_I and E_{II} in the numerical solution $f_i = f(x_i) + h^4 \varphi_o\ E_{I,II} + \ldots$

3.2. Method II : Implicit formula for second derivatives.

The method has been studied in particular in [3] - [6]. It makes use of the formula (3.5) for the first derivatives, and of the formula H_i [5/3,0,1,0,0] = 0, i.e.

$$f''_{i+1}+10f''_i + f''_{i-1} - \frac{12}{h^2}(f_{i+1}-2f_i+f_{i-1}) = 0 \qquad (3.16)$$

for the second derivatives. This formula leads to a fourth-order approximation of the second derivative, and more precisely

$$f''_i = f''(x_i) - \frac{h^4}{240} f^{vi}(x_i) + O(h^6). \qquad (3.17)$$

In the present method, the system is closed by considering : (1) the boundary conditions (3.4), (2) the Hermitian formulas (3.10), (3) the equation (3.1.a) itself written at i=0 and i=I. Consequently the algebraic system to be solved is :

$$\left.\begin{array}{ll} L f_i = 0 & \\ H_i\,[0,0,0,1,0] = 0 & \\ H_i\,[5/3,0,1,0,0] = 0 & \end{array}\right\} i=1,\ldots,I-1 \qquad \begin{array}{r}(3.18a)\\(3.18b)\\(3.18c)\end{array}$$

$$f_o = \varphi_o, \quad Lf_o = 0, \quad H_1\,[0,0,0,1,-1] = 0 \qquad (3.18d)$$

$$f_I = \varphi_1, \quad Lf_I = 0, \quad H_{I-1}\,[0,0,0,1,1] = 0. \qquad (3.18e)$$

The unknowns are vectors (f_i, f'_i, f''_i) and their determination requires the inversion of a 3x3 block tridiagonal matrix, However, it is possible [5] to reduce the dimension of the blocks by using the linear combination $Lf_{i+1}+10Lf_i+Lf_{i-1}$ in order to eliminate the second derivatives thanks to formula (3.18c). In such a case, the unknowns are vectors (f_i, f'_i).

In the case of the example (3.12), the truncation error of the equation analogous to (3.13a) is $-h^4 af^v/720$; this error is $O(h^3)$ for the boundary equations analogous to (3.13b, d). The error E_i of the numerical solution is

$$E_i = f_i - f(x_i) = h^4 \varphi_o \Phi + O(h^5) \qquad (3.19)$$

with Φ defined by (3.15). The error term $E_{II} = \Phi$ is shown in figure 1.

Remark on boundary conditions. In some problems, it can be awkward to use the differential equation itself at a boundary in order to close the system. For example, in the case of the Navier-Stokes equations in primitive variables the use of the momentum equation at a boundary involves the value of the pressure gradient at this boundary. The evaluation of such a quantity can induce difficulties, mainly in the accuracy. This problem can be avoided if the equations $Lf_o=0$ and $Lf_I=0$ in (3.18d.e) are replaced by Hermitian formulas for second derivatives. So, let us consider for the boundary x=0 the general formula :

$$H_1\,[\alpha_o,\beta_o,\gamma_o,\rho,\Theta] = 0 \tag{3.20}$$

where the parameters $\alpha_o,\beta_o,\gamma_o$ are arbitrary and ρ,Θ are those of the relation

$$H_i\,[0,0,0,\rho,\Theta] = 0$$

used at inner point for the first derivatives. Taking into account of this last relation, the formula (3.20) becomes :

$$H_1\,[\alpha_o,\beta_o,\gamma_o,0,0] = 0. \tag{3.21}$$

Now we assume that (3.18c) is used at inner points ; therefore we can eliminate f_2'' from (3.21) and (3.18c) considered at i=1 and we obtain the following fourth-order accurate formula :

$$\begin{aligned} C_1^o\,[\mu_o] \equiv\ & (5-3\mu_o)\,f_1'' + f_o'' \\ & + [\,(7-3\mu_o)f_2'+16f_1'+(7+3\mu_o)f_o'\,]\ /(2h) \\ & + [\,(-27+12\mu_o)f_2 + 24(1-\mu_o)f_1+3(1+4\mu_o)f_o\,]/(2h^2)=0 \end{aligned} \tag{3.22}$$

depending on the single arbitrary parameter $\mu_o=(\alpha_o-5\gamma_o/3)/\beta_o$. For the other boundary, we get

$$C_{I-1}^o\,[\mu_I] = 0 \tag{3.23}$$

which can be deduced formally from (3.22) by changing μ_o in μ_I, h in -h, and the subscripts 0,1,2 in I,I-1,I-2 ,respectively.

When the equations $Lf_o=0$ and $Lf_I=0$ in (3.18d,e) are replaced by (3.22) and (3.23), the truncation error of the resulting 4-point difference equations analogous to (3.13,b,d) is $O\,[(\mu_j+1)h^3]$ with j=0,I. Finally, the error E_i is again given by (3.19) ; at fourth-order, E_i does not depend on μ_o,μ_I. On the other hand, if $\mu_o=\mu_I=-1$, the term $O(h^5)$ in (3.9) is replaced by a term $O(h^6)$.

3.3. Conclusions and Remarks.

(a) The numerical error $E_i=f_i-f(x_i)$ is $O(h^4)$ whatever the truncation near a boundary is $O(h^3)$ (Method II) or $O(h^2)$ (Method I). However, the error (3.19) becomes $O(h^5)$ near a boundary, while the error (3.14) remains $O(h^4)$.

(b) From the analytical and numerical results (fig.1), it is seen that Method II gives more accurate results that Method I, although the error of approximation (3.9) is smaller than (3.17). This is due to two facts : (i) the different order of magnitude of the errors near a boundary as mentionned in (a) ; (ii) the errors in (3.9) and (3.17) have different sign, so that the combination of the errors associated with the respective approximations of the first and second derivations in the differential equation leads to a final truncation error larger in Method I than in Method II. This seems to be effective in the simple linear case considered here, but this superiority of Method II will not be constated (Section 5) in the more complex case of the Navier-Stokes equations.

(c) Only fourth-order accurate formulas have been considered here. It is interesting to point out that sixth-order accurate approximations can be obtained from :

$$H_i\,[2/3,0,1,0,0] = 0, \quad H_i\,[0,1,0,0,0] = 0. \qquad (3.24)$$

These formulas have been used in [17] and [39].

(d) The fourth-order accurate "Mehrstellen method" considered in [9] can be constructed by using special Hermitian formulas. More precisely, this method allows the elimination of the derivatives f'_{i-1}, f'_i, f'_{i+1} and f''_{i-1}, f''_i, f''_{i+1} thanks to the following system of equations :

$$Lf_{i-1} = 0, \quad Lf_i = 0, \quad Lf_{i+1} = 0 \qquad (3.25)$$

$$H_i\,[0,0,0,1,0] = 0 \qquad (3.26)$$

$$H_i\,[0,-1,1,0,0] = 0, \quad H_i\,[1,0,1,0,0] = 0, \quad H_i[0,1,1,0,0] = 0. \qquad (3.27)$$

Therefore, there results a finite-difference equation involving only the values f_{i-1}, f_i, f_{i+1} which, along with the boundary conditions (3.4), forms a simple tridiagonal system. Note that an identical system is obtained from the "standard O.C.I. method" [10], [13].

(e) The "Hermitian type" method developed in [40] is related to the above Mehrstellen method in the sense that it makes use of formulas (3.27). For each value of i, the derivatives f''_{i-1}, f''_i and f''_{i+1} are eliminated respectively, from Lf_{i-1}, Lf_i and Lf_{i+1} thanks to eqs (3.27). We denote the result of this elimination by L^+f_{i-1}, L^0f_i and L^-f_{i+1}. The truncation errors associated with such expressions are defined by :

$$L^{+}f_{i-1} = Lf(x_i) + h^4\,\varepsilon_i + O(h^5),$$
$$L^{o}f_i = Lf(x_i) + \frac{h^4}{4}\,\varepsilon_i + O(h^5), \qquad (3.28)$$
$$L^{-}f_{i+1} = Lf(x_i) + h^4\,\varepsilon_i + O(h^5).$$

consequently, the linear combinations

$$\left.\begin{aligned} L^{+}f_{i-1} - 4\,L^{o}f_i &= 0 \\ 4\,L^{o}f_i - L^{-}f_{i+1} &= 0 \end{aligned}\right\}\; i = 2,\ldots,I-2 \qquad (3.29)$$

leads to a fifth-order approximation. The system (3.29) is closed by adding the boundary conditions (3.4) and the supplementary equations :

$$\begin{aligned} &L^{+}f_o = 0,\ L^{o}f_1 = 0,\ L^{-}f_2 = 0 \\ &L^{+}f_{I-2}=0,\ L^{o}f_{I-1} = 0,\ L^{-}f_I = 0. \end{aligned} \qquad (3.30)$$

The resulting hexa-diagonal system for the unknowns f_i, f'_i, is solved by means of a Gaussian elimination. It must be pointed out that the tridiagonal character of the matrices associated with the Hermitian methods described previously is lost in the present case.

4. HERMITIAN METHODS FOR THE NAVIER-STOKES EQUATIONS.

The steady Navier-Stokes equations in Cartesian coordinates x,y are written

$$u\,\vec{V}_x + v\,\vec{V}_y - \frac{1}{Re}\nabla^2\vec{V} + \nabla p = 0 \qquad (4.1.a)$$

$$\nabla.\vec{V} = 0 \qquad (4.1.b)$$

where $\vec{V} = (u,v)$ is the velocity vector, p in the pressure and Re is the Reynolds number. In eq. (4.1) the subscripts x,y denote derivatives. We look for the solution of (4.1), in a rectangular domain Ω , satisfying the boundary conditions :

$$\vec{V} = \vec{V}^{\Gamma} = (u^{\Gamma}, v^{\Gamma}) \qquad (x,y) \in \Gamma = \partial\Omega \qquad (4.2)$$

where $\vec{V}^{\Gamma}$ is a given vector with zero total flux across Γ .

At each discretization point $x_i = ih$, $i=0,\ldots,I$, $y_j = jh$, $j=0,\ldots,J$, and for any function f(x,y), the unknowns are the approximated values of the function $f_{i,j}$ and those of its derivatives.

The solution of the problem (4.1)-(4.2) makes use of Hermitian relations of fourth-order accuracy. The resulting nonlinear algebraic system is solved by an iterative procedure based on the combination of artificial compressibility and

A.D.I. methods.

The two methods exposed in the previous section will be used here. The method I makes use of formulas of type (3.5) and (3.6) :

$$F_i(f_x, f) \equiv f_{xi+1,j} + 4\, f_{xi,j} + f_{xi-1,j} - \frac{3}{h}(f_{i+1,j} - f_{i-1,j}) = 0 \quad (4.3)$$

$$E_i(f_{xx}, f_x, f) \equiv f_{xxi,j} + \frac{1}{2h}(f_{xi+1,j} - f_{xi-1,j})$$
$$- \frac{2}{h^2}(f_{i+1,j} - 2f_{i,j} + f_{i-1,j}) = 0 \quad (4.4)$$

where $f_{xi,j}$ and $f_{xx\;i,j}$ are respectively the approximation of $f_x(x_i, y_j)$ and $f_{xx}(x_i, y_j)$. Analogous formulas for the y-derivatives f_y and f_{yy} will be denoted by

$$F_j(f_y, f) = 0 \quad (4.5)$$

$$E_j(f_{yy}, f_y, f) = 0 \quad (4.6)$$

The method II is based on the same relations (4.3), (4.5) and on formulas of type (3.16) :

$$I_i(f_{xx}, f) \equiv f_{xx\;i+1,j} + 10 f_{xx\;i,j} + f_{xx\;i-1,j}$$
$$- \frac{12}{h^2}(f_{i+1,j} - 2\, f_{i,j} + f_{i-1,j}) = 0, \quad (4.7)$$

with an analogous relation for the y-derivative :

$$I_j(f_{yy}, f) = 0 \quad (4.8)$$

Moreover, for each of these methods, we shall consider two variants (A and B) according to the degree of implicite-ness introduced in the calculation of the pressure.

4.1. Description of methods.

Let n be the index of iteration, a complete cycle allows to pass from the known values $f^{n}_{i,j}, \ldots,$ to the final values $f^{n+2}_{i,j}, \ldots,$ through the intermediates values $f^{n+1}_{i,j}, \ldots$.

(a) Method I.A.

1st step : $n \longrightarrow n+1$.

$$\vec{V}^{n+1}_{i,j} - \vec{V}^{n}_{i,j} + \kappa\,[u^{n}_{i,j}\,\vec{V}^{n+1}_{x\,i,j} + v^{n}_{i,j}\,\vec{V}^{n}_{y\,i,j}$$

$$- \frac{1}{Re}(\vec{V}^{n+1}_{xx\,i,j} + \vec{V}^{n}_{yy\,i,j}) + \nabla p^{n}_{i,j}\,] = 0 \qquad (4.9.a)$$

$$F_i(\vec{V}^{n+1}_{x}, \vec{V}^{n+1}) = 0, \quad E_i(\vec{V}^{n+1}_{xx}, \vec{V}^{n+1}_{x}, \vec{V}^{n+1}) = 0 \qquad (4.9.b,c)$$

$$p^{n+1}_{i,j} - p^{n}_{i,j} + \lambda\,(u^{n+1}_{x\,i,j} + v^{n}_{y\,i,j}) = 0 \qquad (4.9.d)$$

$$F_i(p^{n+1}_{x}, p^{n+1}) = 0, \; F_j(p^{n+1}_{y}, p^{n+1}) = 0 \qquad (4.9.e,f)$$

2nd step : n+1 → n+2.

$$\vec{V}^{n+2}_{i,j} - \vec{V}^{n+1}_{i,j} + \kappa\big[u^{n+1}_{i,j}\,\vec{V}^{n+1}_{x\,i,j} + v^{n+1}_{i,j}\,\vec{V}^{n+2}_{y\,i,j}$$

$$- \frac{1}{Re}(\vec{V}^{n+1}_{xx\,i,j} + \vec{V}^{n+2}_{yy\,i,j}) + \nabla p^{n+1}_{i,j}\big] = 0 \qquad (4.10.a)$$

$$F_j(\vec{V}^{n+2}_{y}, \vec{V}^{n+2}) = 0, \quad E_j(\vec{V}^{n+2}_{yy}, \vec{V}^{n+2}_{y}, \vec{V}^{n+2}) = 0 \qquad (4.10.b,c)$$

$$p^{n+2}_{i,j} - p^{n+1}_{i,j} + \lambda(u^{n+1}_{x\,i,j} + v^{n+2}_{y\,i,j}) = 0 \qquad (4.10.d)$$

$$F_i(p^{n+2}_{x}, p^{n+2}) = 0, \quad F_j(p^{n+2}_{y}, p^{n+2}) = 0 \qquad (4.10.e,f)$$

In the above equations κ and λ are parameters of convergence. When the boundary conditions (4.2) and supplementary equations as it will be described in section 4.3 are added, the equations (4.9), (4.10) allow the determination of the solution in the following manner :

(i) The equations (4.9a,b,c) lead to the solution of two uncoupled 2x2 block tridiagonal systems, one for $u^{n+1}_{i,j}$, $u^{n+1}_{x\,i,j}$) and the other for ($v^{n+1}_{i,j}$, $v^{n+1}_{x\,i,j}$). The second derivatives $u^{n+1}_{xx\,i,j}$ and $v^{n+1}_{xx\,i,j}$ being eliminated by using (4.9c)

(ii) The pressure $p^{n+1}_{i,j}$ is determined explicitely by eq. (4.9d).

(iii) The equations (4.9e,f) lead to two simple tridiagonal systems respectively for $p^{n+1}_{x\,i,j}$ and $p^{n+1}_{y\,i,j}$.

The calculation of the second step is similar. All these systems are solved by factorisation.

(b) Method II.A. This method differs from I.A. by the use of the Hermitian formulas of type (4.7). More precisely, the equations (4.9c) and (4.10c) are respectively replaced by

$$I_i(\vec{V}_{xx}^{n+1}, \vec{V}^{n+1}) = 0, \quad I_j(\vec{V}_{yy}^{n+2}, \vec{V}^{n+2}) = 0. \qquad (4.11.a,b)$$

Consequently, the elimination of $V_{xx\ i,j}$ and $V_{yy\ i,j}$ is no longer explicit and in both systems given by (4.9a,b) and (4.11a) the blocks are 3x3, the unknowns being the vectors $(u_{i,j}^{n+1}, u_{x\ i,j}^{n+1}, u_{xx\ i,j}^{n+1})$ for the first system and the vectors $(v_{i,j}^{n+1}, v_{x\ i,j}^{n+1}, v_{xx\ i,j}^{n+1})$ for the second one. However, as it was explained in Section 3.2, the derivatives $u_{xx\ i,j}^{n+1}$ and $v_{xx\ i,j}^{n+1}$ are eliminated thanks to a linear combination of (4.9a) by taking into account of (4.11a), so that the dimension of the blocks is reduced to 2. It is the same for the second step.

(c) Method I.B. and II.B. These methods differ from the previous ones by the calculation of the pressure : the term $\nabla.\vec{V}$ is evaluated entirely at the level under consideration. So, the equations (4.9d) and (4.10d) are respectively replaced by

$$p_{i,j}^{n+1} - p_{i,j}^{n} + \lambda(u_{x\ i,j}^{n+1} + v_{y\ i,j}^{n+1}) = 0 \qquad (4.9d)'$$

$$p_{i,j}^{n+2} - p_{i,j}^{n+1} + \lambda(u_{x\ i,j}^{n+2} + v_{y\ i,j}^{n+2}) = 0 \qquad (4.10d)'$$

Because they are more implicit, the resulting methods have much better properties than methods IA and IIA, as it will be seen later. However, they necessitate the solution of two supplementary tridiagonal systems

$$F_j(v_y^{n+1}, v^{n+1}) = 0 , \quad F_i(u_x^{n+2}, u^{n+2}) = 0$$

to calculate $v_{y\ i,j}^{n+1}$ involved in (4.9d)' and $u_{x\ i,j}^{n+2}$ needed in (4.10d)'.

4.2. CONVERGENCE.

Necessary conditions of convergence of the iterative methods I and II were obtained by carrying out a Von Neumann stability analysis. The convective terms are linearized (Oseen's approximation) ; then by combining analytical and numerical studies of the eigenvalues of the amplification matrix, we have obtained the following conditions

$$\kappa>0, \lambda>0, \ \kappa\lambda/h^2 \leq 2/3, \ \mathrm{Re}\ \lambda \leq 2 \qquad (4.12)$$

valid for both methods IA and IIA. It must be noticed that

these conditions do not depend on velocity.

In the case of methods IB and IIB, the first three conditions (4.11) are necessary, but the last one $Re\lambda \leq 2$ is no longer needed. This property makes the interest of the methods of type B, even if they need a little more computations.

4.3. Boundary conditions.

As it was already developed in section 3, the boundary conditions

$$\vec{V}_{i,j} = \vec{V}^{\Gamma}_{i,j} = (u^{\Gamma}_{i,j}, v^{\Gamma}_{i,j}), \quad (x_i, y_j) \in \Gamma$$

are not sufficient to close the various systems. The techniques of closure described in section 3 can be extended to the case of the Navier-Stokes equations. However, the usual problem of the determination of the pressure at a boundary is amplified here by the fact that the pressure gradient is also an unknown. In classical finite-difference method with second-order accuracy, the use of a staggered mesh allows to do not define the pressure at a boundary and the artificial compressibility method does not require boundary conditions for the pressure [41] . In the present case of fourth-order Hermitian methods, the utilization of a staggered mesh, although conceivable, would lead to a so complicated differencing that it has been rejected. Various procedures have been devised and experimented in [28]-[31] .

Procedure a. The pressure on the boundary Γ ($i=0,I$ and $j=0,J$) is calculated from the equations (4.9d) and (4.10d) applied up to the boundary. Such a procedure has been tested in [28], [29] associated with method IA. The convergence of the iterative procedure and the accuracy of the results of pressure are relatively poor (see figs 3 and 4). Consequently, it is preferable to use procedures which do not include the values of the pressure at a boundary.

Procedure b. We describe in detail this procedure in the case of method IA. It is characterized by the fact that the pressure at the boundary Γ does not play any role in the solution. The equations (4.9d) and (4.10d) are applied at inner points, $i=1,\dots,I-1$, $j=1,\dots,J-1$. The pressure gradient is also needed at inner points only and the corresponding systems (4.9e,f), (4.10e,f) are closed by adding non-centered third-order accurate Hermitian relations :

$$B^{o}_{2}(p_x, p) = 0 \qquad\qquad B^{1}_{I-2}(p_x, p) = 0 \qquad (4.13)$$

$$B^{o}_{2}(p_y, p) = 0 \qquad\qquad B^{1}_{J-2}(p_y, p) = 0 \qquad (4.14)$$

where

$$B_i^o(f_x,f) \equiv 2\, f_{x\,i,j} + f_{x\,i-1,j} - \frac{1}{2h}(f_{i+1,j} + 4\, f_{i,j} - 5\, f_{i-1,j})$$

$$B_i^1(f_x,f) \equiv f_{x\,i+1,j} + 2\, f_{x\,i,j} - \frac{1}{2h}(5\, f_{i+1,j} - 4 f_{i,j} - f_{i-1,j})$$

and similar definitions for $B_j^o(f_y,f)$ and $B_j^1(f_y,f)$.

The boundary relations (4.13) and (4.14) apply for n+1 as well as n+2.

For the closure of the systems obtained from the equations (4.9a,b,c) written for i=1,...,I-1, j=1,...,J-1 and defining $(\vec{V}^{n+1}_{i,j}, \vec{V}^{n+1}_{x\,i,j})$ we use, as previously, the relations :

$$B_1^o(\vec{V}_x^{n+1}, \vec{V}^{n+1}) = 0,\quad B_{I-1}^1(\vec{V}_x^{n+1}, \vec{V}^{n+1}) = 0 \tag{4.15}$$

Finally, the similar systems obtained from (4.10a,b,c) for $(\vec{V}^{n+2}_{i,j}, \vec{V}^{n+2}_{y\,i,j})$ are closed with

$$B_1^o(\vec{V}_y^{n+2}, \vec{V}^{n+2}) = 0,\quad B_{J-1}^1(\vec{V}_y^{n+2}, \vec{V}^{n+2}) = 0. \tag{4.16}$$

In the case of <u>method IIA</u>, it is necessary to consider supplementary relations for the second order derivatives of $\vec{V}$ in the normal direction to Γ . These relations are the two-dimensional extensions of formulas (3.22) and (3.23), let respectively :

$$C_1^o(\vec{V}_{xx}^{n+1}, \vec{V}_x^{n+1}, \vec{V}^{n+1};\ \mu_o) = 0,\quad C_{I-1}^1(\vec{V}_{xx}^{n+1}, \vec{V}_x^{n+1}, \vec{V}^{n+1},\ \mu_I) = 0 \tag{4.17}$$

for the first step and similar relations for the y-derivatives at the second step.

In this Procedure b, neither the value of p on Γ nor its gradient plays a role in the solution. If the pressure p on Γ is wanted, it can be calculated at convergence from relations of the type :

$$H_1\,[0,0,0,1,1] = 0,\quad H_{I-1}[0,0,0,1,-1] = 0.$$

<u>Procedure c.</u> This procedure differs from <u>Procedure b</u> only by the use of the incompressibility equation $u_x + v_y = 0$ imposed at a boundary. Consequently, the formulas (4.15) associated with u_x^{n+1} are replaced by $u^{n+1}_{x\,i,j} = -v^{\Gamma}_{y\,i,j}$ for the boundaries i=0 and i=I ; and the analogous formulas (4.16) for v_y^{n+2} are replaced by $v^{n+2}_{y\,i,j} = -u^{\Gamma}_{x\,i,j}$ for the boundaries j=0 and j=J.

4.4. REMARK ON THE COMPUTATION OF THE PRESSURE.

As for the artificial compressibility method associated with classical finite-difference of second order accuracy [41], the present method is equivalent to the iterative solution of a Poisson equation for the pressure. In order to simplify the analysis, let us consider the case of Oseen's equations with a forcing term $\vec{F} = (F^x, F^y)$

$$\vec{V}_o . \nabla\vec{V} + \nabla p - \frac{1}{Re} \nabla^2 \vec{V} = \vec{F} \qquad (4.18\ a)$$

$$\nabla . \vec{V} = 0 \qquad (4.18\ b)$$

where $\vec{V}_o$ is a constant vector. It is found that the pressure p satisfies at convergence, and whatever the method I or II, to the finite-difference equation

$$(D_y^2\ \delta_x^2 + D_x^2\ \delta_y^2) p_{i,j} = D_x\ D_y (D_y\ \delta_x\ F^x_{i,j} + D_x\ \delta_y F^y_{i,j}) \qquad (4.19)$$

where

$$D_x\ f_{i,j} = f_{i+1,j} + 4f_{i,j} + f_{i-1,j} = 6\ f_{i,j} + h^2 \delta_{xx}\ f_{i,j}$$

and analogous definition for D_y ; $\delta_x, \delta_{xx}, \ldots$, are the classical finite-difference operators introduced in section 3.1.

The equation (4.19) is a fourth-order approximation of the Poisson equation

$$\nabla^2 p = \nabla . \vec{F} \qquad (4.20)$$

derived from eqs. (4.18). The discretization of the Laplacien operator involves 21 points (fig.2). It must be noticed that such a discretization does not lead to the existence of two uncoupled pressure fields as it appears in classical centered second order schemes when velocity and pressure are defined at the same points. In the case of second order schemes this phenomenon, which can induce oscillations in the pressure, is generally avoided by using a staggered mesh. Here, the use of such a mesh is not necessary.

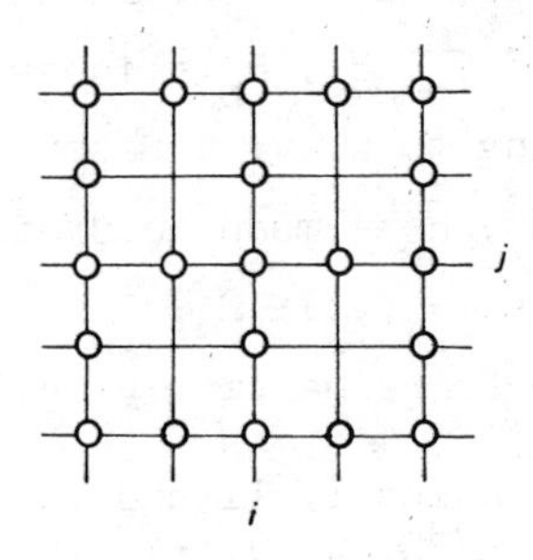

Fig.2 - Grid points involved in the solution of the Poisson equation for pressure.

The equation (4.19) was derived without taking boundaries into account. It could be possible, though very complicated, to derive the form taken by eq.(4.19) near a boundary and the boundary conditions (of Neumann's type) associated with it as it was done for second-order schemes in [41].

5. NUMERICAL APPLICATIONS.

The Hermitian methods described in the previous section have been tested on various examples : (1) An exact solution, (2) flows in a square cavity, (3) jet flows in a channel.

5.1. An exact solution of the Navier-Stokes equations.

The solution considered here is :

$u=-2(1+y)F$, $v=2(1+x)F$, $p=-2F$ (+ const.=2.2).

$F=1/[(1+x)^2+(1+y)^2]$, $0 \leq x \leq 1$, $0 \leq y \leq 1$.

The iterative procedure is begun with $\vec{V}^o_{i,j}=0$, $p^o_{i,j}=1$ and all the derivatives equal to zero. The boundary conditions $\vec{V}=\vec{V}^\Gamma$ deduced from the above exact solution are applied progressively. More precisely $\vec{V}^n_{i,j} = \psi\, \vec{V}^\Gamma_{i,j}$, $(x_i,y_j) \in \Gamma$, with $\psi=n/\mathcal{N}$ if $n \leq \mathcal{N}$ and $\psi=1$ if $n > \mathcal{N}$. This technique improves the speed of convergence of the iterative procedure. The value $\mathcal{N}=100$ has been used here (see [28] for results in the case $\mathcal{N}=1$). The convergence is assumed to be obtained if

$$\max_{i,j;\ u,v,p} \left\{ \frac{1}{\kappa} |u^N_{i,j} - u^{N-1}_{i,j}|, \frac{1}{\kappa}|v^N_{i,j}-v^{N-1}_{i,j}| , \frac{1}{\lambda}|p^N_{i,j}-p^{N-1}_{i,j}| \right\} < \varepsilon$$

where N refers to one of the steps n+1, n+2.

The accuracy of the results is measured by the following errors : E_u, E_v, E_p = inner maximum errors, $(x_i,y_j)\in\Omega$, corresponding to u, v, p respectively ; E^Γ_p = maximum error on Γ ; $\bar{E}_u$, $\bar{E}_v$, $\bar{E}_p$ = mean quadratic errors computed on the inner points $(x_i,y_j)\in\Omega$; $\bar{E}^\Gamma_p$ and $\bar{E}^T_p$ are the analogous errors on Γ and $\Omega+\Gamma$ respectively. All these errors are absolute errors.

Tables I, II and III present results for Re=50, h=1/10 and h=1/20, $\mu_o=\mu_I=-1$. The results are compared to those obtained using a second-order accurate method (explicit artificial compressibility method with a staggered mesh [41]).

The table IV shows the effect of the choice of the parameters μ_o and μ_I (Eqs. 4.17) on results given by Method II. This effect is negligible in the case where Procedure b is used for boundary conditions, but it is much marked in the

Table I. Re = 50 , h = 1/10					
	4th - Order Hermitian Methods, $\varepsilon = 10^{-5}$				2nd-Order Method
Meth.	I.A.b	I.A.c	II.A.b	II.A.c	$\varepsilon = 10^{-4}$
κ	8×10^{-1}	8×10^{-1}	8×10^{-1}	8×10^{-1}	5×10^{-2}
λ	8×10^{-3}	8×10^{-3}	8×10^{-3}	8×10^{-3}	5×10^{-2}
$\mu_o = \mu_I$	-	-	-1	-1	-
N	596	578	601	561	578
E_u	3.84×10^{-5}	$9{,}52 \times 10^{-6}$	3.30×10^{-5}	8.05×10^{-6}	1.32×10^{-4}
E_v	6.56×10^{-5}	9.67×10^{-6}	$6{,}83 \times 10^{-5}$	$9{,}71 \times 10^{-6}$	1.11×10^{-4}
E_p	5.12×10^{-5}	$1{,}94 \times 10^{-5}$	$4{,}95 \times 10^{-5}$	1.69×10^{-5}	1.39×10^{-3}
E_p^{Γ}	$7{,}10 \times 10^{-5}$	$1{,}71 \times 10^{-4}$	$8{,}09 \times 10^{-5}$	1.64×10^{-4}	-
$\overline{E}_u$	$1{,}24 \times 10^{-5}$	3.17×10^{-6}	$1{,}19 \times 10^{-5}$	$2{,}69 \times 10^{-6}$	6.89×10^{-5}
$\overline{E}_v$	2.36×10^{-5}	3.33×10^{-6}	$2{,}50 \times 10^{-5}$	$2{,}83 \times 10^{-6}$	5.62×10^{-5}
$\overline{E}_p$	$1{,}27 \times 10^{-5}$	7.30×10^{-6}	$1{,}17 \times 10^{-5}$	$5{,}82 \times 10^{-6}$	$4{,}08 \times 10^{-4}$
$\overline{E}_p^{\Gamma}$	$2{,}96 \times 10^{-5}$	$5{,}32 \times 10^{-5}$	$3{,}21 \times 10^{-5}$	$5{,}18 \times 10^{-5}$	-
$\overline{E}_p^{T}$	$1{,}96 \times 10^{-5}$	$3{,}01 \times 10^{-5}$	$2{,}03 \times 10^{-5}$	$2{,}91 \times 10^{-5}$	$4{,}08 \times 10^{-4}$

Table II - Re = 50 , h = 1/20					
	4 th - Order Hermitian Methods, $\varepsilon = 10^{-6}$				2nd-Order Method
Meth.	I.A.b	I.A.c	II.A.b	II.A.c	$\varepsilon = 10^{-5}$
κ	5×10^{-2}	4×10^{-2}	5×10^{-2}	$3{,}75 \times 10^{-2}$	1.5×10^{-2}
λ	3×10^{-2}	$2{,}4 \times 10^{-2}$	3×10^{-2}	$2{,}25 \times 10^{-2}$	1.5×10^{-2}
$\mu_o = \mu_I$	-	-	-1	-1	-
N	917	1161	921	1230	1134
E_u	3.65×10^{-6}	1.05×10^{-6}	3.61×10^{-6}	7.10×10^{-7}	2.99×10^{-5}
E_v	4.55×10^{-6}	$9{,}71 \times 10^{-7}$	4.78×10^{-6}	5.94×10^{-7}	3.01×10^{-5}
E_p	$7{,}50 \times 10^{-6}$	$2{,}66 \times 10^{-6}$	7.79×10^{-6}	$1{,}93 \times 10^{-6}$	4.20×10^{-4}
E_p^{Γ}	5.16×10^{-6}	$1{,}60 \times 10^{-5}$	$6{,}40 \times 10^{-6}$	1.46×10^{-5}	-
$\overline{E}_u$	1.13×10^{-6}	$1{,}69 \times 10^{-7}$	$1{,}08 \times 10^{-6}$	$1{,}14 \times 10^{-7}$	$1{,}55 \times 10^{-5}$
$\overline{E}_v$	$1{,}36 \times 10^{-6}$	$2{,}27 \times 10^{-7}$	1.34×10^{-6}	1.49×10^{-7}	$1{,}57 \times 10^{-5}$
$\overline{E}_p$	1.49×10^{-6}	5.44×10^{-7}	$1{,}44 \times 10^{-6}$	4.33×10^{-7}	$1{,}05 \times 10^{-4}$
$\overline{E}_p^{\Gamma}$	1.76×10^{-6}	$4{,}00 \times 10^{-6}$	$2{,}19 \times 10^{-6}$	3.83×10^{-6}	-
$\overline{E}_p^{T}$	1.54×10^{-6}	$1{,}74 \times 10^{-6}$	$1{,}60 \times 10^{-6}$	$1{,}65 \times 10^{-6}$	1.05×10^{-4}

Table III, Re=50, $h=1/10$, $\varepsilon=10^{-5}$		
	$\bar{E}_u \times 10^6$	
$\mu_o=\mu_I$	II-A-b	II-A-c
-5	12.05	3.28
-3	12.09	3.44
-2	12.33	4.14
-1	11.89	2.69
0	11.96	2.98
1	11.99	3.06
5/3	12.00	3.08
3	12.00	3.11
5	12.01	3.12

Table III. Effect of parameters μ_o and μ_I.

Table IV, Re=50, Method I.A.a		
h	1/10	1/20
κ	3.5×10^{2}	5×10^{-2}
λ	2.1×10^{-2}	3×10^{-2}
ε	2.5×10^{-4}	10^{-5}
N	750	2147
E_u	3.26×10^{-5}	2.54×10^{-6}
E_v	8.00×10^{-5}	2.02×10^{-6}
E_p	3.38×10^{-4}	7.36×10^{-4}
E_p^{Γ}	1.64×10^{-3}	3.54×10^{-4}
$\bar{E}_u$	1.06×10^{-5}	6.88×10^{-7}
$\bar{E}_v$	2.82×10^{-5}	6.73×10^{-7}
$\bar{E}_p$	1.36×10^{-4}	6.14×10^{-4}
$\bar{E}_p^{\Gamma}$	8.27×10^{-4}	2.32×10^{-4}
$\bar{E}_p^{T}$	3.31×10^{-4}	5.67×10^{-4}

Table IV. Results given by Method I-A-a.

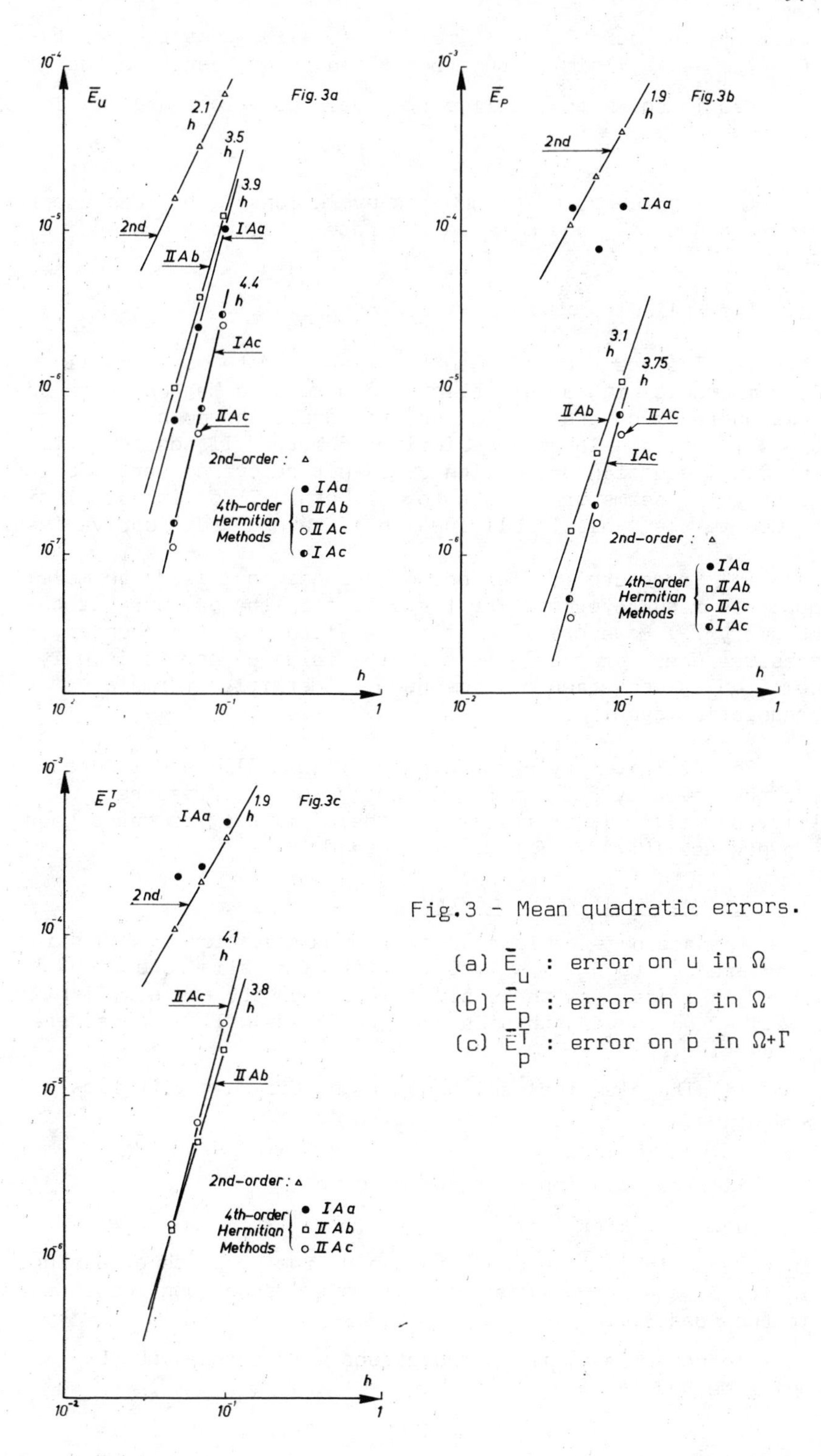

Fig.3 - Mean quadratic errors.

(a) $\bar{E}_u$: error on u in Ω

(b) $\bar{E}_p$: error on p in Ω

(c) $\bar{E}_p^T$: error on p in $\Omega+\Gamma$

case of Procedure c. It must be recalled that the values $\mu_o = \mu_I = -1$ minimize the truncation error near a boundary associated with the linear model equation studied in section 3.2.

The figure 3a, b, c show the evolution of the mean quadratic errors $\bar{E}_u$, $\bar{E}_p$ and $\bar{E}_p^\Gamma$ with the rate of convergence indicated.

The following conclusions can be drawn :

(a) The accuracy of results given by Methods I and II is comparable with a very slight advantage to Method II in the case where Procedure c is used to handle the boundary conditions ; but, the large superiority of Method II constated in the model equation of Section 3. is not recovered here. Consequently, it seems preferable to use Method I for its simplicity due to the explicit elimination of second-order derivatives.

(b) Procedure c for boundary conditions leads to more accurate results except for the values of the pressure at the boundary Γ. However, it must be recalled that the boundary pressure does not play a role in the solution and it is only computed at convergence by means of a Hermitian formula of extrapolation type.

Results given by Procedure a (Table III) are accurate for the velocity but not sufficiently for the pressure (fig. 3.b,c). This is related to the difficulty to get a good convergence (N=2147 if h=1/20 for $\varepsilon=10^{-5}$).

5.2. Flows in a square cavity.

The second example concerns the computation of flows in a square cavity $0 < x < 1$, $0 < y < 1$. A no-slip condition $u=v=0$ is imposed on three sides of the cavity and $u(x,1)=F(x)$, $v(x,1)=0$ on the fourth side $y=1$. Two cases will be considered successively.

(i) The classical solution of the driven cavity flow where

$$u(x,1) = -1 \tag{5.1}$$

(ii) The solution corresponding to

$$u(x,1) = -16x^2(1-x)^2 \tag{5.2}$$

introduced in [43] in order to avoid the presence of a singularity at the corners where the velocity is not continuous as in the case (i).

In both cases, the computations were carried out by using Method IA.

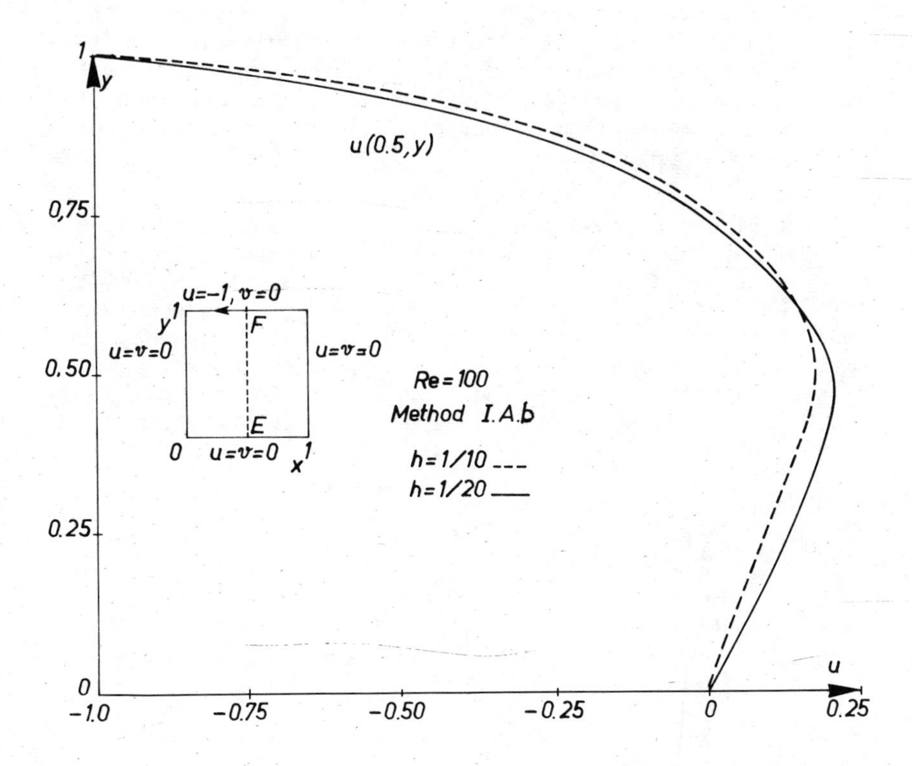

Fig. 4. - Profile of the velocity u on midline EF.

The figures 4 and 5 show results obtained in the case where the velocity distribution on y=1 is given by (5.1), Re=100 and two mesh sizes, h=1/10 and h=1/20. Procedure b was used for boundary conditions. The parameters κ and λ were defined by $\kappa = c_1 h$, $\lambda = c_2 \kappa$ with $c_1 = 0.9$, $c_2 = 0.45$ for h=1/10, and $c_1=6$, $c_2=0.01$ for h=1/20. The iterative procedure begun with all quantities taken equal to zero and the boundary conditions were imposed progressively with $\mathcal{N} = 100$ as explained in the previous section. The number N of iterations needed to reach a convergence characterized by $\varepsilon = 10^{-6}$ was N = 944 in the case h = 1/20.

The examination of Fig.4 shows relatively large differences between the velocity profiles u(0.5, y) obtained for h=1/10 and h=1/20 respectively. This fact can be attributed to the effect of the singularity at the corners A and B which leads to large variations in the truncation error. However, it must be pointed out that the smoothness of the results is not affected by such singularities and no oscillations appear in the numerical solution. This is illustrated by the evolution of the pressure $p(x,1)$ along the side AB as shown in Fig.5.

The figures 6 to 8 display results corresponding to the regular solution obtained with the velocity distribution (5.2). The computations were carried out using the same values κ, and λ as above. Procedure c was used for the treatment of boundary conditions. The number of iterations needed to reach a convergence characterized by $\varepsilon = 10^{-5}$ was N=603 in the case Re=100, h=1/10.

The figure 6 shows the u velocity profiles on the vertical midline x=0.5, given by the Hermitian Method with h=1/10 and by a second-order accurate method (see Section 5.1) with h=1/10 and h=1/20, in the case Re=100. The profile obtained using the Hermitian method and h=1/20 is not represented because the differences with the case h=1/10 are not discernible at the scale of the figure. More precisely, the maximal value of $u(0.5,y)$ is 0.159 for h=1/10 and 0.162 for h=1/20.

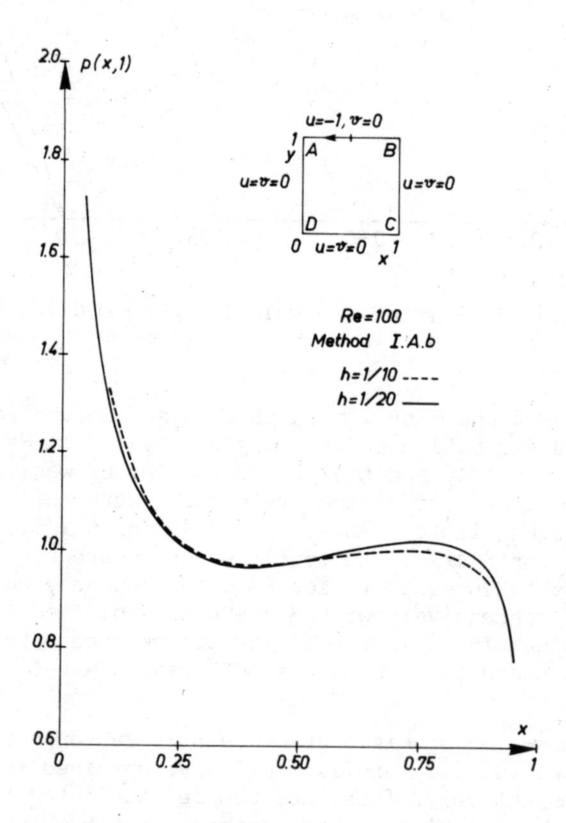

Fig.5. Pressure distribution along AB.

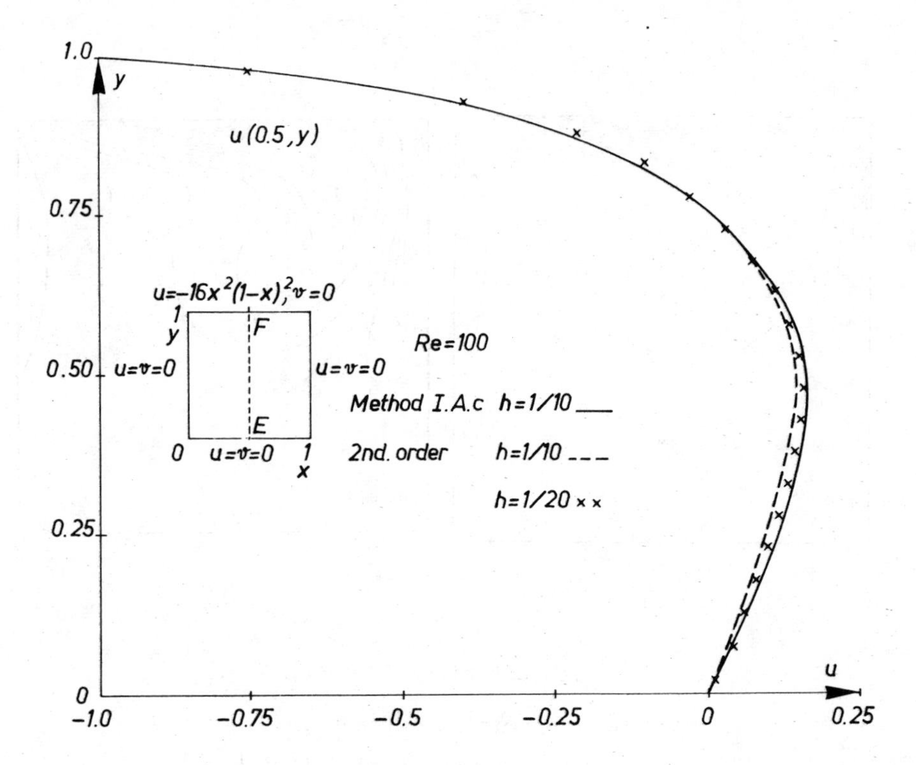

Fig.6 - Profile of the velocity u on midline EF.

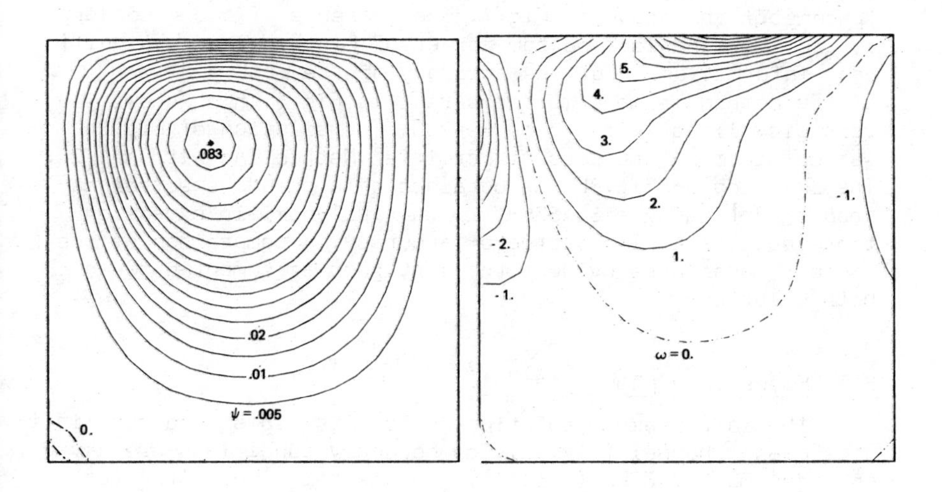

Fig.7a - Streamlines

Fig.7b - Isovorticity lines

(Re=100, Method I.A.c., h=1/20)

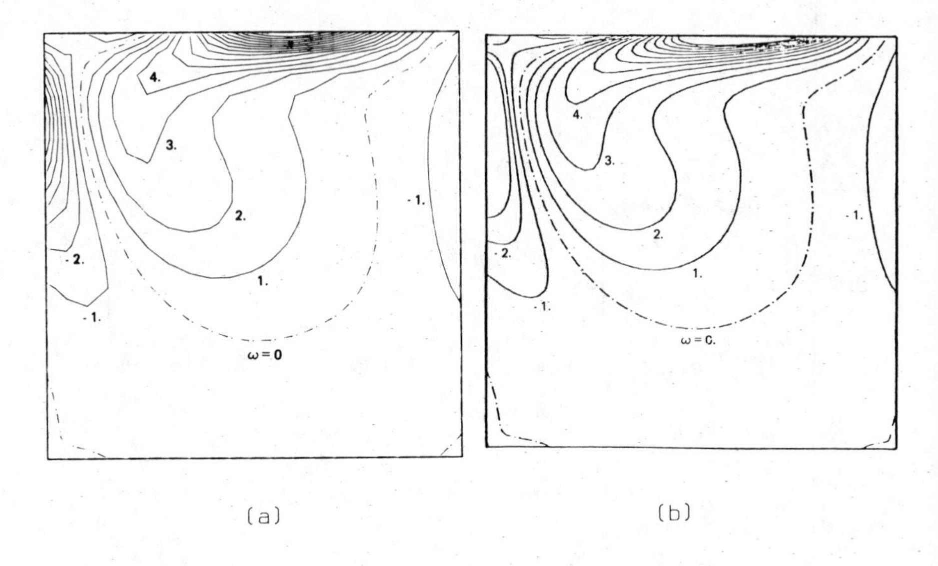

(a) (b)

Fig. 8. - Isovorticity lines, Re=200.

(a) - Hermitian Method I.A.c, h=1/20.

(b) - Pseudo-spectral Method [44].

The isovorticity lines (ω=const.) and the streamlines* (ψ=const.) are shown in Fig.7. The presence of two secondary vortices in the left bottom corner of Fig.7a gives a significant information about the accuracy of the method. The accuracy of the method is made more apparent in Fig.8 which shows isovorticity lines in the case Re=200. The solution of Fig.8a was obtained by the present Hermitian Method IAc with h=1/20. The solution of Fig.8b was obtained by a pseudo-spectral method [44] using a 33x33 Chebyshev polynomials expansion that leads to a high degree of accuracy. A comparison of the contour maps makes evident the very good agreement between both solutions.

5.3. Flows in a plane channel.

The last example solution deals with flows in a semi-infinite plane channel (Fig.9). The boundary conditions are u=v=0 on sides BC and B'C' (y=±1/2), and the entry velocity (at x=0) is given by :

* Computed from a 4th-order Hermitian approximation of $\nabla^2\psi = -\omega$

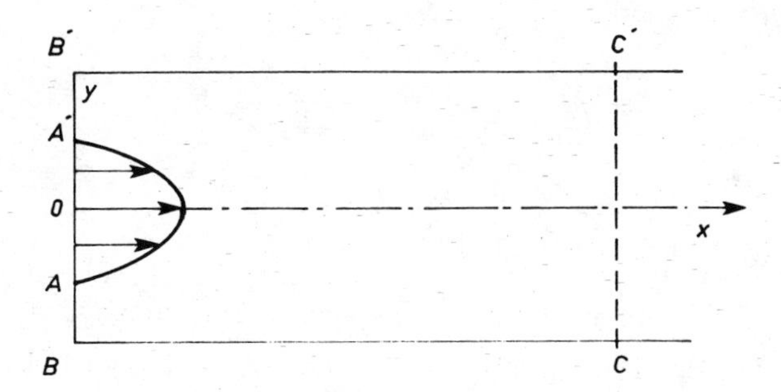

Fig.9. - Geometrical configuration of channel flow.

$$u(0,y) = U_o(y), \quad v(0,y) = 0. \tag{5.3}$$

At the downstream boundary CC' (located at $x = x_e$), the flow is assumed to be of Poiseuille-type. More precisely, we consider two types of boundary conditions*. In the first case, the two components of the velocity are given :

$$u(x_e,y) = U_e(y) = A(1-4y^2), \quad v(x_e,y) = 0 \tag{5.4}$$

where the constant A is determined by the condition :

$$\int_{-1/2}^{1/2} U_o(y)dy = \int_{-1/2}^{1/2} U_e(y)\, dy = \frac{2}{3} A. \tag{5.5}$$

In the second case, the v component and the pressure are given

$$v(x_e,y) = 0, \quad p(x_e,y) = p_o = \text{const.} \tag{5.6}$$

The differences in results obtained by using (5.4) or (5.6) appear only near the downstream boundary.

In some cases where (5.4) was used (mainly with a coarse mesh $h > 1/20$), a blocking in the convergence was experimented: the value of $D_{i,j} = u_{x\,i,j} + v_{y\,i,j}$ was tending toward a steady limit uniformly constant ($\sim 10^{-2}$) in the whole domain of computation. The phenomenon was attributed to a lack of numerical verification of the condition

$$\int_\Omega \nabla.\vec{V}\, d\sigma = \int_\Gamma \vec{V}.\vec{n}\, ds = 0 \tag{5.7}$$

The remedy was to modify slightly the constant A in order to force the discrete analogous of

$\int_\Omega \nabla.\vec{V}\, d\sigma$ to be zero ultimately. That was done by using a

* Other treatments of the boundary CC' are considered in [31]

Newton-type method

$$A_{m+1} = A_m - \frac{S(A_m)}{S'(A_m)}, \quad S'(A_m) = \frac{dS}{dA}(A_m). \qquad (5.8)$$

where

$$S(A_m) = \sum_{i,j=1}^{I,J} (u_{x\,i,j} + v_{y\,i,j}) \qquad (5.9)$$

and $S'(A_m)$ is replaced by a constant value deduced from Green's identity (5.7), i.e. $S'(A_m) = \frac{2}{3}$. The process (5.8) is initiated with $A_0 = A$ as given by Eq. (5.5). It is included in the general iterative procedure (4.9),(4.10)) only in the last phase

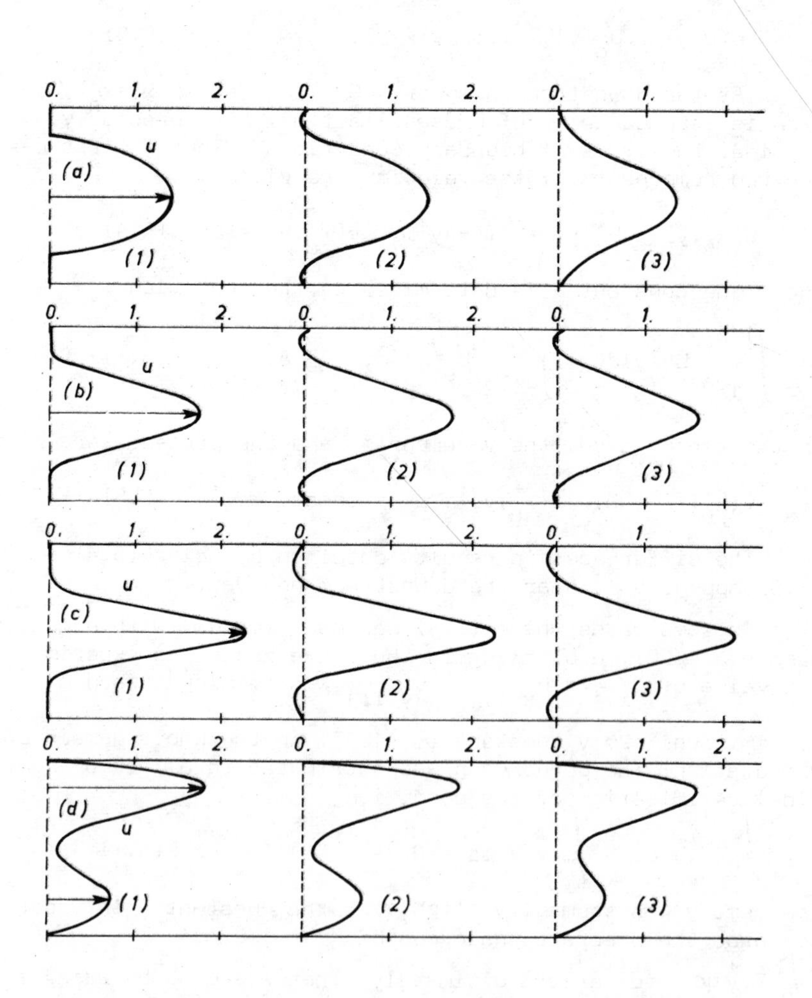

Fig. 10. - Profiles of the velocity u at various sections : (1) x=0, (2) x=0.1, (3) x=0.25 (Re=50, h=1/20).

of the convergence, and it is effectively done 4 to 5 times (A_m is kept constant during a number M=100-200 of iterations n). In the case (d) of Figure 10, a convergence characterized by $D_{i,j} \leq 10^{-5}$ was obtained after 1600 global iterations. The change in A was 1.3%.

Figures 10 and 11 compare results given by various entry velocity profiles. The entry profile is a polynomial of degree r, with r=2 in (a), r=4 in (b), r=8 in (c) and r=5 in (d). In all cases Re=50, the velocity of reference being chosen in order to have the entry flow rate equal to 2/3 in cases (a,b,c) and 0.715 in (d). The downstream boundary is taken at a distance x_e =2 and the mesh size is h=1/20. Boundary conditions (5.4) were used in cases (a) and (d), and conditions (5.6) were used in (b) and (c).

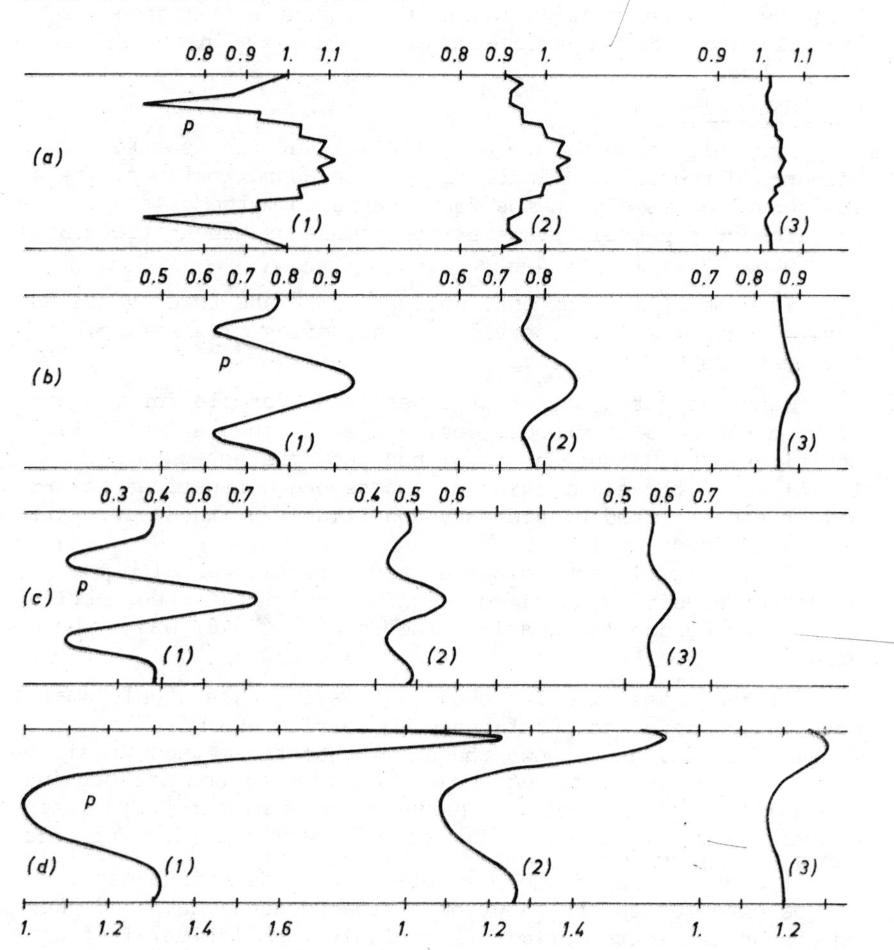

Fig. 11. - Profiles of the pressure p at various sections : (1) x=0, (2) x=0.1, (3) x=0.25 (Re=50, h=1/20).

All the results are obtained by using Method I.B associated with Procedure (c) for the boundary conditions. This procedure is slightly modified when boundary conditions (5.6) are considered. The parameters of convergence are $c_1=0.8$, $c_2=0.45$. When the conditions (5.6) are used, a convergence characterized by $\varepsilon=5\times10^{-5}$ (case b) and $\varepsilon=6\times10^{-5}$ (case c) is obtained after 1000 iterations.

The pressure profiles in case (a) exhibit large oscillations (the v velocity profiles are slightly oscillating). As it was shown in Section 4.4, the presence of such oscillations cannot be attributed to the existence of two uncoupled pressure fields like in second-order methods. The reason for these oscillations lies in the discontinuity of the derivative of $U_o(y)$. This is confirmed by comparing the results of case (a) with those of cases (b,c) for which dU_o/dy is continuous. It must be remarked that no oscillations were encountered when the singular point was located at a corner as in the driven cavity flow.

6. CONCLUSION.

Hermitian methods are an efficient way for constructing higher-order accurate finite-difference approximations. The efficiency lies mainly in the fact that such methods lead to the solution of algebraic systems with simple or block-tridiagonal matrices easily inverted.

In case of multidimensional problems, the tridiagonal nature of matrices is preserved by considering Alternating Direction Methods [41].

Concerning the choice of a Hermitian formula for the approximation of second-order derivatives in the Navier-Stokes equations, the use of Eq. (3.6) has been preferred to Eq. (3.16) since the elimination of the second derivatives is explicit and the need of boundary conditions for these derivatives is avoided.

Obviously, in some cases as the Dirichlet problem for a second-order differential equation without first-order derivative, like Poisson's equation, the formula (3.16) has to be used.

Although the examples above displayed concern only small Reynolds numbers, calculations [31] carried out with larger values (Re=400) have shown the ability of the method in its variant (B), in order to avoid the condition of convergence involving Re. Results with Re=1000 were obtained in [32] by using a method based on the solution of a Poisson equation for calculating the pressure.

The principle of the Hermitian method described here, can be extended to the solution of the unsteady Navier-Stokes equations within the primitive-variable formulation. This can be done by using either a Poisson equation like in [32] or the general method developed in [42] for classical second-order finite-differences. This method makes use of an implicit, second-order accurate, time-discretization scheme (Crank-Nicolson or three-level scheme). Then, the resulting equations are solved, at each time step, by means of an algorithm analogous to (4.9) -(4.10).

Finally, it must be noted that the use of Hermitian methods is not restricted to Cartesian configurations. An Alternating Direction-Artificial Compressibility method has been applied with success to the solution of the Navier-Stokes equations in primitive variables, using a boundary-fitted coordinate transformation method [33].

REFERENCES.

1. COLLATZ, L. - The Numerical Treatment of Differential Equations, Springer Verlag, 1966.

2. ORSZAG, S.A. and ISRAELI, M. - Numerical simulation of viscous incompressible flows. Annual Review of Fluid Mechanics, vol. 6, pp. 281-318, 1974.

3. CIMENT, M. and LEVENTHAL, S.H. - Higher Order Compact Implicit Schemes for the wave equation. Math. Comput., vol. 29, pp. 985-994, 1975.

4. HIRSH, R.S. - Higher-Order Accurate Difference solution of Fluid Mechanics Problems by a Compact Differencing Technique. J. Comput. Phys., vol.19 pp. 90-109, 1975.

5. ADAM, Y. - A Hermitian finite difference method for the solution of parabolic equations. Comp. and Maths. with Appls., vol. 1, pp. 393-406, 1975.

6. ADAM Y. - Highly accurate compact implicit methods and boundary conditions. J. Comput. Phys., vol. 24, pp.10-22, 1977.

7. KRAUSE, E. - Mehrstellenverfahren zur Integration der Grenzschichtgleichungen, DLR Mitt., vol. 71, pp. 109-138, 1971.

8. WIRZ, H.J. - Eine Erweiterung des Verfahrens der Zwischenschritte auf allgemeinere parabolische und elliptische Differentialgleichungen,ZAMM, vol. 52, pp. 329-336, 1972.

9. KRAUSE, E., HIRSCHEL, E.H. and KORDULLA, W. - Fourth order "Mehrstellen" integration for three-dimensional turbulent boundary layers. Computers and Fluids, vol.4, pp. 77-92, 1976.

10. SWARTZ, B.K. - The Construction and Comparison of Finite Difference Analogs of some Finite Element schemes, Mathematical Aspects of Finite Elements in Partial Differential Equations, C.De Boor, Ed., pp. 279-312, Academic Press, 1974.

11. COLLATZ, L. - Hermitean Methods for Initial value Problems in Partial Differential Equations,. Topics in Numerical Analysis, Proc. Royal Irish Academy Conf. Numerical Analysis, J.J.H. Miller, Ed., pp. 41-61, Academic Press, 1972.

12. WEINBERG, B.C., LEVENTHAL, S.H. and CIMENT, M. - The Operator Compact Implicit Scheme for Viscous Flow Problems, AIAA Paper N° 77-638, 1977.

13. CIMENT, M., LEVENTHAL, S.H. and WEINBERG, B.C. - The Operator Compact Implicit Method for Parabolic Equations, J. Comput. Phys. vol. 28, pp. 135-166, 1978.

14. BERGER, A.E., SOLOMON, J.M., CIMENT, M., LEVENTHAL, S.H. and WEINBERG, B.C. - Generalized OCI Schemes for Boundary Layer Problems. Math. Comput., vol. 35, pp. 695-731, 1980.

15. MERCIER, P. and DEVILLE, M., - A multidimensional Compact higher order scheme for 3-D Poisson's Equation, J. Comput. Phys., vol.39, pp. 443-455, 1981.

16. RUBIN, S.G. and KHOSLA, P.K. - Higher-Order Numerical solutions using Cubic splines, AIAA J., vol. 14, pp.851-858 (1976).

17. RUBIN, S.G. and KHOSLA, P.K. - Polynomial Interpolation Methods for viscous Flow Calculations. J. Comput. Phys., vol. 24, pp. 217-244, 1977.

18. RUBIN, S.G. and KHOSLA, P.K. - Turbulent Boundary Layers with and without Mass Injection, Computers and Fluids, vol. 5, pp. 241-259, 1977.

19. RUBIN, S.G. and KHOSLA, P.K. - Navier-Stokes Calculations with a Coupled Strongly Implicit Method, Part II : Spline Deferred - Corrector Solutions. Lecture Notes in Mathematics vol. 771, pp. 469-488, Springer Verlag, 1980.

20. PETERS, N. - Boundary Layer Calculation by a Hermitian Finite Difference Method, Lecture Notes in Physics, vol. 35, pp. 313-318. 1975.

21. DAUBE, O. et TA PHUOC LOC, Etude numérique d'écoulements instationnaires de fluide visqueux incompressible autour de corps profilés par une méthode combinée d'ordre $O(h^2)$ et $O(h^4)$. - J. Méca., vol. 17 pp. 651-678, 1978.

22. TA PHUOC LOC and DAUBE.O. - Higher Order Accurate Numerical Solution of Unsteady Viscous Flow generated by a Transversely oscillating Elliptic cylinder, Vortex Flow, W.L. SWIFT and P.S. BARNA, Eds., pp. 155-171, ASME, 1981.

23. BONTOUX, P., FORESTIER, B. et ROUX, B. - Analyse et optimisation d'une méthode de haute précision pour la résolution des équations de Navier-Stokes instationnaires. J. Méca. Appl., vol.2, pp. 291-316, 1978.

24. ROUX, B., BONTOUX, P., TA PHUOC LOC and DAUBE O. - Optimisation of Hermitian Methods for Navier-Stokes Equations in the Vorticity and Stream Function Formulation, Lecture Notes in Mathematics, vol. 771, pp. 450-468, Springer Verlag, 1980.

25. MEHTA, U.B. - Dynamic Stall of an Oscillating Airfoil, AGARD Fluid Dynamics Panel Symposium on Unsteady Aerodynamics, Paper n° 23, Ottawa, Canada, Sept. 1977.

26. LECOINTE, Y. and PIQUET, J. On the Numerical Solution of Some Types of Unsteady Incompressible Viscous Flow, 2nd Intern. Conf. on Numerical Methods in Laminar and Turbulent Flow, Venice, Italy, July 15-16 th. 1981.

27. LECOINTE, Y., PIQUET, J. and VISONNEAU, M. - "Mehrstellen" Techniques for the Numerical Solution of Unsteady Incompressible Viscous Flow in Enclosures. 2nd. Intern. Symposium on Numerical Methods in Heat Transfer, University of Maryland,CollegePark, Maryland, sept. 28-30th, 1981.

28. PEYRET, R. - Une méthode aux différences finies de haute précision pour la résolution des équations de Navier-Stokes stationnaires, Comptes-Rendus Acad. Sci., vol. 286, A. pp. 59-62, 1978.

29. PEYRET, R. - A Hermitian Finite-Difference Method for the Solution of the Navier-Stokes Equations. Proc. 1st Intern. Conf. on Numerical Methods in Laminar and Turbulent flow, C. Taylor, K. Morgan and C.A. Brebbia, Eds., pp. 43-54, Pentech Press, Swansea, 1978.

30. ELSAESSER, E. and PEYRET, R. - Méthodes hermitiennes pour la résolution numérique des équations de Navier-Stokes, Méthodes Numériques dans les Sciences de l'Ingénieur, E. Absi et R. Glowinski, Eds., pp.249-258, Dunod, Paris, 1979.

31. ELSAESSER, E. - Etudes de méthodes hermitiennes pour la résolution des équations de Navier-Stokes pour un fluide incompressible en régime stationnaire. Application au calcul de jets. Thèse Doctorat 3ième Cycle, Mécanique des Fluides, Université Pierre et Marie Curie, Paris, 1980.

32. GHIA, K.N., SHIN, C.T. and GHIA, U. - Use of Spline Approximations for Higher-order Accurate solutions of Navier-Stokes Equations in Primitive Variables. Proc. 4-th AIAA Computational Fluid Dynamics Conference, pp. 284-291, Williamsburg, Va, July 1979.

33. AUBERT, X. and DEVILLE, M. Steady Viscous Flows by Compact Differences in Boundary-Fitted Coordinates, to be published.

34. RUBIN, S.G. and KHOSLA, P.K. - Higher-order Numerical Methods from Three-Point Polynomial Interpolation, NASA CR-2735, August 1976.

35. OUTREBON, P. - Les Méthodes hermitiennes pour la résolution numérique des équations aux dérivées partielles. Projet DEA, Mécanique Théorique, Université Pierre et Marie-Curie, Paris 1979.

36. ISAACSON, E. and KELLER, H.B. Analysis of Numerical Methods, J. Wiley & sons, New-York 1966.

37. BRAMBLE, J.H. and HUBBARD, B.E. - On the Formulation of Finite Difference Analogues of the Dirichlet Problem for Poisson's Equation, Numer. Math., vol. 4, pp. 313-327, 1962.

38. KREISS, H.O. - Difference approximations for boundary and eigenvalue problems for ordinary differential equations. Math. Comput., vol. 26, pp. 605-624, 1972.

39. ROACHE, P.J. - A Sixth-Order Accurate Direct Solver for the Poisson and Helmholtz Equations, AIAA J., vol. 17, pp. 524-526, 1979.

40. THIELE, F. - Accurate Numerical Solution of Boundary Layer Flows by the Finite-Difference Methods of Hermitian Type. J. Comput. Phys., vol. 27, pp. 138-159, 1978.

41. PEYRET, R. and TAYLOR, T.D. - Computational Methods for Fluid Flow, Springer Verlag, 1982.

42. PEYRET, R. - Numerical Studies of Non homogeneous Fluid Flows, Lecture Notes in Physics, vol. 148, pp. 330-361, 1981.

43. BOURCIER, M. et FRANCOIS, C. - Intégration numérique des équations de Navier-Stokes, Rech. Aérosp., N°131, pp. 22-33, 1969.

44. MORCHOISNE, Y. - Résolution des équations de Navier-Stokes par une méthode pseudo-spectrale en espace-temps., Rech. Aérosp. N° 1979-5, pp. 293-306, 1979.

SOME ASPECTS OF FINITE ELEMENT APPROXIMATIONS OF INCOMPRESSIBLE VISCOUS FLOWS*

M.D. Gunzburger** and R.A. Nicolaides**

ABSTRACT

Several topics are considered which arise in the finite element solution of the incompressible Navier-Stokes equations. Specifically, the question of choosing finite element velocity/pressure spaces is addressed, particularly from the viewpoint of achieving stable discretizations leading to convergent pressure approximations. Following this, the role of artificial viscosity in viscous flow calculations is studied, emphasizing recent work by several researchers for the anisotropic case. The last section treats the problem of solving the nonlinear systems of equations which arise from the discretization.

1. DISCRETIZATION

1.1. Continuous Problem

Let Ω be a bounded region of $\mathbb{R}^2$ or $\mathbb{R}^3$, the flow region, and let $\underline{u}$ and p denote the velocity and pressure fields, respectively and ν the kinematic viscosity. Normalizing the pressure by the constant density, the stationary Navier-Stokes equations take the form

$$\underline{u}\cdot\nabla\underline{u} + \nabla p = \nu\Delta\underline{u} + \underline{f} \quad \text{in} \quad \Omega \tag{1}$$

$$\text{div}\underline{u} = 0 \quad \text{in} \quad \Omega \tag{2}$$

$$\underline{u} = \underline{g} \quad \text{on} \quad \partial\Omega \tag{3}$$

---

*This work was supported by the Air Force Office of Scientific Research under grants nos. AFOSR-83-0101 and AFOSR-82-0213.

**Professor of Mathematics, Carnegie-Mellon University, Pittsburgh, PA 15213 where $\underline{f}$ and $\underline{g}$ are given functions and $\partial\Omega$ denotes the boundary of Ω.

In conservation form (1) may be written

$$\mathrm{Div}(\underline{u}\,\underline{u}^T) + \nabla p = \nu\Delta\underline{u} + \underline{f} \tag{4}$$

where Div denotes the tensor divergence operator and the equivalence of (1) and (4) is shown using (2). Equation (2)-(4) are the equations to be solved.

The standard weak form of (2)-(4) is: find $\underline{u} \in \overline{H}^1(\Omega)$ satisfying (3) and $p \in L_0^2(\Omega)$ such that

$$\nu\int_\Omega \nabla\underline{u}:\nabla\underline{v} - \int_\Omega (\underline{u}\,\underline{u}^T):\nabla\underline{v} \tag{5}$$

$$- \int_\Omega p\,\mathrm{div}\underline{v} = \int_\Omega \underline{f}\cdot\underline{v} \quad \forall \quad \underline{v} \in \overline{H}_0^1(\Omega)$$

$$\int_\Omega q\,\mathrm{div}\underline{u} = 0 \qquad \forall \quad q \in L_0^2(\Omega). \tag{6}$$

In the above formulation, $\overline{H}^1(\Omega)$ is the usual Sobolev space of functions with one square integrable derivative, $\overline{H}_0^1(\Omega)$ is that subspace of $\overline{H}^1(\Omega)$ whose elements are zero on $\partial\Omega$ and $L_0^2(\Omega)$ consists of square integrable functions with zero mean over Ω.

A slightly different treatment of the convection term is given in the following weak form of the momentum equation [1]

$$\nu\int_\Omega \nabla\underline{u}:\nabla\underline{v} - 1/2\int_\Omega (\underline{u}\cdot\nabla\underline{u}\cdot\underline{v} - \underline{u}\cdot\nabla\underline{v}\cdot\underline{u})$$

$$- \int_\Omega p\,\mathrm{div}\underline{v} = \int_\Omega \underline{f}\cdot\underline{v} \quad \forall \quad \underline{v} \in \overline{H}_0^1(\Omega). \tag{7}$$

Again, in view of (2), this formulation is equivalent to (5); however for computations, the form (7) possesses certain advantages [1], and thus is recommended.

1.2. Discrete Problem

Proceeding in the usual way, one chooses finite dimensional subspaces $V^h \subset \overline{H}^1(\Omega)$, $S^h \subset L_C^2(\Omega)$, and seeks $\underline{u}^h \in V^h$ satisfying

$$\underline{u}^h = \underline{g}^h \quad \text{on} \quad \partial\Omega \tag{8}$$

and $p^h \in S^h$ such that

$$\nu \int_\Omega \nabla \underline{u}^h : \nabla \underline{v}^h - 1/2 \int_\Omega (\underline{u}^h \cdot \nabla \underline{u}^h \cdot \underline{v}^h - \underline{u}^h \cdot \nabla \underline{v}^h \cdot \underline{u}^h)$$
$$- \int_\Omega p^h \mathrm{div} \underline{v}^h = \int_\Omega \underline{f} \cdot \underline{v}^h \quad \forall \underline{v}^h \in V_0^h \tag{9}$$

$$\int_\Omega q^h \mathrm{div} \underline{u}^h = 0 \quad \forall q^h \in S^h. \tag{10}$$

In (8) $\underline{g}^h$ denotes an approximation, in V^h restricted to $\partial\Omega$, to $\underline{g}$ (e.g., an interpolant). V_0^h is the subspace of V^h of trial functions which are zero on $\partial\Omega$. Strictly speaking, this formulation is valid for polygonal domains but may be easily extended, e.g., by isoparametric techniques, to more general domains.

Unlike the standard positive definite elliptic case [2,3] mere inclusion of $V^h \subset \overline{H}^1(\Omega)$ and $S^h \subset L_0^2(\Omega)$ is not sufficient to ensure convergence (or even existence) of the discrete solutions. In fact, the spaces V^h and S^h cannot be chosen independently of each other. Mathematically, the following condition is required [4]

$$\sup_{\substack{\underline{v}^h \in V_0^h \\ |\underline{v}^h| = 1}} \int_\Omega q^h \mathrm{div} \underline{v}^h \geq \gamma \|q^h\| \quad \forall q^h \in S^h \tag{11}$$

where $\gamma > 0$ is independent of the discretization parameter h. Here, the norms used in (11) are defined by

$$|\underline{v}|^2 = \int_\Omega \nabla \underline{v} : \nabla \underline{v} \quad \forall \underline{v} \in H_0^1(\Omega)$$

$$\|q\|^2 = \int_\Omega q^2 \qquad \forall q \in L_0^2(\Omega).$$

There are other mathematical conditions on the discrete spaces which must be imposed in order to guarantee convergence [4], however it is (11) which can fail to hold in general, and anyway offers substantial difficulties in its verification. Equation (11) is referred to as the condition of "div-stability" [5,6].

Note that the analogous condition

$$\sup_{\substack{\underline{v} \in \overline{H}_0^1(\Omega) \\ |\underline{v}| = 1}} \int_\Omega q \mathrm{div} \underline{v} \geq \gamma \|q\| \quad \forall q \in L_0^2(\Omega)$$

is necessary to guarantee the existence and stability of the solution of the continuous problem, i.e., the Navier-Stokes equations. This condition is easily verified since it is well known that the problem

$$\operatorname{div}\hat{\underline{v}} = q \quad \text{in} \quad \Omega$$

$$\hat{\underline{v}} = 0 \quad \text{on} \quad \partial\Omega$$

has a solution $\hat{\underline{v}} \in \bar{H}_0^1(\Omega)$ for any $q \in L_0^2(\Omega)$ which satisfies

$$\gamma|\hat{\underline{v}}| \leq \|q\|$$

for some constant $\gamma > 0$. Then

$$\sup_{\substack{v \in \bar{H}_0^1(\Omega) \\ |\underline{v}| = 1}} \int_\Omega q \operatorname{div}\underline{v} \geq \int_\Omega q \operatorname{div}\hat{\underline{v}} / |\hat{\underline{v}}|$$

$$= \int_\Omega q^2 / |\hat{\underline{v}}| \geq \gamma\|q\|.$$

1.3. div-Stable Elements

In practice, one would like to use low degree piecewise polynomial trial spaces V^h, S^h. Unfortunately these are the spaces which encounter the most difficulty in satisfying (11). For the higher order cases, a reasonably efficient test for showing that an element pair is div-stable is known [5]. No such simple test is known for the low order cases. This is discussed further below.

Perhaps the best known low order pair is the piecewise bilinear velocity/piecewise constant pressure combination defined on a quadrilateral subdivision of Ω. This pair has been extensively used in engineering computations and been the object of much theoretical work. In particular, the checkerboard pressure mode is well documented [7]. The effect of this mode is to make γ in (11) zero, so (11) is caused to fail in this case. Less well known [8] is the existence of a mode such that $\gamma = O(h)$. This causes the pressure approximation to fail to converge in general, even when the standard checkerboard mode is filtered out. This is proved in Boland and Nicolaides [6], where also a filtering technique applicable to many cases is given. Use of this filter enables the pressures to converge optimally.

Unfortunately, for arbitrary elemental subdivisions of Ω, the appropriate filters are not known. Therefore, it seems advisable to use element pairs known a priori to be div-stable.

Here are presented two element pairs which have been used in practical caluclations and have been shown rigorously [6] to be stable.

For the first of these, let $\Omega = \{(x,y), 0<x,y<1\}$ and subdivide Ω by lines $x = ih$, $y = jh$, $i,j = 0,1,...,2n$, so that $h = 1/2n$. For V^h one may use continuous piecewise bilinear vector fields, bilinear in each subsquare. The pressure field is a subspace of all piecewise constants with zero mean on Ω, defined as follows: on the coarser grid $h' = 1/n$ whose subsquares each contain four of the smaller subsquares, one merely imposes one constraint on the four scalars. For example, referring to Fig. 1, one imposes the

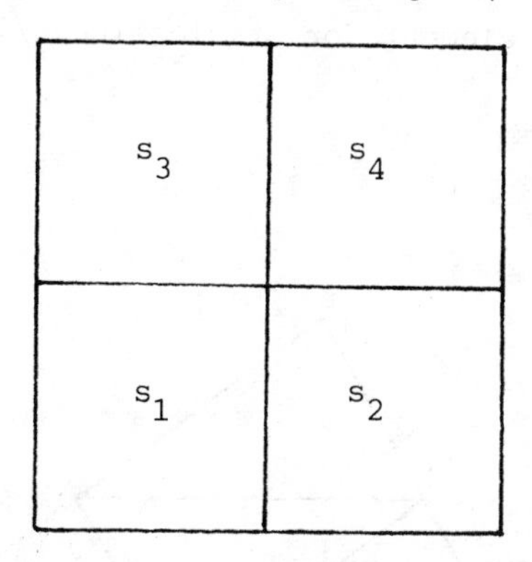

Figure 1. Stable bilinear/constant element pair.

constraint $s_1 + s_2 = s_3 + s_4$. For the general case, isoparametric mapping onto this macroelement is used. The resulting pair is div-stable, requires no filtering and gives convergent approximations [6].

Another stable pair, not requiring the use of isoparametric methods is the following. Let Ω be triangulated regularly and let τ_i be an arbitrary element of the triangulation. In Ω, p^h is taken to be a continuous piecewise linear field with respect to the triangulation. To define v^h, τ_i is further subdivided into four similar triangles and V^h is then defined to be all continuous piecewise linear vector valued fields on the finer triangulation. This element pair is also div-stable and convergent without filtering. Fig. 2a shows a typical τ_i in the pressure triangulation and Fig. 2b depicts the resulting velocity elements derived from τ_i.

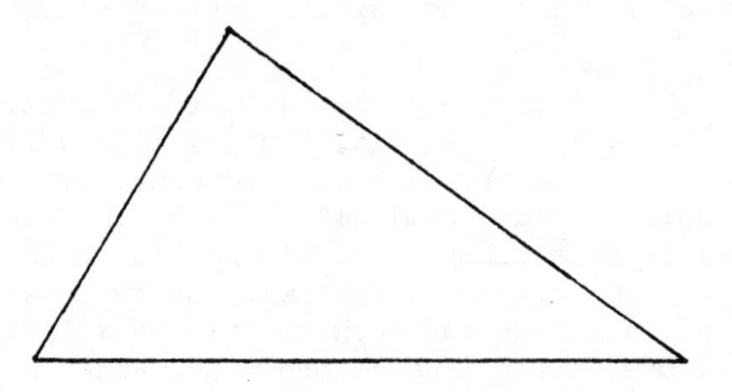

Figure 2a. Pressure element for stable linear/linear element pair.

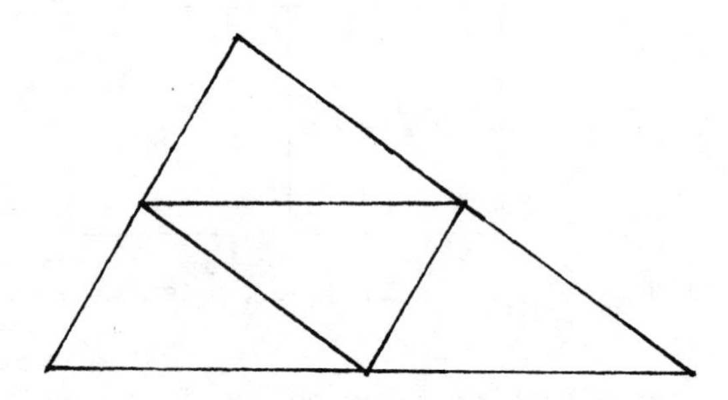

Figure 2b. Velocity elements for stable linear/linear element pair.

Since it is known that these elements are div-stable, i.e., (11) holds, then one may use finite element theory [1,9] to obtain the following estimates:

$$\|\underline{u}-\underline{u}^h\|_{H^1(\Omega)} \leq c_1 h$$

$$\|p-p^h\|_{L^2(\Omega)} \leq c_2 h \tag{12}$$

where c_1 and c_2 do not depend on h. The duality method then shows that

$$\|\underline{u}-\underline{u}^h\|_{L^2(\Omega)} \leq c_3 h^2. \tag{13}$$

Thus, the methods are second order, in the root mean square sense, for the velocities and first order for the pressure. These are the best possible rates obtainable with these elements.

1.4 Upwinding

Since V^h and S^h are used as both test and trial spaces in (8)-(10), the discrete equations will be of centered type. Thus, for fixed h as $\nu \to 0$ one has to expect the numerical solution to develop increasingly large oscillations [10] ("wiggles"). These can of course be eliminated by use of any of the numerous upwinding/artificial diffusion methods [11]. However, the very idea of letting $\nu \to 0$ while keeping h fixed precludes the possibility of accurate computation of viscous effects. For this one must, at the very least, permit h to go to zero with ν. The precise dependence of h on ν depends on what viscous phenomena are to be accurately simulated.

For example, simulation of shear or tangential boundary layers certainly will require $h = 0(\nu^{1/2})$ in the neighborhood of these layers. Conceivably, especially at outflows, there may be other kinds of layers, maybe induced by numerical boundary conditions, requiring the more onerous (and less practical) restriction $h = 0(\nu)$. Usually the latter layers appear not to be of physical interest and may therefore be smoothed by use of anisotropic artificial diffusion, chosen so that tangential layers are not smoothed. Such methods have recently come into prominence [12,13]. Specifically these methods attempt to add diffusion only in the streamwise direction. One must emphasize that $h = 0(\nu^{1/2})$ is still an essential requirement for resolution of physical layers and hence, convergence.

One streamwise artificial diffusion method is now presented. First, observe that (4) may be written in the form

$$\mathrm{Div}[\underline{u}\,\underline{u}^T + pI - \nu\nabla\underline{u}] = \underline{f} \tag{14}$$

where I is the identity matrix. The quantity in the brackets is the momentum flux density. To this one adds the term

$$-\lambda\nabla\underline{u}(\underline{u}\,\underline{u}^T) \tag{15}$$

where λ is a mesh dependent parameter tending to zero with h. Then (14) becomes

$$\mathrm{Div}[\underline{u}\,\underline{u}^T + pI - \nabla\underline{u}(\nu I+\lambda\underline{u}\,\underline{u}^T)] = \underline{f}. \tag{16}$$

In the form (16) one may interpret (15) as an anisotropic perturbation of the viscosity tensor (νI). Clearly the effect of this perturbation vanishes in directions normal to the velocity vector, and hence to streamlines. Alternately, by associating the perturbation with the convection term a connection may be made, in transient cases, with the method of characteristics [14].

This is not pursued further here. Equation (16), along with (2)-(3), may now be discretized in the usual manner using the same test and trial spaces. Generally, in order to achieve the desired stabilizing effect, λ should be $O(h)$. The effect of such a choice of λ on the estimates (12)-(13) is not fully understood. When using higher order elements it is reasonable to expect that the accuracy would be degraded by this crude approach. Therefore it is of interest to note that a similar effect may be achieved by appropriate choices of distinct test and trial spaces [13].

Normally, for internal flows, e.g., cavity flows, and for good choices for numerical outflow conditions [15], $O(\nu)$ layers are not present in the flow field. In such cases, the use of artificial streamwise diffusion methods is not necessary.

One is still left with the task of resolving tangential layers. If a uniform mesh is used throughout the flow field, the $h = O(\nu^{1/2})$ restriction results in an unacceptable number of degrees of freedom. However such a small h is needed only in the neighborhood of the layers themselves. Thus it is advantageous, in the sense of reducing the number of degrees of freedom, to use nonuniform grids. As an example consider the driven cavity problem. Here Ω is a unit square, the upper end of the cavity moves with velocity $\underline{u} = (1,0)$, and $\nu = 1/1000$. The results of three calculations are reported. The first [16] uses an upwind finite difference technique, on a uniform mesh, for a streamfunction - vorticity formulation of (1)-(3). The second [17] again solves the streamfunction - vorticity formulation by a finite element technique using a nonuniform grid. The grid spacing is determined by the functions

$$\begin{aligned} x &= \sin^2(\pi\bar{x}/2) \\ y &= \sin^2(\pi\bar{y}/2) \end{aligned} \tag{17}$$

with a uniform grid spacing in the $\bar{x}$ and $\bar{y}$ coordinates. The third calculation uses the element pair of Fig. 2 in conjunction with the primitive variable formulation and a nonuniform grid again determined by a means similar to (17). In neither finite element calculation were artificial diffusion/upwinding techniques used. In Fig. 3 are given the x-component of velocity u at $x = 1/2$ as a function of y and the y-component of velocity v at $y = 1/2$ as a function of x. The uniform grid calculation used a (129×129) grid so that $h = 0.0078$. The nonuniform grid calculations used a (19×19) grid. With the spacing determined by (17), the minimum $h = 0.0076$ which is comparable to the uniform h of the calculation of Ghia, et. al. [16]. Clearly, one can achieve the same accuracy at a greatly reduced cost by using nonuniform grids.

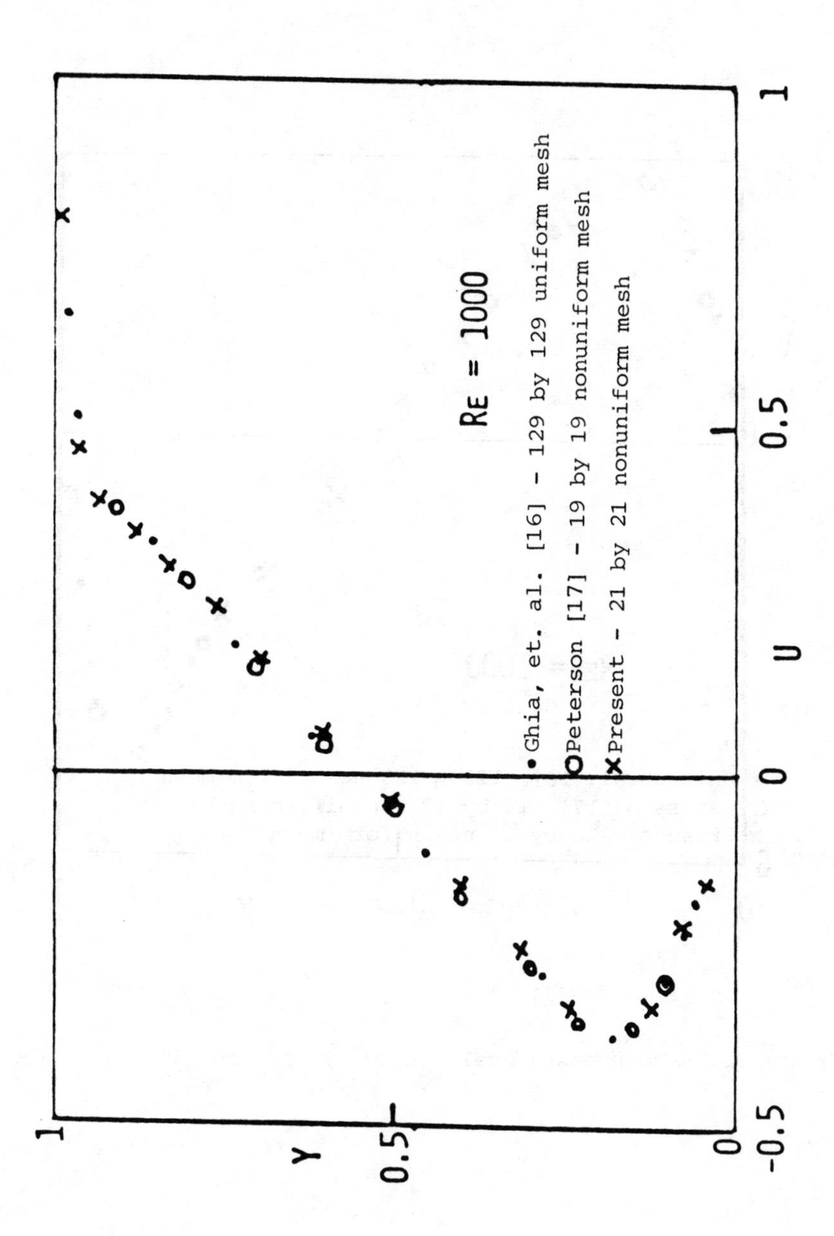

Figure 3a. x-component of velocity at $x = 1/2$ for driven cavity.

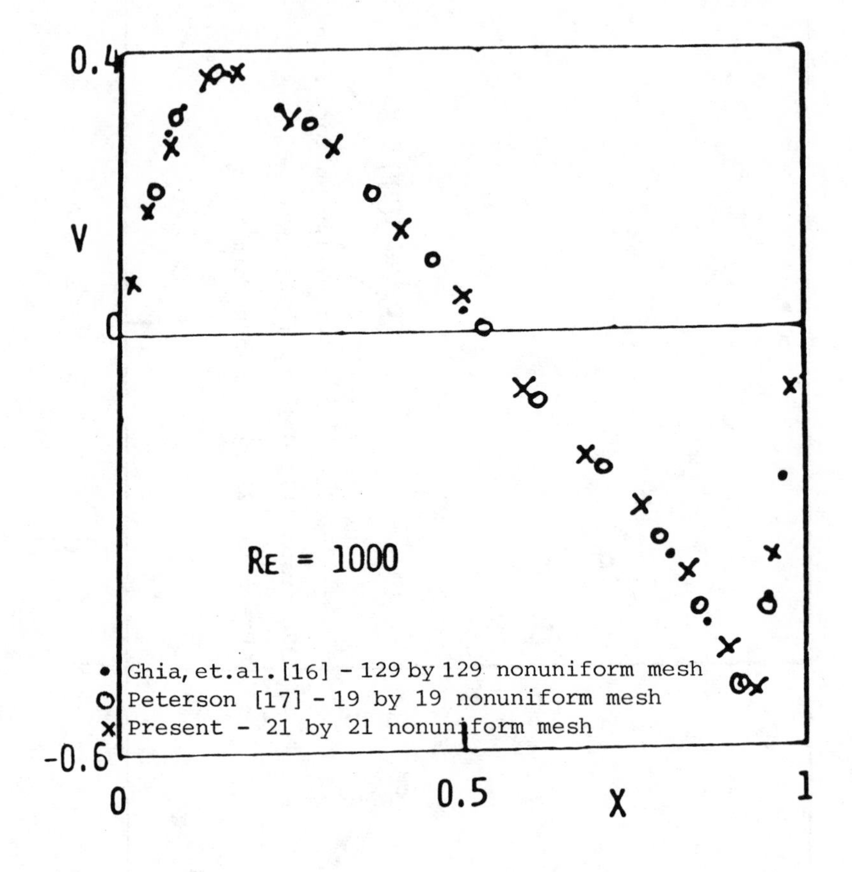

Figure 3b. y-component of velocity at $y = 1/2$ for driven cavity.

2. SOLUTION TECHNIQUES

Whatever method is employed for discretization, the outcome is a large system of nonlinear equations which must be solved for the approximate solution. Concerning the solution of this system by some classical technique, such as Newton's method, one must first observe that the Jacobian of the system requires a remarkably large amount of storage, if the usual band storage scheme is adopted. For example, it is easy to verify that a driven cavity calculation on an $n \times n$ grid requires $27n^3$ words of storage (for the bilinear/constant element). Thus, even with $n = 20$, a rather large storage requirement is apparent. In 3D, the situation is catastrophic. Hence, the classical techniques may be of limited value (although some fairly fine mesh calculations have been reported [11]). Anyway, it is evident that alternative techniques are of interest. Of course many have been proposed. They fall into the two natural classes of "false transient" or time marching algorithms on the one hand and general nonlinear equation solvers on the other.

Here is an example of a time marching technique used successfully by the authors. Let U, P denote the unknowns and U^k, P^k the k-th approximations to U and P, and consider

$$\frac{1}{\Delta t}(U^{k+1} - U^k) + N(U^k) + B\bar{P}^{k+1} = F \qquad k = 0,1,\ldots \tag{18}$$

$$B^T U^{k+1} = G \qquad k = 0,1,\ldots \tag{19}$$

where $N(U)$ represents an approximate Laplacian plus convection terms and B an approximate gradient operator, generated by the particular finite elements used. The first term in (18) is a discretization of $\frac{\partial u}{\partial t}$, forward in time from the k-th time level. U^0 is arbitrary, and is not required to satisfy (19). In the steady state limit, $U^{k+1} = U^k$ and one recovers the solution of the original problem. To get U^{k+1}, given U^k, multiply (18) by B^T and use (19) to get

$$B^T B P^{k+1} = B^T F - B^T N(U^k) \qquad k \geq 1 \tag{20}$$

which must be solved for BP^{k+1}, so that then

$$U^{k+1} = U^k + \Delta t[F - BP^{k+1} - N(U^k)]. \tag{21}$$

If $k = 0$, a term

$$-\frac{1}{\Delta t}(G - B^T U^0)$$

is added to the right hand side of (20). Notice that (20) is the usual "Poisson" equation for the pressure, but with the crucial feature that boundary conditions do not need to be supplied externally. They are automatically present in $B^T B$. Notice also that in (21) not P^{k+1} but BP^{k+1} is required. This is the least squares solution of the equation

$$B^T V^{k+1} = B^T F - B^T N(U^k) \tag{22}$$

where

$$V^{k+1} = BP^{k+1}$$

which may be found by any of the numerous techniques, direct or iterative. The following iterative method, a generalization of a method of Kacmarcz which may be called the row projections method, was used by the authors. For solving

$$B^T V = L$$

this method is the following:

$$V^{\ell+1} = V^{\ell} + \omega \alpha B_{\ell}$$

$$V^0 = B\xi, \quad \xi \text{ arbitrary}, \quad \ell = 1,2,...$$

where B_{ℓ} denotes the ℓth column of B (taken cyclically), ω is a relaxation parameter and α the number chosen so that when $\omega = 1$, the ℓth component of the residual $r^{\ell+1} = L - B^T V^{\ell+1}$ is zero, thus,

$$\alpha = \frac{r_{\ell}^{\ell}}{\rho_{\ell}^2}$$

where ρ_{ℓ}^2 is the sum of the squares of the elements in row ℓ of B^T. Geometrically, the error $(V - V^{\ell+1})$ is projected orthogonally in succession onto the planes whose normals are the rows of B^T. ω effects a kind of overprojection. Generalization to projection onto the planes normal to several rows (corresponding to line, etc., relaxation) is straightforward

Returning to (20), note that only few iterations of (22) are required for each time step, due to the availability of the solution from the earlier time level.

The problem with time marching based on first order time derivatives is the large number of time steps (regardless of the size of ν) needed to reach the steady state. Generally, several thousand steps are needed for each digit of accuracy required, for the kind of grid sizes encountered in practice, and the number of steps rapidly increases as the latter approach zero. Using higher order time derivatives would give considerable improvement, but does not appear to be a common idea. Among the other methods, great promise is shown by the reduced basis - continuation techniques [17,18]. A nonlinear conjugate gradient method is used by Glowinski, et.al. [19], although it does not appear to function well in all cases.

A good example of a nonlinear equation solving technique is furnished by the application of Newton's method to the discrete Navier-Stokes equations (8)-(10). This can be done in the usual way, but at the cost of dealing with the very large equation systems mentioned at the beginning of this section. These systems arise from the necessity of solving the Jacobian equation whose solution is the correction at the current step. In practice, the correction equation is not explicitly formed. Instead the equation

$$-\nu\Delta_h \underline{u}^{(n+1)} + \mathrm{Div}_h(\underline{u}^{(n)}\underline{u}^{(n+1)T} + \theta\underline{u}^{(n+1)}\underline{u}^{(n)T}) + \nabla_h p^{(n+1)} = \underline{f} + \theta\mathrm{Div}_h(\underline{u}^{(n)}\underline{u}^{(n)T}) \tag{23}$$

$$\mathrm{div}_h \underline{u}^{(n+1)} = 0 \qquad \theta = 1$$

is assembled and solved for $\underline{u}^{(n+1)}$, $p^{(n+1)}$. It is interesting to note that with $\theta = 1$ $\underline{u}^{(n+1)}$ is the Newton approximation and with $\theta = 0$, the resulting method is the simple iteration method for which it is known that convergence to a unique solution of the discrete Navier-Stokes system is obtained from any initial $\underline{u}^{(0)}$, even one which does not satisfy (10) and (8). This suggests an obvious type of solution algorithm.

There is still the problem in (23) of solving the linear equations, avoiding the storage problems at the same time. A promising approach is that of "substructuring", which simply refers to a method for ordering the nodes of the triangulation to obtain a favorable structure to the discrete system. Specifically, the nodes are ordered to achieve a matrix structure of the form

$$M = \begin{bmatrix} A_1 & & & B_2 \\ & A_2 & & B_2 \\ & & A_m & B_m \\ C_1 & C_2 & C_m & D \end{bmatrix} \tag{24}$$

where each A_i is a square matrix of order n_i, and D is square and of order n. The standard way to achieve this form is to divide the computational domain into subdomains (substructures) by means of closed sequences of edges (for the 2D case) of the triangulation, and to number nodes within each subdomain sequentially and then to number the nodes on the edges. In (24) the first m block rows are associated with the m subdomains into which Ω has been divided, and express the couplings of interior unknowns of a subdomain with its own boundary. The last row expresses the couplings of the subdomain boundary nodes with the interior nodes and, since generally the boundary nodes are common to more than one subdomain, couples the subdomain together.

To solve an equation

$$Mx = b$$

with m given by (24) and $b = (b_1^T, b_2^T, \ldots, b_{m+1}^T)^T$ conformably partitioned, multiply equation i by $C_i A_i^{-1}$ $i = 1,2,\ldots,m$ and subtract from the $(m+1)$th equation to obtain

$$D_1 x_{m+1} = f \tag{25}$$

say, where

$$D_1 = D - \sum_{i=1}^{m} C_i A_i^{-1} B_i \tag{26}$$

and

$$f = b_{m+1} - \sum_{i=1}^{m} C_i A_i^{-1} b_i . \tag{27}$$

(25) may now be solved for x_{m+1} and then the remaining x_i found from the given system. In practical implementations, only data for a single subdomain needs to be in main memory at any given time and thus the storage problem may be alleviated.

In many Navier-Stokes cases, the diagonal matrices A_i are singular, reflecting the fact that in the continuous problem itself, the pressure is determined only up to an additive constant. In the numerical approximations, this will happen whenever the trial space for the pressure consists of discontinuous piecewise polynomial functions. This situation occurs sufficiently often to warrant an extension of the substructuring algorithm to handle it. The required generalization is presented in [20]. It seems very reasonable that since both the assembly and elimination phases are independent between subdomains, some kind of parallel implementation of this algorithm will be valuable.

CONCLUSIONS

In connection with the incompressible Navier-Stokes equations discretization techniques via finite elements, accuracy questions, and methods for solving the algebraic systems of nonlinear equations, all for the $\underline{u}$,p (primitive) variable case, have been discussed. It is fair to say that the first topic is by now reasonably well understood. The main difficulty concerned the discretization of the incompressibility condition, and this topic now has a variety of theoretical analyses and tests whereby the stability of the discretization can be verified before calculations are performed.

Less satisfactory is the state of affairs relating to solving the systems of nonlinear equations. Here, there are many problems still to be overcome, mostly concerning the efficiency of the solution procedures. The time marching method discussed, along with similar methods based on first order time derivatives, really is too inefficient, requiring of the order of $10^3 - 10^4$ time steps to obtain each digit of the solution. On the other hand, the direct solution methods encounter problems caused by the large storage resources necessary for carrying out the matrix manipulations as well as domain of convergence problems as $\nu \to 0$. In the latter case, the starting approximation has to be closer and closer to the exact solution being computed, as ν becomes smaller. Continuation techniques are naturally suggested for dealing with this latter issue, as stated in the text.

REFERENCES

1. V. GIRAULT and P.-A. RAVIART, Finite element approximation of the Navier-Stokes equations, Lecture Notes in Mathematics 749, Springer, New York (1979).

2. G. STRANG and G. FIX, An Analysis of the Finite Element Method, Prentice-Hall, Englewood Cliffs (1972).

3. J. ODEN and J. REDDY, An Introduction to the Mathematical Theory of Finite Element, Wiley, New York (1976).

4. F. BREZZI, On the existence, uniqueness and approximation of saddle point problems arising from Lagrange multipliers, RAIRO, R2, p. 129 (1974).

5. J. BOLAND and R. NICOLAIDES, Stability of finite element discretizations of incompressibility conditions, SIAM J. Numer. Anal., to appear (1983).

6. J. BOLAND and R. NICOLAIDES, Stable and semi-stable low order finite element schemes for solving Navier-Stokes equations, SIAM J. Numer. Anal., to appear.

7. R. SANI, P. GRESHO, R. LEE and D. GRIFFITHS, The cause and cure (?) of the spurious pressures generated by certain FEM solutions of the incompressible Navier-Stokes equations: Part 1, Int. J. Numer. Meth. Fluids 1, p. 17 (1981).

8. J. BOLAND and R. NICOLAIDES, Counterexample to the uniform div-stability of bilinear-constant velocity-pressure finite elements for viscous flows, Numer. Math., to appear.

9. M. GUNZBURGER and J. PETERSON, On conforming finite element methods for the inhomogeneous stationary Navier-Stokes equations, Numer. Math., to appear.

10. P. GRESHO and R. LEE, Don't suppress the wiggles - they're telling you something, Comput. & Fluids 9, p. 223 (1981).

11. F. THOMASSET, Implementation of Finite Element Methods for Navier-Stokes Equations, Springer, New York (1981).

12. J. DUKOWICZ and J. RAMSHAW, Tensor viscosity method for convection in numerical fluid dynamics, J. Comput. Phys. 32, p. 71 (1979).

13. A. BROOKS and T. HUGHES, Streamwise upwind/Petrov Galerkin formulation for convection-dominated flows with particular emphasis on the incompressible Navier-Stokes equations, Comput. Methods Appl. Mech. Eng. 32, p. 199 (1982).

14. J. DOUGLAS and T. RUSSELL, Numerical methods for convection-dominated diffusion problems based on combining the method of characteristics with finite element or finite difference procedures, SIAM J. Numer. Anal. 19, p. 871 (1982).

15. G. FIX and M. GUNZBURGER, Downstream boundary conditions for viscous flow problems, Comput. Math. Appl. 3, p. 53 (1977).

16. U. GHIA, K. GHIA and C. SHIN, High-Re solutions for incompressible flow using the Navier-Stokes equations and a multigrid method, J. Comput. Phys. 48, p. 387 (1982).

17. J. PETERSON, High Reynolds number solutions for incompressible viscous flow using the reduced basis technique, J. Comput. Phys., to appear.

18. H. KELLER, Continuation methods in computational fluid dynamics, in Conf. Proc. on Numerical and Physical Aspects of Aerodynamic Flows, California State University, Long Beach, CA, T. Cebeci, ed., pp. 3-34, Springer-Verlag (1982).

19. R. GLOWINSKI, B. MANTEL, J. PERIAUX, O. PIRONNEAU, and G. POINIEN, An efficient preconditioned conjugate gradient method applied to nonlinear problems in fluid dynamics, Computing Methods in Applied Science and Engineering, North Holland, Amsterdam (1980).

20. M. GUNZBURGER and R. NICOLAIDES, Gauss elimination with noninvertible pivots, Linear Algebra Applic., to appear.

RESPONSE ANALYSIS OF SOME FINITE ELEMENT APPROXIMATIONS TO THE INCOMPRESSIBLE NAVIER-STOKES EQUATIONS

L. QUARTAPELLE*, J. DONEA**, K. MORGAN***

*Istituto di Fisica, Politecnico di Milano, Italy, **Applied Mechanics Division, Joint Research Centre, Ispra, Italy,
***University College of Swansea, Singleton Park, Swansea, U.K.

1. INTRODUCTION

In spite of the great number of investigations on the approximate solution to the incompressible Navier-Stokes equations, studies aimed at investigating the discretization properties of numerical schemes for the fully multidimensional and incompressible equations are few. A remarkable exception is provided by the work of Arakawa [1] who developed finite difference approximations to the advection terms which satisfy physically relevant conservation laws of the inviscid equations and thus eliminate some nonlinear numerical instabilities (see also [2]). The lack of detailed analyses of the incompressible discretized equations can be traced back, in the case of finite differences, to a certain difficulty in dealing with the discrete representation of the condition of incompressibility viewed either as a solenoidal constraint for the velocity field or as a Poisson equation for the pressure. On the other hand, some recent works on finite element approximations to time-dependent problems have discussed the properties of the semi- and fully-discretized equations on the basis of simple heuristic methods which allow a semi-quantitative inter-comparison of various approximations. Among these studies we can mention: the determination of proper combinations of mass matrix approximations and time integration schemes for the one-dimensional wave equation [3] and the one-dimensional advection equation [4,5]; the discussion of collocation approximations to the advection equation [6]; and the comparison of different kinds of polynomial interpolations for the tidal equation [7]. Since the presence of the incompressibility constraint prevents a direct application of more rigorous modern methods for the analysis of truncation errors as employed in [8], it seems that an analysis of the discretized incompressible Navier-Stokes equations can be attempted on the basis of the aforementioned heuristic approaches.

The aim of this paper is to investigate the discretization properties of some finite element approximations to the equations governing the multidimensional advection and diffusion of a divergenceless vector field, such as the velocity of an incompressible viscous fluid. To this purpose, a linearized model of the unsteady Navier-Stokes equations is introduced (Sec. 2). The equations are discretized in time by means of the finite difference method using a fractional step approach (Sec. 3), and in space by the finite element method (Sec. 4). Explicit, Crank-Nicolson and implicit time integration schemes are considered. The spatial discretization is achieved by means of the 4-noded quadrilateral element with equal-order bilinear interpolation for the velocity and pressure fields in two dimensions. An iterative method to evaluate explicitly the consistent mass matrix is introduced (Sec. 5). It provides an accurate representation of the consistent mass matrix, whence the name of quasi-consistent mass matrix approximation. By using the standard analysis in Fourier modes, dispersion relationships in the reciprocal space of the propagation wave vectors are introduced for the frequency and damping responses of the incompressible equations (Sec.6). The properties of the various discrete approximations are discussed and evaluated by comparing the response of the fully-discretized equations against the exact response of the continuum model equations. Consistent, lumped diagonal and quasi-consistent representations of the mass matrix are considered. Some aspects of the numerical stability of the linear equations subjected to the incompressibility constraint are briefly examined in Section 7 with special attention paid to the case of the explicit scheme. The validity of the quasi-consistent mass matrix approximation is numerically investigated by some two-dimensional nonlinear calculations reported in Section 8. The last section is devoted to a few concluding remarks.

2. LINEARIZED INCOMPRESSIBLE NAVIER-STOKES EQUATIONS

As a model of the Navier-Stokes equations for incompressible viscous flows, let us consider the linear problem

$$\frac{\partial \underline{u}}{\partial t} + (\underline{v}.\underline{\nabla})\underline{u} = -\underline{\nabla}p + 2\nu\underline{\nabla}.\ \underline{\underline{D}}(\underline{u}), \tag{2.1}$$

$$\underline{\nabla}.\underline{u} = 0, \tag{2.2}$$

where

$$D_{ij}(\underline{u}) = \frac{1}{2}\left(\frac{\partial u_j}{\partial x_i} + \frac{\partial u_i}{\partial x_j}\right). \tag{2.3}$$

The advection velocity $\underline{v}$ and the coefficient of kinematic viscosity ν are assumed to be constant and independent of the position vector $\underline{x}$. In the case $\underline{v} \equiv \underline{0}$, equations (2.1) - (2.3) give the evolution equations for an incompressible fluid of high viscosity (unsteady creeping flows), whereas, in the limit $\nu \to 0$, they become a linearized version of the incompressible Euler equations for an ideal fluid.

The analysis in Fourier modes is now employed to describe the spatial dependence of the solution of the linear equations (2.1) and (2.2) over a rectangular domain of finite extent with spatially periodic boundary conditions. This means that numerical phenomena such as the propagation of a surfacial wave along the boundary and the reflection of ordinary waves by the boundary will not be considered in the present discussion. We assume a solution in the form of a spatial Fourier wave with propagation wave vector $\underline{k}$,

$$\underline{u}(\underline{x},t) = \underline{u}(\underline{k},t)\, e^{i\underline{k}.\underline{x}}, \qquad (2.4)$$

$$p(\underline{x},t) = p(\underline{k},t)\, e^{i\underline{k}.\underline{x}}. \qquad (2.5)$$

Then equations (2.1) and (2.2) produce a system of equations

$$\frac{d\underline{u}}{dt} + i(\underline{v}.\underline{k})\underline{u} = - i\underline{k}p + 2\nu i\underline{k}.\,[i(\underline{k}\,\underline{u})_s], \qquad (2.6)$$

$$\underline{k}.\underline{u} = 0, \qquad (2.7)$$

for the Fourier coefficients $\underline{u}(\underline{k},t)$ and $p(\underline{k},t)$, where

$$(\underline{k}\,\underline{u})_s \equiv \frac{1}{2}\,(\underline{k}\,\underline{u} + \underline{u}\,\underline{k}). \qquad (2.8)$$

By eliminating p between the two equations, we can obtain a single equation for $\underline{u}$ in the form

$$\frac{d\underline{u}}{dt} + (i\,\underline{v}.\underline{k} + \nu k^2)\,[\underline{u}-\underline{k}(\underline{k}.\underline{u})/k^2] = 0 \qquad (2.9)$$

(In the absence of the second term within the square brackets, equation (2.9) is the Fourier transform of the advection-diffusion equation for the vector unknown $\underline{u}$ without the incompressibility constraint). A solution to equation (2.9) can be sought in the form

$$\underline{u}(\underline{k},t) = \underline{u}_o\, e^{-\gamma t}, \qquad (2.10)$$

and then $\gamma = \gamma(\underline{k})$ and $\underline{u}_o = \underline{u}_o(\underline{k})$ are the eigenvalues and eigenvectors of the problem

$$-\gamma \underline{u}_o + (i\underline{v}.\underline{k} + \nu k^2)\,[\underline{u}_o - \underline{k}(\underline{k}.\underline{u}_o)/k^2] = 0. \qquad (2.11)$$

This eigenproblem can be conveniently rewritten as

$$\underline{S}\,\underline{u}_o = \sigma\,\underline{u}_o, \qquad (2.12)$$

where $\underline{S} = \underline{S}(\underline{k})$ has typical elements defined by

$$S_{ij} = \delta_{ij} - k_i k_j/k^2 \qquad (2.13)$$

and the eigenvalue $\sigma = \sigma(\underline{k})$ is related to $\gamma = \gamma(\underline{k})$ by relationship

$$\sigma = \gamma/(i\underline{v}.\underline{k} + \nu k^2). \qquad (2.14)$$

It can be seen that operator $\underline{S}$ is idempotent, i.e., $\underline{S}^2 = \underline{S}$ and symmetric, so that it is an orthogonal projection operator. In fact, $\underline{S}(\underline{k})$ projects any vector $\underline{u}(\underline{k},t)$ orthogonally onto the plane perpendicular to the wave vector $\underline{k}$. Since the condition $\underline{k}.\underline{u} = 0$ is simply the Fourier transform of the incompressibility condition(2.2), the projection operator $\underline{S}(\underline{k})$ makes solenoidal the Fourier component with wave vector $\underline{k}$. Therefore $\underline{S}$ will be named the solenoidalization operator in the Fourier transform space.

2.1 Two-dimensional Equations

In this case the two eigenvalues and the associated normalized eigenvectors of $\underline{S}(\underline{k})$ are

$$\sigma_1 = 0, \qquad \underline{u}_1 = \hat{\underline{k}}, \qquad (2.15)$$

$$\sigma_2 = 1, \qquad \underline{u}_2 \perp \hat{\underline{k}}, \qquad (2.16)$$

where $\underline{k} \equiv (k_1,k_2)$ and $\hat{\underline{k}} \equiv \underline{k}/k$ (see figure 1).

The eigenvalues are independent of $\underline{k}$. The eigenvector $\underline{u}_1$ associated with the first eigenvalue $\sigma_1 = 0$ is longitudinal and corresponds to the nonsolenoidal or compressible mode. The eigenvector $\underline{u}_2$ associated with $\sigma_2 = 1$ is transversal and corresponds to the solenoidal or incompressible mode. By the definition of equation (2.14), the values of γ associated with the two modes are $\gamma_1 = 0$ and $\gamma_2 = \gamma_2(\underline{k}) = \nu k^2 + i\underline{v}.\underline{k}$, respectively. Thus, in the continuum, the solution of equation (2.9) satisfies the condition of incompressibility exactly at all wavelengths, as requested.

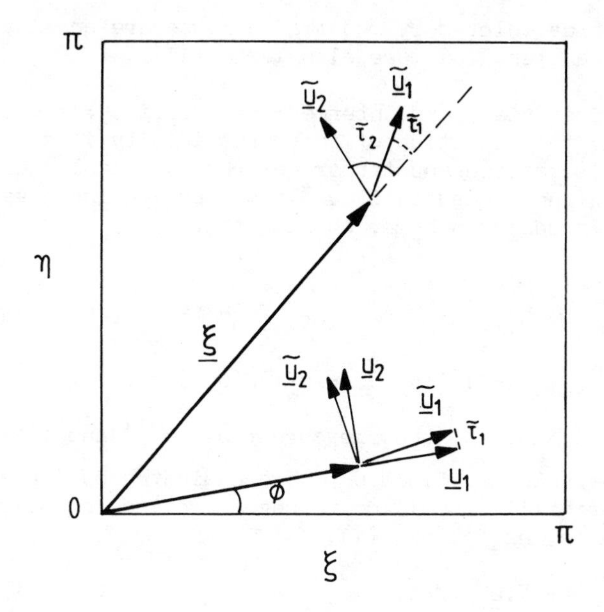

FIGURE 1

Fundamental zone of the dimensionless wave-vector space $\underline{\xi} = \underline{k}h$ for the response analysis of the two dimensional discretized equations.

2.2 Three-dimensional Equations

In the case of the fully three-dimensional equations, the solution of the problem defined by equation (2.12) is

$$\sigma_1 = 0, \qquad \underline{u}_1 = \underline{\hat{k}}, \tag{2.17}$$

$$\sigma_2 = \sigma_3 = 1, \qquad \underline{u}_2, \underline{u}_3 \perp \underline{\hat{k}}, \tag{2.18}$$

where now $\underline{k} \equiv (k_1,k_2,k_3)$. There are two solenoidal modes which are degenerate. The normalized eigenvectors $\underline{u}_2$ and $\underline{u}_3$ can be chosen along any two mutually orthogonal directions perpendicular to $\underline{k}$.

3. TIME DISCRETIZED EQUATIONS

In this section equations (2.1)- (2.3) are discretized in time according to the fractional step method originally proposed by Chorin [9,10] and suitably modified for the case of the linear equations considered here. Let $\underline{u}^n$ and p^n denote the velocity and pressure, respectively, at time $t^n \equiv n\Delta t$. With the solution $\underline{u}^n$ known, the fields $\underline{u}^{n+1}$ and p^{n+1} at the next time level are calculated by means of a procedure consisting of two separate steps [9] (see also Temam [11]).

In the first step, and intermediate velocity field $\underline{u}^{n+\frac{1}{2}}$ not satisfying the condition of incompressibility is calculated from a time discretized counterpart of the dynamical equation (2.1) omitting the pressure term. Thus, considering a general two-level differencing scheme, we have that

$$\frac{\underline{u}^{n+\frac{1}{2}} - \underline{u}^{n}}{\Delta t} + \theta[(\underline{v}.\underline{\nabla})\underline{u}^{n+\frac{1}{2}} - 2\nu\underline{\nabla}.\ \underline{\underline{D}}(\underline{u}^{n+\frac{1}{2}})] + (1-\theta)\ [(\underline{v}.\underline{\nabla})\underline{u}^{n} - 2\nu\underline{\nabla}.\ \underline{\underline{D}}(\underline{u}^{n})] = 0 \qquad (3.1)$$

where $0 \leqq \theta \leqq 1$. Then, the pressure field p^{n+1} and the final divergenceless velocity field $\underline{u}^{n+1}$ are evaluated by imposing the incompressibility condition through a Poisson equation for the pressure, as follows [11]

$$-\nabla^2\ p^{n+1} = -\ \underline{\nabla}.\underline{u}^{n+\frac{1}{2}}/\Delta t, \qquad (3.2)$$

$$\frac{\underline{u}^{n+1} - \underline{u}^{n+\frac{1}{2}}}{\Delta t} + \underline{\nabla}\ p^{n+1} = 0. \qquad (3.3)$$

To employ the scheme defined by equations (3.1) - (3.3) in association with a finite element spatial discretization, the equations must be rewritten in a weak variational form. This can be accomplished according to standard methods [12], and the result is

$$\left(\frac{\underline{u}^{n+\frac{1}{2}} - \underline{u}^{n}}{\Delta t},\ \underline{w}\right) + \theta[((\underline{v}.\underline{\nabla})\underline{u}^{n+\frac{1}{2}}, \underline{w}) + 2\nu(\underline{\underline{D}}(\underline{u}^{n+\frac{1}{2}}),\ \underline{\underline{D}}(\underline{w}))] + (1-\theta)\ [((\underline{v}.\underline{\nabla})\underline{u}^{n}, \underline{w}) + 2\nu(\underline{\underline{D}}(\underline{u}^{n}),\ \underline{\underline{D}}(\underline{w}))] = 0 \qquad (3.4)$$

$$(\underline{\nabla}p^{n+1},\ \underline{\nabla}w) + (\underline{\nabla}.\underline{u}^{n+\frac{1}{2}}, w)/\Delta t = 0, \qquad (3.5)$$

$$\left(\frac{\underline{u}^{n+1} - \underline{u}^{n+\frac{1}{2}}}{\Delta t},\ \underline{w}\right) - (p^{n+1},\ \underline{\nabla}.\underline{w}) = 0. \qquad (3.6)$$

Here we have adopted the standard notation of Hilbert scalar products, so that for instance $(\underline{\underline{D}}(\underline{u}),\ \underline{\underline{D}}(\underline{w})) = \sum_{i,j=1}^{d} (D_{ij}(\underline{u}),\ D_{ij}(\underline{w}))$ where $(.,.)$ denotes the scalar product of square integrable functions. Furthermore, it is implicitly assumed that equations (3.4) and (3.6) and (3.5) must be satisfied for any vector function $\underline{w} \in (H^1)^d$, d=2,3 and any scalar function $w \in H^1$, respectively.

4. SPATIAL DISCRETIZATION BY QUADRILATERAL FINITE ELEMENTS

Limiting our attention to the equations for two-dimensional flows, let us consider a uniform array of 4-noded square elements of side length equal to h and let us denote by (ℓ,m) a typical node of the mesh. We approximate the fields $\underline{u}$ and p at a given time by bilinear polynomials over each element, so that the semidiscrete equations (3.4) - (3.6) can be replaced by their fully discretized counterparts, consisting of purely algebraic equations for the nodal values

$\underline{u}_{\ell m}^{n+\frac{1}{2}}$, $\underline{u}_{\ell m}^{n+1}$ and $p_{\ell m}^{n+1}$, $\ell,m = \ldots 0,1,2,\ldots$

By means of simple calculations we obtain at the representative node (ℓ,m)

$$\underline{M}(\underline{u}_{\ell m}^{n+\frac{1}{2}} - \underline{u}_{\ell m}^{n})/\Delta t + \theta\left[\frac{v}{h}\underline{A}(\hat{\underline{v}}) + \frac{\nu}{h^2}\underline{D}\right]\underline{u}_{\ell m}^{n+\frac{1}{2}}$$

$$+ (1-\theta)\left[\frac{v}{h}\underline{A}(\hat{\underline{v}}) + \frac{\nu}{h^2}\underline{D}\right]\underline{u}_{\ell m}^{n} = 0, \qquad (4.1)$$

$$\frac{1}{h^2} K\, p_{\ell m}^{n+1} + \frac{1}{h}\underline{C}^T\, \underline{u}_{\ell m}^{n+\frac{1}{2}}/\Delta t = 0, \qquad (4.2)$$

$$\underline{M}\,(\underline{u}_{\ell m}^{n+1} - \underline{u}_{\ell m}^{n+\frac{1}{2}})/\Delta t - \frac{1}{h}\underline{C}\, p_{\ell m}^{n+1} = 0, \qquad (4.3)$$

where $\hat{\underline{v}} = \underline{v}/v$ is the unit vector in the direction of the advection velocity and $\underline{M}$, $\underline{A}(\hat{\underline{v}})$, $\underline{C}$, $\underline{C}^T$, $\underline{D}$ and K denote matrices of difference operators acting on the nodal subscript indices ℓ and m. The explicit representation of these operators is

$$\underline{M} = \underline{I}\, M, \qquad (4.4)$$

$$\underline{A}(\hat{\underline{v}}) = \underline{I}\, A(\hat{\underline{v}}) \qquad (4.5)$$

$$M = 1 + r(D_x + D_y + \frac{1}{4}D_{xy} + \frac{1}{4}D_{yx}),\ r = \frac{1}{9}, \qquad (4.6)$$

$$A(\hat{\underline{v}}) = \hat{v}_x C_x + \hat{v}_y C_y, \qquad (4.7)$$

$$\underline{C}^T = (C_x, C_y), \qquad (4.8)$$

$$C_x = \frac{2}{3}(A_x + \frac{1}{4}A_{xy} + \frac{1}{4}A_{yx}) \qquad (4.8_1)$$

$$C_y = \frac{2}{3}(A_y + \frac{1}{4}A_{xy} - \frac{1}{4}A_{yx}); \qquad (4.8_2)$$

$$\underline{D} = \begin{bmatrix} -(D_x + \frac{1}{2}D_{xy} + \frac{1}{2}D_{yx}) & -\frac{1}{4}(D_{xy} - D_{yx}) \\ -\frac{1}{4}(D_{xy} - D_{yx}) & -(D_y + \frac{1}{2}D_{xy} + \frac{1}{2}D_{yx}) \end{bmatrix} \qquad (4.9)$$

$$K = -\frac{1}{3}(D_x + D_y + D_{xy} + D_{yx}) \qquad (4.10)$$

where $\underline{I}$ is the 2x2 identity matrix. In the above expressions A_x, A_y, A_{xy}, A_{yx} and D_x, D_y, D_{xy}, D_{yx} are the basic operators, respectively, of first-order and second-order spatial difference defined in the Appendix. By comparing the expressions contained in equations (4.6) and (4.10), we notice that, in contrast to the one-dimensional case, in two-dimensional problems the consistent mass matrix $\underline{M}$ is not obtained from the lumped diagonal mass matrix (equation (4.6) with $r = 0$) by adding to it the stiffness matrix suitably scaled.

Let us now consider a solution of equations (4.1) - (4.3) in the form of a plane wave with wave vector $\underline{k} \equiv (k_1, k_2)$. Then, introducing the dimensionless wave vector $\underline{\xi} \equiv (\xi, \eta) = \underline{k}\, h \equiv (k_1, k_2)h$, we can write

$$\underline{u}_{\ell m}^{n+1} = \underline{\tilde{u}}^{n+1}\, e^{i(\ell k_1 + m k_2)h} = \underline{\tilde{u}}^{n+1}\, e^{i(\ell\xi + m\eta)}$$

and similarly for $\underline{u}_{\ell m}^{n+\frac{1}{2}}$ and $p_{\ell m}^{n+1}$. Upon substitution into equations (4.1) - (4.3), we obtain

$$\underline{M}(\underline{\tilde{u}}^{n+\frac{1}{2}} - \underline{\tilde{u}}^{n})/\Delta t + \theta[a\,\underline{A}(\underline{\hat{v}}) + d\,\underline{D}]\underline{\tilde{u}}^{n+\frac{1}{2}} \qquad (4.11)$$

$$+ (1-\theta)[a\,\underline{A}(\underline{\hat{v}}) + d\,\underline{D}]\underline{\tilde{u}}^{n} = 0,$$

$$\frac{1}{h^2} K\, \tilde{p}^{n+1} + \frac{1}{h}\underline{C}^{+}\, \underline{\tilde{u}}^{n+\frac{1}{2}}/\Delta t = 0, \qquad (4.12)$$

$$\underline{M}(\underline{\tilde{u}}^{n+1} - \underline{\tilde{u}}^{n+\frac{1}{2}})/\Delta t - \frac{1}{h}\underline{C}\, \tilde{p}^{n+1} = 0, \qquad (4.13)$$

where the Courant number $a \equiv v\,\Delta t/h$ and the diffusion number $d \equiv \nu\Delta t/h^2$ have been introduced. In equations (4.11) - (4.13) $\underline{M}$, $\underline{A}(\underline{\hat{v}})$, $\underline{C}$, $\underline{C}^{+}$ (the conjugate transpose of $\underline{C}$), $\underline{D}$ and K designate matrices of complex numbers dependent on ξ and η, and are given by

$$M = 1 + r(2\cos\xi + 2\cos\eta + \cos\xi\ \cos\eta - 5), \quad r = \frac{1}{9}, \tag{4.14}$$

$$A(\underline{\hat{v}}) = \frac{i}{3}\,[\hat{v}_x \sin\xi\ (\cos\eta+2) + \hat{v}_y \sin\eta\ (\cos\xi+2)], \tag{4.15}$$

$$\underline{C}^T = \frac{i}{3}\,(\sin\xi\ (\cos\eta+2),\ \sin\eta(\cos\xi+2)], \tag{4.16}$$

$$\underline{D} = \begin{bmatrix} -2[\cos\xi(\cos\eta+2)] & \sin\xi\ \sin\eta \\ \sin\xi\ \sin\eta & -2[\cos\eta(\cos\xi+2)] \end{bmatrix}, \tag{4.17}$$

$$K = -\frac{2}{3}\,(\cos\xi + \cos\eta + 2\cos\xi\ \cos\eta-4). \tag{4.18}$$

By simple matrix manipulation it is possible to derive from equations (4.11) - (4.13) the expression

$$\underline{\tilde{u}}^{n+1} = [\underline{I} - \underline{M}^{-1}\ \underline{C}\ K^{-1}\ \underline{C}^{+}]\ [\underline{I} + \theta\underline{M}^{-1}\ (a\ \underline{A} + d\ \underline{D})]^{-1}$$
$$\times\ [\underline{I}-(1-\theta)\underline{M}^{-1}\ (a\ \underline{A} + d\ \underline{D})]\underline{\tilde{u}}^{n}, \tag{4.19}$$

which gives the new velocity $\underline{\tilde{u}}^{n+1}$ in terms only of the previous velocity $\underline{\tilde{u}}^{n}$.

5. QUASI-CONSISTENT MASS MATRIX APPROXIMATION

In this section we introduce an iterative method to take into account approximately the effect of the consistent mass matrix during the time integration without having to solve the involved linear system of coupled equations. This method allows the superior response properties of the consistent mass matrix to be exploited in a purely explicit iterative procedure and therefore lends itself to be combined with explicit time integration schemes in a very simple and efficient method of solution [13].

Let us consider the model linear problem

$$\underline{M}\ \underline{\dot{U}} + \underline{B}\ \underline{U} = 0, \tag{5.1}$$

where the dot indicates time derivative, $\underline{U} \equiv \{U_\ell,\ \ell=1,2,...\}$ is the vector of time-dependent nodal values of a finite element approximation of a scalar field U, $\underline{M}$ is the consistent mass matrix and $\underline{B}$ is a constant matrix. Equation (5.1) results from applying the Galerkin method to a linear partial differential equation of hyperbolic, parabolic or mixed type. Let us introduce the standard lumped diagonal matrix $\underline{L}$ obtained from $\underline{M}$ by the row-sum technique, namely,

$$L_{\ell\ell} = \sum_{\ell'} M_{\ell\ell'},$$
$$L_{\ell\ell'} = 0 \text{ for any } \ell' \neq \ell. \tag{5.2}$$

To avoid the solution of the linear system of algebraic equations implied by the presence of $\underline{M}$ in equation (5.1), we define the following iterative procedure.

Start from $\dot{\underline{U}}_o = 0$. Then for $g=1,2,..G-1$ with $\dot{\underline{U}}_g$ known, determine $\dot{\underline{U}}_{g+1}$ as the solution of the linear equations

$$\underline{L}\,\dot{\underline{U}}_{g+1} + (\underline{M}-\underline{L})\dot{\underline{U}}_g + \underline{B}\,\underline{U} = 0 \tag{5.3}$$

where $\underline{L}$ is the lumped diagonal mass matrix defined by equations (5.2). At the end, assume $\dot{\underline{U}} = \dot{\underline{U}}_G$.

When applied to the equations discretized in time, this iterative method is equivalent to a special case of the general integration algorithm considered in [14]. For $G = 1$ we have immediately

$$\dot{\underline{U}} + \underline{L}^{-1}\,\underline{B}\,\underline{U} = 0, \tag{5.4}$$

that is, the procedure provides the usual lumped diagonal approximation. For $G = 2$ we obtain

$$\dot{\underline{U}} + 2\underline{L}^{-1}\,(\underline{I} - \tfrac{1}{2}\underline{M}\,\underline{L}^{-1})\underline{B}\,\underline{U} = 0, \tag{5.5}$$

and therefore $\underline{M}^{-1}$ is approximated by

$$\underline{Q}^{-1} \equiv \underline{Q}(2)^{-1} = 2\underline{L}^{-1}\,(\underline{I} - \tfrac{1}{2}\underline{M}\,\underline{L}^{-1}). \tag{5.6}$$

In the case of hyperbolic equations, this approximation is coincident with the two-stage method proposed in [13] and assures the fourth-order accuracy in the discretized representation of the first-order spatial derivative.

For $G=3$ it results that $\underline{M}^{-1}$ is approximated by matrix

$$\underline{Q}(3)^{-1} = 3\underline{L}^{-1}\,(\underline{I}-\underline{M}\,\underline{L}^{-1} + \tfrac{1}{3}\underline{M}\,\underline{L}^{-1}\,\underline{M}\,\underline{L}^{-1}), \tag{5.7}$$

and so on. In Figure 2 we report the ratios M/L, M/Q and M/Q(3) of the Fourier transform of the mass matrices for the piecewise linear approximation in one dimension, or, equivalently, for the bilinear approximation in two dimensions along $\underline{\xi} = (\xi,0)$. The iterative method produces approximations to $\underline{M}$ whose accuracy increases with the number of iterations: it is more convenient than standard relaxation procedures since it has no free parameter and converges

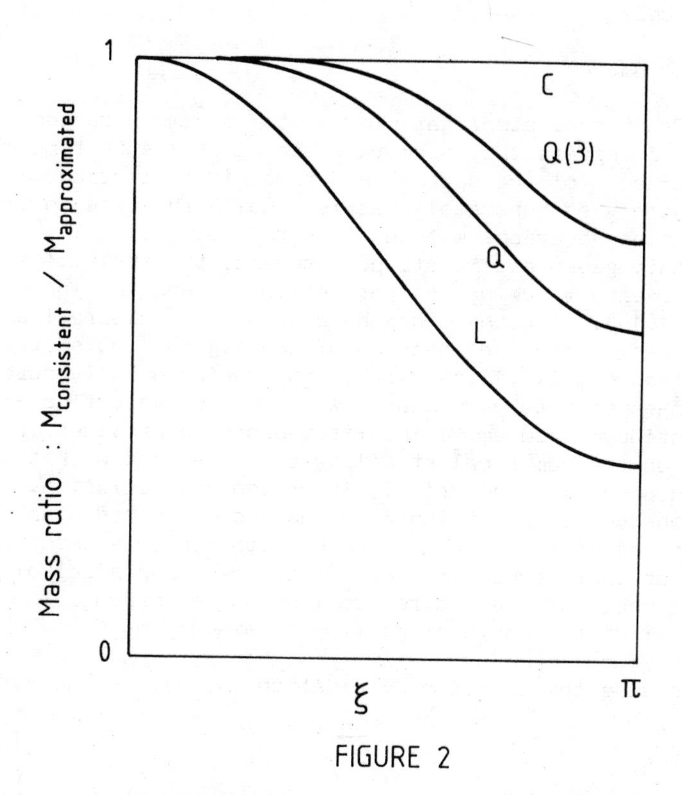

FIGURE 2

Comparison of different mass matrix approximations: consistent (C), lumped diagonal (L), quasi-consistent (Q) and three iteration approximations (Q(3)).

selectively the earlier at the longer wavelength.

An appreciable reduction of the distortion with respect to the lumped diagonal approximation in the range of long and intermediate wavelengths is obtained by means of the procedure with two iterations. This approximation will be named quasi-consistent mass matrix approximation and will be denoted hereafter by Q. It is immediate to see that for $\xi \to 0$ it results $Q/M \to 1 + \frac{1}{36}\xi^4$. Therefore the use of the quasi-consistent mass matrix preserves the fourth-order spatial accuracy of the Galerkin approximation to the advection equation. The effect of this approximate method when applied to the incompressible advection-diffusion equation is investigated in the next section.

6. RESPONSE ANALYSIS OF THE DISCRETIZED INCOMPRESSIBLE EQUATIONS

6.1 Preliminaries

The fractional step character of the scheme here considered to integrate the incompressible equations in time means that the effects of the spatial and time discretizations cannot be analyzed separately, as is usually the case for finite element equations without constraints [3-5]. It follows that, given a time stepping scheme, the analysis must be done for each value of the parameters a and d in equation (4.19). On the other hand, the fully discretized equations can be studied in terms of the eigenvalues of the amplification matrix [15,16] or by introducing a fully discrete response parameter to be compared with the equivalent parameter of the continuum problem. The first approach provides information on the numerical stability of the scheme with the associated aspects of numerical dispersion and diffusion. It will be considered in Section 7. The second method, that has been sometimes prefered in connection with finite elements and that will be considered here, allows a direct appraisal of the consequences of the discretizations on the propagation and diffusion of disturbances of different wavelengths.

Expressing the solution of equation (4.19), in the form

$$\underline{\tilde{u}}^n = \underline{\tilde{u}}_o \, e^{-\tilde{\gamma} n \Delta t}, \tag{6.1}$$

we obtain that $\tilde{\gamma}$ and $\underline{\tilde{u}}_o$ are the eigenvalues and eigenvectors of problem

$$e^{-\tilde{\gamma}\Delta t} \underline{\tilde{u}}_o = [\underline{I}-\underline{M}^{-1} \underline{C} K^{-1} \underline{C}^+] [\underline{I}+\theta\underline{M}^{-1}(a \underline{A} + d \underline{D})]^{-1}$$

$$\times [\underline{I}-(1-\theta)\underline{M}^{-1} (a \underline{A} + d \underline{D})]\underline{\tilde{u}}_o. \tag{6.2}$$

Equation (6.2) is the fully discretized counterpart of equation (2.11). By comparing the eigenvalues and the directions of the associated eigenvectors of the discretized equations with the corresponding quantities of the exact equations, it is possible to investigate the discretization properties of the finite dimensional representation of the basic continuum equations. In the mixed case of advection-diffusion, by equations (4.14) - (4.18), the matrix to be diagonalized is in general complex and not hermitian so that the solution of eigenproblem (6.2) requires calculations in the complex field. However, this can be avoided if we examine in separate phases firstly the effect of the spatial discretization on the condition of incompressibility, and subsequently the combined effects of both discretizations in the case of pure advection and pure diffusion, as will be shown below.

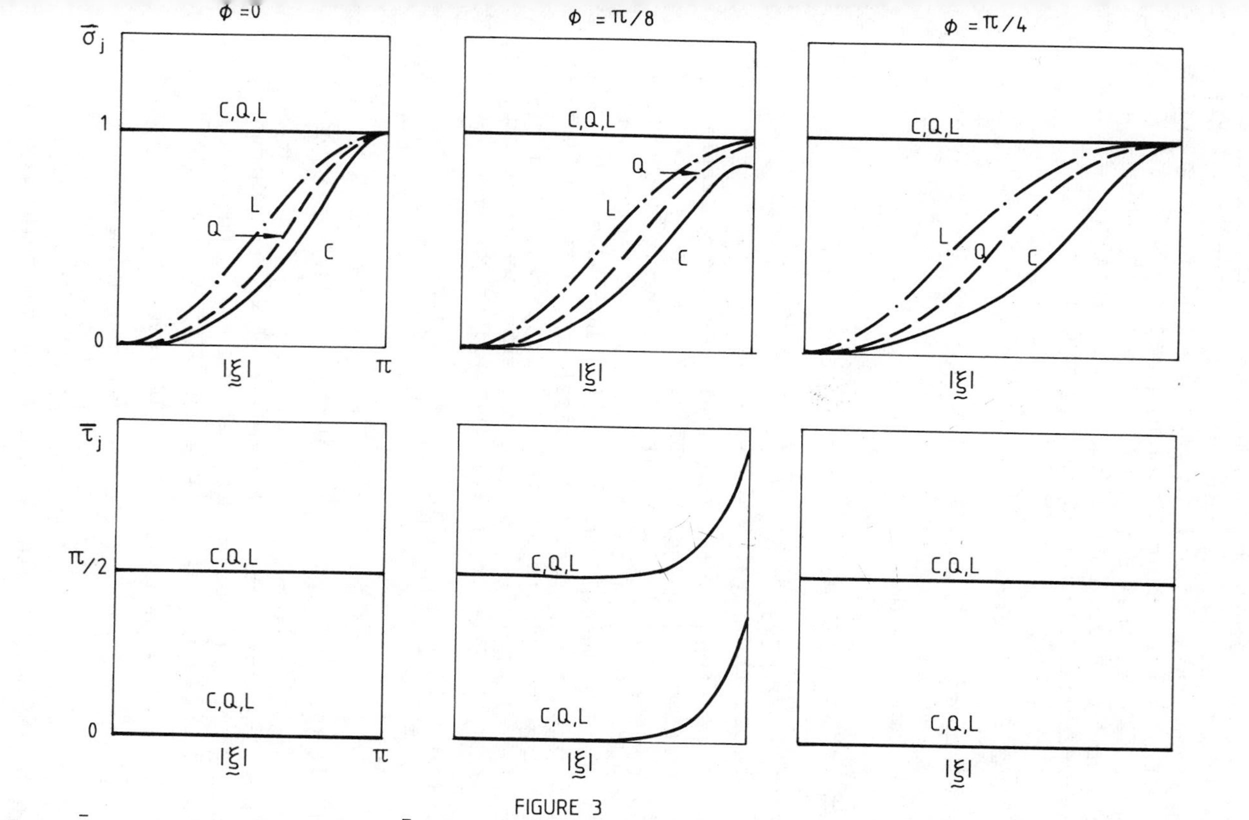

FIGURE 3

Eigenvalues $\bar{\sigma}_j$ and eigendirections $\bar{\tau}_j$ (with respect to $\underline{k}$) of the discretized solenoidalization operator $\underline{\underline{S}}$ along the straight lines of the wavevector space $\phi = 0$, $\phi = \pi/8$, and $\phi = \pi/4$. Comparison of different mass matrix approximations. Solenoidal mode: upper curves; Nonsolenoidal mode: lower curves.

6.2 Spatial Discretization and Incompressibility Condition

The first bracketed operator in the right hand side of equation (6.2) is a consequence of the incompressible character of the basic equations and depends only on the spatial discretization that has been adopted. It is therefore legitimate to regard this operator as the (spatially) semi-discrete counterpart of the solenoidalization operator $\underline{S}$ defined by equation (2.13) for the continuum problem. Using an overbar to denote semi-discrete quantities, i.e., quantities affected only by the spatial discretization, we can write

$$\bar{\underline{S}} \equiv \underline{I} - \underline{M}^{-1} \underline{C} \, \underline{K}^{-1} \underline{C}^{+} = \underline{I} - \underline{C} \, \underline{C}^{+}/(MK), \tag{6.3}$$

where the property $\underline{M} = \underline{I}\, M$ has been used. Let us denote the eigenvalues and eigenvectors of $\bar{\underline{S}} = \bar{\underline{S}}(\underline{k})$ by $\bar{\sigma} = \bar{\sigma}(\underline{k})$ and $\bar{\underline{u}}_o = \bar{\underline{u}}_o(\underline{k})$, respectively, namely,

$$\bar{\underline{S}} \, \bar{\underline{u}}_o = \bar{\sigma} \, \bar{\underline{u}}_o \; . \tag{6.4}$$

By comparing the semidiscrete quantities $\bar{\sigma}_j$ and $\bar{\underline{u}}_j$, j=1,2, with the corresponding exact quantities σ_j and $\underline{u}_j$ discussed in Section 2, the effect of the finite element discretization on the satisfaction of the incompressibility condition for Fourier components of different wavelengths is examined. The problem being two-dimensional, this comparison must be made at each point of the fundamental zone of the (dimensionless) wavevector space depicted in Figure 1. As typical loci we choose straight lines passing through the origin and forming an angle ϕ with the ξ-axis. The direction of the semi-discrete eigenvectors $\bar{\underline{u}}_j$ is characterized by the values of the angle $\bar{\tau}_j$ between $\bar{\underline{u}}_j$ and the first exact eigenvector $\underline{u}_1 \equiv \hat{\underline{k}}$, i.e.,

$$\bar{\tau}_j \equiv \cos^{-1} (\bar{\underline{u}}_j \cdot \underline{u}_1) = \cos^{-1} (\bar{\underline{u}}_j \cdot \hat{\underline{k}}). \tag{6.5}$$

In Figure 3 we report the values of $\bar{\sigma}_j$ and $\bar{\tau}_j$, j=1,2, along three typical straightlines $\phi = 0, \frac{\pi}{8}, \frac{\pi}{4}$ and for three different representations of the mass matrix: consistent (C), lumped diagonal (L) (obtained by setting r=0 in equation (4.14)) and quasi-consistent (Q). Over the entire zone $\bar{\sigma}_2 = \sigma_2 \equiv 1$ for all of the mass approximations, whereas $\bar{\sigma}_1 \neq \sigma_1 \equiv 0$, with the consistent mass giving a better approximation of the exact solution and the quasi-consistent mass being in the middle between the consistent and the lumped diagonal mass approximations. Since $\bar{\sigma}_1 \neq 0$ the semi-discrete solenoidalization operator $\bar{\underline{S}}$ is not able to suppress the nonsolenoidal mode and this happens for Fourier components of all wavelengths. The curves of $\bar{\sigma}_1$ show that the intensity of the nonsolenoidal mode is less than, but comparable to, the intensity of the solenoidal mode and that it is greater for shorter wavelengths. On the other hand, the curves of $\bar{\tau}_j$, j=1,2, give an indication of how much the two semi-discrete modes deviate from the purely nonsolenoidal and solenoidal character, in the continuum sense. Since matrix $\underline{\bar{S}}$ is real and symmetric, the two eigendirections are mutually orthogonal for all $\underline{k}$. On the general line $\phi = \pi/8$ both modes differ from the exact ones appreciably only in the last third of the wavelength range and on high symmetry lines $\phi = 0$ and $\phi = \pi/4$ the modes coincide with their exact counterparts for all of the wavelengths. Furthermore, the mass matrix approximation adopted does not affect the direction of the eigenvectors, a consequence of the structure of the operator $\underline{\bar{S}}$ defined by equation (6.3). It is important to point out that the spurious nonsolenoidal mode leaking through the semi-discrete operator $\underline{\bar{S}}$ may eventually be damped out because of the discrete time integration (see below).

6.3 Advection Equation

In the case of pure advection ($\nu = d = 0$), the eigenvalue problem of equation (6.2) becomes, remembering that $\underline{M}$ and $\underline{A}$ are block-diagonal,

$$e^{-\tilde{\gamma}\Delta t}\, \underline{\tilde{u}}_o = \frac{1-(1-\theta)\ a\ A/M}{1 + \theta\ a\ A/M}\, \underline{\bar{S}}\, \underline{\tilde{u}}_o, \tag{6.6}$$

where $\underline{\bar{S}}$ is defined by equation (6.3). Since $\underline{\bar{S}}$ is the only matrix in problem (6.6), the fully-discrete eigenvalues $\tilde{\gamma}_j$ and eigenvectors $\underline{\tilde{u}}_j$, j=1,2, of the advection equation can be expressed simply in terms of those of the semi-discrete eigenproblem (6.4) as follows

$$e^{-\tilde{\gamma}_j\ \Delta t} = \frac{1-(1-\theta)\ a\ A/M}{1 + \theta\ a\ A/M}\ \bar{\sigma}_j, \tag{6.7}$$

$$\underline{\tilde{u}}_j = \underline{\bar{u}}_j.$$

In other words, in the case of the advection equation, the spatial discretization process has independent effects on the condition of incompressibility and on the dynamical equation.

By equation (4.5), we can write $A = i\alpha$ where

$$\alpha = [\hat{v}_x \sin\xi\ (\cos\eta+2) + \hat{v}_y \sin\eta\ (\cos\xi+2)]/3.$$

Thus equation (6.7) can be rewritten in the form

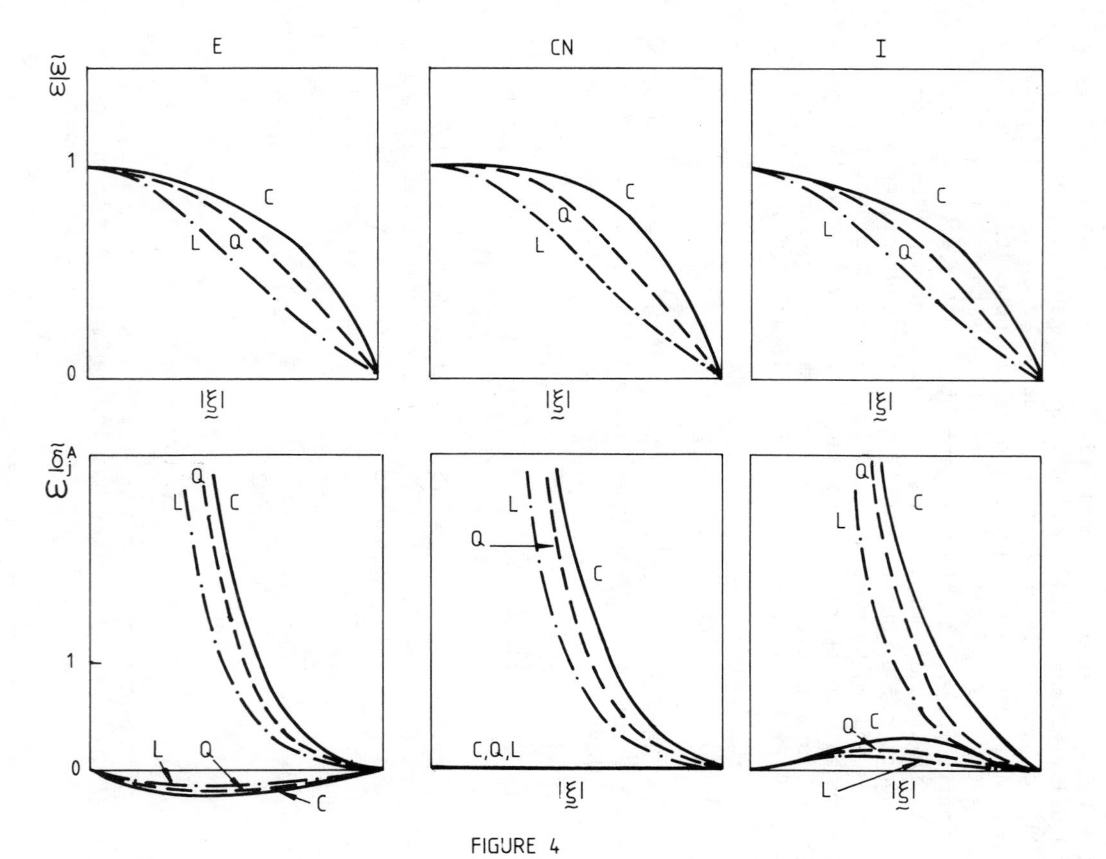

FIGURE 4

Frequency ratio $\tilde{\omega}/\omega$ of the fully-discrete advection equation and associated damping parameters $\tilde{\delta}_j^A/\omega$ of the two modes of the incompressible advection equation. Comparison of different time integration scheme: explicit, Crank-Nicolson, and implicit. $\phi = \pi/8$, $a = \frac{1}{2}$ and $\gamma = 0$.

$$e^{-\tilde{\gamma}_j \Delta t} = \frac{1 - i\,\alpha'}{1 + i\,\alpha''}\;\bar{\sigma}_j \qquad (6.9)$$

where $\alpha' \equiv (1-\theta)\, a\, \alpha/M$ and $\alpha'' \equiv \theta\, a\, \alpha/M$ are both real quantities. Setting $\tilde{\gamma} = \tilde{\delta}^A + i\tilde{\omega}$ into equation (6.9), we can separate the real part from the imaginary part, so obtaining

$$\tilde{\omega}\Delta t = \tan^{-1}\left(\frac{\alpha' + \alpha''}{1-\alpha'\;\alpha''}\right), \qquad (6.10)$$

$$\tilde{\delta}^A_j \Delta t = -\frac{1}{2}\,\ell n\left(\frac{1 + \alpha'^2}{1 + \alpha''^2}\;|\bar{\sigma}_j|^2\right) \qquad (6.11)$$

Notice that the fully discrete frequency response $\tilde{\omega}$ is independent of $\bar{\sigma}_j$, whereas two damping parameters $\tilde{\delta}^A_j$ are obtained for the incompressible advection equation: the first one is associated with the nonsolenoidal mode and the second one with the solenoidal mode. To compare these response parameters of the fully-discrete equations with the exact frequency response $\omega = \underline{v}.\underline{k}$ of the continuum advection equation, we introduce the frequency ratio $\tilde{\omega}/\omega$ and the damping parameters $\tilde{\delta}^A_j/\omega$, namely

$$\frac{\tilde{\omega}}{\omega} = \frac{1}{a\;\underline{\hat{v}}.\underline{\xi}}\,[\tan^{-1}(\alpha') + \tan^{-1}(\alpha'')], \qquad (6.12)$$

$$\frac{\tilde{\delta}^A_j}{\omega} = -\frac{1}{2}\,\frac{1}{a\;\underline{\hat{v}}.\;\underline{\xi}}\;\ell n\left(\frac{1 + \alpha'^2}{1 + \alpha''^2}\;|\bar{\sigma}_j|^2\right) \qquad (6.13)$$

where $\underline{\xi} = (\xi,\eta) = \underline{k}\,h$. In Figure 4 we report the frequency ratio $\tilde{\omega}/\omega$ and the damping parameters $\tilde{\delta}^A_j/\omega$, j=1,2, along the general straight line $\phi = \pi/8$ for the three integration schemes: explicit (θ=0), Crank-Nicolson ($\theta=\frac{1}{2}$) and implicit (θ=1). The Courant number $a = \frac{1}{2}$ and the angle χ between the advection velocity $\underline{v}$ and the x-axis is zero. The frequency curves clearly illustrate the well-known fact that disturbances of short wavelength propagate with a velocity smaller than the exact one [5], and that the mass lumping technique emphasizes this effect [4]. The use of the quasi-consistent approximation assures a much better frequency response particularly in the region of great wavelengths. The lower part of the figure shows that the damping $\tilde{\delta}^A_1$ associated to the nonsolenoidal mode is positive and very large at intermediate and long wavelengths for all the schemes and all the mass matrix approximations here considered. On the contrary, the value of the damping associated to the solenoidal mode is small negative for the explicit scheme, zero for the Crank-Nicolson scheme, and small positive for the implicit scheme over the entire range of wavevector values. This corresponds to the well-known fact that these

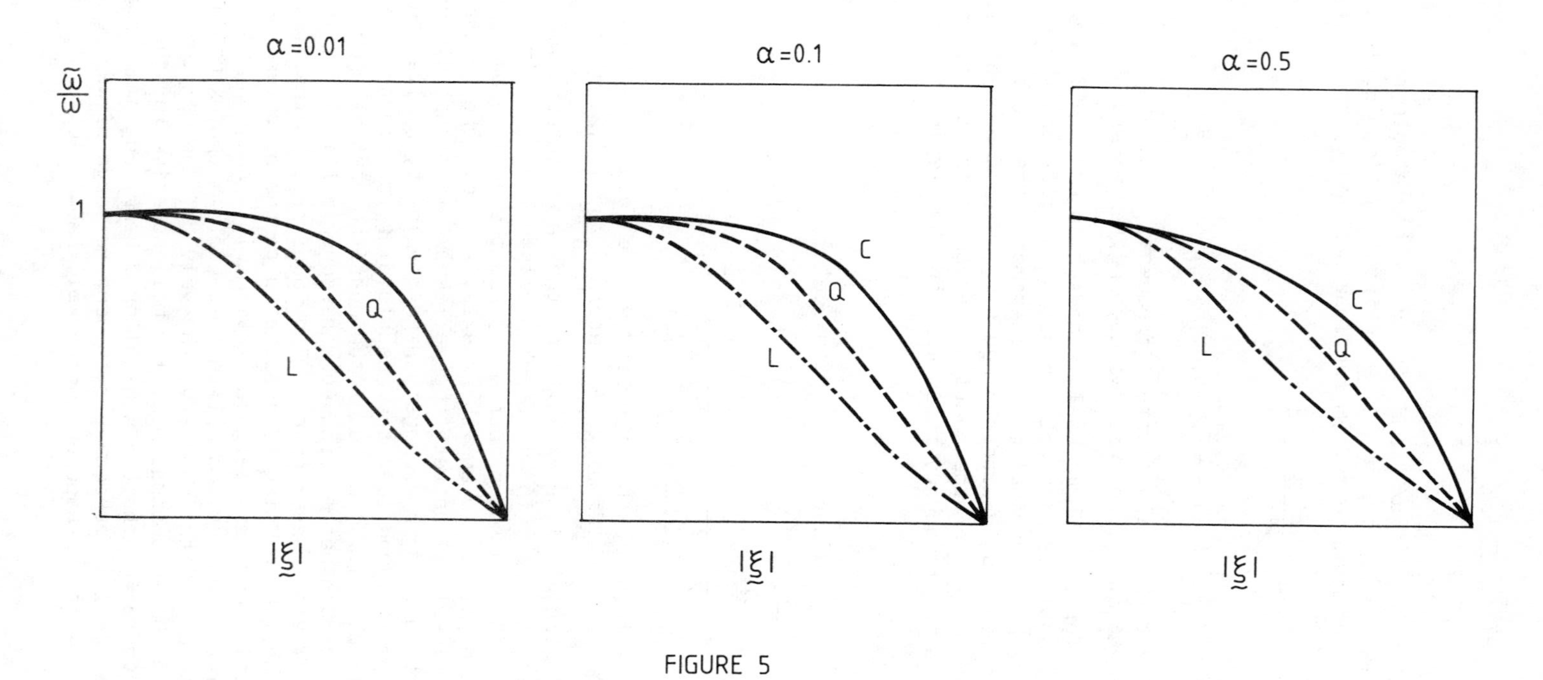

FIGURE 5

Frequency response $\tilde{\omega}/\omega$ of the fully-discrete advection equation for different values of Courant number a = 0.01, 0.1, and 0.5. Explicit scheme, $\phi = \pi/8$, $\chi = 0$.

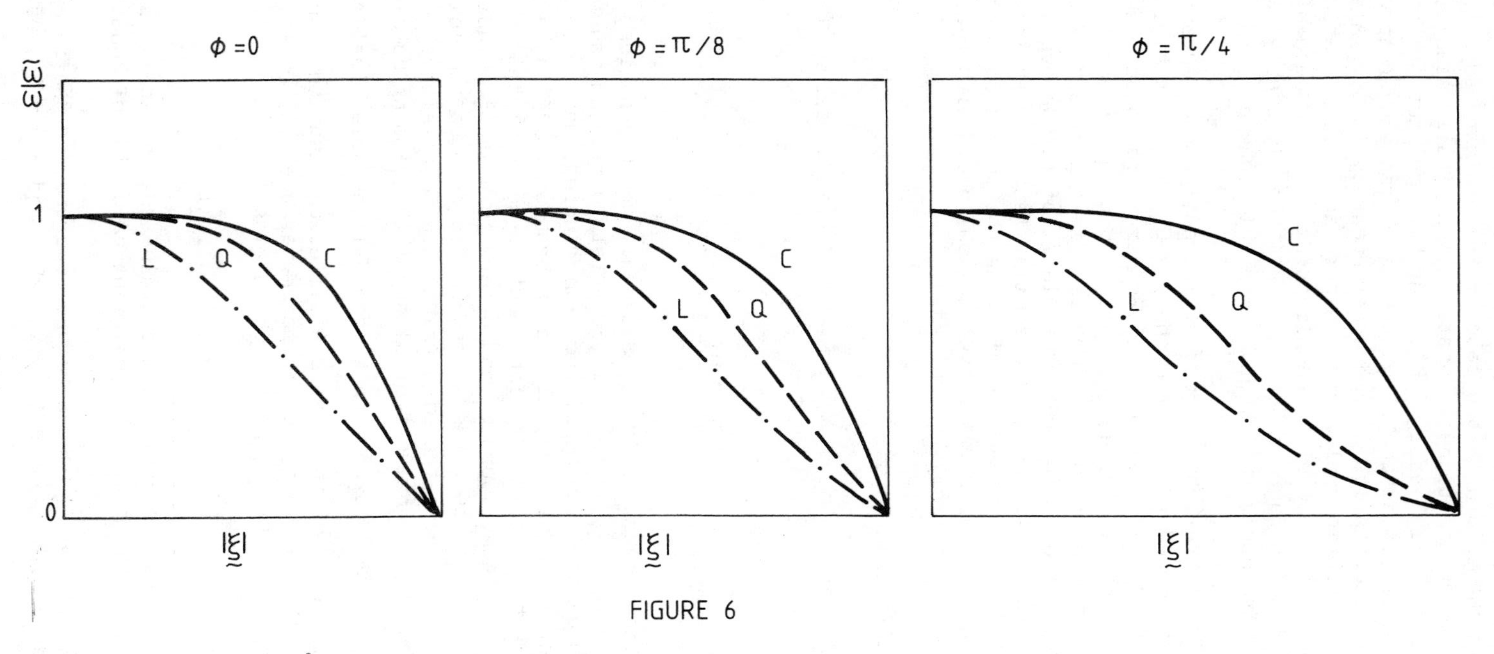

FIGURE 6

Frequency response $\tilde{\omega}/\omega$ of the fully-discrete advection equation along different straight lines of the wavevector space: $\phi = 0$, $\phi = \pi/8$, and $\phi = \pi/4$. Explicit scheme, $a = 0.1$, $\chi = 0$.

three schemes are in the case of the ordinary advection equation unconditionally unstable, marginally stable, and unconditionally stable [16]. Furthermore, notice that the Crank-Nicolson scheme gives the best frequency response in accordance with its second-order accuracy in Δt. Although the explicit scheme applied to the pure advection equation is unstable, in view of its application to mixed case of advection and diffusion where it became conditionally stable, we will report frequency ratio curves for this scheme.

In Figure 5 the effect of increasing the Courant number is illustrated for $\phi = \pi/8$ and $\chi = 0$. Notice that in the range $0.01 < \alpha < 0.1$ the fully-discrete response does not change appreciably, an indication that the effect of the time discretization is not as important as the effect of the spatial discretization. At different places of the wavevector space the frequency response have similar features, as shown by Figure 6. The advantage of using the consistent mass matrix particularly for waves travelling obliquely with respect to the mesh is clearly recognized.

6.4 Diffusion Equation

As far as the diffusion equation is concerned ($\underline{v}$=a=0), from equation (6.2) we have

$$e^{-\tilde{\gamma}\Delta t}\ \underline{\tilde{u}}_o = \bar{\underline{S}}\ [\underline{I}+\theta d/M\ \underline{D}]^{-1}\ [\underline{I}-(1-\theta)d/M\ \underline{D}]\underline{\tilde{u}}_o \qquad (6.14)$$

Since matrix $\underline{D}$ is not diagonal, the eigenvalues $\tilde{\sigma}_j$ and eigenvectors $\underline{\tilde{u}}_j$ of the fully-discrete diffusion equation cannot be expressed in terms of the semi-discrete ones. Moreover, the matrix to be diagonalized in problem (6.14) at a general point $\underline{k}$ has real coefficients but is not symmetric. It follows that the eigenvectors $\underline{\tilde{u}}_1$ and $\underline{\tilde{u}}_2$ are in general mutually non-orthogonal and that the eigenvalues $\tilde{\sigma}_j$ can be complex. As a matter of fact, the eigenvalues are found always to be real and thus they provide the values of the damping parameters $\tilde{\delta}_j \equiv \tilde{\gamma}_j$, j=1,2, of the nonsolenoidal and solenoidal modes, respectively, of the discrete diffusion equation. To compare them with the exact damping parameter $\delta = \nu k^2$, we introduce the damping ratios $\tilde{\delta}_j/\delta$ and the angles

$$\tilde{\tau}_j = \cos^{-1}(\underline{\tilde{u}}_j \cdot \underline{u}_1) = \cos^{-1}(\underline{\tilde{u}}_j \cdot \hat{\underline{k}}). \qquad (6.15)$$

All these quantities are evaluated numerically by diagonalizing the matrix in the right hand side of equation (6.14).

In figure 7 these data are reported for the integration schemes $\theta=0$, $\theta=\frac{1}{2}$ and $\theta=1$, with d=0.1 and $\phi = \pi/8$. As far as

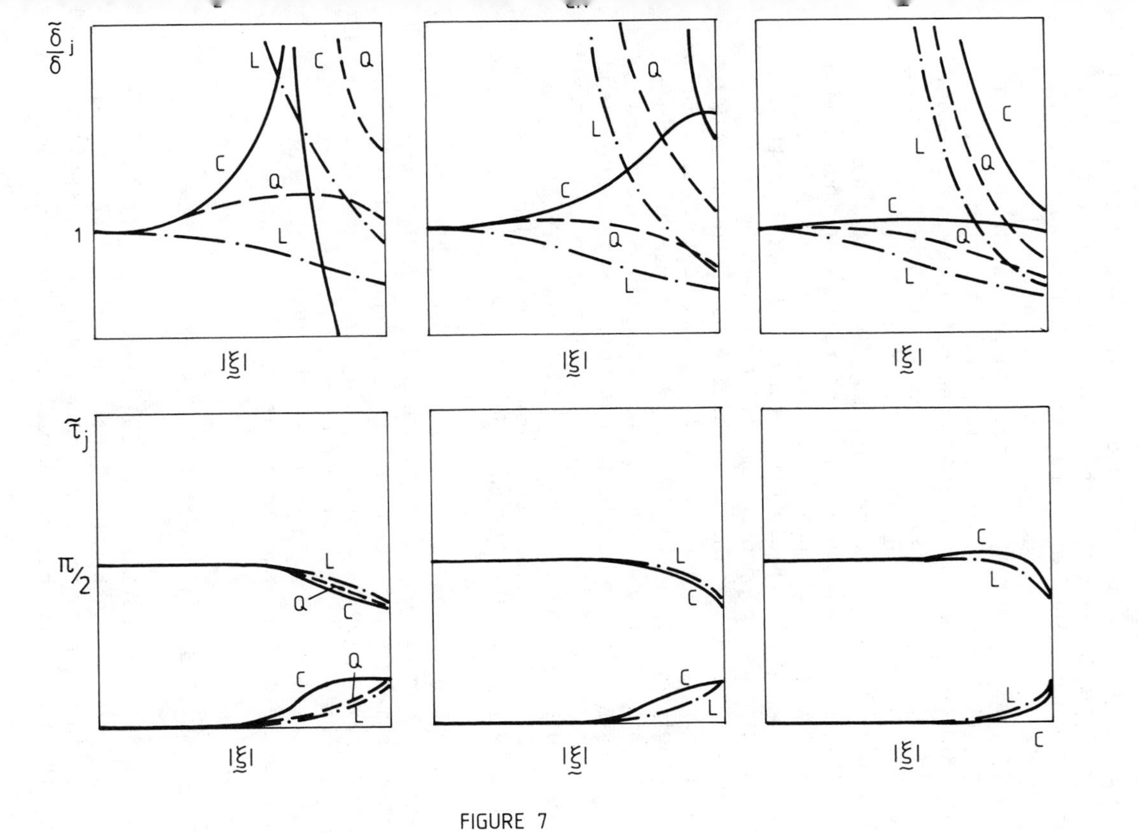

FIGURE 7

Damping ratios $\tilde{\delta}_j/\delta$ and eigendirections $\tilde{\tau}_j$ (with respect to $\underline{k}$) of the two modes of the fully-discrete incompressible diffusion equation. Comparison of different time integration schemes: explicit, Crank-Nicolson, and implicit. $\phi = \pi/8$, $d = 0.1$.

the solenoidal mode is concerned, the diagonal mass matrix approximation attenuates the damping at intermediate and short wavelengths. On the contrary, the consistent mass matrix has the opposite effect of increasing the damping except in the case of the implicit scheme which models the diffusion very accurately at all wavelengths for the considered value of the diffusion coefficient d. The quasi-consistent approximation is intermediate between the other two methods and particularly effective when combined with the Crank-Nicolson scheme. The nonsolenoidal mode has a damping ratio $\tilde{\delta}_1/\delta \rightarrow \infty$ when $|\underline{\xi}| \rightarrow 0$ and always positive for the implicit schemes with any of the mass representations. The nonsolenoidal component of the velocity field is then strongly damped out by these schemes. On the contrary, in the case of the explicit scheme, $\tilde{\delta}_1/\delta$ is always positive for the lumped and quasi-consistent mass approximations, whereas it becomes negative at the shorter wavelengths for the consistent mass matrix: when the explicit scheme is used with the consistent mass matrix at d = 0.1 a nonsolenoidal velocity field containing short wavelength components can develop and is amplified, resulting in a numerical instability. This instability has been actually encountered in computations and will be discussed later. In the lower part of Figure 7, the curves of $\tilde{\tau}_j$ are reported. The eigenvectors $\tilde{\underline{u}}_j$, j=1,2 of the diffusion equation along the general straight line $\phi = \pi/8$ are not mutually orthogonal and deviate from those of the continuum problem appreciably only in the domain of high spatial frequency. In the diffusion case there are differences among the curves corresponding to different representations of the mass matrix.

In Figure 8 the damping ratios $\tilde{\delta}_j/\delta$ along different lines are reported for the explicit scheme and d = 0.01. The consistent mass matrix provides better responses for directions along the ξ-axis and the bisecting diagonal than for the general oblique direction $\phi = \pi/8$, an effect of the anisotropy of the spatial discretization. The quasi-consistent approximation is more uniform over the wavevector domain.

7. NUMERICAL STABILITY ANALYSIS OF THE EXPLICIT SCHEME

The response analysis has shown that the explicit scheme is characterized by an unconditional instability of the solenoidal mode in the pure advection case and by a conditional instability of the nonsolenoidal mode in the pure diffusion case when the consistent mass matrix is employed. These occurrences demand further investigation on the numerical stability of the explicit time integration scheme applied to the incompressible equations. We examine the problem by considering the mixed case of advection-diffusion [17] and by evaluating the eigenvalues only on the ξ-axis of the wavevector space. The last assumption is expected not to limit the general applicability

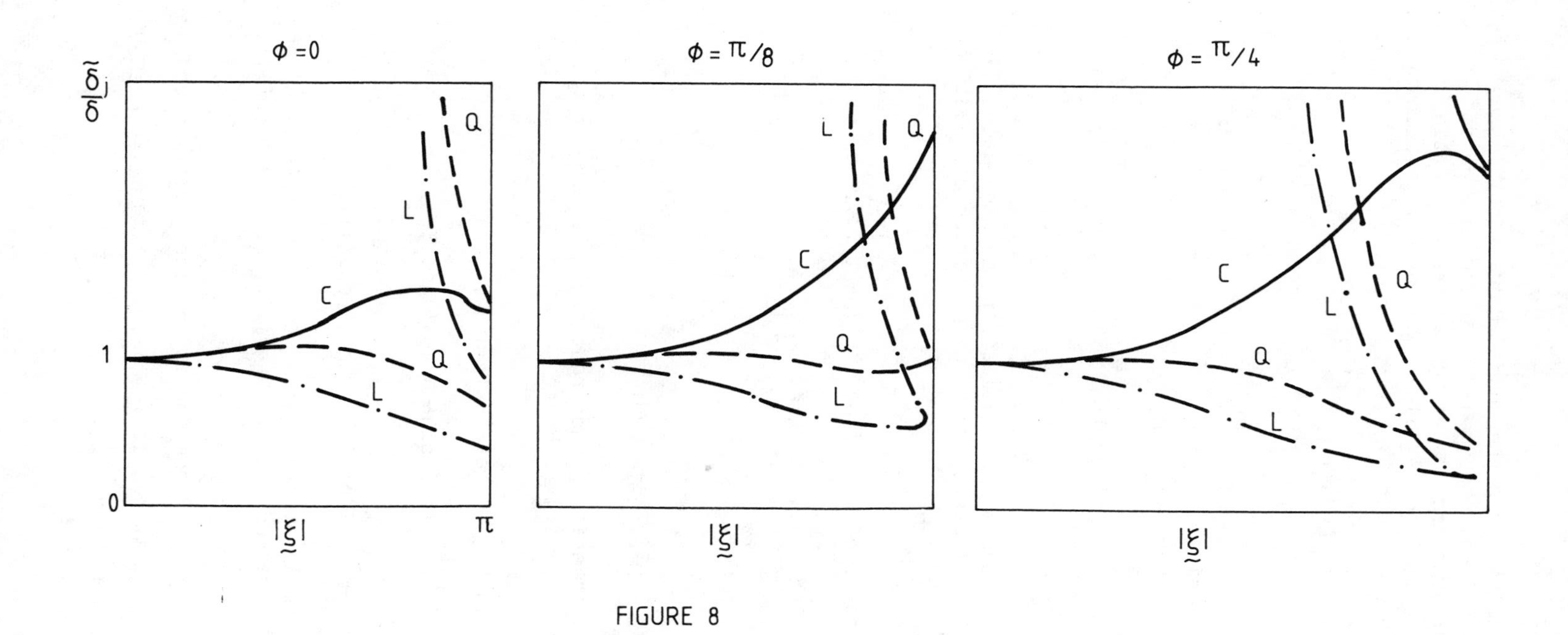

FIGURE 8

Damping ratios $\tilde{\delta}_j/\delta$ of the incompressible diffusion equation along different straight lines of the wavevector space: $\phi = 0$, $\phi = \pi/8$, and $\phi = \pi/4$. Explicit scheme, $d = 0.01$.

of the results since the onset of numerical instability is typically a local phenomena and along the coordinate axis the spacing of nodal points is the shortest. Setting $\eta=0$ in equations (4.14) to (4.18) and taking for simplicity $\hat{\underline{v}}$ in the x-direction, we obtain

$$M = \frac{1}{3}(2 + \cos\xi), \tag{7.1}$$

$$A = i \sin\xi, \tag{7.2}$$

$$\underline{C}^T = i \sin\xi \, (1, 0), \tag{7.3}$$

$$\underline{D} = 2(1-\cos\xi) \begin{bmatrix} 2 & 0 \\ 0 & 1 \end{bmatrix}, \tag{7.4}$$

$$K = 2(1-\cos\xi). \tag{7.5}$$

In the case of the explicit time integration ($\theta=0$), equation (4.19) can be written, by virtue of equation (6.3),

$$\underline{\tilde{u}}^{n+1} = \underline{\bar{S}}\,\underline{H}\,\underline{\tilde{u}}^{n}, \tag{7.6}$$

where, because of equations (7.1) - (7.5), it is

$$\underline{\bar{S}} = \begin{bmatrix} 1 - M^{-1}(1+\cos\xi)/2 & 0 \\ 0 & 1 \end{bmatrix} \tag{7.7}$$

and

$$\underline{H} = \underline{I} - M^{-1}(aA\underline{I} + d\underline{D}). \tag{7.8}$$

It is immediately apparent that the eigenvalues of the amplification matrix $\underline{S}\,\underline{H}$ are

$$\lambda_1 = [1-M^{-1}(1+\cos\xi)/2] \times [1-M^{-1}(i\,a \sin\xi + 4d\,(1-\cos\xi))], \tag{7.9}$$

$$\lambda_2 = 1-M^{-1}(i\,a \sin\xi + 2d\,(1-\cos\xi))], \tag{7.10}$$

where λ_1 and λ_2 are associated to the nonsolenoidal and solenoidal mode, respectively. Stability is assured if $|\lambda_j| \leqq 1$ for both modes.

In the case of the consistent mass matrix given by equation (7.1) it results: $|\lambda_2| \leqq 1$ if $a^2 \leqq 2d \leqq \frac{1}{3}$ and $|\lambda_1| \leqq 1$ if $a^2 \leqq 4d \leqq \frac{1}{3}$.

By taking the most restrictive of both conditions we have the stability condition $a^2 \leq 2d \leq \frac{1}{6}$. In Figure 9 we depict in the plane (d,a) the domains of stability of the two distinct modes and the intersection domain for the global stability of the explicit scheme when use is made of the consistent mass matrix.

Similarly, if the mass matrix is approximated quasi-consistently so that $M^{-1} \cong Q^{-1} = 1/3\ (4-\cos\xi)$, it results $|\lambda_2| \leq 1$ if $a^2 \leq 2d \leq \frac{3}{5}$ and $|\lambda_1| \leq 1$ if $a^2 \leq 4d \leq \frac{3}{5}$. Thus the condition of numerical stability in this case is $a^2 \leq 2d \leq \frac{3}{10}$ and the corresponding domain is still represented in Figure 9. Finally, when the lumped diagonal approximation is employed, $M \cong L = 1$ and the stability condition becomes $a^2 \leq 2d \leq \frac{1}{2}$. Two points deserve attention about these results. Firstly, the size of the stability domain is half with respect to the usual (unconstrained) diffusion equation in consequence of the so-called coupled approach to the diffusion term in the momentum equation (2.1). Secondly, the use of the consistent mass matrix, in spite of the increased implicitness of the resulting numerical scheme, destabilizes the explicit integration of the diffusion equation giving a stability limit lower than the lumped diagonal matrix approximation. The performance of the quasi-consistent mass matrix approximation is intermediate between the consistent and lumped approximations.

8. NUMERICAL TESTS

The effect of the different mass matrix representations on the accuracy of the numerical solutions is assessed against the following analytical test problem [9]. The domain is the square $\Omega = [0,\pi] \times [0,\pi]$ and the exact solution of the problem is given, in dimensionless form, by

$$u_1 = -\cos x_1 \sin x_2\, e^{-2t}, \tag{8.1}$$

$$u_2 = +\sin x_1 \cos x_2\, e^{-2t}, \tag{8.2}$$

$$p = -Re\, \frac{1}{4}\, (\cos 2x_1 + \cos 2x_2) e^{-4t}, \tag{8.3}$$

where Re is the Reynolds number. Boundary and initial conditions are derivable from equations (8.1) - (8.2) in an obvious way. Finite element solutions to the time-dependent problem with Re=20 have been calculated by means of the explicit Euler scheme (θ=0) using Δt = 0.00324 and two uniform meshes of 4x4 and 8x8 4-noded quadrilaterals, respectively. In Table 1 we display the numerical results at some selected time steps: n=1,3,5,7,9 and 20. $e(u_j)$, j=1,2, are the maxima over the

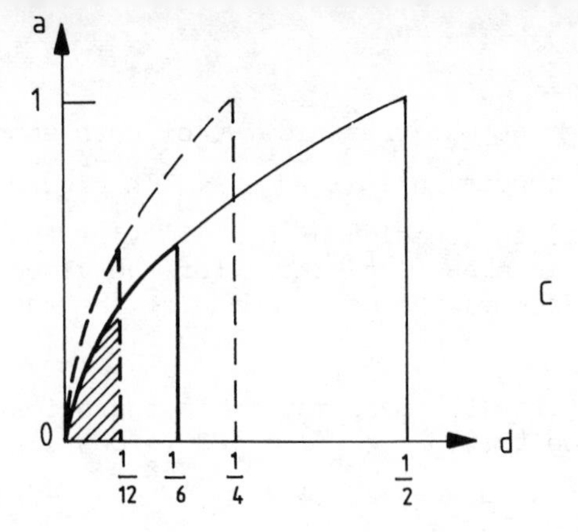

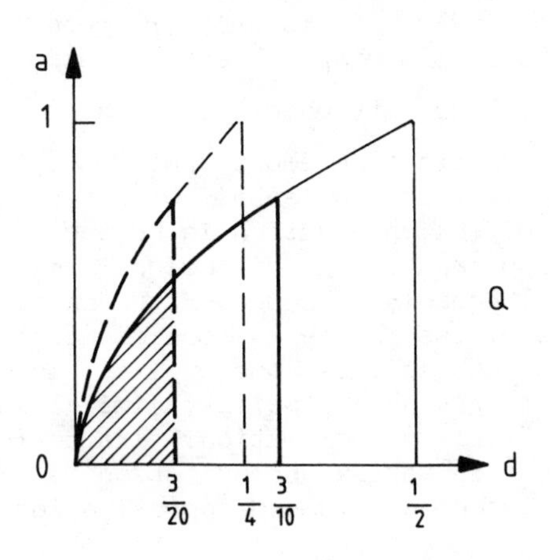

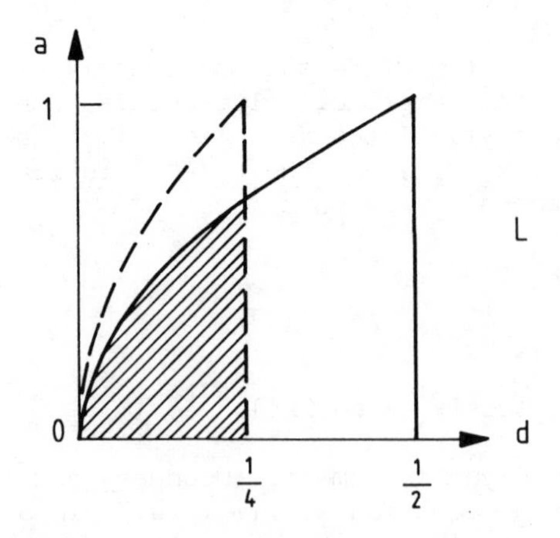

FIGURE 9

Numerical stability limits of the explicit integration scheme for the two-dimensional incompressible advection-diffusion equation with different approximations of the mass matrix (———— : solenoidal mode; - - - - : nonsolenoidal mode).

	n	$e(u_1)$, $e(u_2)$	$e(p)$
4x4 elements; Δt = 0.00324; Re = 20			
LUMPED	1	1.1×10^{-2}	0.2351
	3	2.0	0.0417
	5	2.2	0.0075
	7	2.3	0.0152
	9	2.4	0.0175
	20	2.5	0.0180
QUASI-CONSISTENT	1	0.9	0.1490
	3	1.4	0.0034
	5	1.5	0.0142
	7	1.6	0.0140
	9	1.6	0.0130
	20	1.9	0.0081
CONSISTENT	1	0.8	0.1059
	3	1.1	0.0110
	5	1.2	0.0160
	7	1.3	0.0147
	9	1.4	0.0132
	20	1.9	0.0061
8x8 elements; Δt = 0.00324; Re = 20			
LUMPED	1	4.4×10^{-3}	0.0609
	3	5.3	0.0050
	5	5.5	0.0063
	7	5.5	0.0063
	9	5.6	0.0063
	20	5.7	0.0059
QUASI-CONSISTENT	1	2.1	0.0243
	3	2.5	0.0062
	5	2.8	0.0059
	7	3.0	0.0054
	9	3.4	0.0050
	20	5.1	0.0027
CONSISTENT	1	1.9	0.0198
	3	2.2	0.0063
	5	2.5	0.0059
	7	2.7	0.0055
	9	3.1	0.0050
	20	4.8	0.0027

TABLE 1 COMPARISON OF DIFFERENT MASS MATRIX APPROXIMATIONS IN THE FINITE ELEMENT SOLUTION TO THE ANALYTICAL TEST PROBLEM

nodes of the difference between the exact and the approximate solutions. Following Chorin, the error for the pressure is conventionally defined as "the maximum over the nodes of the difference between the exact and the computed pressure divided by Re". For comparative purpose, in Table 2 we report the numerical results for the same problem calculated by finite differences using a 40x40 uniform grid [9].

The comparison of the FE solutions on the finer grid with the FD solution shows that the velocity error are of the same order of magnitude whereas for the pressure the FE error is from 5 to 10 times smaller than the FD error depending on the used mass matrix approximations. This result gives an indication of the superiority of the finite element based spatial discretization, particularly in the representation of the pressure field. Concerning the effect of the different mass matrix approximations, the scheme using the consistent mass matrix is the most accurate both for the velocity and for the pressure. The lumped mass matrix approximation provides numerical results less accurate than the other two methods, the error being greater in the initial stage for the velocity and at later times for the pressure. The results obtained by the quasi-consistent mass matrix approximation using 8x8 finite elements show a pressure error almost identical to, and a velocity error only marginally greater than, the corresponding errors associated with the consistent scheme. For the considered mesh, the quasi-consistent approximation guarantees an accuracy very near to the one provided by the consistent method without requiring to solve 4 banded symmetric linear systems for the mass matrix at each time step.

The results obtained by using the coarser mesh of 4x4 finite elements illustrates the behaviour of the different mass approximations at shorter wavelengths. In this range the lumped and quasi-consistent approximations are expected to be less accurate than before. This is confirmed by the numerical tests which show that the error of the pressure is greater in both cases whereas the error of the velocity is substantially greater only for the lumped mass approximation. The quasi-consistent method provides therefore more accurate numerical results than the lumped approximation also in the domain of short wavelengths.

9. CONCLUSION

This work has analyzed some consequences of the spatial and temporal discretization processes of the linearized incompressibl Navier-Stokes equations for the primitive variables, velocity and pressure. A method based on the determination of response parameters for the advection and the diffusion parts of the equations has been employed. This approach makes possible to describe in some detail how the incompressibility condition is represented in the case of the equations spatially discretized by finite elements through the introduction of the new concept of solenoidal and nonsolenoidal modes for the incompressible equations. It has been found that the well known results obtained from simpler model equations, such as the one-dimensional equation for a scalar unconstrained variable, are substantially confirmed by the present analysis based on the fully multidimensional incompressible equations. Possible generalizations and extensions of the method of analysis considered in this paper are in the following directions: finite element approximations with unequal-order interpolation for the velocity and the pressure; finite elements of different shape (triangle) and/or with higher-order interpolations (biquadratic etc.); examination of the tridimensional incompressible equations; consideration of three- or multi-level time integration schemes (leap-frog, etc.).

n	$e(u_1)$	$e(u_2)$	$e(p)$
1	1.1×10^{-3}	1.2×10^{-3}	0.0217
3	1.9	2.1	0.0234
5	2.5	2.8	0.0242
7	3.3	3.2	0.0249
9	4.0	3.5	0.0253
20	5.8	3.9	0.0258

TABLE 2. FINITE DIFFERENCE SOLUTION TO THE ANALYTICAL TEST PROBLEM (CHORIN [9]). 40 x 40 MESH; $\Delta t = 0.00324$; Re= 20.

ACKNOWLEDGEMENTS

The authors are grateful to H. Laval for a critical reading of the manuscript and to Cornelis L. van den Muyzenberg who assisted in the design and development of the general FEM computer program used for the numerical tests. A copy of the program for performing the response analysis of the two-dimensional advection-diffusion equation can be obtained from the authors upon request.

One of the authors (K.Morgan) wishes to thank the S.E.R.C. for their financial support in the form of research grant GR/B/69906.

REFERENCES

1. A. ARAKAWA, J. Comput. Phys. 1, 119-143, 1966.

2. A. ARAKAWA and V.R. LAMB, in "Methods in Computational Physics" Vol.17 (J. Chang, Ed.), Academic Press, New York, 1977.

3. R.D. KRIEG and S.W. KEY, Int. J. Num. Meth. Engng. 7, 273-286, 1973.

4. P.M. GRESHO, R.L. LEE and R. SANI, in "Finite Elements in Fluids", Vol.3 (R.H. Gallagher et al., Eds.), John Wiley and Sons, Chichester, England, 1978.

5. J. DONEA, S. GIULIANI and H. LAVAL, in "Finite Element Methods for Convection Dominated Flows", (T.J.R. Hughes Ed.), ADM, Vol. 34, New York. 1979.

6. A. SHAPIRO and G.F. PINDER, J. Comput. Phys. 39, 46-71, 1981.

7. G.W. PLATZMAN, J. Comput. Phys. 40, 36-63, 1981.

8. M.J.P. CULLEN and K.W. MORTON, J. Comput. Phys. 34, 245-267, 1980.

9. A.J. CHORIN, Math. Comp. 22, 745-762, 1968.

10. A.J. CHORIN, Math. Comp. 23, 341-353, 1969.

11. R. TEMAM, "On the Theory and Numerical Analysis of the Navier-Stokes Equations", North-Holland, Amsterdam, 1977.

12. G. STRANG and G.J. FIX, "An analysis of the Finite Element Method", Prentice-Hall, Englewood Cliffs, New Jersey, 1973.

13. J. DONEA and S. GIULIANI, Int. J. Num. Meth. Fluids, 1, 63-79, 1981.

14. A.N. BROOKS, "A Petrov-Galerkin finite-element formulation for convection dominated flows", Ph.D. Thesis, California Institute of Technology, Pasadena, California, May 1981.

15. R.D. RICHTMYER and K.W. MORTON, "Difference Methods for Initial-Value Problems", 2nd Edition, Interscience Publishers, New York, 1967.

16. P.J. ROACHE, "Computational Fluid Dynamics", 2nd Edition, Hermosa Publishers, Albuquerque, 1976.

17. B.P. LEONARD, Appl. Math. Mod. 4, 401-403, 1980.

APPENDIX : SPATIAL DIFFERENCE OPERATORS

$$A_x\, u_{\ell m} \equiv \frac{1}{2}\,(u_{\ell+1,m} - u_{\ell-1,m}) = i\,\sin\xi,$$

$$A_y\, u_{\ell m} \equiv \frac{1}{2}\,(u_{\ell,m+1} - u_{\ell,m-1}) = i\,\sin\eta,$$

$$A_{xy}\, u_{\ell m} \equiv \frac{1}{2}\,(u_{\ell+1,m+1} - u_{\ell-1,m-1}) = i\,\sin(\xi+\eta),$$

$$A_{yx}\, u_{\ell m} \equiv \frac{1}{2}\,(u_{\ell+1,m-1} - u_{\ell-1,m+1}) = i\,\sin(\xi-\eta);$$

$$D_x\, u_{\ell m} \equiv u_{\ell+1,m} - 2u_{\ell m} + u_{\ell-1,m} = 2(\cos\xi-1),$$

$$D_y\, u_{\ell m} \equiv u_{\ell,m+1} - 2u_{\ell m} + u_{\ell,m-1} = 2(\cos\eta-1),$$

$$D_{xy}\, u_{\ell m} \equiv u_{\ell+1,m+1} - 2u_{\ell m} + u_{\ell-1,m-1} = 2[\cos(\xi+\eta)-1],$$

$$D_{yx}\, u_{\ell m} \equiv u_{\ell+1,m-1} - 2u_{\ell m} + u_{\ell-1,m+1} = 2[\cos(\xi-\eta)-1].$$

Section 2:

NAVIER-STOKES EQUATIONS, COMPRESSIBLE

THE NUMERICAL SOLUTION OF THE COMPRESSIBLE VISCOUS FLOW FIELD ABOUT A COMPLETE AIRCRAFT IN FLIGHT

R.W. MacCormack
Department of Aeronautics and Astronautics,
University of Washington, Seattle, WA., U.S.A.

SUMMARY

In 1979 D.R. Chapman [1] estimated the number of mesh points and computer memory words required to calculate the viscous flow field about a complete aircraft at flight Reynolds numbers. Within the next year or two the required computer capability will become available. The complex topological problems of fitting a mesh about an aircraft can be solved and the numerical procedures required for such a flow calculation can also be developed within the same time frame. This paper considers the problems to be solved and outlines some approaches toward their solution. Furthermore, it argues that the Reynolds-averaged Navier-Stokes equations can be solved about a complete aircraft at cruise conditions with the coming computer resources to the same degree of accuracy and cost as present high Reynolds number calculations for flows past relatively simple aerodynamic shapes.

1. INTRODUCTION

We have witnessed extraordinary progress in the field of computational fluid dynamics during the past two decades. We have progressed from linear theory for slender body flow calculations to nonlinear inviscid theory for flows about aircraft-like configurations [2-5]. During the last decade there has also been much activity in calculating three dimensional compressible viscous flow using the Reynolds-averaged Navier-Stokes equations with turbulence modeling. These calculations, for flows past relatively simple aerodynamic shapes, represent the present stage of development. They require only the space-time resolution of the gross turbulence effects and leave the representation of the remaining, though highly significant, turbulence effects to modeling. The computer storage and speed requirements of this stage are much less than those of the next, and perhaps, final stage that represents by both mesh and time step resolution all sizes of the significant energy bearing turbulent eddies. With the present and very near future advances in computer technology and numerical method development, we are now on the threshold of extending the Reynolds averaged calculations to full aircrat configurations at flight conditions. To pass over this threshold into the practical use of such calculations for aircraft design requires: (1) the solution of several topological problems in fitting a system of mesh points about a geometric shape as complex as an aircraft configuration, and (2) the development of numerical procedures with improved reliability and efficiency for solving the equations of compressible viscous flow. These two computational requirements will be discussed in this paper following dicussions on the present status of compressible viscous flow calculations and computer memory and speed requirements for three dimensional aircraft flow field calculations.

One final comment on turbulence modeling before continuing. Much progress is required before the presently available models will be able to account for the turbulence effects of strongly interacting flowfields with moderate or large amounts of separation. There is also no assurance that such capability will be forthcoming in the near or even distant future. However, present models can predict to engineering accuracy turbulent boundary layer interactions with little or no regions of separation. The development of the numerical procedures required to extend present three dimensional viscous flow calculations to complete aircraft simulations without waiting for further improvements in turbulence modeling is still a logical next step. These calculations can also be of engineering accuracy for flows near design cruise conditions and can be used to predict incipient separation, shock and vortex boundary layer interactions, buffet, reduced lift, and interference phenomena. Improvements in turbulence modeling will further extend the range of numerical simulation, eventually to aircraft in complex maneuver.

2. COMPUTER REQUIREMENTS

The shaded region in the lower left hand corner of Figure 1 [6] represents the present range of computational requirements for solving the Reynolds-averaged Navier Stokes equations in three dimensions. The region consists of data points each representing a single flow calculation in terms of the number of grid points and total computer time required to solve a given problem. Although the calculations were performed on several different computers, each computational time was converted to the equivalent time required if the problem were solved on a CDC 7600 computer.

The broken line passing through the center of the shaded region represents an estimate of the computer time required versus the number of mesh points used for a three dimensional Reynolds-averaged Navier-Stokes calculation. The slope of the line is one and implies that the required computer time varies linearly only with the number of mesh points used. This behavior was observed in full Navier-Stokes calculations [6] and should hold true also for the Reynolds-averaged calculations because present computer program execution times, using modern implicit methods, depend primarily on the number of mesh points used and not, as was formerly true, on such factors as the fineness of the mesh and magnitudes of the characteristic speeds and kinematic viscosities of the flow being calculated.

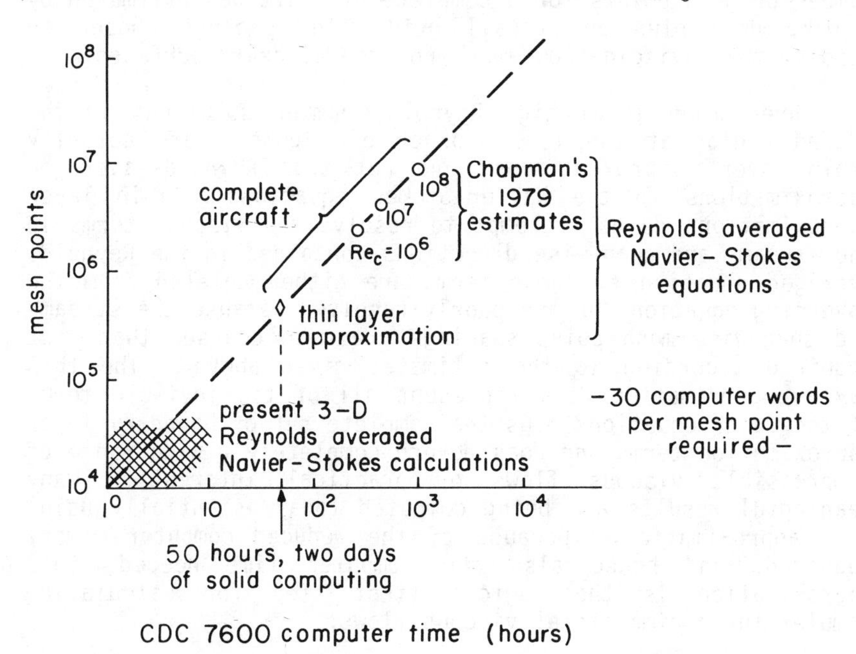

Figure 1: Computer memory and speed requirements for three-dimensional Reynolds-averaged Navier-Stokes calculations.

The three circular points plotted on the broken line of Figure 1 correspond to D.R. Chapman's 1979 estimates for the number of grid points required to calculate the flow about a complete aircraft (wing-body-tail and nacelles) at cruise Reynolds numbers of 10^6, 10^7 and 10^8 using the Reynolds averaged equations. The number of grid points corresponding to these three calculations are 2×10^6, 4×10^6, and 9×10^6 respectively. The total number of grid points in just the upper and lower surface boundary layer of a wing of chord c and aspect ratio

AR is given by $N_x \times N_y \times N_z$ where N_x, N_y and N_z are the number of points along the chord, across the boundary layer, and along the span directions, respectively, and are given by

$$N_x = 4.5\ Re_c^{0.2}\ , \quad N_y = 20\ , \quad \text{and} \quad N_z = 2.25\ AR\ Re_c^{0.2}$$

The estimates for N_x and N_z correspond to grid points spaced approximately one and two boundary layer thicknesses apart along the chord and span directions. This spacing should be sufficient to resolve the Reynolds-averaged Navier-Stokes equations with the effects of subgrid scale motions accounted for by turbulence modeling. Similar estimates were made for the fuselage, tail, nacelle, and pylon surfaces. The total number of grid points for a complete aircraft was estimated by a wake grid plus an overall grid with spacings chosen to resolve the inviscid flow away from the aircraft surfaces.

The three dimensional high Reynolds number solutions in the shaded region at the lower corner of Figure 1 are actually "thin layer" approximations, and not true Reynolds-averaged approximations to the Navier-Stokes equations. Thin layer approximations do not attempt to resolve the viscous terms in the stream- and span-wise directions contained in the Reynolds averaged equations. These terms are either deleted from the governing equations or are poorly resolved because the stream- and span-wise mesh point spacings are much coarser than that required according to the estimates given above. The thin layer approximation does represent all of the inviscid terms of the Euler equations plus the complete set of boundary layer approximation terms and does govern completely a wide range of compressible viscous flows of practical interest. Many meaningful results are being computed today essentially using this approximation. Because of the reduced computer memory required, and hence also the computer time needed, this approximation is the logical first step for calculating complex three-dimensional viscous flows.

For a wing at $Re_c = 10^6$ with AR = 4, Chapman's estimates correspond to an upper surface boundary layer grid of 71 points along the chord, 20 across the boundary layer, and 141 along the span, or 2×10^5 total grid points. For thin layer theory, using estimates by South and Thames [7], with 60 points along the chord, 20 across the boundary layer, and 40 across the span, a total of only 5×10^4 are required, or one fourth as many as are needed to solve the Reynolds-averaged full Navier-Stokes equations for a wing at $Re_c = 10^6$. Assuming that estimates similar to that for the wing can be made for the fuselage, tail, nacelle, and pylon surfaces, the total number of grid points needed to solve the Reynolds-averaged thin-layer Navier-Stokes equations is $N_{T.L.N.S.} = 5 \times 10^5$ grid points, or again approximately one fourth the number required to solve the Reynolds-averaged full Navier-Stokes equations for a complete aircraft at $Re_c = 10^6$. This data point is also plotted on the broken line of Figure 1 as the diamond symbol and indicates that in theory a solution for the flow about a complete aircraft configuration could be obtained in 50 hours,/two days of solid computing/, on a machine with the speed of a CDC 7600.

Because the mesh point spacings in the stream- and spanwise directions of the flow are chosen in the thin layer approximation independent of viscous effects and hence Re_c, and the number of points used across the boundary layer, 20 in the above example, is also chosen independent of Re_c, the total number of mesh points required for a given flow configuration does not depend on Reynolds number. Thus the diamond symbol representing the thin layer approximation in Figure 1 represents all calculations at high values of Re_c and should, according to boundary layer theory, actually improve as Re_c increases.

The number of words of computer memory required for solving the thin layer or complete Reynolds-averaged Navier-Stokes equations is estimated to be approximately 30 times the number of grid points. This factor includes storage for the dependent flow variables, metric coefficients, and turbulence quantities associated with each mesh point. Thus, in addition to the computing time shown in Figure 1, a thin-layer Reynolds-averaged Navier-Stokes calculation for the flow about an aircraft would require approximately 15 million words of computer memory.

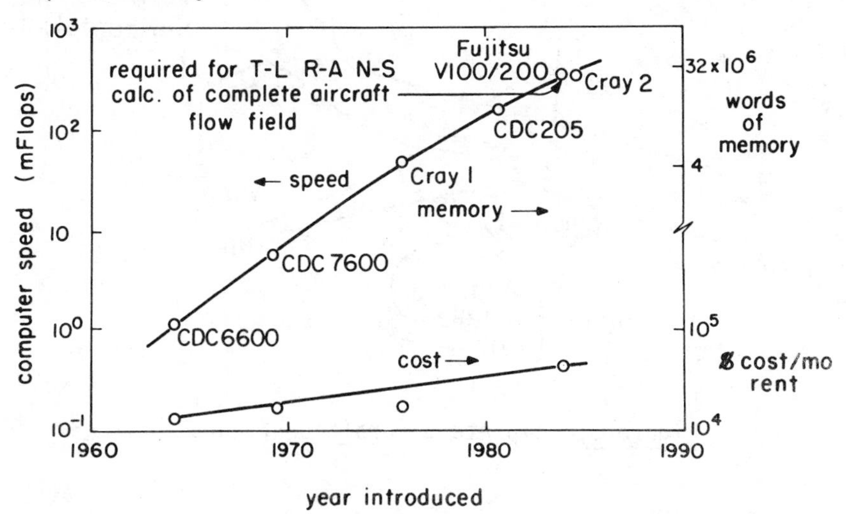

Figure 2: Computer speed, memory, and rental costs versus year introduced.

Figure 2 shows the speed, memory and cost of several scientific computers versus the year introduced or expected to be introduced. The computers include the CDC 6600, 7600, and 205, the Cray 1, and 2, and the Fujitsu V100/200. This figure shows that within the next two years computers with memories of 16 million words and more and with speeds two orders of magnitude and more faster than a CDC 7600 will be available at costs only approximately twice as much. Assuming thin layer Navier-Stokes calculations can make full use of the hardware potential of the coming computers, full aircraft calculations can be made in approximately the same time and at the same cost as many present research calculations past relatively simple aerodynamic shapes. However, before these calculations can be made we will first have to solve some difficult topological problems associated with nesting grids about aircraft component elements and interfacing each with its neighboring grids. Also before such calculations can become practical for aircraft design, the convergence of our present numerical methods must be accelerated to reduce total computation times and costs still further. The first of these subjects, a computational mesh topology for an aircraft will now be discussed. Improved reliability and convergence acceleration of numerical methods and the adaptation of numerical methods to forthcoming computer hardware, will be discussed subsequently.

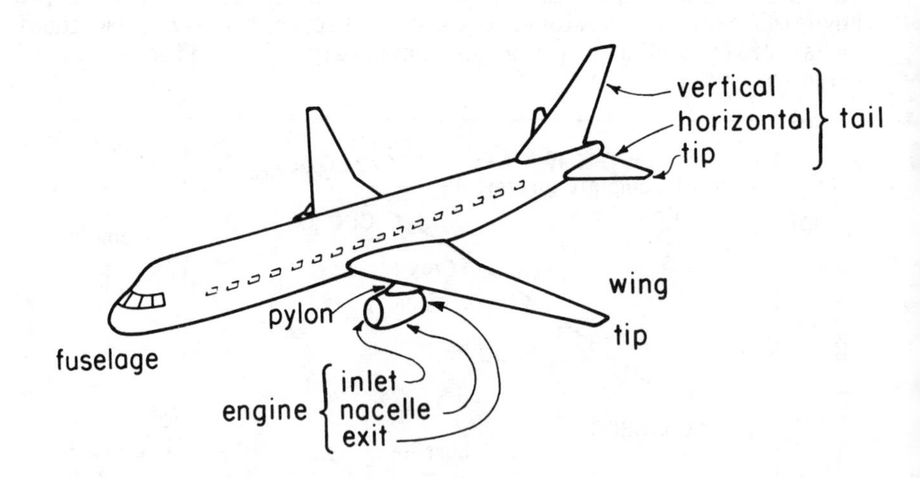

Figure 3: Complete aircraft configuration.

3. COMPUTATIONAL MESH

Figure 3 shows a typical aircraft configuration containing a fuselage, wings, tail, pylons and nacelles. Its geometry is complex and the nesting of a single body fitted mesh about it appears hopeless. The surface geometry includes several distinct shapes: a spherical cap at the fuselage nose, cylindrical sections along the fuselage and nacelles, and rectangular shaped regions along the wing and tail surfaces. Nevertheless, it is possible, if not fairly straightforward, to construct a system of simple meshes, each nested about an aircraft component surface, that completely descretizes the entire flow field surrounding an aircraft. Each component mesh when unwrapped is topologically equivalent to a cube so that the dependent flow variables and metric quantities can each be stored computationally as a regular three dimensional array, i.e. RHO(I,J,K), for I=1,2,3...,I_{MAX}, J=1,...,J_{MAX}, and K=1,...,K_{MAX}. Also each mesh can be aligned at the interface with its neighbor so that the boundary points of one are interior points of the other[1]. We will now design such a system containing nine component meshes for descretizing the flow about the aircraft shown in Figure 3. We will assume that:

1) the leading edges of the wing and stabilizer sections are blunt and their trailing edges are sharp,

2) the leading and trailing edges of the rudder and pylons are sharp, and

3) engine performance conditions can be supplied so that, together with the computed flow field near the engines, boundary conditions can be determined at the engine inlet and exit surfaces.

Alternate geometries could be considered, even to include the extension of flaps and other control surfaces, with the use of additional component meshes, but, as a start we will limit ourselves to the above assumptions.

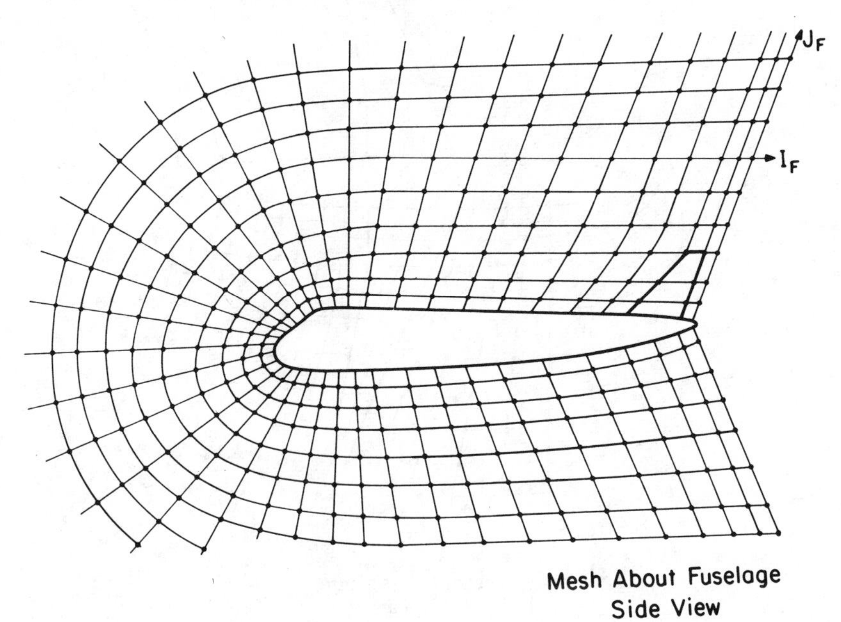

Figure 4a: Mesh about fuselage-side view.

[1]This requirement will be relaxed later.

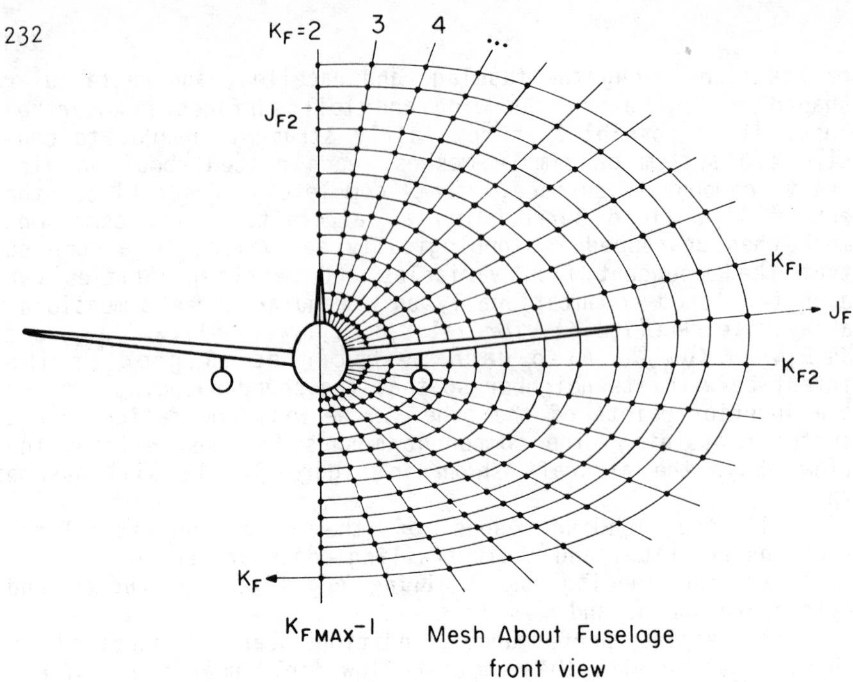

Figure 4b: Mesh about fuselage-front view.

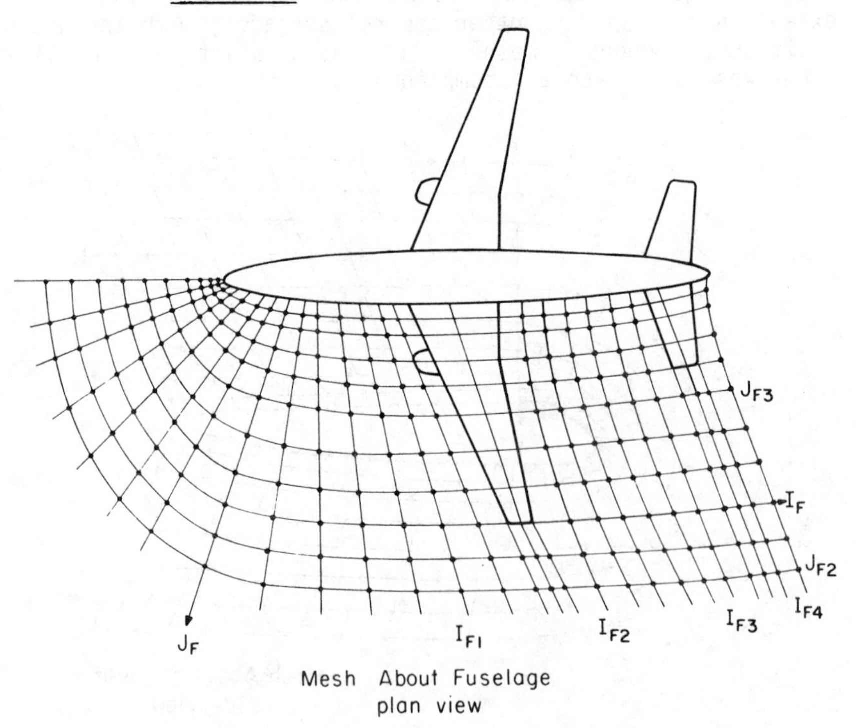

Figure 4c: Mesh about fuselage-plan view.

a. Fuselage mesh

A mesh for a blunt nosed fuselage and sharp edged rudder is sketched in Figure 4a-c. The mesh ordered by the indices I_F, J_F, and K_F is polar-like about the nose and cylindrical-like over the rest of the fuselage. Mesh surfaces of constant K_F are deformed meridional surfaces. The rudder surface lies on a small rectangular section of one such surface for which $K_F=2$. The aircraft geometry is assumed also to be symmetric so that we need only specify the mesh on one side of the aircraft plane of symmetry, from surfaces $K_F=2$ to $K_{FMAX}-1$. Surfaces $K_F=1$ and $K_F=K_{FMAX}$ are used as storage locations for implementing boundary conditions. Surfaces of constant I_F represent deformed conical surfaces about an axis passing through the aircraft nose and along the center of the fuselage. These surfaces are swept forward about the aircraft nose and backward in the mid fuselage and tail regions with sweep angles approximately equal to the leading and trailing edges of the wing and tail sections respectively. Surfaces of constant J_F wrap about the fuselage body, one of which for $J_F=2$ is the fuselage surface itself. The generation of the fuselage mesh requires the geometric specification of the fuselage and rudder surfaces plus the sweep angles of the wing and tail surface edges to be used as mesh constraints. Any one of several mesh generation procedures in use today [8-10] can then be used to numerically determine a set of mesh point coordinate locations (points of intersection of the I-J-K surfaces).

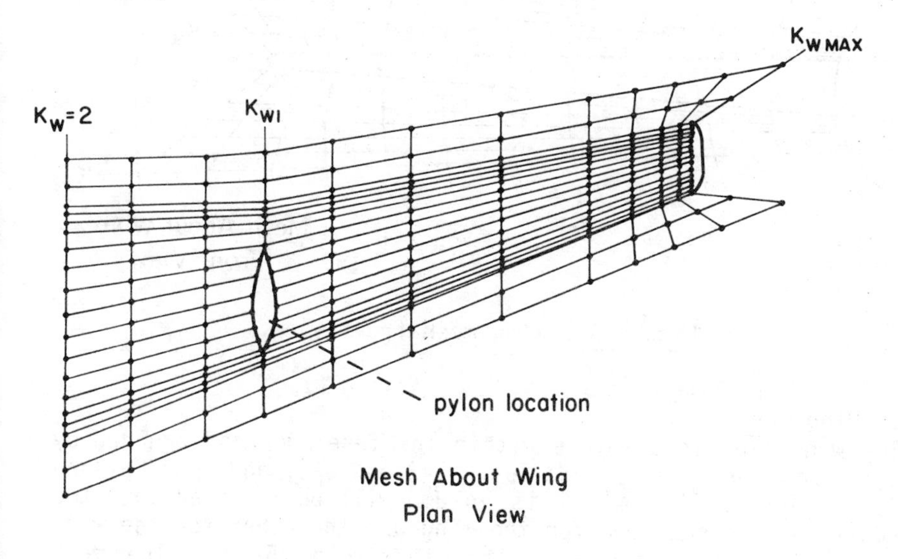

Figure 5a: Wing mesh-plan view.

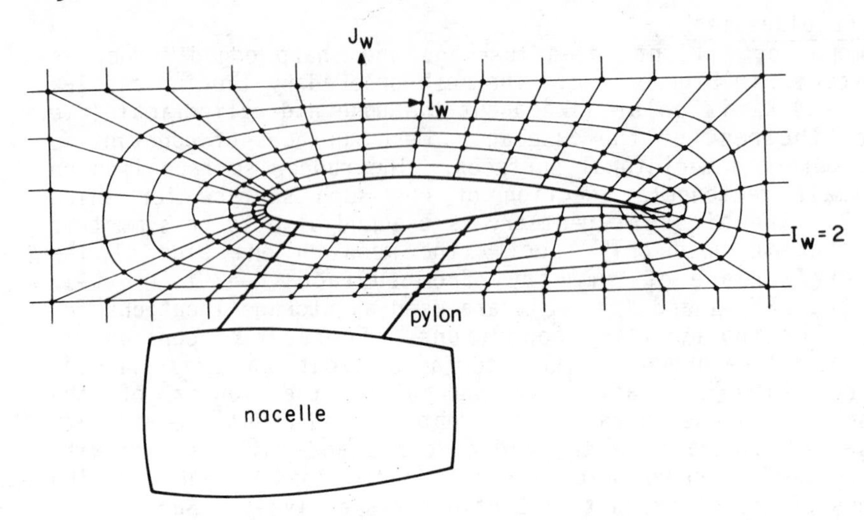

Figure 5b: Wing mesh-side view.

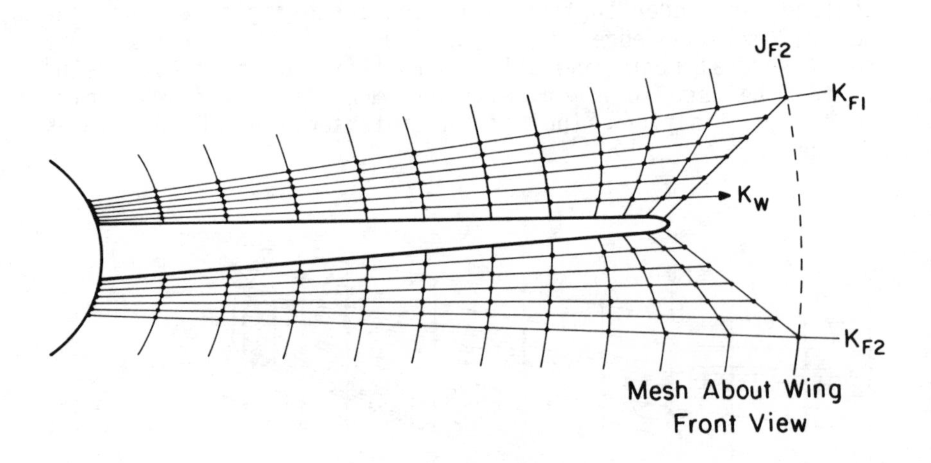

Figure 5c: Wing mesh-front view.

b. Wing mesh
The wing lies in a volume within the fuselage mesh bounded by the surfaces $I_F = I_{F1}$ and I_{F2}, $J_F=2$ and J_{F2}, and $K_F = K_{F1}$ and K_{F2} shown in Fig. 4. This volume will be remeshed into two additional meshes: one for the wing and the other for the wing tip. The wing mesh is sketched in Fig. 5a-c. The mesh surface indices are denoted by $I_W = 1,\ldots, I_{WMAX}$, $J_W=1,\ldots, J_{WMAX}$, and $K_W = 1,\ldots, K_{WMAX}$ with directions as indicated in Fig. 5. Surfaces of constant J_W wrap about the wing with $J_W=2$ coincident with the wing surface and $J_W = J_{WMAX}$ coincient with sections of the fuselage mesh surfaces $I_F = I_{F1}$ and I_{F2}, and $K_F = K_{F1}$ and K_{F2}. Surfaces of constant K_W are spanwise

surfaces with $K_W=2$ coincident with a section of the fuselage surface $J_F=2$ and $K_W = K_{WMAX}-1$ representing the outermost spanwise surface near the wing tip. The I_W and J_W surface intersections with a spanwise surface is shown in Figure 5b. The mesh wraps around both the leading and trailing edges with surfaces $I_W=2$ and $I_{WMAX}-1$ coincident. Note that the wing mesh is aligned with the fuselage-rudder mesh at each common interface. The wing mesh boundary points stored at $J_W = J_{WMAX}$ can be set from values obtained at interior points of the fuselage mesh and vice versa. The location of the pylon passing through the wing mesh is shown in Figure 5a and its treatment will be discussed later.

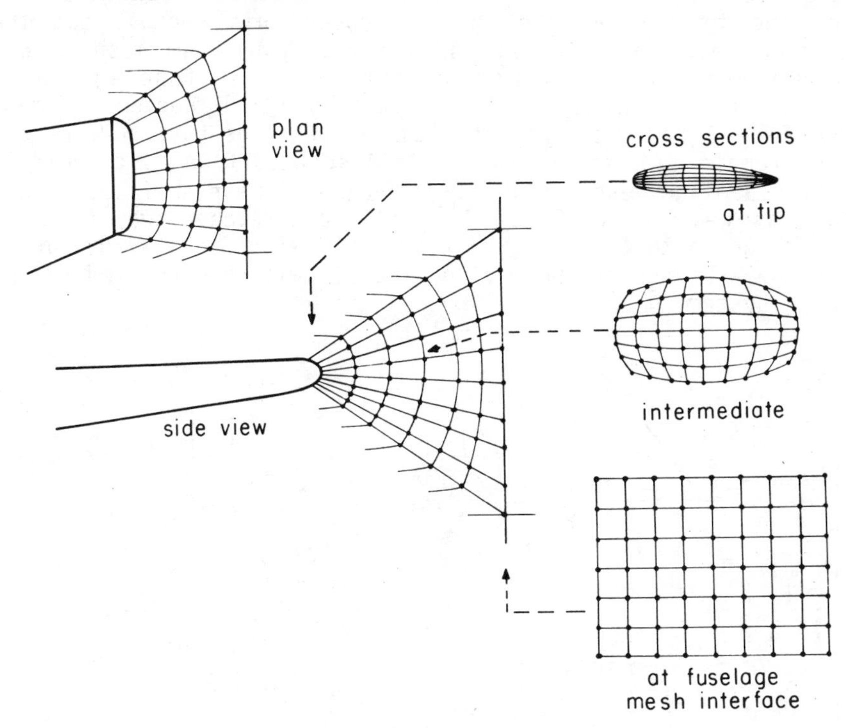

Figure 6: Wing tip mesh.

c. Wing tip mesh

The mesh about the wing tip is sketched in Figure 6. The essential purpose of this mesh is to link the cylindrical-like wing mesh surfaces of constant J_W at the wing tip to the rectangular-like cross section of the fuselage mesh on surface $J_F = J_{F2}$ as shown in Figure 6. Again this component mesh is topologically equivalent to a cube with boundary surfaces in regular alignment with both the fuselage and wing meshes at common interfaces.

d. Stabilizer mesh and stabilizer tip mesh

The stabilizer lies within the volume enclosed by the

following surfaces of the fuselage mesh $I_F = I_{F3}$ and I_{F4}, $J_F=2$ and J_{F3}, and $K_F = K_{F1}$ and K_{F2} shown in Figure 4. This volume and the stabilizer surface are topologically equivalent to that for the wing just treated, and in a similar procedure two meshes can be constructed, one for the stabilizer and the other for the stabilizer tip.

e. Downstream mesh

A separate cylindrical-like mesh can be constructed to discretize the flow field aft of the fuselage with indices I_D, J_D, and K_D directed like those of the fuselage mesh. The fuselage mesh itself could have been extended about the base of the tail as at the nose to cover this region, but the choice made here for an independent downstream mesh offers additional flexibility to fit various fuselage base geometries. In our case, a small central region of mesh surface $I_D=2$ is coincident with the fuselage base surface and the remaining outer region of this mesh surface is coincident with fuselage mesh surface $I_F = I_{FMAX}-1$. Mesh surfaces I_D and J_D, as well as surfaces J_F, will be stretched away from the fuselage so that the entire flow field significantly disturbed by the presence of the moving aircraft will be enclosed by the mesh system.

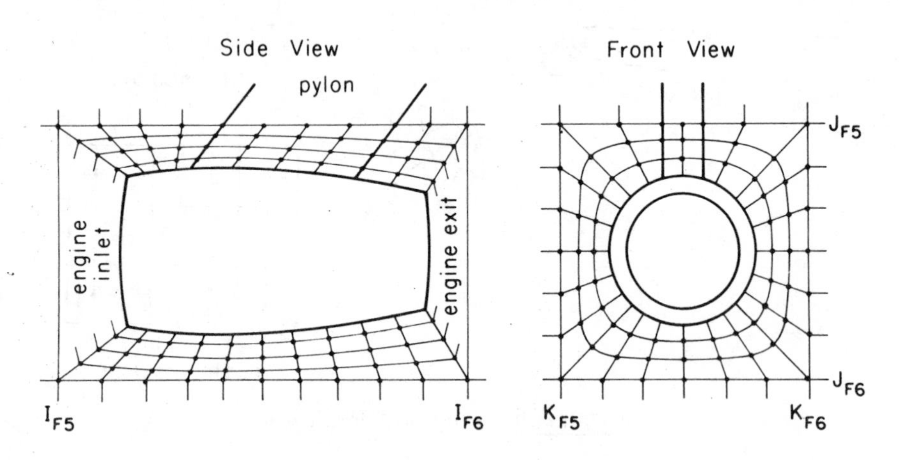

Figure 7: Pylon and nacelle mesh.

f. Pylon and nacelle mesh

The nacelle lies within the volume bounded by mesh surfaces I_{F5}, I_{F6}, J_{F5}, J_{F6}, K_{F5} and K_{F6} shown in Figure 7. The pylon passes through the wing mesh into this enclosed volume along surface K_{W1} (see Figure 5c). This surface within the wing mesh must be modified to represent the pylon thickness. This can be accomplished for the assumed sharp edged pylon by splitting mesh surface K_{W1} at the pylon into two surfaces, each wetting a side of the pylon. The mesh point coordinate locations for one side will be stored in computer memory with the rest of the points of surface K_{W1} and that for the other pylon side surface will be stored separately. During program execution an internal boundary condition procedure can be used to choose the proper mesh coordinates when needed in calculating the flow near the pylon. A more exact treatment of the flow about the pylon, required perhaps for a blunt leading edged pylon, would require an additional separate mesh constructed in an analogous manner to that for the wing.

The pylon-nacelle mesh within the bounded volume shown in Figure 7 is topologically similar to the fuselage mesh given earlier, although here the mesh wraps completely around the nacelle. Surfaces $K_{PN}=2$ and $K_{PNMAX}-1$ are coincident away from the pylon and split at the pylon to represent both the inboard and outboard pylon surfaces. Surfaces J_{PN} wrap about the nacelle with $J_{PN}=2$ coincident with the nacelle surface.

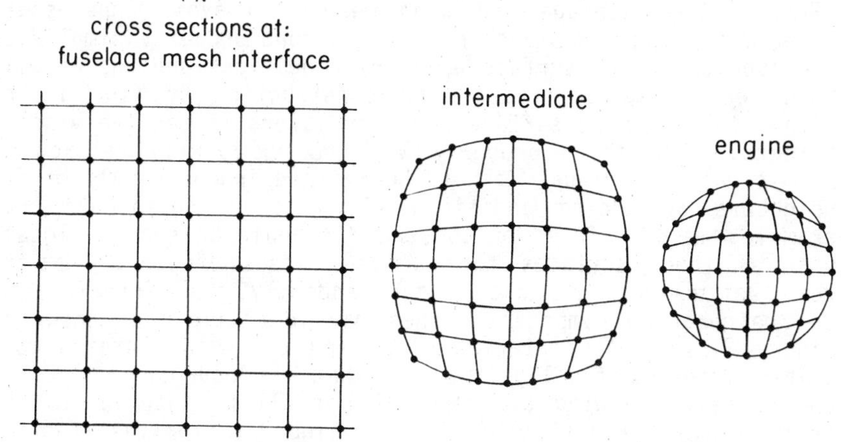

Figure 8: Engine inlet or exit mesh cross sections.

g. Engine inlet mesh and engine exit mesh

As before for the wing tip mesh we can construct an inlet mesh and an exit mesh to link the cylindrical-like nacelle-pylon mesh to the rectangular-like interfaces of the fuselage mesh as shown in Figure 8. Inlet mesh surfaces I_{EI} gradually deform from rectangular-like to circular-like surfaces at the engine entrance boundary. Exit mesh surfaces behave similarly. In addition to aircraft engine entrance and exit geometry, engine performance data will need to be specified at these mesh boundaries.

We have thus outlined the construction of a system of nine meshes to resolve the flow field about a complete aircraft configuration. Each component mesh is no more complicated than those used today to calculate three dimensional flows and standard numerical procedures [8-10] can be used to generate them. The mesh system has three basic properites:

1) each component mesh is topologically equivalent to a cube,
2) each aircraft component surface is coincident with a mesh surface,

and

3) each component mesh is aligned with each of its neighbors at common interfaces so that the boundary points of one can be interior points of the other.

The first property allows each dependent variable to be stored in a regular rectangular-like three dimensional array. This property eliminates the costly logic of index testing and program branching required for irregularly shaped meshes. Even the treatment, to be discussed later, of the "holes" in the fuselage mesh containing the wing, stabilizer, pylon and nacelle and the double-sided wing mesh surface K_{W1} can be done external to the computational loops that determine the flow within the fuselage and wing meshes to avoid unnecessary program testing and branching. The second property simplifies the application of surface boundary conditions and, as we will soon see, also facilitates the construction of fine meshes within the aircraft surface boundary layers to resolve viscous effects. The third property will now be partly relaxed to permit each individual mesh to be refined independently of its neighbors by adding additional I, J or K surfaces. For example, mesh surfaces of constant J should be spaced closely near all body surfaces to resolve viscous effects. If the mesh refinements are chosen independently, the interface at two neighboring component meshes may look like that shown in Figure 9. At this interface the wing tip mesh contains more points than the fuselage mesh so that the boundary points of one, namely the wing tip mesh, cannot all be interior points of the other. Additional boundary value data for the wing tip mesh can be obtained by interpolation of fuselage mesh point values. The interpolation procedure will be fairly simple because the two meshes are still in alignment at their interface.

Table 1 gives estimates of the dimensions of the nine meshes forming a system of nearly half a million grid points for solving a thin layer approximation of the Reynolds averaged Navier-Stokes equations about an aircraft in flight. This system provides better flow resolution than that generally in use today to calculate three-dimensional high Reynolds number flow fields about relatively simple aerodynamic shapes. For a symmetric flow about an aircraft the entire flow field can be obtained by using just these nine meshes. For an asymmetric flow, including the effect of yaw, two such calculations would need to be performed together and coupled along the aircraft plane of symmetry. Both halves of an asymmetric calculation need not reside in computer memory simultaneously nor is a second "image" copy of the matrices of the nine mesh system required unless the airplane geometry is also asymmetric as would occur during lateral maneuver.

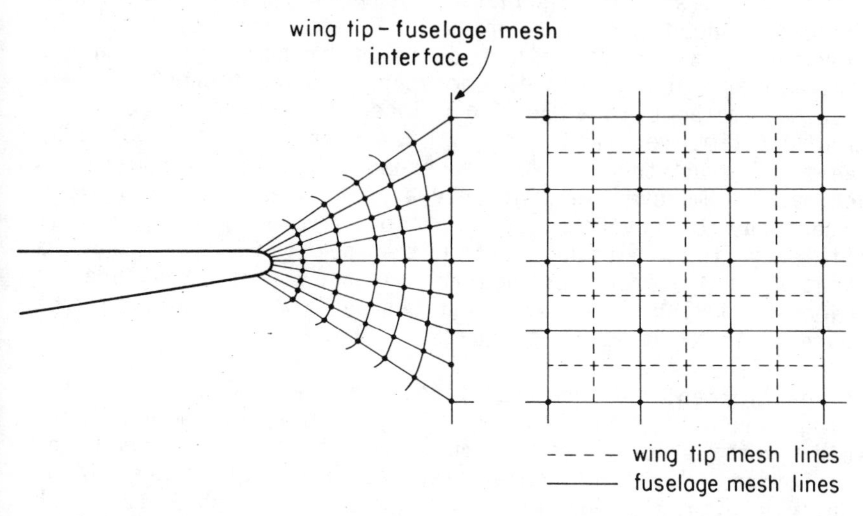

Figure 9: Independent mesh refinement and wing tip-fuselage mesh interface.

MESH	I_{max}	J_{max}	K_{max}	No. Points
1. Fuselage	150	40	40	240,000
2. Wing	120	20	40	96,000
3. Wing-Tip	30	20	20	12,000
4. Horizontal Tail	60	20	30	36,000
5. Horizontal Tail-Tip	20	20	10	4,000
6. Downstream-AFT of Fuselage	40	40	40	64,000
7. Nacelle-Pylon	30	20	40	24,000
8. Engine Inlet	10	10	10	1,000
9. Engine Exit	10	10	10	1,000

Total mesh points 478,000
Total words of mem. ~ 16×10^6

TABLE 1: Estimate of component mesh dimensions for thin-layer Reynolds-averaged Navier-Stokes approximation.

4. NUMERICAL METHODS

During the last two decades we have witnessed remarkable progress, indeed a revolution, in our capability to solve the governing equations of fluid dynamics by numerical methods. Complementary to advances in computer hardware technology, new numerical methods have been developed that have reduced the computing time required formerly to solve a given problem by orders of magnitude. As we expect computer hardware to continue to improve, we similarly expect numerical methodology to continue to advance. A five fold increase in numerical efficiency is predicted during the next five years and a potential improvement by nearly two orders of magnitude is projected during the next fifteen years for solving the compressible Navier-Stokes equations [6].

The development of noniterative implicit methods [11-13] during the last decade has been a key element in computational fluid dynamics progress. Implicit methods, free from the restrictions limiting time step advances of conventional stability criteria for explicit methods, are able to march their solutions through time rapidly. This new capability has been used to solve more difficult problems, particularly those of high Reynolds number viscous flow. Mesh points can now be spaced extremely close together near body surfaces to resolve thin separated boundary layers without restricting the time step size. Time step size is theoretically only limited by accuracy considerations, temporal truncation error, and if permitted is chosen by how far we wish the mean flow to travel during each time step frame so that a complete history of an evolving flow field can be obtained.

Let the following equation represent the equations governing unsteady compressible viscous flow:

$$\frac{\partial U}{\partial t} = -\frac{\partial F}{\partial x} - \frac{\partial G}{\partial y} - \frac{\partial H}{\partial z} \tag{1}$$

where

$$U = \begin{pmatrix} \rho \\ \rho u \\ \rho v \\ \rho w \\ e \end{pmatrix} \tag{2}$$

$$F = \begin{pmatrix} \rho u \\ \rho u^2 + \sigma_x \\ \rho v u + \tau_{xy} \\ \rho w u + \tau_{xz} \\ (e+\sigma_x)u + \tau_{xy}v + \tau_{xz}w - k\,\dfrac{\partial T}{\partial x} \end{pmatrix} \tag{3}$$

$$\sigma_x = p - \lambda(\frac{\partial u}{\partial x} + \frac{\partial v}{\partial y} + \frac{\partial w}{\partial z}) - 2\mu \frac{\partial u}{\partial x} \tag{4}$$

$$\tau_{xy} = -\mu(\frac{\partial u}{\partial y} + \frac{\partial v}{\partial x}) \tag{5}$$

$$\tau_{xz} = -\mu(\frac{\partial u}{\partial z} + \frac{\partial w}{\partial x}) \tag{6}$$

etc..... .

The modern implicit methods applied to the above equation have the following form in general

$$\{I + \Delta t \frac{\Delta A\cdot}{\Delta x}\}\{I + \Delta t \frac{\Delta B\cdot}{\Delta y}\}\{I + \Delta t \frac{\Delta C\cdot}{\Delta z}\}\delta U_{i,j,k} = \Delta U_{i,j,k} \tag{7}$$

where $\Delta U_{i,j,k} = -\Delta t(\frac{\Delta F}{\Delta x} + \frac{\Delta G}{\Delta y} + \frac{\Delta H}{\Delta z})^n_{i,j,k}$ (8)

$\delta U_{i,j,k}$ represents the solution change, $\Delta t \frac{\partial U}{\partial t}$, and A,B, and C are related to the Jacobian matrices of F, G and H with respect to U. The dots following the matrices A,B and C in the above equation indicate that the difference operators with respect to x,y and z apply also to all factors to the right. The subscripts represent mesh point locations and the superscript represents time $t = n\,\Delta t$. The quantity $\Delta U_{i,j,k}$ is an explicit numerical approximation to the right hand side of the governing equation at each mesh point at time $t = n\,\Delta t$. If Δt is small this explicitly calculated change in the solution is a fair approximation. However, if Δt is large this change is valid only locally during the initial part of the time step interval and global information is required to calculate a fair approximation to the solution change over the remainder of the time step. The factors contained in brackets on the left hand side of the numerical approximation to the equation are implicit operators that bring in the additional information required for a good approximation to $\delta U = \Delta t \partial U/\partial t$ over the entire interval. The new updated solution is obtained by first inverting the bracketed factors to determine the solution change δU at each grid point and then by adding it to the present solution.

$$U^{n+1}_{i,j,k} = U^n_{i,j,k} + \delta U_{i,j,k} \tag{9}$$

The implicit factors are in general either block tridiagonal or block bidiagonal matrix operators and represent an approximate factorization of the following operator:

$$\left\{I + \Delta t \frac{\Delta A \cdot}{\Delta x} + \Delta t \frac{\Delta B \cdot}{\Delta y} + \Delta t \frac{\Delta C \cdot}{\Delta z}\right\} \quad (10)$$

Although the above operator is fairly difficult to invert, its factored approximation consisting of a sequence of bi-or tri-diagonal matrix operators is fairly simple.

Although the modern implicit methods for Navier-Stokes calculations have reduced computing times by several orders of magnitude, thousands of time steps are still required to calculate many flows of practical interest. Many flow problems that were intractable earlier are now possible to solve in research studies, but their solution largely remains impractical for design studies. Time step size, unfortunately, remains severely limited, if not by linear stability theory, by practical experience. Large time step sizes produce inaccuracies during the transient phase of the calculation, which can perturb the solution beyond the bounds of linear analysis and cause negative nonphysical densities, internal energies, or pressures and eventual program blow up. Attempts to control perturbations by adding numerical dissipation, "smoothing", terms usually mask the cause of the inaccuracies, destroy precision, and seldom lead to a practical "robust" numerical procedure. Nevertheless, improvements in numerical procedures are being made that promise to significantly reduce the number of time step iterations required to solve compressible viscous flow problems. One approach is to eliminate the sources of inaccuracies that lead to the instability of present methods. The sources of inaccuracy can be found in the finite difference approximation of the terms of the governing equations, the approximate factorization of the implicit matrix operator, and the treatment of boundary conditions. Another approach is the development of convergence acceleration procedures. The next four subsections discuss these approaches.

4.1 Flux splitting

Interest in flux splitting procedures has grown rapidly in recent years following the pioneering work of Steger and Warming [14]. Like Method of Characteristic procedures and the more recent Lambda scheme of Moretti [15], flux split procedures take into consideration characteristic directions of information travel and thereby can approximate the physics of the governing equations more realistically.

To illustrate flux splitting, consider the approximation of the right hand side of Eq. (1) given in Eq. (8). To focus our attention further, let's consider the approximation of the term $\partial F/\partial X$ represented by $\Delta F^n/\Delta x_{i,j,k}$. The flux F contains both inviscid and viscous terms. Let

$$F = F' + F_v \tag{11}$$

where

$$F' = \begin{pmatrix} \rho u \\ \rho u^2 + p \\ \rho vu \\ \rho wu \\ (e+p)u \end{pmatrix} \tag{12}$$

and F_v contains the remaining viscous terms of F. Focusing our attention still further, let's consider the approximation to $\partial F'/\partial X$. If this term is approximated, for example, as in the explicit predictor-corrector method of MacCormack [16], it could be given in the predictor step as a forward difference

$$\frac{\Delta F'}{\Delta x_{i,j,k}} = \frac{F'_{i+1,j,k} - F'_{i,j,k}}{\Delta x} \tag{13}$$

and in the corrector step as a backward difference

$$\frac{\Delta F'}{\Delta x_{i,j,k}} = \frac{F'_{i,j,k} - F'_{i-1,j,k}}{\Delta x} \tag{14}$$

The flux vector F' can be written as [14]

$$F' = A'U \tag{15}$$

where A' is the Jacobian of F' with respect to U. The Jacobian A' can be diagonalized by a similarity transformation S as follows

$$A' = S^{-1}\Lambda S, \tag{16}$$

where

$$\Lambda = \begin{pmatrix} u & 0 & 0 & 0 & 0 \\ 0 & u+c & 0 & 0 & 0 \\ 0 & 0 & u & 0 & 0 \\ 0 & 0 & 0 & u & 0 \\ 0 & 0 & 0 & 0 & u-c \end{pmatrix} \tag{17}$$

c is the speed of sound, and the diagonal elements of Λ represent characteristic signal speeds. Let Λ_+ and Λ_- be diagonal matrices containing, respectively, the positive and negative elements of Λ. We can now split the inviscid flux vector F' into two parts, $F'_+ = S^{-1}\Lambda_+ SU$ which carries information to the right (increasing X direction) and $F'_- = S^{-1}\Lambda_- SU$ which carries information to the left (decreasing x direction) according to the signs of the characteristic signal speeds. Instead of the approximations given earlier for $\partial F'/\partial x$, a more physically realistic approximation is

$$\frac{\Delta F'}{\Delta x}_{i,j,k} = \frac{F'_{+\,i,j,k} - F'_{+\,i-1,j,k}}{\Delta x} + \frac{F'_{-\,i+1,j,k} - F'_{-\,i,j,k}}{\Delta x} , \tag{18}$$

in which information traveling to the right is backward differenced and information traveling to the left is forward differenced. The numerical domains of dependence in the above approximation more closely match the physical domains of dependence and this can eliminate a significant source of inaccuracy.

It is worthwhile to note that the difference between the approximation given above and that given earlier for the predictor step is a term equal to

$$\frac{F'_{+\,i+1,j,k} - 2F'_{+\,i,j,k} + F'_{+\,i-1,j,k}}{\Delta x} \tag{19}$$

Similarly, the difference, for the corrector step is a term equal to

$$\frac{F'_{-\,i+1,j,k} - 2F'_{-\,i,j,k} + F'_{-\,i-1,j,k}}{\Delta x} \tag{20}$$

Both terms have the appearance of, and in fact are, dissipative terms and can be thought of as representing an intelligent way to add dissipation to the explicit predictor-corrector method when needed to control instabilities caused by insufficient consideration of physical domains of dependence.

The above example of flux splitting is only one of many possible procedures. There is a raft of papers currently appearing in the literature using flux split procedures and demonstrating surprisingly precise results.

4.2 Approximate Factorization

According to linear theory, the error introduced by the approximate factorization of Eq. (10) represented by the left hand side of Eq. (7) is of the order of:

$$\left\{\Delta t^2 AB \frac{\Delta^2 \bullet}{\Delta x \Delta y} + \Delta t^2 AC \frac{\Delta^2 \bullet}{\Delta x \Delta z} + \Delta t^2 BC \frac{\Delta^2 \bullet}{\Delta y \Delta z}\right\} \delta U \tag{21}$$

The coefficients in the term above are each proportional to the product of the CFL numbers in two spacial directions. As Δt increases, and hence the CFL numbers increase, the magnitude of the approximate factorization error during the transient phase of the calculation can increase to the point of introducing unacceptable error into the flow field simulation. Though not generally acknowledged, this is probably the primary reason why high Reynolds number flow calculations require thousands of time steps for completion.

We now introduce a different factorization that should significantly reduce error and still retain simple inversion properties. First, let's examine the term $\Delta t \frac{\Delta(A\delta U)}{\Delta x}$ which we rewrite as $\frac{\Delta t}{S_x} \frac{\Delta(AS_x\delta U)}{\Delta x}$ where $S_x = \Delta y \Delta z$. The term represents the flux of δU passing into and out of a finite volume of size $\Delta vol = S_x \Delta x = \Delta x \Delta y \Delta z$ through two sides of area $S_x = \Delta y \Delta z$ with outer unit normals in the negative and positive x directions. Similarly rewriting the other terms, we obtain for the implicit operator

$$\left\{ I + \Delta t \frac{\Delta AS_x \cdot}{S_x \Delta x} + \Delta t \frac{\Delta BS_y \cdot}{S_y \Delta y} + \Delta t \frac{\Delta CS_z \cdot}{S_z \Delta z} \right\} \qquad (22)$$

where $S_y = \Delta x \Delta z$ and $S_z = \Delta x \Delta y$.

We now define

$$\vec{P} = A\vec{i}_x + B\vec{i}_y + C\vec{i}_z \qquad (23)$$

$$\vec{S} = S_x\vec{i}_x + S_y\vec{i}_y + S_z\vec{i}_z \qquad (24)$$

and

$$Q = \vec{P}\cdot\vec{S} = AS_x + BS_y + CS_z \qquad (25)$$

where $\vec{i}_x$, $\vec{i}_y$ and $\vec{i}_z$ are unit vectors in the x,y and z directions. We now write the implicit operator as follows

$$\left\{ I + \frac{\Delta t}{\Delta vol} \Delta Q \cdot \right\} \qquad (26)$$

where $\Delta Q \delta U$ represents the flux of δU passing into and out of volume $\Delta vol = \Delta x \Delta y \Delta z$ through sides

$$\vec{S}_1 = S_x\vec{i}_x + S_y\vec{i}_y + S_z\vec{i}_z \qquad (27)$$

and

$$\vec{S}_2 = -S_x\vec{i}_x - S_y\vec{i}_y - S_z\vec{i}_z. \qquad (28)$$

Side $\vec{S}_1$ is composed of three orthogonally intersecting surfaces partly enclosing volume Δvol with outer unit normals in x,y, and z directions. Side $\vec{S}_2$ contains the opposite sides of the volume to $\vec{S}_1$ and together they completely enclose volume Δvol. The choices of $\vec{S}_1$ and $\vec{S}_2$ are not unique. There are two other pairs of surfaces,

$$\vec{S}_3 = S_x\vec{i}_x - S_y\vec{i}_y + S_z\vec{i}_z \tag{29}$$

and

$$\vec{S}_4 = -S_x\vec{i}_x + S_y\vec{i}_y - S_z\vec{i}_z \tag{30}$$

and also

$$\vec{S}_5 = S_x\vec{i}_x + S_y\vec{i}_y - S_z\vec{i}_z \tag{31}$$

and

$$\vec{S}_6 = -S_x\vec{i}_x - S_y\vec{i}_y + S_z\vec{i}_z \tag{32}$$

The matrix Q contains both inviscid and viscous elements.

Let

$$Q = Q' + Q_v \tag{33}$$

where Q' contains the inviscid elements and Q_v contains the remaining viscous elements. As was done earlier for the matrix A', the matrix Q' can be split into the "positive" and "negative" parts according to the signs of its eigenvalues, characteristic signal speeds, relative to surface $\vec{S}$. The eigenvalues of Q' are u', $u'+c'$, u', u', and $u'-c'$ where $u' = uS_x + vS_y + wS_z$ and $c' = c\sqrt{S_x^2 + S_y^2 + S_z^2}$. Let T represent the similarity transformation that diagonalizes Q'

$$Q'_+ = T^{-1}\Lambda'_+T \tag{34}$$

and

$$Q'_- = T^{-1}\Lambda'_-T \tag{35}$$

where Λ'_+ and Λ'_- contain respectively the positive and negative elements of Λ'.

The matrix Q_v represents viscous diffusion across surface $\vec{S}$ and can be represented by a block tridiagonal matrix. The exact representation of Q_v is fairly complex. Fortunately, for most high Reynolds number flow calculations the matrix norm of Q_v is far smaller than that of Q' and we may take some liberties with its representation. To simplify its numerical treatment, and hopefully not introduce significant new errors, we approximate Q_v with matrix Q'_v defined as follows[13]

$$Q'_v = \frac{\nu}{\rho} I \frac{\Delta\bullet}{\Delta\eta} \tag{37}$$

where

$$\nu = \max\{2\mu + \lambda, \mu, k/c_v\}, \tag{38}$$

$$\frac{1}{\Delta\eta} = \frac{\|\vec{S}\|}{\Delta vol} = \sqrt{\frac{1}{\Delta x^2} + \frac{1}{\Delta y^2} + \frac{1}{\Delta z^2}}, \tag{39}$$

and I is the identity matrix. We now define I_+ as a matrix with "ones" in the same positions as the nonzero elements of matrix $\Lambda_+^{'}$ and similarly I_- with "ones" in the same positions as the nonzero elements of matrix $\Lambda_-^{'}$, so that

$$I = I_+ + I_- \tag{40}$$

Putting the inviscid and viscous approximations together we define

$$Q_+ = Q_+^{'} + T^{-1}I_+T \frac{\nu}{\rho} \frac{\Delta \cdot}{\Delta \eta} \tag{41}$$

and

$$Q_- = Q_-^{'} + T^{-1}I_-T \frac{\nu}{\rho} \frac{\Delta \cdot}{\Delta \eta} \tag{42}$$

We now propose as an approximate factorization instead of the left hand side of Eq (7) the following

$$\left\{I + \frac{\Delta t}{\Delta vol} \Delta Q_+ \cdot\right\} \left\{I + \frac{\Delta t}{\Delta vol} \Delta Q_- \cdot\right\} \tag{43}$$

According to linear theory, the error introduced by the above factorization is of the order $\Delta t^2/\Delta vol^2\ Q_+Q_-\Delta^2\delta U$. But Q_+Q_- is null and hence the above error term has zero norm.

Each of the above bracketed factors represents a block tridiagonal matrix operator which can be inverted fairly easily. Each consists of an inviscid part that sweeps information along characteristic paths and uses one sided difference (bidiagonal) operators and a viscous part that uses central difference (tridiagonal) operators. Because each of the above two factors is tridiagonal, it appears that four sweeps through the flow field data arrays, a forward elimination and a backward substitution for each operator, are required. However, because each factor is operating separately or independently of the other (i.e. $Q_+Q_- = 0$), the two forward eliminations can be performed together on the same sweep and, similarly, the two backward substitutions can be performed together on a single back sweep through the data, for a total of only two sweeps per time step. Finally, to avoid a directional preference, each pair of sides, $\vec{S}_1$ and $\vec{S}_2$, $\vec{S}_3$ and $\vec{S}_4$, and $\vec{S}_5$ and $\vec{S}_6$, should be used sequentially to define Q.

4.3 Boundary conditions

The importance of implicit boundary conditions is clearly recognized and excellent contributions have been reported [17]. For a complete aircraft calculation using the mesh system outlined earlier there are three types of boundaries located at the outer edge of the flow field, at intermesh surfaces, and at aircraft surfaces.

The farfield boundaries are the easiest of the three to treat. They should be located outside the flow region significantly disturbed by the presence of the aircraft. The boundary values can be held fixed or nonreflecting conditions can be imposed if there is a chance that signals can reflect and return to falsely disturb the aircraft flowfield.

A finite volume method can be used to update the interior of the flow field on the non-orthogonal aircraft mesh system. Each finite volume is bounded by six sides and represents a deformed cube. Fluxes are determined at each side and the rate change of mass, momentum, and energy within each finite volume of the mesh system is determined by the net flux transferred across the bounding sides. Conservation can be strictly observed by requiring that the same flux leaving a finite volume is received by a neighbor through their common side. The fluxes are transported by both the explicit (right hand side of Eq. (7)) and implicit (left hand side of Eq. (7)) parts of the calculation. The implicit updating of the interior during each time step should proceed by sweeping from the outer flow field boundary toward the aircraft body. The second sweep of the two sweep implicit procedure outlined in the preceeding subsection should then begin at the body and sweep back towards the far field boundary.

At intermesh boundaries, the flux exiting one component mesh should be received by the corresponding finite volumes lying across the interface. For the mesh interface illustrated in Figure 9 the flux leaving each finite volume of the fuselage mesh should be partitioned into the four receiving finite volumes in the wing tip mesh. Similarly, the fluxes from sets of four finite volumes of the wing tip mesh should be summed and received by the corresponding finite volume of the fuselage mesh.

At aircraft body surfaces the momentum and energy fluxes carried on incoming characteristics during the first sweep through the data at each time step are reflected back out on outgoing characteristics during the second sweep. Characteristic boundary conditions are natural to implement for the flux split and approximate factorization procedures presented earlier. At engine inlet and exit boundary surfaces characteristic boundary condition procedures can also be applied with part of the data being supplied by engine performance specifications.

4.4 Convergence acceleration

An initial condition for a flow field is required to start the calculation off. For simplicity let's assume that at time t=0 a complete aircraft with engines running is suddenly immersed in a uniform flow field. Strong disturbances created at the aircraft body surfaces will then propagate away from the body to signal the surrounding flow to adjust itself to the aircraft's presence. Eventually the region of significantly disturbed flow may extend many body lengths away. The computational mesh, to span this region and still provide fine resolution near the body, will need to be highly stretched.

A choice must be made on the size of the time step that the solution is advanced during each step. In earlier years, when explicit methods prevailed, the choice was limited to the shortest signal propagation time between any two points of the mesh. In our case this would be

$$\Delta t_{explicit} \simeq \frac{\Delta y_{min}}{c} \tag{44}$$

where Δy_{min} is the distance between the two closest points in the boundary layer and c is the signal propagation speed. For information traveling at this pace, approximately Δy_{min} per time step, to reach distances several body lengths away would be highly impractical. Present implicit methods, free of the severe stability conditions of explicit methods, can take time steps orders of magnitude larger.

$$\Delta t_{implicit} \simeq \frac{\delta}{c} \tag{45}$$

where δ represents a small characteristic length. Though many times larger than Δy_{min}, δ is typically of the order of the boundary layer thickness, but, preferably if permitted, is chosen of the order of a few percent of the body length. This choice permits good temporal resolution of the flow near the body, but, unfortunately still represents a small pace for information traveling in the highly stretched mesh regions of the far field. Many time steps would be required for information to travel even between nearest neighbors in the far field. The requirements of good temporal resolution near the body and practical overall flow field convergence times appear at odds with one another. However, the pioneering research of Brandt [18] and more recently of Ni [19], Johnson [20], and Denton [21] on multi-grid procedures offer a solution that promises to accelerate convergence and yet retain near field resolution.

The multigrid procedure presented by Denton for finite volume methods could be applied to an aircraft flow field calculation. Although Denton has reported only its application to explicit methods, the present author [22] has used it successfully with an implicit finite volume method. Basically the procedure is as follows: An implicit finite volume method, as described earlier, calculates the solution change for every finite volume in the mesh system surrounding the aircraft. During this calculation every inviscid and viscous flux term of the governing equations needs to be computed at every finite volume surface. A coarse mesh is then formed by deleting a set of mesh surfaces from the original mesh system. For example, if every other surface is removed, the three dimensional mesh would contain only one eighth as many Finite volumes, each composed of eight finite volumes of the original mesh. For an implicit calculation with the time step chosen such that $c\Delta t \approx \delta$, many more surfaces than just every other one need to be removed in the boundary layer regions of the mesh where the surface spacing is much less than δ. Enough volumes need to be joined together so that the new coarse mesh finite volumes have linear dimensions of at least 2δ, creating volumes greater than $8\delta^3$ each. The time step for a coarse mesh calculation can be chosen as large as $2\Delta t$. Instead of calculating each inviscid and viscous flux term as before, all that is required to determine the coarse mesh solution change is a volume weighted averaging of the solution changes already calculated on the original mesh.

$$\delta U_{i^{(2)},j^{(2)},k^{(2)}} = 2 \frac{\Sigma\, \Delta vol_{i,j,k}\ \delta U_{i,j,k}}{\Sigma\, \Delta vol_{i,j,k}} \tag{46}$$

where the summation extends over all original finite volumes indexed i,j,k now contained in coarse mesh finite volume indexed $i^{(2)}$, $j^{(2)}$, $k^{(2)}$. Though not apparent, the volume weighting and the strong conservation properties of the finite volume method combine to preserve strong conservation form in the coarse mesh calculation. The solution change given above actually represents the net transport of flux across the surfaces bounding each new coarse mesh volume during a time step of size $2\Delta t$. The process can be continued until the entire flow volume is spanned by a coarse mesh containing only a few finite volumes with a correspondingly large time step. Only a simple weighted averaging is required to determine a new set of solution changes for each new coarse mesh. Once determined, the new coarse grid solution changes then need to be partitioned back to the set of original finite volumes. If M-1 additional coarse grids are formed, the new updated solution on the fine mesh, in place of that given by Eq. (9), becomes

$$U^{n+1}_{i,j,k} = U^{n}_{i,j,k} + \delta U_{i,j,k} + \frac{1}{4} \sum_{m=2}^{M} \delta U^{(m)}_{i^{(m)},j^{(m)},k^{(m)}} \tag{47}$$

where each of the changes indexed $i^{(m)},j^{(m)},k^{(m)}$ corresponds to a coarse grid volume containing the original finite volume indexed i,j,k. The time the solution is advanced is equal to

$$\Delta T = \Delta t + \frac{1}{4} \sum_{m=2}^{M} 2^{m-1}\ \Delta t = (1 + \frac{1}{4}\,(2^M - 2))\ \Delta t \tag{48}$$

The factors of one fourth appearing in the above equations have been chosen to insure that the total time step advance multiplied by the signal propagation speed does not exceed the linear dimensions of the flow field.

The above multigrid procedure has the potential to move information rapidly throughout the flow field and, because the terms of the governing equations are approximated only on the finest grid, preserve the resolution of the original mesh. The present author has found this to be true[22] for an implicit calculation of the interaction of a shock wave and a laminar boundary layer. Only one additional subroutine needed to be written to apply the multigrid procedure to an existing implicit program and no additional storage was required for data. For a flow calculation with a uniform far field mesh, the rate of convergence was increased by a factor of from two to four with only an increase in computer time per step of fifteen percent. Larger convergence acceleration factors are expected for the highly stretched mesh system required for an aircraft calculation.

4.5 Vectorization and computer hardware

The described numerical procedures should adapt to the vector architectures of the new generation of scientific computers. The explicit stage represented by Eq. (8) is readily vectorized. The implicit stage, basically consisting at each time step of two bidiagonal sweeps through the flow field, can be highly vectorized with a little more effort. Even the "holes" in the fuselage mesh containing the locations of the wing and stabilizer sections, or the double valued surface in the wing mesh enclosing the pylon, should not greatly impede vectorization. The calculation can proceed as on any regular three dimensional mesh with "dummy" data at "hole" locations that is either ignored entirely or corrected later, external to the vectorized loop. The vectorization of the multigrid procedure would require considerably more effort, but because the basic operations performed are simple summation in ascending from fine to coarse grids and partitioning in returning to the fine grid, there is also a good chance that much of this procedure can be vectorized.

Much of the calculation on each component mesh is independent of that of the others and can be performed simultaneously. Each is coupled to its neighbors at common mesh boundaries but, once the boundary values are supplied, the calculation can then proceed independently. New generation computers with more than one independent processor can be used efficiently much of the time with each processor calculating flow quantities on separate component meshes.

5. CONCLUSIONS

The numerical solution of the compressible viscous flow field about a complete aircraft in flight is not very far off in time. The logical first step is a flow simulation using the thin-layer Reynolds-averaged Navier-Stokes equations. A mesh containing approximately half a million grid points, or finite volumes, and computer storage for fifteen million words of memory is required. If performed on a computer with the speed of a CDC 7600 computer, the calculation would require about fifty hours of solid computing. New computers will become available within the next year or two with the required storage, computing speeds two orders of magnitude faster than a CDC 7600, and costs of only about twice as much per hour. The topological problems of nesting a system of meshes about an aircraft configuration are solvable. More accurate, reliable, and efficient numerical methods, perhaps an order of magnitude faster than present procedures, can be developed in the near future that should adapt to the architectures of the coming new generation of scientific computers. Computation times and costs can perhaps be reduced enough to permit the use of compressible viscous flow simulations for aircraft configuration design studies.

Considerable development is still required in turbulence modeling. Even with no further modeling improvements, realistic calculations will still be able to be performed for flow fields about aircraft near design cruise conditions. Numerical simulations of the off design cases, of military aircraft in maneuver, or of other flows containing moderate or large regions of separation, must await further advances in turbulence modeling.

I would like to make a few final comments to the reader before closing. I have tried to put forth an argument containing both fact and speculation to convince you that it is now reasonable to begin the attempt to solve for the compressible viscous flow field about a complete aircraft at flight conditions. Such a calculation represents more than a quantum leap in computational fluid dynamics and there are many who can point out convincing reasons why this attempt should be delayed. We are taught, sometimes unfortunately, to proceed step by step satisfying all present questions before moving on from our present location at point A toward our goal at point B. However, progress is often made by those who attempt to leap directly from point A to point B. If they land safely they can build a bridge back to point A for others to follow. If they fail we all see more clearly what preparations need to be worked on for the next attempt. We win either way.

ACKNOWLEDGEMENT

This work was supported by the Air Force Office of Scientific Research under Contract AFOSR-83-0057.

REFERENCES

1. CHAPMAN, D.R. (1979), "Computational Aerodynamics Development and Outlook". AIAA Journal, Vol. 17, No. 12.

2. BOPPE, C.W. and STERN, M.A. (1980), "Simulated Transonic Flows for Aircraft with Nacelles, Pylons, and Winglets". AIAA Paper No. 80-0130.

3. YU, N.J. (1980), "Grid Generation and Transonic Flow Calculations for Three-Dimensional Configurations". AIAA Paper No. 80-1391.

4. RIZZI, A. (1982), "Damped Euler-Equation Method to Compute Transonic Flow Around Wing-Body Combinations". AIAA Journal, Vol. 20, No. 10.

5. JAMESON, A., RIZZI, A., SCHMIDT, W. and WHITFIELD, W. (1981), "Finite Volume Solution for the Euler Equations for Transonic Flow Over Airfoils and Wings Including Viscous Effects". AIAA Paper No. 81-1265.

6. National Academy of Science Report (1983), "The Influence of Computational Fluid Dynamics on Experimental Aerospace Facilities-A Fifteen Year Projection". National Academy Press, Washington, D.C.

7. SOUTH, J. and THAMES, F. (1982): Private Communication of an Internal NASA Langley presentation.

8. SORENSON, R.L. (1982), "Grid Generation by Elliptic Partial Differential Equations for a Tri-Element Augmentor-Wing Airfoil". "Numerical Grid Generation", Ed. J. F. Thompson, North-Holland.

9. EISEMAN, P.R. (1982), "Orthogonal Grid Generation", "Numerical Grid Generation". Ed. J. F. Thompson, North-Holland.

10. THOMPSON, J.F., "A Survey of Grid Generation Techniques in Computational Fluid Dynamics". AIAA Paper No. 83-0447.

11. BRILEY, W.R. and MCDONALD, H. (1975), "Solution of the Three-Dimensional Compressible Navier-Stokes Equations by an Implicit Technique". Proceedings of the 4th International Conference on Numerical Methods in Fluid Dynamics, Springer-Verlag.

12. BEAM, R.M. and WARMING, R.F. (1977), "An Implicit Factored Scheme for the Compressible Navier-Stokes Equations". AIAA Paper No. 77-645.

13. MACCORMACK, R.W. (1981), "A Numerical Method for Solving the Equations of Compressible Viscous Flow". AIAA Paper No. 81-0110.

14. STEGER, J. and WARMING, R.F. (1979), "Flux Vector Splitting of the Inviscid Gasdynamics Equations with Application to Finite Difference Methods". NASA TM-78605.

15. MORETTI, G. (1979), "The λ-Scheme". Journal of Computers and Fluids, Vol. 7.

16. MACCORMACK, R.W. (1969), "The Effect of Viscosity in Hypervelocity Impact Cratering". AIAA Paper No. 69-354.

17. YEE, H.C., BEAM, R.M. and WARMING, R.F. (1981), "Stable Boundary Approximations for a Class of Implicit Schemes for the One-Dimensional Inviscid Equations of Gas Dynamics". AIAA Paper No. 81-1009.

18. BRANDT, A. (1981), "Multigrid Solutions to Steady-State Compressible Navier-Stokes Equations". Proceedings of the 5th International Symposium on Computing Methods in Applied Science and Engineering, Eds. Glowinski, R. and Lions, J.L., Versailles, France.

19. NI, R.H. (1981), "A Multiple Grid Scheme for Solving the Euler Equations". AIAA Paper No. 81-1025.

20. JOHNSON, G.M. (1982), "Multiple Grid Acceleration of Lax-Wendroff Algorithms". NASA TM-821843.

21. DENTON, J.D. (1982), "An Improved Time Marching Method for Turbomachinery Flow Calculations". ASME Paper No. 82-GT-239.

22. MACCORMACK, R.W. (1983), "Acceleration of Convergence of Navier-Stokes Calcuations". Proceedings of the Conference on Large Scale Scientific Computing, University of Wisconsin-Madison, May 17-19.

APPROXIMATE-FACTORIZATION ALGORITHMS: THEORY AND APPLICATIONS IN VISCOUS-FLOW COMPUTATIONS

Denny S. Chaussee

NASA Ames Research Center, Moffett Field, California

1. INTRODUCTION

Numerical computations based on the compressible, Reynolds-averaged Navier-Stokes equations have become fairly commonplace since the advent of the new generation of computers (e.g., Cray-1S, CDC 205). Explicit and implicit numerical methods are employed in these calculations. This chapter is restricted to implicit methods, specifically to those types of noniterative algorithms advanced by Lindemuth and Killeen [1], Briley and McDonald [2], and Beam and Warming [3,4]. The remainder of this work follows the Warming and Beam development [5].

Both the unsteady and steady three-dimensional forms of the thin-layer Navier-Stokes equations are presented in conservation-law form. Pulliam and Steger [6] originally developed the finite-difference procedure for the unsteady form of the equations. This procedure improved stability and was simple, flexible, and cast in the delta form. If a steady-state solution is reached, the delta form ensures that it will be independent of the step size. Many solutions have been calculated using this form of the algorithm. These include transonic flow past a hemisphere-cylinder body [6], the nose region of the Convair 990 research aircraft [7], and the front half of the Jetstar aircraft. In addition, the supersonic flow past blunt bodies, both spherical and ablated [8-10], and regions of embedded subsonic flow [11] have been calculated.

For steady, viscous, supersonic high-Reynolds-number flows, a substantial reduction in both computational effort and storage is achieved by using a simplified form of the equations, which is termed the parabolized Navier-Stokes (PNS) equations. These equations are obtained by (1) neglecting the unsteady terms and the streamwise viscous diffusion terms and (2) modifying the streamwise convection flux vector to permit stable marching downstream from an initial set of data. A finite-difference implicit marching algorithm in delta form was first developed by Vigneron et al. [12]. The form that is presented here was developed by Schiff and Steger [13]; it differs from that of Vigneron et al. in the way in which the streamwise convection flux vector is treated. Applying this approximate-factored algorithm in a marching mode requires that (1) the inviscid part of the shock layer be supersonic and (2) the streamwise (marching direction) velocity remains positive. This rules out cases in which axial separation occurs, but permits the correct resolution of crossflow separation. Body shapes that have been marched over by the present algorithm include pointed cones [14], sphere-cones [13], three-dimensional reentry vehicles [15-17], an ogive-cylinder-boattail [18], a delta wing configuration [12,19], the X-24 lifting body [15], the Space Shuttle [20], and, most recently, finned configurations [14,21].

In the following sections a systematic development of the implicit, approximate-factorization algorithm in the delta form is presented. The governing gas-dynamic equations and the unsteady and steady finite-difference algorithms are presented first. Boundary conditions and the development of an implicit formulation are presented in Section 4 and appropriate initial conditions are discussed briefly in Section 5. Results, which cover typical and challenging applications of the steady algorithm, are presented in the final section.

2. GAS DYNAMIC EQUATIONS

2.1 Navier-Stokes Equations

A computational procedure for a wide variety of physical geometries and grid systems can be developed by transforming the Reynolds-averaged Navier-Stokes equations from the Cartesian coordinates (x,y,z,t) to general curvilinear coordinates (ξ,η,ζ,τ) where

$$\left.\begin{aligned} \tau &= t \\ \xi &= \xi(x,y,z,t) \\ \eta &= \eta(x,y,z,t) \\ \zeta &= \zeta(x,y,z,t) \end{aligned}\right\} \qquad (1)$$

The transformations are chosen so that the grid spacing in the computational space (ξ,η,ζ) is uniform and of unit length. This produces a computational region that is a rectangular parallelepiped and has a regular uniform mesh. The generalized coordinate transformation produces a system of equations that can be applied to any regular and nonsingular geometry or grid system. The advantages of this form are as follows.

1. Since spacing in the computational domain can be chosen to be uniform, simple difference formulas can be used in the numerical scheme. This results in computer codes that can be applied to a wide variety of problems without modification of the equations and numerical scheme.

2. Physical boundary surfaces are easily mapped onto coordinate surfaces, which makes the application of boundary conditions easier and aids in the formulation of the thin-layer approximation.

3. The transformation permits the unsteady motion of the grid points, so that moving meshes and distorting surfaces can be treated with ease.

The region in Cartesian space will be referred to as the physical domain, and the region in computational space will be called the computational domain. A one-to-one correspondence is set up between points in the physical space and points in the computational space. With this construction, the Navier-Stokes equations are transformed from Cartesian coordinates to general curvilinear coordinates as described below.

In most cases the transformation from physical space to computational space is not known analytically; instead, it is generated numerically. That is, the x,y,z coordinates of the grid points are provided and the inverse metrics (e.g., x_ξ, y_ξ, z_ξ) are numerically generated using finite-difference approximations. Simple finite-difference approximations can be used to calculate these inverse metrics since the grid points are equally spaced in computational space, that is, $\Delta\xi = 1$. The metrics are then obtained from the inverse metrics as follows:

$$\left.\begin{array}{ll}
\xi_x = J(y_\eta z_\zeta - y_\zeta z_\eta) & \eta_x = J(z_\xi y_\zeta - y_\xi z_\zeta) \\
\xi_y = J(z_\eta x_\zeta - x_\eta z_\zeta) & \eta_y = J(x_\xi z_\zeta - x_\zeta z_\xi) \\
\xi_z = J(x_\eta y_\zeta - y_\eta x_\zeta) & \eta_z = J(y_\xi x_\zeta - x_\xi y_\zeta) \\
\zeta_x = J(y_\xi z_\eta - z_\xi y_\eta) & \xi_t = -x_\tau \xi_x - y_\tau \xi_y - z_\tau \xi_z \\
\zeta_y = J(x_\eta z_\xi - x_\xi z_\eta) & \eta_t = -x_\tau \eta_x - y_\tau \eta_y - z_\tau \eta_z \\
\zeta_z = J(x_\xi y_\eta - y_\xi x_\eta) & \zeta_t = -x_\tau \zeta_x - y_\tau \zeta_y - z_\tau \zeta_z
\end{array}\right\} \quad (2)$$

where

$$J = (x_\xi y_\eta z_\zeta + x_\zeta y_\xi z_\eta + x_\eta y_\zeta z_\xi - x_\xi y_\zeta z_\eta - x_\eta y_\xi z_\zeta - x_\zeta y_\eta z_\xi)^{-1}$$

and J is defined as the metric Jacobian.

Applying the curvilinear transformation to the Cartesian form of the Navier-Stokes equations, the following form is obtained:

$$\partial_\tau Q + \xi_t \partial_\xi Q + \eta_t \partial_\eta Q + \zeta_t \partial_\zeta Q + \xi_x \partial_\xi E + \eta_x \partial_\eta E + \zeta_x \partial_\zeta E$$

$$+ \zeta_y \partial_\zeta F + \xi_y \partial_\xi F + \eta_y \partial_\eta F + \xi_z \partial_\xi G + \eta_z \partial_\eta G + \zeta_z \partial_\zeta G$$

$$= Re^{-1}[\xi_x \partial_\xi R + \eta_x \partial_\eta R + \zeta_x \partial_\zeta R + \zeta_y \partial_\zeta S$$

$$+ \xi_y \partial_\xi S + \eta_y \partial_\eta S + \xi_z \partial_\xi T + \eta_z \partial_\eta T + \zeta_z \partial_\zeta T] \quad (3)$$

where

$$Q = \begin{vmatrix} \rho \\ \rho u \\ \rho v \\ \rho w \\ e \end{vmatrix}, \quad E = \begin{vmatrix} \rho u \\ \rho u^2 + p \\ \rho uv \\ \rho uw \\ u(e + p) \end{vmatrix}, \quad F = \begin{vmatrix} \rho v \\ \rho uv \\ \rho v^2 + p \\ \rho vw \\ v(e + p) \end{vmatrix}, \quad G = \begin{vmatrix} \rho w \\ \rho uw \\ \rho vw \\ \rho w^2 + p \\ w(e + p) \end{vmatrix}$$

$$R = \begin{vmatrix} 0 \\ \tau_{xx} \\ \tau_{xy} \\ \tau_{xz} \\ R_4 \end{vmatrix}, \quad S = \begin{vmatrix} 0 \\ \tau_{xy} \\ \tau_{yy} \\ \tau_{zy} \\ S_4 \end{vmatrix}, \quad T = \begin{vmatrix} 0 \\ \tau_{xz} \\ \tau_{yz} \\ \tau_{zz} \\ T_4 \end{vmatrix} \quad (4)$$

with

$$\left.\begin{aligned} \tau_{xx} &= (\lambda + 2\mu)u_x + \lambda(v_y + w_z), \quad \tau_{xy} = \mu(u_y + v_x), \quad \tau_{xz} = \mu(u_z + w_x) \\ \tau_{yy} &= (\lambda + 2\mu)v_y + \lambda(u_x + w_z), \quad \tau_{yz} = \mu(v_z + w_y) \\ \tau_{zz} &= (\lambda + 2\mu)w_z + \lambda(u_x + v_y) \end{aligned}\right\} \quad (5)$$

$$\left.\begin{aligned} R_4 &= u\tau_{xx} + v\tau_{xy} + w\tau_{xz} + \kappa Pr^{-1}(\gamma - 1)^{-1} \partial_x a^2 \\ S_4 &= u\tau_{xy} + v\tau_{yy} + w\tau_{yz} + \kappa Pr^{-1}(\gamma - 1)^{-1} \partial_y a^2 \\ T_4 &= u\tau_{xz} + v\tau_{zy} + w\tau_{zz} + \kappa Pr^{-1}(\gamma - 1)^{-1} \partial_z a^2 \end{aligned}\right\} \quad (6)$$

The Cartesian velocity components are u, v, and w, the density is ρ, and the total energy is e and is related to the pressure p by

$$p = (\gamma - 1)[e - 0.5\rho(u^2 + v^2 + w^2)] \tag{7}$$

The ratio of specific heats is γ, K is the coefficient of thermal conductivity, μ is the dynamic viscosity, and λ is obtained from Stokes' hypothesis as $-(2/3)\mu$. The Reynolds number is Re and the Prandtl number is Pr. The equations have been nondimensionalized as in [6] and [13].

Note that Eqs. (3) are in the chain-rule form; that is, even though none of the flow variables (or more appropriately functions of the flow variables) occurs as coefficients to the differentials, the metrics do. There is some recent research which advocates the use of this form, since it should still have good shock-capturing properties and in some ways have a simpler form. However, the following development will be restricted to the strong conservation-law form.

To produce the strong conservation-law form, first multiply Eqs. (3) by J^{-1} and use the chain-rule on all the terms, such as

$$J^{-1}\xi_x \partial_\xi E = \partial_\xi[(\xi_x/J)E] - E\,\partial_\xi(\xi_x/J) \tag{8}$$

Collect all terms into two groups,

$$T_1 + T_2 = 0 \tag{9}$$

where

$$\begin{aligned} T_1 = {} & \partial_\tau(Q/J) + \partial_\xi\{[\xi_t Q + \xi_x E + \xi_y F + \xi_z G - Re^{-1}(\xi_x R \\ & + \xi_y S + \xi_z T)]/J\} + \partial_\eta\{[\eta_t Q + \eta_x E + \eta_y F + \eta_x G \\ & - Re^{-1}(\eta_x R + \eta_y S + \eta_z T)]/J\} + \partial_\zeta\{[\zeta_t Q + \zeta_x E \\ & + \zeta_y F + \zeta_z G - Re^{-1}(\zeta_x R + \zeta_y S + \zeta_z T)]/J\} \end{aligned} \tag{10}$$

and

$$\begin{aligned} T_2 = {} & -(E - Re^{-1}R)[\partial_\xi(\xi_x/J) + \partial_\eta(\eta_x/J) + \partial_\zeta(\zeta_x/J)] \\ & - (F - Re^{-1}S)[\partial_\xi(\xi_y/J) + \partial_\eta(\eta_y/J) + \partial_\zeta(\zeta_y/J)] \\ & - (G - Re^{-1}T)[\partial_\xi(\xi_z/J) + \partial_\eta(\eta_z/J) + \partial_\zeta(\zeta_z/J)] \\ & - Q[\partial_\tau(J^{-1}) + \partial_\xi(\xi_t/J) + \partial_\eta(\eta_t/J) + \partial_\zeta(\zeta_t/J)] \end{aligned} \tag{11}$$

When the terms T_2 are eliminated, a strong conservation-law form of the equations results. The following four terms,

$$\left.\begin{aligned}
&\partial_\tau(J^{-1}) + \partial_\xi(\xi_t/J) + \partial_\eta(\eta_t/J) + \partial_\zeta(\zeta_t/J) \,, \\
&\partial_\xi(\xi_x/J) + \partial_\eta(\eta_x/J) + \partial_\zeta(\zeta_x/J) \,, \\
&\partial_\xi(\xi_y/J) + \partial_\eta(\eta_y/J) + \partial_\zeta(\zeta_y/J) \,, \\
&\partial_\xi(\xi_z/J) + \partial_\eta(\eta_z/J) + \partial_\zeta(\zeta_z/J)
\end{aligned}\right\} \qquad (12)$$

are defined as invariants of the transformation and can be shown to be zero analytically. Applying the metric definitions, Eq. (2), to the second invariant term in Eq. (12), for example, yields

$$\begin{aligned}
&\partial_\xi(y_\eta z_\zeta - y_\zeta z_\eta) + \partial_\eta(z_\xi y_\zeta - y_\xi z_\zeta) + \partial_\zeta(y_\xi z_\eta - z_\xi y_\eta) \\
&\quad = y_{\eta\xi} z_\zeta + y_\eta z_{\zeta\xi} - y_{\zeta\xi} z_\eta - y_\zeta z_{\eta\xi} + z_{\xi\eta} y_\zeta \\
&\quad\quad + z_\xi y_{\zeta\eta} - y_{\xi\eta} z_\zeta - y_\xi z_{\eta\zeta} + y_{\xi\zeta} z_\eta + y_\xi z_{\eta\zeta} \\
&\quad\quad - z_{\xi\zeta} y_\eta - z_\xi y_{\zeta\eta} = 0
\end{aligned} \qquad (13)$$

Since analytical differentiation is commutative, each of the terms then sums to zero as above. This eliminates the T_2 term of Eq. (9), and the resulting equations are in a strong conservation-law form. Capsulizing, the strong conservation-law form of the Navier-Stokes is written as

$$\begin{aligned}
\partial_\tau\hat{Q} + \partial_\xi\hat{E} + \partial_\eta\hat{F} + \partial_\zeta\hat{G} = Re^{-1}\{&\partial_\xi[J^{-1}(\xi_x R + \xi_y S + \xi_z T)] \\
&+ \partial_\eta[J^{-1}(\eta_x R + \eta_y S + \eta_z T)] \\
&+ \partial_\zeta[J^{-1}(\zeta_x R + \zeta_y S + \zeta_z T)]\}
\end{aligned} \qquad (14)$$

where

$$\hat{Q} = J^{-1}Q \,, \quad \hat{E} = J^{-1}\begin{vmatrix} \rho U \\ \rho u U + \xi_x p \\ \rho v U + \xi_y p \\ \rho w U + \xi_z p \\ (e + p)U - \xi_t p \end{vmatrix} \,, \quad \hat{F} = J^{-1}\begin{vmatrix} \rho V \\ \rho u V + \eta_x p \\ \rho v V + \eta_y p \\ \rho w V + \eta_z p \\ (e + p)V - \eta_t p \end{vmatrix} \qquad (15)$$

$$\hat{G} = J^{-1} \begin{vmatrix} \rho W \\ \rho uW + \zeta_x p \\ \rho vW + \zeta_y p \\ \rho wW + \zeta_z p \\ (e + p)W - \zeta_t p \end{vmatrix} ,$$

$$\hat{R} = J^{-1}(\xi_x R + \xi_y S + \xi_z T) = \frac{1}{J} \begin{vmatrix} 0 \\ \xi_x \tau_{xx} + \xi_y \tau_{xy} + \xi_z \tau_{xz} \\ \xi_x \tau_{xy} + \xi_y \tau_{yy} + \xi_z \tau_{zy} \\ \xi_x \tau_{xz} + \xi_y \tau_{yz} + \xi_z \tau_{zz} \\ \xi_x R_4 + \xi_y S_4 + \xi_z T_4 \end{vmatrix} \tag{16}$$

and

$$\hat{S} = J^{-1}(\eta_x R + \eta_y S + \eta_z T) = \frac{1}{J} \begin{vmatrix} 0 \\ \eta_x \tau_{xx} + \eta_y \tau_{yx} + \eta_z \tau_{zx} \\ \eta_x \tau_{xy} + \eta_y \tau_{yy} + \eta_z \tau_{zy} \\ \eta_x \tau_{xz} + \eta_y \tau_{yz} + \eta_z \tau_{zz} \\ \eta_x R_4 + \eta_y S_4 + \eta_z T_4 \end{vmatrix}$$

$$\hat{T} = J^{-1}(\zeta_x R + \zeta_y S + \zeta_z T) = \frac{1}{J} \begin{vmatrix} 0 \\ \zeta_x \tau_{xx} + \zeta_y \tau_{yx} + \zeta_z \tau_{zx} \\ \zeta_x \tau_{xy} + \zeta_y \tau_{yy} + \zeta_z \tau_{xz} \\ \zeta_x \tau_{xz} + \zeta_y \tau_{yz} + \zeta_z \tau_{zz} \\ \zeta_x R_4 + \zeta_y S_4 + \zeta_z T_4 \end{vmatrix} \tag{17}$$

The contravariant velocities (U,V,W) are defined to be

$$\left.\begin{aligned} U &= \xi_t + \xi_x u + \xi_y v + \xi_z w \\ V &= \eta_t + \eta_x u + \eta_y v + \eta_z w \\ W &= \zeta_t + \zeta_x u + \zeta_y v + \zeta_z w \end{aligned}\right\} \tag{18}$$

This form permits the algorithm to predict the proper location and strength of discontinuities, that is, shock waves and slip surfaces.

2.2 Thin-Layer Approximation

In high-Reynolds-number flows, the effects of viscosity are concentrated near rigid boundaries and in wake regions. Typically, in three-dimensional, finite-difference computations there are only enough grid points available (because of computer storage limits) to concentrate grid lines near rigid surfaces. The resulting grid system usually has fine grid spacing in directions nearly normal to the surfaces and coarse grid spacing along the surface (see Fig. 1).

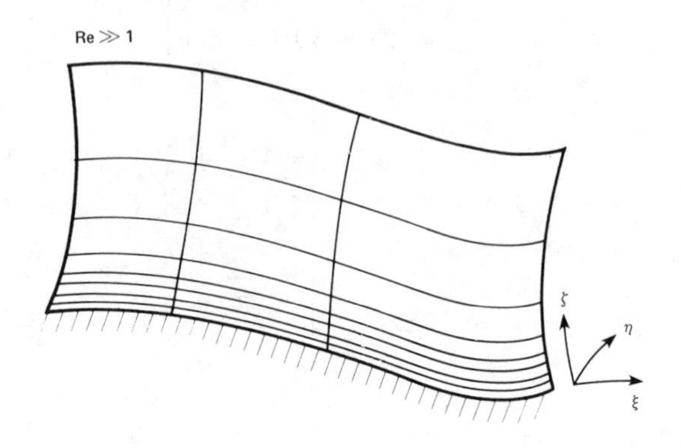

Figure 1. Grid system.

Even though the full Navier-Stokes equations may be used, the viscous terms associated with derivatives along the body will not be computationally resolved, because of grid coarseness; in most cases for attached and mildly separated flows, these terms are negligible. The terms in the normal direction will be resolved for sufficiently fine grid spacing, and these terms are substantial in magnitude.

In boundary-layer theory, appropriate scaling arguments show that streamwise components of the viscous terms can be neglected relative to the normal terms. These arguments are relied upon as a justification for the thin-layer approximation. The application of the thin-layer approximation requires the following:

1. That all body surfaces be mapped onto coordinate surfaces, for example, ξ-coordinate surfaces, as in Fig. 1.

2. That all the viscous derivatives in the ξ- and η-direction be neglected, and the terms in the ζ-direction be retained. Grid spacing in the ζ-direction should be sufficiently small to resolve the ζ-viscous terms (at least one or two grid points in the laminar sublayer). However, all of the convective terms are retained.

The thin-layer approximation is similar in philosophy but not the same as the boundary-layer theory, because the normal momentum equation is retained and the pressure can vary through the boundary layer. This approximation can break down for low Reynolds numbers and in regions of massive flow separation, and it should be stressed that it is not a necessary step in the development of the equations and numerical algorithm.

Applying the thin-layer approximation to Eqs. (14), where all the viscous terms associated with the ξ and η derivatives are neglected, the final form of the equations is obtained

$$\partial_\tau \hat{Q} + \partial_\xi \hat{E} + \partial_\eta \hat{F} + \partial_\zeta \hat{G} = Re^{-1} \partial_\zeta \hat{T} \tag{19}$$

where the variables are defined in Sec. 2.1.

2.3 Parabolized Form of the Navier-Stokes Equations

To obtain the parabolized Navier-Stokes equations from the steady Navier-Stokes equations, the streamwise viscous terms must be neglected and the convective flux vector $\hat{E}$ must be modified such that $\hat{A} = \partial\hat{E}/\partial\hat{Q}$ has positive eigenvalues. The above modifications change the nature of the steady Navier-Stokes equations from elliptic to parabolic. The parabolic nature of the new set of equations permits the stable marching of these equations in the streamwise direction. It can be shown that the PNS equations retain their parabolic nature as long as the streamwise velocity u is positive.

To obtain the parabolic approximation, neglect the streamwise derivatives $\partial/\partial\xi$ of the flow variables in the viscous term of Eq. (14), thus yielding for the term

$$\frac{\partial\hat{R}}{\partial\xi} = R\,\frac{\partial(\xi_x/J)}{\partial\xi} + S\,\frac{\partial(\xi_y/J)}{\partial\xi} + T\,\frac{\partial(\xi_z/J)}{\partial\xi} \cong 0 \tag{20a}$$

and additionally, with the thin-layer approximation

$$\frac{\partial\hat{S}}{\partial\eta} = R\,\frac{\partial(\eta_x/J)}{\partial\eta} + S\,\frac{\partial(\eta_y/J)}{\partial\eta} + T\,\frac{\partial(\eta_z/J)}{\partial\eta} \cong 0 \tag{20b}$$

Hence the right-hand side (RHS) of Eq. (14) becomes

$$RHS = Re^{-1}\left[\frac{\partial\hat{T}}{\partial\zeta}\right] \tag{21}$$

Two important points are now made. The first is that if pressure is specified in the flux vector E, that is, if the given p is not a function of Q, then the sound speed contribution to the eigenvalues of $\partial E/\partial Q = A$ is removed. In this way, the eigenvalues of A remain positive as long as $u > 0$. The second point is that for high-Reynolds-number viscous flow, the pressure is approximately constant through the thin, subsonic viscous sublayer near the wall. Moreover, according to boundary-layer theory for high-Reynolds-number flow, the approximation $\partial p/\partial n = 0$ is valid over the entire thickness of the viscous layer. Thus, this approximation is even more apropos over just the subsonic portion of the viscous layer. These points constitute the basis of the sublayer model as developed by Schiff and Steger [13].

By introducing the sublayer approximation into Eqs. (14) and (21), the parabolized Navier-Stokes equations can be written as

$$\frac{\partial \hat{E}_s}{\partial \xi} + \frac{\partial \hat{F}}{\partial \eta} + \frac{\partial \hat{G}}{\partial \zeta} = Re^{-1} \frac{\partial \hat{T}}{\partial \zeta} \tag{22}$$

where the streamwise flux vector, $\hat{E}_s$, is defined in [13].

3. IMPLICIT FINITE-DIFFERENCE ALGORITHMS

The finite-difference scheme discussed in the following sections is the implicit approximate-factorization algorithm in the delta form, as developed by Beam and Warming [4]. The calculation of viscous flows requires the use of fine grid spacing close to the body in order to resolve the boundary layer. These fine grids result in exceedingly small allowable integration steps in the case of explicit methods (because of stability restrictions) and, hence, long convergence times. The implicit method discussed here is not constrained by such restrictive stability limits and permits the use of large integration steps.

The basic unsteady algorithm is first- or second-order accurate in time and is noniterative. The equations are spatially factored, which reduces the solution process to solving three one-dimensional problems at a given time-level. Since central differences are used, the algorithm produces a block-tridiagonal system for each spatial coordinate. The stability and accuracy of the unsteady numerical algorithm have been described in detail by Warming and Beam [5].

The basic steady algorithm (parabolized Navier-Stokes) is conservative, can be first- or second-order accurate in the marching direction, is second-order accurate in the other spatial directions, and is noniterative. The equations are spatially factored, which reduces the solution process to solving two one-dimensional problems at a given marching step. Since central differences are used, the PNS algorithm produces a block-tridiagonal system for each direction. The stability and accuracy have been discussed in some detail by Schiff and Steger [13].

In order to obtain a finite-difference scheme, the various derivatives in the governing equations must be approximated by difference formulas. For the time derivatives and time-like derivatives, approximations are made as in [6] and [13].

The next step is to take the continuous differential operators $\partial\eta$ and $\partial\zeta$ and approximate them with finite-difference operators on a discrete mesh. The transformations are chosen such that the grid spacing in the computational domain is uniform and of unit length; $\Delta\xi = 1$, $\Delta\eta = 1$, and $\Delta\zeta = 1$. Let j be the index in the ξ-direction, k for the η-direction, and ℓ for the ζ-direction. Second-order central differences are used for the right-hand-side approximations.

To avoid solving a nonlinear system of equations at each step, the flux vectors at the new level of the steady marching algorithm, for example, j + 1, are replaced by local linearizations about the old level, that is, j. These are obtained by applying a Taylor series expansion. Note that the linearizations are second-order accurate and so if a second-order time-scheme is chosen, the linearizations would not degrade the time-accuracy. A more detailed presentation can be found in [6] and [13].

3.1 Unsteady Formulation

The finite-difference form of Eq. (19) results in the following approximate-factorization algorithm:

$$(I + h\delta_\xi \hat{A}^n - \varepsilon_I J^{-1}\nabla_\xi\Delta_\xi J)(I + h\delta_\eta \hat{B}^n - \varepsilon_I J^{-1}\nabla_\eta\Delta_\eta J)$$

$$(I + h\delta_\zeta \hat{C}^n - hRe^{-1}\delta_\zeta J^{-1}\hat{M}^n J - \varepsilon_I J^{-1}\nabla_\zeta\Delta_\zeta J)(\hat{Q}^{n+1} - \hat{Q}^n)$$

$$= -\Delta t(\delta_\xi \hat{E}^n + \delta_\eta \hat{F}^n + \delta_\zeta \hat{G}^n - Re^{-1}\delta_\zeta \hat{T}^n)$$

$$- \varepsilon_E J^{-1}[(\nabla_\xi\Delta_\xi)^2 + (\nabla_\eta\Delta_\eta)^2 + (\nabla_\zeta\Delta_\zeta)^2]J\hat{Q}^n \qquad (23)$$

where the δ's are central-difference operators and Δ and ∇ are forward and backward-difference operators, respectively. $\hat{A}^n$, $\hat{B}^n$, and $\hat{C}^n$ are the Jacobian matrices and are presented in detail in [6]. The smoothing parameters ε_E and ε_I are discussed in Sec. 3.3. The coefficient matrix $\hat{M}^n$, which is defined in [6], is the result of linearizing the viscous vector $\hat{T}^n$.

3.2 Steady Formulation

The steady three-dimensional algorithm written in delta form is

$$[\tilde{A}^j_s + (1 - \alpha)\Delta\xi(\delta_\eta \tilde{B}^j + \delta_\zeta \tilde{C}^j - Re^{-1}\bar{\delta}_\zeta \tilde{M}^j)](\hat{Q}^{j+1} - \hat{Q}^j)$$

$$= -(\tilde{A}^j_s - \tilde{A}^{j-1}_s)\hat{Q}^j + \alpha(\hat{E}^j_s - \hat{E}^{j-1}_s) - (1 - \alpha)\Delta\xi\{\delta_\eta[\eta^{j+1}_x (E/J)^j$$

$$+ \eta^{j+1}_y (F/J)^j + \eta^{j+1}_z (G/J)^j] + \delta_\zeta[\zeta^{j+1}_x (E/J)^j + \zeta^{j+1}_y (F/J)^j$$

$$+ \zeta^{j+1}_z (G/J)^j] - Re^{-1}\bar{\delta}_\zeta \tilde{T}^j\} - [(\xi_x/J)^{j+1} E^j_p$$

$$- (\xi_x/J)^j E^{j-1}_p] + \mathscr{D} \tag{24}$$

where δ_η and δ_ζ are central differences and the smoothing term $\mathscr{D}$ is defined by

$$\mathscr{D} = \varepsilon_e \tilde{A}^j_{s_{k,\ell}} (1/J)^j [(\nabla_\eta \Delta_\eta)^2 (J\hat{Q})^j + (\nabla_\zeta \Delta_\zeta)^2 (J\hat{Q})^j] \tag{25}$$

Here ε_E must be less than 1/16 for stability.

As in the unsteady case, an approximately factored form of Eq. (24), which retains the same order of accuracy in ξ, can be obtained if we note that

$$[\tilde{A}^j_s + (1 - \alpha)\Delta\xi(\delta_\eta \tilde{B}^j)](\tilde{A}^j_s)^{-1}\left[\tilde{A}^j_s + (1 - \alpha)\Delta\xi\left(\delta_\zeta \tilde{C}^j - \frac{1}{Re}\bar{\delta}_\zeta \tilde{M}^j\right)\right]\Delta\hat{Q}^j$$

$$= \text{LHS}(24) + O(\Delta\xi)^3 \tag{26}$$

Unlike the factored unsteady equation, Eq. (26) has the term $(\tilde{A}^j_s)^{-1}$ between the two factors. The factorization error also contains this term. The term $(\tilde{A}^j_s)^{-1}$ becomes large when u is small and hence can give rise to large errors and may even result in an instability if u is sufficiently small (near surface boundaries). Upon substituting the left-hand side of Eq. (24) for the left-hand side of Eq. (26), the factored algorithm is obtained. The algorithm is solved by the sequence of implicit inversions

$$\left.\begin{aligned} [\tilde{A}^j_s + (1 - \alpha)\Delta\xi(\delta_\eta \tilde{B}^j)]\Delta\hat{Q}_* &= \text{RHS}(24) \\ \left[\tilde{A}^j + (1 - \alpha)\Delta\xi\left(\delta_\zeta \tilde{C}^j - \frac{1}{Re}\bar{\delta}_\zeta \tilde{M}^j\right)\right]\Delta\hat{Q}^j &= \tilde{A}^j_s \,\Delta\hat{Q}_* \end{aligned}\right\} \tag{27}$$

For an excellent discussion on necessary and sufficient conditions for a stable marching procedure using the parabolized Navier-Stokes approximation refer to the work by Schiff and Steger [13].

3.3 Explicit and Implicit Numerical Dissipation

Fourth-order dissipation terms such as $\varepsilon_E J^{-1}(\nabla_\xi \Delta_\xi)^2 J\hat{Q}$ are added explicitly to the unsteady equations to damp high-frequency growth; thus, they serve to control nonlinear instability. Linear-stability analysis shows that the coefficient ε_E must be less than 1/24. The addition of the implicit second-order difference terms operating on $\Delta\hat{Q}^n$ with ε_I extends the linear stability bound of the fourth-order terms. The usual procedure of adding dissipation is to set $\varepsilon_E = \Delta t$ and $\varepsilon_I = 2\varepsilon_E$; this results in a consistent definition of ε_E. As the time-step is increased, the amount of artificial dissipation added relative to the spatial derivatives of convection and diffusion remains constant.

To enhance stability in the steady case, some additional smoothing terms have also been included within the marching algorithm. They are referred to as the implicit smoothing, ε_I, as in the unsteady case, and ε_A to allow for smoothing near the body surface where the u velocity goes to zero.

Adding implicit smoothing to Eq. (27):

$$\{\tilde{A}_s^j(1 - \varepsilon_I[-J_{j+1}^{-1}(\nabla\Delta)_\eta J_{j+1}]) + (1 - \alpha)\Delta\xi(\delta_\eta \tilde{B}^j)\}\Delta\hat{Q}_*$$

$$= \text{RHS}(24) + \varepsilon_I[J_{j+1}^{-1}(\nabla\Delta)_\eta J_{j+1} - J_j^{-1}(\nabla\Delta)_\eta J_j]\hat{Q}_j \tag{28}$$

where if $\varepsilon_I \neq 0$, ε_I is usually 2 or 3 times ε_E and then ε_E in the right-hand side of Eq. (24) is not restricted. Otherwise, if $\varepsilon_I = 0$, then $\varepsilon_E < 1/16$ by stability theory.

Since the explicit smoothing term contains the A-matrix, an additional stabilizing term is needed because the axial velocity u appears along the diagonal of this matrix. At and near the body surface and, at times, as a result of geometry changes, the axial velocity can approach zero. If this happens, the smoothing term has a negligible effect unless the stabilizing term is added. To be consistent with the equations, the term has to be of the form $\tilde{A}_s + \varepsilon_A I$ where ε_A is some parameter and I is the identity matrix. The term is to be added such that it goes to zero when the flow is conical.

Therefore, it is of the form

$$\varepsilon_A I[J^{j+1}]^{-1}(\hat{Q}^{j+1} - \hat{Q}^j) = 0 \tag{29}$$

Equation (29) can be rewritten as

$$\varepsilon_A I(\Delta\hat{Q})^{j+1} - \varepsilon_A I\hat{Q}^j(1 - J^j/J^{j+1}) = 0 \tag{30}$$

Equation (30) is then incorporated into the algorithm by adding $\varepsilon_A I(\Delta\hat{Q})^{j+1}$ to the left-hand side and by adding $\varepsilon_A I\hat{Q}^j[1 - J^j/J^{j+1}]$ to the right-hand side. On the left-hand side, the $\tilde{A}_s^j$ matrix is redefined as $\tilde{A}_s + \varepsilon_A I$. The value of ε_A varies between zero and one. These additional stabilizing terms do not have an adverse effect on the marching solutions, but tend only to enhance the solution procedure.

4. BOUNDARY CONDITIONS

Probably the single most important aspect in the successful application of any numerical technique to the solution of fluid-flow problems is the proper treatment of the impermeable and permeable boundaries that encompass the given region. An impermeable boundary is one across which no mass can flow, such as the solid surface of a body or a plane of symmetry. A permeable boundary, such as a shock wave or inflow or outflow boundary, allows mass to flow through its surface. Of all these boundaries, the impermeable surface of a solid body has probably received the most attention by the numerical modelist, and rightfully so. It is this boundary, in conjunction with the free-stream conditions, that generates the behavior of the surrounding flow field.

The increased use of implicit schemes for the solution of the Euler- and Reynolds-averaged Navier-Stokes equations has generated considerable interest in the development of stable, accurate, implicit boundary-condition procedures. The application of viscous boundary conditions at the surface of a body in an implicit manner is a straightforward procedure. However, the application of accurate implicit boundary conditions for inviscid flows requires more detailed analysis. The body boundary conditions for viscous flow are addressed in the remainder of this section; a more detailed analysis of the implicit shock-boundary conditions is contained in [14].

The condition to be satisfied at a surface boundary in viscous flow is the no-slip condition. This condition gives the equations

$$\left.\begin{aligned} \rho u &= 0 \\ \rho v &= 0 \\ \rho w &= 0 \end{aligned}\right\} \tag{31}$$

at the wall. Equations (31) are valid when the slip velocity at the wall is negligible (this is true for high-Reynolds-number flows). The assumption that the flow Reynolds number is large also gives the equation

$$\frac{\partial p}{\partial n} = 0 \tag{32}$$

where n is the direction normal to the surface of the wall. Since

$$e = \frac{p}{\gamma - 1} + \frac{\rho}{2}(u^2 + v^2 + w^2) \tag{33}$$

then differentiating Eq. (33) and making use of Eqs. (31) and (32),

$$\frac{\partial e}{\partial n} = 0 \tag{34}$$

In the transformed coordinate system this equation can be written as

$$[\xi_x\zeta_x + \xi_y\zeta_y + \xi_z\zeta_z]\frac{\partial e}{\partial \xi} + [\eta_x\zeta_x + \eta_y\zeta_y + \eta_z\zeta_z]\frac{\partial e}{\partial \eta} + [\zeta_x^2 + \zeta_y^2 + \zeta_z^2]\frac{\partial e}{\partial \zeta} = 0 \tag{35}$$

where the body is a constant ζ-surface. When the ζ-coordinate lines are orthogonal to the body, Eq. (35) reduces to

$$\frac{\partial e}{\partial \zeta} = 0 \tag{36}$$

This equation can be written using a first-order accurate, one-sided finite difference as

$$e_2 - e_1 = 0 \tag{37}$$

where the grid points 1 and 2 are as shown in Fig. 2.

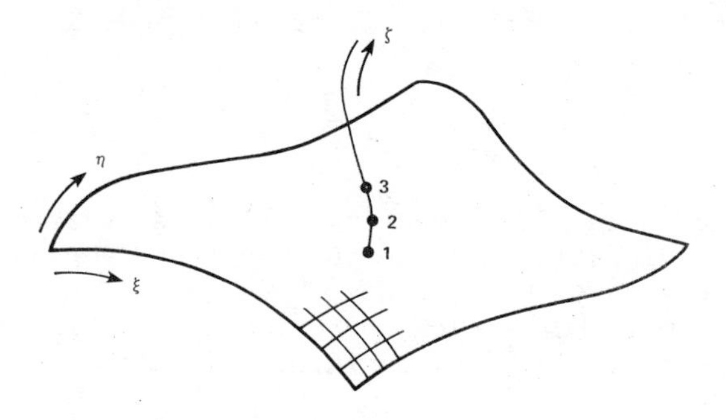

Figure 2. Grid surface.

From Eq. (14) it can be seen that the number of dependent variables at any grid point is five. Hence, in order to update all the five variables at a surface grid point, five equations are required. Equations (31) and (37) furnish a total of four equations. The fifth equation is obtained when either a constant-temperature-wall or adiabatic-wall condition is specified. For an adiabatic wall the condition to be satisfied at the wall is

$$\frac{\partial T}{\partial n} = 0 \tag{38}$$

Assuming a perfect gas, the following relationship is obtained

$$p = \rho RT \tag{39}$$

Differentiating Eq. (39) with respect to n and making use of Eq. (38), the following condition is obtained

$$\frac{\partial \rho}{\partial n} = 0 \tag{40}$$

In the transformed coordinate system this equation can be written as

$$[\xi_x\zeta_x + \xi_y\zeta_y + \xi_z\zeta_z] \frac{\partial \rho}{\partial \xi} + [\eta_x\zeta_x + \eta_y\zeta_y + \eta_z\zeta_z] \frac{\partial \rho}{\partial \eta} + [\zeta_x^2 + \zeta_y^2 + \zeta_z^2] \frac{\partial \rho}{\partial \zeta} = 0 \tag{41}$$

and, when the ζ coordinate lines are orthogonal to the body, Eq. (41) can be written using a first-order accurate, one-sided finite difference as

$$\rho_2 - \rho_1 = 0 \tag{42}$$

In the constant-temperature-wall case, the equation of state can be written as

$$\rho_1 = p_1/RT_{wall} \tag{43}$$

There now exist five equations at any surface grid point for both the adiabatic-wall and the constant-temperature-wall cases. It remains to be seen how these equations are implemented in schemes that employ approximate factorization. Consider the unsteady algorithm in three dimensions,

$$[I + \Delta\tau\delta_\xi\tilde{A}][I + \Delta\tau\delta_\eta\tilde{B}][I + \Delta\tau\delta_\zeta\tilde{C}]\Delta Q = P \tag{44}$$

where P denotes the right-hand-side vector whose elements are five-element column vectors. Carrying out two of the inversions we obtain

$$[I + \Delta\tau\delta_\zeta\tilde{C}]\Delta Q = [I + \Delta\tau\delta_\eta\tilde{B}]^{-1}[I + \Delta\tau\delta_\xi\tilde{A}]^{-1}P = T \tag{45}$$

Equation (45) is a system of equations that can be written in the matrix form as

$$\begin{vmatrix} D_1 & R_1 & & & \\ L_2 & D_2 & R_2 & & \\ & L_3 & D_3 & R_3 & \\ & & \cdot & & \\ & & & \cdot & \\ & & & & \cdot \end{vmatrix} \begin{vmatrix} \Delta\vec{Q}_1 \\ \Delta\vec{Q}_2 \\ \Delta\vec{Q}_3 \\ \cdot \\ \cdot \\ \cdot \end{vmatrix} = \begin{vmatrix} \vec{t}_1 \\ \vec{t}_2 \\ \vec{t}_3 \\ \cdot \\ \cdot \\ \cdot \end{vmatrix} \tag{46}$$

where

$$\Delta Q = \begin{vmatrix} \Delta\vec{Q}_1 \\ \Delta\vec{Q}_2 \\ \Delta\vec{Q}_3 \\ \cdot \\ \cdot \\ \cdot \end{vmatrix}, \qquad T = \begin{vmatrix} \vec{t}_1 \\ \vec{t}_2 \\ \vec{t}_3 \\ \cdot \\ \cdot \\ \cdot \end{vmatrix} \tag{47}$$

and

$$\left.\begin{aligned} L_2 &= -\frac{\Delta\tau}{2}\tilde{C}_1 \\ D_2 &= I \\ R_2 &= \frac{\Delta\tau}{2}\tilde{C}_3 \end{aligned}\right\} \tag{48}$$

The adiabatic-wall boundary condition is implemented by defining

$$\left.\begin{aligned} D_1 &= \begin{vmatrix} -J_1/J_2 & 0 & 0 & 0 & 0 \\ 0 & 1 & 0 & 0 & 0 \\ 0 & 0 & 1 & 0 & 0 \\ 0 & 0 & 0 & 1 & 0 \\ 0 & 0 & 0 & 0 & -J_1/J_2 \end{vmatrix}, \\ R_1 &= \begin{vmatrix} 1 & 0 & 0 & 0 & 0 \\ 0 & 0 & 0 & 0 & 0 \\ 0 & 0 & 0 & 0 & 0 \\ 0 & 0 & 0 & 0 & 0 \\ 0 & 0 & 0 & 0 & 1 \end{vmatrix}, \quad t_1 = \begin{vmatrix} 0 \\ 0 \\ 0 \\ 0 \\ 0 \end{vmatrix} \end{aligned}\right\} \tag{49}$$

Note that Eq. (49) is merely a matrix representation of Eqs. (31), (37), and (42). The constant-temperature-wall boundary condition can be implemented with the following definitions for D_1, R_1, and t_1:

$$\left.\begin{aligned} D_1 &= \begin{vmatrix} 1 & 0 & 0 & 0 & -1/RT_{wall} \\ 0 & 1 & 0 & 0 & 0 \\ 0 & 0 & 1 & 0 & 0 \\ 0 & 0 & 0 & 1 & 0 \\ 0 & 0 & 0 & 0 & -J_1/J_2 \end{vmatrix}, \\ R_1 &= \begin{vmatrix} 0 & 0 & 0 & 0 & 0 \\ 0 & 0 & 0 & 0 & 0 \\ 0 & 0 & 0 & 0 & 0 \\ 0 & 0 & 0 & 0 & 0 \\ 0 & 0 & 0 & 0 & 1 \end{vmatrix}, \quad t_1 = \begin{vmatrix} 0 \\ 0 \\ 0 \\ 0 \\ 0 \end{vmatrix} \end{aligned}\right\} \quad (50)$$

The preceding implementation of surface boundary conditions assumes that the coordinate lines intersect the surface of the body orthogonally. A simple way of implementing surface boundary conditions in cases in which the coordinate lines are not orthogonal to the body surface is to initially use Eqs. (49) or (50) to update all the dependent variables on the surface and then to correct the total energy at the surface using Eq. (35) [and the density for an adiabatic wall using Eq. (41)]. Although the viscous surface boundary conditions have been developed for the unsteady Navier-Stokes equations, a very similar approach can be used for the steady form of these equations (such as the parabolized Navier-Stokes equations). These boundary conditions can also be made more accurate by using three-point differences instead of two-point differences, as in Eqs. (37) and (42). Additional algebraic manipulation will then be required to restore the tridiagonal nature of the global matrix.

5. INITIAL CONDITIONS

To initiate a given calculation, the values of the dependent variables at each grid point in the computational plane must be specified. This initialization procedure depends on whether a marching technique or an iterative approach is being used to obtain the solution.

For an iterative approach in which one is interested only in the converged solution, the initial data, in most instances, can be a rather rough approximation of what the actual solution is believed to be. In using a purely shock-capturing approach, for example, it is common practice to set the flow variables at the interior points equal to the free-stream variables. In doing this, imposition of the tangency condition at the body creates a compression wave that propagates into the field and eventually settles down to its correct location as the solution converges.

Initialization for the blunt-body problem in which the bow shock is treated as a discontinuity is somewhat more complicated. In addition to initializing the dependent variables at the interior points, one must estimate the shock position and slope. Usually, this simply requires that the shock be specified analytically, thus yielding its slope and the flow variables behind it. In conjunction with a simple approximation for the body variables — for example, modified Newtonian flow followed by a linear interpolation for the flow variables between the body and shock — the initialization procedure is complete.

The initial data required when using a marching procedure in which the solution at each step is of interest are usually determined from some other source and supplied to the marching code. For example, when solving for the supersonic flow over a three-dimensional, blunt-nose configuration, a blunt-body computer program supplies the data in the starting plane. This includes all flow variables between the body and the shock, the shock position, and the shock slopes.

6. COMPUTATIONAL RESULTS

This section presents the computational results for different problems ranging from a simple sharp-cone flow to a multiple-shocked three-dimensional flow. These calculations were performed on the Ames Research Center's Cray-1S computer. For state-of-the-art applications using the present algorithm refer to [7,11,21,22].

6.1 Supersonic Cone Flow

To show the advantage of the implicit boundary conditions, both the present ones and those of [12], in the marching mode, the supersonic flow past a pointed cone was calculated. The flow consisted of an $M_\infty = 4$ inviscid flow past a 5° cone at 0° angle of attack. Three points in the circumferential direction and 23 equispaced points in the radial direction were used for the calculation. It was run with a cylindrical coordinate option. Figure 3, where iterations to convergence versus Courant number (ν) are presented, shows the advantage of using the implicit boundary conditions of [12]. With explicit boundary conditions, ν had to be less than 0.7 for stability and required 525 iterations for convergence. However, with the new implicit boundary conditions, ν was 63, and only 11 iterations were needed. The error in the surface density was less than 0.5% at all values of ν, as long as the

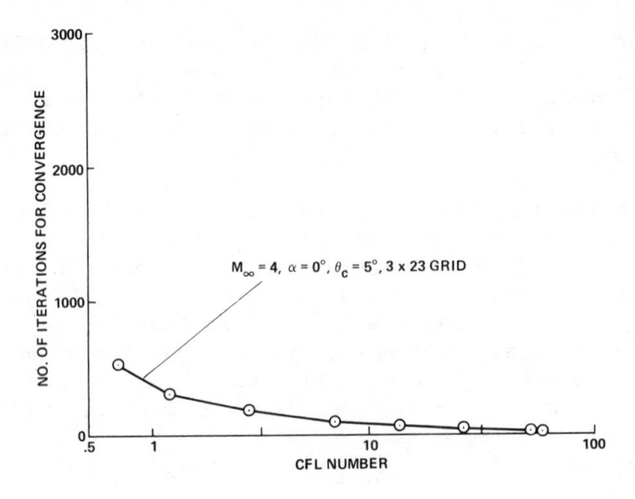

Figure 3. Convergence rate for inviscid supersonic flow over a cone.

calculations were done in a cylindrical frame of reference. In a typical viscous case, the present viscous boundary conditions would be slowly imposed over the first 100 iterations and the the above analysis (or speed) would hold for convergence.

Next, the laminar hypersonic flow around a blunt slender cone with a 4.7° cone half angle is calculated using the steady form of the Navier-Stokes equations (PNS). The free-stream Mach number is 10, the Reynolds number based on the cone length is 2.3×10^6, and the angle of attack is 20°, which corresponds to the ratio $\alpha/\theta_c = 4.25$.

At different axial-plane locations, a comparison of measured surface shear-stress direction angle and the calculated limiting streamline angle is presented in Fig. 4. There is a good agreement between the experiment and the calculation in predicting the circumferential angle at which the shear-stress angle changes sign (primary separation point).

The tangential conical components (with rays originating from an equivalent sharp cone) of the flow velocity at the first set of grid points off the surface was used to simulate the lines of shear stress. These are presented in a top and side view in Fig. 5. The primary separation line is clearly visible as the coalescence of the skin-friction lines. The particle trace also shows a qualitative picture of the development of the secondary separation between the lee attachment line and the primary separation line.

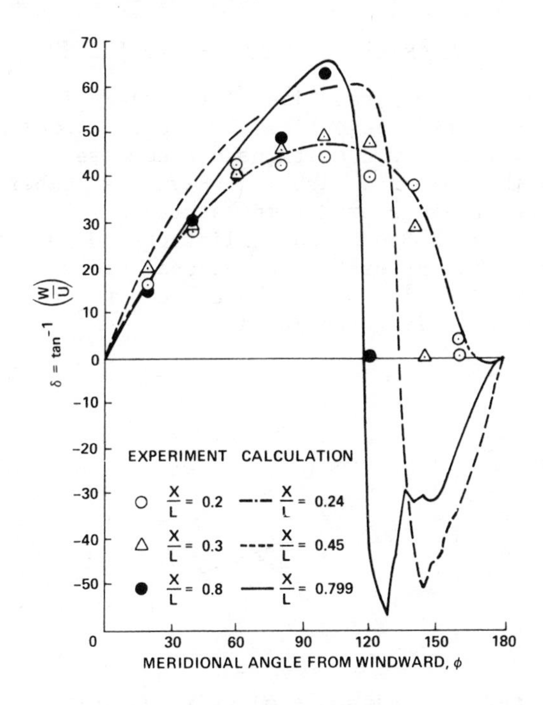

Figure 4. Comparison of the calculated and experimental shear-stress angle: $\theta_c = 4.7°$; $M_\infty = 10$; $Re_L = 2.3 \times 10^6$; $\alpha = 20°$.

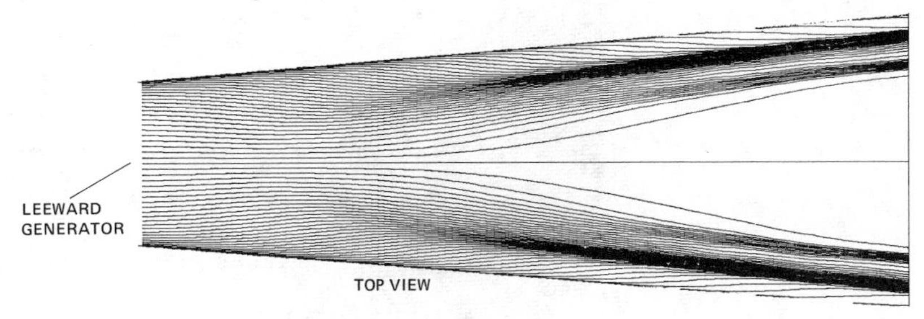

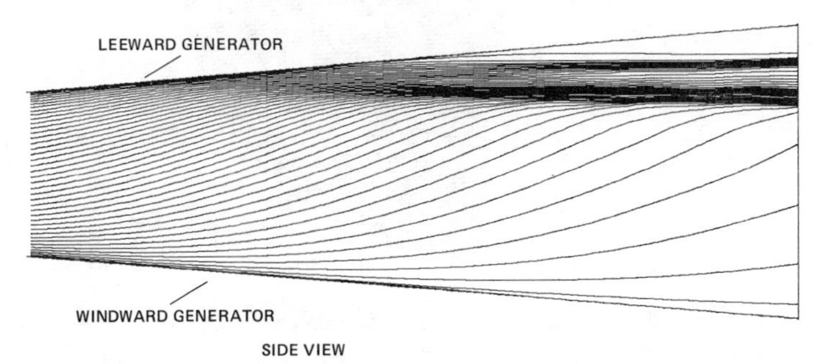

Figure 5. Calculated limiting streamlines: $\theta_c = 4.7°$; $M_\infty = 10$; $Re_L = 2.3 \times 10^6$; $\alpha = 20°$. a) Top view. b) Side view.

6.2 Supersonic Flow About a Finned Configuration

Results using the parabolized (steady) Navier-Stokes equations for a six-finned projectile at an angle of attack of 2° are presented. The flow conditions are composed of a free-stream Mach number of 4, a turbulent Reynolds number of 10^5/in., a free-stream static temperature of 100°R, and a wall static temperature of 540°R. An elliptic grid generator was used to create grids at each computational cross section of the finned projectile in Fig. 6. A typical example of such a cross-sectional grid is given in Fig. 7.

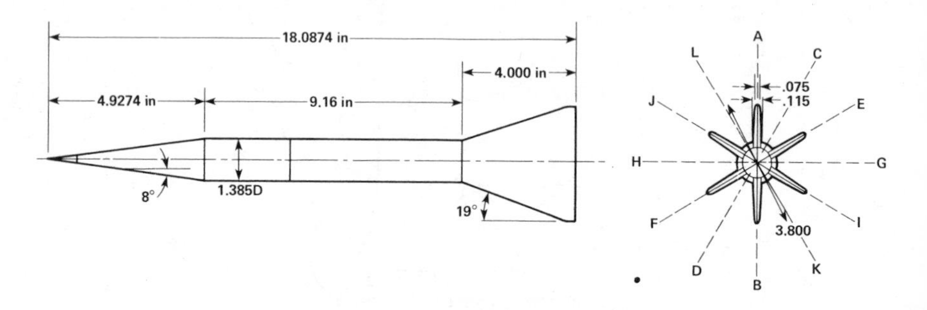

Figure 6. Finned projectile geometry.

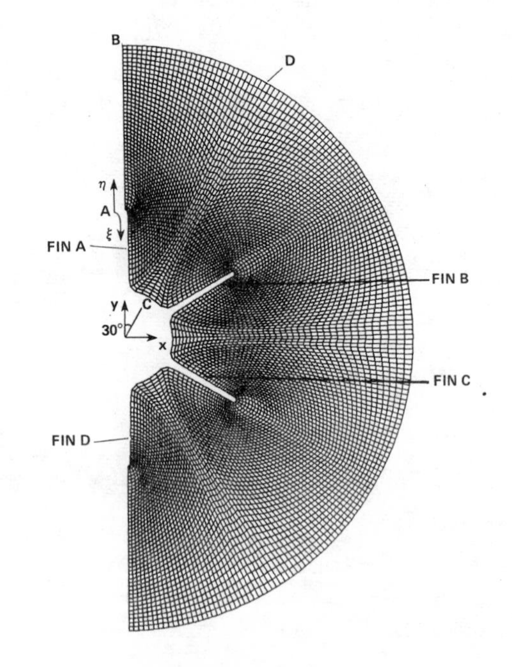

Figure 7. Grid for finned projectile at x = 17.55 in.

In Fig. 8, which is a crossflow velocity vector plot, a strong crossflow-separation region is evident over much of the body surface. The fin shocks associated with fins C and D (see Fig. 6 for fin nomenclature) can be clearly seen as a sharp change in velocity direction.

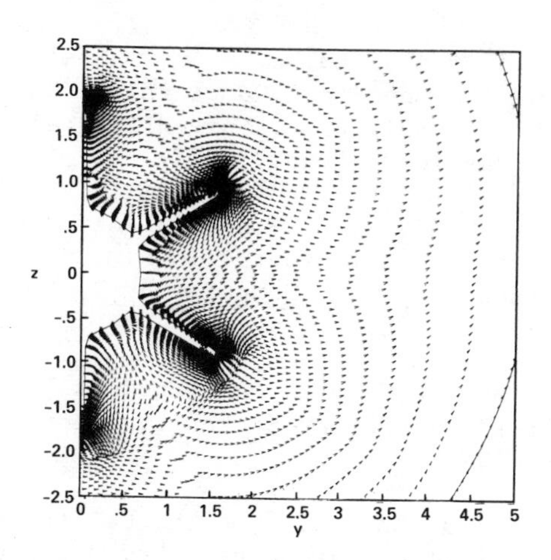

Figure 8. Crossflow velocity vectors for the projectile at $x = 17.55$ in. for $\alpha = 2°$, $M_\infty = 4$, and a turbulent $Re_L = 10^5$/in.

7. CONCLUDING REMARKS

The process of developing approximate-factorization algorithms in delta form for both unsteady and steady viscous flows has been presented. These algorithms are cast in conservation-law form and, further, are simplified by using a thin-layer approximation to the governing equations.

The implementation of implicit surface viscous boundary conditions is discussed in detail. The complete development of inviscid implicit boundary conditions has been referenced. An example showing the advantage of using these implicit boundary conditions has been included.

Three-dimensional results from the steady form of the algorithm are presented and compared with experiment whenever possible. A number of references to other results for both the unsteady and the steady computer codes show that the delta form of the approximate-factorization algorithm is robust and applicable to a wide variety of configurations.

REFERENCES

1. LINDEMUTH, I. AND KILLEEN, J. - Alternating Direction Implicit Techniques for Two-Dimensional Magnetohydrodynamic Calculations. J. Comput. Phys., Vol. 13, pp. 181-208, 1973.

2. BRILEY, W. F. AND McDONALD, E. - An Implicit Numerical Method for Multi-Dimensional Compressible Navier-Stokes Equations. Report M911363-6, United Aircraft Research Laboratories, Hartford, Conn., 1973.

3. BEAM, R. AND WARMING, R. F. - An Implicit Finite-Difference Algorithm for Hyperbolic Systems in Conservation-Law Form. J. Comput. Phys., Vol. 22, pp. 87-110, Sept. 1976.

4. BEAM, R. AND WARMING, R. F. - An Implicit Factored Scheme for the Compressible Navier-Stokes Equations. AIAA J., Vol. 16, Apr. 1978.

5. WARMING, R. F. AND BEAM, R. - On the Construction and Application of Implicit Factored Schemes for Conservation Laws. SIAM-AMS Proceedings, Proceedings of the Symposium on Computational Fluid Mechanics, New York, 1977.

6. PULLIAM, T. H. AND STEGER, J. L. - Implicit Finite-Difference Simulations of Three-Dimensional Compressible Flow. AIAA J., Vol. 18, pp. 159-167, Feb. 1980.

7. CHAUSSEE, D. S., BUNING, P. G. AND KIRK, D. B. - Convair 990 Transonic Flowfield Simulation. AIAA Paper 83-1785, 1983.

8. KUTLER, P., PEDELTY, J. A., AND PULLIAM, T. H. - Supersonic Flow Over Three-Dimensional Ablated Nosetips Using an Implicit Numerical Procedure. AIAA Paper 80-0063, Pasadena, Calif., 1980.

9. KUTLER, P., CHAKRAVARTHY, S. R., AND LOMBARD, C. P. - Supersonic Flow over Ablated Nosetips Using an Unsteady Implicit Numerical Procedure. AIAA Paper 78-213, Huntsville, Ala., 1978.

10. RIZK, Y. M., CHAUSSEE, D. S., AND McRAE, D. S. - Numerical Simulation of Viscous-Inviscid Interactions on Indented Nose Tips. AIAA Paper 82-0290, 1982.

11. RIZK, Y. M. AND CHAUSSEE, D. S. - Three-Dimensional Viscous Flow Computations Using a Directionally Hybrid Implicit-Explicit Procedure. AIAA Paper 83-1910, 1983.

12. VIGNERON, Y. C., RAKICH, J. V., AND TANNEHILL, J. C. - Calculation of Supersonic Viscous Flow over Delta Wings with Sharp Subsonic Leading Edges. NASA TM-78500, 1978.

13. SCHIFF, L. B. AND STEGER, J. L. - Numerical Simulation of Steady Supersonic Viscous Flow. AIAA Paper 79-130, New Orleans, La., 1979.

14. RAI, M. M. AND CHAUSSEE, D. S. - New Implicit Boundary Procedures Theory and Applications. AIAA Paper 83-0123, 1983.

15. CHAUSSEE, D. S., PATTERSON, J. L., KUTLER, P., PULLIAM, T. H., AND STEGER, J. L. - A Numerical Simulation of Hypersonic Viscous Flows Over Arbitrary Geometries at High Angle of Attack. AIAA Paper 81-0050, St. Louis, Mo., 1981.

16. RIZK, Y. M., CHAUSSEE, D. S., AND McRAE, D. S. - Computation of Hypersonic Flow around Three-Dimensional Bodies at High Angles of Attack. AIAA Paper 81-1261, Palo Alto, Calif., 1981.

17. CHAUSSEE, D. S. AND RIZK, Y. M. - Computation of Viscous Hypersonic Flow Over Control Surfaces. AIAA Paper 82-0291, Jan. 1982.

18. SCHIFF, L. B. AND STUREK, W. B. - Numerical Simulation of Steady Supersonic Flow over an Ogive-Cylinder-Boattail Body. AIAA Paper 80-0066, Jan. 1980.

19. TANNEHILL, J. C., VENKAPTAPATHY, E., AND RAKICH, J. V. - Numerical Solution of Supersonic Viscous Flow over Blunt Delta Wings. AIAA Paper 81-0049, Jan. 1981.

20. VENKATAPATHY, E., RAKICH, J. V., AND TANNEHILL, J. C. - Numerical Solution of the Space Shuttle Orbiter Flow Field. AIAA Paper 82-0028, 1982.

21. RAI, M. M., CHAUSSEE, D. S., AND RIZK, Y. M. - Calculation of Viscous Supersonic Flows over Finned Bodies. AIAA Paper 83-1667, 1983.

22. FUJII, K. AND KUTLER, P. - Numerical Simulation of the Leading-Edge Separation Vortex for a Wing and Strake-Wing Configuration. AIAA Paper 83-1098, July 1983.

FINITE DIFFERENCE SIMULATION OF UNSTEADY INTERACTIVE FLOWS

GEORGE S. DEIWERT

Research Scientist, CFD Branch
NASA Ames Research Center, Moffett Field, CA 94035

§1. INTRODUCTION

During the past decade advances in computer hardware and numerical methods have permitted the development of computer programs capable of simulating unsteady interactive flows. The physical realism of the simulated unsteady flows has been validated by comparison with experimental measurements. The promising results obtained so far, coupled with continued improvements in both computer performance and algorithm efficiency, encourage further development of these methods and their implementation in the study of unsteady interactive aerodynamic flows.

In this chapter the development of time dependent numerical simulations of unsteady interactive flows of an aerodynamic nature is reviewed. The focus is primarily on compressible flows at flight Reynolds numbers and noniterative schemes based on Navier-Stokes equations. In the following sections the governing equations are outlined, time and length scales are discussed, and numerical methods currently in use are reviewed. A selection of computed results and their comparison with experiment are presented followed by concluding remarks.

§2. GOVERNING EQUATIONS

The equations of motion for continuum fluid mechanics are the Navier-Stokes equations. For many flows of aerodynamic interest these equations can be greatly simplified such as for inviscid flows (Euler equations), or irrotational flows (potential equations), or simple, thin shear layers (boundary layer equations). For unsteady interactive flows, however, such simple uncouplings are not possible and the full form of the equations are generally considered. Interactions between the inviscid external flow and the viscous wall-bounded flow are typified by rapid thickening of the shear layer with strong streamline curvature. This is often accompanied by separation or flow reversal. Streamwise pressure gradients can be quite large and shock waves may exist that penetrate the shear layer. In addition, interaction between two viscous dominated flows occurs at the trailing edge of bodies or between elements of multielement configurations. Here two (or more) shear layers, with different upstream histories, interact and form a complex shear layer. Further complexity exists at flight Reynolds numbers in the form of turbulence which extends the range of length and time scales that require consideration.

For unsteady interactive flows of aerodynamic interest some simplifications to the full Navier-Stokes equations for compressible flow can be made. One is to time average the equations over a time scale which is small when compared to the aerodynamic time scale for unsteady flow yet large when compared to the time scale of the turbulent eddies. This results in the Reynolds-averaged form of the Navier-Stokes equations which contain Reynolds stress terms that must be modeled empirically. Another simplification that is sometimes used is the thin- shear-layer approximation. Here all streamwise- and cross-derivatives of the viscous as well as the turbulent stress terms are neglected. The momentum equation across the shear layer is still retained, however, so that the critical coupling between the wall-bounded shear flow and the inviscid, external flow is not lost.

It is generally convenient to cast the equations in conservation-law form to facilitate the capture of discontinuities and so that global conservation of the dependent variables can be easily maintained. The conservative form of the differential equations avoids fictitious sources along discontinuities and permits the numerical attainment of the weak solution to the equations. It is highly desirable to write the equations for a body-oriented coordinate system so that the description and modeling of Reynolds stress terms in the wall-bounded shear layers are not unnecessarily complex, and so that empirical models developed for thin shear layers can be easily used and modified. There are two possible ways to write the equations for generalized geometries while maintaining strong conservation-law form. One is to write the equations in integral form using Cartesian momentum components and space coordinates, and contravariant velocity components. These equations are applied to volume elements of arbitrary shape and are commonly referred to as the finite volume formulation [1 - 10]. They are written below as

$$\partial_t \int_{vol} Q \, dvol \; + \; \int_s F \cdot \vec{n} \, ds \; = \; 0 \tag{1}$$

where for Cartesian momentum components

$$Q = \begin{pmatrix} \rho \\ \rho u \\ \rho v \\ e \end{pmatrix}, \quad F = \begin{pmatrix} \rho\vec{q} \\ \rho u\vec{q} + \tau \cdot \vec{e}_x \\ \rho v\vec{q} + \tau \cdot \vec{e}_y \\ e\vec{q} + \tau \cdot \vec{q} - K_e \nabla T \end{pmatrix} \tag{2}$$

Although the momentum components, u and v, are written in the Cartesian coordinate system, the velocity and heat flux vectors can be written in the curvilinear mesh coordinate system so that

$$\vec{q} = u\vec{e}_x + v\vec{e}_y = U\vec{g}_\xi + V\vec{g}_\eta \tag{3}$$

$$\nabla T = \partial_x T\vec{e}_x + \partial_y T\vec{e}_y = \partial_\xi T\vec{g}_\xi + \partial_\eta T\vec{g}_\eta \tag{4}$$

where U and V are the contravariant velocity components.

The other way to write the equations is to use Cartesian momentum components and contravariant velocity components and transform the space coordinates to a generalized system. When the Navier-Stokes equations in conservation-law form are transformed from the Cartesian coordinates to arbitrary curvilinear coordinates they do not generally retain conservation-law form. Following the method proposed by Viviand [11] they can be put in conservation law form and are written below as:

$$\partial_t(JQ) + \partial_\xi(JF\cdot\vec{g}^\xi) + \partial_\eta(JF\cdot\vec{g}^\eta) = 0 \tag{5}$$

In both formulations (eq. (1) and eq. (5)) the Reynolds stresses and turbulent heat flux terms have been included in the stress tensor and heat flux vector by using the eddy viscosity and eddy conductivity concept, whereby the coefficients of viscosity and thermal conductivity are the sum of the molecular (laminar) part and an eddy (turbulent) part. Eddy viscosity models incorporate turbulent transport into the molecular transport stress tensor by adding the scalar eddy-viscosity transport coefficient μ_T thereby relating turbulent transport directly to gradients of the mean flow variables. In a Cartesian coordinate system, the two-dimensional molecular stress tensor can be written as

$$\tau_\ell = (p+\sigma_x)\vec{e}_x\vec{e}_x + \tau_{xy}\vec{e}_x\vec{e}_y + \tau_{yx}\vec{e}_y\vec{e}_x + (p+\sigma_y)\vec{e}_y\vec{e}_y \tag{6}$$

where the components are defined by

$$\sigma_x = -2\mu u_x - \lambda(\partial_x u + \partial_y v) \tag{7}$$

$$\sigma_y = -2\mu v_y - \lambda(\partial_x u + \partial_y v) \tag{8}$$

$$\tau_{xy} = \tau_{yx} = -\mu(\partial_y u + \partial_x v) \tag{9}$$

$$\lambda = -2\mu/3 \tag{10}$$

The total shear (molecular plus turbulent) is written as

$$\tau = \tau_\ell + \tau_T = (p+\tilde{\sigma}_x)\vec{e}_x\vec{e}_x + \tilde{\tau}_{xy}\vec{e}_x\vec{e}_y + \tilde{\tau}_{yx}\vec{e}_y\vec{e}_x + (p+\tilde{\sigma}_y)\vec{e}_y\vec{e}_y \tag{11}$$

where

$$\tilde{\sigma}_x = -2(\mu+\mu_T)u_x - \lambda(\partial_x u + \partial_y v) \tag{12}$$

$$\tilde{\sigma}_y = -2(\mu + \mu_T)v_y - \lambda(\partial_x u + \partial_y v) \quad (13)$$

$$\tilde{\tau}_{xy} = \tilde{\tau}_{yx} = -(\mu + \mu_T)(\partial_y u + \partial_x v) \quad (14)$$

In a similar manner, turbulent heat transport is defined in terms of mean energy gradients and an eddy conductivity coefficient K_e so that

$$K_e = K + K_T\,, \qquad \mu_e = \mu + \mu_T \quad (15)$$

Typically, the eddy conductivity coefficient is related to the eddy viscosity coefficient via a turbulent Prandtl number Pr_T where

$$Pr_T = C_p\,\mu_T / K_T \quad (16)$$

The simplest eddy viscosity models are algebraic. In many cases algebraic models of the type suggested by Smith and Cebeci [12] are used where in the boundary layer near the solid surface we have

$$\mu_T = \rho \ell_p^2 \left[(\partial_y u)^2 + (\partial_x v)^2\right]^{1/2} \quad (17)$$

$$\ell_p = 0.4\,\eta\,[1 - exp(-\eta/A)] \quad (18)$$

$$A = 26\mu_s / \sqrt{\rho_s \tau_s} \quad (19)$$

and in the outer part of the boundary layer and in wakes we have

$$\mu_T = 0.0168\,U_\delta \delta_i^* \Big/ \left[1 + 5.5[(\eta - \eta_{DS})/\overline{\delta}]^6\right] \quad (20)$$

$$\delta_i^* = \int_{\eta_{DS}}^{\overline{\delta}} (1 - U/U_\delta)\,d\eta \quad (21)$$

To approximate the influence of upstream history, a simple relaxation procedure can be used such that

$$\mu_T(\xi, \eta) = \alpha\,\mu_T(\xi - \Delta\xi, \eta) + (1 - \alpha)\mu_{T_{eq}} \quad (22)$$

where $\mu_{T_{eq}}$ is defined by equations (17) and (20), α is a relaxation parameter with value between zero and one and $\Delta\xi$ is the local streamwise computational mesh spacing. For $\alpha = 0$, there is no relaxation and for $\alpha = 1$, the eddy viscosity is frozen. A value of 0.3 is commonly used for α and the turbulent Prandtl number is assumed constant at 0.90.

An attractive alternative to the Smith-Cebeci model is the algebraic model proposed by Baldwin and Lomax [13]. This model is particularly well suited to complex flows that contain regions in which the length scales are not clearly defined. It is described briefly as follows: For wall-bounded shear layers, a two-layer formulation is used such that

$$\mu_T = (\mu_T)_{inner} \qquad \eta < \eta_{crossover} \quad (23)$$

$$\mu_T = (\mu_T)_{outer} \qquad \eta > \eta_{crossover} \tag{24}$$

where η is the normal distance from the wall and $\eta_{crossover}$ is the smallest value of η at which values from the inner and outer formulas are equal. The Prandtl-Van Driest formulation is used in the inner (or wall) region.

$$(\mu_T)_{inner} = \rho \ell_p^2 |\varsigma| \tag{25}$$

$$|\varsigma| = |\partial_y u - \partial_x v| \tag{26}$$

The formulation for the outer region is given by

$$(\mu_T)_{outer} = 0.0168\, C_{cp}\, F_{wake}\, F_{Kleb}(\eta) \tag{27}$$

$$F_{wake} = \begin{pmatrix} \eta_{max} F_{max} \\ C_{wk}\, \eta_{max}\, q_{dif}^2 / F_{max} \end{pmatrix} \text{ the smaller} \tag{28}$$

The quantities η_{max} and F_{max} are determined from the function

$$F(\eta) = \eta\, |\varsigma|\, [1 - exp(-\eta/A)]$$

where F_{max} is the maximum value of $F(\eta)$, and η_{max} is the value of η at which it occurs. The function $F_{Kleb}(\eta)$ is the Klebanoff intermittency function given by

$$F_{Kleb}(\eta) = [1 + 5.5(C_{Kleb}\, \eta/\eta_{max})^6]^{-1} \tag{29}$$

The quantity q_{dif}^2 is the difference between the maximum and minimum total velocity squared in the profile (along a η coordinate line),

$$q_{dif}^2 = q_{max}^2 - q_{min}^2 \tag{30}$$

and for boundary layers, the minimum is defined as zero.

$$C_{cp} = 1.6 \quad , \quad C_{wk} = 0.25 \quad , \quad C_{Kleb} = 0.3 \tag{31}$$

The advantage of this model for boundary-layer flows are as follows: (1) for the inner region, the velocity and length scales are always well defined, and the model is consistent with the "law of the wall"; (2) in the outer region for well-behaved (simple) boundary layers, where there is a well-defined length scale (η_{max}), the velocity scale is determined by F_{max}, which is a length scale times a vorticity scale; (3) in the outer region of complex boundary layers where the length from a wall becomes meaningless, a new length scale is determined from a velocity (q_{dif}) divided by a velocity gradient ($|\varsigma|$), and the velocity scale is q_{dif}.

It is also possible to model the Reynolds stresses directly rather than relating them to the mean field gradients via an eddy viscosity concept, but this has received little attention to date for unsteady interactive flows. In using this direct modeling approach it is convenient to write the Reynolds stress tensor in terms of the coordinate system aligned with the shear layer. This results in considerable simplification of empirical descriptions and permits the direct use of models developed for thin shear layers.

$$\tau_T = (-\rho\overline{U'^2})\mathfrak{I}_\xi\mathfrak{I}_\xi + (-\rho\overline{U'V'})\mathfrak{I}_\xi\mathfrak{I}_\eta + (-\rho\overline{V'U'})\mathfrak{I}_\eta\mathfrak{I}_\xi + (-\rho\overline{V'^2})\mathfrak{I}_\eta\mathfrak{I}_\eta \tag{32}$$

The accuracy of numerical simulations with the Reynolds-averaged Navier-Stokes equations depends principally upon the accuracy of the turbulence modeling. The eddy coefficients are given by empirical expressions which can range from fairly simple algebraic expressions based on mixing-length concepts to fairly complex expressions based on empirical transport equations to determine length and velocity scales. Most unsteady interactive computations to date (except some used only to approach a steady state in a timewise manner) have relied on the simpler algebraic expressions for eddy viscosity and a constant turbulent Prandtl number to determine eddy conductivity. These algebraic models are developed from boundary layer concepts and in general have not been validated for other than thin shear layers. In recent work by Shamroth [14] a differential expression for turbulent kinetic energy combined with an algebraic length scale to describe the eddy viscosity was used to study subsonic flow over an oscillating airfoil where the influence of viscous/inviscid interactions is small.

§3. TIME AND LENGTH SCALES

To simulate unsteady flows it is necessary to know what time and length scales are important and thus require resolution. Time scales exist that range from the very short periods associated with the high-frequency dissipative turbulence structure (the Kolomogorov micro-scale) to the very long times associated with slow moving signals which are propagated along upstream characteristic paths in a transonic flow field. (For example, pressure waves propagate upstream at a speed equal to $(1-M_\infty)a$, which for Mach numbers close to unity can be quite slow.) Length scales exist that range from the very small micro scale structure of the dissipative turbulent eddies to the very long scales that extend from the aerodynamic body to the outer boundaries of the computational control volume. Many unsteady flows of aerodynamic interest have important time and length scales that are somewhere in the middle of this vast range. Numerical schemes can be selected that neglect the very short time and length scales and yet are sufficient so that the scales of concern can be resolved.

Strong interactive effects occur when the range of flow conditions and airfoil motion parameters produce unsteady shock-induced boundary layer separation, trailing-edge separation or various combinations of interactions which result in separation-induced transonic flutter, buffet, aileron buzz and dynamic stall. In the absence of forced motions the characteristic speed that drives the unsteady behavior is the free-stream velocity (U_∞) and the characteristic length is the body-length scale (L), or separation scale (S). A non dimensional frequency parameter ($\overline{\Omega}$), can be defined that describes the characteristic time scale as $\overline{\Omega} = fL/U_\infty$ where f is the dimensional frequency of unsteady motion. For forced frequencies, such as occur with propellers and helicopter rotors, the dominant driven frequency is sinusoidal with higher harmonics becoming important as the blades pass through the trailing vortices of the preceding blade(s). These flows are characterized by a nondimensional frequency parameter defined as $\overline{k} = \omega c/2\pi$ where ω is the circular frequency and c is the chord of the airfoil section.

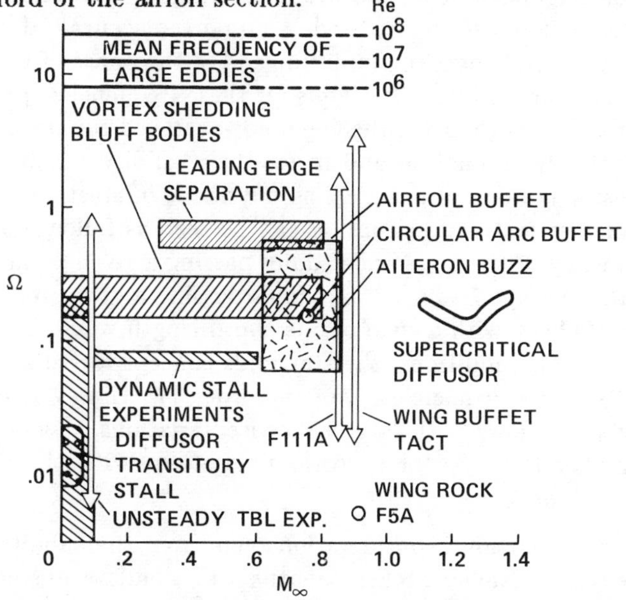

Fig. 1 Comparison of frequency range of unsteady flows with mean frequency of large-scale turbulent eddies.

An important consideration concerns how high the frequencies, f or $\overline{k}/c$, can be relative to the mean frequency, f_T, of the turbulent eddies for realistic simulations with the Reynolds-averaged equations. For the concept to be valid, the averaging time interval must be long when compared to the characteristic time f_T^{-1} of the principal turbulent eddies and short when compared to the characteristic time f^{-1} of the unsteady mean flow. Hence, f should be much smaller than f_T. To obtain a perspective on this question, Chapman [15] assembled some relevant data for unsteady aerodynamic flows. These are reproduced in Fig. 1 which maps typical unsteady aerodynamic flow and turbulent eddy domains as a function of non dimensional frequency parameter, $\overline{\Omega}$, and flight Mach number, M_∞.

The lines representing the mean frequency of the turbulent eddies are based on flat plate experiments and correspond to $f\bar{\delta}/U_\infty = 0.2$, the experimentally observed mean turbulent burst period. Also shown are domains representative of airfoil buffet, wing buffet, leading-edge separation, vortex shedding behind bluff bodies, supercritical diffusor stall, low-speed diffusor transitory stall, dynamic stall, transonic wing rock, and unsteady boundary layer experiments [16]. Almost all of the frequencies of these unsteady aerodynamic flows are one to two orders of magnitude smaller than f_T. The two open circle points in Fig. 1 represent airfoil buffeting and aileron buzz for which frequencies the Reynolds-averaged equations have provided good simulations when using turbulence models developed for steady flows. The unsteady frequencies in these cases are two orders of magnitude less than f_T. At the highest frequencies tested, the usual steady flow turbulence models supported accurate descriptions of the time-varying changes in amplitude and in the phase of the velocity profiles and turbulence intensity. Thus for these frequencies, just one order of magnitude less than f_T, the Reynolds-averaged equations are adequate for unsteady simulation and hence their validity can be expected for many unsteady flows of practical aerodynamic interest. A partial explanation for this fortunate situation can be seen as follows: While the average frequency of the large scale eddies passing a given point on a surface is f_T, the average frequency of eddies passing a given spanwise station on an airfoil say with a span of one chord length would be the order of $100 f_T$. For such conditions, the Reynolds concept for time averaging may be realistic for frequencies f, of the order f_T. However, for highly three-dimensional flows with large spanwise variations, f, may need to be much smaller then f_T for realistic simulations with the Reynolds-averaged equations.

Another important consideration concerns the ability of the Reynolds-averaged Navier-Stokes equations to simulate unsteady flows with a wide range of frequency spectra such as can occur in rotating machinery or in helicopter rotors where the multiple elements, each of which generates and interacts with vortices, induce higher harmonics. Applications to date have been conducted for two-dimensional flows without the complications of three-dimensional effects, free-stream turbulence, airfoil vibrations or structural oscillations. These have resulted in essentially cyclic unsteadiness with a single narrow-band frequency. It is well known from experimental observations that many flows have complexities resulting in broader band distributions of frequencies. A capability to simulate these types of flow would permit the study, for example, of unsteady inlet flows feeding into compressors, compressor stall, certain flutter problems, gust loading and wing buffet. Such unsteady flow simulations would probably necessitate removal of the Reynolds averaging restriction and would use instead a large eddy simulation scheme in which only the fine scale turbulence associated with dissipation is empirically modeled. Since such computations are not yet feasible with todays algorithms and computers, the simulations must await a later generation of computational power and sophistication.

To determine finite difference solutions to the Reynolds-averaged equations, a computational grid must be constructed about the aerodynamic shape of interest. The grid must be capable of resolving all the essential length scales and at the same time be efficient so as not to over resolve the flow field and saturate computer storage systems and processing times. Stretching and clustering of grid points are used extensively and dynamic remeshing during transient phases of the solution is desirable to assure adequate resolution of high gradient regions.

The primary variable determining the required minimum number of grid points is the boundary layer thickness, $\overline{\delta}$. This thickness can be estimated from flat plate boundary turbulent layer behavior as $\overline{\delta} \approx 0.37L/Re_L^{0.2}$. For steady attached turbulent boundary layers the well-known "law of the wall" describes the boundary layer behavior near the body surface. In interactive regions, however, the log-law region of the turbulent boundary layer can be annihilated and it is necessary to resolve the boundary layer to the scale of the viscous sublayer if accurate simulations of separation and surface shear are to be expected. To assure this resolution, the first grid line off the surface should lie within the sublayer where the velocity varies linearly with distance from the surface (i.e. $u^+ = y^+$, where $u^+ = u/u_\tau$, $y^+ = \eta u_\tau/\nu$, and $u_\tau = (\tau_s/\rho_s)^{1/2}$.) This occurs for values of $y^+ \approx 8$ and can be estimated from the free-stream Reynolds number and body length scale by $\Delta\eta_{min} \approx 0.08L/(Re)^{1/2}$, where $\Delta\eta_{min}$ is the distance away from the body surface to $y^+ \approx 8$. From this first point additional grid lines can be distributed away from the body with exponentially increased spacings to a distance somewhere just outside the boundary layer. An external grid can be further constructed to extend the computational field to the outer edge of the computational control volume, again using either geometrical or algebraic progressions to increase grid spacings away from the body.

Unsteady flows typically contain regions of high gradients that move about in space; shock waves and shear layers, for example. Resolution of these high gradient regions requires a tight clustering of grid lines and efficient use of grid lines is best achieved by moving or adapting the clustered grid to the moving region. Fortunately, in many instances, this can be realized by adapting just one family of grid lines, for example lines of constant η for shear layers and lines of constant ξ for shocks normal to the stream-wise direction. When there are interacting shocks that result in structures not aligned with a principal coordinate the problem of adaptive meshing becomes much more cumbersome and complex. Unsteady interactive computations with completely general adaptive meshing of this kind have not yet been attempted. Examples of one coordinate adaptive meshing for shocks are given by MacCormack [2], Schiff [17], and Deiwert [9], and examples for near wake-flows (moving shear layers) by Deiwert [10]. Note that in the equations presented in the previous section, the time-varying metrics have been included to facilitate adaptive meshing.

§4. NUMERICAL METHODS

Finite difference methods for solving the Reynolds-averaged Navier-Stokes equations can be classified by type: either explicit, implicit, or some hybrid combination of the two. Explicit methods offer the advantage of low cost per step and ease of formulation and computer programming. Associated with them are time-step stability constraints based on convection of signals (the Courant condition) and on diffusion of signals (the viscous stability condition). The Courant condition restricts the time step, Δt, to values less than $\Delta\xi/(|U|+a)$ where $\Delta\xi$ is the mesh spacing and $(|U|+a)$ is the local convection speed in the ξ-direction plus the local speed of sound. A similar restriction exists for the η-direction. The viscous stability condition restricts the time step to values less than $\Delta\xi^2/2\nu$ (or $\Delta\eta^2/2\nu$). If these time-step constraints are compatible with the unsteady frequency of the flow being computed (i.e., if Δt from stability considerations is not orders less than f^{-1}) then explicit methods are a good choice. An example of an explicit method used to compute unsteady transonic flow over an airfoil section is the MacCormack [1-3] method which is of the Lax- Wendroff [18] type and solves the equations in the finite volume formulation.

Equation (1) can be solved for arbitrary geometries with computational meshes for arbitrary configuration. For simplicity in treating the Reynolds-stress equations, the coordinate system should be orthogonal and body oriented in the viscous dominated region.

Constructing the transformation between Cartesian space and the computational space we have

$$\xi = \xi(x,y,t) \quad , \quad \eta = \eta(x,y,t) \tag{33}$$

with a Jacobian of the transformation

$$J = \partial_\xi x\, \partial_\eta y - \partial_\xi y\, \partial_\eta x_\eta \tag{34}$$

The covariant base vectors can be written as

$$\vec{g}_\xi = \partial_\xi x\, \vec{e}_x + \partial_\xi y\, \vec{e}_y \quad , \quad \vec{g}_\eta = \partial_\eta x\, \vec{e}_x + \partial_\eta y\, \vec{e}_y \tag{35}$$

and the contravariant base vectors as

$$\vec{g}^\xi = \partial_x \xi\, \vec{e}_x + \partial_y \xi\, \vec{e}_y \quad , \quad \vec{g}^\eta = \partial_x \eta\, \vec{e}_x + \partial_y \eta\, \vec{e}_y \tag{36}$$

The transformation metrics are

$$g_{\xi\xi} = \vec{g}_\xi \cdot \vec{g}_\xi = (\partial_\xi x)^2 + (\partial_\xi y)^2 \tag{37}$$

$$g_{\eta\eta} = \vec{g}_\eta \cdot \vec{g}_\eta = (\partial_\eta x)^2 + (\partial_\eta y)^2 \tag{38}$$

$$g_{\xi\eta} = g_{\eta\xi} = \partial_\xi x\, \partial_\eta x + \partial_\xi y\, \partial_\eta y \tag{39}$$

and

$$g^{\xi\xi} = \vec{g}^\xi \cdot \vec{g}^\xi = (\partial_x \xi)^2 + (\partial_y \xi)^2 \tag{40}$$

$$g^{\eta\eta} = \vec{g}^\eta \cdot \vec{g}^\eta = (\partial_x \eta)^2 + (\partial_y \eta)^2 \tag{41}$$

$$g^{\xi\eta} = g^{\eta\xi} = \partial_x \xi\, \partial_x \eta + \partial_y \xi\, \partial_y \eta \tag{42}$$

For the computational coordinate system we can write equation (1) in diffferential form as

$$\partial_t Q + \frac{1}{L_i}\partial_\xi(F \cdot \vec{n}_\xi\, \Delta s_\xi) + \frac{1}{L_j}\partial_\eta(F \cdot \vec{n}_\eta\, \Delta s_\eta) = 0 \tag{43}$$

where $L_i = vol_{i,j}/\Delta\xi$ and $L_j = vol_{i,j}/\Delta\eta$. The flux components can be written as

$$(F \cdot \vec{n}_\xi \Delta s_\xi) = J\Delta\eta \begin{pmatrix} \rho U \\ \rho u U + \tau \cdot \vec{e}_x \cdot \vec{g}^\xi \\ \rho v U + \tau \cdot \vec{e}_y \cdot \vec{g}^\xi \\ eU + \tau \cdot \vec{q} \cdot \vec{g}^\xi - K_e \nabla T \cdot \vec{g}^\xi \end{pmatrix} \tag{44}$$

$$(F \cdot \vec{n}_\eta \Delta s_\eta) = J\Delta\xi \begin{pmatrix} \rho V \\ \rho u V + \tau \cdot \vec{e}_x \cdot \vec{g}^\eta \\ \rho v V + \tau \cdot \vec{e}_y \cdot \vec{g}^\eta \\ eV + \tau \cdot \vec{q} \cdot \vec{g}^\eta - K_e \nabla T \cdot \vec{g}^\eta \end{pmatrix} \tag{45}$$

where

$$\vec{n}_\xi \Delta s_\xi = (\vec{g}^\xi / \sqrt{g^{\xi\xi}})(\sqrt{g_{\eta\eta}}\Delta\eta) = J\Delta\eta\, \vec{g}^\xi \tag{46}$$

and

$$\vec{n}_\eta \Delta s_\eta = (\vec{g}^\eta / \sqrt{g^{\eta\eta}})(\sqrt{g_{\xi\xi}}\Delta\xi) = J\Delta\xi\, \vec{g}^\eta \tag{47}$$

If the solution $Q^n_{i,j}$ is known at time $t = n\Delta t$ at each mesh point (i, j), the solution at time $t = (n+1)\Delta t$ is calculated by

$$Q^{n+1}_{i,j} = \mathcal{L}(\Delta t)\, Q^n_{i,j} \tag{48}$$

where $\mathcal{L}(\Delta t)$ is a symmetric sequence of time split, one-dimensional difference operators $\mathcal{L}_\xi(\Delta t_\xi)$ and $\mathcal{L}_\eta(\Delta t_\eta)$. For example,

$$\mathcal{L}(\Delta t) = \mathcal{L}_\eta(\Delta t/2)\mathcal{L}_\xi(\Delta t)\mathcal{L}_\eta(\Delta t/2) \tag{49}$$

In this sequence the $\mathcal{L}_\eta$ operator is called twice, each time advancing the solution in time by $\Delta t/2$ by accounting only for the effect of the η-derivative in equation (43) on the solution. Similarly, the $\mathcal{L}_\xi$ operator advances the solution by Δt once by accounting only for the effect of the ξ-derivative on the solution. The $\mathcal{L}_\eta$ operator solves the time-split differential "equation"

$$\partial_t Q = \frac{1}{L_j}\partial_\eta (F \cdot \vec{n}_\eta \, \Delta s_\eta) = 0 \tag{50}$$

by first predicting a new value $\overline{Q_{i,j}}$ from the current solution value and then by correcting the predicted value. The two-step operations for $\mathcal{L}_\eta$ and $\mathcal{L}_\xi$ are written below as

$$\overline{Q_{i,j}^{n+1/2}} = Q_{i,j}^n - \frac{\Delta t}{vol_{i,j}} \nabla_\eta (F \cdot \vec{n}_\eta \, \Delta s_\eta)_{i,j}^n \tag{51}$$

$$Q_{i,j}^{n+1/2} = \frac{1}{2}\left[Q_{i,j}^n + \overline{Q_{i,j}^{n+1/2}} - \frac{\Delta t}{vol_{i,j}} \Delta_\eta \overline{(F \cdot \vec{n}_\eta \, \Delta s_\eta)_{i,j+1}^{n+1/2}}\right] \tag{52}$$

$$\overline{Q_{i,j}^{n+1}} = Q_{i,j}^{n+1/2} - \frac{\Delta t}{vol_{i,j}} \nabla_\xi (F \cdot \vec{n}_\xi \, \Delta s_\xi)_{i,j}^{n+1/2} \tag{53}$$

$$Q_{i,j}^{n+1} = \frac{1}{2}\left[Q_{i,j}^{n+1/2} + \overline{Q_{i,j}^{n+1}} - \frac{\Delta t}{vol_{i,j}} \Delta_\xi \overline{(F \cdot \vec{n}_\xi \, \Delta s_\xi)_{i+1,j}^{n+1}}\right] \tag{54}$$

In regions where stability constraints require very small time steps relative to the undisturbed external flow conditions (in the thin shear layers for example) the operator sequence can be repeated m times for every one time step in the external flow-field as follows

$$\mathcal{L}(\Delta t) = \left[\mathcal{L}_\eta\left(\frac{\Delta t}{2m}\right)\mathcal{L}_\xi\left(\frac{\Delta t}{m}\right)\mathcal{L}_\eta\left(\frac{\Delta t}{2m}\right)\right]^m \tag{55}$$

where the number of repeat times, m is determined by

$$\frac{\Delta t}{m} < \min_{i,j} [\, \Delta t_{\xi_{max}} \, , \, 2\,\Delta t_{\eta_{max}}] \tag{56}$$

and Δt is the time step determined from the external flow conditions.

Generally, however, at the high Reynolds numbers associated with flight conditions, the shear layers are so thin and require such finely spaced meshes, that the time constraints imposed by both the Courant condition and the viscous stability condition are prohibitively small. To circumvent this problem, either semi-implicit schemes are used (e.g., MacCormack [4,5]) whereby the diffusion dominated regions are treated implicitly and the convective dominated regions explicitly, or fully implicit schemes are used for the entire flow field (e.g., Beam and Warming

[19] and Briley and McDonald [20]). The MacCormack hybrid scheme requires the solution of simple tridiagonal matrices for the viscous terms and characteristic equations for some of the convective terms in the diffusion dominated regions, and retains an explicit formulation for convective dominated regions. To achieve this computational efficiency in the viscous region, the term $(\mathcal{L}_\eta)$ is split into a parabolic part, which is treated implicitly, and a hyperbolic part, which is treated explicitly. Thus, for the hyperbolic part we have

$$F_H = \begin{pmatrix} \rho\vec{q} \\ \rho u\vec{q} + p\vec{e}_x \\ \rho v\vec{q} + p\vec{e}_y \\ (e+p)\vec{q} \end{pmatrix} \tag{57}$$

and the parabolic part

$$F_P = F - F_H \tag{58}$$

Similarly, the operator $\mathcal{L}_\eta$ is split into two parts as

$$\mathcal{L}_\eta(\Delta t) \leftarrow \mathcal{L}_{\eta_H}(\Delta t)\mathcal{L}_{\eta_P}(\Delta t) \tag{59}$$

$$\mathcal{L}(\Delta t) = \left[\mathcal{L}_{\eta_H}\left(\frac{\Delta t}{2N}\right)\mathcal{L}_{\eta_P}\left(\frac{\Delta t}{2N}\right)\mathcal{L}_\xi\left(\frac{\Delta t}{N}\right)\mathcal{L}_{\eta_P}\left(\frac{\Delta t}{2N}\right)\mathcal{L}_{\eta_H}\left(\frac{\Delta t}{2N}\right)\right]^N \tag{60}$$

Details of the mixed MacCormack method can be found in [4] and [5]. The MacCormack hybrid method still requires satisfaction of a Courant condition for stability. Since this condition is based on the convective dominated flow regime where mesh spacings are large relative to the diffusion dominated regime, the time steps are not restrictive for unsteady flow computations.

Because of the programming logic required to hybridize the method it is difficult to vectorize this procedure for modern array processors. The fully implicit method of Beam and Warming requires the solution of block tridiagonal matrices, and hence requires more computation per grid point per time step than the hybrid method does on the average, but it is readily vectorized when approximate factorization of the differencing operators is used. Linear analysis indicates that the method is neutrally stable and has no formal time-step constraint. In actual practice, however, stability time constraints dependent on both the coordinate transformation and the mean flow variation occur. Because there is not yet a straight-forward procedure to estimate these constraints, the solutions must be monitored for violation of the stability condition. This is generally done by tracking the development of the residuals at each step of the solution.

To facilitate the approximate factorization of the implicit differencing operators so that the algorithm can be vectorized, it is convenient to make the "thin shear layer" approximation. Generally, this is consistent with thin shear layer approximations inherent in the turbulent transport models as well and poses no additional restriction on the generality of the scheme. The thin layer approximation requires that critical body surfaces be mapped onto η = constant planes, and that Re >> 1. The variation of the viscous terms along the body surface (here, taken as ξ) are neglected, and the terms in the η, or near-normal direction to the body surface are retained. To solve equation (5), the Beam and Warming implicit finite-difference algorithm is used. Written in operator notation, and using approximate factorization to facilitate efficient use of vector processors, the difference equation is

$$\overline{\mathcal{L}}_\xi \overline{\mathcal{L}}_\eta \Delta_t(JQ) = \mathcal{R}_\xi + \mathcal{R}_\eta \tag{61}$$

$$\overline{\mathcal{L}}_\xi = \left(I + \Delta t\, \delta_\xi\, A^n - \epsilon_I\, J \nabla_\xi \Delta_\xi J^{-1}\right) \tag{62}$$

$$\overline{\mathcal{L}}_\eta = \left(I + \Delta t\, \delta_\eta\, B^n - \epsilon_I\, J \nabla_\eta \Delta_\eta J^{-1} - \Delta t \delta_\eta\, J M^n J^{-1}\right) \tag{63}$$

$$\Delta_t(JQ) = (JQ)^{n+1} - (JQ)^n \tag{64}$$

$$\mathcal{R}_\xi = -\Delta t\, \delta_\xi \big(JF_H \cdot \mathcal{J}^\xi\big)^n - \epsilon_E\, J\, (\nabla_\xi \Delta_\xi)^2 J^{-1} (JQ)^n \tag{65}$$

$$\mathcal{R}_\eta = -\Delta t\, \delta_\eta (JF \cdot \mathcal{J}^\eta)^n - \epsilon_E\, J\, (\nabla_\eta \Delta_\eta)^2 J^{-1} (JQ)^n \tag{66}$$

where the δ_ξ and δ_η are central difference operators, ∇_ξ and ∇_η are backward- and Δ_ξ and Δ_η are forward- difference operators in the ξ and η directions respectively. For example, $\Delta_\eta Q = Q(\xi, \eta + \Delta\eta) - Q(\xi, \eta)$. Indices denoting spatial location have been suppressed for convenience. The Jacobian matrices

$$A = \partial_Q(JF_H \cdot \mathcal{J}^\xi) \quad \text{and} \quad B = \partial_Q(JF_H \cdot \mathcal{J}^\eta),$$

along with the coefficient matrix

$$M = \partial_Q(JF_P \cdot \mathcal{J}^\eta)$$

are described in Steger [21]. Fourth-order explicit (preceded by ϵ_E) and second-order implicit (preceded by ϵ_I) terms have been added to control nonlinear instabilities. The subscript H refers to the hyperbolic part of equation (2) (as described in eq. (57) in the mixed algorithm) and for the ξ-direction, the parabolic part has been neglected consistent with the thin-layer approximation.

An important consideration concerns the frequency range supported by a finite difference solution method. Finite grids can support only a finite number of frequencies in a discrete Fourier series. Higher frequencies than the grid will support are aliased to lower frequencies. For example, on an equispaced grid of n points, $n/2$ harmonics of the form

$e^{inx/2}$ can be accurately supported. Frequencies higher than $n/2$ reappear as lower frequencies. In unsteady flows with moving shocks, high frequencies are continually generated by nonlinear convective interactions. For example, the product of waves $e^{imx}e^{inx}$ arise due to terms such as uU and vU, and these produce two harmonics, a lower one proportional to $(m - n)$ and a higher one proportional to $(m + n)$. The numerical problem that occurs in this situation has been described by Mehta and Lomax [22] and is illustrated in Fig. 2 which is taken from their paper. Schematically, amplitude is shown as a function of wave number. The frequencies to the right of the mesh cutoff line are subgrid frequencies that alias back to the low frequency range and introduce numerical error. This error can be sufficient to cause numerical instability. Artificial numerical dissipation is generally introduced to remove the high frequency terms before significant aliasing occurs. Additionally, mesh clustering is generally used in the vicinity of shocks where high frequency terms are generated and this reduces the physical extent of aliasing problems that would otherwise occur on a coarse grid.

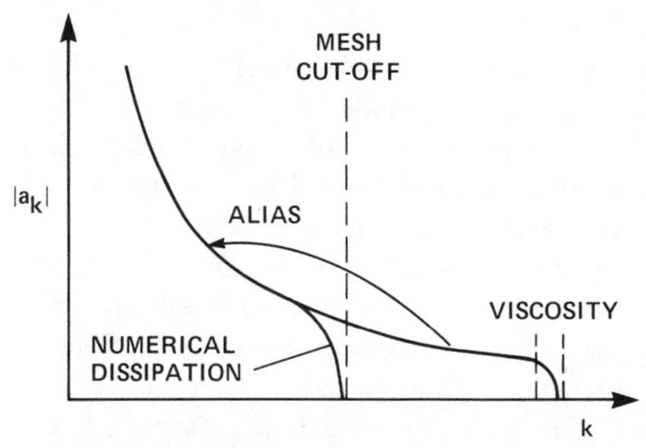

Fig. 2 Numerical dissipation of subgrid amplitudes.

A second consideration concerns the use of shock-capturing methods to describe a discontinuity as it moves about in the mesh. When such a technique is used the shock profile is "smeared" over a few mesh points. For most applications the shock strength and location are adequately represented and, in fact, for shock/boundary layer interactions, the smeared shock structure in the interaction region is preferred over a discontinuous representation. Alternatively, shock fitting may be esthetically more attractive in the inviscid regions. However, shock capturing is practically more advantageous, resulting in no real loss of accuracy near the body. It is recognized, of course, that shocks really are essentially discontinuities in the exterior flowfield.

A third consideration concerns the computation of turbulence or, rather, the effect of turbulence. No real attempt is made to resolve the full range of scales inherent in a turbulent shear flow, nor is there any attempt to account for the inherent three-dimensional structure of turbulent fields.

There is, however, an interaction between the numerical procedures and the computation of turbulence effects. Two different issues are involved. One is the manner in which the subgrid scales are accommodated and the other is the manner in which the turbulence model is implemented. Subgrid scales are continually generated by the larger scale structure by means of nonlinear wave interactions in the convective terms. Numerical control of the subgrid energy production is achieved by the addition of dissipation either through approximations to spatial derivatives, or by artificial terms. In either instance the dissipation is arbitrary to the extent that it must lie within the error band of the large scale resolution and it must prevent the accumulation of energy in the highest frequencies supported by the mesh. This artificial dissipation is not related to the eddy viscosity that is empirically modeled and must not be of comparable order. However, even though the detailed form of this dissipation is somewhat arbitrary, its presence is essential to prevent the flow of subgrid scale energy to the large scale terms where it would not have physical meaning.

The second interaction is more subtle and is related to the manner in which the turbulence model is incorporated into the computer code. While the analytical form of a given eddy viscosity model is well described, its implementation and the means by which certain key parameters (particularly length scales) are evaluated are not clear. Every code developer practices the art of model implementation. The numerical effect of the complete model is the sum of all its parts, including grid distribution effects, metric evaluation techniques, difference approximations, the adaptation of "thin-shear-layer" models to describe complex shear layers, etc. Because the models are empirical, a wide degree of freedom is often exercized in their implementation. The accuracy of the models with all these ingredients is difficult to evaluate and the final assesment of the method must be based on comparisons with experiments and benchmark computations.

§5. RESULTS

We mention five examples of unsteady interactive flow which have been computed. In each case the flow is transonic and there are in each instance some experimental results with which comparisons are made. In each of the five cases, the time scale of interest is narrow-banded and is long when compared with the mean frequency of the turbulent eddies.

§5.1 biconvex airfoil

First, the experiments of McDevitt et al. [23] are considered; the transonic flow past an 18% thick biconvex circular arc airfoil at zero angle of incidence was investigated. The circular arc was placed in a high-Reynolds number channel with walls contoured to match streamlines predicted from a transonic Navier-Stokes code [6]. Both Mach number and Reynolds number were varied. At a Mach number of 0.72, the flow was steady, the viscous/inviscid interaction was somewhat weak, and flow separation occurred just ahead of the trailing edge. At a Mach number of 0.783, the flow was quasisteady, the viscous/inviscid interaction was strong and resulted in shock-induced separation well ahead of the trailing edge with reattachment in the wake of the airfoil. At Mach numbers in between these values the flow was observed to be highly unsteady with shock waves and separation points oscillating fore and aft with a dimensionless frequency of $\overline{\Omega} = 0.49$ [24].

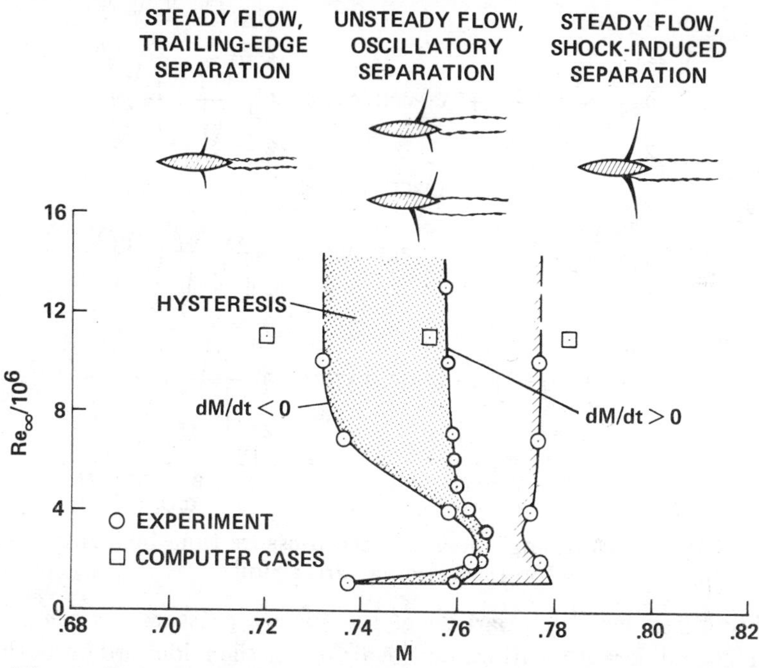

Fig. 3 Experimental flow domains for 18% circular arc airfoil.

Levy [25], using a code written by Deiwert [8] based initially on the explicit MacCormack method and subsequently the hybrid MacCormack method [9], simulated the same flow conditions and observed the same cyclic behavior for the same Mach number range as was observed experimentally. The computed frequency of the oscillations and the magnitude of the shock excursions agreed remarkably well with experimentally measured values. Shown in Fig. 3 are the Reynolds number and Mach number domains for which the flow was observed to be either weakly-interacting steady, stongly-interacting quasisteady, or strongly- interacting unsteady. Experimentally, as Mach number is increased the onset of the unsteady flow regime occurs for Mach numbers near 0.76 and terminates at values near 0.78. As Mach number is decreased the unsteady regime persists down to values near 0.73. The hysteresis region showing this difference in the unsteady flow domain with increasing and decreasing Mach numbers is shaded in Fig. 3. The computations of Levy are for fixed Mach numbers of 0.72, 0.754, and 0.783, all for a Reynolds number of 11 x 10^6. These are denoted by the square symbols in the figure. For the Mach number 0.72, the computed flow is steady as observed in the experiment; for 0.754 it is unsteady, and for 0.783 it is quasisteady, again, as observed experimentally.

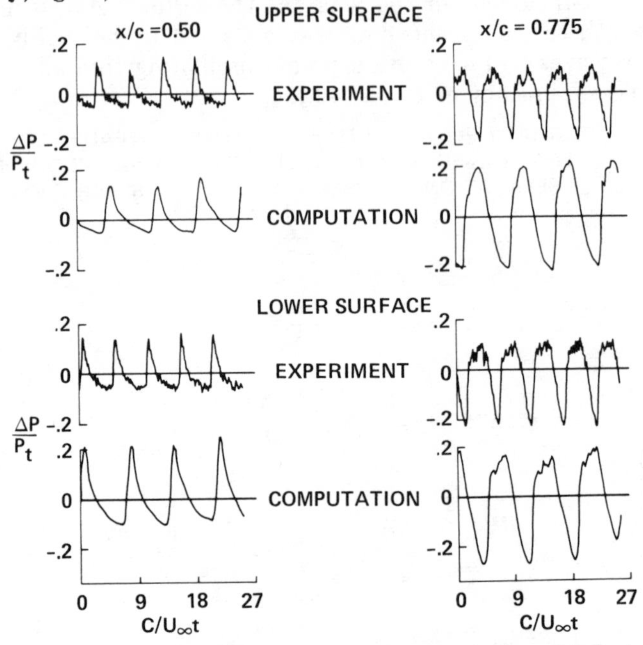

Fig. 4 "Buffeting" flow, surface-pressure time histories for 18% circular-arc airfoil.

Fig. 4 shows a comparison of the surface pressure variation with time at four different locations on the airfoil: at the midchord on both the upper and lower surface and at 77.5%-chord on both the upper and the lower surface. The oscillations on the upper and lower surfaces are a half period out of phase with each other. The frequency of the computed and measured oscillations agrees to within 20%. It is of special interest to note the agreement between some of the details of the pressure oscillations. For example, at the midchord position, both experimental and computed pressures show a very rapid rise and then a slower, almost exponential-like, decay. At the 77.5%-position, the pressure rises, is less rapid, and there is almost a pressure plateau with a fine scale structure before there is a decay to a pressure minimum again. Some of this fine scale structure is indicated in the computed results as well as in the experiments.

§5.2 buffet boundaries

Another unsteady flow application of the same code used above was the determination of buffet boundaries for the Korn 1 airfoil [26]. Fig. 5

shows a lift-drag polar and a lift curve for the Korn 1 supercritical airfoil for a nominal Mach number of 0.75. The computations, performed for angles of incidence ranging from -1.54° to 4.34°, are compared with the experimental data of Kacprzynski et al. [27]. The computed drag polar indicates the onset of buffet somewhat after maximum lift has been realized and is illustrated by two different C_C vs C_D branches for angles of 3.25° and 4.34°.

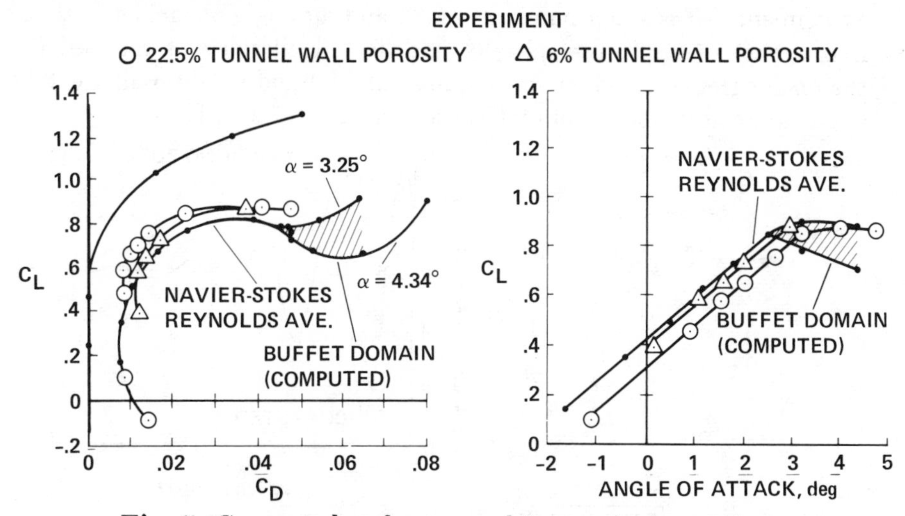

Fig. 5 Computed and measured transonic drag polar and lift curve for a supercritical airfoil.

The lift and drag vary periodically along the branch corresponding to the particular angle of incidence. Other angles of incidence greater than 3° (not shown) would exhibit different paths of periodic variation. The lift curve indicates that the onset of buffet occurs at an angle of incidence of nearly three degrees to the free stream. Here, for a given incidence, the minimum and maximum lift values define a buffet envelope. The computations were performed assuming free boundaries at the nominal wind-tunnel test conditions and no adjustments were made to account for Mach number or flow angularity corrections because of wall interference. Neglecting Mach number corrections, comparisons with the lift-curve data suggest equivalent angle of attack corrections of roughly -0.3° and -1.3° for the 6% and 20.5% wall porosity experiments, respectively. This compares with suggested corrections of -0.89° suggested by the experimental investigators in an earlier study

§5.3 aileron buzz

Another unsteady phenomenon, this time associated with a moving boundary, is represented by the performance characteristics of the aileron of a P-80 aircraft. This configuration was exhaustively investigated experimentally in the mid 1940s by Erikson and Stephenson [28]. This flow has been simulated by Steger and Bailey [29] using the fully implicit algorithm of Beam and Warming and a method that couples the solution of an ordinary differential equation describing the motion of the aileron with the flow field solution. The interrupted shock-wave motion over the aft portion of the airfoil causes a shift in phase of the aerodynamic hinge moment with respect to the movement of the aileron, thereby exciting an oscillation of the aileron (buzz) in one degree of freedom. In the experiment, for a Mach number of 0.82 and an angle of incidence to the free stream of -1.0°, the initially undeflected aileron was released (i.e. the one degree of freedom was made available) and would oscillate 22.2° about a mean incidence of -1.1° at a frequency of 22.2 Hz.

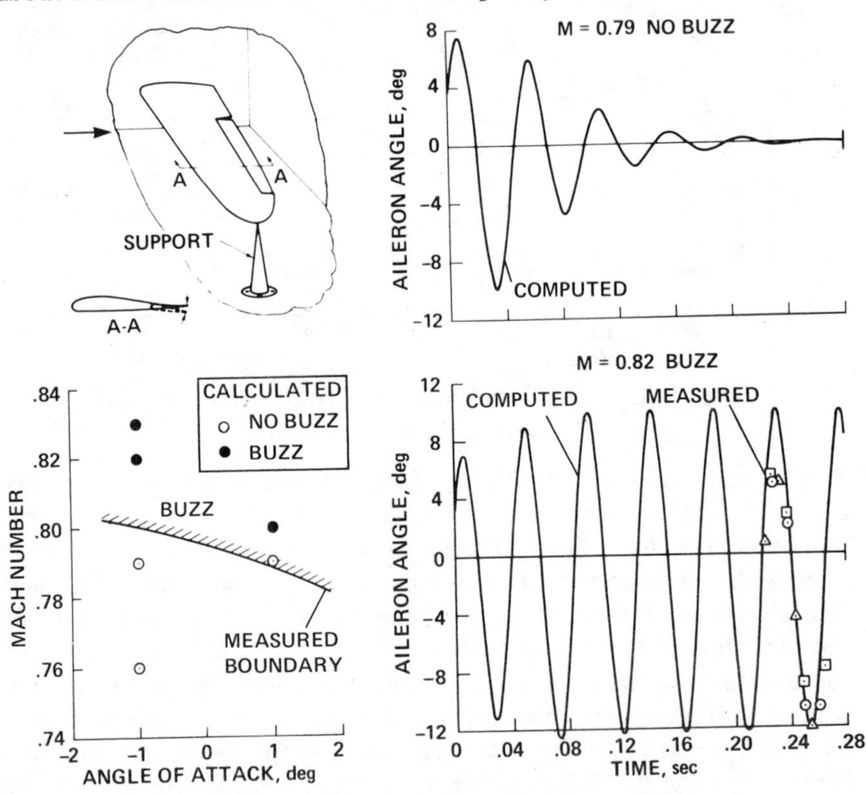

Fig. 6 Computed and measured characteristics of transonic aileron buzz.

Computationally, this unsteady behavior was not obtained for an initially undeflected aileron, but when the aileron was released from an initial position of 4° it experienced oscillations of 18.4° about a mean incidence of -3.0° at a frequency of 21.2 Hz. Similar computations for an airfoil angle of incidence of -1.0° were made for a free-stream Mach number of 0.79 and showed that even with an initial deflection of 4°, the oscillations would damp out in a few cycles. The results of these computations are compared with experiment in Fig. 6 which shows both the buzz boundary as a function of free-stream Mach number and airfoil angle of incidence, and aileron deflection angles as a function of time for free-stream Mach numbers of 0.79 and 0.82 for the airfoil at incidence of -1.0°.

§5.4 stall boundaries

A fourth example is the computation of the stall boundary of a given airfoil. Levy and Bailey [30], using both the hybrid MacCormack algorithm and the fully implicit Beam and Warming algorithm, performed a series of computations for a wide range of Mach numbers and angles of incidence for both a NACA 65-213 airfoil (the P-80 airfoil section) and the Korn 1 supercritical airfoil. Shown in Fig. 7 is the boundary for the onset of unsteady flow as a function of lift coefficient and free-stream Mach number for the Korn 1 section. Comparison with experimental data of Ohman et al. [31] for the same configuration show generally good agreement, especially for the higher values of Mach number and lower lift coefficients where the shock-wave system determines a buffet-onset boundary.

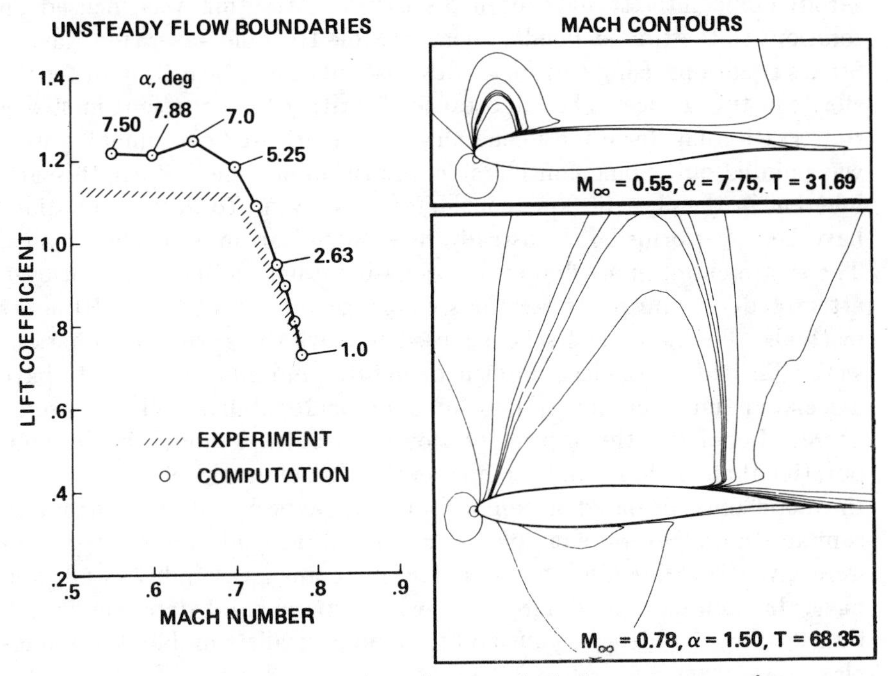

Fig. 7 A performance characteristic in maneuver of Korn-1 airfoil at Re = 21 x 10^6.

For the lower Mach numbers and high-lift coefficients (corresponding to high angles of incidence) the stall boundary is similar to that of classical low-speed trailing-edge separation as opposed to shock-wave/boundary-layer interactions. This is illustrated by the computed Mach contours for both a high incidence and low incidence configuration in the figure.

§5.5 dynamic stall

Several studies of dynamic stall have been performed using the compressible Navier-Stokes equations. In most instances the implicit method of Briley and McDonald was used to study flows with free-stream Mach numbers low enough so that there were no regions of supersonic flow and hence no shock/boundary-layer interactions. Included in these works are the laminar dynamic stall studies of Gibeling et al. [32] and Sankar and Tassa [33] and turbulent dynamic stall studies of Shamroth and Gibeling [34] and Tassa and Sankar [35]. Shamroth[14] also used a one equation differential eddy viscosity model to study an oscillating airfoil without stall.

§6. CONCLUDING REMARKS

In the preceding sections some of the considerations and problems associated with numerically simulating unsteady interactive flows of aerodynamic interest have been discussed. Attention was focused on solutions to the time-dependent compressible Reynolds-averaged Navier-Stokes equations using empirical eddy viscosity models to account for the effects of turbulence. The importance of writing the equations in strong conservation-law form for a generalized body-oriented coordinate system was pointed out. Some considerable discussion of time and length scales inherent in the class of flows considered was given. To date, simulations have been performed for unsteady flows with narrow frequency bands. The treatment of many flows with broad-band unsteadiness has not been attempted yet and poses a serious challenge to current state of the art methods. The numerical schemes used to solve the governing equations were classified as explicit, implicit or hybrid, and all are known to have associated time-step constraints for numerical stability. There is some degree of choice in the implementation of turbulence models in the computational algorithms, and the final evaluation of an algorithm including the model must be based on comparisons with experiment and benchmark computations. Several examples of simulated unsteady interacting flows were given covering such aerodynamic phenomena as buffet, stall, and buzz. In each case good agreement with experimental data was found, thus lending confidence to our ability to numerically simulate these complex flows.

References

1. MACCORMACK, R. W. - The Effect of Viscosity in Hypervelocity Impact Cratering. AIAA Paper No. 69-354, Jan. 1969.

2. MACCORMACK, R. W. and BALDWIN, B. S. - A Numerical Method for Solving the Navier-Stokes Equations with Application to Shock-Boundary Layer Interactions. AIAA Paper No. 75-1, Jan. 1975.

3. BALDWIN, B. S., MACCORMACK, R. W. and DEIWERT, G. S. - Numerical Techniques for the Solution of the Compressible Navier-Stokes Equations and Implementation of Turbulence Models. AGARD-LS-73, Computational Methods for Inviscid and Viscous Two- and Three-dimensional Flow Fields, 1975.

4. MACCORMACK, R. W. - A Rapid Solver for Hyperbolic Systems of Equations. A. I. van de Vooren and P. J. Zandbergen, eds., Lecture Notes in Physics, 59, Springer-Verlag, New York, pp. 307-317, 1976.

5. MACCORMACK, R. W. - An Efficient Explicit-Implicit-Characteristic Method for Solving the Compressible Navier-Stokes Equations. SIAM-AMS Proc., 11, pp. 130-155, 1978.

6. DEIWERT, G. S. - Numerical Simulation of High Reynolds Number Transonic Flows. AIAA Jour., 13, pp. 1354-1359, 1975.

7. DEIWERT, G. S. - Computation of Separated Transonic Turbulent Flows. AIAA Jour. 14, pp. 735-740, 1976.

8. DEIWERT, G. S. - On the Prediction of Viscous Phenomena in Transonic Flows. Adamson, T. C. and Platzer, M. F. eds., Transonic Flow Problems in Turbomachinery, Hemisphere Publishing Corp. pp. 371-391, 1977.

9. DEIWERT, G. S. - Recent Computation of Viscous Effects in Transonic Flow. A. I. van de Vooren and P. J. Zandbergen, eds., Lecture Notes in Physics, 59, Springer-Verlag, pp. 159-164, 1976.

10. DEIWERT, G. S. - Computation of Turbulent Near Wakes for Asymmetric Airfoils. K. H. Bronau and H. U. Meier, eds., Viscous and Interacting Flow Field Effects, Proceedings of the 4th U.S. Air Force and the Federal Republic of Germany Data Exchange Agreement Meeting BMVg-FBWT 79-31, pp. 455-467, 1979. Also NASA TM-78581, March 1979.

11. VIVIAND, H. - Conservation Forms of the Gas Dynamics Equations. Recherche Aerospatiale, 1 pp. 153-158, 1974.

12. SMITH, A. M. O. and CEBECI, T. - Numerical Solution of the Turbulent Boundary-Layer Equations. Douglas Aircraft Div. Report, DAC 33735, 1967.

13. BALDWIN, B. S. and LOMAX, H. - Thin Layer Approximation

and Algebraic Model for Separated Turbulent Flows. AIAA Paper 78-257, 1978.

14. SHAMROTH, S. J. - A Turbulent Flow Navier-Stokes Analysis for an Airfoil Oscillating in Pitch. Michel, R., Cousteix, J. and Houdeville, R. eds., Unsteady Turbulent Shear Flows Springer-Verlag, pp. 185-196, 1981.

15. CHAPMAN, D. R. - Computational Aerodynamics Development and Outlook. AIAA Journal 17, pp. 1293-1313, 1979.

16. MICHEL, R., COUSTEIX, J. and HOUDEVILLE, R. eds., - Unsteady Turbulent Shear Flows. Springer-Verlag, 1981.

17. SCHIFF, L. B. - A Numerical Solution of the Axisymmetric Jet Counterflow. A. I. van de Vooren and P. J. Zandbergen, eds., Lecture Notes in Physics, 59, Springer-Verlag, New York, pp. 391-397, 1976.

18. LAX, P. and WENDROFF, B. - Systems of Conservation Laws. Comm. Pure and Appl. Math. 13, pp. 217-237, 1960.

19. BEAM, R. M. and WARMING, R. F. - An Implicit Factored Scheme for the Compressible Navier-Stokes Equation. AIAA Jour. 16, pp. 393-402, 1978.

20. BRILEY, W. R. and MCDONALD, H. - Solution of the Multidimensional Compressible Navier-Stokes Equations by a Generalized Implicit Method. J. Comp. Physics, Vol. 24, No. 4, ,p 372, August 1977.

21. STEGER, J. L. - Implicit Finite-Difference Simulation of Flow about Arbitrary Two-Dimensional Geometries. AIAA Jour. 16, pp. 679-686, 1978.

22. MEHTA, U. and LOMAX, H. - Reynolds Averaged Navier-Stokes Computations of Transonic Flows- The State-of-the-Art. D. Nixon ed., Transonic Aerodynamics, M. Summerfield ed. Progress in Astronautics and Aeronautics 81, AIAA, New York, pp. 297-375, 1982.

23. MCDEVITT, J. B., LEVY, L. L., and DEIWERT, G. S. - Transonic Flow about a Thick Circular-arc Airfoil. AIAA Jour. 14, pp. 606-613, 1976.

24. SEEGMILLER, H. L., MARVIN, J. G. and LEVY, L. L. - Steady and Unsteady Transonic Flows. AIAA Jour. 16, pp. 1262-70, 1978.

25. LEVY, L. L. - Experimental and Computational Steady and Unsteady Transonic Flows about a Thick Airfoil. AIAA Jour. 16, pp. 564-570, 1978.

26. DEIWERT, G. S. and BAILEY, H. E. - Prospects for Computing Airfoil Aerodynamics with Reynolds Averaged Navier-Stokes Codes. NASA CP 2045, 1978.

27. KACPRZYNSKI, J. J. and OHMAN, L. H. - Wind Tunnel Tests of a Shockless Lifting Airfoil No. 1. National Research Council of Canada, NAE Project Report. ,5x5/0054, 1972.

28. ERIKSON, A. L. and STEPHENSON, J. D. - A Suggested Method of Analyzing for Transonic Flutter of Control Surfaces Based on Available Experimental Evidence. NACA RM A7F30, 1947.

29. STEGER, J. L. and BAILEY, H. E. - Calculation of Transonic Aileron Buzz. AIAA Jour. 18, pp. 249-55, 1980.

30. LEVY, L. L. and BAILEY, H. E. - Computation of Airfoil Buffet Boundaries. AIAA Jour. 19, pp. 1488-90, 1981.

31. OHMAN, L. H., KACPRZYNSKI, J. J. and BROWN, D. - Some Results from Tests in the NAE High Reynolds Number Two-dimensional Test Facility on Shockless and Other Airfoils. Canadian Aero. and Space Jour. 19, pp. 297-312, 1973.

32. GIBELING, H. J., SHAMROTH, S. J. and EISEMAN, P. R. - Analysis of Strong-Interaction Dynamic Stall for Laminar Flow on Airfoils. NASA CR-2969, 1978.

33. SANKAR, N. L. and TASSA, Y. - Reynolds Number and Compressibility Effects on Dynamic Stall of a NACA 0012 Airfoil. AIAA Paper 80-0010, Jan. 1980.

34. SHAMROTH, S. J. and GIBELING, H. J. - Analysis of Turbulent Flow about an Isolated Airfoil Using a Time-Dependent Navier-Stokes Procedure, AGARD Specialists Meeting on Boundary Layer Effects on Unsteady Flows, Aix-en-Provence, Sept. 1980.

35. TASSA, Y. and SANKAR, N. L. - Dynamic Stall of an Oscillating Airfoil in Turbulent Flow Using Time Dependent Navier-Stokes Solver. Unsteady Turbulent Shear Flows, R. Michel, J. Cousteix, and R. Houdeville, Eds., Springer-Verlag, pp 185-196, 1981.

Additional Symbols

A^n, B^n	Jacobian matrices
c	cord length
C_{CP}	turbulence model constant
C_{Kleb}	turbulence model constant
C_{WK}	turbulence model constant
$\vec{e}_x$, $\vec{e}_y$	unit vectors, Cartesian space
F	flux vector
F_{Kleb}	turbulence model function
F_{wake}	turbulence model function
f	dimensionless frequency
$\vec{g}_\xi$, $\vec{g}_\eta$	covariant base vectors
$\vec{g}^\xi$, $\vec{g}^\eta$	contravariant base vectors
$g_{\xi\xi}$, $g_{\eta\eta}$	transformation metrics
$g_{\xi\eta}$, $g_{\eta\xi}$	transformation metrics
$g^{\xi\xi}$, $g^{\eta\eta}$	transformation metrics
$g^{\xi\eta}$, $g^{\eta\xi}$	transformation metrics
J	Jacobian of transformation
K_e	effective thermal conductivity
L	length scale
L_i, L_j	volumetric metrics
$\mathcal{L}$	symmetric sequence of difference operators
$\mathcal{L}_\xi$, $\mathcal{L}_\eta$	one-dimensional difference operators
M^n	coefficient matrix
m, n	wave numbers
N	time step fraction

$\vec{n}$	unit normal vector
$\vec{n}_\xi$, $\vec{n}_\eta$	unit normal vectors in ξ and η directions
Q	solution vector
s	surface
U, V	contravariant velocity components
vol	volume element
α	turbulence model relaxation parameter
$\bar{\delta}$	boundary layer thickness
δ_i^*	kinetic displacement thickness
δ_ξ, δ_η	central difference opertors in ξ and η directions
Δ_ξ, Δ_η	forward difference opertors in ξ and η directions
∇_ξ, ∇_η	backward difference opertors in ξ and η directions
$\Delta\xi$, $\Delta\eta$	mesh spacing in ξ and η directions
Δs_η, Δs_ξ	mesh spacing in ξ and η directions
∂_t, ∂_ξ, ∂_η	partial differentiation w.r.t. time and ξ and η directions
ϵ_E, ϵ_I	explicit and implicit smoothing coefficients
η, ξ	curvilinear coordinate directions
η_{DS}	dividing streamline distance off surface
λ	bulk viscosity coefficient
μ	molecular viscosity coefficient
$\mu_{T_{eq}}$	equilibrium turbulent dynamic viscosity
σ_x, σ_y	normal components of stress tensor
τ_{xy}, τ_{yx}	shear components of stress tensor
τ_l	molecular stress tensor
τ_T	turbulent stress tensor
$\bar{\omega}$	circular frequency
$\bar{\Omega}$	dimensionless frequency parameter

Subscripts

x, y	differentiation w.r.t x and y
ξ, η	differentiation w.r.t ξ and η
t	differentiation w.r.t. time
∞	free stream condition
T	turbulent component
P	parabolic part
H	hyperbolic part
i, j	mesh point i,j
δ	evaluated at boundary layer edge
max	maximum value
$inner$	inner layer of turbulence model
$outer$	outer layer of turbulence model
$crossover$	boundary between inner and outer layer

A FINITE ELEMENT ALGORITHM FOR THE NAVIER-STOKES EQUATIONS

A. J. Baker

Professor of Engineering Science and Mechanics
University of Tennessee, USA

SUMMARY

Finite element approximation theory is employed to develop a numerical solution algorithm for the three-dimensional Navier-Stokes equations governing inviscid, viscous and/or turbulent flow of a compressible fluid. The fundamental theoretical statement requires the semi-discrete approximation error to be orthogonal to the space of functions employed in forming the approximation. Various additional constraints can be appended to the basic statement as necessary. For example, the additional constraint requiring the semi-discrete approximation error in the substantial time derivative terms of the Navier-Stokes equations to be orthogonal to a modified function space yields a highly phase selective artificial dissipation mechanism for control of non-linearly induced instabilities. An idealized von Neumann analysis is employed to estimate the functional form of the constraint, and numerical experiments for shocked ducted flows are employed to refine these estimates. Results of several basic computational experiments confirm additional theoretical and practical aspects of the algorithm.

1. INTRODUCTION

The assessment of the potential impact of finite element theory and practice applied to construction of numerical solution algorithms for computational fluid dynamics (CFD) is underway. The formal elegance of finite element methodology has produced a sound theoretical basis for the very comprehensive numerical simulation capabilities now in existence throughout structural mechanics [1]. The potential for such an impact in fluid dynamics remains to be verified.

This past decade has witnessed a greatly expanded interest in derivation and evaluation of numerical solution algorithms for a broad problem class in CFD, in particular aerodynamics. The study of computational aerodynamics was given initial impetus by MacCormack, who in 1969 published [2] an explicit, predictor-corrector finite difference algorithm for the Euler equations. Using split-operator techniques, this truly elementary construction yields a second-order accurate (in space and time) approximation to the Euler equations. The addition of artificial diffusion permits prediction of shocked flows, and the basic theoretical formulation has enjoyed many refinements and world-wide use.

For viscous or turbulent flows, an explicit algorithm becomes severely penalized by the "stiffness" associated with the discrete Navier-Stokes approximation. Modifications have been formulated for the MacCormack algorithm [3,4], but interest has generally shifted to implicit formulations, as exemplified for example by the finite difference algorithms of Beam and Warming [5] and Briley and McDonald [6]. These approximate factorization (AF) algorithms are specifically addressed to the steady-state, compressible flow equation system, but can be extended to a time-accurate solution for the transient problem. The addition of artificial diffusion is required for stability, and the algorithms can simulate shocked flows. The extension of this AF finite difference algorithm concept to a generalized coordinates description was reported by Steger and Pulliam [7], for use with numerical grid generation procedures, cf., Thompson and co-workers [8, 9].

Most recently, the formality of finite element function theoretic concepts has been applied to algorithm constructions in CFD [10-13]. While in its infancy compared to the extensive development of difference algorithms, the methodology does appear to yield robust algorithm formulations. In its elementary interpretation, the finite element algorithm concept returns calculus and vector field theory, within the framework of classical mechanics, to the construction of discrete approximations for any branch of mechanics. Of necessity, using Taylor series expansions, one must be able to verify the equivalent (finite difference) order-of-accuracy for elementary derivatives within the governing equation system. Specifically upon dissection, one can always relate a finite element generated approximation to an appropriate Taylor series expansion. The important issue is that the finite element theory produces the discrete analog expressions completely independent of the **a posteriori** ability to construct an equivalent difference representation.

In fluid mechanics, confidence in the theoretical statement occurs only through examination of detailed numerical assessment of progressively more complicated and pertinent differential equation descriptions. Strict adherence to convergence theory has been verified for scalar parabolic partial differential equations [10], using linear, quadratic, and cubic Lagrange and Hermite finite element bases. Agreement with linear convergence theory concepts is documented for solutions of the mildly non-linear, parabolic laminar boundary layer equations [11], for both linear and quadratic bases. Similar results for the consequentially non-linear parabolic turbulent boundary layer equations are also reported [12]. Of primary theoretical interest here, the Sobolev norm used to quantize convergence was a strongly non-linear function, yet the theory accurately predicted algorithm performance. Similar assessments are reported [13] for the complete Euler equations.

In this chapter, the implicit finite element numerical solution algorithm of reference 13 is presented for the unsteady three-dimensional Navier-Stokes equations expressed in generalized coordinates. The solution statement is applicable to inviscid, viscous and/or turbulent flowfield descriptions, upon specification of the stress tensor. The theoretical statement utilizes a weighted residuals formulation, and extremization of semi-discrete approximation error within an augmented Galerkin criterion. A von Neumann linearized stability analysis is employed to estimate the set of parameters introduced by the augmented error constraint statement. The results of numerical experiments for shocked, ducted flows, quantizing solution algorithm accuracy and convergence with discretization refinement, are highlighted.

2. PROBLEM STATEMENT

The partial differential equation set governing transient, three-dimensional aerodynamic flows is the familiar and very non-linear Navier-Stokes system. In non-dimensional conservation form, using Cartesian tensor summation notation, the equation system for a compressible, viscous, heat-conducting fluid is

$$L(\rho) = \frac{\partial \rho}{\partial t} + \frac{\partial}{\partial x_j}\left[u_j \rho\right] = 0 \tag{1}$$

$$L(\rho u_i) = \frac{\partial(\rho u_i)}{\partial t} + \frac{\partial}{\partial x_j}\left[u_j \rho u_i + p\delta_{ij} - \sigma_{ij}\right] = 0 \tag{2}$$

$$L(\rho e) = \frac{\partial(\rho e)}{\partial t} + \frac{\partial}{\partial x_j}\left[u_j \rho e + u_j p - \sigma_{ij} u_i - q_j\right] = 0 \tag{3}$$

In equations 1-3, ρ is density, ρu_i is the momentum vector, p is pressure, and e is mass specific total energy. Assuming a polytropic gas, $p = (\gamma-1)\rho\varepsilon$ and the equation of state is

$$p = (\gamma-1)\left[\rho e - \tfrac{1}{2}\rho u_j u_j\right] \tag{4}$$

The Stokes stress tensor σ_{ij}, heat flux vector q_j, and specific internal energy ε, are defined as,

$$\sigma_{ij} = \frac{\mu}{Re}\left[\frac{\partial u_i}{\partial x_j} + \frac{\partial u_j}{\partial x_i}\right] - \frac{2\mu}{3Re}\frac{\partial u_k}{\partial x_k}\delta_{ij} \tag{5}$$

$$q_j = -\kappa \frac{\partial e}{\partial x_j} \tag{6}$$

$$\varepsilon = e - \tfrac{1}{2} u_i u_i \qquad (7)$$

where μ is the absolute viscosity, κ is the coefficient of heat conductivity, and δ_{ij} is the Kronecker delta.

The Euler equations are contained within equations 1-4 upon specification that equations 5-6 vanish identically. The form of equations 1-4 is also representative of the mass-weighted formulation of the time-averaged Navier-Stokes equations for a turbulent flow [14]. In this instance, the variables are interpreted as appropriate descriptors of the mean flow, and σ_{ij} and q_j are generalized to include non-vanishing correlations of sub-grid scale phenomena. In this formulation, for example, the total stress tensor becomes,

$$\sigma_{ij} \equiv \bar{\sigma}_{ij} - \overline{\rho u_i' u_j'} \qquad (8)$$

where $-\overline{\rho u_i' u_j'}$ is the Reynolds stress tensor and $\bar{\sigma}_{ij}$ denotes the time averaged form for equation 5.

For many aerodynamic flows of practical interest, it is sufficient to assume that

$$\overline{\rho u_i' u_j'} \approx \bar{\rho}\, \overline{u_i' u_j'} \qquad (9)$$

The equation set 1-9 must be closed by additional equations defining the six components of the symmetric kinematic Reynolds stress tensor $-\overline{u_i' u_j'}$. One approach is to employ the Reynolds stress transport equations. An alternative is to employ an algebraic Reynolds stress model [15,16], in concert with solution of the transport equations for the turbulent kinetic energy k and isotropic dissipation function ε, where

$$k \equiv \tfrac{1}{2}\, \overline{u_i' u_i'} \qquad (10)$$

$$2/3\, \delta_{ij}\, \varepsilon \equiv 2\bar{\nu} \overline{\left(\frac{\partial u_i'}{\partial x_k} \frac{\partial u_j'}{\partial x_k} \right)} \qquad (11)$$

Equations 10-11 constitute the dependent variable definitions for the so-called two-equation turbulence model, with governing equations.

$$L(k) = \frac{\partial k}{\partial t} + \frac{\partial}{\partial x_j}\left[\bar{u}_j k + \left(C_k\, \overline{u_i' u_j'} \frac{k}{\varepsilon} - \bar{\nu} \delta_{ij} \right) \frac{\partial k}{\partial x_i} \right] + \overline{u_i' u_j'} \frac{\partial \bar{u}_i}{\partial x_j} + \varepsilon = 0 \qquad (12)$$

$$L(\varepsilon) = \frac{\partial \varepsilon}{\partial t} + \frac{\partial}{\partial x_j}\left[u_j \varepsilon + \left(C_\varepsilon \overline{u_i' u_j'} \frac{k}{\varepsilon} \right) \frac{\partial \varepsilon}{\partial x_i} \right]$$

$$+ C_\varepsilon^1 \overline{u_i' u_j'} \frac{\varepsilon}{k} \frac{\partial u_i}{\partial x_j} + C_\varepsilon^2 \frac{\varepsilon^2}{k} = 0 \qquad (13)$$

Standard values for the model constants C_k and C_ε^i are given in reference 15.

3. NUMERICAL SOLUTION ALGORITHM

Finite Element Formulation

The three-dimensional Navier-Stokes equation system is identified for the unsteady description. The direct steady-state algorithm will emerge as a special case of the transient finite element algorithm formulation. Denoting $\rho u_i \equiv m_i$ and $\rho e \equiv g$, equations 1-13 describe evolution of the vector dependent variable **q**, with elements $q_\alpha \Rightarrow \{q\}$ of the form

$$\{q\}^T \equiv \{\rho, m_i, g, p, \sigma_{ij}, q_j, k, \varepsilon\} \qquad (14)$$

The initial-valued, partial differential equations 1-3, 12 and 13 are of the form

$$L(\mathbf{q}) = \frac{\partial \mathbf{q}}{\partial t} + \frac{\partial}{\partial x_j}\left[u_j \mathbf{q} + \mathbf{f}_j \right] + \mathbf{s} = 0 \qquad (15)$$

In equation 15, $\mathbf{f}_j(\mathbf{q})$ and $\mathbf{s}(\mathbf{q})$ are specified non-linear functions of their argument. The form for the remaining algebraic and partial differential equations 4-9 is

$$L(\mathbf{q}) = \mathbf{q} + \mathbf{f}(\mathbf{q}) = 0 \qquad (16)$$

The n-dimensional equation system 15-16 is defined on the Euclidean space R^n spanned by the **x** coordinate system with scalar components x_i, $1 \leq i \leq n$. The solution domain Ω is,

$$\Omega \equiv R^n \times t = \{(\mathbf{x},t): \quad \mathbf{x} \in R^n \text{ and } t \in [t_o, t)\} \qquad (17)$$

The domain boundary is $\partial\Omega = \partial R \times t$, whereupon, the general form for the boundary condition constraint is.

$$\ell(\mathbf{q}) = a_1^{\alpha}\mathbf{q} + a_2^{\alpha}\frac{\partial}{\partial x_j}\mathbf{q}\,\hat{n}_j + a_3^{\alpha} = 0 \qquad (18)$$

In equation 18, the a_i^{α} are specified coefficients and $\hat{n}_j$ is the unit normal vector. Finally, an initial distribution for $\mathbf{q}$ on $\Omega_0 = R^n \times t_0$ is required.

$$\mathbf{q}(\mathbf{x},t_0) = \mathbf{q}_0(\mathbf{x}) \qquad (19)$$

The functional requirement of any numerical solution algorithm for equations 15-19 is to construct a suitable approximation $\mathbf{q}^h(\mathbf{x},t)$ to the dependent variable set $\mathbf{q}(\mathbf{x},t)$, and to extremize the error in this approximation in some norm. A finite element algorithm is constructed as the formal enforcement of these two basic requirements. The approximation $\mathbf{q}^h(\mathbf{x},t)$ is constructed from members of a finite-dimensional subspace of $H_0^1(\Omega)$, the Hilbert space of all functions possessing square integrable first derivatives and satisfying the boundary conditions, equation 18. While extremely flexible in theory, the usual practice is to employ polynomials, truncated at degree k, and defined on disjoint interior subdomains Ω_e, the union of which forms the discretization of Ω, ie., $\Omega \equiv \cup\Omega_e$. Hence, by these definitions,

$$\mathbf{q}(\mathbf{x},t) \approx \mathbf{q}^h(\mathbf{x},t) \equiv \bigcup_{e=1}^{M} \mathbf{q}_e(\mathbf{x},t) \qquad (20)$$

and the elemental semi-discrete approximation $\mathbf{q}_e$ is

$$\mathbf{q}_e(\mathbf{x},t) \equiv \{N_k(\mathbf{x})\}^T\{Q(t)\}_e \qquad (21)$$

In equations 20-21, the elements of $\{Q(t)\}_e$ denote members of $\mathbf{q}^h$ evaluated at the nodes of Ω_e, and subscript e signifies pertaining to the eth finite element domain, $\Omega_e \equiv R_e^n \times t$. The elements of the row matrix $\{N_k(\mathbf{x})\}^T$ are polynomials on $\mathbf{x}$ with components x_i, $1 \leq i \leq n$, which are complete to degree k and constructed to form a cardinal basis [17].

The principal requirement of the numerical solution algorithm is to render the error in $\mathbf{q}^h$ extremum in some norm. This is accomplished within finite element theory by requiring the semi-discrete approximation error in equations 15, 16, and 18, i.e., $L(\mathbf{q}^h)$ and $\ell(\mathbf{q}^h)$, to be orthogonal to the function space defining $\mathbf{q}^h$. Non-linearly induced instability is controlled by additionally requiring the semi-discrete approximation error in the substantial derivative portions of equations 1-3, i.e., $L^c(\cdot) = \frac{\partial(\cdot)}{\partial t} + \frac{\partial}{\partial x_j}[u(\cdot)]$, to be orthogonal to the modified basis $\beta\cdot\nabla\{N_k\}$. These linearly independent constraints are combined to form the theoretical statement of the finite element solution algorithm as

$$\int_{R^n} \{N_k\} L(q^h) dx + \vec{\beta}_1 \cdot \int_{R^n} \nabla \{N_k\} L^c(q^h) dx$$

$$+ \beta_2 \int_{\partial R} \{N_k\} \ell(q^h) dx \equiv \{0\} \tag{22}$$

Upon definition of k in equation 21, equation 22 represents a coupled system of algebraic equations, and non-linear ordinary differential equations on t. The form of the latter is

$$[C] \frac{d\{Q\}}{dt} + \{G(Q)\} = \{0\} \tag{23}$$

which is transformed to a non-linear algebraic system using a Taylor series as, for example,

$$\{\mathbf{F}\} \equiv \{Q\}_{j+1} - \{Q\}_j - \Delta t\{Q\}'_{j+\theta} + \ldots \equiv \{0\} \tag{24}$$

The solution of equation 24 yields the fully discrete approximate solution $\{\mathbf{Q}(n\Delta t)\}$, and the trapezoidal rule results upon setting $\Theta = ½$ in the derivative evaluation $\{\cdot\}'$.

The solution algorithm for equation 24 draws on standard procedures, eg., the Newton iteration algorithm is.

$$[\mathbf{J}(\mathbf{Q})]^p_{j+1} \{\delta Q\}^{p+1}_{j+1} = - \{\mathbf{F}\}^p_{j+1} \tag{25}$$

The dependent variable is the iteration vector, related to the solution in the conventional manner,

$$\{Q\}^{p+1}_{j+1} \equiv \{Q\}^p_{j+1} + \{\delta Q\}^{p+1}_{j+1} \tag{26}$$

where superscript p denotes the iteration index. The Jacobian for equation 25 is defined as

$$[\mathbf{J}(\mathbf{Q})] \equiv \frac{\partial\{\mathbf{F}\}}{\partial\{\mathbf{Q}\}} \tag{27}$$

Generalized Coordinates

A principal requirement in computational aerodynamics is to accurately interpolate solution domain boundaries ∂R that are non-regular and perhaps non-smooth. The term "generalized coordinates" has gained acceptance to describe an algorithm statement appropriate for use with a regularizing, boundary-fitted coordinate transformation. Many procedures are available to construct such transformations [18]. The generated coordinate transformation is the mapping

$$\mathbf{x} = \mathbf{x}(\eta_j) \tag{28}$$

Since equation 21 defines a general interpolation, the local form for equation 28 is

$$\mathbf{x} = \{N_k(\vec{\eta})\}^T \{\mathbf{X}\}_e, \quad \mathbf{x} \in R_e^n \tag{29}$$

The elements of $\{\mathbf{X}\}_e$ are the coordinates of the nodes of R_e^n, $1 \leq i \leq n$, hence $\cup R_e$, the global discretization. The elements of $\{N_k(\vec{\eta})\}$ for R_e^2 and R_e^3 are well known [1] for $1 \leq k \leq 3$.

The numerical algorithm requirement is the transformation of the divergence operator in equation 15, i.e.,

$$\frac{\partial(\cdot)}{\partial x_j} = \frac{\partial(\cdot)}{\partial \eta_k} \frac{\partial \eta_k}{\partial x_j} \tag{30}$$

The elements of the inverse Jacobian, $J_e^{-1} \equiv [\partial \eta_k / \partial x_j]_e$, are,

$$\left[\frac{\partial \eta_k}{\partial x_j}\right]_e \equiv J_e^{-1} = \frac{1}{\det[J]_e} \left[\text{transformed J cofactor}\right]_e \tag{31}$$

where $[J]_e = [\partial x_i / \partial \eta_j]_e$ is directly evaluated from equation 29. The differential element for equation 22 is

$$d\mathbf{x} = \det[J]_e d\vec{\eta} \tag{32}$$

The consequence of this transformation is best illustrated by an example. Using the divergence theorem for the second term in equation 15, the first error constraint in equation 22 for $L(\rho u_i)$, see equation 2, becomes

$$\int_{R^n} \{N\} L(\rho u_i^h) d\mathbf{x} = \int_{R^n} \{N\} \frac{\partial \rho u_i^h}{\partial t} \det[J] d\vec{\eta}$$

$$+ \oint_{\partial R} \{N\} \left[u_j^h \rho u_i^h + p^h \delta_{ij} - \sigma_{ij} \right] \cdot \hat{n}_j \det[J] d\vec{\eta}$$
$$- \int_{R^n} \frac{\partial \{N\}}{\partial \eta_k} \left(\frac{\partial \eta_k}{\partial x_j} \right) \left[u_j^h \rho u_i^h + p^h \delta_{ij} - \sigma_{ij}^h \right] \det[J] \, d\vec{\eta} \qquad (33)$$

Define the contravariant components of convection velocity on R_e as

$$\bar{u}_k^e = \det[J]_e \left(\frac{\partial \eta_k}{\partial x_j} \right) u_j^e = [\,\text{Cof. } J\,]_e \, u_j^e \qquad (34)$$

where $[\text{Cof } J]_e$ is the transformed co-factor matrix of $[J]_e$. Using equation 21 to interpolate $\bar{u}_k^e$ on R_e^n, and noting the surface integral in equation 33 vanishes identically on all interior elemental boundaries ∂R_e, equation 33 becomes the matrix statement.

$$\int_{R^n} \{N\} L(\rho u_i^h) dx \Rightarrow$$

$$S_e \Bigg[\int_{R_e^n} \Big(\{DET\}_e^T \{N\}\{N\}\{N\}^T \{RHOUI\}_e' - \{UBARK\}_e^T \{N\} \frac{\partial}{\partial \eta_k} \{N\}\{N\}^T \{RHOUI\}_e$$
$$- \{ETAKI\}_e^T \{N\} \frac{\partial}{\partial \eta_k} \{N\}\{N\}^T \{P\}_e + \{ETAKJ\}_e^T \{N\} \frac{\partial}{\partial \eta_k} \{N\}\{N\}^T \{SIGIJ\}_e \Big) d\vec{\eta}$$
$$+ \oint_{\partial R_e \cap \partial R} \Big(\{UBARK\}_e^T \{N\}\{N\}\{N\}^T \{RHOUI\}_e + \{DET\}_e^T \{N\}\{N\}\{N\}^T \{P\}_e \delta_{IK}$$
$$- \{DET\}_e^T \{N\}\{N\}\{N\}^T \{SIGIK\}_e \Big) \cdot \hat{\eta}_k \, d\vec{\eta} \Bigg] \qquad (35)$$

In equation 35, since the $\{N_k\}$ for q^h are orthonormalized on R_e^n, the limits for each integral now become ± 1. S_e is the matrix operator [16] projecting the element contributions, computed from equation 35, into the corresponding global matrix statement, equation 23. The e-subscripted terms in equation 35 are independent of η_k and can be extracted from the integrand, thus leaving only products of the polynomials, and their derivatives, constituting the elements of $\{N_k\}$. These are directly evaluable using numerical quadrature [16], yielding equation 35 as the matrix statement.

$$\int_{R^n} \{N\} L(\rho u_i^h) dx \Rightarrow S_e \Big[\{DET\}^T [M3000] \{RHOUI\}'_e$$

$$- \{UBARK\}_e^T \left[M30K0\right] - \hat{n}_K [N3000] \Big) \{RHOUI\}_e$$

$$- \left(\{ETAKI\}_e^T \left([M30K0] - \hat{n}_I \{DET\}^T [N3000] \right) \{P\}_e \right.$$

$$+ \{ETAKL\}_e^T [M30K0] - \hat{n}_L \{DET\}^T [N3000] \Big) \{SIGIL\}_e \Big] \qquad (36)$$

In equation 36, the indices K and L obey the tensor summation rule, $1 \leq (K,L) \leq n$, I is the free index (for ρu_I^h), and [M30K0] is the hypermatrix [16] equivalent of $\partial/\partial\eta_k$ (transformed), contracted with corresponding element distributions. The matrix [N3000], premultiplied by the unit outward pointing normal $\hat{n}_i$ to ∂R, results from the integration of the last term in equation 35. $\{DET\}_e$ is the nodal distribution of $\det[J]_e$ on R_e^n, while $\{ETAKI\}_e$ and $\{ETAKL\}_e$ are corresponding nodal distributions of components of J_e^{-1} on R_e^n. Within the generalized coordinates framework, therefore, the grid and metric data required for the finite element algorithm reduce to specification of the nodal distributions of $J_e^{-1} = [\partial\eta_k/\partial x_j]_e$ and $\det[J]_e = \det[\partial x_i/\partial\eta_k]_e$. The remaining equation defined in equations 15-16, and the additional terms in equation 22 are evaluated in an identical manner. The complete formulation for two-dimensional flow is given later.

Tensor Matrix Product Jacobian

The concept of the generalized coordinates transformation is applicable to definition and use of any cardinal basis $\{N_k\}$ in equation 21. For efficiency, however, a tensor product approximation to the Newton algorithm Jacobian, can be constructed by specifying use of cardinal bases $\{N_k(\vec{\eta})\}$ spanning quadrilateral and hexahedron element domains on R^2 and R^3 respectively. In this instance, the Newton algorithm Jacobian $[J(F)]$ can be approximated by the tensor (outer) matrix product [19]

$$[J(F)] \Rightarrow [J_1] \otimes [J_2] \otimes [J_3] \qquad (37)$$

Each component $[J_\eta]$ is constructed from its definition, equation 27, assuming interpolation and differentiation are one-dimensional. The solution statement, equation 25, then becomes

$$[J_1] \otimes [J_2] \otimes [J_3] \quad \{\delta Q\}^{p+1}_{j+1} = -\{F\}^{p}_{j+1} \tag{38}$$

Defining,

$$\{P1\}^{p+1}_{j+1} \equiv [J_2] \otimes [J_3]\{\delta Q\}^{p+1}_{j+1}$$

$$\{P2\}^{p+1}_{j+1} \equiv [J_3]\{\delta Q\}^{p+1}_{j+1} \tag{39}$$

the operational sequence for equation 25 becomes

$$[J_1]\{P1\}^{p+1}_{j+1} = -\{F\}^{p}_{j+1}$$

$$[J_2]\{P2\}^{p+1}_{j+1} = \{P1\}^{p+1}_{j+1}$$

$$[J_3]\{\delta Q\}^{p+1}_{j+1} = \{P2\}^{p+1}_{j+1} \tag{40}$$

Other permutations of the index sequence for $[J_\eta]$ could be utilized. The key aspect is replacement of the very large sparse matrix $[J]$, equation 27, with α-block-tridiagonal (or pentadiagonal, for k=2 in equation 21) matrices $[J_\eta]$. This yields a significant reduction in both storage for the Jacobian and the CPU required to construct the LU decomposition and to perform the back substitution. The compromise occurs with decay of the Newton algorithm convergence rate to less than quadratic. This procedure in no way affects the formation of $\{F\}$, to which the accuracy features intrinsic to the finite element algorithm statement are attributed. Compromises in the construction of $\{F\}$ will invariably produce inferior results. A complete description of the tensor matrix approximation construction, equation 37-40, is given in references 13 and 16.

4. THEORETICAL ANALYSIS

General Concepts

The important theoretical issue is quantization of algorithm accuracy and convergence in suitably defined Sobolev norms. At large Reynolds number, away from aerodynamic surfaces, equations 1-3 are principally hyperbolic. In an analysis [20] for a linear scalar hyperbolic equation, the total approximation error is constructed in terms of semi-discrete, discrete and temporal truncation error contributions. The total approximation error $\{\mathbf{e}(n\Delta t)\}$ is the difference between the exact solution $\mathbf{q}(\mathbf{x},t)$, at the nodes of $\cup R_e^n$, and $\{Q(n\Delta t)\}$. This is constituted in part by the semi-discrete approximation error, $\mathbf{e}^h(\mathbf{x},t) = \mathbf{q}(\mathbf{x},t) - \mathbf{q}^h(\mathbf{x},t)$, and the temporal approximation error introduced by selection of a specific value for Θ, equation 24. The principal result is quantization of convergence of the semi-discrete approximation error, which for smooth solutions of the linear hyperbolic equation satisfies the inequality [20],

$$\| \mathbf{e}^h(\cdot\,,t)\|_{H^1(\Omega)} < C_1\Delta^{k+1} \| \mathbf{q}\|_{H^{k+1}(\Omega)} + C_2\Delta\int_0^t \mathbf{q}(\tau)\|_{H^{k+1}(\Omega)} d\tau \tag{41}$$

In equation 41, C_1 and C_2 are constants independent of the mesh, Δ is the extremum mesh measure, k is the degree of the semi-discrete approximation, and $\|\cdot\|_{H^P}$ denotes the H^P Sobolev norm. Under discretization refinement, convergence is dominated by the second term, hence, is independent of the degree of the finite element basis $\{N_k(x)\}$. By extension of the analysis, for $\Theta = \frac{1}{2}$ in equation 24, the temporal approximation error is bounded by $C_3\Delta t\|Q^0\|_{H^0}$, where C_3 is a constant and Q^0 is the interpolation on $\cup R_e^1$ of the initial data $\mathbf{q}_0(x)$. The validity of equation 41 for smooth solutions of a linear hyperbolic equation, eg., equation 1 with constant velocity, is verified [10].

It is also required to determine the functional form for the constraint parameter set $\vec{\beta}_1$, equation 22, and to expose the role this term plays in the algorithm. Considering equation 1, linearized by $\mathbf{u} = u\hat{\imath} = U_0\hat{\imath}$ = constant, the Fourier decomposition of the corresponding semi-discrete approximate solution, on a uniform mesh of measure Δx, is

$$q^h(j\Delta x,t) = Q_0 \exp[\, i\omega(j\Delta x - \Gamma U_0 t)] \tag{42}$$

In equation 42, $\Gamma = \sigma + i\delta$, and σ and δ are real numbers, $i = \sqrt{-1}$, $\omega = 2\pi/\lambda$ is the wave number for Fourier mode of wavelength λ, and $x = j\Delta x$, $j = 0, 1, 2,\ldots$. For the definition $\vec{\beta}_1 = \nu\Delta x\hat{\imath}$, where $\nu > 0$ is a scalar parameter, the k=1 algorithm expansions for σ and δ are [21],

$$\sigma = 1 + \left(\frac{-1}{180} + \frac{\nu^2}{12}\right) d^4 + 0(d^6) \tag{43}$$

$$\delta = -\frac{\nu}{12} d^3 + 0(d^5) \tag{44}$$

where $d = \omega\Delta x = 2\pi/n$, and n is the discrete Fourier mode index, $\lambda_n = n\Delta x$, and $0(\cdot)$ indicates order. The semi-discrete approximation q^h can be made sixth-order accurate by eliminating the $0(d^4)$ term in equation 43, yielding the "optimum" estimate $\nu_0 = (15)^{-\frac{1}{2}}$. Correspondingly, $\delta < 0$ in equation 44, and an artificial dissipation mechanism is thereby introduced.

The original analysis has been expanded [13] for equations 1-3, by redefining the dissipation parameter $\vec{\beta}_1$ in the form,

$$\vec{\beta}_1 \equiv \Delta x(\nu^1 \delta_t + \nu^2 \delta_x)\hat{i} \tag{45}$$

where δ_t and δ_x are Kronecker delta-type switches, yielding ν^1 operating only on the time derivative, and ν^2 operating solely on the spatial derivative term in $L^c(\cdot)$. Proceeding through the substitutions yields, for the k=1 algorithm,

$$\sigma = 1 - d^2(\nu^1 - \nu^2) + d^4\left[-\frac{1}{180} + \frac{\nu^1\nu^2}{12} + (\nu^1 - \nu^2)(\nu^1)^3\right] + O(d^6) \tag{46}$$

$$\delta = d(\nu^1 - \nu^2) - d^3\left[\frac{\nu^2}{12} - (\nu^1 - \nu^2)(\nu^1)^2\right] + O(d^5) \tag{47}$$

The quadratic algorithm construction (k=2 in equation 21) is also reported [13]. Setting $\nu^1 = \nu = \nu^2$ in equations 46-47 yields the results of the original analysis. Alternately, enforcing sixth-order accuracy in equations 46-47 produces the constraint

$$\nu^2 = \frac{d^2\,[1/180 - (\nu^1)^4] + (\nu^1)^2}{d^2\,[\nu^1/12 - (\nu^1)^3] + \nu^1} \tag{48}$$

Figure 1 is a plot of equation 48, with n as a parameter. Sixth-order accuracy can be achieved only for $\nu^1 > 0$; for any level,

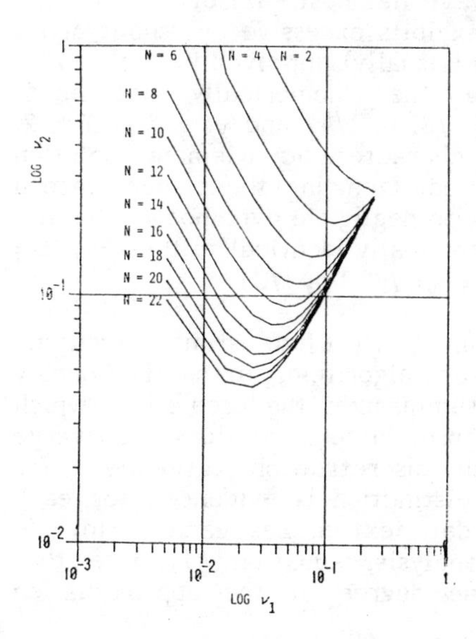

Figure 1. Solution to Equation 48.

ranges over an order of magnitude dependent upon n, with the largest levels required for the shortest wavelengths. All data coverage at the point $\nu^1 = (15)^{-1/2} = \nu^2$. Defining $\nu^1 = 0$ renders the k=1 semi-discrete approximation q^h a second-order accurate representation of the analytical solution for all $\nu^2 \neq 0$. The form of equation 45 for the multi-dimensional algorithm in generalized coordinates is [16]

$$\boldsymbol{\beta}_1^{\alpha} = (\det J)(\boldsymbol{\nu}_{\alpha}^1 \, \delta_t + \boldsymbol{\nu}_{\alpha}^2 \, \delta_{\mathbf{x}}) \tag{49}$$

Accuracy and Convergence

Assessment of solution accuracy, and convergence with discretization refinement, of the developed finite element algorithm is reported [13], for inviscid one-dimensional flow in the Riemann shock tube. This is a well-suited problem definition, since the resultant flow structure is richly endowed with discontinuities and sharp field gradients. The case with unique stagnation sound speeds, in the two chambers initially separated by the diaphragm, has been exhaustively examined in the finite-difference literature [22-24]. The diaphragm is placed midway within a duct of uniform cross section, with unit length discretized into $12 \leq M \leq 400$ finite elements R_e^1 of uniform measure Δ. The initial condition specification is $u(x) = 0$ and $p = 1 = \rho$ on $0 \leq x \leq 0.5$, and $p = 0.1$ and $\rho = 0.125$ on $0.5 < x \leq 1.0$, with $\gamma = 1.4$.

Figure 2 graphs the $M = 400$, $k = 1$ solution discrete approximation $\{Q(n\Delta t)\}$, at $n\Delta t = 0.14154$s, as obtained using the elementary order-of-accuracy optimized parameter set $\nu_{\alpha}^1 = \nu_0 = \nu_{\alpha}^2$, $\nu_m^1 = 0$, and $\nu_0 = (15)^{-1/2}$. Each symbol is located at a nodal coordinate of $\cup R_e^1$, and the dashed lines denote the initial conditions. The shock is centered at $x = 0.75$, the contact discontinuity is centered at $x = 0.62$, and the rarefaction wave lies upstream of $x = 0.5$, the diaphragm location. This solution exhibits excessive overshoot about the shock, and Figure 3 graphs the (visually) improved k=1, M = 200, discrete solution obtained using the "numerically optimized" dissipation parameter set $\nu_{\alpha}^1 = \nu_0\{3/8, 0, 1/4\}$ and $\nu_{\alpha}^2 = \nu_0\{3/4, 2, 1\}$, $\nu_0 = (15)^{-1/2}$. Each of the characteristic Riemann solution features appears accurately predicted, including the planar plateau regions and a crisply defined shock with negligible overshoot. The M = 100, k=2 algorithm performance is nominally identical to the k=1, M = 200 prediction, using $\nu_{\alpha}^1 = 0$ and $\nu_{\alpha}^2 = \nu_0\{1/4, 3/4, 1/2\}$.

Data of this type have been employed [16] to estimate accuracy and convergence of the finite element algorithm, in the H^1 Sobolev and energy (E) norms. Figure 4 summarizes the prediction, which confirms the semi-discrete approximate solutions converge monotonically in H^1 and E with discretization refinement for $10 < M \leq 400$. A modest slope distinction is evidenced for each dependent variable, and the k=2 data extremizes each norm. In confirmation of the elementary analysis, equation 41, note that convergence is independent of the degree of the approximation subspace $\{N_k\}$.

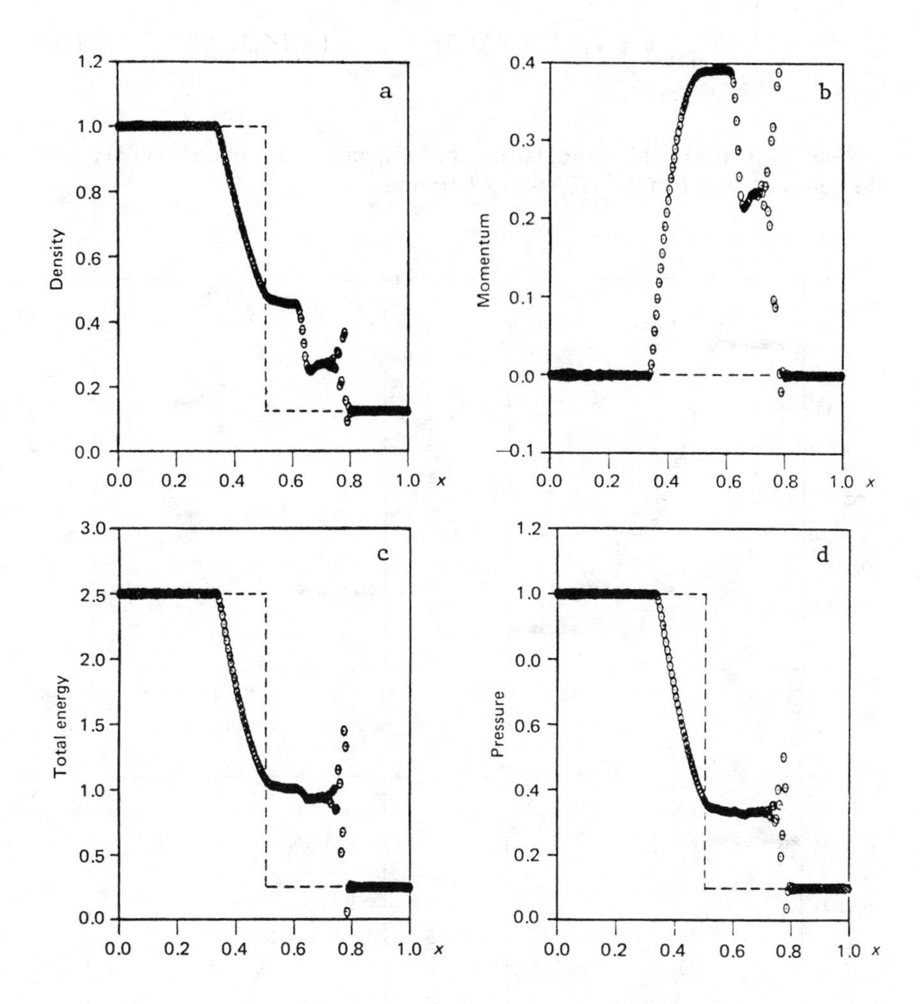

Figure 2. M = 400, k=1 Finite Element Algorithm Solution, Riemann Shock Tube, t = 0.14154s, $\nu^1_\alpha \equiv \nu \equiv \nu^2_\alpha$, $\nu^1_m \equiv 0$. (---) Denotes Initial Conditions.

5. TWO-DIMENSIONAL, BI-LINEAR ALGORITHM CONSTRUCTION

Metric Definitions

Consider problems defined on the two-dimensional space R^2, and restrict attention to k=1 in equation 21 i.e., the bi-linear, tensor product cardinal basis, $\{N_1(\vec{\eta})\}$. In this case, R^2_e exhibits only vertex nodes, hence,

$$\left(\frac{\partial \eta_k}{\partial x_j}\right)_e \equiv \frac{1}{\det J_e}\{N_1\}^T\{ETAKJ\}_e \qquad 1 \leq (K,J) \leq 2 \qquad (50)$$

Noting that $(\det J)_e^{-1}$ cancels in the algebra, see equation 31, the elements of two times $\{ETAKJ\}_e$ become

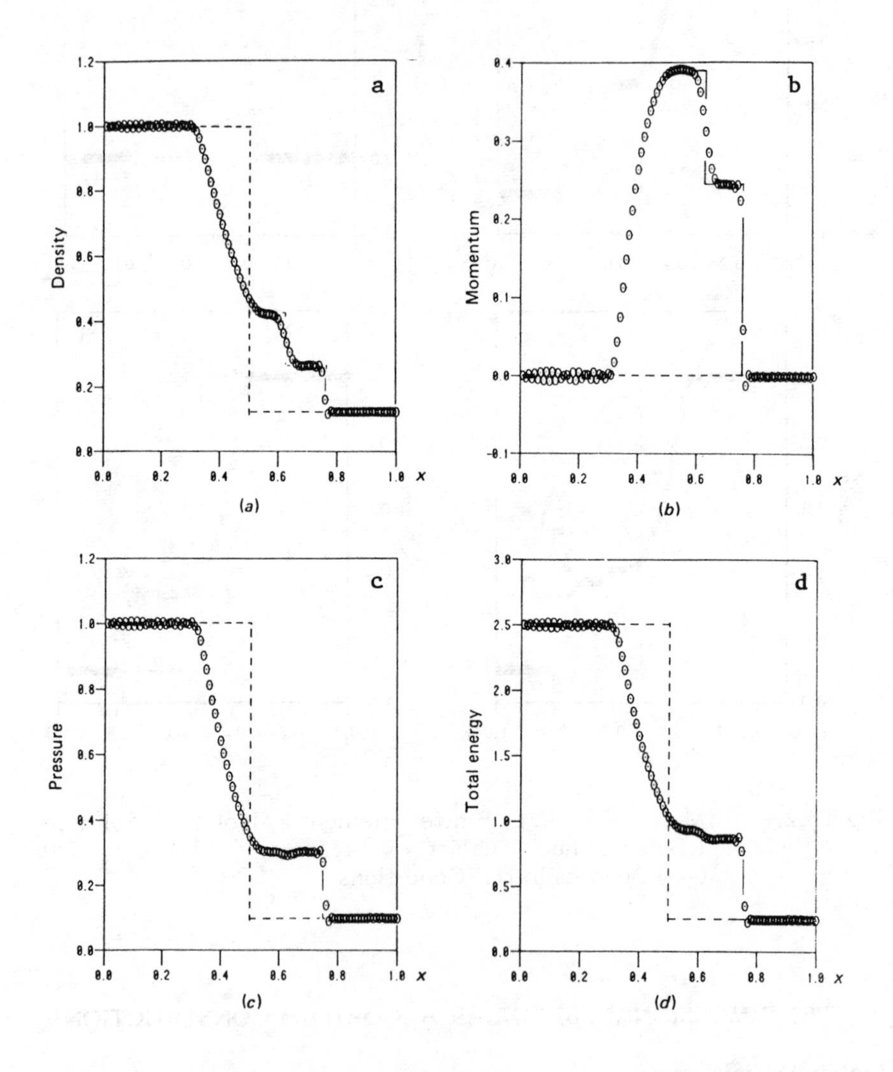

Figure 3. Finite Element Algorithm Solution, Riemann Shock Tube, t = 0.14154s, M = 200, k=1, ν_α^1 = {3/8, 0, 1/4 }, ν_α^2 = {3/4, 2, 1 }, a) Density, b) Momentum, c) Energy, d) Pressure.

$$2\,\{ETAKJ\}_e = \begin{Bmatrix} Y4-Y1 \\ Y3-Y2 \\ Y3-Y2 \\ Y4-Y1 \\ (1,1) \end{Bmatrix}_e, \begin{Bmatrix} X1-X4 \\ X2-X3 \\ X2-X3 \\ X1-X4 \\ (1,2) \end{Bmatrix}_e, \begin{Bmatrix} Y1-Y2 \\ Y1-Y2 \\ Y4-Y3 \\ Y4-Y3 \\ (2,1) \end{Bmatrix}_e, \begin{Bmatrix} X2-X1 \\ X2-X1 \\ X3-X4 \\ X3-X4 \\ (2,2) \end{Bmatrix}_e \quad (51)$$

In equation 51, the indices in parenthesis indicate (K,J), and each array contains only two distinct entries. The corresponding definition of interpolation for $\det[J]_e$ yields [13]

$$\{DET\}_e = \frac{1}{4}\begin{Bmatrix} (X2-X1)\;(Y4-Y1) - (X4-X1)\;(Y2-Y1) \\ (X2-X1)\;(Y3-Y2) - (X3-X2)\;(Y2-Y1) \\ (X3-X4)\;(Y3-Y2) - (X3-X2)\;(Y3-Y4) \\ (X3-X4)\;(Y4-Y1) - (X4-X1)\;(Y3-Y4) \end{Bmatrix}_e \quad (52)$$

In equations 51-52, the element nodal coordinates are $\{\mathbf{X}\}_e = \{XJ, YJ, 1 \le J \le 4\}_e$.

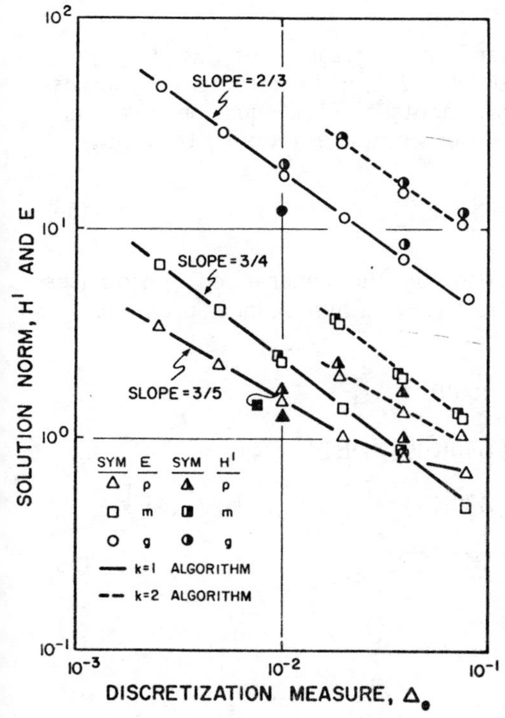

Figure 4. Semi-discrete Approximation Accuracy and Convergence in $\|q^h\|^2_{H^1}$ and $\|q^h\|^2_E$, Finite Element Algorithm Solution for Riemann Shock Tube.

The remaining metric operation is computation of the contravariant convection velocity approximation $\bar{u}_k^e$. From equation 34, and for the definition,

$$\bar{u}_k^e \equiv \{N_1\}^T\{UBARK\}_e \tag{53}$$

completion of the algebra yields.

$$\{UBARK\}_e = \{ETAKJ\}_e\left(\{N_1\}^T\{UJ\}_e\right) \tag{54}$$

The term in parenthesis in equation 54 is a scalar, and the evaluation is on a nodal basis, whereupon $\{N_1\}$ reduces to the Kronecker delta. Summation is implied over the index J, and at the nodes UJ $\equiv$ MJ/R, i.e., $u_j \equiv m_j/\rho$.

Generalized Coordinates Algorithm

The dependent variable set $\mathbf{q}$ includes ρ, $m_i = \rho u_i$, i = 1, 2, p, and $g = \rho e$, and the solution parameters σ_{ij} and q_j. The algorithm parameter set $\boldsymbol{\beta}_1$ is a two-dimensional vector with elemental scalar components equal to $\det[J]_e \nu^i_{\alpha j}$. The subscript α signifies the appropriate dependent variable, the superscript i = 1, 2 indicates the two distinct components, equation 49, and the subscript j = 1, 2 is a tensor index.

Order the discrete dependent variable set as $\{\mathbf{Q}\}^T$ = {R,MI,G,P,SIJ, QJ}, where I and J (=1,2) are discrete tensor indices. The respective algorithm statements $\{\mathbf{F}\}$, equation 24, are constructed as an assembly of the elemental evaluations [16], i.e.,

$$\{\mathbf{F}\} \equiv \mathbf{S}_e\{\mathbf{F}\}_e \tag{55}$$

The element column matrices $\{\mathbf{F}\}_e$, in the generalized coordinates description, are thus formed as the following matrix inner products.

$$\begin{aligned}\{FR\}_e = &\Big(\{DET\}^T[B3000] \\ &+\nu^1_{RJ}\{ETAKJ\}^T[B40K00]\{DET\}\Big)\{R\}'_{j+1} \\ &+\frac{\Delta t}{2}\Big(-\{ETAKI\}^T[B30K0]\{MI\} + \nu^2_{RJ}\{UBARL\}^T \\ &[B40KL0]\{ETAKJ\}\{R\}\Big)_{j+1,j}\end{aligned} \tag{56}$$

$$\{FMI\}_e = \Big(\{DET\}^T[B3000]$$
$$+ \nu^1_{IJ}\{ETAKJ\}^T[B40K00]\{DET\}\Big)\{MI\}'_{j+1}$$
$$+ \frac{\Delta t}{2}\Big(-\{UBARK\}^T[B30K0]\{MI\}$$
$$- \{ETAKI\}^T[B30K0]\{P\} + \{ETAKJ\}^T[B30K0]\{SIJ\}$$
$$+ \nu^2_{IJ}\{UBARL\}^T[B40KL0]\{ETAKJ\}\{MI\}\Big)_{j+1,j} \qquad (57)$$

$$\{FG\}_e = \Big(\{DET\}^T[B3000]$$
$$+ \nu^1_{GJ}\{ETAKJ\}^T[B40K00]\{DET\}\Big)\{G\}'_{j+1}$$
$$+ \frac{\Delta t}{2}\Big(-\{UBARK\}^T[B30K0](\{G\} + \{P\})$$
$$+ \{ETAKJ\}^T[B30K0]\{QJ\}$$
$$+ \{ETAKJ\}^T[B40K00]\{UL\}\{SLJ\}$$
$$+ \nu^2_{GJ}\{UBARL\}^T[B40KL0]\{ETAKJ\}\{G\}\Big)_{j+1,j} \qquad (58)$$

$$\{FP\}_e = \{DET\}^T[B3000]\{P\}$$
$$- (\gamma - 1)\Big(\{DET\}^T[B3000]\{G\}$$
$$+ \tfrac{1}{2}\{DET\}^T[B40000]\{UJ\}\{MJ\}\Big)_{j+1} \qquad (59)$$

$$\{FSIJ\}_e = \{DET\}^T[B3000]\{SIJ\}$$
$$- \mu\Big(\{ETAKJ\}^T[B30K0]\{UI\}$$
$$+ \{ETAKI\}^T[B30K0]\{UJ\}$$
$$- 2/3\delta_{IJ}\{ETAKL\}^T[B30K0]\{UL\}\Big)_{j+1} \qquad (60)$$

$$\{FQJ\}_e = \{DET\}^T [B3000] \{QJ\} - \kappa \{ETAKI\}^T [B30K0] \left(\{GSR\} - \{ULUL\} \right)_{j+1} \quad (61)$$

The generalized elemental hypermatrices [M...], see equation 36, are named [B...] in equations 56-61 for the two-dimensional problem. The matrices [B30K0], premultiplied by $\{UBARK\}^T$ and $\{ETAKJ\}^T$, are augmented on $\partial R_e \cap \partial R$ by the surface integral term, see equation 36. Each matrix [B...] is evaluated only once, by integration over a master element domain R_e^2 using numerical quadrature, and each is listed in reference 16, Appendix B. The Boolean indices (0,K) in [B30K0] indicate that the corresponding basis elements in $\{N_k\}$ are not differentiated, or are differentiated once into scalar components parallel to the η_k coordinate system. In equations 56-59, the discrete indices J, K, and L, occurring in both matrix and variable names, are tensor summation indices with range $1 \leq (J,K,L) \leq 2$. The free index I in equation 57 denotes Cartesian scalar components of m_i, i.e., {MI}, and both I and J are free indices in equations 60-61.

The subscript e has been deleted for clarity, in equations 56-61. However, all matrices other than [B...] contain element dependent entries. The matrices {DET} and {ETAKJ} contain elements equal to the nodal values of det J_e and the entries of $[\partial \eta_k / \partial x_j]_e$. The nodal values of the scalar components of $\bar{u}_k$ are denoted { UBARK }. The elements of {SIJ } are nodal values of the stress tensor, computed in principal coordinates in terms of $u_i = m_i/\rho$. In equation 60, δ_{IJ} is the Kronecker delta; in equation 61, the elements of { GSR } are nodal values of $e \equiv g/\rho$, while those of { ULUL} are nodal values of specific kinetic energy $\frac{1}{2} u_\ell u_\ell$. In all equations, the notation $\{\cdot\}_{j+1}^{\prime}$ denotes $\{\cdot\}_{j+1}^{p} - \{\cdot\}_j$, and $(\tilde{\cdot})_{j+1,j}$ indicates evaluation of the argument $(\cdot)$ at both t_{j+1} and t_j, followed by addition after multiplication by Θ and $(1 - \Theta)$, see equation 24. Finally, note that equations 56-61 completely express the three-dimensional form of the algorithm, by the exchange of B-prefix standard matrices with those of C-prefix, and extension of the discrete tensor index range to $1 \leq (I,J,K,L) \leq 3$.

Tensor Matrix Product Jacobian

Construction of the tensor matrix product approximation to the Newton algorithm Jacobian, see equations 27, 38-40, is a calculus operation, since equations 56-61 are in a functional form, and

$[J] \equiv S_e[J]_e$. The lead term in each equation $\{F\}$ is $\{DET\}^T[B3000]\{Q\}_{j+1}$, which serves to illustrate the construction. The first term in $[J(F)]_e$, which accounts for self-coupling, is computed as

$$\frac{\partial\{FI\}_e}{\partial\{QJ\}_e}^e \equiv [JQQ]_e\,\delta_{IJ} = \{DET\}^T[B3000]\delta_{IJ} \tag{62}$$

Referring to equation 35 for the definition, the elemental operation in forming equation 62 is

$$[JQQ]_e \equiv \{DET\}_e^T[B3000] \equiv \int_{R_e^2} \{DET\}_e^T\{N_k\}\{N_k\}\{N_k\}^T d\vec{\eta} \tag{63}$$

Assuming a rectangular element domain R_e^2, described by measures ℓ and ω and spanned by $\{N_1\}$, the evaluation of equation 63 yields

$$[JQQ]_e \equiv \{DET\}^T[B3000] = \Delta_e[B200] = \ell\omega[B200] = \frac{\ell\omega}{36}\begin{bmatrix} 4 & 2 & 1 & 2 \\ & 4 & 2 & 1 \\ & & 4 & 2 \\ (sym) & & & 4 \end{bmatrix} \tag{64}$$

The tensor matrix product approximation to equation 64 involves evaluation of equation 63 on a one-dimensional element. Denoting $M \equiv A$ for $n = 1$,

$$[JQQ_\eta]_e \equiv \Delta_\eta[A200] = \int_{R_e^1} \{N_1\}\{N_1\}^T dx \tag{65}$$

Evaluating the integrals defined in equation 65 yields,

$$[JQQ_1]_e = \frac{\ell}{6}\begin{bmatrix} 2 & 1 \\ 1 & 2 \end{bmatrix} \qquad [JQQ_2]_e = \frac{\omega}{6}\begin{bmatrix} 2 & 1 \\ 1 & 2 \end{bmatrix} \tag{66}$$

assuming $\Delta_1 = \ell$ and $\Delta_2 = \omega$. Accounting for entry locations in $[JQQ]_e$, as determined by the node numbering convention, and letting $\otimes$ denote the tensor matrix product [19], it is readily verified that

$$[JQQ]_e = [JQQ_1]_e \otimes [JQQ_2]_e \tag{67}$$

The operations defined by equations 64-67 can be extended to all terms in the Newton algorithm Jacobian, cf. reference 16. The construction of certain terms also involves differentiation with respect to the parameter $\bar{u}_k = \bar{m}_k/\rho$, using the chain rule. Recalling that operations are performed on the element matrices, letting $\bar{D}_{KI}$ denote the one-dimensional element average of det $J_e(\partial\eta_k/\partial x_i)_e$, and defining K to signify the discrete free index corresponding to $\partial/\partial\eta_k$, the non-empty tensor matrix product Newton algorithm Jacobians for equations 56-61 are.

$$
\begin{aligned}
[JRR]_e &= \{DETK\}^T[A3000] \\
&+ \nu^1_{RJ}\{ETAKJ\}^T[A40K00]\{DETK\} \\
&+ \frac{\Delta t}{2}\nu^2_{RJ}\Big[\{ETAKJ\}^T[M40KL0]\{UBARL\} \\
&- \left(\frac{\bar{m}_L}{\bar{\rho}^2}\right)\{ETAKJ\}^T[M40K0L]\{R\}\Big] \\
[JRMI]_e &= \frac{\Delta t}{2}\Big[-\{ETAKI\}^T[A30K0] \\
&+ \nu^2_{RJ}\bar{D}_{KI}\{ETAKJ\}^T[A30KK]\Big]
\end{aligned}
\tag{68}
$$

$$
\begin{aligned}
[JMIR]_e &= \frac{\Delta t}{2}\left(\frac{\bar{m}_k}{\bar{\rho}^2}\right)\Big[\{MI\}^T[A30K0] \\
&+ \nu^2_{IJ}\{MI\}^T[A4KK00]\{ETAKJ\}\Big] \\
[JMIMI]_e &= \{DETK\}^T[A3000] \\
&+ \nu^1_{IJ}\{ETAKJ\}^T[A40K00]\{DETK\} \\
&+ \frac{\Delta t}{2}\Big[-\{UBARK\}^T[A30K0] - \bar{D}_{KI}\{MI\}^T[A30K0]\frac{1}{\bar{\rho}} \\
&+ \nu^2_{IJ}(\{UBARK\}^T[A40KK0]\{ETAKJ\} \\
&+ \frac{1}{\bar{\rho}}\{MI\}^T[A4KK00]\{ETAKJ\})\Big] \\
[JMIMJ]_e &= \frac{\Delta t}{2}\bar{D}_{KJ}\left(\frac{1}{\bar{\rho}}\right)\Big[-\{MJ\}^T[A30K0] \\
&+ \nu^2_{JL}\{MJ\}^T[A4KK00]\{ETAKL\}\Big] \\
[JMIP]_e &= \frac{-\Delta t}{2}\{ETAKI\}^T[A30K0] \\
[JMISIJ]_e &= \frac{\Delta t}{2}\{ETAKJ\}^T[A40K00]\{DETK\}
\end{aligned}
\tag{69}
$$

$$[JGR]_e = \frac{\Delta t}{2}\left(\frac{\bar{m}_k}{\bar{\rho}^2}\right)\Big[\{G{+}P\}^T[A30K0] + \nu^2_{GJ}\{G\}^T[A4KK00]\{ETAKJ\}\Big]$$

$$[JGMI]_e = \frac{\Delta t}{2}\bar{D}_{KI}\left(\frac{1}{\bar{\rho}}\right)\Big[-\{G{+}P\}^T[A30K0] + \nu^2_{GL}\{G\}^T[A4KK00]\{ETAKJ\}\Big]$$

$$\begin{aligned}[JGG]_e = {} & \{DETK\}^T[A3000] \\ & + \nu^1_{GJ}\{ETAKJ\}^T[A40K00]\{DETK\} \\ & + \frac{\Delta t}{2}\Big[-\{UBARK\}^T[A30K0] \\ & + \nu^2_{GJ}\{UBARK\}^T[A40KK0]\{ETAKJ\}\Big]\end{aligned}$$

$$[JGP]_e = -\frac{\Delta t}{2}\{UBARK\}^T[A30K0]$$

$$[JGSIJ]_e = \frac{\Delta t}{2}\{ETAKJ\}^T[A40K00]\{UI\}$$

$$[JGQJ]_e = \frac{\Delta t}{2}\bar{D}_{LJ}\{ETAKL\}^T[A30K0] \qquad (70)$$

$$[JPR]_e = -\left(\frac{\gamma-1}{2}\right)\left(\frac{\bar{m}_k}{\bar{\rho}^2}\right)\{DETK\}^T([A40000]\{MK\})$$

$$[JPMK]_e = \left(\frac{\gamma-1}{2}\right)\Big[\{DETK\}^T([A40000]\{UK\}) + \left(\frac{1}{\bar{\rho}}\right)\{DETK\}^T([A40000]\{MK\})\Big]$$

$$[JPG]_e = -(\gamma-1)\{DETK\}^T[A3000]$$

$$[JPP]_e = \{DETK\}^T[A3000] \qquad (71)$$

$$\begin{aligned}[JSIJR]_e = {} & \frac{\mu}{\bar{\rho}^2}\Big[\bar{m}_i\{ETAKJ\}^T[A30K0] \\ & - \bar{m}_j\{ETAKI\}^T[A30K0] \\ & - 2/3\,\delta_{IJ}\bar{m}_\ell\{ETAKL\}^T[A30K0]\Big]\end{aligned}$$

$$[\mathrm{JSIJMI}]_e = -\frac{\mu}{\bar{\rho}}\Big[\{\mathrm{ETAKJ}\}^T[\mathrm{A30K0}] - 2/3\,\delta_{IJ}\{\mathrm{ETAKL}\}^T[\mathrm{A30K0}]\Big]$$

$$[\mathrm{JSIJSIJ}]_e = \{\mathrm{DETK}\}^T[\mathrm{A3000}] \qquad (72)$$

$$[\mathrm{JQIR}]_e = \frac{\kappa}{\bar{\rho}^2}\Big[\bar{g}\{\mathrm{ETAKI}\}^T[\mathrm{A30K0}] - \frac{\bar{m}_i^2}{\bar{\rho}}\{\mathrm{ETAKI}\}^T[\mathrm{A30K0}]\Big]$$

$$[\mathrm{JQIMI}]_e = \frac{\kappa\bar{m}_i}{\bar{\rho}^2}\{\mathrm{ETAKI}\}^T[\mathrm{A30K0}]$$

$$[\mathrm{JQIG}]_e = -\frac{\kappa}{\bar{\rho}}\{\mathrm{ETAKI}\}^T[\mathrm{A30K0}]$$

$$[\mathrm{JQIQI}]_e = \{\mathrm{DETK}\}^T[\mathrm{A3000}] \qquad (73)$$

As occurred in equations 56-61, the subscript e has been deleted for clarity in equations 68-73. However, all matrices other than those denoted [A...] contain element-dependent data. All algorithm standard hypermatrices [A...] defined in equations 68-73 are listed in Appendix B of reference 16. The scalars $\bar{m}_i$, $\bar{\rho}$ and $\bar{g}$ are algebraic averages of the entries in the element matrices {MI}, {R}, and {G}, respectively, on R_e^1. All evaluations in equations 68-73 employ the most recent solution approximation $\{\mathbf{Q}\}_{j+1}^p$.

Equations 38-40 define the matrix solution sequence of the Newton algorithm, and a two-dimensional problem involves only a single intermediate determination {**P**}. The specific form of the solution steps defined in equation 40 is

$$S_e\begin{bmatrix} [\mathrm{JRR}]_e & [\mathrm{JRM1}]_e & 0 & 0 & 0 & 0 & 0 \\ [\mathrm{JM1R}]_e & [\mathrm{JM1M1}]_e & 0 & 0 & [\mathrm{JM1P}]_e & [\mathrm{JM1SIJ}]_e & 0 \\ [\mathrm{JM2R}]_e & [\mathrm{JM2M1}]_e & [\mathrm{JM2M2}]_e & 0 & 0 & 0 & 0 \\ [\mathrm{JGR}]_e & [\mathrm{JGM1}]_e & 0 & [\mathrm{JGG}]_e & [\mathrm{JGP}]_e & [\mathrm{JGSIJ}]_e & [\mathrm{JGQI}]_e \\ [\mathrm{JPR}]_e & [\mathrm{JPM1}]_e & 0 & [\mathrm{JPG}]_e & [\mathrm{JPP}]_e & 0 & 0 \\ [\mathrm{JSIJR}]_e & [\mathrm{JSIJM1}]_e & 0 & 0 & 0 & [\mathrm{JSIJSIJ}]_e & 0 \\ [\mathrm{JQIR}]_e & [\mathrm{JQIM1}]_e & 0 & [\mathrm{JQIG}]_e & 0 & 0 & [\mathrm{JQIQI}]_e \end{bmatrix}\{\mathbf{P}\} = -\{\mathbf{F}\} \qquad (74)$$

$$S_e \begin{bmatrix} [JRR]_e & 0 & [JRM2]_e & 0 & 0 & 0 & 0 \\ [JM1R]_e & [JM1M1]_e & [JM1M2]_e & 0 & 0 & 0 & 0 \\ [JM2R]_e & 0 & [JM2M2]_e & 0 & [JM2P]_e & [JM2SIJ]_e & 0 \\ [JGR]_e & 0 & [JGM2]_e & [JGG]_e & [JGP]_e & [JGSIJ]_e & [JGQI]_e \\ [JPR]_e & 0 & [JPM2]_e & [JPG]_e & [JPP]_e & 0 & 0 \\ [JSIJR]_e & 0 & [JSIJM2]_e & 0 & 0 & [JSIJSIJ]_e & 0 \\ [JQIR]_e & 0 & [JQIM2]_e & [JQIG]_e & 0 & 0 & [JQIQI]_e \end{bmatrix} \{\delta Q\} \equiv \{P\} \qquad (75)$$

Non-Iterative and Direct Steady-State Solution

The non-iterative and direct steady-state algorithms are special cases of the developed formulation. The non-iterative form simply constitutes acceptance of the first solution, $\{\delta \mathbf{Q}\}^2_{j+1}$, from the Newton algorithm, using only the single evaluation $\{\mathbf{F}\}^1_{j+1} \equiv \{\mathbf{F}\}_j$. By definition, $\{\mathbf{Q}\}^1_{j+1} \equiv \{\mathbf{Q}\}_j$, the dependent variable matrix at time step t_j. Therefore $\{\mathbf{Q}\}_{j+1} = \{0\}$ in equations 56-58, and the corresponding expression in brackets is not evaluated. Furthermore, for the second bracketed expressions, $\Delta t/2 \Rightarrow \Delta t$ and the evaluation $(\cdot)_{j+1,j}$ reduces to $(\cdot)_j$. This procedure eliminates the error control parameter $\nu^1_{\alpha j}$ from the algorithm. The tensor matrix product Jacobian formulation is unaltered, except that since the terms involving ν^1 have been eliminated from $\{\mathbf{F}\}$, the corresponding terms in $[J_\alpha]_1$ are omitted. The multiplier $\Delta t/2$ remains appropriate, and $\{\mathbf{Q}\}_{j+1} \equiv \{\mathbf{Q}\}_j$ in equations 59-61. The direct steady-state algorithm is identical to the non-iterative formulation, except that $\Delta t/2 \Rightarrow \Delta t$ in the Jacobians. This multiplier, common to all elements of $\{\mathbf{F}\}$, can be removed by placing Δt^{-1} as a scalar multiplier on $\{DET\}^T[A3000]$ in the Jacobians $[JQQ]$ of the differential equations exhibiting initial-value character, ie., ρ, m_i, and g.

Documentary Results

The Riemann shock tube problem can be expanded to a multi-dimensional specification, and results compared to the exact solution. From the standpoint of computer cost, multi-dimensional problem solutions are usually conducted on discretizations that are significantly coarser than those illustrated in Figures 2-3, for example. However, the developed finite element algorithm is documented [16] to retain its essential accuracy features on the progression to coarser discretizations for the Riemann problem. Sensitive "eyeball norm" measures of accuracy are convection velocity $\bar{u}_1 = m_1/\rho$ and temperature. Figures 5-6 summarize the k=1 algorithm fully discrete solutions, for these variables at t = 0.14154s, for three discretizations, M = 200, M = 50 and M = 25. The solid lines in Figures 5 b),c) and 6 b),c) are traces of the M = 200 solution. The coarser discretization has certainly degraded solution accuracy; however, the shock and trailing high temperature plateau remain quite sharply defined over what amounts to very few nodal variables.

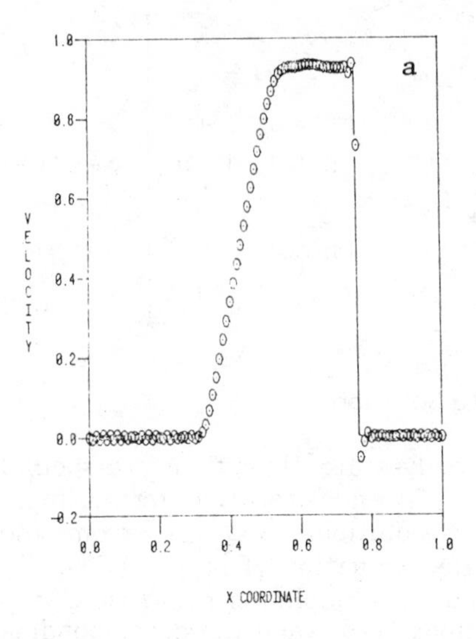

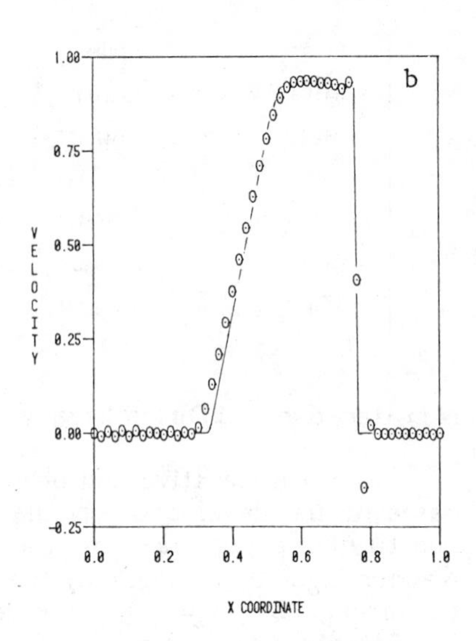

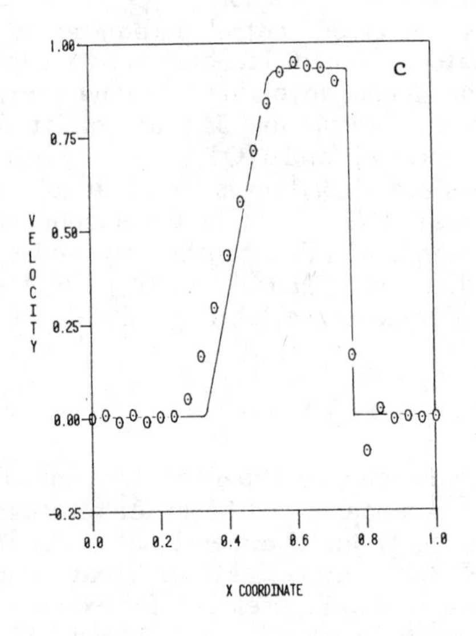

Figure 5. Riemann Shock Tube Solution Convection Velocity Distribution, k=1 Algorithm, t = 0.14154s, $\vec{\nu}_{\alpha}^{1}$ optimal, a) M = 200, b) M = 50, c) M = 25.

Additional results [16] indicate that the dissipation parameter set $\vec{\nu}_{\alpha}^{1}$, the multiplier of the second term in parenthesis modifying $\{\cdot\}'$ in equations 56-58, becomes progressively ineffective in solutions generated on coarser grids. The alternate dissipation parameter set remains required, and the definition $\vec{\nu}_{\alpha}^{2} = \nu_{0}\{\,3/4,\,2,\,1\,\}$, as obtained in the Riemann optimization study, was employed for the multi-dimensional problem solutions. A two-dimensional inviscid simulation of the Riemann problem defines $m_i = 0$ as the initial condition, and applies a vanishing normal derivative in the x_2 direction as the boundary condition for $\{Q\}$, $1 \leq I \leq 5$. This is the default specification for the finite element algorithm and no difference approximations are required for its enforcement.

Figure 7 summarizes the k=1 algorithm discrete solution for $\{Q\} = \{MI,P,R,G\}$ at t = 0.14154s, as obtained using a $M = 32 \times 6$ uniform discretization of R^2 and the standard definition of $\vec{v}_2$. This solution is in exact agreement with the one-dimensional solution, and the shock is sharply defined across the span of two elements in the x_1 direction. Solutions have been generated for various orientations of the shock tube coordinate system $(\vec{\eta})$ in the fixed Cartesian reference frame **x**. Figure 8 summarizes the k=1 discrete solution for a 26°

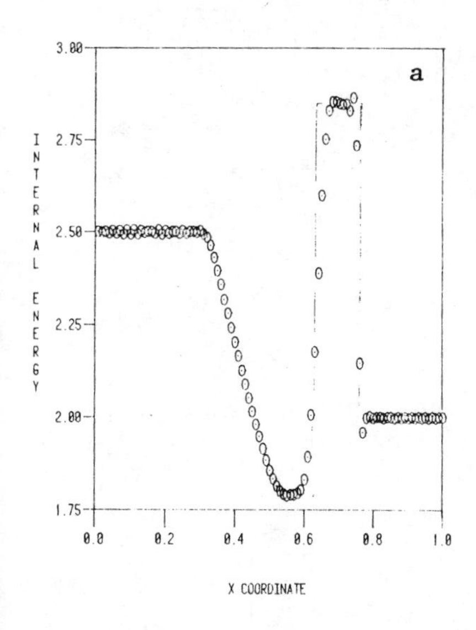

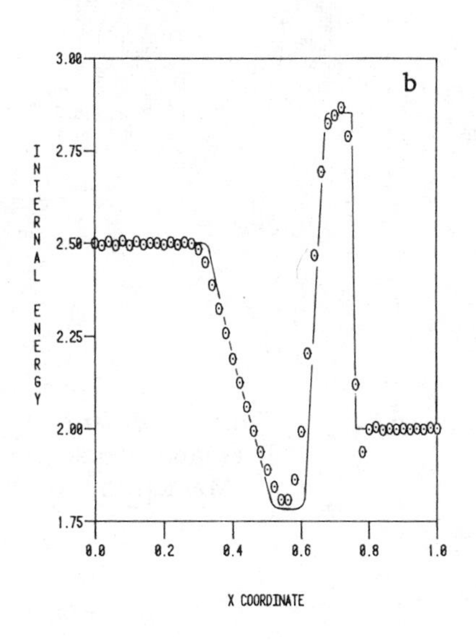

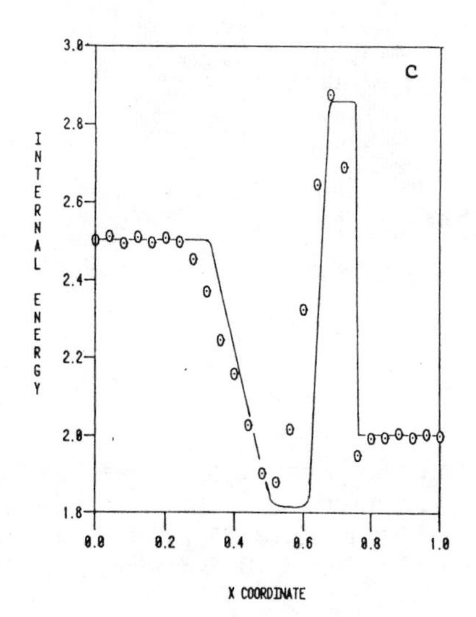

Figure 6. Riemann Shock Tube Solution Temperature Distribution, k=1 Algorithm, t = 0.14154s, $\vec{v}^{i}_{\alpha}$ optimal, a) M = 200, b) M = 50, c) M = 25.

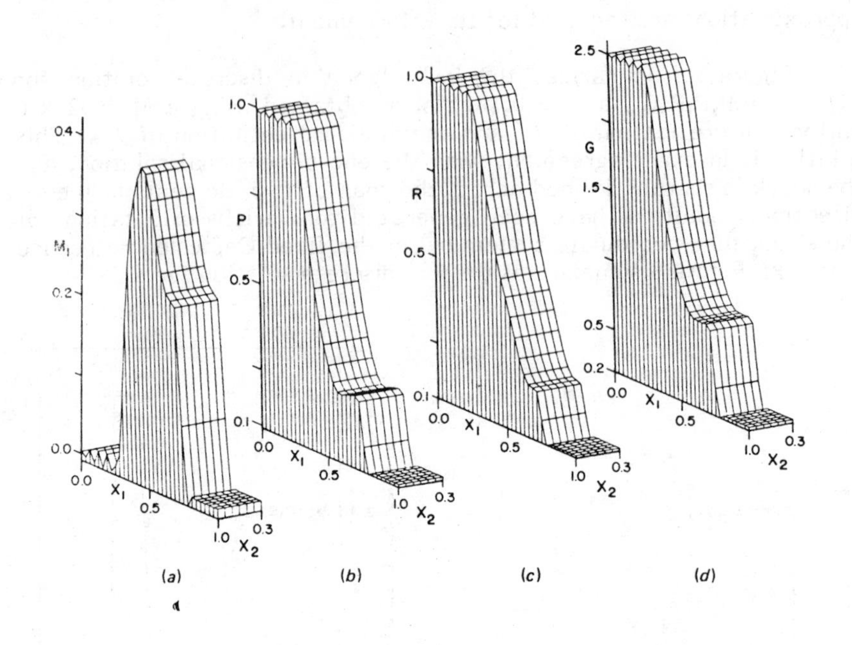

Figure 7. Finite Element Algorithm Solution, Two-Dimensional Riemann Shock Tube, M = 32 x 16, k=1, t = 0.14154s, a) Momentum, b) Pressure, c) Density, d) Energy.

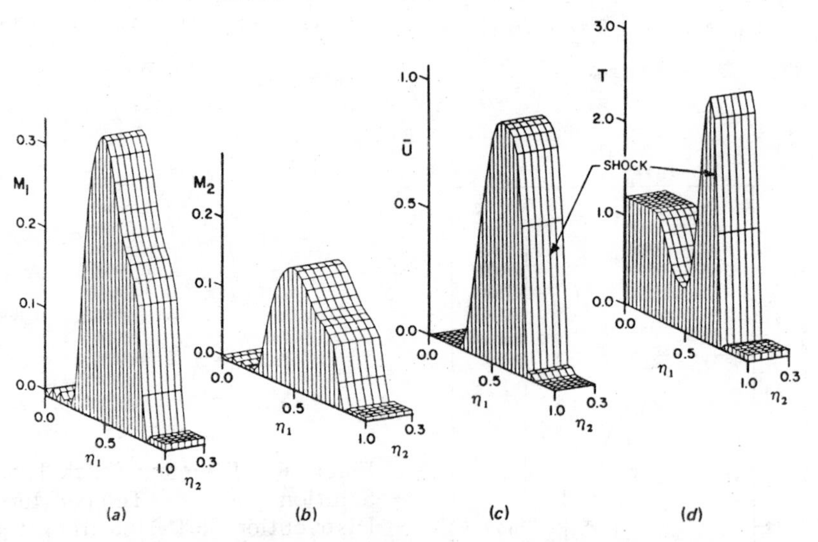

Figure 8. Finite Element Algorithm Solution, Two-Dimensional Riemann Shock Tube Oriented 26° to Cartesian Reference, M = 32 x 6, k=1, t = 0.14154s, a) Momentum m_1, b) Momentum m_2, c) Speed, d) Temperature.

rotation, as obtained on a M = 32 x 6 discretization for the standard $\vec{\nu}^2_\alpha$. The computed extremum of {M2} is one-third that of {M1}; squaring and adding to form $\bar{u}_k$ produces essentially exact agreement with the one-dimensional solution, see Figure 5c), also Figure 6c). Figure 9 summarizes the solution generated assuming the diaphragm oriented at an angle (45°) to the mesh. The result is propagation of an oblique shock with large cross-mesh gradients. The shock position is sharply defined in all variables, as is the trailing plateau. (The solution trashiness evidenced near the upstream right corner is the result of an inadequate domain extent [16].)

The two-dimensional Riemann problem can be redefined as viscous by setting $m_i \equiv 0$ at the walls, and employing a cold wall boundary condition for the energy equation. For a Reynolds number of Re = 10^5, and using a uniform M = 32 x 20 grid aligned with principal coordinates, Figure 10 summarizes the k=1 algorithm solution at t = 0.14154s, obtained using the standard dissipation set $\vec{\nu}^2_\alpha$. The shock is sharply defined in the pressure solution, Figure 10a, and the growth of a boundary layer is just visible in m_1 behind the shock, Figure 10b. Away from the walls, the centroidal nodal distributions of p and m_1 are in essentially exact agreement with the inviscid solution. In the solution for the dominant shear stress σ_{12}, Figure 10c, the shock intersection with the wall is well defined, and its passage has induced a residual large local level with steep gradients. The viscous solution employed the same time-step as the inviscid solution, and averaged three iterations/time-step for the convergence requirement $\varepsilon = 10^{-3}$.

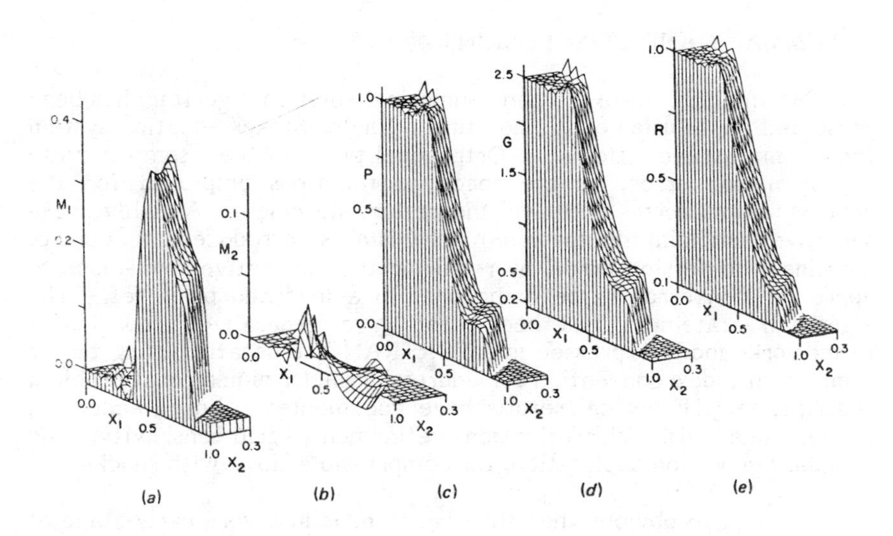

Figure 9. Finite Element Algorithm Solution, Two-Dimensional Riemann Shock Tube, Oblique Diaphragm and Shock, M = 32 x 6, k=1, t = 0.14154s, a) Momentum m_1, b) Momentum m_2, c) Pressure, d) Energy, e) Density.

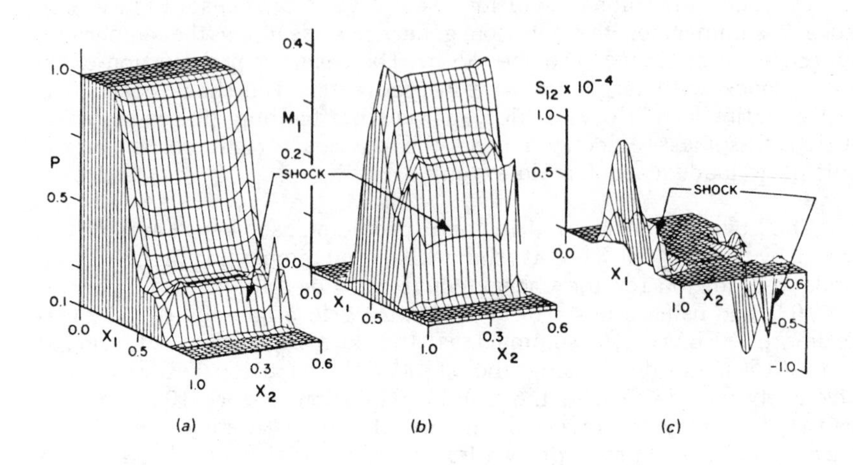

Figure 10. Finite Element Algorithm Solution, Two-Dimensional Viscous Riemann Problem, Re = 10^5, k=1, M = 32 x 20, t = 0.14154s, a) Pressure, b) Momentum m_1, c) Shear Stress σ_{12}.

6. SUMMARY AND CONCLUSIONS

An implicit finite element numerical solution algorithm has been presented for solution of the non-linear Navier-Stokes equation system for compressible flow. Orthogonality of the semi-discrete approximation error, to the space of functions employed for the approximation, forms the basic theoretical statement. A highly phase selective artificial dissipation mechanism is introduced, to control non-linear dispersion error, by requiring the convective semi-discrete approximation error to be orthogonal to a modified basis set. The algorithm statement has been cast in a generalized coordinates framework, and is expressed in a FORTRAN-like notation using tensor summation index convention for addressing multi-dimensional problem descriptions. Numerical results have documented solution accuracy, convergence with discretization refinement, grid-sensitivity, and boundary condition application, for compressible flows with shocks.

It is quite obvious that the algorithm is in a very early stage of development and evaluation. However, the theoretical and formalized programmatic structures of the finite element construction do appear to be well suited to eventual maturation of an accurate, efficient and robust procedure for application to flow prediction at large Reynolds numbers. Further confirmation of the projection is of course required, and is underway.

7. ACKNOWLEDGMENTS

The research projects which have resulted in development and evaluation of this finite element algorithm have been principally supported by the U.S. Air Force, Wright-Patterson Air Force Base, and the National Aeronautics and Space Administration, Langley Research Center. In addition, the author wishes to acknowledge the generosity of the University of Tennessee Computer Center in providing consequential computer support. This chapter is a modest expansion on a paper of a similar title appearing in the AIAA Journal [25].

8. REFERENCES

1. ZIENKIEWICZ, O.C., The Finite Element Method, McGraw-Hill, London, 1977.
2. MacCORMACK, R.W., "The Effect of Viscosity in Hypervelocity Impact Cratering," AIAA Paper No. 69-354, 1969.
3. MacCORMACK, R.W., "An Efficient Explicit-Implicit-Characteristic Method for Solving the Compressible Navier-Stokes Equations," Proc. SIAM-AMS Sym. Computational Fluid Dynamics, New York, 1977.
4. MacCORMACK, R.W., "A Numerical Method For Solving the Equations of Compressible Viscous Flow," AIAA Paper No. 81-0110, 1981.
5. BEAM, R.M. and WARMING, R.F., "An Implicit Factored Scheme for the Compressible Navier-Stokes Equations," AIAA J., Vol. 16, pp. 393-402, 1978.
6. BRILEY, W.R. and McDONALD, H., "Solution of the Multi-Dimensional Compressible Navier-Stokes Equations by a Generalized Implicit Method," J. Comp. Phys., Vol. 24, p. 372, 1977.
7. STEGER, J.L. and PULLIAM, T.H., "An Implicit Finite Difference Code for Inviscid and Viscous Cascade Flow," AIAA Paper No. 80-1427, 1980.
8. THAMES, F.C., THOMPSON, J.F., MASTIN, C.W. and WALKER, R.L., "Numerical Solutions for Viscous and Potential Flow About Arbitrary Two-Dimensional Bodies Using Body-Fitted Coordinate Systems," J. Comp. Phys., Vol. 24, No. 1, pp. 245-273, 1977.
9. THOMPSON, J.F. and MASTIN, C.W., "Grid Generation Using Differential Systems Techniques," in Numerical Grid Generation Techniques, NASA Report CP 2166, pp. 37-72, 1980.
10. BAKER, A.J. and SOLIMAN, M.O., "Utility of a Finite Element Solution Algorithm for Initial-Value Problems," J. Comp. Phys., Vol. 32, No. 3, pp. 289-324, 1979.
11. SOLIMAN, M.O. and BAKER, A.J., "Accuracy and Convergence of a Finite Element Algorithm for Laminar Boundary Layer Flow," J. Comp. and Fluids, Vol. 9, pp. 43-62, 1981.
12. SOLIMAN, M.O. and BAKER, A.J. "Accuracy and Convergence of a Finite Element Algorithm for Turbulent Boundary Layer Flow," Comp. Mtd. Appl. Mech. & Engr., V. 28, pp. 81-102, 1981.

13. BAKER, A.J., "Research On A Finite Element Numerical Algorithm for the Three-Dimensional Navier-Stokes Equations," USAF Report AFWAL-TR-82-3012, 1982.
14. CEBECI, T. and SMITH, A.M.O., Analysis of Turbulent Boundary Layers, Academic Press, New York, 1974.
15. RODI, W., Turbulence Models and Their Application in Hydraulics, Int. Assoc. Hydraulic Research, Delft, The Netherlands, 1980.
16. BAKER, A.J., Finite Element Computational Fluid Mechanics, McGraw-Hill/Hemisphere, New York, 1983.
17. PRENTER, P.M., Splines and Variational Methods, John Wiley, New York, 1975.
18. Proceedings, NASA Workshop On Numerical Grid Generation Techniques for Partial Differential Equations, NASA CP-2166, 1980.
19. HALMOS, P.R., Finite Dimensional Vector Spaces, Van Nostrand, Princeton, NJ, 1958.
20. ODEN, J.T., and REDDY, J.N., An Introduction to the Mathematical Theory of Finite Elements, John Wiley, New York, 1976.
21. RAYMOND, W.H., and GARDER, A., "Selective Damping in a Galerkin Method For Solving Wave Problems with Variable Grids," Monthly Weather Rev., V. 194, pp. 1583-1590, 1976.
22. SOD, G.A., "A Survey of Several Finite Difference Methods for Systems of Non-Linear Hyperbolic Conservation Laws," J. Comp. Phys., Vol. 27, pp. 1-31, 1978.
23. VAN LEER, B., "Towards the Ultimate Conservative Difference Scheme. V. A Second-Order Sequel to Godunov's Method," J. Comp. Phys., Vol. 32, pp. 101-136, 1979.
24. ZALESAK, S.T., "High Order ZIP Differencing of Convective Terms," J. Comp. Phys., Vol. 40, pp. 497-508, 1981.
25. BAKER, A.J., and SOLIMAN, M.O., "A Finite Element Algorithm For Computational Fluid Dynamics," AIAA Journal, Vol. 21, 1983, in press.

Section 3:

VISCOUS-INVISCID INTERACTION, INCOMPRESSIBLE

A NUMERICAL VIEW ON STRONG VISCOUS-INVISCID INTERACTION

A.E.P. Veldman

National Aerospace Laboratory NLR
Amsterdam, The Netherlands

SUMMARY

For about half a century attempts to calculate separated boundary layers have been frustrated by a Goldstein singularity, with which every calculation method in vogue in those days came to an untimely end. In this chapter the origin of this singularity will be elucidated from a numerical point of view. It is shown that the singularity is due to the mathematical way in which the flow equations are treated. Several strategies to avoid the singular behaviour are discussed. Analysis of the iterative behaviour of these methods reveals a common underlying condition governing their convergence.

1 INTRODUCTION

Viscous effects can have a substantial influence on the aerodynamic properties of airplanes. Therefore it is important that numerical prediction methods take this influence into account. Fortunately, in many cases viscosity plays its role only in a thin region consisting of boundary layer and wake, whereas the remaining part of the flow field is inviscid in character. This subdivision of the flow field reduces the task of solving the flow equations, in comparison with the efforts required for solving the viscous flow equations (Navier-Stokes) in the complete flow domain.

However, a subdivision requires some kind of coupling mechanism with which the interaction between both parts of the flow field is accounted for. Special care has to be taken in designing the numerical coupling algorithm, since an "unlucky" choice may lead to difficulties which are of purely mathematical nature, and have little to do with the physics of the problem. An example hereof is the well-known Goldstein singularity [1] in a point of flow separation: it occurs because prescribing the pressure as a boundary condition to the viscous flow equations leads to an ill-posed problem near separation (even the existence of a solution with prescribed pressure is uncertain then).

When the interaction between the viscous and the inviscid part of the flow field is weak, the numerical coupling can proceed in the classical way (called "direct"). Herein the two parts are treated alternately: the inviscid flow equations are solved with prescribed displacement thickness, whereas the viscous flow equations are solved with prescribed pressure. But when the two parts of the flow field are interacting strongly (which is the case near separation, trailing edges and shock waves) the classical procedure leads to numerical difficulties (of which the Goldstein singularity is an example). These can be circumvented when the numerical coupling algorithm is changed; for instance prescription of the displacement thickness to the viscous flow equations (the "inverse" method) removes the singular behaviour in separation. The first observation hereof has been presented by Catherall and Mangler [2]. After this demonstration of numerical feasibility, the literature on viscous-inviscid interaction has been growing vastly. Several numerical coupling algorithms have been proposed; we will discuss them in the sequel of this chapter.

In addition much attention has been paid to the modelling of the physics of the flow, especially in those regions where a strong interaction exists between the viscous and inviscid parts of the flow field. Asymptotic theories, based on singular perturbation theory, have been used to gain insight into the relative importance of the various physical phenomena in the flow.

For laminar flow the so-called triple deck describes a number of situations (see the review papers by Stewartson [3] and Smith [4]). One of the main conclusions is that the boundary-layer equations are sufficient to describe the viscous part of the flow. The latter conclusion has been confirmed, amongst others, by comparisons between numerical solutions for separation bubbles using boundary-layer equations and Navier-Stokes equations which have been performed by Briley and McDonald [5] and Heidsieck [6].

For turbulent flow also asymptotic theories are available. Melnik, et al [7] have presented a model for strong trailing-edge interaction in case of attached flow. A theory for turbulent separation is proposed by Sychev and Sychev [8]. A comprehensive review of models for shock-wave/boundary-layer interaction can be found in a paper by Adamson and Messiter [9].

Insight in the importance of flow phenomena can also be obtained by means of experiments. For instance, strong interaction near trailing edges has been studied by Cleary, et al [10]. For more literature on the modelling of viscous-inviscid interaction we refer to the proceedings of two recent conferences [11, 12].

2 NUMERICAL COUPLING ALGORITHMS

The coupling algorithm between the viscous and the inviscid part of the flow field plays a central role in a numerical method for calculating viscous-inviscid interaction. To focus attention to the coupling algorithm we will start from a schematic description of the flow equations. For ease of presentation the equations governing the interaction are written in terms of u_e (the velocity at the edge of the viscous region) and δ^* (the displacement thickness). The actual situation is more complicated, but this does not influence the principle of the interaction.

The two regions in which the flow field is divided each yield a relation between u_e and δ^*. In symbolic notation we denote these relations by

$$\text{inviscid flow} : u_e = P[\delta^*], \quad (1)$$

$$\text{viscous flow} : u_e = B[\delta^*]. \quad (2)$$

In order not to make the situation too difficult it is assumed that the system (1) + (2) possesses a unique solution. In view of the complexity of the flow equations, the solution is usually pursued via iterative procedures. In this section we will describe a number of possibilities.

The classical direct iterative method to solve (1) + (2) reads (the subscript n denotes the iteration count)

$$\text{direct method: } \begin{aligned} u_e^{(n)} &= P\ [\delta^{*(n-1)}], \\ \delta^{*(n)} &= B^{-1}[u_e^{(n)}]. \end{aligned} \quad (3)$$

The numerical difficulties encountered by the direct method can be explained from the properties of B^{-1} (the symbolic inverse of B). It will be discussed in the next section that B^{-1} possesses a singular character in regions of strong interaction.

The difficulties associated with B^{-1} can be circumvented by reversing the order of the iteration process (3); thus the inverse method is created

$$\text{inverse method: } \begin{aligned} \delta^{*(n)} &= P^{-1}[u_e^{(n-1)}], \\ u_e^{(n)} &= B\ [\delta^{*(n)}]. \end{aligned} \quad (4)$$

The operators P^{-1} and B do not give rise to specific difficulties. However, the iteration process requires a large amount of underrelaxation in order to obtain (slow) convergence, as will be explained in section 4.

A better method is created when the direct and inverse approaches are combined into a semi-inverse method, which can be schematized by

$$u_e^P = P[\delta^{*(n-1)}],$$

$$u_e^B = B[\delta^{*(n-1)}],$$

$$\delta^{*(n)} = R[u_e^P, u_e^B, \delta^{*(n-1)}],$$

where R is a relaxation formula. The methods of Le Balleur [13] and Carter [14] belong to this type.

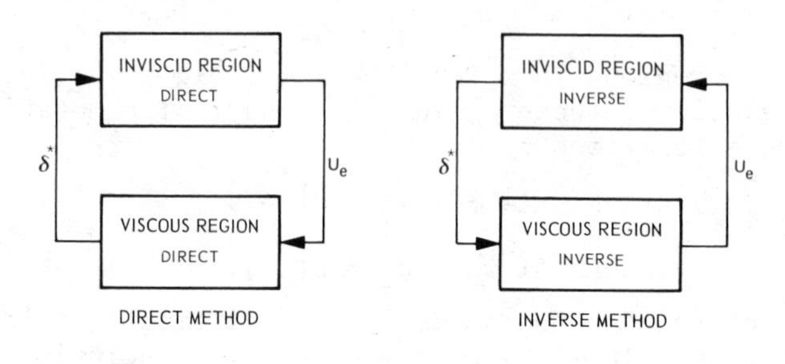

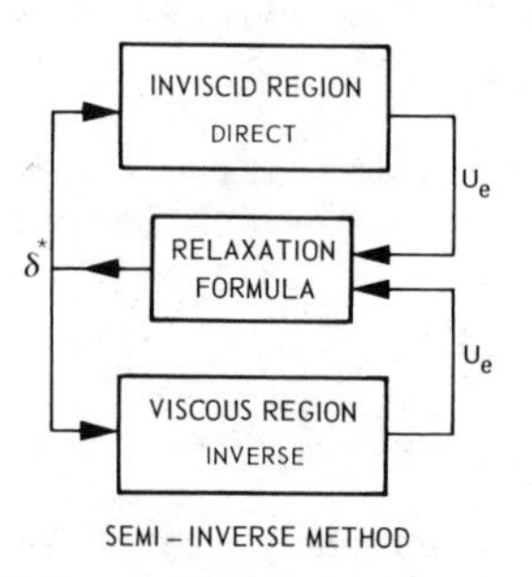

Fig. 1 Global organization of methods for calculating viscous-inviscid interaction

Next to the above methods, in which both parts of the flow field are treated separately (see also Fig. 1), another approach exists in which it is tried to solve the equations (1) + (2) simultaneously. When one of the equations is simple such an approach becomes feasible. Two strategies can be distinguished. One possibility is used by various authors; we mention Moses, et al [15], Wai and Yoshihara [16] and Ghose and Kline [17]. They describe the viscous region by means of an integral method. In this way the viscous equations can be implemented in the numerical code as a boundary condition to the inviscid flow equations.

The other possibility is presented in [18-21]. The inviscid flow equation (1) is replaced (temporarily and only if required for numerical convenience) by a much simpler equation, which is called the interaction law

$$u_e = I[\delta^*]. \qquad (1')$$

This equation is chosen such that:

i) it is simple enough to be treated as a boundary condition to the viscous flow equations;

ii) it avoids difficulties of purely numerical character;

iii) it gives a reasonable (i.e. at least locally adequate) description of the inviscid contribution to the viscous-inviscid interaction.

When, from a physical point of view, the interaction law is considered to be not accurate enough an outer iterative process can be added,

$$u_e^{(n)} - I[\delta^{*(n)}] = P[\delta^{*(n-1)}] - I[\delta^{*(n-1)}],$$

$$u_e^{(n)} - B[\delta^{*(n)}] = 0, \qquad (6)$$

such that after convergence ($n \to \infty$) the solution again satisfies (1) + (2). The method denoted in (6) is called <u>quasi-simultaneous</u>, since the most important part of the interaction (given by the interaction law) is treated simultaneously with the viscous flow equations (Fig. 2).

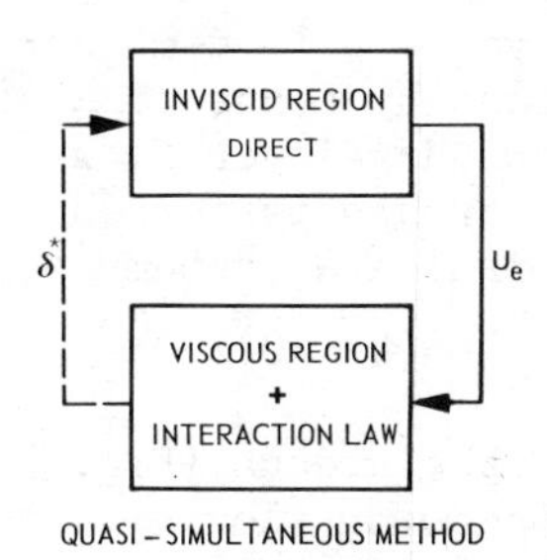

Fig. 2 Global organization of quasi-simultaneous method

3 A SEPARATION BUBBLE: AN EXAMPLE OF STRONG VISCOUS-INVISCID INTERACTION

The character of strong viscous-inviscid interaction can be understood from a study of the equations describing the viscous and inviscid parts of the flow field. Especially the viscous equation plays an important role since its character is essentially different in weak and strong interaction. A number of ways exist in which the equations can be studied.

One way is to treat the flow problem as unsteady [22]. Then the theory of characteristics for hyperbolic equations can be applied to investigate the direction in which information flows through the viscous and inviscid regions.

A second way is to study the equations in an asymptotic sence (for vanishing viscosity), see for instance [3,4].

The asymptotic structure obtained will show a number of regions. The way in which these regions are coupled reveals the character of the interaction between the viscous and inviscid regions. Also, the equations describing the various regions reflect which physical phenomena are important.

A third way is to look at the flow equations with a numerical eye. The matrices obtained after discretization of the equations can be investigated. In this section we will pursue this type of approach in more detail.

From a number of investigations which have been reported in the literature it appears that the essential properties of strong viscous-inviscid interaction are independent of the choice of the viscous flow equations. It makes no difference whether boundary-layer equations are used in integral formulation or in differential formulation [18,23]; neither makes it a difference whether the viscous layer is modelled laminar or turbulent, incompressible or compressible [24]. A fortiori, when the viscous region is thin, even the use of the full Navier-Stokes equations does not essentially change the character of the interaction [6].

3.1 Definition of a separated flow problem

In view of the above we will demonstrate the character of the viscous flow equations by means of a simple example: laminar, incompressible boundary-layer flow over the Carter and Wornom trough [25]. The trough is formed by an indented plate which in a Cartesian coordinate system (x,y) is given by

$$y_B = -t \operatorname{sech}(4x-10),\ 0 \leqq x < \infty,$$

where t is the depth of the trough (Fig. 3). The plate is placed parallel to an oncoming flow with unit velocity. The Reynolds number of the flow, based on unit length, is defined as $Re = 1/\nu$, where ν is the kinematic viscosity.

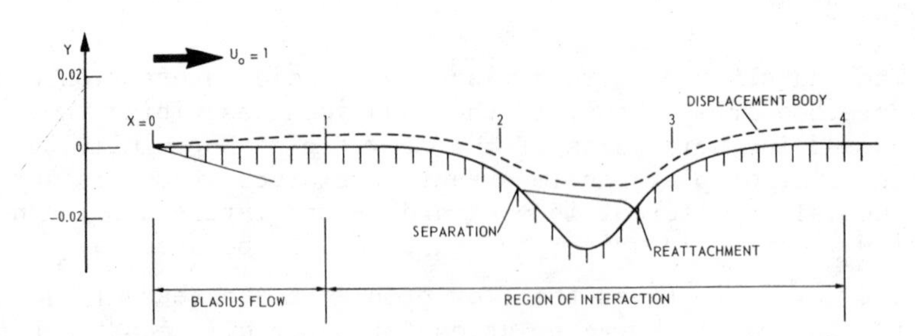

Fig. 3 Dented plate for calculation of separation bubbles (separation streamline and displacement body are indicated for Re = 36 x 10^4). Note difference in x and y scale.

The description of the outer inviscid flow is also kept simple. The edge velocity u_e is taken to be a superposition of the inviscid velocity U_e over the trough without viscous effects, and a viscous correction $u_{e_{\delta*}}$ which is due to the viscous displacement. Thus

$$u_e(x) = U_e(x) + u_{e_{\delta*}}(x) \ . \tag{7}$$

Thin airfoil integrals have been used by Carter and Wornom [25] to describe both terms in the right-hand side of (7). We will use the same expressions, remarking that the use of more sophisticated expressions of course will lead to somewhat different results; however, this does not change the essential features we are interested in. The inviscid velocity distribution $U_e(x)$ is chosen according to

$$U_e(x) = 1 + \frac{1}{\pi} \oint_0^\infty \frac{dy_B/d\xi}{x-\xi} d\xi, \tag{8}$$

whereas Carter and Wornom have modelled the displacement effect by means of

$$u_{e_{\delta*}} = \frac{1}{\pi} \oint_1^4 \frac{d\xi^*/d\xi}{x-\xi} \xi \ . \tag{9}$$

Thus it is assumed that only the region between x = 1 and x = 4 contributes to the interaction.

The above choice for the inviscid flow operator P is so simple that it can also be used as an interaction law I. Thus the iteration process indicated in (6) can be avoided in this example.

The viscous region is described by the laminar, incompressible boundary-layer equations

$$u \frac{\partial u}{\partial x} - \frac{\partial \tilde{\Psi}}{\partial x} \frac{\partial u}{\partial \tilde{y}} = u_e \frac{d}{dx} u_e + \frac{\partial^2 u}{\partial \tilde{y}^2}, \tag{10a}$$

$$u = \frac{\partial \tilde{\Psi}}{\partial \tilde{y}} , \tag{10b}$$

with boundary conditions

$$\tilde{y} \to \infty: u \to u_e, \ \tilde{\Psi} \to u_e(\tilde{y} - \delta), \tag{11a}$$

$$\tilde{y} = 0: \tilde{\Psi} = 0, \ u = 0. \tag{11b}$$

The usual boundary-layer scaling has been applied, i.e.

$$\tilde{y} = Re^{\frac{1}{2}}(y-y_B), \ \tilde{\Psi} = Re^{\frac{1}{2}} \ \Psi \text{ and } \delta = Re^{\frac{1}{2}} \ \delta^*,$$

and Ψ is the streamfunction. These equations are solved in the region $1 \leqq x \leqq 4$. The initial profile which is required at x = 1 has been set equal to a Blasius profile.

Combining (8) and (9) the outer flow is described by means of the interaction law

$$u_e(x) = U_e(x) + \frac{Re^{-\frac{1}{2}}}{\pi} \oint_1^4 \frac{d\delta/d\xi}{x-\xi}\, d\xi \tag{12}$$

This relation is applied as a boundary condition which closes the above set of equations (10). The boundary condition possesses an elliptic character. This makes it necessary to perform a number of calculation sweeps through the boundary layer.

3.2 Discretization of the equations of motion

In our calculations the following finite-difference approximations have been used. The bounded domain, obtained after setting the outer edge of the boundary layer at $\tilde{y} = 7\,\delta(x)$, has been covered by a (non-equidistant) mesh with grid points $(x_i, \tilde{y}_j)$; $i = 0,1, \ldots, N$; $j = 0,1, \ldots, M$. The x-derivatives in (10a) were replaced by the familiar second-order three-point-backward formula, using information from grid lines i-1 and i-2. In regions of reversed flow the most stable procedure is to switch the direction of discretization, thus using grid points along the lines i+1 and i+2. However, when the mesh size in x-direction is not too small it is our experience that using the three-point-backward formula throughout (thus ignoring the local flow direction) does not necessarily give rise to numerical instabilities [18]. The $\tilde{y}$-derivatives in (10a) have been discretized with second-order formulas centered around $\tilde{y}_j$: $j = 1, \ldots, M-1$; whereas in (10b) they were centered around $\tilde{y}_{j-\frac{1}{2}}$, $j = 1,\ldots M$.

To obtain insight into the character of the discrete equations resulting from the above process (or another one if preferred) it is advisable to linearize them with Newton's method. The discrete equations, including boundary conditions (11a) and (11b) can now be ordered to form a 2x2 block tridiagonal matrix with one additional column (which is due to the presence of u_e). When the elimination process of this matrix is started at $j = 0$ and is proceeded towards the outer edge, it can be arranged such that one equation is left which gives a relation between $u_{e_i} = u_e(x_i)$ and $\delta_i = \delta(x_i)$. This relation plays an important role; due to the Newton linearization it contains information on the local behaviour of the boundary-layer. Let us suppose that this equation reads

$$u_{e_i} + d_i\, \delta_i = rhs_i \quad . \tag{13}$$

The next step is to combine this equation with the remaining boundary condition which gives information about the outer flow. In a direct method it would prescribe u_{e_i}, and in an inverse method δ_i would be given. In our example, defined above, the remaining boundary condition is (12), which can be transformed into a linear combination of u_{e_i} and δ_i. In all three cases the remaining boundary condition fits into the form

$$u_{e_i} + c_i\, \delta_i = RHS_i \quad . \tag{14}$$

Equations (13) and (14) constitute two equations in two unknowns, to be studied in more detail below. When these equations are solved the process of back-substitution can be started. At each station a few of these iteration sweeps have to be made until the Newton process has converged sufficiently.

Equation (13) describes the character of the viscous layer, d_i being the quantity to pay attention to. This quantity describes how u_e reacts on changes in δ. We can write $d_i = -\partial u_e/\partial \delta$ at the fixed station x_i. In our calculations we have been monitoring d_i. Figure 4 shows a typical example; it corresponds to a calculation of the Carter and Wornom through for Re = 36×10^4 and t = 0.03. All of our calculations show a positive d_i when the flow is attached, and a slightly negative d_i for separated flow. When an integral description of the viscous layer is employed this behaviour can be proved analytically [23]. For finite difference calculations such a proof is still lacking, although much numerical evidence is being built up recently [18, 24, 26, 27].

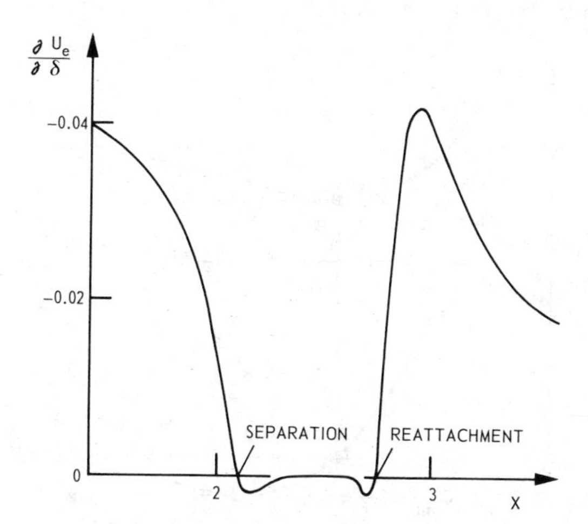

Fig. 4 Separation bubble: derivative of edge velocity with respect to displacement thickness at same boundary layer station (Re = 36 x 10 , Δx = 1/20).

When $d_i = 0$ the boundary-layer equation (13) reduces to $u_{e_i} = rhs_i$. One can say that now the boundary-layer determines the value of u_e, which is in contrast with the "classical" situation in which the inviscid outer flow determines u_e. This change of hierarchy between the viscous and inviscid regions has been noticed already by Lagerstrom [28] who uses the terminology "self-induced pressure"; note that the pressure is directly related to u_e via Bernoulli's law. Also in asymptotic triple-deck theory the change of hierarchy can be observed; for a discussion hereof we refer to [29]. Further, when $|d_i|$ is small the displacement thickness δ reacts violently on changes in u_e. The boundary layer thus plays a much more active role than in the situation of attached flow where d_i is relatively large. This difference in the behaviour of d_i constitutes the difference between strong and weak interaction.

Further information can be obtained when at a given station x_i, with all previous stations fixed, the viscous equations are solved with a varying value of δ_i prescribed. In this way u_e and c_f (the wall stress coefficient) can be studied as a function of δ. Figure 5 shows a typical example corresponding to a station close to separation. It is found that u_e possesses a minimum, which coincides with the vanishing of the shear stress. It will be clear that for values of u_e below this minimum no solution of the boundary-layer equations exists. Note that the slope of the u_e-curve equals $\partial u_e/\partial\delta$. Further, figure 5 shows that also c_f as a function of δ possesses a minimum; it occurs for a situation where the flow is well separated. This implies that also the prescription of the wall shear as a boundary condition to the viscous equations can lead to difficulties. Indeed, such difficulties have been reported already by Carter [30].

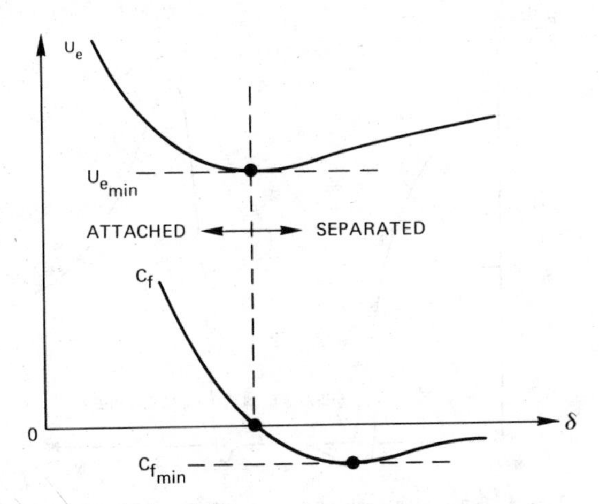

Fig. 5 Behaviour of u_e and c_f as a function of δ at a fixed boundary layer station

Results similar to the ones depicted in figure 5 have been presented by Drela and Thompkins [26] for incompressible flow. But also for compressible flow this type of relations exist, as is apparent from the investigations by Ardonceau, et al [24] and Murphy and King [27].

3.3 Experiments with various boundary conditions

Now that the character of eq. (13) is known this equation can be combined with eq. (14). It is clear that the existence of a solution of (13) + (14) is unlikely when $d_i = c_i$. For a direct boundary condition (u_e prescribed, i.e. $c_i = 0$) this implies difficulties when $d_i = 0$, i.e. at a point of separation! With this observation we have arrived at the root of the

Goldstein singularity.

It is interesting to see what actually happens in the calculations when other values of c_i are used. We have done some numerical experiments. As may be expected we encountered Goldstein-type singularities when c_i was lying in the range of values which d_i takes (see Fig. 4); be aware that the values of d_i in figure 4 are those for the converged solution. There are two possible choices for which there is hope to avoid singularities: select $c_i > \max_j d_j$ or $c_i < \min_j d_j$. In our experiments the first strategy always ran into a singularity. This is due to the fact that during the iterations the values of d_i were varying, and apparently, for this choice of c_i, in each calculation one of the coefficients d_j came close enough near c_i to have the calculation break down. However, experiments with $c_i < \min_j d_j$ were much more satisfactory. It was our experience that during the iteration process d_j did not become much smaller than its converged value (at least in the critical zone of separated flow) and calculations thus proceeded without difficulties.

From the numerical experiments described above we have learned that numerical survival requires the prescription of a boundary condition (14) with c_i sufficiently negative. This requirement has been obtained from a consideration of the calculation at one fixed station x_i. Thus nothing has been said yet on the convergence of the boundary layer sweeps which have to be made. Section 4 will be devoted to this subject.

3.4 Selection and discretization of the interaction law

The next step is to find a boundary condition which, after discretization, leads to a discrete relation of the type (14) with a favourable value of c_i. Our philosophy is to let the numerics follow the physics. In the physical situation the boundary-layer experiences an outer flow. Therefore we select a description of the outer flow, or, more precise, a good approximation hereof as a boundary condition. For the separated flow calculations the thin-airfoil expression (12) is likely to give a reasonable description of the outer flow. Let us see what this expression looks like after discretization.

Define

$$J_i = \oint_1^4 \frac{d\delta/d\xi}{x_i - \xi}\, d\xi, \quad i = 1, \ldots, N.$$

For the integration the segment [1,4] is divided into the intervals $[x_j, x_{j+1}]$. δ is approximated by a quadratic function on the two intervals adjacent to x_i, i.e. on $[x_{i-1}, x_{i+1}]$ (now the Cauchy principal value makes sense); and δ is taken linear on the other intervals. For sake of convenience when $i = N$, extend the segment [1,4] with one more interval $[x_N, x_{N+1}]$ for which $\delta_{N+1} = \delta_N$. Thus the discretization is performed in the following way.

$$J_i = \sum_{\substack{j=0 \\ j \neq i-1,i}}^{N} \int_{x_j}^{x_{j+1}} \frac{d\delta}{d\xi}\bigg|_{j+\frac{1}{2}} \frac{d\xi}{x_i - \xi} + $$

$$+ \oint_{x_{i-1}}^{x_{i+1}} \left\{\frac{d\delta}{d\xi} + (\xi - x_i)\frac{d^2\xi}{d\xi^2}\right\}_i \frac{d\xi}{x_i - \xi}$$

$$= \sum_{\substack{j=0 \\ j \neq i-1,i}}^{N} \frac{d\delta}{d\xi}\bigg|_{j+\frac{1}{2}} \ln\left|\frac{x_i - x_j}{x_i - x_{j+1}}\right|$$

$$+ \frac{d\delta}{d\xi}\bigg|_i \ln\left|\frac{x_i - x_{i-1}}{x_i - x_{i+1}}\right| - \frac{d^2\delta}{d\xi^2}\bigg|_i (x_{i+1} - x_{i-1}).$$

The derivatives are discretized as follows

$$\frac{d\delta}{d\xi}\bigg|_{j+\frac{1}{2}} = \frac{\delta_{j+1} - \delta_j}{x_{j+1} - x_j},$$

$$\frac{d\delta}{d\xi}\bigg|_i = \frac{x_i - x_{i+1}}{(x_{i-1} - x_i)(x_{i-1} - x_{i+1})}\delta_{i-1} + \frac{x_{i-1} - x_{i+1}}{(x_i - x_{i-1})(x_i - x_{i+1})}\delta_i + $$

$$+ \frac{x_{i-1} - x_i}{(x_{i+1} - x_{i-1})(x_{i+1} - x_i)}\delta_{i+1},$$

$$\frac{d^2\delta}{d\xi^2}\bigg|_i = \frac{2}{(x_{i-1} - x_i)(x_{i-1} - x_{i+1})}\delta_{i-1} + \frac{2}{(x_i - x_{i-1})(x_i - x_{i+1})}\delta_i + $$

$$+ \frac{2}{(x_{i+1} - x_{i-1})(x_{i+1} - x_i)}\delta_{i+1}.$$

The above formulas can be used for non-equidistant distributions in x-direction. When an equal mesh spacing h is used the formulas become somewhat simpler. In the latter situation they can be arranged to

$$J_i = \sum_{\substack{j=0 \\ j \neq i-1,i}}^{N} \left\{\frac{1}{h}(\delta_{j+1} - \delta_j)\ln\left|\frac{i-j}{i-j-1}\right|\right\} - \frac{2}{h}(\delta_{i+1} - 2\delta_i + \delta_{i-1}).$$

Writing this in the form

$$J_i = \sum_{j=1}^{N} a_{ij}\delta_j + a_{io}\delta_o + a_{i,N+1}\delta_{N+1},$$

for equidistant x-spacing the matrix (a_{ij}) is symmetric with positive diagonal entries, whereas the off-diagonal elements are negative. Moreover the matrix is diagonally dominant, and hence positive definite.

The above discretization is substituted in the interaction law (12), which then takes the discrete form

$$u_{e_i} = U_{e_i} + \sum_{j=1}^{N} \alpha_{ij}\delta_j + R_i, \tag{15}$$

where $\alpha_{ij} = \frac{1}{\pi} Re^{-\frac{1}{2}} a_{ij}$ and $R_i = \frac{1}{\pi} Re^{-\frac{1}{2}} (a_{io}\delta_o + a_{i,N+1}\delta_{N+1})$. By shifting the contribution of δ_i to the left-hand side we obtain an expression of the type (14), with $c_i = -\alpha_{ii}$ which for equal mesh spacing becomes

$$c_i = -\frac{4}{(\pi h Re^{\frac{1}{2}})} \ . \tag{16}$$

Note that the sign of c_i is negative, which we desired. The above discretization is just an example; other ways can be used which also result in a coefficient c_i comparable with (16) [18].

3.5 The global iteration scheme

To complete the description of the quasi-simultaneous numerical method the organisation of the global iteration scheme is presented. These iterations are required because of the elliptic character of the inviscid flow. At the beginning of the first sweep an initial guess is made for the displacement thickness. Hereafter in each sweep the discrete interaction law (15) is implemented in the following Gauss-Seidel fashion:

$$u_{e_i}^{(n)} - \alpha_{ii}\delta_i^{(n)} = U_{e_i} + \sum_{j=1}^{i-1} \alpha_{ij}\delta_j^{(n)} + \sum_{j=i+1}^{N} \alpha_{ij}\delta_j^{(n-1)} + R_i^{(n-1)}, \tag{17}$$

where the index (n) denotes the number of the iteration sweep. The global iterations are terminated when the difference between $\delta_j^{(n)}$ and $\delta_j^{(n-1)}$ becomes smaller than a prescribed small number.

3.6 Results and remarks

The numerical experiments described above have been made using separation bubbles; we will present some results in this section. Figure 6 shows velocity profiles in the separated region for t = 0.03 and Re = 36 x 10^4. Corresponding skin friction and pressure behaviour can be inferred from figures 7 and 8.

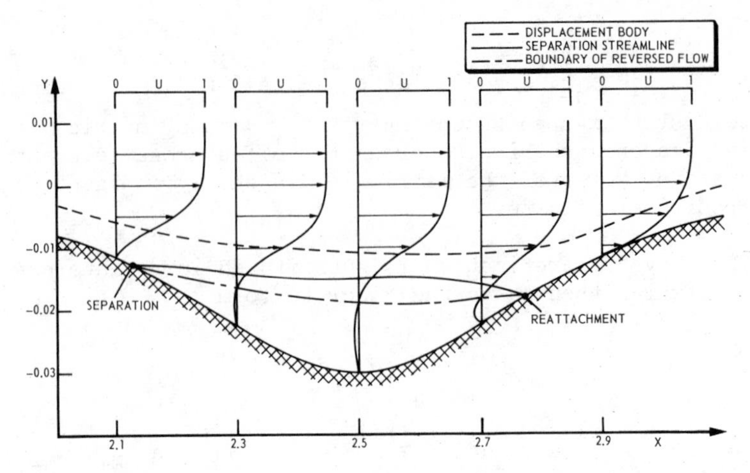

Fig. 6 Separation bubble: velocity profiles for a case with severe separation, Re = 36 x 10^4

The latter figures also display results for a smaller Reynolds number, Re = 8 x 10^4; a case which has already been computed by Carter and Wornom. The best impression of the influence of the viscous-inviscid interaction can be seen in figure 8 where the inviscid pressure distribution without interaction has also been plotted.

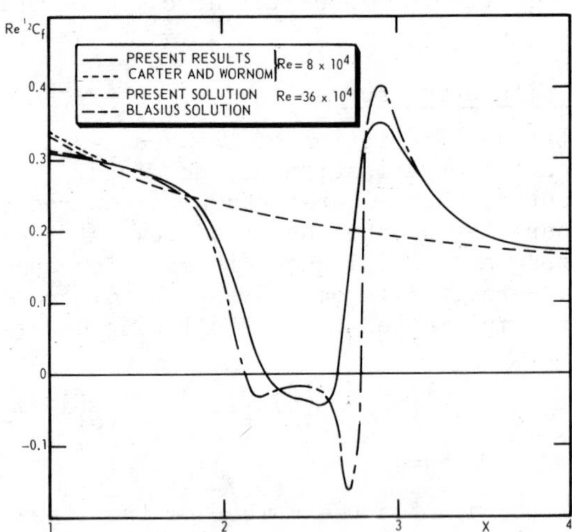

Fig. 7 Separation bubble: skin friction

For separation bubbles of which the thickness is comparable to the oncoming boundary-layer thickness the present method converges easily. For instance, the calculation for Re = 36 x 10^4 on a grid 60 x 40 requires 18 boundary-layer sweeps to reach $\max_i |\delta_i^{(n)} - \delta_i^{(n-1)}| < 10^{-4}$ [18]. However, for larger separation bubbles numerical difficulties show up again. The Newton process, performed at each boundary-layer station, does not converge at stations where large separation occurs.

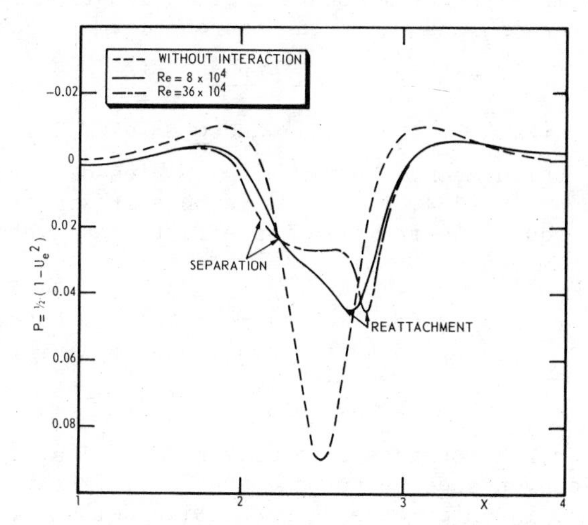

Fig. 8 Separation bubble: pressure distributions with and without separation

Dijkstra [31] has taken a closer look at the Newton matrix at such stations and found it to be singular. His impression is that the discrete set of equations (10a) and (10b) at such a station tend to be a dependent system. This dependency, of course, cannot be solved by prescribing a suitable boundary condition (14).

This phenomenon may be related to a bifurcation of the solution (i.e. multiple solutions exist) or to a non-existence of a solution. In favour of the latter possibility is that unsteady calculations for large separation bubbles fail to converge to a steady limit. It must be remarked that we are now talking about bubbles in which the actual physical flow will be turbulent; hence the laminar calculations do not necessarily have a physical meaning. Nevertheless a further investigation can be worthwhile.

4 NUMERICAL ANALYSIS OF A MODEL PROBLEM

In the previous section we have seen which type of boundary condition is required in order to survive numerically at a fixed station, i.e. to avoid numerical singularities after which the calculation breaks down. A further requirement is that the overall calculation, i.e. the global iteration process, converges. This section is devoted to the latter requirement.

4.1 Definition of a model problem

In order to give a theoretical analysis of the global iteration process we define here a model problem which is a simplification of the full problem described in the previous section. The model problem consists of an equation representing the viscous region, and which in discrete form is taken as

$$\underline{u}_e - \underline{D}\,\underline{\delta} = \underline{rhs}, \tag{18}$$

where $\underline{u}_e = (u_e(x_1), \ldots, u_e(x_N))^t$ and $\underline{\delta} = (\delta(x_1), \ldots, \delta(x_N))^t$. Further $\underline{D}$ is the diagonal matrix with entries $-d_i$, where d_i is the coefficient in (13). Thus (18) is an approximation of the viscous flow equations in which the effect of neighbouring stations is removed.

The outer flow is modelled by means of the thin-airfoil discretization from the previous section, i.e.

$$\underline{u}_e - Re^{-\frac{1}{2}}\,\underline{H}\,\underline{\delta} = \underline{U}_e\,, \tag{19}$$

where the matrix $\underline{H}$ consists of elements $H_{ij} = \frac{1}{\pi}\,a_{ij}$. Again for theoretical convenience we suppose that equidistant x-spacing has been used, in which case $\underline{H}$ is positive definite.

4.2 Analysis

The above model problem (18) and (19) will not only be used to analyze the global convergence of the present quasi-simultaneous method. Also the behaviour of the other techniques, mentioned in section 2, for solving the flow equations will be treated theoretically.

The convergence of the various iterative techniques can be studied from the homogeneous problem

$$\underline{u}_e - \underline{D}\,\underline{\delta} = \underline{0},\; \underline{u}_e - Re^{-\frac{1}{2}}\,\underline{H}\,\underline{\delta} = 0. \tag{20}$$

It will turn out that the following property plays an important role in the analysis:

$$\underline{Q} = Re^{-\frac{1}{2}}\,\underline{H} - \underline{D} \text{ is positive definite.} \tag{21}$$

Note that this property implies that $c_i = -\frac{1}{\pi}\,Re^{-\frac{1}{2}}\,a_{ii} = -Re^{-\frac{1}{2}}\,H_{ii} < d_i$. Hence the requirement, obtained in the previous section, under which the set of equations (13) and (14) can be solved is fulfilled whenever (21) is satisfied. We can already say that (21) is always satisfied for attached flow since then $d_i < 0$. However, for separated flow, where d_i is slightly positive, no guarantee can be given a priori that (21) is satisfied.

(a) The direct method

The direct method to solve (20) can be written as

$$\underline{u}_e^{(n)} = Re^{-\frac{1}{2}}\,\underline{H}\,\underline{\delta}^{(n-1)},\quad \underline{\delta}^{(n)} = \underline{D}^{-1}\,\underline{u}_e^{(n)}\,,$$

or equivalently, eliminating $\underline{u}_e$,

$$\underline{\delta}^{(n)} = Re^{-\frac{1}{2}}\,\underline{D}^{-1}\,\underline{H}\,\underline{\delta}^{(n-1)}$$

A necessary and sufficient condition for the convergence of the direct method is

$$Re^{-\frac{1}{2}} \rho(\underline{D}^{-1} \underline{H}) < 1, \tag{22}$$

where ρ denotes the spectral radius. For attached flow the eigenvalues of $\underline{D}^{-1} \underline{H}$ are negative and hence the method either converges (when Re is large enough) or can be made convergent by underrelaxation of the type

$$\underline{\delta}^{(n)} = [\omega Re^{-\frac{1}{2}} \underline{D}^{-1} \underline{H} + (1-\omega) \underline{E}] \underline{\delta}^{(n-1)} \tag{23}$$

where $\underline{E}$ is the unit matrix. For separated flow $\underline{D}$ is singular or almost singular which causes the direct method to fail.

(b) The inverse method

The inverse method, in which the iteration process is reversed, can be written

$$\underline{\delta}^{(n)} = Re^{\frac{1}{2}} \underline{H}^{-1} \underline{D} \underline{\delta}^{(n-1)} .$$

$\underline{H}$ is invertable, hence the iteration matrix exists for attached as well as for separated flow. Further $\rho(\underline{H}^{-1} \underline{D}) \geq 1/\rho(\underline{D}^{-1} \underline{H})$, hence the inverse method diverges when the direct method converges; but as long as

$$\text{Real part (eigenvalues of } Re^{\frac{1}{2}} \underline{H}^{-1} \underline{D}) < 1$$

is satisfied, the inverse method can be made convergent by applying sufficient underrelaxation of a type similar to (23). For the present model problem the latter requirement is equivalent to (21). Hence (21) is necessary and sufficient for the inverse method to be applicable. In general, the relaxation factor required will have to be found by trial and error; it is proportional to $Re^{-\frac{1}{2}}$ when Re is large.

(c) The quasi-simultaneous method

The present, quasi-simultaneous, method solves equation (20), i.e.

$$(Re^{-\frac{1}{2}} \underline{H} - \underline{D}) \underline{\delta} = \underline{0},$$

by means of the Gauss-Seidel method. A necessary and sufficient condition for its convergence is (21).

(d) A semi-inverse method

A semi-inverse method to solve (20), which is related to the methods used by le Balleur [13] and Carter [14] and which can be analyzed easily, is given by

$$\underline{u}_e^P = Re^{-\frac{1}{2}} \underline{H} \underline{\delta}^{(n-1)} ,$$

$$\underline{u}_e^B = \underline{D} \underline{\delta}^{(n-1)} ,$$

$$\underline{\delta}^{(n)} = \underline{\delta}^{(n-1)} + \omega(\underline{u}_e^B - \underline{u}_e^P).$$

Eliminating $\underline{u}_e^P$ and $\underline{u}_e^B$ we have

$$\underline{\delta}^{(n)} = [\underline{E} - \omega(\underline{H} - Re^{\frac{1}{2}} \underline{D})] \underline{\delta}^{(n-1)} = [\underline{E} - \omega \underline{Q}] \underline{\delta}^{(n-1)}.$$

This iteration process converges if and only if the eigenvalues of the matrix between square brackets lie between -1 and +1. This implies that the eigenvalues of $\omega \underline{Q}$ must lie between 0 and 2. Again (21) is necessary and sufficient for the existence of an ω for which convergence occurs.

Let us now summarize the conclusions which can be drawn from the analysis of this model problem:

(i) The direct method fails at separation.

(ii) The condition (21) is necessary and sufficient for:
- the convergence of the quasi-simultaneous method;
- the existence of an ω for which the inverse method converges;
- the existence of an ω for which the semi-inverse method converges.

It is remarkable to see that the convergence of the latter three methods is governed by one relation (21).

We emphasize that the above analysis is only given for the model problem in which the influence of non-diagonal entries of the discrete boundary-layer operator B has been neglected.

Wigton and Holt [32] have also presented an elaborate stability analysis of the inverse and semi-inverse methods. Their analysis is based on Fourier analysis, which basically is a local consideration of error propagation in which the influence of boundary conditions is neglected. The present analysis makes use of a study of the eigenvalues of the full iteration matrix which gives it a more global character. These two types of stability analyses are very common (compare chapters 2-2 and 2-5 of [33]) and usually yield results which are comparable. The present analysis makes no exception to this rule.

5 CONCLUSION

In his famous paper of 1948 [1] Goldstein discusses the singularity in the solution of the boundary-layer equations, when pressure is prescribed, in a point of vanishing skin friction. He suggests a number of criteria for the occurrence of a singularity. We quote one of his alternatives:

> "... Another possibility is that a singularity will always occur except for certain special pressure variations in the neighbourhood of separation, and that, experimentally, whatever we may do, the pressure

variations near separation will always be such that no singularity will occur".

With the knowledge obtained in the 70-ies we know that this suggestion by Goldstein is correct. The physics have no problem in finding a singularity-free solution; the pressure distribution, as well as other flow quantities, follows from a "combined effort" (= interaction) of viscous and inviscid flow.

However, it is the mathematical treatment of the flow problem which is responsible for the occurrence of singularities. The classical methods are based on a hierarchy between the viscous and inviscid parts of the flow field, in which the inviscid region is given a dominant position. In regions of weak interaction this is a good description of the physics, and in these regions the classical methods work well. In regions of strong interaction both parts of the flow field are equally important; the classical methods fail in these regions. The previous sections show that mathematical methods which reflect this physical lack-of-hierarchy have no problems in finding the solution. Herewith the Goldstein singularity is definitely condemned as a mathematical artifice.

6 REFERENCES

1. GOLDSTEIN, S. - On laminar boundary-layer flow near a position of separation. Quart.J. Mech. Appl. Math. Vol. I, pp. 48-69, 1948.
2. CATHERALL, D. and MANGLER, K.W. - The integration of the two-dimensional laminar boundary-layer equations past the point of vanishing skin friction. J. Fluid Mech. Vol. 26, pp. 163-182, 1966.
3. STEWARTSON, K. - Multi-structured boundary layers on flat plates and related bodies. Advances in Appl. Mech., Vol.14 pp. 145-239, 1974.
4. SMITH, F.T. - On the high Reynolds number theory of laminar flows. IMA J. Appl.Math., Vol. 28, pp. 207-281, 1982.
5. BRILEY, W.R. and McDONALD, H. - Numerical prediction of incompressible separation bubbles. J. Fluid Mech., Vol. 69, pp. 631-656, 1975.
6. HEIDSIECK,R.D. - Navier-Stokes solutions for laminar incompressible boundary layers with strong viscous-inviscid interaction. Report LR-353, Delft University of Technology, 1982.
7. MELNIK, R.E., CHOW, R. and MEAD, H.R. - Theory of viscous transonic flow over airfoils at high Reynolds number. AIAA paper 77-680, June 1977.
8. SYCHEV, V.V. and SYCHEV, VIK.V. - On turbulent separation (in Russian) J. Comp. Math. and Math. Phys. Vol. 20, pp. 1500-1512, 1980.

9. ADAMSON Jr., R.C., and MESSITER, A.F. - Analysis of two-dimensional interactions between shock waves and boundary layers, Ann. Rev. Fluid Mech. Vol. 12, pp. 103-138, 1980.

10. CLEARY, J.W., VISWANATH, P.R., HORSTMAN, C.C. and SEEGMILLER, H.L. - Asymmetric trailing-edge flows at high Reynolds number. AIAA paper 80-1396, 1980.

11. MONNERIE, B. and QUINN, B. (Eds.) - Computation of viscous-inviscid interactions. AGARD Conference Proceedings CP 291, Colorado Springs, 29 Sept.-1 Oct. 1980.

12. CEBECI, T. (Ed.) - Proceedings of the second symposium on numerical and physical aspects of aerodynamic flows, Long Beach, 17-20 Jan. 1983.

13. LE BALLEUR, J.C. - Couplage visqueux-non visqueux: méthode numérique et applications aux écoulements bidimensionnels transsoniques et supersoniques. La Recherche Aérospatiale, Vol. 183, pp. 65-76, 1978.

14. CARTER, J.E. - Viscous-inviscid interaction analysis of transonic turbulent separated flow. AIAA Paper 81-1241, 1981.

15. MOSES, H.L., JONES III, R.R., O'BRIEN, W.F. and PETERSON, R.S. - Simultaneous solution of the boundary layer and freestream with separated flow. AIAA J. Vol. 16, pp. 61-66 1978.

16. WAI, J.C. and YOSHIHARA, H. - Planar transonic airfoil computations with viscous interactions. AGARD-CP-291, paper 9, 1980.

17. GHOSE, S. and KLINE, S.J. - Prediction of transitory stall in two-dimensional diffusers. Report MD-36, Thermosciences Div., Dept. of Mech. Engrg. Stanford University, 1976.

18. VELDMAN, A.E.P. - A numerical method for the calculation of laminar, incompressible boundary layers with strong viscous-inviscid interaction. NLR TR 79023 U, 1979.

19. VELDMAN, A.E.P. - New, quasi-simultaneous method to calculate interacting boundary layers, AIAA J. Vol. 19, pp. 79-85, 1981.

20. VELDMAN, A.E.P. - The calculation of incompressible boundary layers with strong viscous-inviscid interaction. AGARD-CP-291, paper 12, 1980.

21. VELDMAN, A.E.P. and LINDHOUT, J.P.F. - A quasi-simultaneous calculation method for strongly interacting viscous flow around an infinite swept wing. NLR MP 83001 U, 1983.

22. COUSTEIX, J., LE BALLEUR, J.C. and HOUDEVILLE, R. - Calculation of unsteady turbulent boundary layers in direct or inverse mode, including reverse flows: analysis of singularities. La Recherche Aérospatiale, No. 1980-3, pp. 3-13, 1980.

23. HORTON, H.P. - Numerical investigation of regular laminar boundary-layer separation. AGARD-CP-168, paper 7, 1975.

24. ARDONCEAU, P., ALZIARY, Th. and AYMER, D. - Calcul de l'interaction onde choc/couche limite avec decollement. AGARD-CP-291, paper 28, 1980.

25. CARTER, J.E. and WORNOM, S.F. - Solutions for incompressible separated boundary layers including viscous-inviscid interaction. NASA SP-347, pp. 125-150, 1975.

26. DRELA, M. and THOMPKINS Jr., W.T. - A study of non-unique solutions of the two-dimensional boundary-layer equations at laminar separation and reattachment points. In ref. 12, 1983.

27. MURPHY, J.D. and KING, L.S. - Airfoil flow-field calculations with coupled boundary-layer potential codes. In ref. 12, 1983.

28. LAGERSTROM, P. - Solutions of the Navier-Stokes equation at large Reynolds number. SIAM J. Appl. Math. Vol. 28, pp. 202-214, 1975.

29. VELDMAN, A.E.P. - Boundary layers with strong interaction: from asymptotic theory to calculation method. Proceedings BAIL I Conference, Dublin, pp. 149-163, 1980.

30. CARTER, J.E. - Solutions for laminar boundary layers with separation and reattachment. AIAA paper 74-583, 1974.

31. DIJKSTRA, D. - Private communication.

32. WIGTON, L.B. and HOLT, M. - Viscous-inviscid interaction in transonic flow. AIAA paper 81-1003, 1981.

33. AMES, W.F. - Numerical methods for partial differential equations. Academic Press, New York, 1977.

CALCULATION OF SEPARATED FLOWS BY VISCOUS-INVISCID INTERACTION

Richard H. Pletcher
Professor of Mechanical Engineering
Iowa State University

1. INTRODUCTION

The present chapter deals with viscous-inviscid interaction calculation schemes for two-dimensional flows containing limited regions of separation. For such flows, a predominant main flow direction can be identified. Examples include airfoil flows with leading edge or midchord separation bubbles and the subsonic flow over a rearward-facing step. These configurations are illustrated in Figs. 1 and 2.

Until fairly recently it was common practice to utilize the full Navier-Stokes equations to compute such flows. The validity of the boundary-layer approximation was thought to end as the separation point was approached. This was because of the well-known Goldstein [1] singularity of the standard boundary-layer formulation at separation and because the entire boundary-layer approximation is subject to question as the layer thickens and the normal component of velocity becomes somewhat larger (relative to u) than in the usual high Reynolds number flow. The singular behavior appears in finite-difference calculations, for which the pressure gradient is specified, as a tendency for the normal component of velocity, v, to increase without limit near the separation point as the streamwise step size is reduced [2]. A finite v will be obtained for a finite step size, but the solution will not be unique.

It is now known that the singularity can be removed by employing an "inverse" calculation procedure [3]. An inverse calculation method for the boundary-layer equations is a scheme whereby a solution is obtained which satisfies boundary conditions which differ from the standard ones. The usual procedure in an inverse method is to replace the outer boundary condition, $\lim_{y \to \infty} u(x,y) = u_e(x)$, by the specification of a displacement thickness or wall shear stress which must be satisfied by the solution. The pressure gradient [or $u_e(x)$]

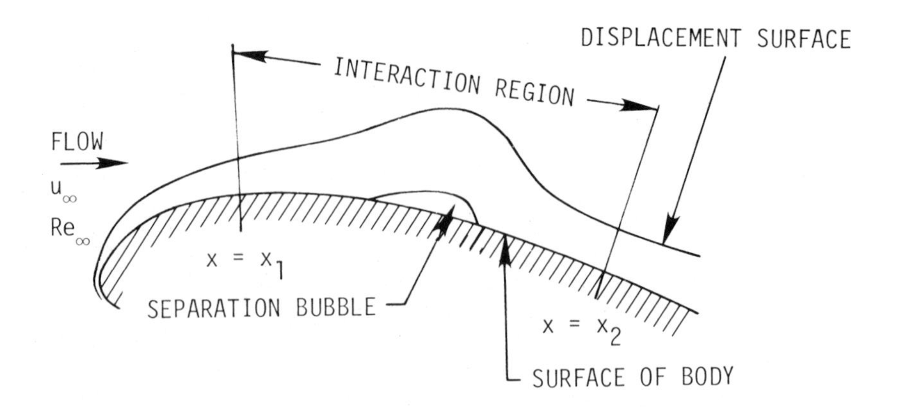

Fig. 1. Schematic diagram of interaction region on a two-dimensional body.

is determined as part of the solution. It should be noted clearly that it is the boundary conditions which differ between the conventional direct methods and the inverse methods. It is perhaps more correct to think of the problem specification as being direct or inverse rather than the method. However, we will yield to convention and refer to the solution method as being direct or inverse. It also appears that the singularity can be removed when conventional direct methods are employed if an auxiliary pressure interaction relationship is used [4,5].

A second concern involves the correct representation of convective terms when flow reversal is present. The difficulty with the convective terms can be viewed as follows: We recall that the steady boundary-layer equations are parabolic. For $u > 0$, the solution can be marched in the positive x (mainstream) direction. Physically, information is carried downstream from the initial plane by the flow. In regions of reversed flow, however, the "downstream" direction is in the negative x direction. Mathematically we observe that when $u < 0$, the boundary-layer momentum equation remains parabolic, but the correct marching direction is in the negative x direction.

A solution procedure can be devised to overcome the problem associated with the "correct" marching direction by making initial guesses or approximations for the velocities in the reversed flow portion of a flow with a separation bubble, storing these velocities, and correcting them by successive iterative calculation sweeps over the entire flowfield. To do this requires using a difference representation which honors the appropriate marching direction, forward or backward, depending on the direction of flow. To follow this iterative procedure means abandoning the once-through simplicity of the usual boundary-layer approach. Computer storage must also be provided for velocities in and near the region of reversed flow. Such multiple-pass procedures have been employed by

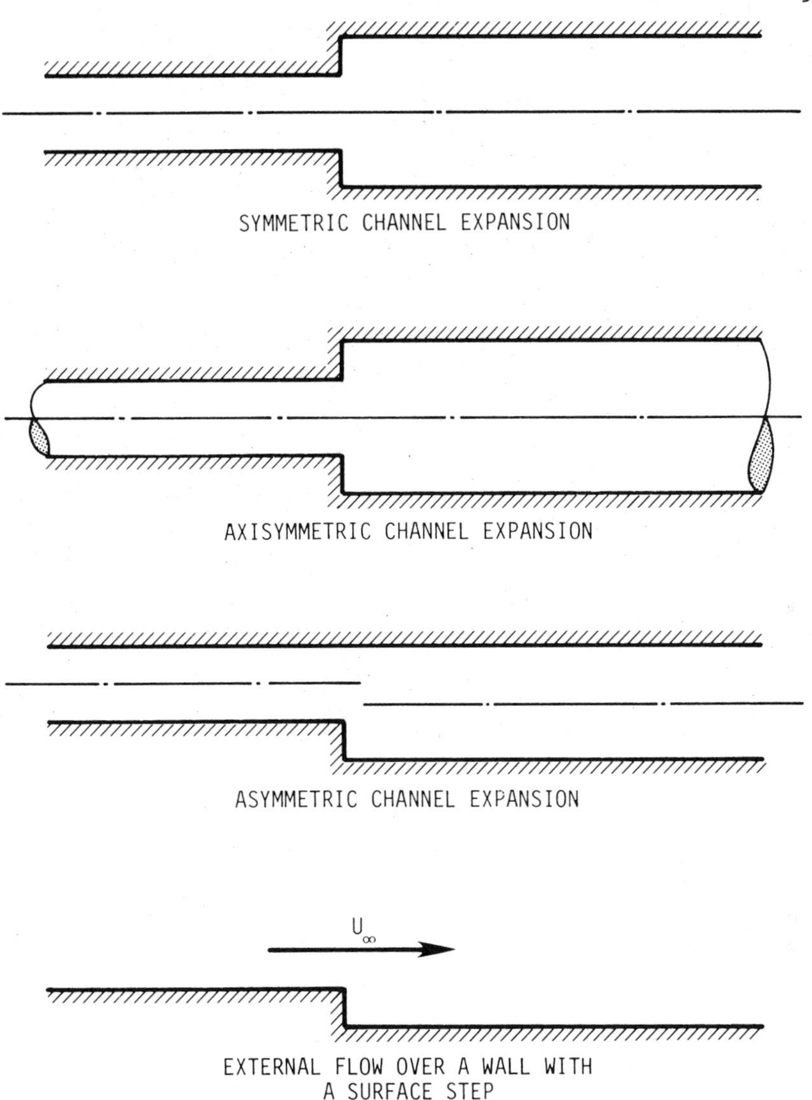

Fig. 2. Rearward-facing step geometries.

Klineberg and Steger [3], Carter and Wornom [6], Cebeci [7], and others.

Reyhner and Flügge-Lotz [4] suggested a simpler alternative to the multiple-pass procedure. Noting that the reversed flow velocities are generally quite small for confined regions of recirculation, they suggested that the convective term $u\partial u/\partial x$ in the boundary-layer momentum equation be represented in the reversed flow regions by $C|u|\partial u/\partial x$, where C is zero or a small positive constant. This representation has become known as the FLARE approximation and permits the boundary-layer solution to proceed through separated regions by a simple forward marching procedure. It should be clear that the FLARE procedure introduces an additional approximation (or assumption) into the boundary-layer formulation, namely that the $u\partial u/\partial x$ term is small relative to other terms in the boundary-layer equation in the region of reversed flow. On the other hand, the FLARE approximation appears to give smooth and plausible solutions for many flows with separation bubbles. Experimental and computational evidence accumulated to date indicates that for naturally occurring separation bubbles, the u component of velocity in reversed flow regions is indeed fairly small in magnitude, usually less than about 10% of the maximum velocity found in the viscous region.

It should be noted that, although ways of satisfactorily treating the $u\partial u/\partial x$ convective term in regions of reversed flow have been presented, it still does not appear possible to obtain a unique, convergent solution of the steady boundary-layer equations alone by a direct marching procedure. Direct calculation procedures reported to date have always employed an interaction relationship whereby the pressure gradient specified becomes dependent upon the displacement thickness (or related parameter) of the viscous regions, usually in a time-dependent manner [5]. On the other hand, we should note that a convergent, unique solution can be obtained for the steady boundary-layer equations using inverse methods without the use of an auxiliary interaction relationship.

We have indicated above that it is now possible to overcome the difficulties associated with obtaining solutions to the boundary-layer equations for flows containing zones of recirculation. The question of the accuracy of the boundary-layer approximation for these flows remains. In support of the validity of the boundary-layer approximation, it is noted that the formation of a separation bubble normally does not cause the thickness of the viscous region to increase by an order of magnitude; that is, the boundary-layer measure of thinness, $\delta/L \ll 1$ is still met. The main analytical justification for use of the boundary-layer approximation is from the "triple deck" theory of Lighthill [8] and Stewartson [9], which is applicable to large Reynolds number flows containing small separated regions. The triple deck theory is discussed in more detail elsewhere in this volume. This theory provides guidance for very large Reynolds numbers, but the exact range of applicability of the boundary-layer approximation for separated flows remains unknown and must be determined by comparisons with experimental data and numerical solutions to more complete governing equations. Although the issue is not conclusively resolved, such comparisons reported to date tend to indicate that the boundary-layer equations provide a useful approximation for many separated flows of practical interest [10,11,12,13].

The essential elements of a viscous-inviscid interaction calculation procedure are the following:

a) A method for obtaining an improved inviscid flow solution which provides a pressure distribution or an edge velocity distribution which accounts for the viscous flow displacement effect. In principle any inviscid flow "solver" could be used but it is also frequently possible to employ a greatly simplified inviscid flow calculation scheme based on a small disturbance approximation.

b) A technique for obtaining a solution to the boundary-layer equations suitable for the problem at hand. For a flow which may separate, an inverse boundary-layer procedure would be appropriate.

c) A procedure for relating the inviscid and viscous flow solutions by matching the pressure gradient at the surface in a manner which will drive the changes from one iterative cycle to the next toward zero.

Some of the problems associated with item b), the boundary-layer solution, have been pointed out above. It is primarily the computational procedures for viscous regions which require special treatment when separation is encountered. The numerical procedures used for the inviscid flow are insensitive to separation in the viscous-inviscid interaction approach, and no special techniques are required to deal with these flows. The presence of separation does, however, influence the inviscid flow solution causing a displacement of the inviscid flow streamlines. Item c) above, the matching of the viscous and inviscid solutions, will be discussed in the next section.

2. ANALYSIS

2.1. General

Over the years numerous viscous-inviscid interaction schemes have been proposed. It will not be possible to discuss all of them here. Instead we will summarize suitable approaches for two types of separation bubble flows. The first is the flow in the neighborhood of a separation bubble on an airfoil (see Fig. 1), and the second is the separated flow formed by an abrupt expansion in channel cross-sectional area (see Fig. 2). In both of these configurations the analysis for the viscous region and the procedure for the matching remains much the same. The most convenient approach for the inviscid analysis differs for these two flows. The equations and specific examples will be for incompressible flow. However, the interaction approach has been extended to compressible flows in the transonic regime [14,15] and examples of this appear elsewhere in this volume. For transonic flows the viscous analysis needs to be changed very little, and the interaction approach has been successfully used with both Euler equation solvers [15] and solutions to the full potential equation [14] for the inviscid flow.

In many cases it is possible to employ a small disturbance approximation to compute the inviscid flow. When this can be done, the computational effort required for the inviscid flow solution is less than for the viscous flow. When the small disturbance approach is not a good approximation, such as in the transonic regime, the inviscid solution may require at least as much computational effort as the viscous solution.

2.2. Numerical procedures for viscous regions

The equations for viscous flow will be presented here in physical coordinates. For some flows, the use of transformed variables is helpful, but since no single transformation appears to work well for all separated flows of interest, the concepts will be illustrated in terms of physical coordinates.

To avoid oscillations in skin friction when reversed flow is present, it is recommended [13] that the boundary-layer momentum and continuity equations be solved simultaneously in a coupled manner. To this end, it is convenient to define a streamfunction, ψ:

$$u = \frac{\partial \psi}{\partial y}$$

and

$$v = -\frac{\partial \psi}{\partial x}$$

For incompressible flow the conservation equations for mass and momentum are written as

$$u = \frac{\partial \psi}{\partial y} \tag{1}$$

$$Cu \frac{\partial u}{\partial x} - \frac{\partial \psi}{\partial x}\frac{\partial u}{\partial y} = u_e \frac{du_e}{dx} + \frac{1}{\rho}\frac{\partial \tau}{\partial y} \tag{2}$$

where $C = 1.0$ when $u > 0$ and $C = 0.0$ when $u \leq 0$ and

$$\tau = \mu \frac{\partial u}{\partial y} - \rho \overline{u'v'} = (\mu + \mu_T)\frac{\partial u}{\partial y} = \bar{\mu}\frac{\partial u}{\partial y} \tag{3}$$

The above equations are in a form applicable to either lamina or turbulent flow. For laminar flow, the primed velocities and μ_T are zero, and for turbulent flow, the unprimed velocities are time-mean quantities.

The boundary conditions for the inverse procedure for both internal and external flows for which an outer inviscid flow is present are given by

$$u(x,0) = \psi(x,0) = 0 \tag{4}$$

and

$$\psi_e = u_e \left[y_e - \delta^*(x)\right] \tag{5}$$

where $\delta^*(x)$ is a prescribed function. The boundary condition for ψ_e follows from the definition of δ^*. A distribution of u is also required at the starting value of x.

Although the emphasis in the present chapter will be on flows with interaction, it is also possible to utilize the boundary-layer approach without interaction to treat an interesting class of flows with separation. These are certain fully developed internal flows undergoing abrupt expansions as illustrated in Fig. 2. No inviscid core region can be identified in these flows. The full range of applicability of the non-interacting boundary-layer approach for internal expansion flows has not yet been determined, but the procedure appears generally useful for symmetric expansions and asymmetric expansions at Reynolds numbers for which a separation bubble only forms on one (the step) wall. In these cases, the boundary layer equations are applied to the entire channel flow. For steady internal flow, the channel mass flow rate is known from the initial velocity profile. This information, expressible as a specified value of stream function at the outer boundary (solid wall or line of symmetry), provides an additional relationship through which the pressure gradient can be determined. No slip or symmetry conditions are imposed on the streamwise component of velocity. That is, when the computation domain is bounded by a second outer solid boundary, the conditions

$$u(x,H) = 0 \tag{6}$$

and

$$\psi(x,H) = \psi_{Tot} \tag{7}$$

are imposed at y = H, where H is the channel height. For symmetric channel flows, the line of symmetry forms the upper computational boundary at which the conditions

$$\left(\frac{\partial u}{\partial y}\right)_{y=\frac{H}{2}} = 0 \tag{8}$$

and

$$\psi\left(x, \frac{H}{2}\right) = \frac{\psi_{Tot}}{2} \tag{9}$$

are imposed.

The boundary-layer equations can, of course, be solved in the conventional direct mode for external flows without separation, in which case the outer boundary condition is the standard one

$$\lim_{y\to\infty} u(x,y) = u_e(x) \tag{10}$$

Details on implementation of the boundary conditions will be given later.

Equations (1) and (2) are represented in finite-difference form as

$$\frac{u_j^{n+1} + u_{j-1}^{n+1}}{2} = \frac{\psi_j^{n+1} - \psi_{j-1}^{n+1}}{\Delta y_-} \tag{11}$$

$$Cu_j^{n+1} \frac{\left(u_j^{n+1} - u_j^n\right)}{\Delta x} - \frac{\left(\psi_j^{n+1} - \psi_j^n\right)}{\Delta x} \frac{\left(u_{j+1}^{n+1} - u_{j-1}^{n+1}\right)}{(\Delta y_+ + \Delta y_-)}$$

$$= \chi^{n+1} + \frac{2}{\rho(\Delta y_+ + \Delta y_-)} \left[\bar{\mu}_{j+1/2} \frac{\left(u_{j+1}^{n+1} - u_j^{n+1}\right)}{\Delta y_+}\right.$$

$$\left. - \bar{\mu}_{j-1/2} \frac{\left(u_j^{n+1} - u_{j-1}^{n+1}\right)}{\Delta y_-}\right] \tag{12}$$

In the above, $C = 1$ when $u_j^{n+1} > 0$ and $C = 0$ when $u_j^{n+1} \leqslant 0$;

$\chi = -\frac{1}{\rho}\frac{dp}{dx}$; $\Delta y_- = y_j - y_{j-1}$; $\Delta y_+ = y_{j+1} - y_j$. Newton linearization is next applied to the above nonlinear convective terms. We let $u_j^{n+1} = \hat{u}_j^{n+1} + \delta_u$ (likewise for u_{j+1}^{n+1} and u_{j-1}^{n+1}) and $\psi_j^{n+1} = \hat{\psi}_j^{n+1} + \delta_\psi$, where the carets indicate provisional values of the variables in an iterative process. The quantities δ_u and δ_ψ are the changes in the variables between two iterative sweeps, i.e., $\delta_\phi = \phi_j^{n+1} - \hat{\phi}_j^{n+1}$ for a general variable ϕ. Terms involving products of δ's are dropped. This results in expressions which are linear in variables at the n+1 level. The resulting difference equations can be written in the form

$$\psi_{j-1}^{n+1} - \psi_j^{n+1} + b_j\left(u_{j-1}^{n+1} + u_j^{n+1}\right) = 0 \tag{13}$$

$$B_j u_{j-1}^{n+1} + D_j u_j^{n+1} + A_j u_{j+1}^{n+1} + E_j \psi_j^{n+1} = H_j \chi^{n+1} + C_j \tag{14}$$

where

$$A_j = -\frac{\left(\hat{\psi}_j^{n+1} - \psi_j^n\right)}{\Delta x(\Delta y_+ + \Delta y_-)} - \frac{2\bar{\mu}_{j+1/2}}{\rho\Delta y_+(\Delta y_+ + \Delta y_-)} \tag{15}$$

$$B_j = \frac{\left(\hat{\psi}_j^{n+1} - \psi_j^n\right)}{\Delta x(\Delta y_+ + \Delta y_-)} - \frac{2\bar{\mu}_{j-1/2}}{\rho\Delta y_-(\Delta y_+ + \Delta y_-)} \tag{16}$$

$$C_j = \frac{c\left(\hat{u}_j^{n+1}\right)^2}{\Delta x} - \frac{\hat{\psi}_j^{n+1}\left(\hat{u}_{j+1}^{n+1} + \hat{u}_{j-1}^{n+1}\right)}{\Delta x(\Delta y_+ + \Delta y_-)} \tag{17}$$

$$D_j = \frac{c\left(2\hat{u}_j^{n+1} - u_j^n\right)}{\Delta x} + \frac{2}{\rho(\Delta y_+ + \Delta y_-)}\left(\frac{\bar{\mu}_{j+1/2}}{\Delta y_+} + \frac{\bar{\mu}_{j-1/2}}{\Delta y_-}\right) \tag{18}$$

$$E_j = - \frac{\left(\hat{u}_{j+1}^{n+1} - \hat{u}_{j-1}^{n+1}\right)}{\Delta x(\Delta y_+ + \Delta y_-)} \tag{19}$$

$$H_j = 1 \tag{20}$$

$$b_j = \frac{\Delta y_-}{2} \tag{21}$$

It is important to note that the pressure gradient parameter χ^{n+1} is one of the unknowns both in the inverse formulation where $\delta^*(x)$ is used as the boundary condition and in the noninteracting boundary-layer approach for internal flows. Equations (13) and (14) are block tridiagonal in form with 2×2 blocks and require the simultaneous solution of 2(NJ) - 2 equations for 2(NJ) - 2 unknowns at each streamwise marching step. The number of grid points across the flow is NJ and includes the boundary points. The elements below the main diagonal can be eliminated and a recursion formula developed for the back substitution. Before the back substitution is carried out, however, the parameter χ^{n+1} must be determined. Further details on the solution procedure are given in the Appendix.

In earlier interaction studies [11] an inverse solution procedure which permitted solution of the boundary-layer equation in an uncoupled manner [16] was applied to predict separation bubbles on airfoils. This procedure predicted unreasonable oscillations in skin-friction coefficient when applied to rearward-facing step flows. No such oscillations appear when the coupled procedure described above is used. For this reason the coupled procedure is believed to be superior and is recommended over the uncoupled scheme described in [16].

2.3. Numerical procedures for the inviscid flow

The best procedure for the inviscid flow depends very much on the specific problem to be solved. Here we will consider two flow situations. The first is the incompressible flow over an airfoil in which a leading edge or midchord separation bubble occurs. Viscous-inviscid interaction is assumed to be important only in the neighborhood of the separation bubble, as illustrated in Fig. 1.

For this case a good estimate of the effect of the displacement correction for the inviscid flow solution can be obtained by the use of a small disturbance approximation. We let $u_{e,o}$ denote the tangential component of velocity for the inviscid flow in the absence of the separation bubble. This inviscid velocity is often computed by standard procedures for flow over the solid body [17], neglecting all effects of the viscous flow. Vatsa and Carter [18] have suggested that at high angles of attack it is desirable to establish a reference inviscid solution, $u_{e,o}$, which includes the viscous effects of the wake. Having established $u_{e,o}$ from an appropriate inviscid solution procedure for the airfoil configurations, we let u_c be the velocity on the displacement surface induced only by the sources and sinks distributed on the surface of the body due to the displacement effect of the viscous flow in the interaction region. Then, the tangential component of velocity of a fluid particle on the displacement surface can be written as

$$u_e = u_{e,o} + u_c \tag{22}$$

Following Lighthill [19], the intensity of the line source or sink displacing a streamline at the displacement surface of the viscous flow can be evaluated as

$$q = \frac{d(u_e \delta^*)}{dx} \tag{23}$$

Using a small disturbance approximation valid for small values of δ^*, u_c can be evaluated from the Hilbert integral

$$u_c(x) = \frac{1}{\pi} \int_{-\infty}^{\infty} \frac{d(u_e \delta^*)}{dx'} \frac{dx'}{(x - x')} \tag{24}$$

Strong interaction is assumed to be limited to the region $x_1 \leq x \leq x_2$ shown in Fig. 1. The viscous and inviscid solutions are only repeated iteratively over this region. However, we note that the range of integration in Eq. (24) extends from $-\infty$ to $+\infty$. Outside the interaction region, the intensity of the source, q, is assumed to approach zero smoothly as x approaches $\pm\infty$. An arbitrary extrapolation of the form

$$q'(x) = \frac{b}{x^2} \tag{25}$$

is often used for the regions $x < x_1$ and $x > x_2$, in order to evaluate the integral in Eq. (24). The constant b is chosen to match the q obtained from the boundary-layer solution at x_1 and x_2. Equation (24) can now be written as

$$u_c(x) = \frac{1}{\pi}\left[\int_{-\infty}^{x_1} \frac{q'(x')}{(x - x')}\, dx' + \int_{x_1}^{x_2} \frac{q(x')}{(x - x')}\, dx' + \int_{x_2}^{\infty} \frac{q'(x')}{(x - x')}\, dx'\right] \tag{26}$$

The first and third integrals can be evaluated analytically. The second integral is evaluated numerically, normally using the trapezoidal rule. The singularity at $x = x'$ can be isolated using the procedure found in Jobe [20]. Some investigators have found it possible to evaluate the integral numerically with no special attention given to the singularity as long as $(x - x')$ remained finite [21].

The second flow configuration for which viscous-inviscid interaction methods will be discussed is illustrated in Fig. 2. As a specific example, we will consider the two-dimensional internal flow over a rearward-facing step. The interaction approach is applicable when viscous effects are confined to regions near the upper and lower boundaries and an inviscid core can be identified.

For the step flow configuration, viscous effects on both walls influence the inviscid flow, and the small disturbance approach is not generally used. Instead, the inviscid flow is assumed to be irrotational, and the Laplace equation for stream-function

$$\frac{\partial^2\psi}{\partial x^2} + \frac{\partial^2\psi}{\partial y^2} = 0 \tag{27}$$

is solved numerically in the effective channel defined by the displacement surfaces of the viscous flow regions (see Fig. 3). The solution is subject to the boundary conditions of a specified inlet velocity, no flow across the displacement surfaces, and a linear variation in the x component of velocity across the channel at the downstream boundary. The mathematical specifications of these boundary conditions are shown in Fig. 3.

Since the inviscid flow domain is irregular with respect

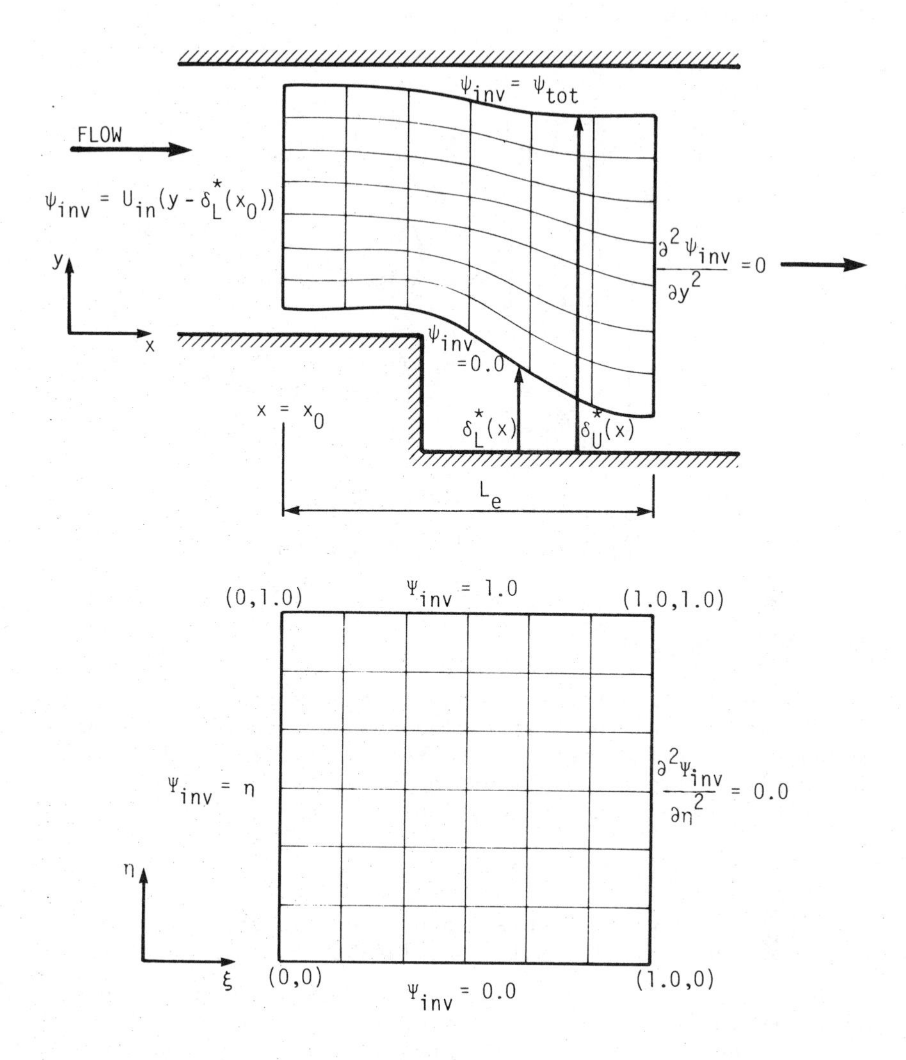

Fig. 3. Effective inviscid flow channel and transformed coordinate system.

to the Cartesian coordinate system, it becomes advantageous to define a new set of independent variables

$$\xi = \frac{x - x_o}{L_e} \tag{28}$$

$$\eta = \frac{y - \delta_L^*(x)}{\delta_U^*(x) - \delta_L^*(x)} \tag{29}$$

and transform Eq. (27) to

$$\frac{\partial^2\psi_{inv}}{\partial\xi^2} - \frac{2}{\delta^*_{UL}}\left(\frac{d\delta^*_L}{d\xi} + \eta\,\frac{d\delta^*_{UL}}{d\xi}\right)\frac{\partial^2\psi_{inv}}{\partial\xi\partial\eta}$$

$$+ \frac{1}{\left(\delta^*_{UL}\right)^2}\left\{L_e^2 + \left(\frac{d\delta^*_L}{d\xi} + \eta\,\frac{d\delta^*_{UL}}{d\xi}\right)^2\right\}\frac{\partial^2\psi_{inv}}{\partial\eta^2}$$

$$+ \left\{\frac{2}{\delta^{*\,2}_{UL}}\,\frac{d\delta^*_{UL}}{d\xi}\left(\frac{d\delta^*_L}{d\xi} + \eta\,\frac{d\delta^*_{UL}}{d\xi}\right)\right.$$

$$\left. - \frac{1}{\delta^*_{UL}}\left(\frac{d^2\delta^*_L}{d\xi^2} + \eta\,\frac{d^2\delta^*_{UL}}{d\xi^2}\right)\right\}\frac{\partial\psi_{inv}}{\partial\eta} = 0 \qquad (30)$$

where $\delta^*_{UL} = \delta^*_U - \delta^*_L$.

The physical aspects of this transformation are shown in Fig. 3, as well as the transformed boundary conditions. After employing central differences to represent terms in Eq. (30), the resulting linear system of equations can be solved by an alternating direction implicit (ADI) scheme [13,22].

2.4. Viscous-inviscid interaction

The interaction calculation proceeds in the following way. For the airfoil bubble calculation, $u_{e,o}$ is obtained for the body of interest, and the viscous flow is computed up to the beginning of the interaction region by a conventional direct boundary-layer method. These two solutions do not change. Similarly, for internal sudden expansion calculations, standard boundary-layer (parabolic) calculation methods for two-dimensional channel flow [13] are used to compute the flow up to the beginning of the interaction region. Next, an initial $\delta^*(x)$distribution is prescribed over the interaction region (over both solid boundaries in internal flow cases). This initial $\delta^*(x)$ distribution is purely arbitrary, but should match the $\delta^*(x)$ computed by standard methods at the beginning of the interaction region. The boundary-layer solution is next obtained by an inverse procedure using this $\delta^*(x)$ as a boundary condition. An edge velocity distribution, $u_{e,BL}(x)$ is obtained as an output.

Using the same $\delta^*(x)$, the inviscid flow solution is obtained by the small disturbance procedure for airfoil bubble calculations and by the numerical solution to Laplace's equation for streamfunction for internal flows. This establishes a second distribution for the edge (surface) velocity distribution, $u_{e,inv}(x)$. The $u_e(x)$ from the two calculations, boundary-layer and inviscid, will not agree until convergence has been achieved. The difference between $u_e(x)$ calculated both ways can be used as a potential to calculate an improved distribution for $\delta^*(x)$. A suitable scheme for computing an improved distribution of $\delta^*(x)$ has been developed [11,12] by noting that a response to small excursions in local u_e tends to preserve the volume flow rate per unit width in the boundary layer, i.e., $u_e\delta^* \cong$ constant. This concept is put into practice by computing the appropriate new distribution of δ^* to use for a new pass through the boundary-layer calculation by [11,12]

$$\delta^*_{k+1} = \delta^*_k \left(\frac{u_{e,BL_k}}{u_{e,inv_k}} \right) \tag{31}$$

where k denotes iteration level. It is important to note that Eq. (31) only serves as a basis for correcting δ^* between iterative passes so that no formal justification for its use is required so long as the iterative process converges. At convergence $u_{e,BL} = u_{e,inv}$; thus, Eq. (31) represents an identity which has no effect on the final solution. In this sense, the use of Eq. (31) is somewhat like the use of an arbitrary over-relaxation factor in the numerical solution of an elliptic equation by successive over-relaxation.

In general, the viscous-inviscid interaction calculation is completed by making successive passes first through the inverse boundary-layer scheme, then through the inviscid flow procedure with δ^* being computed by Eq. (31) prior to each boundary-layer calculation. However, when the small disturbance procedure is being used for airfoil bubble calculations, it was noted in [11] that convergence can usually be accelerated and computation time reduced significantly (a 50% reduction was reported in [17]) by making several (typically three) passes through the inviscid calculation procedure and the $\delta^*(x)$ updating scheme using a "frozen" distribution of $u_{e,BL}$ in Eq. (31). In other words, the viscous flow solution is only recomputed after every three passes through the inviscid calculation procedure. Nothing has been reported on the suitability of this acceleration procedure for internal flows where the inviscid solution is obtained from the solution to Laplace's equation. It is also sometimes [11] possible to speed convergence by using over-relaxation with Eq. (31).

For a sudden expansion in a channel involving an abrupt (step) change in geometry, care must be taken to evaluate $\delta^*(x)$ in Eq. (31) using the same reference axis (the step side wall downstream of the step) in the δ^* computation throughout the flow.

Convergence of the viscous-inviscid interaction scheme is considered to have occurred when

$$\frac{\left|u_{e,BL} - u_{e,inv}\right|}{u_{e,inv}} \leq \varepsilon \qquad (32)$$

where ε is a prescribed tolerance.

3. COMPUTATIONAL EXAMPLES

3.1. Separation bubbles on airfoils

External subsonic flows which give rise to thin separation bubbles invariably separate in the laminar state and undergo transition to turbulent flow prior to or coincidental with reattachment. Unfortunately, there are no known practical procedures for computing the details of laminar-turbulent transition from first principles. Common practice is to evaluate the turbulent viscosity as the product of a viscosity for fully turbulent flow μ_{FT} and an empirical intermittency factor, γ. The accuracy of numerical predictions for these flows is highly dependent upon the modeling used for transition and turbulence.

In [23] three transition models were evaluated for midchord and leading edge separation bubbles on airfoils. The models have been designated as A, B, and C. Model A is based on existing correlations for natural transition on flat plates and airfoils. Model B makes use of correlations for separation bubble transitions. Model C assumes that transition occurs instantaneously at the point at which the nearly constant pressure region characteristic of bubble separation ends. This point is established by the correlation presented in [11]. Details of these models are too lengthy to present here, but they can be found in [23] for all three models and in [11] for models A and B. The fully turbulent viscosity μ_{FT} was evaluated according to model D of [16].

Figure 4, taken from [23], compares the pressure coefficients predicted by models B and C with the measurements of Gault [24] for the NACA 66_3-018 airfoil at two degrees angle of attack and a chord Reynolds number of 2×10^6. Predicted separation and reattachment points are shown in the figure. The bubble is seen to form somewhat downstream of midchord.

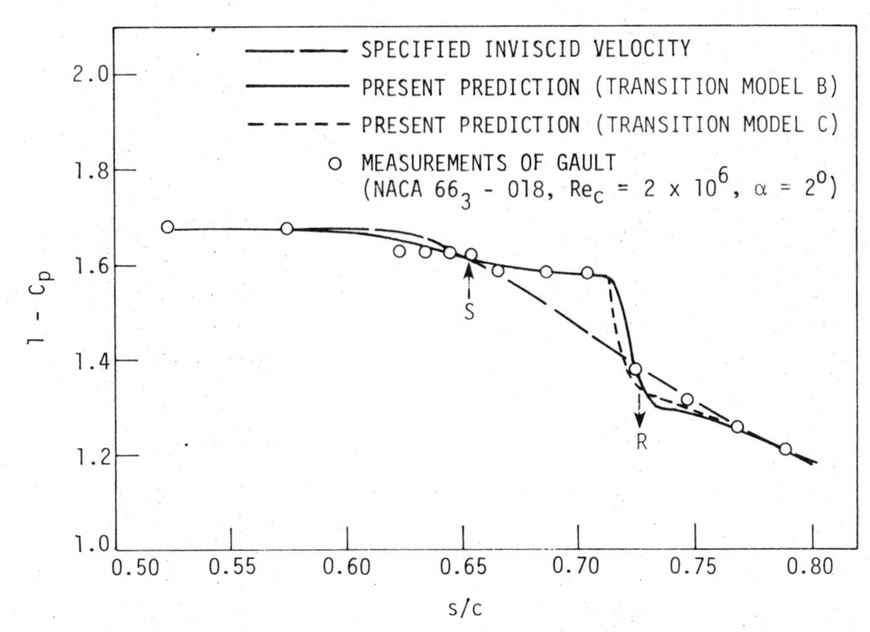

Fig. 4. Comparison of predicted pressure distribution with experimental data on NACA 66_3-018 airfoil, $\alpha = 2°$.

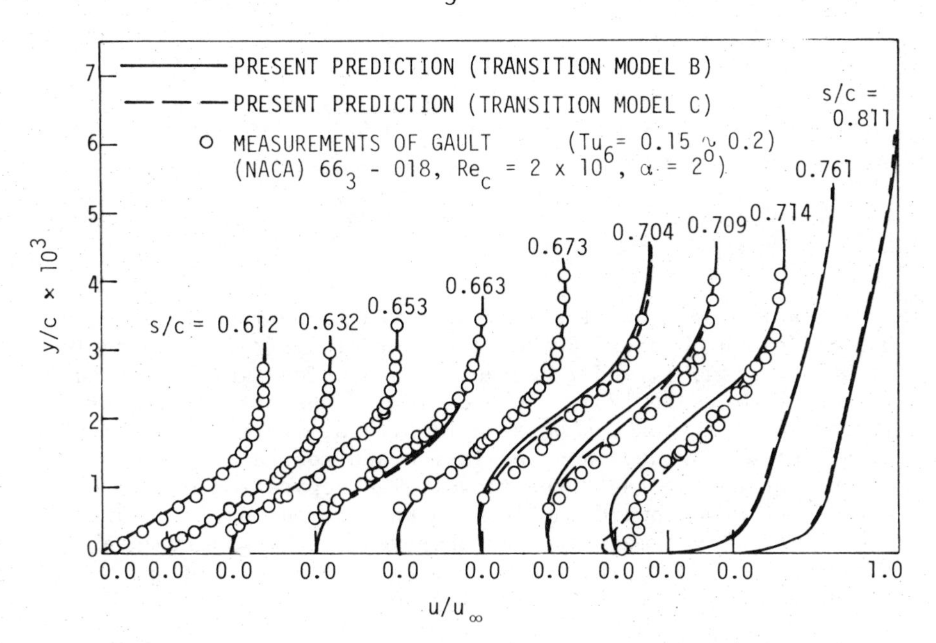

Fig. 5. Comparison of predicted mean velocity profiles with experimental data for NACA 66_3-018 airfoil, $\alpha = 2°$.

Velocity profile comparisons are presented in Fig. 5 for this flow. Both transition models B and C are seen to provide fairly good predictions for this case.

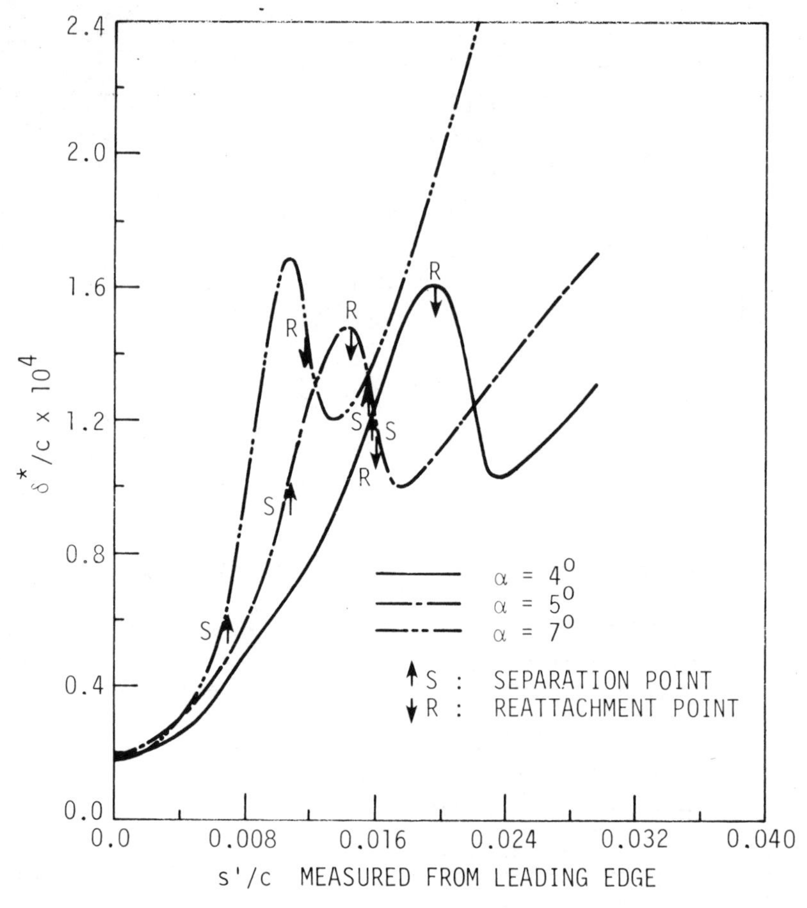

Fig. 6. Variation of displacement thickness distribution with angle of attack on NACA 63-009 airfoil, $Re_c = 5.8 \times 10^6$.

Leading edge bubbles were observed to form on the NACA 63-009 airfoil at a chord Reynolds number of 5.8×10^6 and angles of attack of 4, 5, and 7 degrees. Predictions were compared to measurements in [23]. An overview of the effect of increasing angle of attack can be seen in Fig. 6, where the predicted displacement surfaces and separation and reattachment points are presented for the three flows. The separation bubble can be seen to move toward the leading edge as the angle of attack increases. The same trend was observed in the experimental measurements. A typical comparison of the predicted and measured [25,26] pressure distribution for the NACA 63-009 airfoil is shown in Fig. 7. The prediction of Crimi and Reeves [27] is also shown on the figure. From 10-15 interaction iterations were required to reduce ε in Eq. (32) to 0.006 for these flows.

The results shown in Figs. 4-7 were obtained using the uncoupled inverse procedure of [16] with the small disturbance

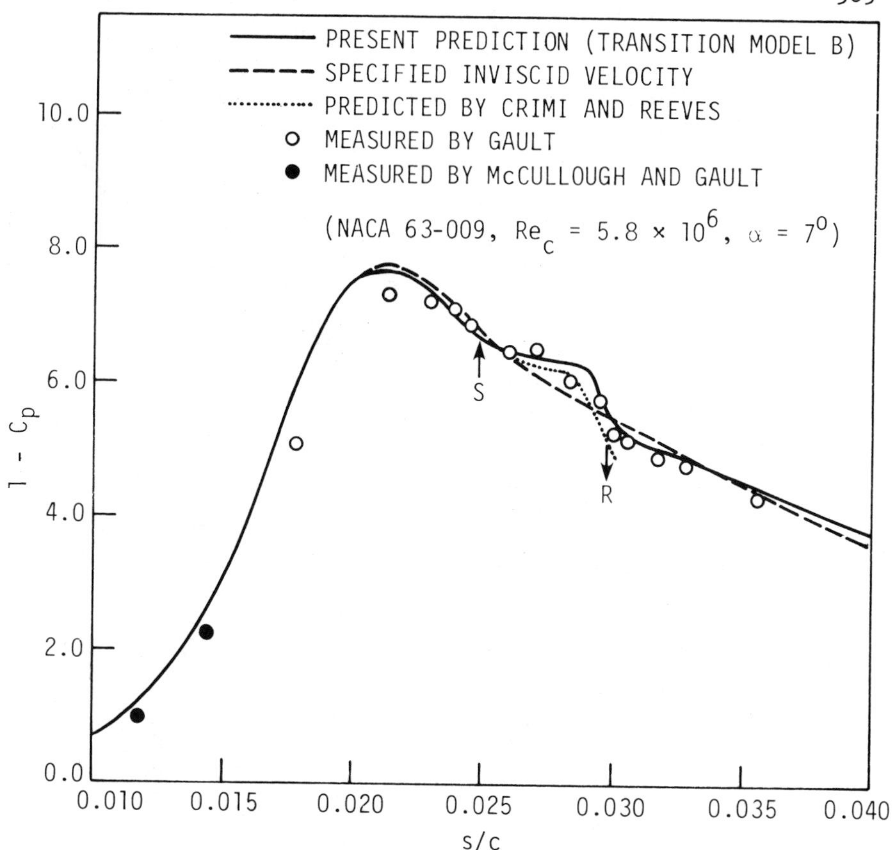

Fig. 7. Comparison of predicted pressure distribution with experimental data on NACA 63-009 airfoil, $\alpha = 7°$.

inviscid formulation. Other airfoil bubble predictions using this approach can be found in [11,23].

3.2. Channel flows with separation

The viscous-inviscid interaction procedure discussed in this chapter has also been applied to predict laminar and turbulent separating flows occurring in channels. These separating flows have included the flow over a rearward-facing step and in a two-dimensional diffuser [13,28].

Comparisons of predicted and measured velocity profiles (from [13]) are shown in Figs. 8 and 9 for laminar flow over a rearward-facing step at a Reynolds number of 412, based on step height and maximum channel velocity. The measurements were obtained by Eriksen [29] for a step height to channel inlet height ratio of 0.0664. The interaction zone ranged from 13.5 step heights upstream of the step to 36 step heights downstream. Figure 8 presents profiles in the zone which includes the recirculating flow downstream of the step. The

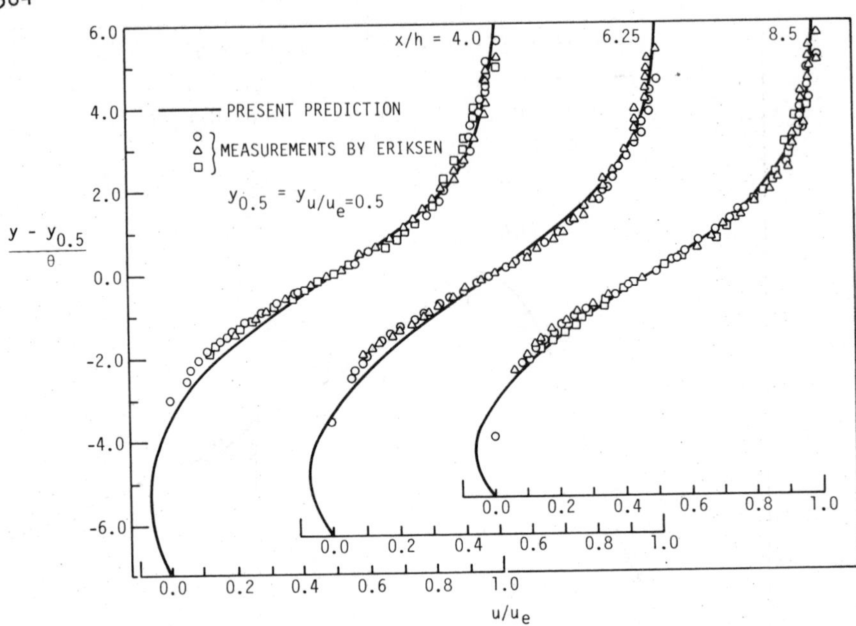

Fig. 8. Velocity profiles for a laminar flow in a sudden channel expansion.

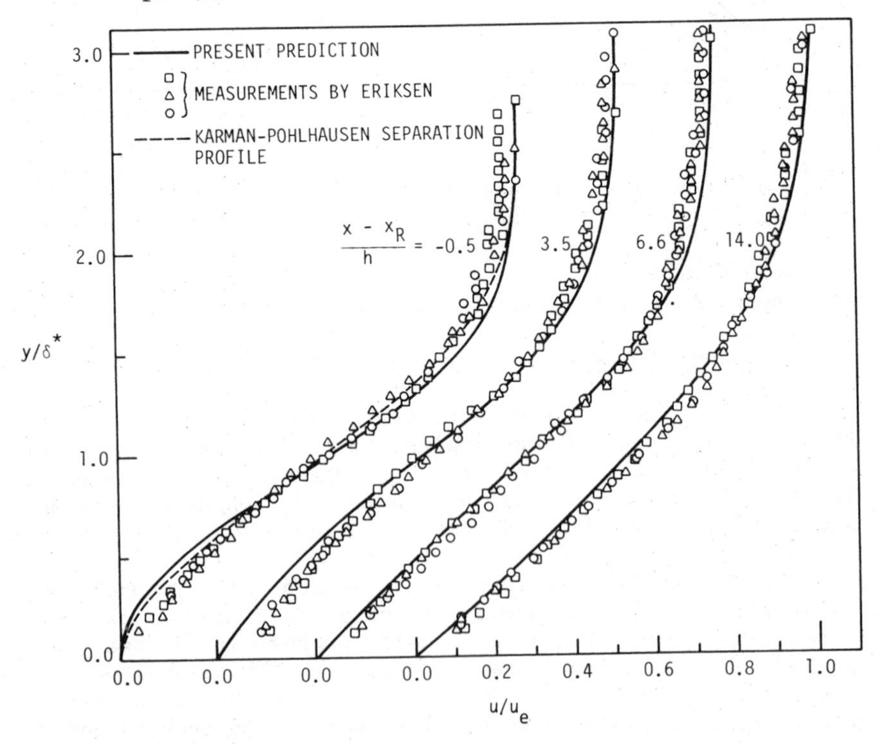

Fig. 9. Velocity profiles for a laminar flow in a sudden channel expansion.

flow in the zone of reattachment and downstream is shown in Fig. 9. Overall, the predicted results agree reasonably well with the measurements, although the predicted reattachment length of 13.5 step heights is slightly longer than the 11 step heights observed experimentally. The interaction procedure converged to $\varepsilon = 0.0007$ (see Eq. 32) in 12 iterations.

In [13,28] comparisons are made with turbulent separating flow measurements obtained by Simpson et al. [30] in a diffusing channel. Turbulence model D of [16] was seen to predict mean flow parameters quite well. It is interesting to note that for the Simpson flow the computational domain ended with flow separation still present; i.e., the computation was not for a bubble separation. Despite this unusual downstream flow condition, the interaction procedure converged to $\varepsilon = 0.005$ in 11 iterations. Comparisons between predictions and measurements for turbulent rearward-facing step flows can also be found in [13,28]. Unfortunately, conventional turbulence modeling has been found to give relatively poor predictions for step flows.

3.3. Predictions of separated channel flows without interaction

When a fully developed flow passes through an abrupt channel expansion as illustrated in Fig. 2, there is no obvious way in which the viscous-inviscid interaction approach can be utilized, because the flow cannot easily be separated into viscous and inviscid regions. Nearly all reported numerical predictions for such flows have employed the full Navier-Stokes equations. It has been recently observed [31] that at least some laminar, fully developed flows undergoing a symmetric expansion can be accurately predicted by a once-through application of the boundary-layer calculation method described in this chapter. This is an interesting development in light of the fact that the once-through boundary-layer procedure requires an order of magnitude less computation time than is required for the numerical solution of the full Navier-Stokes equations. In [31] very good agreement with experimental data and Navier-Stokes solutions was observed for developed laminar flow undergoing 2:1 and 3:1 expansions in a symmetric two-dimensional channel. As an example, Fig. 10 (from [31]) compares predicted velocity profiles based on the boundary-layer equations with the laser anemometer measurements and numerical solutions of the Navier-Stokes equations reported by Durst et al. [32]. The Reynolds number based on step height and maximum velocity upstream of the step for this case was 56, and the channel expansion ratio was 3:1. The range of applicability of the once-through boundary-layer procedure has not been determined, but surely limitations do exist.

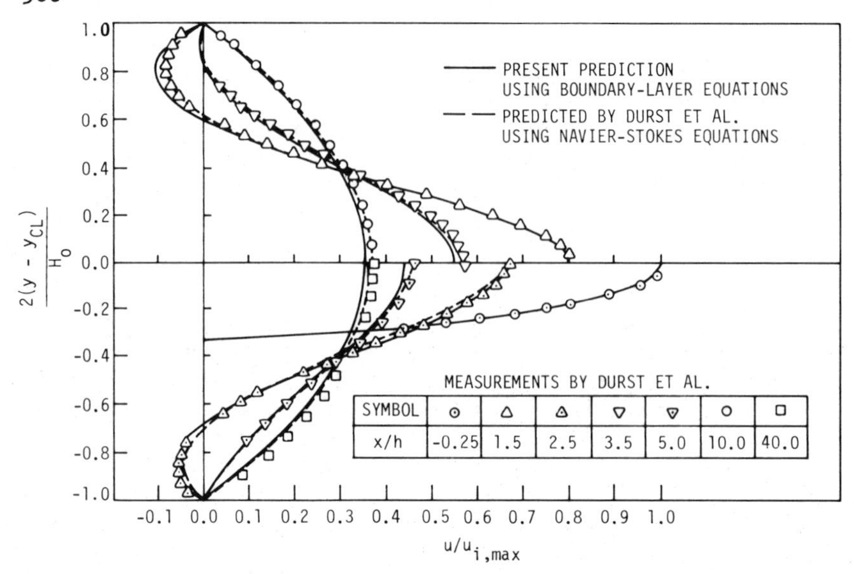

Fig. 10. Velocity profiles for laminar flow in a channel with a symmetric sudden expansion.

4. CONCLUDING REMARKS

The essential elements of a viscous-inviscid interaction calculation procedure for incompressible separated flows has been described. The procedure appears to provide an attractive alternative to the numerical solution of the full Navier-Stokes equations for those flows in which distinct viscous and inviscid regions can be identified. Considerably less computation time is believed to be required for the interaction procedure than for Navier-Stokes solutions when the same level of grid refinement is used in viscous regions, although specific comparisons of computation time have not been reported. Interaction concepts appear applicable to both external and internal flows. It has been observed that some fully developed flows passing through sudden expansions in two-dimensional channels can be accurately predicted by a once-through calculation based on the boundary-layer equations and the FLARE approximation.

5. ACKNOWLEDGMENTS

Many of the results described in this article were obtained in studies sponsored by the National Science Foundation under Grant CEE-781290 and the Army Research Office under Grant DAAG29-76-G-0155. The important contributions of Mr. O. K. Kwon and Mr. E. J. Hall in these studies are gratefully acknowledged.

6. REFERENCES

1. GOLDSTEIN, S. - On Laminar Boundary Layer Flow near a Position of Separation. Q. J. Mech. Appl. Math., Vol. 1, pp. 43-69. 1948.

2. PLETCHER, R. H. and DANCEY, C. L. - A Direct Method of Calculating Through Separated Regions in Boundary Layer Flow. J. Fluids Eng., Vol. 98, pp. 568-572. 1976.

3. KLINEBERG, J. M. and STEGER, J. L. - On Laminar Boundary Layer Separation. AIAA Paper 74-94, Washington, D.C. 1974.

4. REYHNER, T. A. and FLÜGGE-LOTZ, I. - The Interaction of a Shock Wave with a Laminar Boundary Layer. Int. J. Non-Linear Mech., Vol. 3, pp. 173-199. 1968.

5. NAPOLITANO, M., WERLE, M. J., and DAVIS, R. T. A. - Numerical Technique for the Triple-Deck Problem. AIAA Paper 78-1133, Seattle, Wash. 1978.

6. CARTER, J. E. and WORNOM, S. F. - Forward Marching Procedure for Separated Boundary Layer Flows. AIAA J., Vol. 13, pp. 1101-1103. 1975.

7. CEBECI, T. - Separated Flows and Their Representation by Boundary Layer Equations. Report ONR-CR215-234-2, Office of Naval Research, Arlington, Va. 1976.

8. LIGHTHILL, M. J. - On Boundary Layers and Upstream Influence. II. Supersonic Flows Without Separation. Proc. Roy. Soc. London, Ser. A, Vol. 217, pp. 478-507. 1953.

9. STEWARTSON, K. - Multistructured Boundary Layers on Flat Plates and Related Bodies, Adv. Appl. Mech., Vol. 14, Academic Press, New York, pp. 145-239. 1974.

10. WILLIAMS, J. C., - Incompressible Boundary Layer Separation. Annu. Rev. Fluid Mech., Vol. 9, pp. 113-144, Annual Reviews, Inc., Palo Alto, Calif. 1977.

11. KWON, O. K. and PLETCHER, R. H. - Prediction of Incompressible Separated Boundary Layers Including Viscous-Inviscid Interaction. J. Fluids Eng., Vol. 101, pp. 466-472. 1979.

12. CARTER, J. E. - A New Boundary-Layer Interaction Technique for Separated Flows. NASA TM-78690. 1978.

13. KWON, O. K. and PLETCHER, R. H. - Prediction of the Incompressible Flow Over a Rearward-Facing Step, Engineering Research Institute Technical Report 82019/HTL-26, Iowa State University, Ames, Iowa. 1981.

14. CARTER, J. E. - Viscous-Inviscid Interaction Analysis of Transonic Turbulent Separated Flow. AIAA Paper 81-1241, Palo Alto, Calif. 1981.

15. WHITFIELD, D. L., SWAFFORD, T. W. and JACOCKS, J. L. - Calculation of Turbulent Boundary Layers with Separation and Viscous-Inviscid Interaction. AIAA J., Vol. 19, No. 10, pp. 1315-1322. Oct. 1981.

16. PLETCHER, R. H. - Prediction of Incompressible Turbulent Separating Flow. J. Fluids Eng., Vol. 100, pp. 427-433. 1978.

17. HESS, J. L. and SMITH, A. M. O. - Calculation of Potential Flow about Arbitrary Bodies. Prog. Aeronaut. Sci., Vol. 8, pp. 1-138, Pergamon Press, New York. 1967.

18. VATSA, V. N. and CARTER, J. E. - Analysis of Airfoil Leading Edge Separation Bubbles. AIAA Paper 83-0300, Reno, Nev. 1983.

19. LIGHTHILL, M. J. - On Displacement Thickness. J. Fluid Mech., Vol. 4, pp. 383-392. 1958.

20. JOBE, C. E. - The Numerical Solution of the Asymptotic Equations of Trailing Edge Flow. Technical Report AFFDL-TR-74-46, Air Force Flight Dynamics Laboratory. 1974.

21. BRILEY, W. R. and McDONALD, H. - Numerical Prediction of Incompressible Separation Bubbles. J. Fluid Mech., Vol. 69, pp. 631-656. 1975.

22. HALL, E. J. - Application of Viscous-Inviscid Interaction to Separated Flows with Heat Transfer Including Rearward-Facing Step Flows. M.S. Thesis, Iowa State University. 1983.

23. PLETCHER, R. H., KWON, O. K., and CHILUKURI, R. - Prediction of Separating Turbulent Boundary Layers Including Regions of Reversed Flow. Final Technical Report HTL-22, ISU-ERI-Ames-80112. 1980.

24. GAULT, D. E. - An Experimental Investigation of Regions of Separated Laminar Flow. NACA TN-3505. 1955.

25. GAULT, D. E. - Boundary Layer and Stalling Characteristics of the NACA 63-009 Airfoil Section. NACA TN-1894. 1949.

26. McCULLOUGH, G. B. and GAULT, D. E. - Examples of Three Representative Types of Airfoil Section Stall at Low Speed. NACA TN-2502. 1951.

27. CRIMI, P. and REEVES, B. L. - Analysis of Leading-Edge

Separation Bubbles on Airfoils. AIAA J., Vol. 14(11), pp. 1548-1555. 1976.

28. KLINE, S. J., CANTWELL, B. J., LILLEY, G. M. (Ed.) - 1980-1981 AFOSR-HTTM-Stanford Conference on Complex Turbulent Flows: Comparison of Computation and Experiment. Vol. 2, Mech. Engr. Dept., Stanford University. 1982.

29. ERIKSEN, V. L. - An Experimental Investigation of the Laminar Flow of Air over a Downstream-Facing Step. M.S. Thesis, University of Minnesota. 1968.

30. SIMPSON, R. L., CHEW, Y. T., and SHIVAPRASAD, B. G. - Data set prepared for the 1980-1981 AFOSR-HTTM-STANFORD Conference on Complex Turbulent Flows: Comparison of Computation and Experiment. Eds. S. J. Kline, B. J. Cantwell, G. M.Lilley, Case No. 431. Stanford University. 1981.

31. KWON, O. K., and PLETCHER, R. H. - Predicting Sudden Expansion Flows Using the Boundary Layer Equations. ASME Paper No. 83-FE-11, Houston, Texas. 1983.

32. DURST, F., MELLING, A., and WHITELAW, J. H. - Low Reynolds Number Flow over a Plane Symmetric Sudden Expansion. J. Fluid Mech., Vol. 64, pp. 111-128. 1974.

7. APPENDIX

The solution to Eqs. (13) and (14) can be obtained from

$$u_j^{n+1} = A_j' u_{j+1}^{n+1} + H_j' \chi^{n+1} + C_j' \tag{33}$$

$$\psi_j^{n+1} = B_j' u_{j+1}^{n+1} + D_j' \chi^{n+1} + E_j' \tag{34}$$

providing the coefficients A_j', H_j', C_j', B_j', D_j', E_j', and the quantities u_{j+1}^{n+1} and χ^{n+1} are known a priori. The coefficients are given by

$$A_j' = - A_j/R_1$$

$$B_j' = A_j'R_2$$

$$C_j' = [C_j - B_jC_{j-1}' - E_j(b_jC_{j-1}' + E_{j-1}')]/R_1$$

$$D'_j = b_j H'_{j-1} + D'_{j-1} + H'_j R_2$$

$$E'_j = b_j C'_{j-1} + E'_{j-1} + C'_j R_2$$

$$H'_j = [H_j - B_j H'_{j-1} - E_j (b_j H'_{j-1} + D'_{j-1})]/R_1$$

$$R_1 = D_j + (B_j + E_j b_j) A'_{j-1} + E_j (B'_{j-1} + b_j)$$

$$R_2 = b_j (1 + A'_{j-1}) + B'_{j-1}$$

Since the inner ($j = 1$) boundary conditions on u_j^{n+1} and ψ_j^{n+1} are zero, the coefficients A'_1, B'_1, C'_1, D'_1, E'_1, and H'_1 are also zero, and the coefficients above can be computed starting from $j = 2$ and continuing to the outer boundary ($j = NJ$).

The pressure gradient parameter χ^{n+1} is evaluated by simultaneously solving the equations obtained from Eqs. (33) and (34) by replacing j with NJ - 1 and the boundary conditions as follows.

At $j = NJ - 1$, Eqs. (33) and (34) become

$$u_{NJ-1}^{n+1} = A'_{NJ-1} u_{NJ}^{n+1} + H'_{NJ-1} \chi^{n+1} + C'_{NJ-1} \tag{35}$$

$$\psi_{NJ-1}^{n+1} = B'_{NJ-1} u_{NJ}^{n+1} + D'_{NJ-1} \chi^{n+1} + E'_{NJ-1} \tag{36}$$

First, for flow in which Eq. (5) is used as a boundary condition, we can write the condition as

$$\psi_{NJ}^{n+1} = u_{NJ}^{n+1} \left(y_{NJ} - \delta^{*n+1} \right) \tag{37}$$

and the pressure gradient can be expressed in terms of velocities at the outer edge by

$$\chi^{n+1} = \frac{1}{\Delta x} \left[\left(2\hat{u}_{NJ}^{n+1} - u_{NJ}^{n} \right) u_{NJ}^{n+1} - \left(\hat{u}_{NJ}^{n+1} \right)^2 \right] \tag{38}$$

Equation (11) is written as

$$\psi_{NJ}^{n+1} = \psi_{NJ-1}^{n+1} + \frac{\Delta y_-}{2}\left(u_{NJ}^{n+1} + u_{NJ-1}^{n+1}\right) \tag{39}$$

Solving Eqs. (35)-(39) for χ^{n+1} gives

$$\chi^{n+1} = \frac{\left(\frac{F_3}{F_1}\right)\left(2\hat{u}_{NJ}^{n+1} - u_{NJ}^{n}\right) - \left(u_{NJ}^{n+1}\right)^2}{\Delta x - \frac{F_2}{F_1}\left(2\hat{u}_{NJ}^{n+1} - u_{NJ}^{n}\right)} \tag{40}$$

where

$$F_1 = y_{NJ} - \delta^{*n+1} - B'_{NJ-1} - \frac{\Delta y_-}{2}(1 + A'_{NJ-1})$$

$$F_2 = D'_{NJ-1} + \frac{\Delta y_-}{2} H'_{NJ-1}$$

$$F_3 = E'_{NJ-1} + \frac{\Delta y_-}{2} C'_{NJ-1}$$

Once the pressure gradient parameter χ^{n+1} is determined, the edge velocity u_{NJ}^{n+1} can be calculated using Eqs. (35)-(39) as

$$u_{NJ}^{n+1} = \left(\frac{F_2}{F_1}\right)\chi^{n+1} + \frac{F_3}{F_1} \tag{41}$$

Then, ψ_{NJ}^{n+1} can be computed directly from Eq. (37). Now the back substitution process can be initiated using Eqs. (33) and (34) to compute u_j^{n+1} and ψ_j^{n+1} from the outer edge to the wall. The Newton linearization requires that the system of equations be solved iteratively with $\hat{u}_j^{n+1}$ and $\hat{\psi}_j^{n+1}$ being updated between iterations. The iterative process is continued at each streamwise location until the maximum change in u's and ψ's between two successive iterations is less than some predetermined tolerance. The calculation is initiated at each streamwise station by setting $\hat{u}_j^{n+1} = u_j^n$ and $\hat{\psi}_j^{n+1} = \psi_j^n$.

For a symmetrical channel calculation for which Eqs. (8) and (9) are used as boundary conditions, we express $(\partial u/\partial y)_{H/2}$

in terms of a one-sided, second-order accurate difference representation

$$\left(\frac{\partial u}{\partial y}\right)_{NJ}^{n+1} \cong \frac{u_{NJ}^{n+1}}{2}\left(\frac{4}{\Delta y_-} - \frac{1}{\Delta y_{--}}\right) - \frac{2u_{NJ-1}^{n+1}}{\Delta y_-} + \frac{u_{NJ-2}^{n+1}}{2\Delta y_{--}} \tag{42}$$

where

$$\Delta y_- = y_{NJ} - y_{NJ-1}$$

and

$$\Delta y_{--} = y_{NJ-1} - y_{NJ-2}$$

The outer boundary conditions, Eqs. (8) and (9), can now be written as

$$u_{NJ}^{n+1} = c_1 u_{NJ-1}^{n+1} - c_2 u_{NJ-2}^{n+1} \tag{43}$$

and

$$\psi_{NJ}^{n+1} = \frac{\psi_{Tot}}{2} \tag{44}$$

where

$$c_1 = \frac{4}{4 - K}$$

and

$$c_2 = \frac{K}{4 - K}$$

and

$$K = \frac{\Delta y_-}{\Delta y_{--}}$$

Equations (43) and (44) are to be solved with Eqs. (35), (36), and (39). However, one additional relationship is needed since five unknowns appear (u_{NJ}^{n+1}, u_{NJ-1}^{n+1}, u_{NJ-2}^{n+1}, ψ_{NJ-1}^{n+1}, χ^{n+1}) and only four independent relationships among them have been identified thus far [Eqs. (35), (36), (39) and (43)]. The additional equation can be obtained by specializing Eq. (33) for u_{NJ-2}^{n+1} as

$$u_{NJ-2}^{n+1} = A'_{NJ-2}\, u_{NJ-1}^{n+1} + H'_{NJ-2}\, \chi^{n+1} + C'_{NJ-2} \tag{45}$$

This system of equations can be solved for χ^{n+1} by defining

$$\alpha_1 = 1 - A'_{NJ-1}(c_1 - c_2 A'_{NJ-2})$$

$$\alpha_2 = (c_1 - c_2 A'_{NJ-2})H'_{NJ-1} - c_2 H'_{NJ-2}$$

$$\alpha_3 = (c_1 - c_2 A'_{NJ-2})C'_{NJ-1} - c_2 C'_{NJ-2}$$

$$\alpha_4 = 1 + \frac{2}{\Delta y_-} B'_{NJ-1} + A'_{NJ-1}$$

$$\alpha_5 = -\left(H'_{NJ-1} + \frac{2}{\Delta y_-} D'_{NJ-1}\right)$$

$$\alpha_6 = \frac{\psi_{Tot}}{\Delta y_-} - \frac{2}{\Delta y_-} E'_{NJ-1} - C'_{NJ-1}$$

then

$$\chi^{n+1} = \frac{\alpha_1\alpha_6 - \alpha_3\alpha_4}{\alpha_2\alpha_4 - \alpha_1\alpha_5} \tag{46}$$

The axial component of velocity at the line of symmetry can be found from

$$u_{NJ}^{n+1} = \frac{\alpha_2}{\alpha_1}\chi^{n+1} + \frac{\alpha_3}{\alpha_1} \tag{47}$$

At this point the back substitution process can be initiated using Eqs. (33) and (34) to compute u_j^{n+1} and ψ_j^{n+1} from the outer boundary to the wall. The remaining portions of the algorithm are as discussed above.

For a full channel calculation in which Eqs. (6) and (7) are used as boundary conditions, we first express the boundary conditions as

$$u_{NJ}^{n+1} = 0 \tag{48}$$

$$\psi_{NJ}^{n+1} = \psi_{Tot} \tag{49}$$

Utilizing Eqs. (48) and (49) in Eqs. (35), (36), and (39) permits the pressure gradient parameter to be determined by

$$\chi^{n+1} = \frac{\psi_{Tot} - \left(\frac{\Delta y_-}{2}\right) C'_{NJ} - E'_{NJ-1}}{\left(\frac{\Delta y_-}{2}\right) H'_{NJ-1} + D'_{NJ-1}} \tag{50}$$

The back substitution process can now be used to compute u_j^{n+1} and ψ_j^{n+1} from Eqs. (33) and (34). The remaining portions of the algorithm are as discussed above.

CALCULATION OF SEPARATION BUBBLES USING BOUNDARY-LAYER-TYPE EQUATIONS

A. Halim* and M. Hafez**

1. INTRODUCTION

Incompressible laminar flows with small separation bubbles have been successfully simulated using Navier Stokes equations for some years (see Roache[1]). For example, Briley[2] and Leal[3] used stream function-vorticity formulation while Ghia et al[4] used primitive variables. Simplified equations of the first formulation (where the streamwise viscous terms are neglected) were used by Ghia and Davis[5], Werle and Bernstein[6], Ghia, Ghia and Tesch[7], and Inoue[8]. Similarly, partially parabolized, or semi-elliptic equations in terms of primitive variables were used by Mahgoub and Bradshaw[9], and Rubin[10].

In the first approach, careful treatment of the vorticity boundary condition at the wall is required, while conservation of mass is usually difficult to achieve in the second approach (see Orszag and Israeli[11]).

Recently, boundary layer equations have been used to calculate separated flows. There are three main difficulties with classical boundary layer methods. First, it is well known that when the pressure is prescribed, the equations admit a singular solution (see the reviews by Brown and Stewartson[12] and Williams[13]). By differentiating the x-momentum equation twice with respect to y and using the continuity equation, the shear stress at the wall is given by

---

*Research Engineer, The George Washington University, Joint Institute for Advancement of Flight Sciences, Hampton, VA
**Senior Scientist, Computer Dynamics, Inc., Virginia Beach, VA
The authors would like to acknowledge the support of NASA Langley Research Center; Office of Naval Research, Contract No. N00014-80-C-0494 and Air Force Office of Scientific Research Contract F49620-81-C-0041.

$$\frac{1}{2}(\tau^2)_x = C\, u_{yyyy} \qquad (1)$$

Hence, unless, u_{yyyy} vanishes at the separation point, the shear stress has a square root singularity. More precisely, Goldstein[14] showed that the function $\lim_{Re\to\infty}(Re^{1/2}\tau)$ which is generally bounded upstream, is nonanalytic at the point of zero skin friction. There is no contradiction between this theory and Dean's results[15] which show that the Navier-Stokes equations are analytic there. The limiting processes are different.

The second difficulty of direct boundary layer calculations is the possibility of nonunique solutions. This is clear from the similar solution. It has also been identified for non-similar solutions (see Murphy and King[16] and the references therein).

Finally, the marching procedure is not stable for reversed flows. Flugge Lotz and Reyhner[17] neglected the streamwise convective term when u is negative. More accurate solutions can be obtained by either using the appropriate upwind difference schemes and/or integrating the equations in the appropriate direction locally. The stability of such calculations can be improved by solving artificial time-dependent equations where the diagonal dominance is enhanced due to the time-like term.

The numerical breakdown of classical boundary layer methods near a separation point has been demonstrated by many authors (e.g. Wele and Davis[18], Klienberg and Steger[19], and Pletcher and Dancey[20]). The solution becomes singular and it is difficult to proceed the calculations downstream of the separation point. Following Goldstein's analysis, it can be shown that near the separation point x_s,

$$\delta^*(x) = \delta^*(x_s) - \tau(x)/\frac{dp}{dx} + \ldots \qquad (2)$$

So that if $\frac{d\tau}{dx}$ is infinite there, so is $\frac{d\delta^*}{dx}$. Therefore regularity can be ensured by requiring either $\tau(x)$ or $\delta^*(x)$ to be regular. Catherall and Mangler[21] were the first to obtain regular solution by specifying the displacement thickness. Carter[22] demonstrated that results of inverse boundary layer calculations are in good agreement with Navier Stokes solution. Similar success was reported by Cebeci, Keller, and Williams[23] using a nonlinear eigenfunction formulation. On the other hand, Klienberg and Steger[19] and Horton[24] specified the shear stress in their calculations and obtained regular solutions. Indeed, Horton concluded that the condition $u_{yyyy}=0$, at the separation point was satisfied up to the computational accuracy. It should be mentioned that specifying the displacement thickness leads to faster convergence and it seems it avoids the nonuniqueness problem.

The appropriate displacement thickness or wall shear must be obtained as part of the overall problem, from the interaction between the boundary layer and the inviscid flow. Calculation of laminar separation bubbles that includes this viscous inviscid interaction has been given by Carter and Wornom[25]. On the other hand, it is argued that due to the interaction, the mainstream is always able to adjust itself to prevent the singularity from developing. Briley and McDonald[26][27] calculated the leading edge separated flows using a direct procedure, where the unsteady boundary layer equations are repeatedly solved until a steady state solution is obtained; after each time step, the prescribed pressure is updated from thin airfoil theory, thereby accounting for the displacement interaction. The same problem has been also solved by Kwon and Pletcher[28], Cebeci and Schimke[29], and Vatsa and Carter[30]. Davis and Werle[31][32] viewed the interactive boundary layer as an alternative to the asymptotic (triple deck) theory of Stewartson and Messiter. Recently, Carter[33], LeBalleur[34], Wigton[35], and Veldman[36] introduced different coupling procedures to accelerate the convergence of the interaction. For some special cases, inviscid rotational flow (see Taulbee and Robertson[37]) or just boundary layer (see Pletcher[38]) provide reasonable approximate solutions. In general, neither is uniformly valid and the interaction is required. The interaction reflects the boundary value nature of the problem which is important even for supersonic flows where (space-) marching procedures can be used for both the inviscid and the boundary layer parts separately (see Garvine[39]).

In this paper, numerical calculations of separated flows are presented where boundary layer equations in terms of stream function, with different boundary conditions are integrated. For a retarded flow on a flat plate, results are compared with Briley's and Carter's solutions. More accurate models accounting for the pressure variation across the viscous layer are discussed. In many cases, the effect of such a variation is negligible.

For viscous inviscid interaction problems, a patching procedure is used where the unknowns along a vertical line in both domains are solved simultaneously leading to strong coupling. In this method, the grid should represent both the viscous and the inviscid scales, but the equations are simpler than the full Navier Stokes (biharmonic) equation.

2. A MODEL PROBLEM

Briley[2] suggested to calculate retarded flows on a flat plate using Navier Stokes equations and compared the results with Howarth's boundary layer series solution where it breaks down at the separation point as predicted by Goldstein analysis. Later, Carter[22] prescribed the displacement thickness from Briley's solution and demonstrated that the inverse boundary layer solution is in excellent agreement with Briley's results. Recently many authors solved the same problem - Amarante and Cheng[40] and Cebeci and Stewartson[41] used interactive boundary layer methods while Inoue[8] used parabolized Navier Stokes equations. We chose this model problem to test present formulations.

3. BOUNDARY LAYER EQUATIONS

In terms of the tangential and the normal velocity components, the continuity and the tangential momentum equations read:

$$u_x + v_y = 0 \tag{3}$$

$$uu_x + vu_y = - p_x + \nu u_{yy} \tag{4}$$

with the boundary conditions u=v=0 at y=0, and $u=u_e$ at $y=y_e$ (edge of boundary layer), where it is assumed that p_x does not vary with y. Equations (3) and (4) can be written in terms of stream function and vorticity, namely:

$$u\omega_x + v\omega_y = \nu\omega_{yy} \tag{5}$$

$$-\omega = \psi_{yy} \tag{6}$$

with the boundary conditions $\psi=\psi_y=0$ at y=0 and $\psi_y=u_e$ and $\omega=0$ at $y=y_e$.

Another form of the above equations is simply:

$$u\,\psi_{yyx} + v\,\psi_{yyy} = \nu\,\psi_{yyyy} \tag{7}$$

The advantage of form (7) is discussed next.

First, the discretization is straightforward; (for uniform mesh):

$$\psi_{yyyy} \simeq (\psi_{i,j+2} - 4\,\psi_{i,j+1} + 6\psi_{i,j} - 4\,\psi_{i,j-1} + \psi_{i,j-2})/\Delta y^4 \tag{8-1}$$

$$\psi_{yyy} \simeq (\psi_{i,j+2} - 2\psi_{i,j+1} + 2\psi_{i,j-1} - \psi_{i,j-2})/\Delta y^3/2. \tag{8-2}$$

$$\psi_{yyx} \simeq ((\psi_{i,j+1} - 2\psi_{i,j} + \psi_{i,j-1}) - (\psi_{i-1,j+1} - 2\psi_{i-1,j} + \psi_{i-1,j-1}))\Delta y^2/\Delta x$$

$$\text{if } u>0 \tag{8-3}$$

$$\simeq((\psi_{i+1,j+1}-2\psi_{i+1,j}+\psi_{i+1,j-1})-(\psi_{i,j+1}-2\psi_{i,j}+\psi_{i,j-1}))/\Delta y^2/\Delta x$$

$$\text{if } u<0 \tag{8-4}$$

u and v are expressed in terms of ψ:

$$u \simeq (\psi_{i,j+1} - \psi_{i,j-1})/2\Delta y \tag{8-5}$$

$$v_{C\cdot D} \simeq - (\psi_{i+1,j} - \psi_{i-1,j})/2\Delta x \tag{8-6}$$

or $$v_{B\cdot D} \simeq - (\psi_{i,j} - \psi_{i-1,j})/ \Delta x \tag{8-7}$$

At each x=constant line the unknowns form a five diagonal scalar system of equations. Both wall boundary conditions can be easily incorporated implicitly. Similarly, the edge boundary conditions can be implemented by augmenting the system with the extra relations, for example $\psi_y=u_e$ and $\psi_{yy}=0$. In this case, the non-zero coefficients of the matrix are shown in sketch (1).

$$\begin{bmatrix} x & & x & & & & & & & \\ x & x & x & & & & & & & \\ x & x & x & x & x & & & & & \\ & & & & & & & & & \\ & & & & x & x & x & x & x \\ & & & & & & & x & \\ & & & & & & x & & x \end{bmatrix} \begin{bmatrix} \psi_J \\ \psi_{J-1} \\ \psi_{J-2} \\ \\ \psi_2 \\ \psi_1 \\ \psi_0 \end{bmatrix}$$

Sketch (1) - Nonzero Elements of the System Matrix

The resulting equations can be solved efficiently by a Gaussian elimination procedure especialized to five diagonal scalar systems (an extension of Thomas algorithm). For attached flows, assuming an initial profile is given and with a backward difference approximation of v, equation (7) is integrated by marching in the x-direction. Few iterations are required at each station to account for the nonlinearity. For separated flows, centered differences are used for v and a downstream boundary condition seems to be required. Two possibilities are tested, at the last station, a backward difference approximation is used for v or v is set equal to zero. Also, global (sweeps) rather than local iterations are employed.

4. PARTIALLY PARABOLIZED NAVIER STOKES EQUATIONS

In this model, only the streamwise viscous terms in the full equations are neglected, namely,

$$u_x + v_y = 0 \tag{9}$$

$$uu_x + vu_y = - p_x + \nu u_{yy} \tag{10}$$

$$uv_x + vv_y = - p_y + \nu v_{yy} \tag{11}$$

It is argued that the term v_{yy} in equation (11) is also of higher order.

Equations (9-11) can be written in terms of stream function and vorticity:

$$u\omega_x + v\omega_y = \nu\omega_{yy} \tag{12}$$

$$-\omega = \psi_{xx} + \psi_{yy} \tag{13}$$

The difference between equations (12) and (13) and the boundary layer equations (5) and (6) is the term ψ_{xx} in equation (13). Inoue[8] solved equations (12)-(13) for Briley's problem and obtained close results.

Substituting equation (13) into equation (12) leads to a fourth order equation in ψ:

$$u(\psi_{xx} + \psi_{yy})_x + v(\psi_{xx} + \psi_{yy})_y = \nu\,(\psi_{xx} + \psi_{yy})_{yy} \tag{14}$$

Centered differences are always used for $\psi_{xx}+\psi_{yy}$, upwind differences are used for the streamwise convection operator $u\partial_x$. The wall boundary conditions are $\psi=\psi_y=0$ and at the outer edge $\psi_{xx}+\psi_{yy}=0$ and $\psi_y=u_e$.

To solve equation (14), two artificial time dependent terms are added, $\alpha\psi_t$ and $\beta(\psi_{xx}+\psi_{yy})_t$ to enhance diagonal dominance. Again a five diagonal scalar system at each station is solved in the same manner the boundary layer equation (7)is solved. At the downstream boundary, the ψ_{xx} term in equation (14) is neglected and the equation reduces to equation (7). Global iteration is always needed.

The extra terms in equation (14) and not in equation (7) are:

$$-u\psi_{xxx} - v\,\psi_{xxy} + \nu\,\psi_{xxyy}$$

These terms are equal to p_{xy} which has been neglected in boundary layer calculations. For many cases, a boundary layer type equation can be used, where the term p_{xy} is lagged. For example, the pressure can be obtained from a triadiagonal system resulting from the discretization of:

$$p_{yy} = - (u\ v_x + v\ v_y)_y \tag{15}$$

with the boundary conditions:

$$p_y = 0 \quad \text{at the wall}$$

$$\text{and} \quad p = H - \frac{1}{2}(u^2 + v^2) \tag{16}$$

at the outer edge, where H is a constant. In equation (15), the viscous term νv_{yyy} has been neglected. After the pressure is calculated, the term p_{xy} is updated. To ensure stability of calculations, the pressure is underrelaxed.

5. VISCOUS/INVISCID INTERACTION

The viscous terms are negligible far from the wall, and the flow becomes inviscid. There are several methods for boundary layer and inviscid calculations. Four possibilities are considered:

1. using Hilbert integral for inviscid flow and integral boundary layer calculations,

2. using Hilbert integral and finite difference boundary layer calculations,

3. using finite differences for inviscid flows and integral boundary layer calculations,

4. using finite differences for both inviscid and boundary layer flows.

In our opinion, except for class 4, the inviscid and viscous equations should be solved simultaneously (see for example references (42), (36), and (43)). A coupling procedure is needed only for class 4, where at each step, the inviscid and the boundary layer equations are solved separately; the coupling procedure update the boundary conditions for both calculations and the process is repeated until convergence.

An alternative formulation, based on a patching (or zonal) procedure is tested here. One requirement for the present method is the existence of an overlap region, where the equations on both sides are valid. In the viscous region, the partially parabolized Navier Stokes equations are used where the vorticity is convected and diffused. On the other hand, the equation for the inviscid region is

$$\psi_{xx} + \psi_{yy} = - \omega \tag{17}$$

where

$$\omega = \omega(\psi) \tag{18}$$

i.e., the vorticity is only convected. Hence, the patching line should be in a region, where the viscous terms are negligible. If the patching line is chosen far from the wall, the vorticity there vanishes and the flow is irrotational It is clear that this is a special case and the irrotationality is not necessarily required in the present method. Numerical implementation is simple. At each vertical line, the viscous terms are tested (for all the points), when they are relatively small, the five diagonal system is replaced by a tridiagonal system resulting from equation (17). It would be more efficient, to have a dynamic grid generation method, where the grid points are distributed according to the proper scales based on order of magnitude analysis.

If one wishes, for some reason (say to take full advantage of the different scales involved), to solve the viscous and the inviscid problem separately, a coupling procedure is needed. Two methods have been tested, the first is a direct local coupling and the second is a semi-inverse implicit procedure as shown in Sketches (2) and (3) respectively.

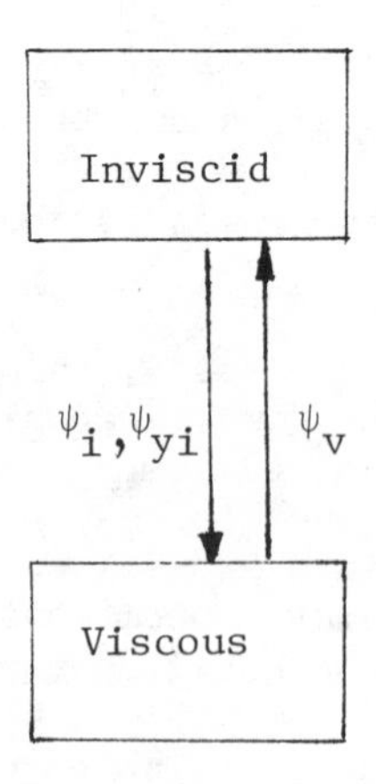

Sketch (2) - A direct local coupling procedure

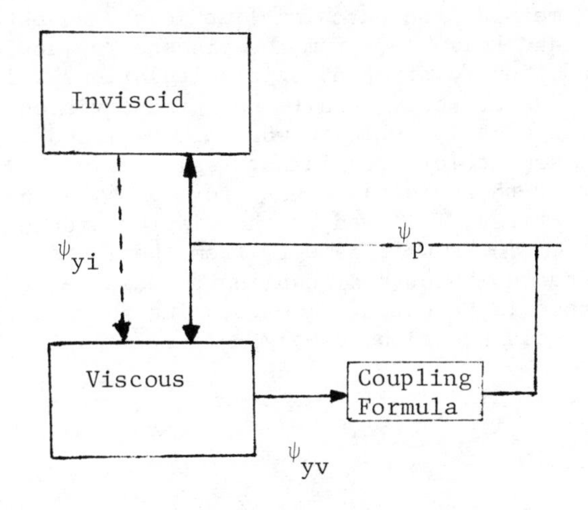

Sketch (3-a) - A semi-inverse implicit coupling procedure.

$$\begin{bmatrix} K_{ii} & 0 & K_{ip} \\ 0 & K_{vv} & K_{vp} \\ K_{pi} & K_{pv} & K_{pp} \end{bmatrix} \qquad \begin{bmatrix} \psi_i \\ \psi_v \\ \psi_p \end{bmatrix}$$

ψ_i = unknowns in Inviscid Region, K_{ii} = matrix of inviscid equations

ψ_v = unknowns in Viscous Region, K_{vv} = matrix of viscous equations

ψ_p = unknowns along the patching line, K_{pp} = matrix of coupling equations

Sketch (3-b) - Partitioning of the total matrix for the implicit coupling procedure

In the first method, the patching line is chosen between grid points say P and P+1. ψ_P, from the viscous region is used as boundary condition for the inviscid calculations which provide ψ_{P+1} and ψ_{P+2} as boundary conditions for the viscous calculations. Convergence is achieved when ψ_P, ψ_{P+1} and ψ_{P+2} are the same (within some tolerance) in both calculations. The second method starts with an initial guess for p. The inviscid problem is solved first, followed by the viscous problem using the same values for ψ_P as well as ψ_{P+1} from the inviscid output. Next ψ_P is updated through a coupling formula, equation (19), and the process is repeated. The coupling formula is consistent with the inviscid equation, namely,

$$(\psi^{n+1}_{i+1,P} - 2\psi^{n+1}_{i,P} + \psi^{n+1}_{i-1,P})/\Delta x^2$$

$$+ \alpha(\psi_{i,P+1} + \psi_{i,P-1})/\Delta y^2 - 2\psi^{n+1}_{i,P}/\Delta y^2 = -\omega(i,P) \tag{19}$$

where α is a relaxation parameter, $\psi_{i,P+1}$ is obtained from the inviscid calculations, while the viscous calculations provide $\psi_{i,P-1}$. A tridiagonal solver is used to obtain $\psi^{n+1}_{i,P}$. If the x-term and ω are neglected, equation (19) reduces to:

$$\psi^{n+1}_{i,P} = \psi^{n}_{i,P} + \alpha\cdot\frac{\Delta y}{2}(u_i - u_v) \tag{20}$$

Equation (20) is similar to Carter's formula[33]. It is found, however, that equation (19) leads to faster convergence.

(As a side remark, we notice that, for transonic calculations, equation (19) should reflect, the local character of the flow, whether subsonic or supersonic and in this case, equation (19) is replaced by:

$$\delta\psi_{xx} - \beta\delta\psi_x - \alpha\delta\psi = -[(\psi_{x}/\tilde{\rho})_x + (\psi_{y}/\tilde{\rho})_y] \tag{19'}$$

where $\delta\psi$ is $\psi^{n+1} - \psi^n$ and β and α are relaxation parameters and $\tilde{\rho}$ is an artificial density to produce dissipation in the supersonic region. Transonic calculations will be reported elsewhere.)

6. NUMERICAL RESULTS

Three uniform meshes (in x and y) are used: Coarse (35, 77), intermediate (76,77), and fine (151,77). Howarth profile is specified at the upstream boundary. Different edge boundary conditions are tested. Five cases are discussed here.

Case 1: $\psi_e = u_e(y_e - \delta^*)$, and $\psi_{yy}=0$, where δ^* is prescribed from Briley's solution.

Case 2: $\psi_e = u_e(y_e - \delta^*)$, and $\psi_y = u_e$.

Case 3: $\psi_y = \bar{u}$, and $\psi_{yy} = 0$, where $\bar{u}$ is the output of Case 1.

Case 4: $\psi_y = u_e$, and $\psi_{yy} = 0$.

The results shown for Cases 1 to 4 are for intermediate mesh.

Case 5: Same as Case 4 but using fine mesh.

Case 1 is the same as Carter's inverse boundary layer calculations. Case 2 has a hybrid boundary condition where both ψ and ψ_y are prescribed. As expected, the solution is regular since by specifying ψ, v is guaranteed to be bounded. We notice that if ψ and ψ_y are prescribed at the wall and at $y=y_e$, an approximate solution, with closed streamlines is obtained using a cubic polynomial ($\psi_{yyyy}=0$). Case 3 is interesting since it demonstrates that the calculations are stable and the solution is unique and regular. The results of Cases 4 and 5 are somewhat surprising since this is similar to the classical direct boundary layer calculations. There are, however, some differences, since centered differences are used for v and a downstream boundary condition is required. These results are, of course, for finite Reynolds number and finite grid size (the scheme is first order accurate). Indeed, if backward differences are used for v, the solution becomes singular as approaching the separation point and the calculation is not stable downstream of the zero skin friction point.

Figure (1) shows the streamlines calculated by Briley and by Carter. In figure (2), the corresponding streamlines of the five cases discussed above are presented. The rates of convergence are plotted in figure (3). We notice that the convergence is faster when ψ is prescribed. Comparing Case 1 to Carter's calculation, it is at least twice as fast. Figures (4) and (5) show the displacement thickness and the wall shear distributions.

The discrepancy between the boundary layer solutions of Cases (4) and (5) and the Navier Stokes solution may be due to the different boundary conditions. In Briley's calculation, $\psi_{xx}+\psi_{yy}=-\omega=0$ is used, while the boundary layer approximation of the vorticity, ψ_{yy} is set equal to zero in Cases 4 and 5. From the streamlines shown in figure (1), it is clear that at $y=y_e$, ψ_{xx} does not vanish.

If viscous inviscid interaction is allowed, the results do not agree with Briley's, since in his study, it was assumed that "along the outer flow boundary u was available from an inviscid solution. In the resulting Navier Stokes solution, the v-component of velocity along the outer-flow boundary generally will not agree with that from the inviscid solution used to prescribe the distribution of u along that boundary, and the mismatch in v is an indication of interaction between the inviscid and viscous solutions which has been neglected. To account for the interaction, it would be necessary to successively recompute the inviscid and Navier-Stokes solutions, allowing in some manner for the influence of one upon the other, until the two solutions no longer change significantly."

The present viscous inviscid interaction method described above, does not produce any separation for the cases Briley solved. In a recent paper, by Cebeci and Stewartson[41], it was shown that, based on interactive boundary layer calculations, the critical value of x_o (a parameter defining the corner point of the external velocity) to induce separation is 0.215 for $Re=10^6/48$. Comparable results, for the displacement thickness are shown for $x_o=0.23$ (corresponding to Briley's solution for $x_o=0.202$). They found that there is a maximum value of x_o, after which the numerical procedure breaks down. Furthermore, this maximum value is a decreasing function of Reynolds number and seems to approach the value 0.12 predicted by Howarth. It was concluded that there is a bound to the usefulness of interactive boundary layer theory, and once it is exceeded the theory in some sense goes sour; possibly the global flow properties then rapidly change over to those corresponding to Kirchoff free-streamline flow.*

Cebeci and Stewartson used Hilbert integral for inviscid flow and Keller's box scheme for the boundary layer calculation. Their results are compared to the present viscous inviscid calculations where finite differences are used for both regions, as shown in figure (6). The agreement is reasonable, taking into consideration the finite domain of integration of the finite difference calculations (130x115 points are used with $x_e=1.2$, and $y_e\sqrt{Re} = 8.11$). The performance of the present method for the cases where the interactive boundary layer calculations diverge is the subject of further investigation.

7. CONCLUDING REMARKS

The boundary layer equations, in terms of a stream function (differentiated to yield a fourth order equation) is integrated, marching in the main stream direction. A special Gaussian elimination procedure is applied to a scalar five diagonal system of equations at each step, which, in a sense, is similar to Davis Coupled Method[44] (where continuity and momentum equations are solved together). The present scheme is not, however, as compact as Cranck-Nicholson's or Keller's box schemes[45]. Nevertheless, for boundary layer type calculations, a smooth stretching transformation is usually employed and accuracy is not sacrificed for the resulting nonuniform mesh.

Solutions of retarded flows are presented for different outer boundary conditions, namely direct, inverse and hybrid.

*Navier Stokes solutions similar to those of Briley (but with higher Re, x_o, and y_e) were successfully calculated by Murphy[46].

Regular solutions at the separation and reattachment points are obtained. Since centered differences are used for $v=-\psi_x$, a downstream boundary condition is required and thus nonuniqueness as well as singularities are avoided, at least for this formulation with the grid size and the Reynolds numbers considered.

To allow for viscous inviscid interaction, a patching procedure is tested where a partially parabolized Navier-Stokes equation for ψ is solved simultaneously with the inviscid Poisson equation via a vertical line relaxation. For each line, a coupled system (tridiagonal in the inviscid region and five diagonal in the viscous region) is solved by efficient Gaussian elimination. Artificial time-dependent terms are added to enhance the stability of calculations.

It seems that the present stream function formulation leading to a single equation is advantageous for discretization and convergence. In this work, only small separation bubbles are calculated. Simulation of large regions of massive separation would be of great interest; for such cases, however, the flow in reality may not remain laminar.

References

1. ROACHE, P. - Computational Fluid Dynamics, Hermosa Pub., 1972.
2. BRILEY, W. R. - A numerical study of laminar separation bubbles using the Navier-Stokes equations, J. Fluid Mech., Vol. 47, pp. 713-736, 1971.
3. LEAL, L. G. - Steady separated flow in a linearly decelerated free stream, J. Fluid Mech., Vol. 59, pp. 513-535, 1973.
4. GHIA, V., GHIA, K. N., RUBIN, S. G. and KHOSLA, P. K. - Study of separated flow in a channel using primitive variables, Computers and Fluids, Vol. 9, pp. 123-142, 1981.
5. GHIA, V. and DAVIS, R. T. - Navier-Stokes solutions for flow past a class of two-dimensional semi-infinite bodies, AIAA Journal, Vol. 12, No. 12, pp. 1659-1665, 1974.
6. WERLE, M. J. and BERNSTEIN, J. M. - A comparative numerical study of models of the Navier-Stokes equations for incompressible separated flows, AIAA Paper 74-48, Jan. 1975.
7. GHIA, K. N., GHIA, V., and TESCH, W. A. - Evaluation of several approximate models for laminar incompressible separation by comparison with complete Navier-Stokes solutions, AGARD Conf. Proc. No. 168, pp. 6-1 to 6-15, 1975.
8. INOUE, O. - Separated boundary layer flows with high Reynolds numbers, Lecture Notes in Physics, Vol. 141, pp. 224-229, 1981.
9. MAHGOUB, H., BRADSHAW, P. - Calculation of Turbulent inviscid flow interactions with large normal pressure gradients, AIAA J., Vol. 17, pp. 1025-1029, 1979.

10. RUBIN, S. - Incompressible Navier-Stokes and parabolized Navier-Stokes formulations and computational techniques, this volume.
11. ORSZAG, S. and ISRAELI, M. - Numerical simulation of viscous incompressible flows, Ann. Rev. Fluid Mech., Vol. 6, pp. 281-318, 1974.
12. BROWN, S. and STEWARTSON, K. - Laminar separation, Ann. Rev. Fluid Mech., Vol. 1, pp. 45-72, 1969.
13. WILLIAMS, J. - Incompressible boundary layer separation, Ann. Rev. Fluid Mech., pp. 113-144, 1979.
14. GOLDSTEIN, S. - On laminar boundary layer flow near a position of separation, Quart. J. Mech. Appl. Math., Vol. 1, pp. 43-69, 1948.
15. DEAN, W. - Note on the motion of liquid near a position of separation, Proc. Camb. Phil. Soc., Vol. 46, pp. 293-306, 1950.
16. MURPHY, J. and KING, L. - Airfoil Flow-field calculations with coupled boundary-layer potential codes, Second Symposuim on Numerical and Physical Aspects of Aerodynamic Flows, Calif. State Univ., Long Beach, CA, 1983.
17. REYHNER, T. A. and FLUGGE-LOTZ, I. - The interaction of a shock wave with a laminar boundary layer, Int. J. on Non-Linear Mech., Vol. 3, No. 2, pp. 173-199, 1968.
18. WERLE, M. and DAVIS, R. - Incompressible laminar boundary layers on a parabola at angle of attack: A study of the separation point, J. of Appl. Math., Vol. 7, 1972.
19. KLIENBERG, J. and STEGER, J. - On laminar boundary-layer separation, AIAA Paper 74-94, 1974.
20. PLETCHER, R. and DANCEY, C. - A direct method of calculating through separated regions in boundary layer flow, J. of Fluids Engineering, pp. 568-572, 1976.
21. CATHERALL, D. and MANGLER, K. - The integration of the two-dimensional laminar boundary-layer equations past a point of vanishing skin friction, J. Fluid Mech., Vol. 26, pp. 163-182, 1966.
22. CARTER, J. - Inverse solutions for laminar boundary layer flows with separation and attachment, NASA TR R-447, 1975.
23. CEBECI, T., KELLER, H. and WILLIAMS, P. - Separating boundary-layer flow calculations, J. of Comp. Physics, Vol. 31, pp. 363-378, 1979.
24. HORTON, H. - Separating laminar boundary layers with prescribed wall shear, AIAA J., Vol. 12, 1974.
25. CARTER, J. and WORNOM, S. - Solutions for incompressible separated boundary layers including viscous-inviscid interaction, NASA SP 347, 1975.
26. BRILEY, W. and McDONALD, H. - Numerical prediction of Incompressible separation bubbles, J. Fluid Mech., Vol. 69, pp. 631-656, 1975.
27. BRILEY, W. and McDONALD, H. - A survey of recent work of interacted boundary layer theory for flow with separation, Second Symposium on Numerical and Physical Aspects of Aerodynamic Flows, Calif. State Univ., Long Beach, CA, 1983.

28. KWON, O. and PLETCHER, R. - Prediction of subsonic separation bubbles on airfoils by viscous-inviscid interaction, Second Symposium on Numerical and Physical Aspects of Aerodynamic Flows, Calif. State Univ., Long Beach, CA, 1983.
29. CEBECI, T. and SCHIMKE, S. - The calculation of separation bubbles in interactive turbulent boundary layers, J. Fluid Mech., Vol. 131, pp. 305-317, 1983.
30. VATSA, V. and CARTER, J. - Analysis of airfoil leading edge separation bubbles, AIAA Paper 83-0300, 1983.
31. DAVIS, R. and WERLE, M. - Numerical methods for interacting boundary layers, Proc. 1976 Heat Transfer Fluid Mech. Institute, Stanford Univ. Press, 1976.
32. DAVIS, R. and WERLE, M. - Progress on Interacting boundary layer computations at high Reynolds number, First Symposium on Numerical and Physical Aspects of Aerodynamic Flows, Calif. State Univ., Long Beach, CA, 1981.
33. CARTER, J. - A new boundary-layer inviscid iteration technique for separated flow, AIAA Paper 79-1450, 1979.
34. LeBALLEUR, J. - Numerical flow calculation and viscous-inviscid interaction techniques, this volume.
35. WIGTON, L. and HOLT, M. - Viscous-Inviscid interaction in transonic flow, AIAA Paper 81-1003, 1982.
36. VELDMAN, A. - New, quasi-simultaneous method to calculate interacting boundary layers, AIAA J., Vol. 19, No. 1, pp. 79-85, 1981.
37. TAULBEE, D. and ROBERTSON, J. - Turbulent separation analysis ahead of a step, J. of Basic Engineering, Vol. 94, pp. 544-550, 1972.
38. PLETCHER, R. - Calculation of separated flows by viscous-inviscid interaction, this volume.
39. GARVINE, R. - Upstream influence in viscous interaction problems, Physics of Fluids, Vol. 11, pp. 1413-1423, 1968.
40. AMARANTE, J. and CHENG, I. - On the viscous-inviscid flow interactions in the vicinity of a laminar separation bubble, AIAA Paper 79-1478, 1979.
41. CEBECI, T. and STEWARTSON, K. - On the calculation of separation bubbles, J. Fluid Mech., Vol. 133, pp. 287-296, 1983.
42. GOHSE, S. and KLINE, S. - The computation of optimum pressure recovery in two dimensional diffusers, J. of Fluids Engineering, Vol. 100, pp. 419-426, 1978.
43. MOSES, H., JONES, R., and O'BRIEN, W. - Simultaneous solution of the boundary layer and freestream with separated flow, AIAA J., Vol. 16, No. 1, pp. 61-66, 1978.
44. BLOTTNER, F. - Investigation of some finite-difference techniques for solving the boundary layer equations, Comp. Meth. Appl. Mech. Eng., Vol. 6, pp. 1-30, 1975.
45. KELLER, H. - Numerical methods in boundary layer theory, Ann. Rev. Fluid Mech., Vol. 10, pp. 417-433, 1978.
46. MURPHY, J. - An efficient solution procedure for the incompressible Navier Stokes equations, AIAA J., Vol. 15, No. 9, pp. 1307-1314, 1977.

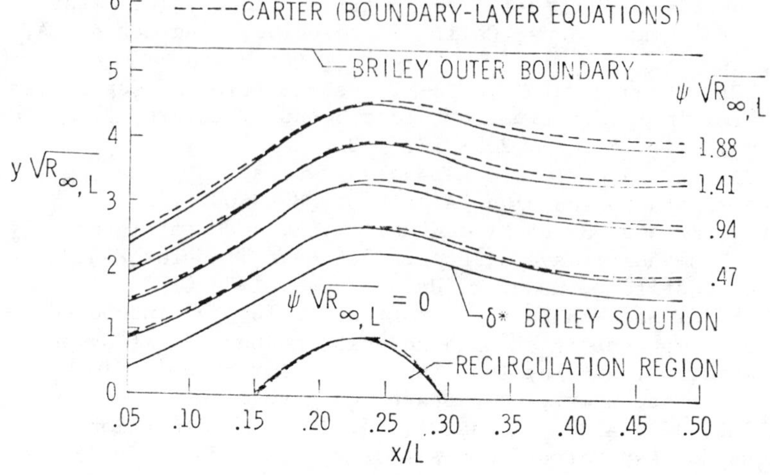

Figure (1) Streamlines from Briley's and Carter's Solution

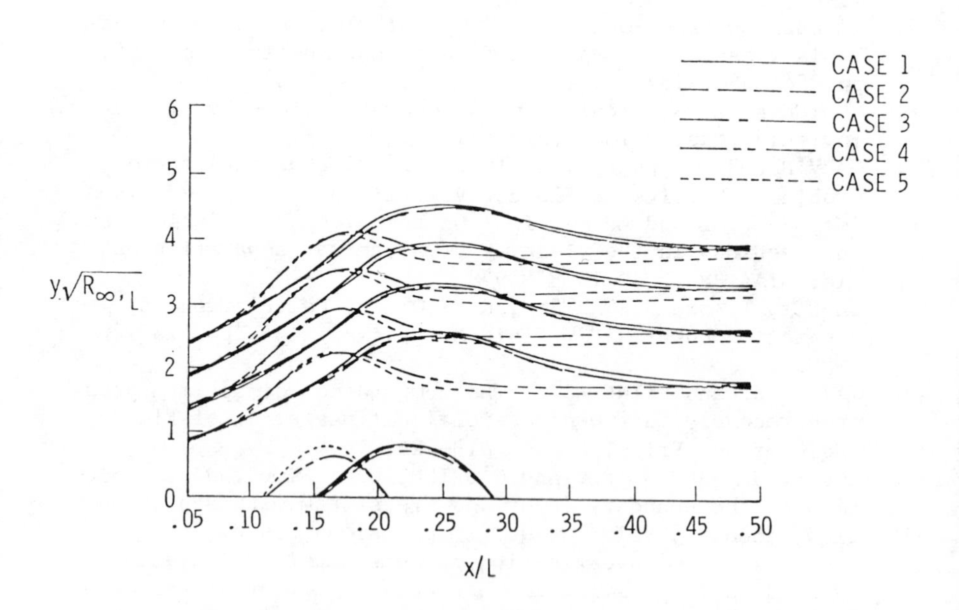

Figure (2) Streamlines from present boundary layer calculations

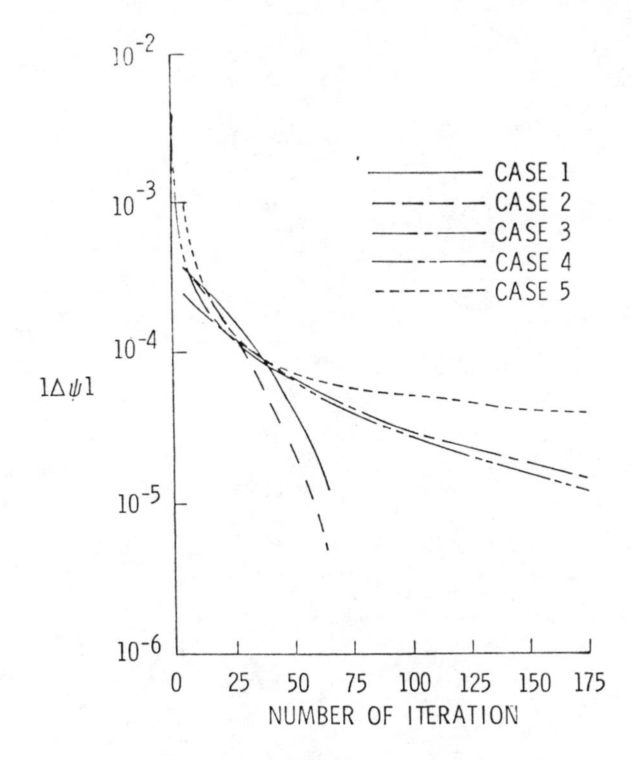

Figure (3) Rates of convergence of present boundary layer calculations

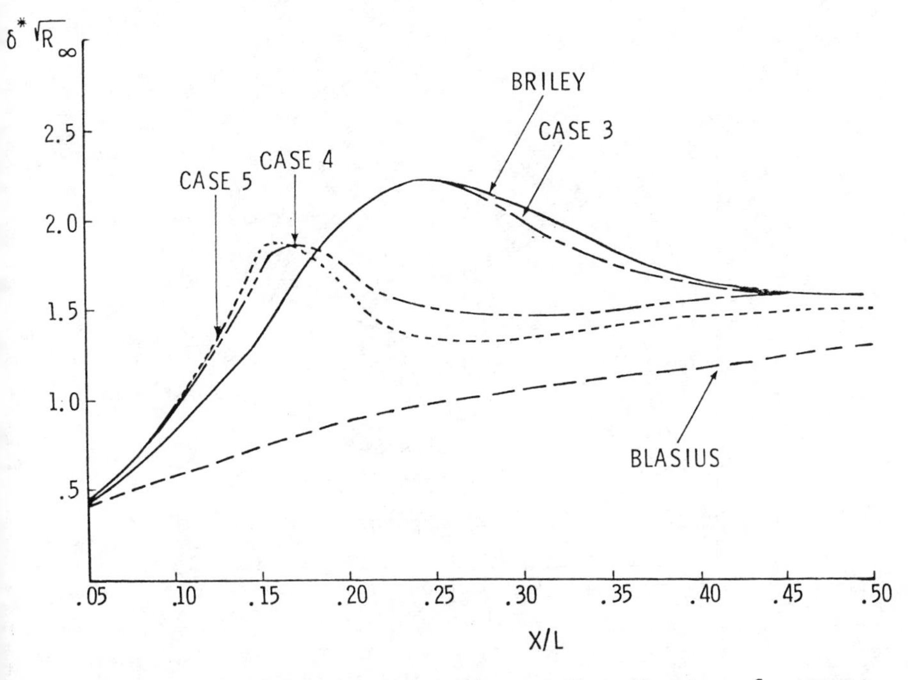

Figure (4) Displacement thickness distribution of present boundary layer calculations

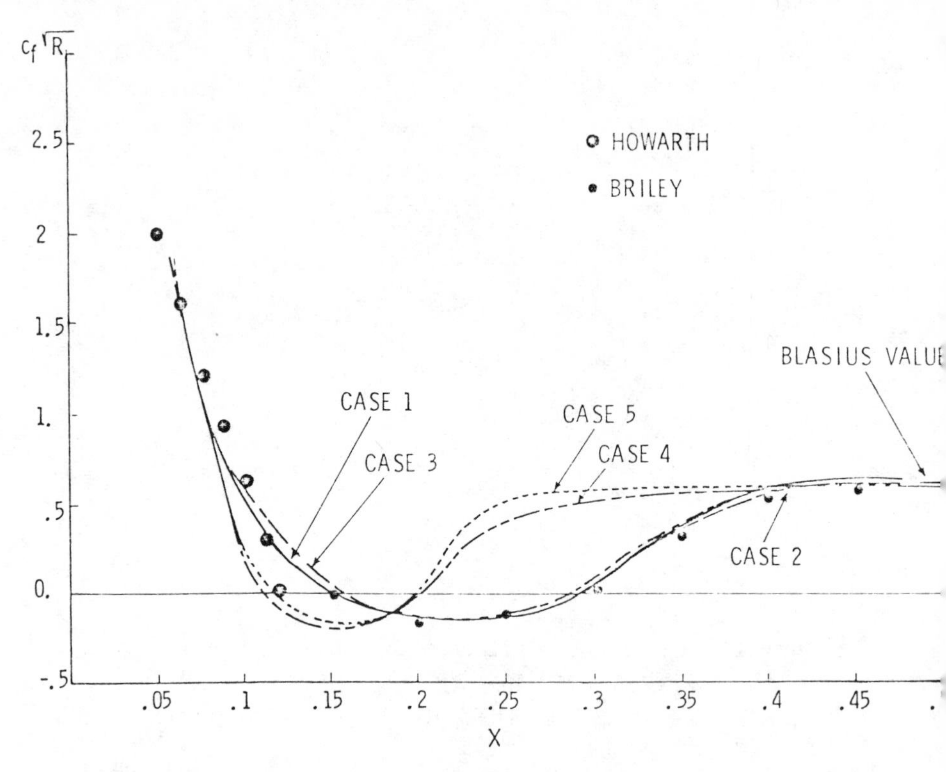

Figure (5) Wall Shear distribution of present boundary layer calculations

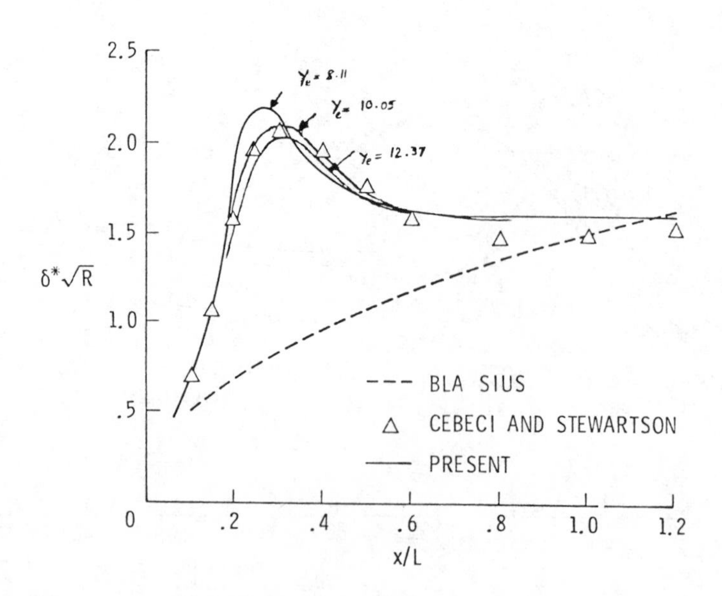

Figure (6) Displacement thickness distribution of present viscous inviscid interactive calculations

APPENDIX

Boundary Layer Calculation of Flow in A Duct With Sudden Expansion.

In the main text, results are shown for a retarded flow over a flat plate. Separation bubbles occur due to pressure gradient. Another example, where separation may occur due to geometry changes, is flow in a duct with sudden expansion. Navier Stokes equations are usually solved to simulate such flows. Pletcher[38] found, however, that boundary layer calculations produce excellent results for a wide range of Reynolds numbers and area changes. In such cases, the upstream influence is completely negligible and the effect of singularity due to the discontinuity of the geometry is very local. A similar calculation is repeated here using the present stream function formulation.

In Pletcher's calculations, the boundary layer equations in terms of the primitive variables u, v, and p are solved marching with the main flow direction. At each station, the pressure is found such that it satisfies a global constraint, namely conservation of mass. On the other hand, if the x-momentum equation is differentiated with respect to y, the pressure term drops and using a stream function, the global constraint is satisfied through the boundary condition. More precisely, a fourth order equation is solved at each station with four boundary conditions: $u=v=o$ at the wall and ψ=constant and $\psi_{yy}=o$ at the axis of symmetry.* The results are shown in figure A and they are in good agreement with Pletcher's and with Navier Stokes solution.

Obviously, such a good agreement of boundary layer and Navier Stokes solutions is not always possible. In general, viscous/inviscid interaction is necessary and boundary layer equations are not adequate to represent the whole flow field. The patching procedure recommended in the paper, consists of an inviscid Poisson's equation and parabolized Navier Stokes in the viscous layer. The latter are solved in a similar way boundary layer equations are solved, except for a pressure term that varies in the lateral direction and is updated using the normal momentum equation.

---

*For asymmetric configurations the boundary conditions are $u=v=0$ at both upper and lower walls. The ψ-distribution at the inlet is always given assuming fully developed flows.

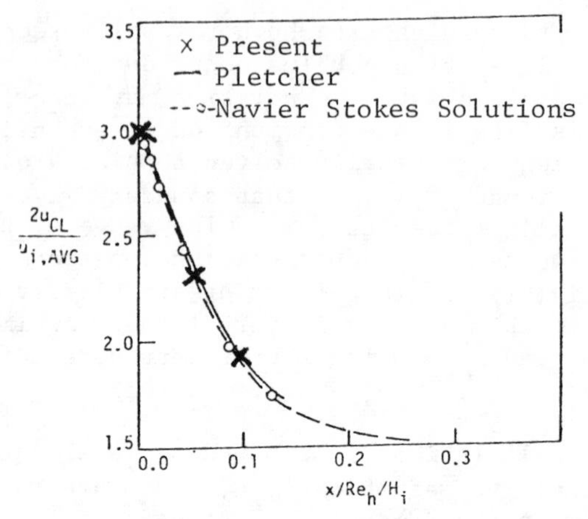

Figure A-1 Centerline velocity distribution for laminar flow in a channel with a symmetric sudden expansion, $h/H_i = 0.5$.

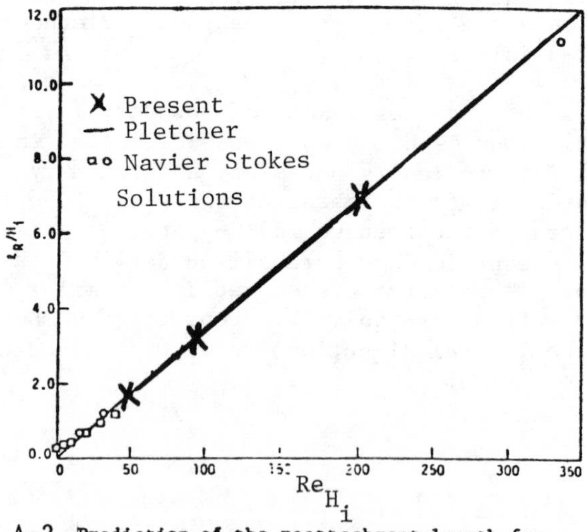

Figure A-2 Prediction of the reattachment length for laminar flow in a channel with a symmetric sudden expansion, $h/H_i = 0.5$.

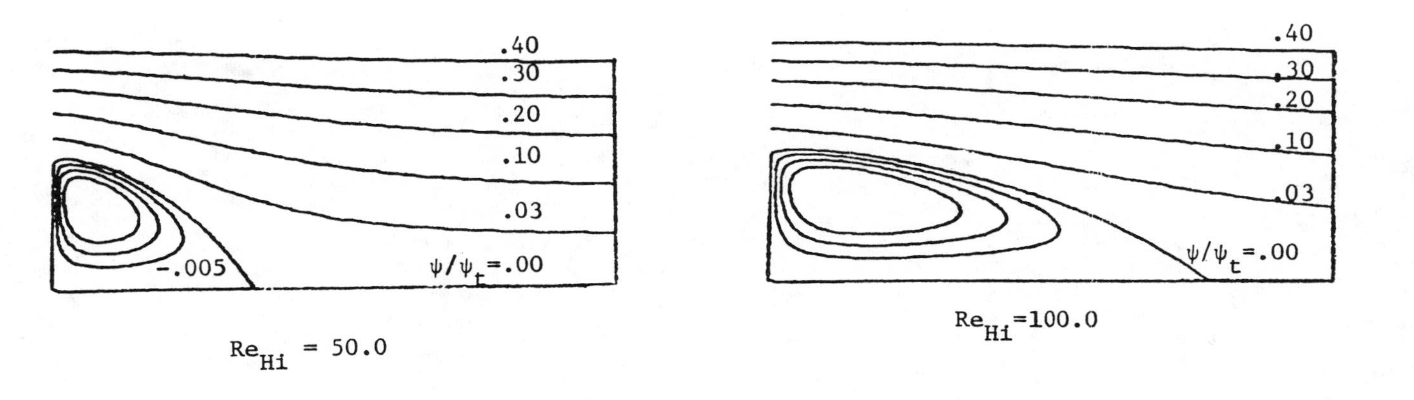

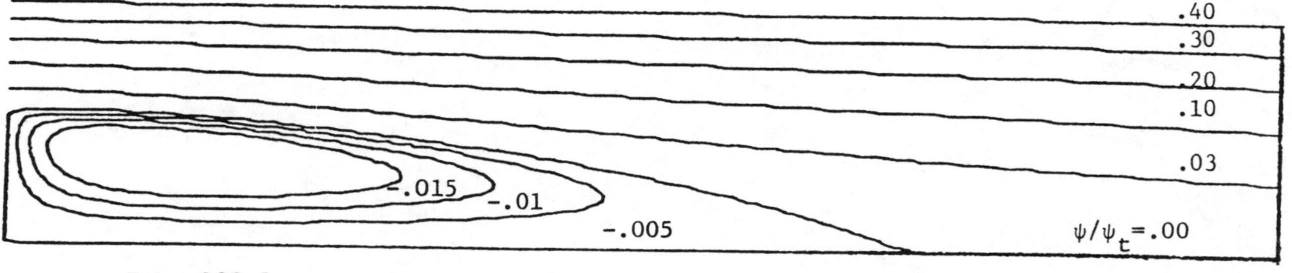

Figure A-3 Streamline contours for laminar flow in a channel with a symmetric sudden expansion, $h/H_i=0.5$.

Section 4:

VISCOUS-INVISCID INTERACTION, COMPRESSIBLE

NUMERICAL FLOW CALCULATION AND VISCOUS-INVISCID INTERACTION TECHNIQUES

J.C. LE BALLEUR*

Office National d'Etudes et de Recherches Aérospatiales (ONERA)
BP 72, 92322 CHATILLON CEDEX

1 - INTRODUCTION

Progress in further computing capability at high Reynolds numbers is now achieved, either with Direct Navier-Stokes (NS) solvers, or with Viscous-Inviscid Solvers (VIS), [1 to 8].

The latter indirect techniques (VIS), which were more or less boundary-layer-like in the early stages of evolution, becomes increasingly NS-like, including separations, shock wave-boundary layer and trailing-edges interactions. These methods provide a capability to generate new Indirect NS-solvers which benefit from the numerical conditioning of the boundary layer techniques[9 to 12]. They also provide the best way to get approximate solvers, which minimize the computer cost or storage by means of some modelling of the momentum equations. Typical developments at this level have been given by Melnik[13 to 15], Lock[16], Werle [17], Carter[18,19], Yoshihara[20], Whitfield[21], Kline[22], Veldman[23], LeBalleur[24 to 28], etc...

After summarizing briefly the different models and developments areas of the VIS-techniques, based on the ideas of zonal computation and of viscous-inviscid splitting, we detail here composite solvers which involve a thin-layer modelling, but which are capable to compute complex compressible flows with multiple separations or strong viscous interactions, at a rather low cost.

These solvers have been generated on the basis of Interacting Defect Integral (IDI) equations[6,29], of an integral turbulent closure which includes an out-of-equilibrium transport[26], and of two explicit relaxation techniques for the numerical viscous-inviscid coupling[25]. These

(*) Senior Scientist, Group Head.

two techniques, called Direct and Semi-Inverse relaxations, insure the stability control with respect to the local flow conditions and mesh size, and allow to switch the viscous solution toward a direct or inverse problem, according to a criterion of local shape parameter in the viscous layer[25, 26,27]. After coupling, the converged discrete solution is fully insensitive to the direct-inverse switch. These algorithms have been extended for the coupling and positioning of the viscous wakes layers, in fully dissymmetrical conditions, and also to solve quasi-3D viscous flows, within the infinite swept wings approximation[39].

Despite of their simplicity, the Interacting Defect Integral (IDI) equations do not involve necessarily the Prandtl's assumptions[6]. The Defect Analysis of the integral continuity equation and of the displacement thickness removes any thin-layer approximation from the wall-transpiration model[5,3,6,29]. The superiority of the formulation is moreover enforced, when compared to an interacting Prandtl layer, because the failures of the supercritical interactions of Crocco[30] are fully removed[24], which is equivalent to insure a full capability in computation of shock waves-boundary layers interactions. At last, with respect to the momentum-balance modelling, the IDI equations are proved[6,29] to be consistent, not only with a Prandtl layer, but also with a first order approximation of the "Defect Formulation" [6] of Thin-layer-NS equations. In this case, the normal pressure gradient is not zero, even though a boundary-layer-like closure is still assumed for modelling a few extra-unknowns in the IDI equations.

The IDI approach minimizes the grid requirement and computer ressources in viscous numerical solvers, or equally opens an access to the computation of a maximum flow complexity. In addition, although involving more empiricism, the turbulent closures of integral methods have been found during the recent AFOSR-Stanford-1981 conference[31] to provide a highly competitive prediction tool, with a range of applications wider than it would have been expected.

Examples are given here in complex flows, involving the problems of airfoils, wings, separation, trailing-edges and shock-waves interactions. The coupling is achieved using either potential or Euler inviscid solvers.

2 - VISCOUS-INVISCID SPLITTING .

The viscous-inviscid splitting is not unique and does not involve necessarily approximations. Recent advances provide now three main development areas for the VIS-methods.

2.1 - Zonal or Subdomains solvers :

Although the idea of zonal solver was probably first connected to the concept of Prandtl, one of the development area of VIS is now to increase the computing efficiency of the Direct NS-solvers, in reducing the extent of the viscous computational domain. The resulting "Zonal NS-solvers" have then to be coupled with an outer inviscid solver along a "patching" boundary, located in the inviscid part of the flow. This approach exhibits some common features with an interacting Prandtl solver, for the analytical questions raised in patching the boundary conditions, and for the multiple coupling algorithms which may be possibly investigated.

On the contrary of the Prandtl solvers, however, the cost reduction in the viscous zone does not result from a fast marching solution, but only from a relative decrease of the number of grid nodes. The difficulty is then to avoid that a coupling relaxation technique of low convergence rate will decrease or cancel the cost reduction[32,33]. Until now, although very few answers are available, a direct and simultaneous solution of the viscous and inviscid unkowns on the patching boundary, at a given node and time-step[34], as detailed in a following chapter by Cambier et al., may eliminate succesfully the coupling algorithm, so long as explicit unsteady solvers are used in both computational domains.

2.2 - Interacting Thin-Layer solvers :

The second development of VIS-methods intends to extrapolate the approximations and techniques of the boundary layer, via viscous-inviscid coupling. Solvers of low cost are then generated, based on approximate momentum equations.

Such solvers are however NS-like if the coupling is "strong" and numerically consistent, which may provide the capability to compute global complex flows with multiple viscous interactions. The simplest usual example is the zonal Prandtl solver, patched with an inviscid flow. This approach extends the idea of Crocco[30] for supersonic separation, and removes the separation singularities[29].

The more recent idea of a viscous "Defect Formulation"[6,26] with respect to a fully overlaying inviscid flow, provides a more satisfactory approach at a similar cost, if we restrict to a first order approximate form, and insures the general capability at shock wave-boundary layer interactions which was missing in the Crocco's model.

The improvements have been to make the wall-transpiration model free of any approximation, to provide always a full upstream influence recovery in supersonic zones by means of a subcritical coupling[24], and to approximate the normal pressure gradient of the layer at first order with the overlaying inviscid pressure field. This last point, which has been proved to be the more rational way to introduce the approximate IDI equations[6,26] (following sections), has now been used also in finite difference techniques designed for the Prandtl's equations[19,35].

Obviously, future progress in Interacting Thin-Layer solvers are strongly connected to the generation of increasingly efficient coupling algorithms, and also to a better understanding of the problem of 3D-separations.

2.3 - Composite solvers :

The third development area in VIS-methods is an extension of the second-one. The approaches intend here to maintain the splitting into an inviscid-like predictor problem, plus a viscous-corrector problem, with overlaying computational domains. They intend to benefit still of fast marching solutions for the viscous part, and to couple the predictor and corrector problems via iterative numerical methods.

Within these constitutive choices, similar to those used in Thin-layer solvers, it may be conceived easily[6,29, 26,9,10,11] to introduce into the viscous-corrector the terms which were neglected in section 2.2, and to generate then indirect NS-solvers, at least for Thin-layer-NS equations. These composite solvers are then based on a viscous-inviscid splitting of the equations, such that the equivalent and coupled sub-systems are designed for numerical techniques which are well-conditioned at high Reynolds numbers.

On the contrary of the zonal NS-solvers, the constitutive predictor and corrector here are subject to fast numerical techniques, inviscid-like and boundary-layer-like. They may then reduce substantially the computing cost, so long as the coupling cost, of course, is not prohibitive. Progress achieved in the coupling techniques of approximate solvers will be used here straightforwardly.

3 - DEFECT FORMULATION - APPROXIMATE DEFECT INTEGRAL EQUATIONS .

From this point, the viscous-inviscid splitting is assumed to be achieved with the composite solver approach.

3.1 - Full Defect Formulation :

The Defect Formulation is the composite solver for which the inviscid-like predictor problem is assumed to satisfy actually the inviscid equations, in such a way that the viscosity controls only the boundary conditions of the predictor.

We consider here for simplicity the steady quasi-3D problem (infinite swept wing). Unsteady or three-dimensional equations are given in[27,29,36]. At a given span station $y=cst$, a contour $z=o$, whose curvature is $K(x)$, which may be a wall or a wake center-line, defines a curvilinear tangent reference system, orthogonal, with metric coefficients

$h=h_1=1+Kz$, $h_2=h_3=1$, We denote $\bar{u},\bar{v},\bar{w}$ the velocity components, $\bar{\rho}$ the density, $\bar{p}$ the pressure, u,v,w,ρ,p the corresponding unknowns in the overlaying inviscid-like predictor, $\bar{V}$ the viscous stress terms. The Navier-Stokes problem may be written, with $\partial/\partial y = 0$:

$$(1) \qquad \frac{\partial \bar{F}}{\partial x} + \frac{\partial \bar{G}h}{\partial z} + K\bar{H} = \bar{V}$$

$$(2) \qquad \bar{F} = \begin{bmatrix} \bar{\rho}\bar{u} \\ \bar{\rho}\bar{u}^2+\bar{p} \\ \bar{\rho}\bar{u}\bar{v} \\ \bar{\rho}\bar{u}\bar{w} \end{bmatrix}, \quad \bar{G} = \begin{bmatrix} \bar{\rho}\bar{w} \\ \bar{\rho}\bar{u}\bar{w} \\ \bar{\rho}\bar{v}\bar{w} \\ \bar{\rho}\bar{w}^2+\bar{p} \end{bmatrix}, \quad \bar{H} = \begin{bmatrix} 0 \\ -\bar{\rho}\bar{u}\bar{w} \\ 0 \\ \bar{\rho}\bar{u}^2+\bar{p} \end{bmatrix}.$$

$$(3) \qquad 0 = \lim_{z\to\infty} (f-\bar{f}) , \qquad f = \{u,v,w,p,\rho\}.$$

In addition to the no-slip condition on the walls, and the inviscid boundary conditions, the coupling relations (3) correspond to a smooth analytical matching with the outer inviscid flow. This condition is free of any assumption and eliminates the influence of a patching boundary. The Defect Formulation splits the system (1)(2)(3) into the following equivalent one, where F,G,H denote the inviscid fluxes :

$$(4) \quad \text{Predictor :} \quad \frac{\partial F}{\partial x} + \frac{\partial Gh}{\partial z} + KH = 0$$

$$(5) \quad \text{Defect :} \quad \frac{\partial}{\partial x}(F-\bar{F}) + \frac{\partial}{\partial z}(G-\bar{G})h + K(H-\bar{H}) = -\bar{V}$$

$$(6) \quad \begin{cases} \text{Wall :} & \rho w_{(x,y,o)} = \dfrac{\partial}{\partial x}\displaystyle\int_o^{\infty}(\rho u-\bar{\rho}\bar{u})_{(x,y,z)}.dz \\ \text{Wake :} & \rho w_{(x,y,o+)} - \rho w_{(x,y,o-)} = \dfrac{\partial}{\partial x}\displaystyle\int_{-\infty}^{+\infty}(\rho u-\bar{\rho}\bar{u}).dz \end{cases}$$

$$(7) \quad \text{Wake :} \quad p_{(x,y,o+)} - p_{(x,y,o-)} = \int_{-\infty}^{+\infty} -\frac{\partial\,(p-\bar{p})}{\partial z}.dz$$

The defect equations (5) control the departure from the inviscid quantities, directly on the fluxes $\bar{F},\bar{G},\bar{H}$. The z-integration of the defect continuity equation provides the exact displacement effect (6). The exact curvature effect (7) corresponds to the continuity of the viscous pressure at the wake center $z=o$, which necessarily introduces a jump on the inviscid quantities between $z=o+$ and $z=o-$.

The basic difference with the boundary layer theories is here to take into account the full z-variations of the terms $\rho,\rho u,p,\bar{p}$ inside the viscous layer.

3.2 - Approximate Defect Integral method :

From this point, the so-called Thin-layer-NS equations (or parabolized equations) are assumed. All viscous terms, excepted the Prandtl-like terms τx and τy , are eliminated. The system is however still an elliptic one.

If, in addition, a first order expansion based on the gauge δ , which scales the viscous layer, is applied to the momentum equations, we get the approximate defect system:

$$
(8) \quad \begin{cases}
\frac{\partial}{\partial x}(\rho u-\bar{\rho}\bar{u}) + \frac{\partial}{\partial z}(\rho w-\bar{\rho}\bar{w}) = 0 \\
\frac{\partial}{\partial x}(\rho u^2-\bar{\rho}\bar{u}^2) + \frac{\partial}{\partial z}(\rho uw-\bar{\rho}\bar{u}\bar{w}) = -\frac{\partial}{\partial x}(p-\bar{p}) + \frac{\partial \tau x}{\partial z} \\
\frac{\partial}{\partial x}(\rho uv-\bar{\rho}\bar{u}\bar{v}) + \frac{\partial}{\partial z}(\rho vw-\bar{\rho}\bar{v}\bar{w}) = \frac{\partial \tau y}{\partial z} \\
0 = \frac{\partial}{\partial z}(p-\bar{p})
\end{cases}
$$

The z-momentum equation, with the matching condition (3), degenerates then at first order into $\bar{p}(x,y,z)=p(x,y,z)$, which vanishes the curvature effect (7), without vanishing however the normal pressure gradient inside the layer. At the same first order approximation ($\bar{p}=p$) , the defect integral equations for continuity, x and y momentum, and the local defect momentum equation at $z=\delta(x,y)$ (entrainment eq.) are, if $\bar{w}(x,y,o)=o$ and $q^2= u^2+ v^2 + w^2$:

$$(9) \quad \left\{ \frac{\partial}{\partial x}\left[\rho q \delta_1\right] = \rho w \right\}_{(x,y,o)}$$

$$(10) \quad \left\{ \frac{\partial}{\partial x}\left[\rho q^2(\theta_{11}+ \frac{u}{q}\delta_1)\right] - \rho uw \right\}_{(x,y,o)} = \rho q^2 \frac{Cfx}{2}$$

$$(11) \quad \left\{ \frac{\partial}{\partial x}\left[\rho q^2(\theta_{21}+ \frac{v}{q}\delta_1)\right] - \rho vw \right\}_{(x,y,o)} = \rho q^2 \frac{Cfy}{2}$$

$$(12) \quad \left\{ \frac{u}{q}\frac{\partial \delta}{\partial x} - \frac{w}{q} \right\}_{(x,y,\delta)} = E \equiv \left\{ \frac{1}{\rho q} \frac{\frac{\partial \tau x}{\partial z}}{\frac{\partial (u-\bar{u})}{\partial z}} \right\}_{(x,y,\delta)}$$

with the following definitions of the thicknesses :

$$
(13) \quad \begin{cases}
\delta_1 \cdot \rho q_{(x,y,o)} = \int_0^\infty \left[\rho u_{(x,y,z)} - \bar{\rho}\bar{u}_{(x,y,o)}\right].dz \\
(\theta_{11}+ \frac{u}{q}\delta_1).\rho q^2_{(x,y,o)} = \int_0^\infty \left[\rho u^2_{(x,y,z)} - \bar{\rho}\bar{u}^2_{(x,y,z)}\right].dz \\
(\theta_{21}+ \frac{v}{q}\delta_1).\rho q^2_{(x,y,o)} = \int_0^\infty \left[\rho uv_{(x,y,z)} - \bar{\rho}\bar{u}\bar{v}_{(x,y,z)}\right].dz
\end{cases}
$$

The set of defect integral equations at first order (9)(10)

(11)(12), which may be solved with the turbulent closure of section 3.3, reduces to the traditional form, simply because it determines unusual "Defect thicknesses" (13), which control only the departure from an overlaying inviscid flow and includes the z-variations of $\rho u, \rho u^2, \rho uv$. This definition of δ_1 eliminates in addition any approximation in (9), which is identical to (6).

It is possible also to introduce a second order estimate of the viscous pressure, in solving a second order modelling of the z-momentum defect equation. This provides $(p-\bar{p})$, and the second order curvature effect (7). For simplicity, this second order modelling of $\bar{p}$ is assumed here as a correction, uncoupled from the x and y momentum balance (10)(11)(12). This modelling has been based on "induced" (and z-averaged) curvatures $K^*_{(x,y,o+)}$ of the interacting inviscid streamlines [6,25,26]:

$$(14) \qquad \frac{\delta}{\delta z}(p-\bar{p}) = K^*_{(x,y,o+)} \cdot \left[\rho u^2_{(x,y,z)} - \overline{\rho u}^2_{(x,y,z)} \right]$$

$$(15) \qquad p_{(x,y,o+)} - p_{(x,y,o-)} = - \left[K^* \rho q^2 (\theta_{11} + \frac{u}{q}\delta_1) \right]_{(x,y,o+)} + \left[K^* \rho q^2 (\theta_{11} + \frac{u}{q}\delta_1) \right]_{(x,y,o-)}$$

The value of the modelling (14)(15) is dependent on a strong viscous coupling of the induced curvature $K^*_{\pm}$, which may be identified for example with the displacement surface curvature. Near a trailing-edge, the accuracy of the modelling requires also to take into account the dissymmetry, not only for the thicknesses $\delta_1^{\pm}$, $\theta_{11}^{\pm}$, but mainly for the induced curvatures $K^*_{(x,y,o+)}$ and $K^*_{(x,y,o-)}$, whose signs are frequently opposite in the near wake.

3.3 - Integral turbulent modelling :

The integral closure relations, detailed in [26,27], are assumed to be boundary-layer-like, and are deduced from an analytical description of the viscous velocity profiles $\vec{q}\,(\delta, C_2, C_3)$, connected with an out-of-equilibrium modelling of the turbulence. δ is the thickness of the layer, C_2 and C_3 are two free shape parameters. We denote $\vec{x}$ and $\vec{y}$ the unit vectors, respectively tangent and normal to the inviscid velocity $\vec{q}$.

a - Velocity profiles : The description of the 3D-velocity profiles assumes a one-sided cross-flow, Fig. 1, and modifies the vectorial description of Coles. The resulting profiles uses a new Wake function and are capable to describe the full range of the shape parameters, including configurations with large reverse flows. The description combines a Wake defect component, plane and parallel to $\vec{W}$, and a Logarithmic shear stress component, non-colinear and parallel to $\vec{\tau}$. If $z = \delta.n$, the velocity profile is :

$$(16) \quad \begin{cases} \vec{q} = \vec{q} - \vec{W}.\tilde{F}(n) + \vec{\tau}.\text{Log } n \\ \vec{W} = q.(C_2\vec{x} + C_3\vec{y}) \\ \vec{\tau} = q\dfrac{C_1}{C_4}.\left[(1-C_2)\vec{x} + C_3\vec{y}\right] \end{cases}$$

$$(17) \quad \begin{cases} C_4 = \sqrt{(1-C_2)^2 + C_3^2}, \quad C_1 = \dfrac{1}{0.41}\cdot\sqrt{\dfrac{Cf}{2}} \\ C_4 = \sqrt{\dfrac{Cf}{2}}\cdot\left[\dfrac{1}{0.41}\text{Log}(R_\delta\sqrt{\dfrac{Cf}{2}}) + 5.25\right] \end{cases}$$

The universal Law of the Wall is assumed to provide the skin-friction and C_1 from (17). The Wake function $\tilde{F}(n)$ has been defined in two-dimensions [26], and involves a rather empiric relation $n^*(\delta_1,\delta)$, in order to dissociate the shear layer from the wall in extensive separations, and to control the maximum reverse flow velocities (n^* is zero for attached flows or incipient separations, see [26]) :

$$\begin{cases} \tilde{F}_{(n)} = F\left[\dfrac{n-n^*}{1-n^*}\right], \qquad F_{(n)} = (1-n^{3/2})^2 . \end{cases}$$

$$(18) \quad \begin{cases} n^* = 0 , & 0 < \delta_1 < .44\delta \\ \delta.n^* = 4.598\,(\delta_1 - .44\delta)^2, & .44\delta < \delta_1 < .69\delta \\ \delta.n^* = 2.299\,(\delta_1 - .565\delta), & .69\delta < \delta_1 < \delta \end{cases}$$

where δ_1 is the incompressible displacement thickness. As a first approximation, the wall shear stress may be assumed colinear with the vector $\vec{\tau}$. At large Reynolds numbers, C_1 decreases, and the polar velocity profile is roughly a triangular one. In compressible flow, the density is deduced from the velocity profile and from the inviscid total enthalpy. The formulae (16) with $C_1=0$ provide at last a modelling of the upper and lower half-wake velocity profiles.

b - Equilibrium turbulence : For an equilibrium turbulence, the modelling of the entrainment E_{eq} and global dissipation ϕ_{eq} are assumed to be deduced from the velocity profile using an algebraic closure, for example a mixing length scaled on the outer shear layer $\tilde{\delta}$, defined as $\tilde{\delta} = (1-n^*).\delta$, smaller than δ in case of extensive separation. A suggested analytical approximation [26] is then, for small cross-flows: :

$$(19) \quad \begin{cases} \tilde{u}_w = 1 - \dfrac{2.22}{1+1.22\,n^*}\dfrac{\delta_1}{\delta} \\ E_{eq} = \left[0.053\,(1-\tilde{u}_w) - 0.182\dfrac{Cf}{\sqrt{|Cf|}}\right]\lambda_1\lambda_2\lambda_3 \\ \phi_{eq} = \left[|Cf|.|\tilde{u}_w| + 0.018\,(1-\tilde{u}_w)^3\right]\lambda_1\lambda_2\lambda_3 \end{cases}$$

The dissipation ϕ_{eq} is the incompressible value. The correction terms are such that $\lambda_1=\lambda_2=\lambda_3=1$ for usual equilibrium boundary layers, $\lambda_1=2$ and $Cf=o$ for usual wakes. The terms λ_2 and λ_3 are curvature and external turbulence corrections see [26].

c - Out-of-equilibrium turbulence : For out-of-equilibrium turbulences (at the present time in two-dimensions), the departure of the shear stress $\tau(x,z)$ from the equilibrium model $\tau_{eq}(x,z)$ is assumed to be only x-dependent. Then :

$$(20) \qquad \begin{bmatrix} \tau(x,z) \\ E(x) \\ \phi(x) \end{bmatrix} = \frac{\tilde{\tau}(x)}{\tilde{\tau}_{eq}(x)} \begin{bmatrix} \tau_{eq}(x,z) \\ E_{eq}(x) \\ \phi_{eq}(x) \end{bmatrix}$$

Denoting $k(x,z)$ the turbulent kinetic energy, $\varepsilon(x,z)$ the unit-dissipation of energy, the additional unknown ratio $\tilde{\tau}/\tilde{\tau}_{eq}(x)$ in (20) is at last approximated with a simplified model $\tilde{\tau}(x),\tilde{k}(x),\tilde{\varepsilon}(x)$ for the turbulent transport of averadged quantities across the shear layer. The Launder, Rodi modelling has been simplified [26] into the integral equations:

$$(22) \qquad \begin{cases} \dfrac{D\tilde{k}}{Dt} = \dfrac{\tilde{\tau}}{\tilde{\tau}_{eq}} \phi_{eq} \dfrac{q^3}{2} - \tilde{\varepsilon}\,\tilde{\delta} \\[2ex] \dfrac{D\tilde{\tau}}{Dt} = 1.5 \dfrac{\tilde{\varepsilon}\,\tilde{\delta}}{\tilde{k}} \left[\left(\dfrac{\tilde{k}}{\tilde{k}_{eq}}\right)^2 \dfrac{\tilde{\varepsilon}_{eq}}{\tilde{\varepsilon}} \tilde{\tau}_{eq} - \tilde{\tau} \right] \\[2ex] \dfrac{\tilde{\varepsilon}}{\tilde{\varepsilon}_{eq}} = \left(\dfrac{\tilde{k}}{\tilde{k}_{eq}}\right)^{3/2} \end{cases}$$

with the following connection of the equilibrium levels $\tilde{k}_{eq}$, $\tilde{\varepsilon}_{eq}$, $\tilde{\tau}_{eq}$ to the mean velocity profiles :

$$(23) \qquad \begin{cases} \tilde{\delta}\,\tilde{\varepsilon}_{eq} = 0.5 \; \phi_{eq} \; q^3 \\[1ex] \tilde{k}_{eq} = \left[0.045 \; \lambda_1\lambda_2\lambda_3(1-\tilde{u}_w) \; \phi_{eq}\right]^{1/2} . q^2 \\[1ex] \tilde{\tau}_{eq} = \lambda_1\lambda_2\lambda_3 \left[0.09 \; (1-\tilde{u}_w)\right]^2 . q^2 \end{cases}$$

3.4 - Viscous Influence Function :

With the previous turbulent closure, the defect integral equations reduce to a quasi-linear differential system, along each boundary layer or half-wake.

When for example the inviscid flow is potential, the density $\rho(x,y,o)$ is connected to the velocity $q(x,y,o)$, and the equations may be summarized :

$$(24) \qquad A^{\pm}\cdot\begin{bmatrix} \frac{\partial\delta}{\partial x} \\ \delta\frac{\partial C_2}{\partial x} \\ \delta\frac{\partial C_3}{\partial x} \\ \frac{\delta}{q}\frac{\partial q}{\partial x} \\ \delta\frac{\partial\tilde{k}}{\partial x} \\ \delta\frac{\partial\tilde{\tau}}{\partial x} \end{bmatrix}_{(x,y,o\pm)} = \begin{bmatrix} \frac{\rho w}{\rho q} - \frac{\bar{\rho}\bar{w}}{\rho q} \\ \frac{Cfx}{2} - \frac{\bar{\rho}\bar{w}}{\rho q}(\frac{u}{q} - \frac{\bar{u}}{q}) \\ \frac{Cfy}{2} - \frac{\bar{\rho}\bar{w}}{\rho q}(\frac{v}{q} - \frac{\bar{v}}{q}) \\ \frac{\tilde{\tau}}{\tilde{\tau}_{eq}}E_{eq} + \frac{\bar{\rho}\bar{w}}{\rho q} \\ b_5 \\ b_6 \end{bmatrix}_{(x,y,o\pm)}$$

The full matrix $A^{\pm}(C_2,C_3,q,\delta,\tilde{k},\tilde{\tau})$ and the terms b_5 , b_6 are very complex and highly non-linear. The non-linearity induces the separation singularity if $q(x,y,o\pm)$ is prescribed. The singularity is removed in inverse problems, such that $[w/q](x,y,o)$ is prescribed in case of a boundary layer, or such that $[w/q(x,y,o+) - w/q(x,y,o-)]$ and $[q(x,y,o+) - q(x,y,o-)]$ are prescribed in case of a wake, see [24,29].

The viscous unknowns $[\delta,C_2,C_3,\tilde{k},\tilde{\tau}]^{\pm}(x,y)$ and the inviscid unknowns $[q,w](x,y,o\pm)$ may be integrated using the system (24) , and the inviscid-field equations which connect $q(x,y,o\pm)$ and $w(x,y,o\pm)$. The different numerical techniques used to intgrate the viscous part (24) are detailed in [26].

The elimination of the viscous unknowns in the quasi linear system (24) provides at last the viscous relations between $q(x,y,o\pm)$ and $w(x,y,o\pm)$, which control the boundary conditions of the pseudo-inviscid flow, see [24,25,29]. In case of a boundary layer, this viscous influence function on the inviscid field may be written :

$$(25) \qquad \left\{ \frac{w}{q} = B^{*}\frac{1}{q}\frac{\partial q}{\partial x} + C \right\}_{(x,y,o)}$$

The non-symmetrical contributions of $w(x,y,o)$ and $q(x,y,o)$ in the formula (25) are connected with the "thin-layer" degeneration of the viscous problem at high Reynolds number, when observed with the inviscid scale. The influence function is highly non-linear because of $B^{*}(x,y)$. This term is negative in attached flow and positive in separated zones, grows to infinity at separation and reattachment stations. It is gauged with the small scale $\delta(x)$, which generates the subcritical branching at viscous-inviscid interaction, and the viscous upstream influence recovery at supersonic stations. In case of a wake, the viscous influence function is a similar rank-2 system :

$$(26) \qquad L^{*}_{(x,y)}\frac{\partial}{\partial x}\begin{bmatrix} q(x,y,o+) \\ q(x,y,o-) \end{bmatrix} = \begin{bmatrix} w/q(x,y,o+) \\ w/q(x,y,o-) \end{bmatrix} + C$$

4 - VISCOUS-INVISCID COUPLING VIA EXPLICIT RELAXATION TECHNIQUES .

Numerically, a direct solution of the coupled discrete problem, including the inviscid solver and the boundary conditions (24), is generally out of scope. Iterative relaxation techniques, alternating uncoupled viscous and inviscid solutions, are here considered. In addition, the relaxation step for coupling is assumed to be performed with "explicit" schemes at each coupling node on the inviscid boundary (in so far as the coupling relaxation is considered as a time-like process). Such explicit coupling methods provide the flexibility to interchange easily the inviscid solver, but require to solve the difficult problem of the stability control.

4.1 - Fixed point explicit techniques :

The simplest of the explicit coupling techniques are only based on viscous and inviscid alternate problems, whose boundary conditions generate a fixed point iteration for the coupling unknowns at cycle n , either $w^n(x,y,o)$, or $q^n(x,y,o)$.

For example, denoting $w^n_{i,k}$ and $q^n_{i,k}$ the discrete inviscid field at y=cst, assuming k=1 at z=o, a direct inviscid problem may be solved with $w^n_{i,1}$ prescribed. The resulting $q^n_{i,1}$ distribution may the be prescribed to a direct viscous problem which solves (24) and provides a new guess of w(x,y,o), denoted $\tilde{w}_i$. The direct fixed point iteration :

$$(27) \qquad w^{n+1}_{i,1} = \tilde{w}_i \left[w^n_{j,1} \right] \qquad i,j=\{1,\dots I\}$$

extrapolates the usual boundary layer theory, which however does not require to converge (27), but only to stop the calculation when n=2 (weak interaction). The convergence of (27) provides the recovery of the strong interaction effects as well as the consistency the triple-deck theory. The convergence is however only insured if all the eigenvalues of the Jacobian matrix $\partial\tilde{w}_i/\partial w^n_{j,1}$ are inside the unit circle in the complex plane [25]. The coupling operator of the direct iteration $\tilde{w}_i(w^n_{j,1})$, which is a product of the direct inviscid operator $q^n_{m,k}(w^n_{j,1})$ with the direct viscous operator $\tilde{w}_i(q^n_{m,1})$, generally does not satisfy this property (next section), when a consistent discretization of $\partial q^n/\partial x(x,y,o)$ is used in the viscous solution of systems (24) or (25)(26). It was proved for example [25] that a numerically consistent solution of (27) becomes always unstable when decreasing the mesh size, even if the flow is simply a flat plate boundary layer flow. A solution of (27) with the addition of a uniform and empirical underrelaxation does not remove the difficulty if the grid becomes fine enough. The sensitivity increases when the flow trends to separate.

The "uncoupled" solution of the viscous system (24), at a given coupling iteration, requires also to solve an inverse viscous problem at separation. It is then possible to solve an inverse inviscid problem with $q^n_{i,1}$-prescribed, the resulting $w^n_{i,1}$ distribution being prescribed to the inverse viscous problem, which generates a new guess of $q(x,y,o)$, denoted $\tilde{q}_i$. The inverse fixed point iteration :

$$q^{n+1}_{i,1} = \tilde{q}_i \left[q^n_{j,1} \right] \qquad i,j=\{1,\ldots 1\} \tag{28}$$

is the product of the inverse inviscid operator $w^n_{m,k}(q^n_{j,1})$ with the inverse viscous-one $\tilde{q}_i(w^n_{m,1})$. The Jacobian matrices of (27) and (28) are inverse, so that (28) is very well conditioned near separation. It was proved however [25] that the stability of the inverse iteration (28), which is not dependent on the local mesh size, is sensitive on the contrary to the length of the computational domain along x (instability at low waves numbers), and that an additional underrelaxation do not necessarily remove the instabilities. A zonal switch between the direct (27) and inverse (28) techniques do not fully avoid this difficulty, and requires an inviscid solver whose complexity may be prohibitive, especially in transonic flows.

4.2 - Stability control - Local relaxation :

Explicit relaxation techniques for coupling have been developed to control the fixed point iterations. The difficulty, even within a linear stability analysis, is that the coupling problem at z=o involves the full inviscid operator, which is z-dependent. An estimate, at least approximate, of the linear stability of the inviscid operator with respect to the boundary conditions at z=o is then necessary.

Satisfactory results for the stability control have been obtained [25] with the rather crude approximation of small linear perturbations around a locally uniform inviscid flow, with $w \ll u$. Denoting with prime the small perturbation of the local solution, the inviscid perturbation operator is then relevant of Prandtl-Glauert-like equations :

$$\left\{ \begin{array}{l} \left[1 - M^2 \left(\frac{u}{q}\right)^2 \right] \varphi_{xx} + \varphi_{zz} = 0 \\ u\,u' = q\,q' \\ u' = u\,\varphi_x \quad , \quad v' = o \quad , \quad w' = u\,\varphi_z \,. \end{array} \right. \tag{29}$$

Particular solutions of (29), that vanish if $z \rightarrow +\infty$:

$$\varphi = \varphi_o \cdot e^{\,i\alpha \left[x + z \sqrt{\left(M \frac{u}{q}\right)^2 - 1} \right]} \tag{30}$$

provide the answer for a Fourier analysis of the inviscid amplification, with respect to the boundary conditions at z=o . Denoting $\hat{u}'$, $\hat{w}'$, $\hat{q}'$ the perturbations at wave number α, we get from (29)(30) :

$$(31)\qquad \begin{cases} \hat{w}'_{(x,y,o)} = i\ \beta\ \hat{q}'_{(x,y,o)} \\ \beta = \sqrt{(\frac{q}{u})^2 - M^2}\ , \qquad i^2 = -1\ . \end{cases}$$

The linear amplification of the viscous operator at wave number α is directly deduced from the viscous influence function (25) :

$$(32)\qquad \hat{w}'_{(x,y,o)} = i\ \alpha\ B^*\ \hat{q}'_{(x,y,o)}$$

Joining (31) and (32), we deduce the amplifications μ_D, μ_I of the direct and inverse fixed-point iterations (27)(28), at wave number α and node i (μ_D,μ_I are complex numbers) :

$$(32)\qquad \begin{cases} \hat{w}'^{\,n+1}_{i,1} = \mu_D\ (x_i)\ .\ \hat{w}'^{\,n}_{i,1} \\ \hat{q}'^{\,n+1}_{i,1} = \mu_I\ (x_i)\ .\ \hat{q}'^{\,n}_{i,1} \\ \mu_D = \dfrac{\alpha\ B^*}{\beta}\ , \qquad \mu_D . \mu_I = 1\ . \end{cases}$$

From the amplification coefficients, an optimal relaxation coefficient may be computed at each node. We get then the "Direct" relaxation technique, overrelaxation-like, whose stability at wave number α_{max} insures the stability at all wave numbers[25] :

$$(33)\qquad \begin{cases} w^{n+1}_{i,1} = \omega^D_i . \left[\tilde{w}_i - w^n_{i,1} \right] \\ \omega^D_i = \omega . \omega_{opt}\left[\mu_D(x_i,\alpha_{max})\right]\ , \qquad \alpha_{max} = \dfrac{\pi}{\Delta x_i} \\ \omega_{opt}(\mu) = \dfrac{1-R}{(1-R)^2 + I^2}\ , \qquad \mu = R + iI \\ 0 < \omega < 2 \end{cases}$$

From the formulae (24) to (33), we notice that the local relaxation rate ω^D_i decreases when the mesh size Δx_i decreases, when the thicknes δ of the viscous layer increases, and also near separation points ($B^* \to \infty$), or sonic-like stations ($\beta \to o$).

In a similar way, an "Inverse" relaxation technique may be defined [25] with the local control :

$$(34)\qquad \begin{cases} q^{n+1}_{i,1} = \omega^I_i \left[\tilde{q}_i - q^n_{i,1} \right] \\ \omega^I_i = \omega . \omega_{opt}\left[\mu_I(x_i,\alpha_{min})\right] \\ 0 < \omega < 2 \end{cases}$$

4.3 - Semi-Inverse relaxation :

In order to remove the difficulty of α_{min} in (34), and also to couple a direct inviscid problem with an inverse viscous solution, the more complex "Semi-Inverse" relaxation has been deduced from the previous stability analysis.

The viscous solution is still marched in free stream direction, usually with an implicit integration scheme, using a $w^n_{i,1}$-prescribed inverse method, which provides a first guess of the inviscid velocity $\tilde{q}_i$. The usual $w^n_{i,1}$-prescribed inviscid solution provides a second guess $q^n_{i,1}$, which is compared to $\tilde{q}_i$, in order to iteratively correct $w^n_{i,1}$.

In the Fourier's space, at wave number α , we get from (31) that the inverse iteration (28) may be simulated, at least for small perturbations, with the following Semi-inverse correction :

$$(35) \qquad \left\{ w^{n+1} - w^n = \frac{\beta}{\alpha}\left[\frac{\partial \tilde{q}}{\partial x} - \frac{\partial q^n}{\partial x}\right] = -\frac{i\beta}{\alpha^2}\left[\frac{\partial^2 \tilde{q}}{\partial x^2} - \frac{\partial^2 q^n}{\partial x^2}\right]\right\}_{i,1}$$

With some additional analysis [25,26] to include the optimal relaxation control of the inverse amplification μ_1 , a stable and overrelaxation-like correction may be written :

$$(36) \qquad \left\{ \begin{aligned} & w^{n+1}_{i,1} - w^n_{i,1} = \omega.\omega^A_i \frac{|\beta|_i}{\alpha}\left[\frac{q^n}{\tilde{q}}\frac{\partial \tilde{q}}{\partial x} - \frac{\partial q^n}{\partial x}\right]_{i,1} \\ & \qquad + \omega.\omega^B_i \frac{|\beta|_i}{\alpha^2}\left[\frac{q^n}{\tilde{q}}\frac{\partial^2 \tilde{q}}{\partial x^2} - \frac{\partial^2 q^n}{\partial x^2}\right]_{i,1} \\ & \omega^C_i = \omega_{opt}\left[\mu_1(x_i,\alpha)\right] , \qquad \alpha = \alpha_{max}(x_i) \\ & 0 < \omega < 2 \end{aligned} \right.$$

As originally suggested [25], we use presently $\omega^A_i = \omega^C_i$ and $\omega^B_i = 0$ at subsonic nodes, $\omega^A_i = 0$ and $\omega^B_i = \omega^C_i$ at supersonic nodes.

At supersonic nodes, one another selection may be $\omega^A_i \cdot \alpha_{max} = -|\beta| \cdot \omega^C_i$ and $\omega^B_i = \omega^C_i$, as more recently suggested by Wigton, Holt [37]. This new selection changes the previous relative optimum of the relaxation control into an absolute optimum, but only at wave number α_{max}. Similar effects are obtained here by discretizing $\partial^2 q^n/\partial x^2$ downwind of $\partial^2 \tilde{q}/\partial x^2$ in (36), and $\partial q^n/\partial x$ downwind of $\partial \tilde{q}/\partial x$ within the direct-relaxation zones. This makes it easier to recover the viscous upstream influence inside supersonic regions. At subsonic nodes at last, it may be shown [6,29] that the semi-inverse method (36) is linearly equivalent to the more recent technique of Carter [18,19], which is approximately a prime integral of (36).

A switch between the Direct relaxation (33) and the

Semi-Inverse relaxation (36) along the contour z=o may be used easily. The switching is controlled by the viscous calculation only, which solves an inverse problem when the incompressible shape parameter H_i of the viscous layer is high ($H_i > 1.8$ for a turbulent flow).

4.4 - Dissymmetrical wakes coupling :

The displacement effect of symmetrical wakes calculations may be coupled using Direct, Inverse, or Semi-inverse relaxation techniques which are very similar to (33)(36), excepted that the normal-velocity jump in the inviscid flow along the cut has here to be considered.

The wake calculation may also be dissymmetrical, either to reach the correct positioning of the cut with respect to the closure of section 3.3 (minimal velocity locus), or because of an inviscid dissymmetry $q(x,y,o+) \neq q(x,y,o-)$, or to get an asymmetrical estimate of the half-wakes thicknesses (turbulent modelling, wake curvature effect).

The mutual interference of the upper-lower viscous-inviscid interactions changes mainly the Viscous influence function into (26), and then the viscous amplification :

$$(37) \quad \begin{cases} L^{*} . \dfrac{\partial}{\partial x} \begin{bmatrix} \hat{q}'(x,y,o+) \\ \hat{q}'(x,y,o-) \end{bmatrix} = \begin{bmatrix} \hat{w}'(x,y,o+) \\ \hat{w}'(x,y,o-) \end{bmatrix} \\ \dfrac{\partial}{\partial x} \left[\hat{q}'(x,y,o+) - \hat{q}'(x,y,o-) \right] = 0 \end{cases}$$

The second relation of (37) assumes that the coupling of the displacement effect is solved with a frozen curvature effect, and provides the upper-lower viscous amplification at wave number α :

$$(38) \quad \hat{w}'_{(x,y,o\pm)} = i \, \alpha \, B^{*\pm} \, \hat{q}'_{(x,y,o\pm)}$$

The coefficients $B^{*\pm}(x_i)$ may be used directly to compute the upper and lower values $\mu_I^{\pm}$, $\omega_i^{A\pm}$, $\omega_i^{B\pm}$ in the Semi-Inverse relaxation, from (32)(36). The formulae (36) may then be applied to the upper and lower sides [26].

The coefficients $B^{*\pm}(x_i)$ provide also upper and lower values $\mu_D^{\pm}$, $\omega_i^{D\pm}$, from (32)(33). The Direct relaxation (33) has however to be changed [26] into :

$$(39) \quad \begin{cases} (w_{i,1}^{n+1} - w_{i,1}^{n})^{\pm} = \omega . \omega_i^{D\pm} . \left[(\tilde{w}_i - w_{i,1}^{n}) + q_{i,1}^{n} . R_i \right]^{\pm} \\ R_i^{\pm} = \pm \dfrac{1}{2} \left[\left(\dfrac{\tilde{w}_i - w_{i,1}^{n}}{q_{i,1}^{n}} \right)^{-} - \left(\dfrac{\tilde{w}_i - w_{i,1}^{n}}{q_{i,1}^{n}} \right)^{+} \right] \\ 0 < \omega < 2 \end{cases}$$

The residual $R_i^{\pm}$ in relation (39) is the local angular error of the wake geometry positioning at cycle n .

The exact positioning of the wake geometry may be achieved iteratively. After converging the displacement coupling with (36)(39) along an approximate geometry, the residual $R_1^{\pm}$ provides an improved update of the wake geometry, which may be converged within a few cycles, without noticeable stability problem [26].

The interacting curvature effect of the wake (7), even with the modelling (14)(15), is more complex than in the boundary layer theory, because the local averadged-curvatures $K^*(x,y,o\pm)$ are no more issued from a preliminary and purely inviscid calculation, but require the "induced" curvature of the coupled inviscid flow (strong interaction). In the same way as for the displacement coupling, an iterative update of the pressure jump across the wake-cut generates a fixed point iteration for the induced curvatures $K^*(x,y,o\pm)$ and raises a stability problem. The stability control depends on the mesh size,see[26], and may become impossible on a fine grid, even with an underrelaxation technique. In such cases, we use here an inconsistent smoothing when discretizing the local curvatures $K^*(x,y,o\pm)$.

5 - INTERACTED POTENTIAL SOLVERS : SEPARATED AIRFOILS, TRAILING-EDGES, WINGS .

5.1 - Two-dimensional subsonic-transonic solver :

In two-dimensions ($u \simeq q$, $v = o$), the interacted solution of the full set of Defect Integral equations (9) (10)(11)(12)(15) and turbulent transport equations (20)(22),

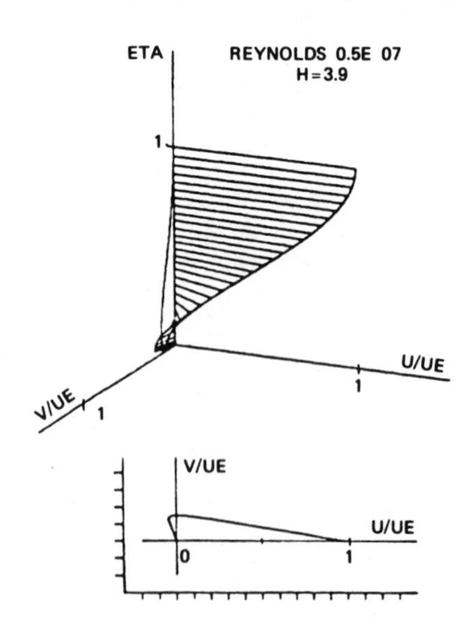

Fig. 1 — Three-dimensional modelling of turbulent velocity profiles.

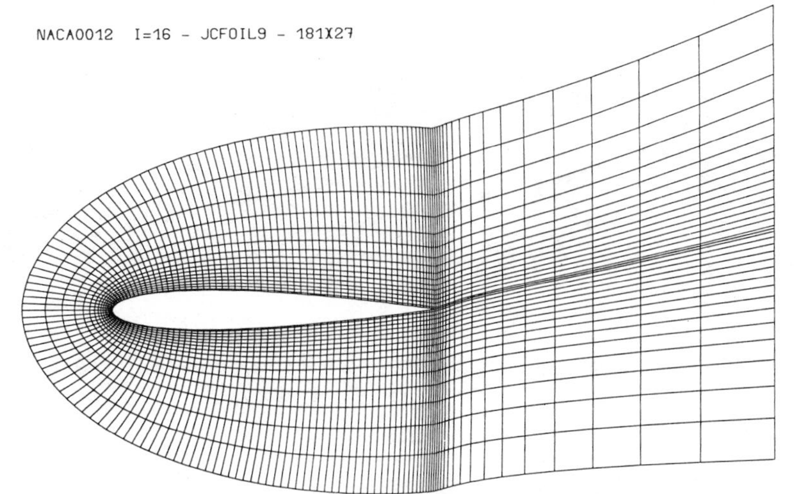

Fig. 2 — Mesh with adaptation to the viscous effects (NACA 0012, $M = 0.116$, $\alpha = 16°$, $R = 2 \times 10^6$).

closed with the relations (13)(16)(17)(18)(19)(23) deduced from the turbulent velocity profiles modelling of Fig. 1 , provides a numerical solver for lifting compressible flows, including separations[26]. Within the areas of laminar or transitional boundary layers, the entrainment equation is replaced by the integral kinetic energy equation of the mean flow, and a smooth intermittency function is used to weight the turbulent closure relations with the laminar closure of the similar solutions, see [26].

The strong viscous-inviscid interaction calculates then the laminar-turbulent boundary layers with smooth transitions, as well as the fully dissymmetrical wakes, and takes into account the displacement, curvature and wake-positioning effects. Separation bubbles, as well as rather extensive trailing-edge separations may be resolved, approximately until the maximum lift. The approximation of the full potential equation is assumed for the inviscid field, which is presently solved with the finite differences relaxation technique of Chattot[38], in conservative or non-conservative form. An adaptable C-grid has been defined for the wake-cut positioning, and for the mesh-clustering at the stations of shock-waves or viscous interactions[26]. The Direct and Semi-Inverse relaxation of the viscous coupling is achieved simultaneously with the potential relaxation (each 5 to 10 cycles), as well as the updates of the mesh (each 200 cycles for example). The non-linearities of the viscous equations are solved only each 20 to 40 cycles. Two or three successive grids may be used.

The Fig. 2 shows the behavior of the mesh after converging, in case of a NACA0012 airfoil at low speed and 16° of incidence (maximum lift). The Fig. 3 displays the stream-

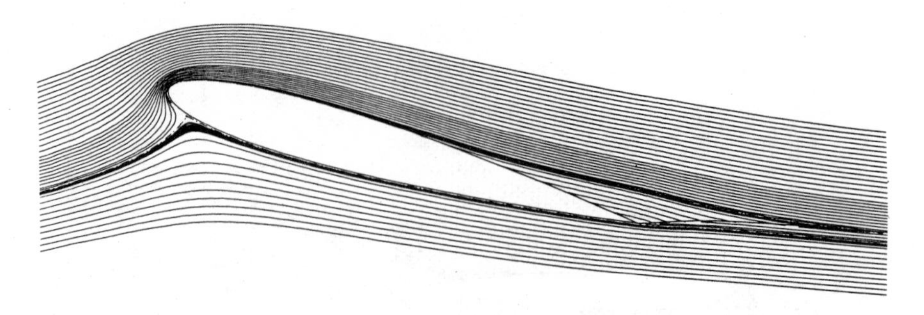

Fig. 3 – Streamlines of the interacted inviscid predictor (NACA 0012, $M = 0.116$, $\alpha = 16°$, $R = 2 \times 10^6$).

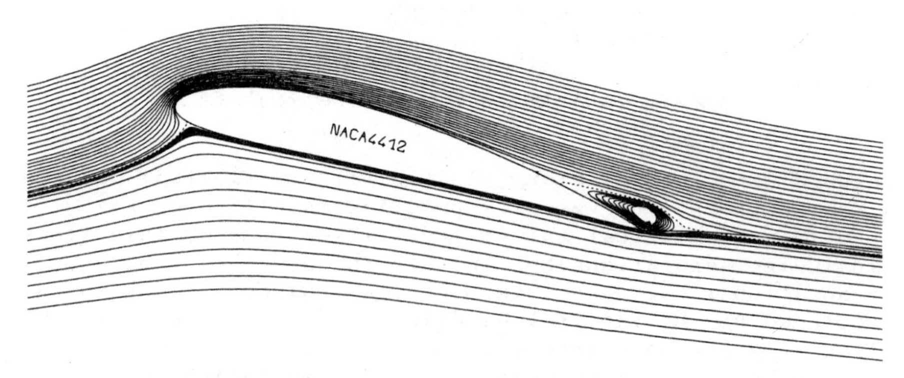

Fig. 4 – Transonic solver at low speed. Viscous-inviscid composite solution (NACA 4412, $V = 20$ m/s, $\alpha = 13.6°$, $R = 1.5 \times 10^6$).

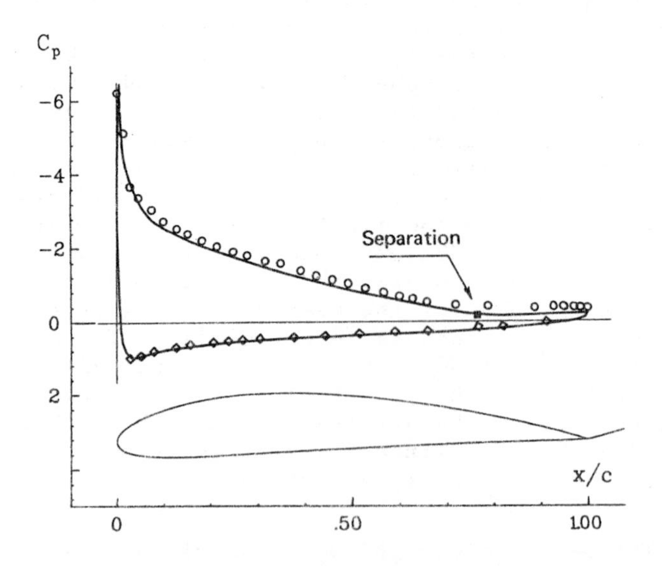

Fig. 5 – Pressure distribution (NACA 4412, $V = 20$ m/s, $\alpha = 13.6°$, $R = 1.5 \times 10^6$).

lines of the interacted inviscid predictor, and shows the large influence of the wall-transpiration effects on the far field. The dotted streamline, where the streamfunction is zero, visualizes indirectly the exact displacement body. The analytical description (16) of the turbulent velocity profiles used for the viscous corrector problem, which describes the full wake dissymmetry, may be combined with the inviscid predictor to visualize the overall composite solution which is computed. The Fig. 4 gives an example of noticeable trailing-edge separation on the NACA4412 section, at low speed and maximum lift. The flow is the "Stalled Airfoil" test-case of the 1980-81-Stanford conference. The corresponding pressure distribution is shown on Fig. 5 .

The Fig. 6 and 7 display typical transonic results on supercritical airfoils, for the test-cases of Stanford-1981 . The rear-loading, with possibly separation, is generally well predicted. The usual grids are however too coarse to resolve correctly the zone of shock wave-boundary layer interaction, the step size being frequently equal to the overall interaction zone. On such grids, the shock-induced separation is not accessible to the calculation. Without separation, a simple overprediction of the compression, restricted to the shock area, is the most often observed when

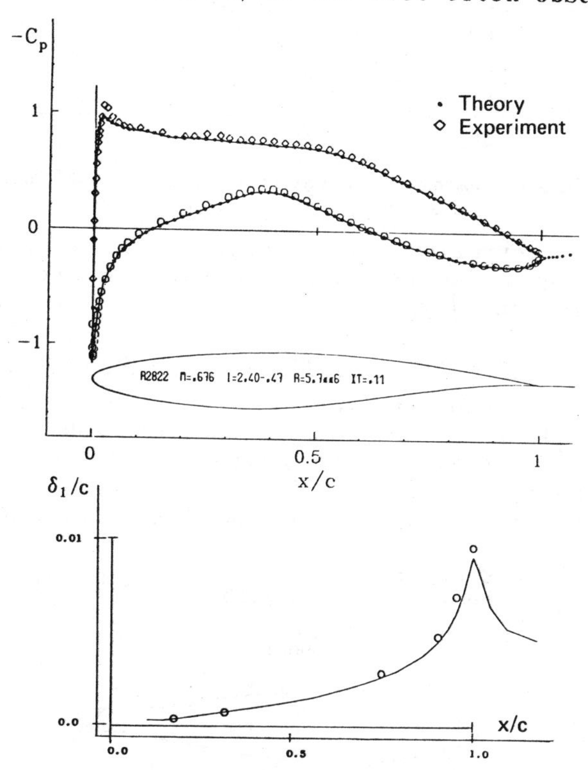

Fig. 6 — Transonic solver at subcritical conditions (RAE 2822, $M = 0.676$, $\alpha = 2.40°$, $R = 5.7 \times 10^6$).

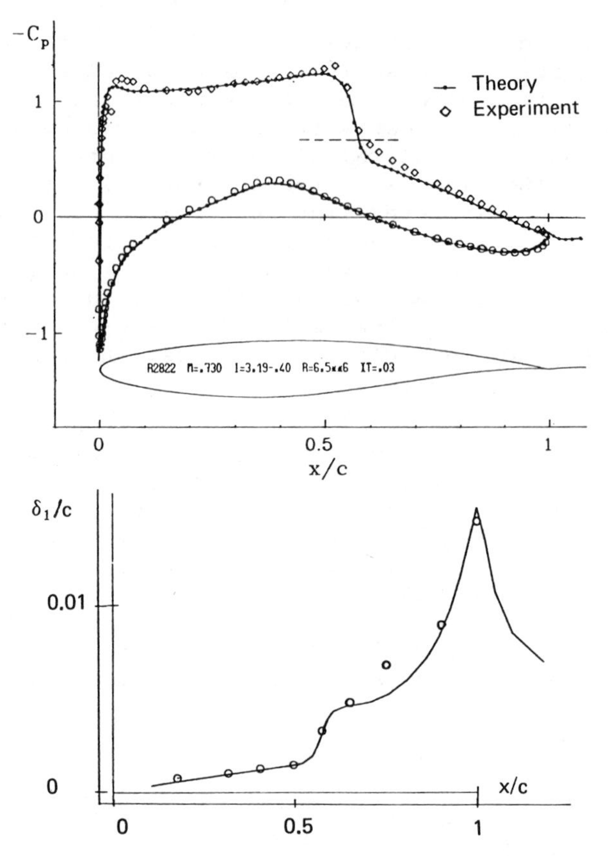

Fig. 7 — Transonic solver at supercritical conditions. (RAE 2822, $M = 0.730$, $\alpha = 3.19°$, $R = 6.5 \times 10^6$).

using the conservative solver, Fig. 7 . Although empirical local techniques, see [26] , may improve this point, the overall solution is yet rather satisfactory in attached flow.

5.2 - Base flows - Spoiler flap modelling :

The transonic solver has been extended with an additional modelling for the base or spoiler problems [28], which are assumed to be similar to backward-facing steps, as shown on the geometry (G), Fig. 8 .

For simplicity, the potential calculation is achie-

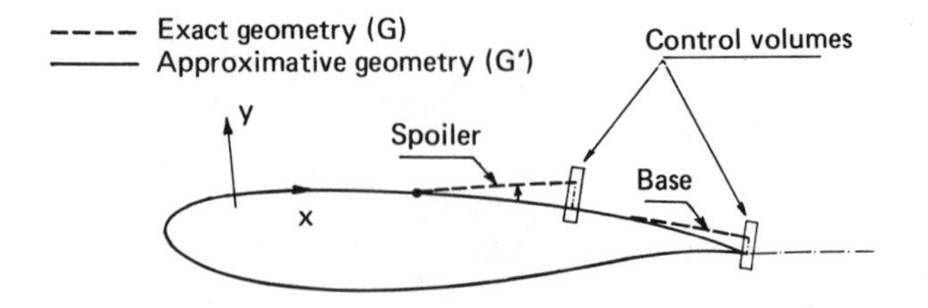

Fig. 8 — Base flows or spoilers modelling.

ved with an approximate geometry (G') where the discontinuities have been removed. The angular error at each station between (G) and (G') is then cancelled with an equivalent blowing at the wall of (G'), which is superimposed to the transpiration velocity deduced from the strong viscous coupling. Global control volumes, at last, estimate the sudden increases of the viscous layer thicknesses at the stations of discontinuities, which do or do not induce separation, Fig. 8 . The reattachment process is fully computed, either along the wall, or inside the wake (rear stagnation point).

The Fig. 9,10,11 show the results obtained on the thick RA16SC1-airfoil, at slightly supercritical conditions. The negative lift, which is only due to the 10°-deflexion of a spoiler flap on the upper surface, is rather well predicted. The flow is near to separate at the spoiler hinge. An extensive separation is calculated downstream of the spoiler. An acceptable plateau pressure is predicted, as well as the shift of the reattachment toward a stagnation

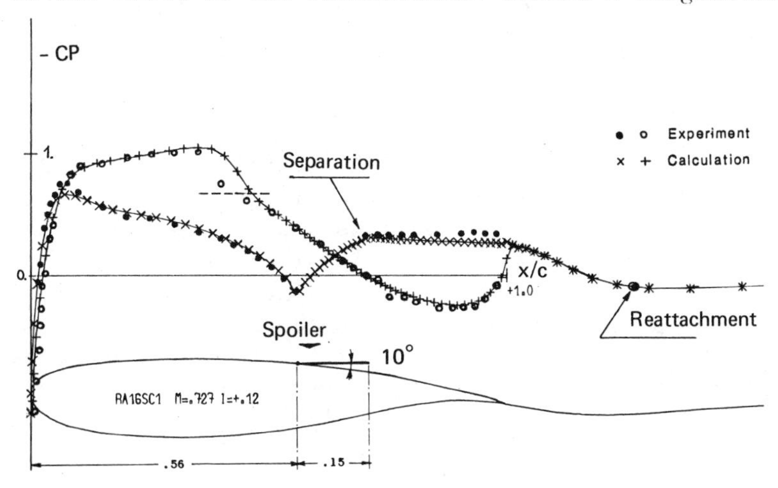

Fig. 9 – Transonic solver with spoiler flap on RA 16SC1 supercritical airfoil ($M=0.727$, $\alpha=0°$, $R=4.2 \times 10^6$, $\delta_{SPOILER}=10°$).

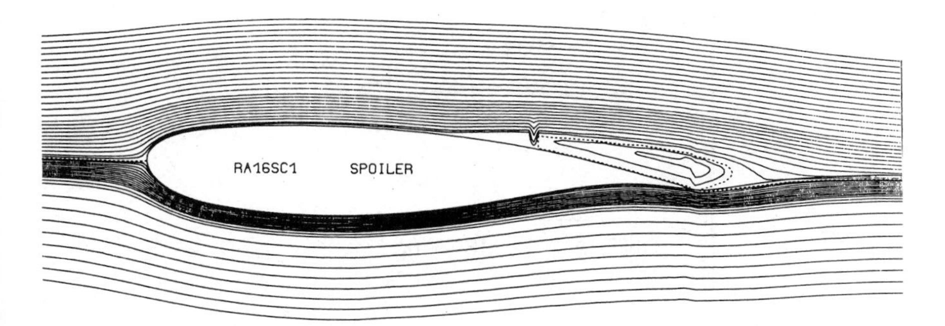

Fig. 10 – Transonic solver with spoiler flap. Composite solution streamlines (RA 16SC1, $M=0.727$, $\alpha=0°$, $R=4.2 \times 10^6$, $\delta_{SPOILER}=10°$).

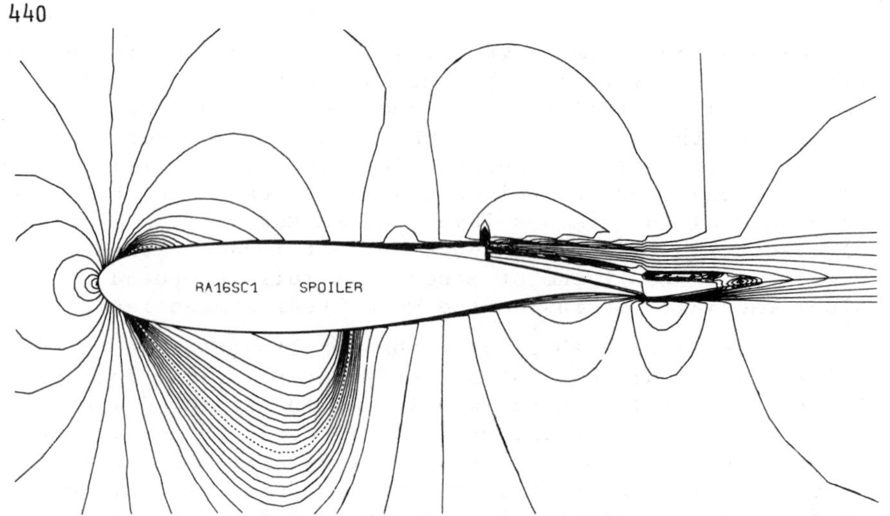

Fig. 11 — Transonic solver with spoiler flap. Iso-Mach contours (RA 16SC1, $M = 0.727$, $\alpha = 0°$, $R = 4.2 \times 10^6$, $\delta_{SPOILER} = 10°$).

point inside the thick wake. The streamlines, Fig. 10 , and the Mach number contours, Fig. 11 , visualize the extensive separation which is calculated.

5.3 - Quasi-3D flows - Infinite swept wings :

The quasi-threedimensional viscous calculation and coupling techniques of sections 3 and 4 have been used to compute transonic flows on infinite swept wings.

Presently, the method is restricted to fully turbulent flows, assumes the equilibrium entrainment closure, and does not resolve the full dissymmetry ot the viscous wake. The integral method uses the full three-dimensional modelling of the viscous velocity profiles (16)(17)(18). The inverse integral method may compute quasi-threedimensional separations

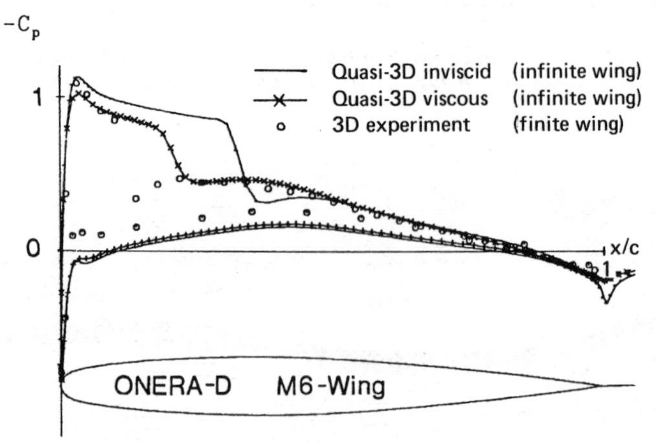

Fig. 12 — Quasi-three-dimensional transonic solver. Infinite swept wing (ONERA-D, $M = 0.84$, $\alpha = 2°$, $R = 2.5 \times 10^6$, $\varphi = 30°$).

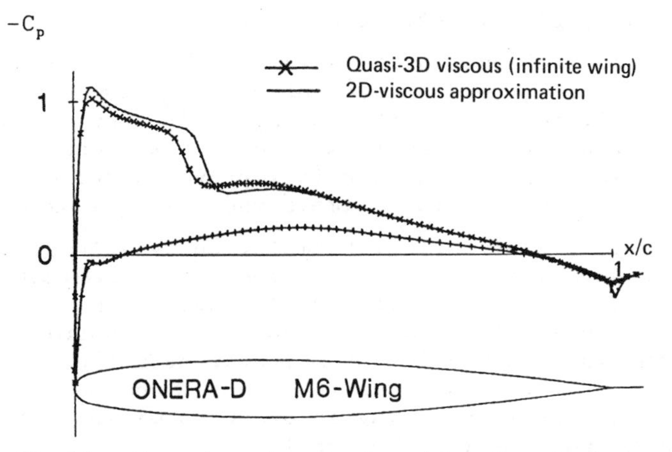

Fig. 13 — Comparison of two- and quasi-three-dimensional viscous calculation. Infinite swept wing (ONERA-D, M = 0.84, $\alpha = 2°$, $R = 2.5 \times 10^6$, $\varphi = 30°$).

marching in the x-direction, either when only the wall streamlines accumulate, or when a full reverse flow domain is developed. The overall numerical technique is however still a quasi-two-dimensional one in the (x,z) plane.

The Fig. 12 and 13 show the results which have been obtained by Blaise[39] for a supercritical wing without separation. If the comparison with the experiment is not significant, because the experimental wing has a finite span, the Fig. 13 indicates that a noticeable difference is present when the two-dimensional and quasi-three-dimensional calculated solutions are compared.

6 - INTERACTED EULER SOLVER : TRANSONIC SHOCK WAVE-BOUNDARY LAYER INTERACTION .

The limitations of the potential assumption may require to use a full Euler solver for the inviscid predictor, for example in order to calculate the internal viscous flows inside a choked transonic channel.

6.1 - Additional boundary conditions :

The Defect Formulation of the NS-equations (4)(5)(6)(7) requires that the pseudo-inviscid flow has a non-zero normal velocity at the walls or at wake-lines. Within the potential approximation, the problem is fully determined if the normal velocity is prescribed. When the steady Euler equations are solved, the normal velocity is still a sufficient boundary condition in the zones where the viscous displacement induces a suction at the wall. Inside the zones of injection, on the contrary, additional boundary conditions have to be prescribed in order to define the entropy and enthalpy, according to the theory of characteristics. This difficulty was noticed for example by Johnston, Sockol [40],

Murman, Bussing [41], Whitfield et al. [21].

The problem, which is equivalent to define a stratified flow below the displacement body, is however only an apparent difficulty. On the contrary, it provides an increased number of free parameters to choose the non-unique viscous-inviscid splitting, and to improve the viscous defect calculation when a rather simple modelling is assumed.

If the full Defect Formulation is solved, the additional boundary conditions are almost indeterminate, any stratification of the inviscid flow being taken into account in the viscous defect calculation. On the contrary, if a modelling is involved in the momentum defect calculation, the entropy at the wall has to be optimized in connection with the momentum modelling in order to minimize the overall approximations. At first order for example, equations (8), the entropy at the wall has to minimize the departure between the viscous and inviscid pressure fields, so that $\bar{p}(x,y,z) \simeq p(x,y,z)$. With the Defect Integral method, an additional criterion of this optimization is the velocity profiles modelling.

First results have been obtained here without a careful optimization, assuming a zero normal derivative of the inviscid entropy at the wall, in analogy with the potential approximation.

6.2 - Shock wave - boundary layer interaction :

The computation method, developed jointly with Blaise, involves exactly the same numerical techniques as the airfoil solver, for the viscous calculation and the coupling. The steady Defect Integral equations, with the two-equations entrainment model, are coupled using the same Direct and Semi-Inverse relaxation (33)(36) as in potential inviscid flow. The Euler solver of the coupling iteration is performed with cycles of 100 time-steps of the explicit pseudo-unsteady method (with constant total enthalpy) of Veuillot, Viviand [42].

The calculated configuration is a symmetrical plane transonic nozzle, providing a choked flow with a compression shock-wave at Mach number $M = 1.3$. The flow has been expe-

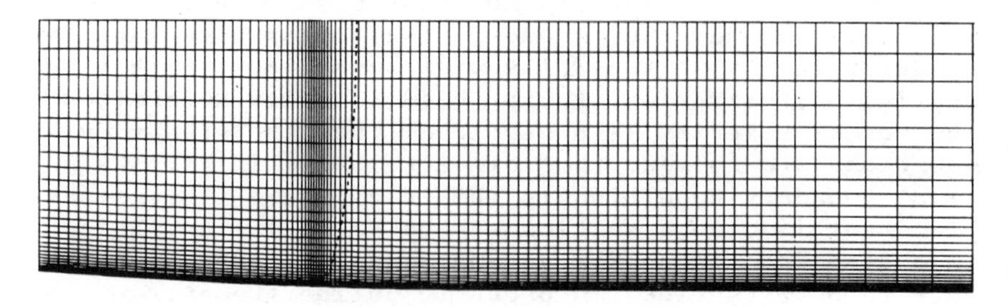

Fig. 14 — Half-mesh for calculation of a shock wave-boundary layer interaction in a transonic channel.

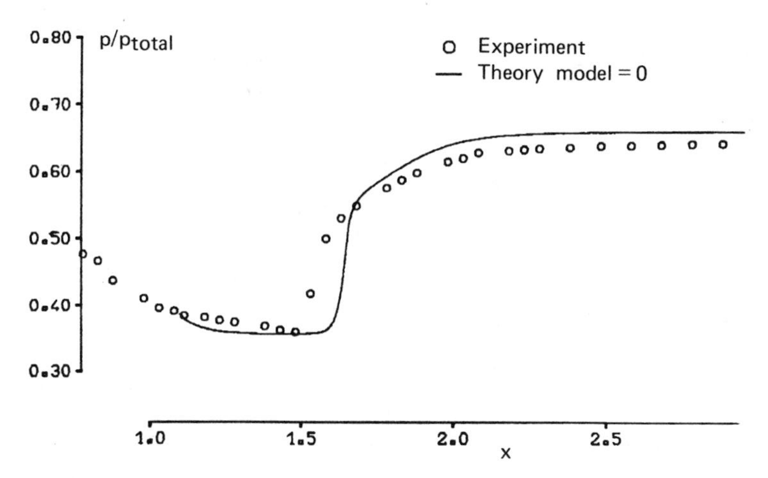

Fig. 15 — Pressure distribution at the wall in the shock wave-boundary layer interaction zone (calculation with equilibrium entrainment).

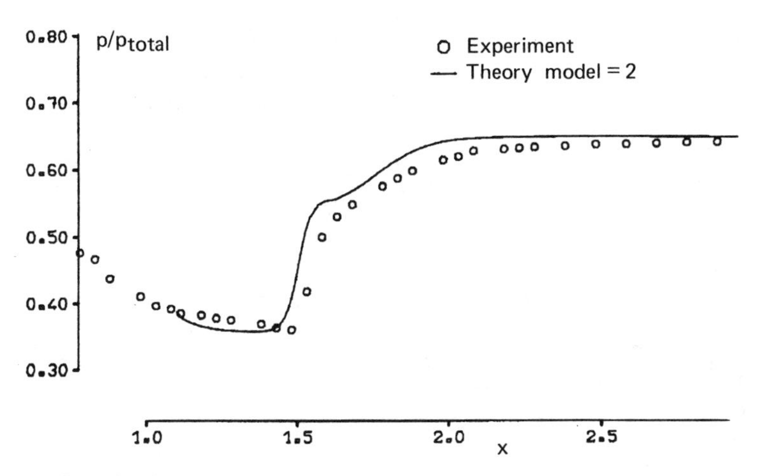

Fig. 16 — Pressure distribution at the wall in the shock wave-boundary layer interaction zone (calculation with a two-equations model for the entrainment).

rimented by Delery[43], and has been used also in the NS-calculations presented by Cambier et al. in a following chapter. The Fig. 14 shows the lower half-domain of the calculation, and the grid clustering necessary to resolve the beginning of the turbulent shock wave-boundary layer interaction. Due the low divergence of the channel and to the boundary layers on the side-walls in the experimental device, the back-pressure of the calculation is not exactly known. It has been tuned imperfectly, with trial and error computations, to get approximately the position of the shock-wave.

The Fig. 15 and 16 show two calculated pressure distributions at the wall. The first one assumes an equilibrium turbulent entrainment, Fig. 15 . The second-one involves the

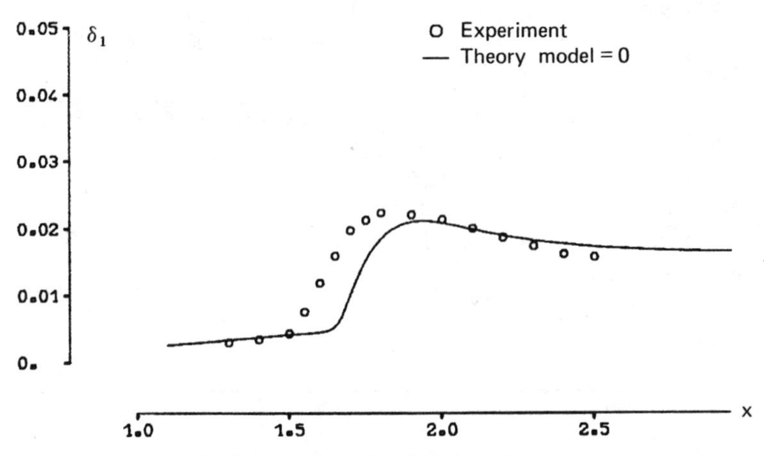

Fig. 17 – Displacement thickness.

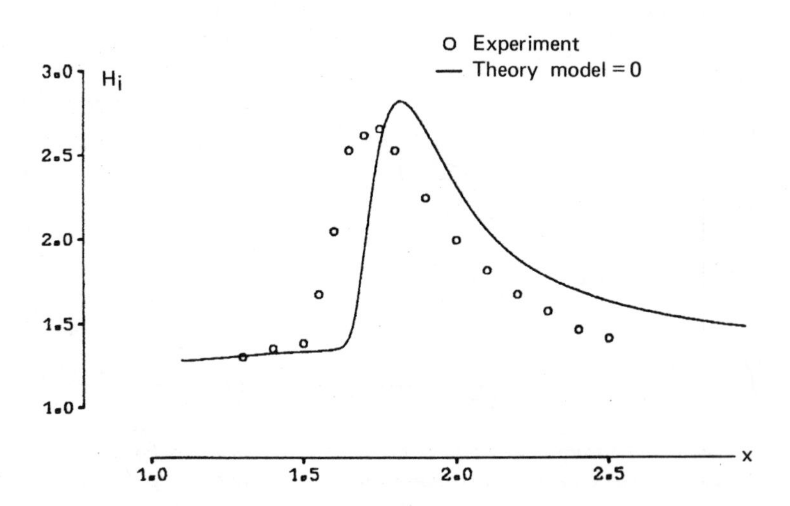

Fig. 18 – Incompressible shape parameter.

Fig. 19 – Pressure contours. Interacted pseudo-inviscid flow.

Fig. 20 – Iso-Mach contours. Composite viscous solution.

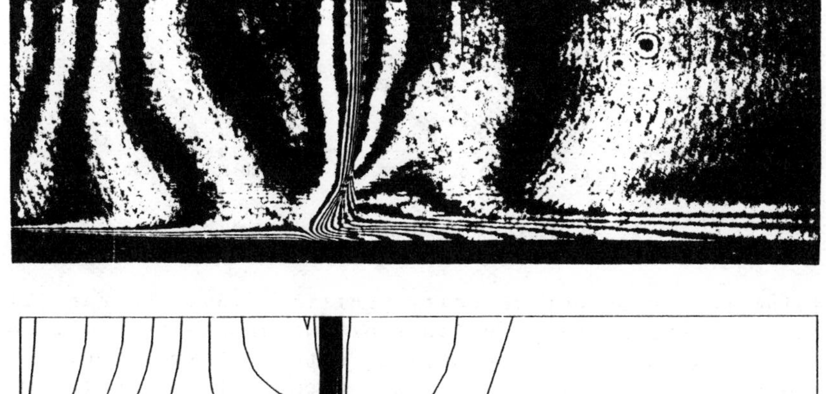

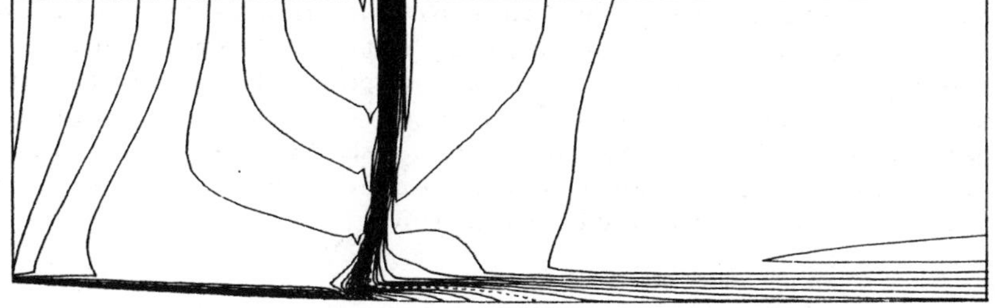

Fig. 21 – Density contours of the composite solution and comparison with an experimental interferogram.

the full two-equations model to calculate the out-of-equilibrium entrainment, Fig. 16 , which appears to improve the agreement with experiment. The Fig. 17 and 18 compare with measurements the integral quantities, displacement thickness and incompressible shape parameter (equilibrium entrainment).

The Fig. 19 shows the pressure contours in the interacted pseudo-inviscid field, which are also a first order approximation of the viscous pressure contours. It may be noticed especially on Fig. 19 that the pattern of the shock wave is curved near the viscous layer, which is connected to the viscous upstream influence recovery. It may be noticed also the spreading of the viscous pressure distribution at the wall, and the non-zero normal pressure gradient insithe viscous layer at the foot of the shock-wave. The Fig. 20 visualizes the Mach-contours of the composite solution, in-

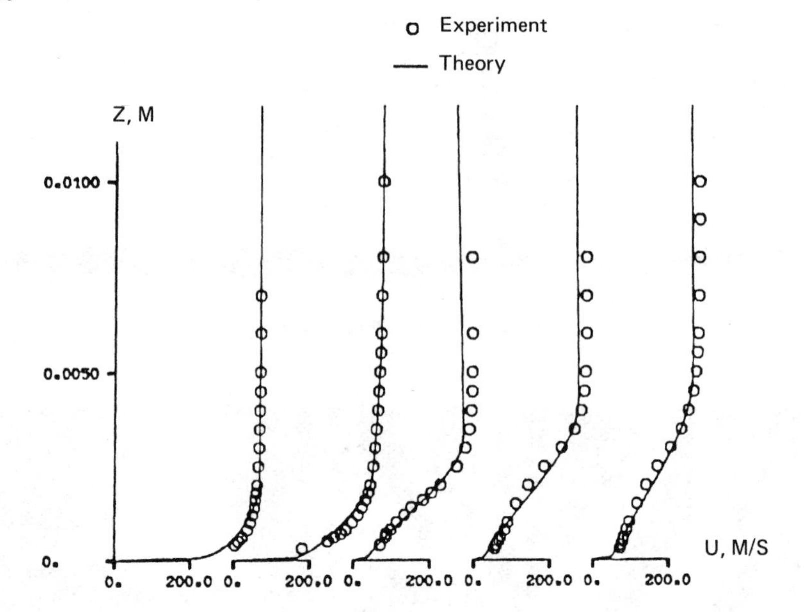

Fig. 22 – Viscous velocity profiles in the interaction zone.

volving the turbulent velocity profiles (16). The Fig. 21 shows that the density contours of the composite numerical solution exhibit a strong analogy with the experimental interferogram. At last the Fig. 22 compares the calculated viscous velocity profiles with measurements. The incipient separation at M = 1.30 is however so small that the reverse flow bubble is imperceptible, excepted for the shape parameter or skin-friction distributions.

7 - CONCLUSION .

Recent advances in Viscous-Inviscid numerical methods (VIS) provide now Indirect solvers, which are capable of an overall calculation (or simulation) in many of the complex flows at high Reynolds numbers, in external or internal aerodynamics. These VIS-methods are mainly developed within three complementary areas :

(i) the Zonal (or subdomains) Coupling techniques, based on zonal viscous solutions with a patching of the boundary conditions, which are capable to incorporate the direct NS-solvers;

(ii) the Interacting Thin-Layers solvers, with overlaying viscous and inviscid domains, based on some approximation of the "Defect Formulation", which are capable to use the boundary layer techniques, and to provide solutions at low cost for separations and shock wave-boundary layer interactions;

(iii) the indirect Composite NS-solvers, based on a splitting into interacting numerical sub-problems, viscous-like and inviscid-like, with overlaying computational domains, which are capable to include the numerical techniques of (ii) and to solve the NS-problem with relaxation methods.

Among the solvers of low cost at level (ii), very competitive results may be derived from strongly Interacting Defect Integral equations, based on the modelling of 3D-turbulent velocity profiles and of an out-of-equilibrium turbulent entrainment. This model is capable to compute complex flows with multiple viscous-inviscid interactions, including the problems of separated flows, trailing-edges, shock wave-boundary layer interactions, base flows or wakes. This model provides the numerical solvers where the discretization is minimal, and makes then accessible the computation of the maximal flow complexity.

Results have been presented in 2D or quasi-3D flows, using either potential or Euler inviscid solvers. The coupling technique is an extension of the previously suggested [25] Direct and Semi-Inverse relaxations, including an automatic stability control. Recent calculations with an Euler solver have been shown to require additional boundary conditions, and promise very encouraging results in transonic channels flows or turbulent shock wave-boundary layer interactions, including small separations.

REFERENCES

1. METHA U., LOMAX H. - Reynolds-averadged Navier-Stokes computations of transonic flows. The state of the art. Proceed. Symp. Transonic Perspective, ed. D. Nixon, Progress in Astronautics, vol. 81 (1982).

2. Mc CROSKEY W. - Unsteady airfoils. Annual Review of Fluid Mech., 14, p 285-311 (1982).

3. LE BALLEUR J.C., PEYRET R., VIVIAND H.- Numerical studies in high Reynolds number aerodynamics. Computers and Fluids vol. 8, n°1, p 1-30 (March 1980).

4. VIVIAND H. - Traitement des problèmes d'interaction fluide parfait-fluide visqueux en écoulement bidimensionnel compressible à partir des équations de Navier-Stokes. AGARD-LS-94 (1978).

5. LE BALLEUR J.C. - Calculs couplés visqueux-non visqueux incluant décollements et ondes de choc en écoulement bidimensionnel. AGARD-LS-94 (1978).

6. LE BALLEUR J.C. - Calcul des écoulements à forte interaction visqueuse au moyen de méthodes de couplage. AGARD-CP-291, General Introduction, paper 1, (1981).

7. LOCK R.C., FIRMIN M.C.P. - Survey of techniques for estimating viscous effects in external aerodynamics. RAE Tech. Memo Aero 1900 (April 1981).

8. McDONALD H., BRILEY W.R. - A survey of recent work on interacted boundary layer theory for flow with separation Proceed. 2nd Symp. Numerical and Physical Aspects of Aerodynamic flows, Calif. St. Univ. Long-Beach, Springer (1983)

9. DODGE P.R. - Numerical method for 2D and 3D flows. AIAA Journal, vol. 15, n° 7, p 961-965 (1977).

10. COSNER R.R. - Relaxation solutions for viscous transonic flow about fighter-type Forebodies and Afterbodies. AIAA Paper 82-0252, Orlando (1982).

11. KHOSLA P.K., RUBIN S.G. - A composite velocity procedure for the compressible Navier-Stokes equations - AIAA Paper n° 82-0099 (Jan 1982).

12. MAHGOUB H.E.H., BRADSHAW P. - Calculation of turbulent-inviscid flow interactions with large normal pressure gradients. AIAA J., vol.17, n°10, p 1025-1029 (Oct 1979).

13. MELNIK R.E., CHOW R., MEAD H.R. - Theory of viscous transonic flow over airfoils at high Reynolds number. AIAA-Paper n° 77-680, Albukerque (1977).

14. MELNIK R.E. - Turbulent interactions on airfoils at transonic speeds. Recent developments. AGARD-CP-291, Lecture 10, Colorado-Springs (1981).

15. MELNIK R.E., MEAD H.R., JAMESON A. - A multi-grid method for the computation of viscid/inviscid interaction on airfoils. AIAA Paper 83-0234, Reno (Jan 1983).

16. LOCK R.C. - A review of methods for predicting viscous effects on aerofoils at transonic speeds. AGARD-CP-291, Lecture 2, (1981).

17. WERLE M.J., VATSA V.N. - New method for supersonic boundary layers. AIAA Journal, vol.12, n°11, (1974).

18. CARTER J.E. - Viscous inviscid interaction analysis of transonic turbulent separated flow. AIAA Paper 81-1241, Palo Alto, (1981).

19. CARTER J.E., VATSA V.N. - Analysis of separated boundary layers flow. Proceed. 8th Int. Conf. Num. Meth. in Fluid Dyna., Aachen, Lecture Notes in Physics, Springer-Verlag (1982).

20. WAI J.C., YOSHIHARA H. - Planar transonic airfoils computations with viscous interactions. AGARD-CP-291, paper n° 9, (1981).

21. WHITFIELD D.L., THOMAS J.J., JAMESON A., SCHMIDT W. - Computation of transonic viscous-inviscid interacting flow. Proceed. 2nd Symp. Numerical and Physical Aspects of Aerodynamic flows, California State Univ. Long-Beach, Springer Verlag to appear, (Jan 1983).

22. GHOSE S., KLINE S.J. - The computation of optimum pressure recovery in two-dimensional diffusers. Journal of Fluid Engin., vol. 100, p 419-426, (1978).

23. VELDMAN A.E.P. - New, quasi-simutaneous method to calculate interacting boundary layers. AIAA Journal, vol. 19, n° 1, p 79-85, (1981).

24. LE BALLEUR J.C. - Couplage visqueux-non visqueux : Analyse du problème incluant décollements et ondes de choc. La Recherche Aerosp. 1977-6, p 349-358, (Nov 1977), or English transl. ESA-TT-476.

25. LE BALLEUR J.C. - Couplage visqueux-non visqueux : Méthode numérique et applications aux écoulements bidimensionnels transsoniques et supersoniques. La Recherche Aerosp. 1978-2, p 67-76, (March 1978), or English transl. ESA-TT-496.

26. LE BALLEUR J.C. - Strong matching method for computing transonic viscous flows including wakes and separations. Lifting airfoils. La Recherche Aerosp. 1981-3, p 21-45, English and French editions, (May 1981).

27. LE BALLEUR J.C. - Numerical viscid-inviscid interaction in steady and unsteady flows. Proceed. 2nd Symp. Numerical and Physical Aspects of Aerodynamic Flows, Californ. State Univ. Long-Beach, (Jan 1983), or ONERA-TP-1983-8, or Springer-Verlag 1983 to appear (reduced version).

28. LE BALLEUR J.C. - Calculation method for transonic separated flows over airfoils including spoilers effects. Proceed. 8th Int. Conf. Num. Meth. in Fluid Dyn., Aachen, Lecture Notes in Physics, Springer-Verlag (1982), or ONERA-TP 1982-66.

29. LE BALLEUR J.C. - Viscid-inviscid coupling calculations for two- and three-dimensional flows. Lectures series 1982-04, Von Karman Institute, Computational Fluids Dynamics, Belgium, (March 1982).

30. CROCCO L., LEES L. - A mixing theory for the interaction between dissipative flows and nearly isentropic streams. Journ. Aero. Sci., vol. 19, n°10, (1952).

31. KLINE S.J. - Editor, Proceed. 1980-81-AFOSR-HTTM-STANFORD Conference on Complex Turbulent Flows, Stanford Univ., California (Oct 1981)

32. BRUNE G.W., RUBBERT P.E., FORESTER C.K. - The analysis of flow fields with separation by numerical matching. AGARD-CP-168 (1975).

33. SEGINER A., ROSE W.C. - An approximate calculation of strong interaction over a transonic airfoil including strong shock-induced flow separation. AIAA Paper 76-333, San Diego, (1976).

34. CAMBIER L., GHAZZI W., VEUILLOT J.P., VIVIAND H. - Une approche par domaines pout le calcul d'écoulements compressibles. Proceed. 5th Symp. Computing Meth. in Appl. Sci. and Engineering, Glowinsky and Lions ed., North-Holland, INRIA (1982).

35. WILMOTH R.G., DASH S.M. - A viscous-inviscid interaction model of jet entrainment. AGARD-CP-291, paper 13, (1981).

36. LAZAREFF M.,LE BALLEUR J.C. - Computation of three-dimensional viscous flows on transonic wings via boundary layer-inviscid flow interaction. La Recherche Aerosp. 1983-3, English and French editions, (May 1983).

37. WIGTON L.B.,HOLT M. - Viscous inviscid interaction in transonic flow. Proceed. 5th Comput. Fluid Dyn. Conf., AIAA Paper n° 81-1003, Palo Alto (1981).

38. CHATTOT J.J.,COULOMBEIX C.,TOME C. - Calcul d'écoulements transsoniques autour d'ailes. La Recherche Aerosp. , n° 1978-4, p 143-159, (July 1978), English trans. ESA-TT-561

39. BLAISE D. - Mise en oeuvre et developpement d'une méthode de couplage fort pour le calcul d'écoulements autour d'ailes en flèche infinies. Doct. Thesis 3°Cycle, Univ. of Lille 1, (1982), Note Tech. ONERA to be published.

40. JOHNSTON W., SOCKOL P. - A viscous-inviscid interactive procedure for rotational flow in cascades of two-dimensional airfoils of arbitrary shape. AIAA Paper 83-0256, Reno, (Jan 1983).

41. MURMAN E.M., BUSSING R.A. - On the coupling of boundary layer and Euler equation solutions. Proceed. 2nd Symp. Numerical and Physical Aspects of Aerodynamic Flows, California State Univ. Long-Beach, (Jan 1983), Springer Verlag to appear.

42. VIVIAND H.,VEUILLOT J.P. - Méthodes Pseudo-Instationnaires pour le calcul d'écoulements transsoniques. ONERA Publication n° 1978-4 , (1978).

43. DELERY J. - Experimental investigation of turbulence properties in transonic shock/boundary layer interaction AIAA Journal, vol. 21, n° 2, p 180-185, (Feb 1983).

TRANSONIC VISCOUS-INVISCID INTERACTION USING EULER AND INVERSE BOUNDARY-LAYER EQUATIONS*

David L. Whitfield
Department of Aerospace Engineering
Mississippi State University
Mississippi State, MS 39762

James L. Thomas
NASA Langley Research Center
Hampton, VA 23665

1. INTRODUCTION

A natural approach to the computation of turbulent transonic viscous-inviscid interaction is to use the Reynolds averaged Navier-Stokes equations as a global solution procedure. Shock/boundary-layer interactions and separated flow can be handled in a straightforward manner. However, there are two major drawbacks to using the Navier-Stokes equations: (1) computer resources and (2) turbulence modeling. The limitation having to do with computer resources involves computer speed and memory. Kutler[1] provides a current summary of the computing power of current and proposed mainframes. According to the present trend [1] (without the Numerical Aerodynamic Simulator (NAS) NASA has proposed), it could be one to two decades before computers will attain the power required to perform computations about realistic aerodynamic configurations using the Navier-Stokes equations. The limitation having to do with turbulence modeling is not unique to the Navier-Stokes equations because any solution methodology must contend with this problem in one way or another.

An alternate approach to the computation of turbulent transonic viscous-inviscid interaction is to use a zonal solution method as opposed to a global solution method. Zonal solution methodology involves using different equation sets and possibly different numerical algorithms for various regions of the flow. An advantage of zonal methods is that simpler equations are used in regions where permissible and, consequently, computational time and storage requirements can be reduced. A disadvantage of zonal methods is that information must be exchanged at zonal boundaries or zonal overlap regions, which can cause convergence and stability problems.

*This research was sponsored by the NASA Langley Research Center, Hampton, VA 23665. Dr. F. Thames was the grant monitor.

One type of zonal method that has been used for viscous-inviscid interacton is to use an inviscid flow solution method for the flow away from a body which is patched or matched to a viscous flow solution method for the flow near the body. Most previous investigations have used a potential flow solution method for the inviscid region, and an attached boundary-layer solution method for the viscous region. Many of these previous investigations have been reviewed by Lock, [2] Melnik, [3] and Le Balleur. [4]

With regard to inviscid flow solution methods, there is a growing tendency in the technical community to solve the Euler equations as opposed to the potential flow equations. The reason for this is that Euler solutions contain much more information than potential flow solutions. Many things are naturally accounted for by the Euler equations that must be either specified or explicitly modeled in the potential flow equations, such as total pressure changes, slip surfaces, entropy gradients, etc.

Transonic flows with shock waves of moderate strength, typically encountered near the design point of long-range transport aircraft, have been accurately and efficiently predicted by potential flow methods. For flows with local Mach numbers exceeding 1.2, it is expected that the entropy and vorticity production effects associated with stronger shocks will become important. Such flows are of practical interest, being encountered at off-design points and in the upper transonic range. An accurate viscous-inviscid interaction method for such flows requires the use of an inviscid method without irrotational isentropic assumptions.

A disadvantage of using the Euler equations in place of the potential flow equations would be computer resources. However, Kutler[1] points out that the computer time required to obtain steady-state Euler solutions is becoming competitive with full-potential solutions.

Perhaps the first truly three-dimensional time-dependent Euler code was developed by Jacocks and Kneile[5] (not published until 1981). Solutions of many difficult three-dimensional problems are given in Ref. [5], and many additional three-dimensional subsonic, transonic, and supersonic flow problems continue to be solved using improved versions of this code (all referred to as ARO-1). The general applicability of the ARO-1 code to practical problems, and the obvious fact that its utility could be further extended by including viscous effects, motivated the work on which this Chapter is based.

With regard to viscous flow solution methods for interactive computations, there is a growing tendency in the technical community to solve the inverse boundary-layer equations (where pressure distribution is obtained as part of the solution) as opposed to the direct boundary-layer equations (where the pressure distribution is prescribed). The reason for using the inverse method is that the Goldstein singularity in the vicinity of separation is avoided for steady flow. An inverse method, however, requires that something other than the pressure distribution be prescribed, such as (although not restricted to) the boundary-layer displacement thickness distribution. It then becomes necessary to use a rational means of prescribing the displacement thickness distribution, for example, to insure convergence.

Inverse boundary-layer computational methods are either finite difference or integral methods. The particular choice is, to a large measure, personal preference rather than one approach demonstrating consistently superior results over the other. It is interesting to note that a result of the 1968 Stanford Conference on the computation of turbulent boundary layers[6] was that finite difference methods showed more promise than integral methods for the future. However, 13 years later at the 1981 Stanford Conference,[7] integral methods made a strong showing. A result of the 1982 conference was that integral methods should not be shelved but rather should receive further development.

Two of the reasons that probably contribute to a favorable view of integral methods are: (1) turbulence modeling and (2) the type of flow regimes considered at the 1968 and 1981 conferences. While it is true that integral methods require some sort of turbulence modeling, they also require some sort of velocity profile modeling. This velocity profile modeling damps somewhat the sensitivity of turbulence modeling on the solution of some integral methods compared to the sensitivity of turbulence modeling on the solution of finite difference methods. Because it is easier to model velocity profiles from many complex turbulent flows than it is to model turbulence, integral methods have held their own and, in many cases, provided even better results than finite difference methods.

With regard to flow regimes, had the 1968 and 1981 conferences been predominantly concerned with flows having large amounts of heat and/or mass transfer at the wall, for example, it is unlikely that integral methods would have faired very well and probably few would have been entered in the competition. However, for turbulent flow over adiabatic surfaces, integral methods have proved to be powerful tools, providing good results while requiring almost negligible computer resources.

This Chapter is concerned with the use of a zonal method for the computation of transonic viscous-inviscid interacting flow about airfoils. An Euler equation solution method is used for the inviscid portion of the flow and this method is discussed in the following section. An inverse integral compressible turbulent boundary-layer solution method is used for the viscous portion of the flow and a discussion of this method follows the section on the Euler method. Because the Euler equations are used, more is required with regard to matching the inviscid and viscous solutions than would be required had the potential equations been used, and a section is devoted to this topic. The Chapter concludes with numerical results, including numerical and experimental comparisons.

2. EULER EQUATION METHOD

The explicit, finite volume formulation developed by Jameson, Schmidt, and Turkel[8], elements of which are discussed elsewhere in this volume, is used to solve the two-dimensional Euler equations. A finite volume spatial discretization, corresponding to a second-order, central difference finite difference formulation, is applied directly to the integral form of the time-dependent Euler equations. The spatial discretization is decoupled from the time discretization through the method of lines and the resulting system of equations is solved using a Runge-Kutta time integration method. The Runge-Kutta method has the advantage of allowing explicit time steps greater than a Courant number of one at the expense of additional function evaluations incurred in the stages. A class of second-order Runge-Kutta schemes is given in Ref. [9] which allows Courant numbers from 2 to 4.5 by varying the number of stages from 3 to 6. The increase in work associated with the increase in stages reaches a point of diminishing returns and numerical experiments have indicated that the four-stage scheme with a Courant number of 2.8 is a reasonable compromise.

As with any central difference scheme applied to the Euler equations, some dissipation is required to suppress the appearance of oscillations in the neighborhood of shock waves and stagnation points. The dissipative terms are composed of a blend of second and fourth difference terms which are formally of third order in smooth regions of the flow and locally first order in the region of a shock[8]. To minimize the computational work, the dissipative terms are only evaluated in the first stage of each time step.

Since only steady state solutions are of interest, convergence to steady state is accelerated by: (1) using a local time step determined by the local Courant number limitation and (2) addition of a forcing term that depends on the difference between local and freestream values of total enthalpy. The fourth difference dissipative terms are required to eliminate nonlinear instabilities when accelerating convergence using a local time step.

Far field boundary conditions are based on a characteristic combination of variables and the wall pressure is determined from the normal momentum equation. The lift is sensitive to the location of the far field boundary and, for the calculations presented in Section 5, the boundary location was extended until no appreciable change was found, corresponding to a location approximately 50 chords from the airfoil. Extensions to the normal momentum relations developed by Rizzi[10] were made to include the effect of porosity at the wall when simulating viscous effects as discussed in Section 4. The effects of the modifications on the resulting pressure distribution were small, as shown in Ref. [9], but were included in the computations to be presented in Section 5. Several alternate methods of determining wall pressure by extrapolation were tried but the normal momentum equation approach was least sensitive to the grid spacing away from the airfoil.

3. INVERSE BOUNDARY-LAYER METHOD

Various finite difference and integral inverse boundary-layer methods have been developed and most of these are pointed out and briefly discussed in Ref. [11]. With regard to integral methods, which are of interest here, Kuhn and Neilsen[12] have developed a moment of momentum method which involves the momentum integral equation and an equation obtained by integrating in the direction normal to the surface the product of the coordinate normal to the surface and the differential momentum equation. Another inverse integral method is the lag-entrainment method of East, Smith, and Merryman.[13] Methods of the lag-entrainment type were used, for example, by Le Balleur,[14] Melnik, Mead, and Jameson,[15] and Vatsa and Carter.[16] The inverse integral method used here involves the momentum and mean flow kinetic energy equation, where the latter equation is obtained by integrating in the direction normal to the surface the product of the velocity parallel to the surface and the differential momentum equation. This approach was also used by Thiede.[17] The momentum and mean flow kinetic energy inverse integral compressible turbulent boundary-layer method used here is explained in detail in Ref. [18]. Elements of the method as pertaining to separated flow calculations are discussed below along with certain recent refinements to the shape factor and turbulence modeling relations for separated flows.

The analytical velocity profile expression in Ref. [11] was developed for attached or separated flow over the entire domain $0 \leq y < \infty$, and depends on the skin friction c_f, displacement thickness shape factor $\bar{H}$, and Reynolds number based on the boundary-layer momentum thickness Re_θ. That the same profile expression can be used for flows with or without separation makes it particularly useful in inverse methods for computing flows with or without separation. The extent to which this single simple closed form velocity profile expression represents experimental data for attached, separated, or reattached turbulent flow is demonstrated in Fig. 1 of Ref.[11]. Further demonstrations of this profile family for strictly attached flow are given in Ref.[19]. Included in Fig. 2 of Ref. [19] is a case where the logarithmic portion of an experimental velocity profile was annihilated due to a strong adverse pressure gradient. The velocity profile family, however, represented this flow quite well for all values of y^+.

It is assumed that the compressible velocity profile can be described by the same expression as a constant property (incompressible) velocity profile. Compressible integral thicknesses are related to incompressible thicknesses by means of velocity-temperature relations developed in Ref.[20] for an adiabatic wall. The relations between compressible and constant property variables are given in Ref.[18] and the discussion below is limited to incompressible variables. A similar approximation is made in the formulation of Le Balleur.[14]

The foundation of the velocity profile is an expression for the inner near-wall region of the boundary layer. The inner region solution was obtained in Ref. [21] from analytical solutions of the differential momentum and turbulent kinetic energy equations in the wall region. The inner region solution, expressed in terms of the inner variable $y^+ = yu_\tau/\nu$, is combined in composite form with an outer region solution, expressed in terms of the outer variable y/θ, as:

$$\frac{u}{u_e} = \frac{u_i}{u_e} + \frac{u_o}{u_e}$$

$$= \frac{u_\tau}{u_e}\frac{s}{0.09}\tan^{-1}[0.09y^+] + [1 - \frac{u_\tau}{u_e}\frac{s\pi}{0.18}]\tanh^{\frac{1}{2}}[a(y/\theta)^b] \quad (3.1)$$

The subscripts e, i, and o denote edge, inner, and outer locations in the boundary layer, respectively, and

$$\frac{u_\tau}{u_e} = \sqrt{\left|\frac{c_f}{2}\right|} \quad (3.2)$$

$$s = \mathrm{sgn}(c_f) \quad (3.3)$$

For the purposes of defining integral thicknesses at high Reynolds number, the inner solution can be evaluated at its outer limit defined as:

$$\frac{(u_i)_\infty}{u_e} = \lim_{y^+\to\infty}\frac{u_i}{u_e} = \sqrt{\left|\frac{c_f}{2}\right|}\;\frac{s\pi}{0.18} \quad (3.4)$$

The velocity profile expression (3.1) then becomes

$$\frac{u}{u_e} = \frac{(u_i)_\infty}{u_e} + [1 - \frac{(u_i)_\infty}{u_e}]\tanh^{1/2}[a(y/\theta)^b] \quad (3.5)$$

from which the analogy with Caluser's representation of the turbulent boundary layer, Ref. [22], becomes clear in that the function of the outer solution is to return the effective slip velocity in the near wall region to freestream conditions.

Using a skin friction relation of the form

$$c_f = c_f(Re_\theta, H) \quad (3.6)$$

enables the profile to be described as a function of the parameters y/θ, H, a, b, and the local unit Reynolds number. Note several skin friction relations are reviewed in the 1981 Stanford Conference with the conclusion that little difference exists between the various correlations, at least for attached high Reynolds number flows. A skin friction correlation which extends an attached flow correlation to separated flow based on the limited experimental data available is given in Ref. [11].

The parameters a and b appearing in the velocity profile expression can be eliminated in either of two ways: (1) fitting experimental velocity profiles versus Re_θ and H at two points in the boundary layer, Ref. [11]; or (2) using the consistency conditions defined by the definitions of δ^* and θ and the velocity profile expression of Eq. (3.5) to solve nonlinear equations for a and b as a function of $(u_i)_\infty/u_e$ and H. The two approaches give very similar results for attached flows. For separated flows, the second approach has an advantage in that $(u_i)_\infty/u_e$, representing the maximum reverse flow velocity, can be correlated more easily than the skin friction, which for separated flows is very small and difficult to measure experimentally. Shown in Fig. 1 is a summary of separated flow experimental data from Refs. [23 - 25] and the present correlation which for attached flow ($H < 4$) represents Eqs. (3.4) and (3.6) evaluated at $Re_\theta = 50,000$. Note the experimental data exhibit a rather broad minimum in the region near separation which causes the Goldstein singularity to appear when integrating the direct integral equations as shown in Ref. [11]. The experimental data exhibit a maximum reverse flow velocity in separated flow of approximately 0.2. The wake velocity predicted from Coles law-of-the-wake, Ref. [6], is also shown which agrees very closely with the data in the region of separation but asymptotes to a maximum reverse flow velocity of one-third as the displacement thickness shape factor tends to infinity.

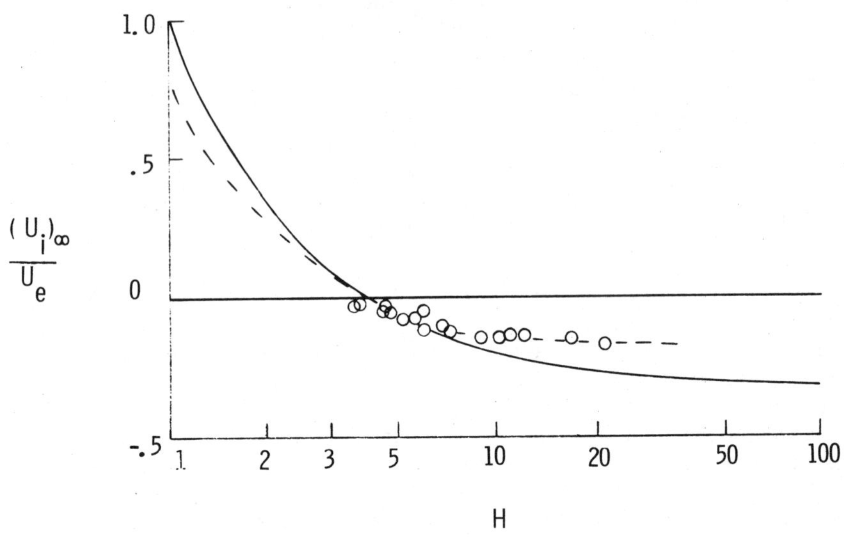

Figure 1. Correlation of maximum reverse flow velocities in separated flow with experimental data of Refs. [23-25].

Shown in Fig. 2 are the corresponding energy integral thickness H_{θ^*} correlations compared to a summary of separated flow experimental data from Refs. [23-26], where H_{θ^*} is defined as

$$H_{\theta^*} = \frac{\theta^*}{\theta} = \int_0^\infty \frac{u}{u_e}\left[1 - \left(\frac{u}{u_e}\right)^2\right] d(y/\theta) \tag{3.7}$$

The present correlation results can be represented as

$$\begin{aligned} H_{\theta^*} &= \frac{1}{9}\left(H + 7 + \frac{10}{H}\right) & H \leq 5 \\ &= \frac{1}{18}\left(H + 24 - \frac{5}{H}\right) & H > 5 \end{aligned} \tag{3.8}$$

where the expression for $H \leq 5$ corresponds to Coles law-of-the-wake result.

Le Balleur [14] has modified Cole's wake function to include a region of constant reverse flow velocity next to the

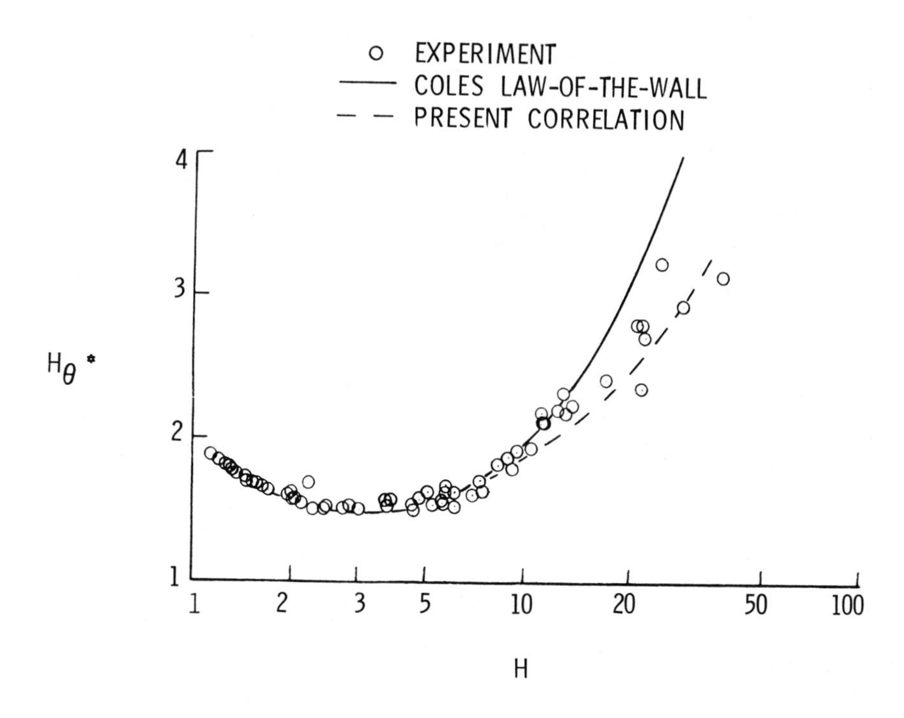

Figure 2. Correlation of energy integral thickness for attached and separated flows with experimental data of Refs.[23-26].

wall which enables the wake velocity to be correlated with experiment as in Fig. 1. Such a modification to the wake profile provides close agreement with the present correlation shown in Fig. 2.

Using Eq. (3.4) and the correlation shown in Fig. 1 enables the skin friction to be deduced in the separated flow region. The resulting skin friction asymptotes to a maximum negative value of approximately 0.0002. Analogous to similar solutions of the Falkner-Skan equation in separated zones, Ref. [13], it is expected that the skin friction and maximum reverse flow velocities will asymptote to zero in the limit as shape factor tends to infinity. As shown in Fig. 1, little experimental data exist for shape factors beyond $H > 20$ and hence the present correlation for such high shape factors is tentative. For typical transonic shock/boundary layer interactions, with small separation bubbles, the shape factor is less than 10.

The moment of momentum, lag-entrainment, and mean flow kinetic energy integral methods all involve a specification of velocity profiles and contain a term involving the shear stress distribution. However, most integral methods, for example that of Thiede, [17] do not explicitly evaluate terms involving the shear stress distribution. Rather such terms were determined empirically using experimental data. A distinguishing feature of the present method, however, is that the dissipation integral, defined for a mean flow kinetic energy method as

$$D = \frac{1}{\rho_e u_e^2} \int_0^\infty \tau \frac{\partial(u/u_e)}{\partial y} dy \tag{3.9}$$

is obtained by numerical evaluation rather than using an empirical dissipation relation. The shear stress distribution used in the integrand of Eq. (3.9) is a constant laminar plus turbulent shear stress in the region just at the wall, and a Cebeci-Smith type turbulence model in the inner and outer regions of the boundary layer.[27] The derivative of the velocity distribution in Eq. (3.9) is obtained by differentiating the analytical velocity profile expression given in Ref. [11]. The numerical results so obtained can be reproduced very closely by a two-term approximation of Eq. (3.9) as the sum of inner and outer contributions. In the inner region, the shear is constant and taken to be the wall value, and in the outer region Clauser's eddy viscosity model [22]

$$\tau = K \rho_e u_e \delta^* \frac{\partial u}{\partial y} \tag{3.10}$$

is applied where K = 0.0168. The integral thus becomes

$$D = D_i + D_o$$
$$= \left|\frac{C_f}{2} \frac{(u_i)_\infty}{u_e}\right| + KH \int_0^\infty \left[\frac{\partial(u/u_e)}{\partial(y/\theta)}\right]^2 d(y/\theta) \tag{3.11}$$

which can be approximated as

$$D = \left|\frac{C_f}{2}\right|^{3/2} \frac{\pi}{0.18} + K\left(\frac{2}{3}\frac{H-1}{H}\right)^3 \frac{\pi^2}{2} \tag{3.12}$$

The above turbulent dissipation correlation is an excellent approximation for flows near equilibrium, defined here as flows for which the shape factor is nearly constant. For flows in which the shape factor is changing rapidly, i.e., separated flows, Eq. (3.12) is not a particularly good approximation as shown in the experiment results in Refs. [23-26]. Entering and within separation, the turbulent shear stresses are below equilibrium values while near and downstream of reattachment, the turbulent shear stresses are above equilibrium values.

Several ways to account for this turbulent lag or history effect have been proposed. A relaxation equation used in several methods, and first proposed by Goldberg,[28] is

$$\frac{\delta}{D}\frac{dD}{dx} = \lambda[(D)_{EQ} - (D)] \qquad (3.13)$$

where subscript EQ denotes a value calculated by means of Eq. (3.12) and λ is an empirical constant. Green, et. al.[29] derive a lag equation for the maximum turbulent shear stress from the modeled turbulent kinetic energy equation which they convert into a lag equation for the entrainment. For equilibrium flows, a relation can be obtained between maximum turbulent shear stress and the outer part of the turbulent dissipation defined in Eq. (3.11) and is given by

$$C_{\tau} = \frac{\tau_{max}}{\rho_e u_e^2} = \left(\frac{4K}{\pi}\right)^{1/3} (D_o)^{2/3} \qquad (3.14)$$

Using the above relation, the lag equation for turbulent shear stress derived in Ref. [29] can be converted into an equation for the outer turbulent dissipation as

$$\frac{\delta}{(D_o)}\frac{d(D_o)}{dx} = \lambda[(D_o)_{EQ}^{1/3} - (D_o)^{1/3}] \qquad (3.15)$$

The term $(D_o)_{EQ}$ corresponds to the outer dissipation term in Eq. (3.11) and the constant $\lambda \simeq 5$ follows directly from Ref. [29] and Eq. (3.14). Equation (3.15) has been used in the interacted results presented below and in later sections. A pressure gradient term used in Ref. [29] has not been included in Eq. (3.15) because its influence was found to be small and in practice, for highly separated flows such as in Ref. [25] for which higher relaxation rates are expected, it has been useful to increase λ by a factor of two, i.e.,

$$\lambda = 5.0 \left[1 + \left(\frac{H-1}{H}\right)^3\right] \qquad (3.16)$$

The trends in the turbulent dissipation history for separated flows can be reproduced by either of Eq. (3.15) or (3.13). An alternative method is to include a rate equation based on integrated forms of a two-equation turbulence model as in Ref. [14].

Note that the inner near wall region is known to respond much more rapidly than the outer region, and the inner contribution to the turbulent dissipation is evaluated using equilibrium values given in Eq. (3.11).

Predictions of the inverse boundary-layer method, both with and without relaxation of the turbulent dissipation, are compared with the separated flow experimental data of Simpson[23] in Fig. 3. The displacement thickness distribution was prescribed from the experiment; the calculated edge Mach number, shape factor, and skin friction agree reasonably well with the experimentally measured values. The influence of the turbulent dissipation relaxation Eq. (3.15) is to improve agreement with experiment in the separated flow region, most noticeably in the Mach number and shape factor gradients.

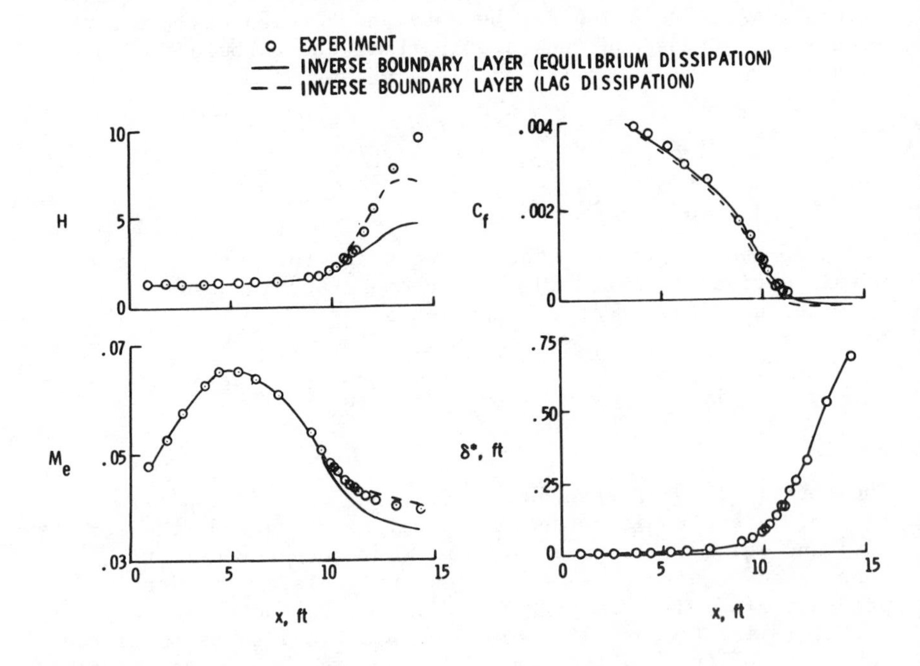

Figure 3. Comparison of inverse integral boundary-layer method with separated flow measurements of Simpson.[24]

Comparison is made in Fig. 4 with the 6° conical diffuser data of Pozzorini [30] which was one of the test cases for the 1981 Stanford Conference and is near separation. For these calculations, a simple 1-D isentropic core model was integrated simultaneously with the inverse boundary-layer model so that all quantities shown were predicted downstream of the initial conditions at the inlet. The pressure coefficient is referenced to a station ahead of the conical diffuser in a region for which no calculations were made. The agreement with experiment is excellent throughout the conical diffuser.

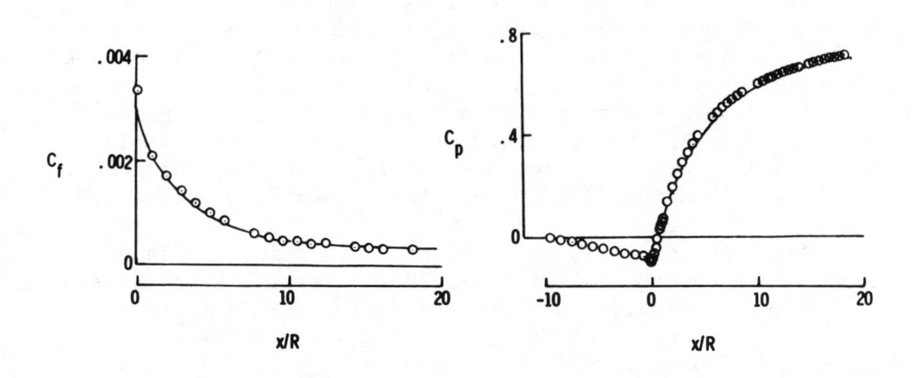

Figure 4. Comparison of inverse integral boundary-layer method with experimental measurements of Pozzorini[30] made in a 6° conical diffuser.

4. VISCOUS-INVISCID INTERACTION

Lighthill [31] proposed four ways of treating the influence of boundary layers and wakes on the flow outside these viscous regions. Lighthill, as most investigators since, considered inviscid potential flow. If the Euler equations are used in place of the potential flow equations, then additional considerations are necessary because, in addition to the continuity equation, two momentum equations and an energy equation are involved for two-dimensional flow. The matching of Euler and boundary-layer solutions has been considered by Johnston and Sockol [32] and Murman and Bussing [33]. Also, Le Balleur [4] has considered the general problem of matching although his interactive computations have involved the potential flow equations. This section summarizes the various techniques that can be used for matching Euler and boundary-layer solutions, and the specific approach used in this work.

Murman and Bussing [33] follow a development similar to that of Le Balleur [4] and discuss the coupling of Euler and boundary-layer equations at three locations: $n = \delta$(boundary-layer edge), $n = \delta^*$ (boundary-layer displacement thickness), and $n = 0$ (body surface). Selection of $n = \delta$ involves placing the boundary of the equivalent inviscid flow at a location just outside the viscous region. Relations for matching (or actually patching), and a discussion of how the boundary conditions might be handled are given in Ref. [33].

Selection of $n = \delta^*$, and neglecting gradients in the equivalent inviscid flow for $\delta^* \leq n \leq \delta$, results in $n = \delta^*$ as a streamline of the equivalent inviscid flow [33] and the Euler equations can be solved using solid wall boundary conditions at $n = \delta^*$. This normally requires remeshing for the Euler solver after each viscous solution, as was done to obtain the displacement thickness interaction results presented in Ref. [9]. If the gradients of the equivalent inviscid flow for $\delta^* \leq n \leq \delta$ are not negligible, then the boundary at $n = \delta^*$ should be handled as an inflow or outflow boundary.[33]

Selection of $n = 0$ corresponds to locating the inner boundary of the equivalent inviscid flow at the same location as the inner boundary of the viscous flow. This location, $n = 0$, is used here and it is also the one discussed by Johnston and Sockol.[32] The choice of $n = 0$ produces a region where the equivalent inviscid flow and the viscous flow overlap. How this overlap region is treated influences the interaction. Hence, it is necessary to make the treatment clear, and helpful to point out its consequences.

The following development was used in Ref. [9] and it follows the general approach presented by Johnston and Sockol.[32] Consider the flow in the neighborhood of a surface and write the steady two-dimensional Navier-Stokes equations as

$$\frac{\partial \vec{F}}{\partial x} + \frac{\partial \vec{G}}{\partial y} = 0 \tag{4.1}$$

and the steady two-dimensional Euler equations as

$$\frac{\partial \vec{f}}{\partial x} + \frac{\partial \vec{g}}{\partial y} = 0 \tag{4.2}$$

where

$$\vec{f} = \begin{bmatrix} \rho u \\ \rho uu + p \\ \rho uv \\ (e + p)u \end{bmatrix} \qquad \vec{g} = \begin{bmatrix} \rho v \\ \rho uv \\ \rho vv + p \\ (e + p)v \end{bmatrix}$$

$$e = \frac{p}{\gamma - 1} + \tfrac{1}{2}\rho(u^2 + v^2)$$

and u, v are velocity components in the x, y directions, and p, ρ, and e are the pressure, density, and total energy per unit volume. An explicit description of the elements of $\vec{F}$ and $\vec{G}$ is not needed. Integrating Eqs. (4.1) and (4.2) with respect to y over $o \leq y \leq h$, and considering the solution vectors $\vec{g}$ and $\vec{G}$ to coincide for $y > h$ (where h is taken outside the viscous region), the two integrals can be combined to obtain [32]

$$\vec{g}_o = \vec{G}_o + \frac{\partial}{\partial x}\int_o^h (\vec{f} - \vec{F})dy \tag{4.3}$$

where the subscript o indicates y = 0.

Equation (4.3) provides a relation for the elements of the $\vec{g}$ vector at the surface for use in the Euler solver as needed. Solutions to the Navier-Stokes or thin-layer Navier-Stokes equations could be used for $\vec{F}$ in Eq. (4.3) for the evaluation of g_o. However, it is undesirable to solve the Navier-Stokes equations and desirable to use some sort of boundary-layer approximation in evaluating g_o. If $\vec{F}$ is simply approximated by the boundary-layer solution, $\vec{\bar{f}}$, then the integrand of Eq. (4.3) is similar to that existing in the formulation of Le Balleur, [14] i.e., $(\vec{f} - \vec{\bar{f}})$, and, of course, both $\vec{f}$ and $\vec{\bar{f}}$ are variable through the viscous region. An alternate approach was suggested by Johnston, Sockol, and Reshotko [34] whereby the exact solution $\vec{F}$ is represented by a composite function $\vec{F}_c$ where

$$\vec{F}_c = \vec{F} = \vec{f} + \vec{\bar{f}} - \vec{f}_o \tag{4.4}$$

The function $\vec{F}_c$ is stated [32] to be constructed in the spirit of a matched asymptotic expansion and is graphically represented in Ref. [32]. Evaluation of the composite function at y = 0 simply yields $(\vec{F}_c)_o = \vec{\bar{f}}$. Evaluation at y = h indicates that the edge properties in the boundary-layer solution should be taken to be those of the Euler solution evaluated at y = 0 and not those of the Euler solution evaluated at y = h as assumed in a conventional matching procedure, such as that of Le Balleur.[4] Using $\vec{F}_c$ for $\vec{F}$, and a similar function for $\vec{G}$, Eq. (4.3) becomes [32]

$$\vec{g}_o = \vec{\bar{g}}_o + \frac{\partial}{\partial x}\int_o^h (\vec{f}_o - \vec{\bar{f}})dy \tag{4.5}$$

where $\vec{\bar{f}}$ and $\vec{\bar{g}}$ are elements of the boundary-layer equations given by

$$\frac{\partial \vec{\bar{f}}}{\partial x} + \frac{\partial \vec{\bar{g}}}{\partial y} = 0 \tag{4.6}$$

where

$$\vec{\bar{f}} = \begin{bmatrix} \bar{\rho}\bar{u} \\ \bar{\rho}\bar{u}\bar{u} + p \\ 0 \\ (\bar{e} + \bar{p})\bar{u} \end{bmatrix} \qquad \vec{\bar{g}} = \begin{bmatrix} \bar{\rho}\bar{v} \\ \bar{\rho}\bar{u}\bar{v} - \tau \\ \bar{p} \\ (\bar{e} + \bar{p})\bar{v} - \bar{u}\tau - q \end{bmatrix}$$

and τ is the shear stress and q is the heat flux.

Note that the integrand in Eq. (4.5) involves $\vec{f}$ at the surface, i.e. $\vec{f}_o$, and the variation of the Euler solution through the viscous region is not explicity involved in Eq. (4.5). This is a consequnce of the composite function for $\vec{F}$ given by Eq. (4.4) which involves the variation of both $\vec{f}$ and $\vec{\bar{f}}$ through the viscous region. Using the definitions of $\vec{f}$, $\vec{g}$, $\vec{\bar{f}}$ and $\vec{\bar{g}}$, the four elements of $\vec{g}_o$ are determined as follows.

The first term $(\rho v)_o$ is given by

$$(\rho v)_o = (\bar{\rho}\bar{v})_o + \frac{\partial}{\partial x}\int_o^h [(\rho u)_o - \bar{\rho}\bar{u}]dy \tag{4.7}$$

For no porosity in the boundary-layer solution $[(\bar{\rho}\bar{v})_o = 0]$

$$(\rho v)_o = \frac{d}{dx}[(\rho u)_o \delta^*] \tag{4.8}$$

where δ^* is defined as

$$(\rho u)_o \delta^* = \int_o^h [(\rho u)_o - \bar{\rho}\bar{u}]dy \tag{4.9}$$

The second term $(\rho uv)_o$ is given by

$$(\rho uv)_o = (\bar{\rho}\bar{u}\bar{v} - \tau)_o + \frac{\partial}{\partial x}\int_o^h [(\rho u^2 + p)_o - (\bar{\rho}\bar{u}^2 + \bar{p})]dy \tag{4.10}$$

For no-slip boundary conditions for the boundary-layer solution ($\bar{u}_o = 0$), and taking the boundary-layer pressure equal to the pressure from the Euler solution at the surface

$$(\rho uv)_o = -\tau_o + \frac{d}{dx} [(\rho u^2)_o(\delta^* + \theta)] \tag{4.11}$$

where θ is defined as

$$(\rho u^2)_o(\delta^* + \theta) = \int_0^h [(\rho u^2)_o - \bar{\rho}\bar{u}^2]dy \tag{4.12}$$

As pointed out in Ref. [32], this approach will not provide the information necessary to obtain the pressure, and a specific approach to obtain the third element of $\vec{g}_o$ is not given in Ref. [32]. The pressure is obtained here through an extension of the work of Rizzi [10] by including a surface porosity term in Rizzi's normal momentum relation. The third term $(\rho v^2 + p)$, therefore, is obtained by determining p_o as mentioned, and determining (ρv^2) by Eq. (4.8) where the density is obtained from the previous time step.

The last term $[(e + p)v]_o$ is given by

$$[(e + p)v]_o = [(\bar{e} + \bar{p})\bar{v} - \bar{u}\tau - q]_o + \frac{\partial}{\partial x}\int_0^h \{[(e + p)u]_o - [(\bar{e} + \bar{p})\bar{u}]\}dy \tag{4.13}$$

Using no-slip and no porosity boundary conditions for the boundary-layer solution ($\bar{u}_o = \bar{v}_o = 0$), an adiabatic surface $[(q)_o = 0]$, and the definition of total enthalpy ($\rho H = e + p$), Eq. (4.13) becomes

$$[(e + p)v]_o = (\rho vH)_o = \frac{d}{dx}\int_0^h [(\rho uH)_o - \bar{\rho}\bar{u}\bar{H}]dy \tag{4.14}$$

The boundary-layer method [19] was developed for an adiabatic surface with variable total enthalpy across the boundary layer that takes into account total enthalpy overshoot and nonunity Prandtl number. [20] A correlation for the integral in Eq. (4.14) has not been developed as yet; hence the approximation $H_o = \bar{H}$ is taken to prevent having to numerically evaluate Eq. (4.14) at each point. This approximation yields

$$[(e + p)v]_o = (\rho vH)_o = H_o \frac{d}{dx} [(\rho u)_o \delta^*] \tag{4.15}$$

which, by Eq. (4.8), is now simply an identity.

The method used for the displacement thickness iteration for the viscous-inviscid interaction results was that of Carter. [35] A relaxation of unity was used in Carter's method for the transonic interaction results, and a relaxation of up to two was used for the low-speed interaction results.

5. VISCOUS-INVISCID INTERACTION RESULTS

Comparisons are made with experimental data [36] obtained for the RAE 2822 airfoil, a supercritical 12-percent thick airfoil, at transonic Mach numbers. Three cases, corresponding to Cases 6, 9, and 10 (using the numbering system of Ref. [36]) are considered ranging from a fully attached shock/boundary layer interaction for Case 6 to an interaction with separation occuring at the foot of the shock for Case 10. As with all transonic wind-tunnel tests, corrections to Mach number and angle-of-attack due to wind tunnel wall effects are required and are difficult to determine accurately. For the calculations presented, the angle-of-attack correction suggested in Ref. [36] was applied and no correction was applied to the Mach number. Table 1 lists the geometric angle-of-attack α_g, corrected angle-of-attack α_c, freestream Mach number M_∞, and Reynolds number based on chord Re_c for each of the cases. An alternate procedure would have been to match the lift coefficient by varying angle-of-attack and Mach number such as in Ref. [15].

The calculations presented were performed using a C-mesh with 128 points along the airfoil and wake surface and 30 points outward from the airfoil. The grid was highly stretched away from the airfoil and, as noted in Section 2, was extended outward until the lift coefficient exhibited no appreciable change. Some numerical experiments were conducted for Case 9 by varying the mesh density from 128 x 30 up to 200 x 50 with only small differences noted from the interacted results shown below. The interaction code was run for 1200 time steps corresponding to more than a four order of magnitude reduction in the maximum residual, defined as the maximum time rate of change of density. The boundary-layer computation was updated every 20 time steps and typically a solution is sufficiently converged for engineering applications after 600 cycles.

The pressure coefficient and boundary-layer properties for Cases 6, 9, and 10 are shown in Figs. 5, 6, and 7, respectively. For the attached flow solutions, Cases 6 and 9, the features of the shock/boundary layer interaction are predicted accurately, corresponding to a rapid thickening of the boundary layer displacement and momentum thicknesses, a sharp decrease in skin friction, and a forward movement of the shock location as compared to an inviscid calculation. Examination of the pressure coefficients near the leading edge indicates an additional angle-of-attack and Mach number correction which, as can be noted from results presented in Ref. [9], would improve the agreement in that region. However, numerical experiments indicate that, for this airfoil, the shock location is relatively independent of the angle-of-attack. The insensitivity of shock location to angle-of-attack can also be seen by examining the experimental data shown in Figs. 5, 6, and 7.

CASE	M_∞	α_g	α_c	Re_c $\times 10^{-6}$
6	0.725	2.92°	2.54°	6.5
9	0.730	3.19°	2.78°	6.5
10	0.750	3.19°	2.81°	6.2

Table 1. Test conditions for RAE 2822 airfoil.

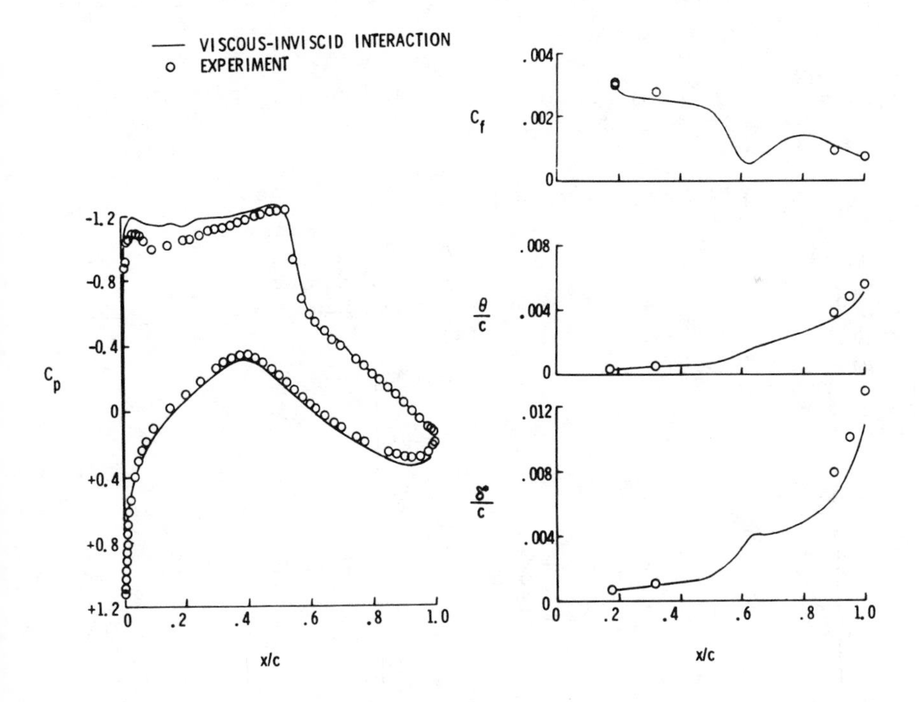

Figure 5. Viscous-inviscid interaction results for RAE 2822 airfoil, Case 6.

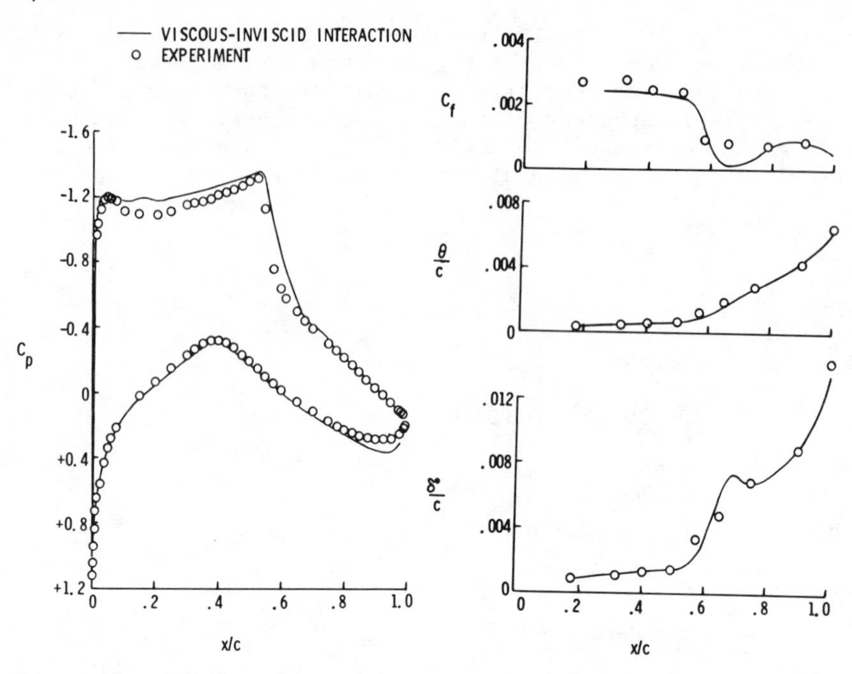

Figure 6. Viscous-inviscid interaction results for RAE 2822 airfoil, Case 9.

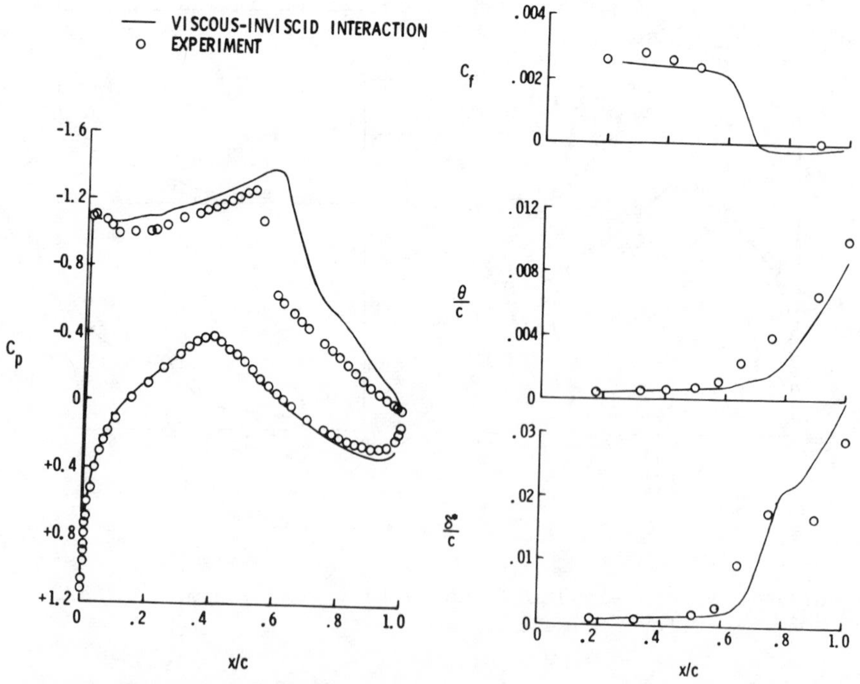

Figure 7. Viscous-inviscid interaction results for RAE 2822 airfoil, Case 10.

For Case 6, the shock location is predicted accurately but the integral thicknesses are slightly underpredicted at the trailing edge. Since the constant-property displacement-thickness shape factor is below 3 for this flow, it can be predicted with a direct boundary-layer method. The boundary-layer properties have been computed using the experimentally measured pressures with both a direct version of the present integral method and a direct version of the lag-entrainment integral method.[29] Little difference between the results of either of these two procedures and the integral method results in Fig. 5 were seen. Hence, the differences shown in Fig. 5 are consistent with boundary-layer approximations to the experimental flow field and cannot be attributed to any errors incurred in the viscous-inviscid interaction. Note that the integral thicknesses are, on the other hand, predicted quite well for Case 9 in Fig. 6.

The separated flow results shown in Fig. 7 indicate a shock position approximately 12 percent chord behind the experimental shock position. Separation is predicted from the foot of the shock rearwards to the trailing edge whereas experimental oil flow measurements indicate separation to occur from approximately 60 to 70 percent chord.[36] The reason for the disagreement is not fully understood at the present time. Numerical experiments indicate that the extent of separation is sensitive to the angle-of-attack with a region less than 10 percent chord in extent at $\alpha = 2.20°$. However, as noted previously, the shock position is relatively insensitive to angle-of-attack changes. Comparing Figs. 5-7 the shock position is predicted progressively rearward of the experiment correspondingly as the displacement thickness behind the shock is becoming increasingly greater. Thus the disagreement may be attributed to a streamwise blockage effect not accounted for in wind tunnel angle-of-attack and Mach number corrections.

REFERENCES

1. KUTLER, P., - A Perspective of Theoretical and Applied Computational Fluid Dynamics. AIAA Paper No. 83-0037, January 1983.

2. LOCK, R.C., - A Review of Methods for Predicting Viscous Effects on Aerofoils and Wings at Transonic Speeds. AGARD-CPP-291, 1980.

3. MELNIK, R.E., - Turbulent Interactions on Airfoils at Transonic Speeds - Recent Developments," AGARD-CPP-291, 1980.

4. LE BALLEUR, J. C., - Viscid-Inviscid Coupling Calculations for Two and Three Dimensional Flows. Lecture Series 1982-04, von Kármán Institute for Fluid Dynamics, March 29-April 2, 1982.

5. JACOCKS, J. L. and KNEILE, K. R., - Computation of Three-Dimensional Time-Dependent Flow Using the Euler Equations. AECD-TR-80-49, Arnold Air Force Station, TN, July 1981.

6. KLINE, S. J., MORKOVIN, M. V., SOVRAN, G., and COCKRELL, D. S., Editors, Proceedings, Computation of Turbulent Boundary Layers - AFOSR-IFP-Stanford Conference, Vol. I. Stanford University Press, Stanford, California, 1968.

7. KLINE, S. J., et. al. - The 1980-81 AFOSR-HTTM-Stanford Conference on Complex Turbulent Flows: Comparison of Computation and Experiment, Stanford University Press, to be published.

8. JAMESON, A., SCHMIDT, W., and TURKEL, E., - Numerical Solutions of the Euler Equations by Finite Volume Methods Using Runge-Kutta Time-Stepping Schemes. AIAA Paper No. 81-1259, June 1981.

9. WHITFIELD, D. L., THOMAS, J. L., JAMESON, A., and SCHMIDT, W. - Computation of Transonic Viscous-Inviscid Interaction Flow. Second Symposium on Numerical and Physical Aspects of Aerodynamic Flows, Ed. Cebeci, T., Springer-Verlag, New York, 1983.

10. RIZZI, A. - Numerical Implementation of Solid-Body Boundary Conditions for the Euler Equations, ZAMM, Vol. 58, pp. 301-304, 1978.

11. WHITFIELD, D. L., SWAFFORD, T. W., and JACOCKS, J. L. - Calculation of Turbulent Boundary Layers with Separation and Viscous-Inviscid Interaction. AIAA Journal, Vol. 19, No. 10, pp. 1315-1322, October 1981.

12. KUHN, G. D. and NEILSEN, J. N. - Prediction of Turbulent Separated Boundary Layers. AIAA Paper No. 73-663, July 1973.

13. EAST, L. F., SMITH, P. D., and MERRYMAN, P. J. - Prediction of the Development of Separated Turbulent Boundary Layers by the Lag-Entrainment Method. RAE TR 77046, Farnborough, Hants, U. K., 1977.

14. LE BALLEUR, J. C. - Strong Matching Method for Computing Transonic Viscous Flows Including Wakes and Separations. Lifting Airfoils. La Recherche Aérospatiale, No. 1981-3, English Edition, pp. 21-45, 1981-3.

15. MELNIK, R. E., MEAD, H. R. and JAMESON, A. - A Multi-Grid Method for the Computation of Viscid/Inviscid Interaction on Airfoils. AIAA Paper No. 83-2034, January 1983.

16. VATSA, V. N. and CARTER, J. E. - Development of an Integral Boundary-Layer Technique for Separated Turbulent Flow. UTRC 82-28, June 1981.

17. THIEDE, P. G. - Prediction Method for Steady Aerodynamic Loading on Airfoils with Separated Transonic Flow. AGARD CP-204, September 1976.

18. WHITFIELD, D. L., SWAFFORD, T. W. and DONEGAN, T. L. - An Inverse Integral Computational Method for Compressible Turbulent Boundary Layers. Recent Contributions to Fluid Mechanics, Ed. A. Haase, Springer-Verlag, 1982.

19. WHITFIELD, D. L. - Integral Solution of Compressible Turbulent Boundary Layers Using Improved Velocity Profiles. AEDC-TR-78-42, Arnold Air Force Station, TN December 1978.

20. WHITFIELD, D. L. and HIGH, M. D. - Velocity-Temperature Relations in Turbulent Boundary Layers With Nonunity Prandtl Numbers. AIAA Journal, Vol. 15, No. 3, pp. 431-434, March 1977.

21. WHITFIELD, D. L. - Analytical, Numerical, and Experimental Results on Turbulent Boundary Layers. AEDC-TR-76-62, Arnold Air Force Station, TN, July 1976.

22. CLAUSER, F. H. - Turbulent Boundary Layers in Adverse Pressure Gradients - Journal of Aeronautical Sciences, Vol. 21, No. 2, pp. 91-108, February, 1954.

23. SIMPSON, R. L., STRICKLAND, J. H. and BARR, P. W. - Features of a Separating Turbulent Boundary Layer in the Vicinity of Separation, Journal of Fluid Mechanics,

Vol. 79, Part 3, pp. 553-594, March 1977.

24. SIMPSON, R. L., CHEW, Y. T. and SHIVAPRASAND, B. G. - The Structure of a Separating Turbulent Boundary Layer. Journal of Fluid Mechanics, Vol. 113, pp. 23-51, 1981.

25. DELERY, J. M. - Experimental Investigation of Turbulence Properties in Transonic Shock/Boundary Layer Interactions. AIAA Journal, Vol. 21, No. 2, pp. 180-185, February 1983.

26. LE BALLEUR, J. C. and MIRANDE, J. - Etude Expérimentale et Théorique de Recollement Bidimensionnel Turbulent, Incompressible. AGARD CP-168, Göttingen, May 1975.

27. CEBECI, T. and SMITH, A. M. O. - Analysis of Turbulent Boundary Layers, Academic Press, New York, 1974.

28. GOLDBERG, P. - Upstream History and Apparent Stress in Turbulent Boundary Layers. MIT Gas Turbine Laboratory Report No. 85, 1966.

29. GREEN J. E., WEEKS, D. M. and BROOMAN, J. W. F. - Prediction of Turbulent Boundary Layers and Wakes in Incompressible Flow by a Lag Entrainment Method. RAE TR 72231, 1972.

30. POZZORINI, R., - Das Turbulente Stromungsfeld in Einem Langen Kreiskegel-Diffusor. Ph.D. Dissertation 5646, Eidgenossischen Technischen Hochschule, Zurich, Ed. Truninger A. G., Zurich, 1976.

31. LIGHTHILL, M. J. - On Displacement Thickness. Journal of Fluid Mechanics, Vol. 4, Part 4, pp. 383-392, 1958.

32. JOHNSTON, W. and SOCKOL, P. - Matching Procedure for Viscous-Inviscid Interactive Computations. AIAA Journal, Vol. 17, No. 6, pp. 661-663, June 1979.

33. MURMAN, E. M. and BUSSING, T. R. A. - On the Coupling of Boundary Layer and Euler Equation Solutions. Second Symposium on Numerical and Physical Aspects of Aerodynamic Flows, Ed. Cebeci, T., Springer-Verlag, New York, 1983.

34. JOHNSTON, W., SOCKOL, P. and RESHOTKO, E. - A Viscous-Inviscid Interactive Compressor Calculation. AIAA Paper No. 78-1140, 1978.

35. CARTER, J. E. - A New Boundary-Layer Inviscid Iteration Technique for Separated Flow. AIAA Paper No. 79-1450, 1979.

36. COOK, P. H., McDONALD, M. A. and FIRMIN, C. C. P. - Airfoil RAE 2822 - Pressure Distributions, and Boundary Layer and Wake Measurements. Experimental Data Base for Computer Program Assessment. AGARD AR-138, 1979.

THE MODULAR APPLICATION OF A SHOCK/BOUNDARY LAYER INTERACTION SOLUTION TO SUPERCRITICAL VISCOUS-INVISCID FLOW FIELD ANALYSIS

G. R. Inger*

West Virginia University
Morgantown, WV

ABSTRACT

Shock-boundary layer interaction can significantly influence not only the local transonic flow on missiles, wings and turbine blades but its influence can also extend significantly downstream within the boundary layer and thereby alter the global aerodynamic properties of lift, drag and pitching moment. It is therefore important that shock-boundary layer interactions and their Reynolds and Mach number scaling be fundamentally understood and appropriate theoretical tools be developed for their prediction in engineering applications. This paper describes the features of an approximate non-asymptotic triple deck theory of shock-turbulent boundary layer interaction that accurately describes non-separating two dimensional flows over a wide range of practical Reynolds numbers and its application as an element in the overall viscous flow analysis of the body. Two main aspects of the problem are examined: (1) The local interactive thickening and skin friction drop in the shock foot region, including the effects of the incoming boundary layer shape factor, wall curvature, an improved "viscous ramp" model of the interaction and an approximate prediction of incipient separation behavior; (2) The significant influence of such interaction on the subsequent downstream turbulent boundary layer thickening, profile shape and skin friction behavior. Comparisons with experimental data are given and applications presented for both supercritical airfoils and transonic bodies of revolution.

NOMENCLATURE

A	Van Driest-Cebeci wall turbulence damping parameter
C_f	skin friction coefficient, $2\tau_w/\rho_{e_o} U_{e_o}^2$

*Associate Chairman
Department of Mechanical and Aerospace Engineering

C_p	pressure coefficient, $2p'/\rho_{e_o} U_{e_o}^2$
H	boundary layer shape factor, δ^*/Θ^*
H_i	incompressible shape factor
M	Mach number
p	static pressure
p'	interactive pressure perturbation, $p-p_1$
Δp	pressure jump across incident shock
Re_ℓ, Re	Reynolds numbers based on length ℓ and boundary layer thickness, respectively
T	absolute temperature
T	basic interactive wall-turbulence parameter
u', v'	streamwise and normal interactive disturbance velocity components, respectively
U_o	undisturbed incoming boundary layer velocity in x-direction
x, y	streamwise and normal distance coordinates (origin at the inviscid shock intersection with the wall)
$y_{w_{eff}}$	effective wall shift seen by interactive inviscid flow
β	M_1^2-1
γ	specific heat ratio
δ	boundary layer thickness
δ^*	boundary layer displacement thickness
δ_{SL}	inner deck sublayer thickness
ε_T	kinematic turbulent eddy viscosity
ε_T'	interactive perturbation of turbulent eddy viscosity
η	y/δ_o
μ	ordinary coefficient of viscosity
ν	μ/ρ
ω	viscosity-temperature dependence exponent, $\mu \sim T^\omega$
ρ	density
Θ^*	boundary layer momentum thickness
τ	total shear stress
τ'	interactive perturbation of total shear stress

Supscripts

1	undisturbed inviscid values ahead of incident shock
e	conditions at the boundary layer edge
inc	incompressible value
inv	inviscid disturbance solution value
o	undisturbed incoming boundary layer properties

1. INTRODUCTION

Shock-boundary layer interaction can significantly influence the transonic flowfield and aerodynamics of missiles, wings and turbine blades. This influence is not only local but can also extend significantly downstream within the boundary layer and thereby alter the global properties of lift, drag and pitching moment. Some of the important effects that these interactions may exert even in the non-separating case are: (a) the interactive-thickening slightly alters the large-scale local inviscid pressure and both the shock location and obliquity; (b) the interaction zone itself may not scale with the local boundary layer thickness, thereby introducing a kind of "unit Reynolds number" effect; (c) possible incipient local separation at the shock foot if the local shock strength is strong enough (and/or Reynolds number is low), which then drastically changes the entire nature and extent of the interaction to a larger scale one involving a bifurcated - shock interaction pattern; (d) an overall increase of the boundary layer displacement and momentum thicknesses downstream; (e) additional downstream distortion for some considerable distance of the more detailed boundary layer properties such as the shape factor and skin friction.

In view of these effects, it is important that shock-boundary layer interactions and their Reynolds and Mach number scaling be fundamentally understood and appropriate theoretical tools be developed for their prediction in engineering applications. This paper describes the features and applications of an approximate non-asymptotic triple-deck theory of transonic shock-turbulent boundary layer interaction which provides such a tool for non-separating two-dimensional flows over a wide range of practical Reynolds numbers. Section 2 contains a brief description of the foundation and essential features of the theoretical model including validating comparisons with experiment. It is shown how the theory is constructed to treat a wide range of practical Reynolds numbers and incoming turbulent boundary layer profile shapes. Section 3 then describes the results of a comprehensive parametric study showing how the interactive pressure, displacement thickness and local skin friction distributions each depend on the shock strength, Reynolds number and shape factor. These results further provide an improved "viscous wedge" model of the local interaction which embodies proper scaling behavior as well as an approximate account of incipient separation that is in good agreement with experimental trends. We further examine the application of this theory as an element in global viscous flow field analyses of both supercritical air-foils and transonic bodies of revolution. In such problems it will be shown that even in non-separating cases the changes across the interaction may significantly alter the subsequent turbulent boundary layer behavior for appreciable distances, especially when larger downstream adverse pressure gradient are present.

2. OUTLINE OF THE LOCAL INTERACTION THEORY

2.1 Rationale of non-asymptotic triple deck approach

It is well-known [1] that when separation occurs, the disturbance flow pattern associated with a nearly-normal shock-boundary layer interaction is a very complicated one involving a bifurcated shock pattern, whereas the unseparated case pertaining to turbulent boundary layers up to roughly $M_1 \approx 1.3$ has instead a much simpler type of interaction pattern which is more amenable to analytical treatment. With some judicious simplifications, it is possible to construct a fundamentally-based approximate theory of the problem, as documented in detail in Ref. 2-3. For purposes of orientation and completeness, a brief outline of this theory will be given here.

We consider small disturbances of an arbitrary incoming turbulent boundary layer due to a weak external shock and examine the detailed perturbation field within the layer. We purposely employ a non-asymptotic triple-deck flow model patterned in some ways after Lighthill's approach [4] because of its essential soundness and adaptability to further improvement, because of its similarity to related types of multiple-deck approaches that have proven highly successful in treating turbulent boundary layer response to strong known adverse pressure gradients [5], and becuase of the large body of turbulent boundary layer interaction data that supports the predicted results in a variety of specific problems [3]. At high Reynolds numbers it has been established [3,6,7] that the local interaction disturbance field in the neighborhood of the impinging shock organizes itself into three basic layered-regions or "decks" (Figure 1): 1) an outer region of potential

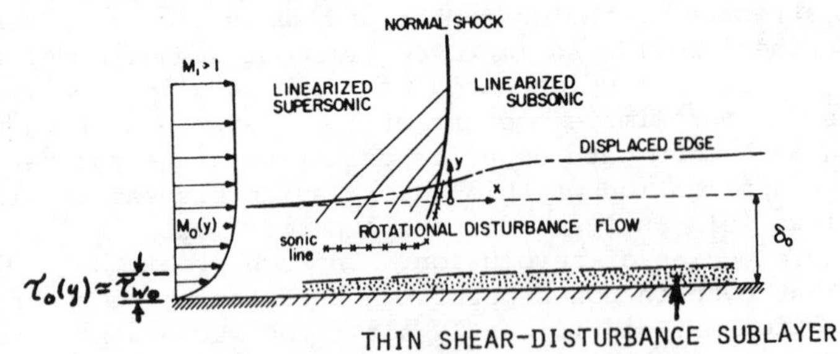

Fig. 1. Triple-Deck Structure of the Local Disturbance Field in a Transonic Interaction.

inviscid flow above the boundary layer, which contains the incident shock and interactive wave systems; 2) an intermediate deck of frozen shear stress-rotational inviscid disturbance flow occupying the outer 90% or more of the incoming boundary layer thickness; 3) an inner shear-disturbance sublayer adjacent to the wall which accounts for the interactive skin friction perturbations (and hence any possible incipient separation) plus most of the upstream influence of the interaction. The "forcing function" of the problem here is thus impressed by the outer deck upon the boundary layer; the middle deck couples this to the response of the inner deck but in so doing can itself modify the disturbance field to some extent, while the slow viscous flow in the thin inner deck reacts very strongly to the pressure gradient disturbances imposed by these overlying decks. This triple deck structure also has been employed in the theoretical studies of Gadd [8], Honda[9] and others, and has been verified by a large body of experimental evidence and recent numerical studies with the full Navier-Stokes equations [10]. The essential correctness of this model is further supported by its success in related boundary layer perturbation problems involving viscous-inviscid interactions, turbulent boundary layer response to sudden changes in surface roughness or pressure gradient, and flow past various kinds of surface distortions including skin friction measuring devices.

While there is general agreement about the validity of the triple deck approach over a wide range of Reynolds number and the well-known qualitative differences in interactive response between laminar and turbulent flow (e.g., the much smaller upstream influence and larger separation-resistance of the later), questions have been raised concerning (a) the relative interactive importance of the inner shear-disturbance deck and (b) the accuracy of deliberately using a non-asymptotic treatment of the details within the boundary layer. Regarding the first, we note that while asymptotic ($Re_\ell \to \infty$) theory predicts an exponentially-small thickness and displacement effect contribution of the inner deck [6,7], this is not apparently true at ordinary Reynolds numbers, where many analytic and experimental studies have firmly established that this deck, although indeed very thin, still contributes significantly to the overlying interaction and its displacement thickness growth [6]. Bolstered by these facts, plus the unanimous conclusion reached in the detailed reviews by Green [11], Rose, Murphy and Watson [12], and Hankey and Holden [13] that the viscous sublayer is an important component of turbulent interaction problems, we take the point of view here that the inner deck is in fact significant at the Reynolds numbers of practical interest. In this regard, we re-emphasize that this deck contains <u>all</u> of the skin friction and incipient separation effects in the interaction, which alone are sufficient reasons to examine it in detail. Regarding (b), it is pointed out that application of $Re_\ell \to \infty$ asymptotic theory results (no matter how rigorous in this limit) to ordinary Reynolds numbers is itself an approximation which may be no more accurate, (indeed perhaps less so) than a physically well constructed non-asymptotic theory. Direct extrapolated-asymptotic versus non-asymptotic theory comparison have definitely shown this to be the case for laminar flows (especially as regards the skin friction aspect [14]) and the situation has been shown to be possibly even worse in turbulent flow [3]. For example, the asymptotic first order theory formally excludes both the streamwise interactive pressure gradient effect on the shear-disturbance deck and both the normal pressure gradient and so-called "streamline divergence" effects on the middle deck; however, physical considerations plus experimental observations and recent comparative numerical studies [15,16] suggest that these effects may in fact be significant at practical Reynolds numbers and should not be neglected. Of course, second order asymptotic corrections can be devised to redress this difficulty but, as Neyfeh and Regab [17] have shown, run the risk of breaking down even worse when extrapolated to ordinary Reynolds numbers. In the present work, we avoid these problems by using a deliberately non-asymptotic triple-deck model appropriate to realistic Reynolds numbers that includes the inner deck pressure gradient terms plus the middle deck $\partial p/\partial y$ and streamline divergence effects, along with some simplifying approximations that render the resulting theory tractible from an engineering standpoint. With this viewpoint in mind, we now examine in more detail the nature of the disturbance flow problem in each of the three basic decks.

2.2 Formulation of the disturbance problem in each deck

Outer Potential Flow Region. Assuming that the incident shock and its reflection system are weak with isentropic non-hypersonic flow, we have here a small disturbance potential inviscid motion imposed upon the undisturbed uniform flow U_{o_e} outside the boundary layer:

$$[M_{o_e}^2 - 1 + (\gamma + 1)\, u' \frac{M_{o_e}^2}{U_{o_e}}] \frac{\partial u'}{\partial x} \simeq \frac{\partial v'}{\partial y} \tag{1}$$

$$\partial v'/\partial x \simeq \partial u'/\partial y \tag{2}$$

$$\frac{\partial^2 p'}{\partial y^2} + [1 - M_{o_e}^2 - 2\frac{u' M_{o_e}^2}{U_{o_e}}] \frac{\partial^2 p'}{\partial x^2} \quad 0 \tag{3}$$

where the third term within the square brackets of Eqs. (1) and (3) is significant in the transonic regime $1 < M_{o_e} < 1.1$ and automatically includes the supersonic-subsonic jump conditions to this order of approximation [18]. Since a variety of efficient analytical and numerical methods are presently available to solve this system in either transonic flow or in purely supersonic flow (in which case Eqs. 1-3 further reduce to an Ackeret-type problem), we assume that such a solution may be carried out for all x on the upper region $y > \delta_o$ subject to the usual far-field conditions as $y \to \infty$. The remaining disturbance boundary condition that must be supplied along $y = \delta_o$ then couples this solution to the underlying double-deck: it requires that the outer disturbance flow pertain to an effective streamline shape (relative to the wall) defined by the total interactive displacement effect of the inner decks. To insure physically-smooth matching along this outer-inner interface, then, we require both v'/U_{o_e} and p' to be continuous with their middle deck counterparts along $y = \delta_o$.

Middle Rotational-Disturbance Flow Deck. This layer contributes to and transmits the displacement effect, contains the boundary layer lateral pressure gradient due to the interaction and carries the significant influence of the incoming boundary layer profile shape. Our analysis of this layer rests on the key simplifying assumption that for non-separating interactions the turbulent Reynolds shear stress changes are small and have a negligible back effect on the mean flow properties along the interaction zone; hence this stress can be taken to be "frozen" along each streamline at its appropriate value in the undisturbed incoming boundary layer. This approximation, likewise adopted by a number of earlier investigators with good results, is supported not only by asymptotic analysis [19] but especially by the results of Rose's detailed experimental studies [20] of a non-separating shock-turbulent boundary layer interaction which showed that over the short-ranged interaction length straddling the shock the pressure gradient and inertial forces outside a thin layer near the wall are at least an order of magnitude larger than the corresponding changes in Reynolds stress. Furthermore, there is a substantial body of related experimental results on turbulent boundary layer response to various kinds of sudden perturbations and rapid pressure gradients which also strongly support this view [3]. These studies unanimously confirm that, at least for non-separating flows, significant local Reynolds shear stress disturbances are essentially confined to a thin sublayer within the Law of the Wall region (see below) where the turbulence rapidly adjusts to the local pressure gradient, while outside in the Law of the Wake region the turbulent stresses respond very slowly and remain nearly frozen at their initial values far out of local equilibrium with the wall stress.

Confining attention, then, to the short range local shock interaction zone where the aforementioned "frozen turbulence" approximation is applicable, the disturbance field caused by a weak shock is one of small rotational inviscid perturbation of the incoming non-uniform turbulent boundary layer profile $M_o(y)$, governed by equations

$$\frac{\partial}{\partial y}\left[\frac{v'(x,y)}{U_o(y)}\right] = \frac{1-M_o^2(y)}{\gamma M_o^2(y)} \cdot \frac{\partial (p'/p_o)}{\partial x} \tag{4}$$

$$\frac{\partial u'}{\partial x} = -\frac{\partial p'/\partial x}{\rho_o(y)\, U_o(y)} - \frac{dU_o}{dy} \cdot \frac{v'}{U_o} \tag{5}$$

$$\frac{\partial^2 p'}{\partial y^2} - \frac{2}{M_o}\frac{dM_o}{dy}\frac{\partial p'}{\partial y} + \left[1-M_o^2 - \frac{2\, u' M_o^2}{U_o}\right]\frac{\partial^2 p'}{\partial x^2} = 0 \tag{6}$$

where Eq. (4) is a result of the combined particle-isentropic continuity, x-momentum and energy equations. It is noted that, consistent with the assumed short range character of the interaction, the streamwise variation of the undisturbed turbulent boundary layer properties that would occur over this range are neglected, taking $U_o(y)$, $\rho_o(y)$ and $M_o(y)$ to be arbitrary functions of y only with δ_o, δ_o^* and τ_{w_o} as constants. Now Eq. (6) is a generalization of Lighthill's well-known pressure perturbation equation for non-uniform flows [4] which includes a non-linear correction term for possible transonic effects within the boundary layer including the diffracted impinging shock above the sonic level of the incoming boundary layer profile. Excluding the hypersonic regime, Eqs. (4)-(6) therefore apply to a wide range of initially supersonic external flow conditions and the complete speed range across the boundary layer except at the singular point $M_o \rightarrow 0$ (which we avoid by consideration of the inner deck as shown below). In particular, use of Eq. (6) provides an account of any lateral pressure gradient that develops across the interacting boundary layer.

As is the case with the outer deck, a variety of analytical or numerical methods may be used to solve this middle deck disturbance problem [2,4]. Whatever the method chosen, we imagine that it provides at each streamwise station x an evaluation of the disturbance pressure distribution $p'(x,y)$; then y-integration of Eq. (4) gives

$$\frac{v'(x,y)}{U_o(y)} = \underbrace{\left[\frac{v'}{U_o}\right](x,y_{w_{eff}})}_{=0} + \frac{\partial}{\partial x}\left\{\int_{y_{w_{eff}}}^{y} \frac{p'}{pe_1}\left[\frac{1-M_o^2(\bar{y})}{M_o^2(\bar{y})}\right]d\bar{y}\right\} \tag{7}$$

where $y_{w_{eff}} > 0$ is the effective wall shift or displacement height associated with the inner deck defined such that the inviscid $v'(x,y_{w_{eff}})$ and hence $\partial p'/\partial y(x,y_{w_{eff}})$ both vanish (see below). Equation (7) provides the disturbance streamline slope distribution across the boundary layer at any streamwise station, and its value at $y = \delta_o$ yields the total streamline displacement effect of the two inner decks (the lower limit on the integral being the inner deck contribution). We then may obtain the corresponding total displacement thickness growth along the interaction by streamwise-quadrature of the perturbation boundary layer continuity equation integral; this yields to first perturbation

$$\Delta\delta^*(x) \simeq \int_{y_{w_{eff}}}^{\delta_o} \frac{p'}{p_{e_1}}\left(\frac{1-M_o^2}{M_o^2}\right)dy + (\delta_o-\delta_o^*)\left[\frac{Me_1^2}{\gamma\, Me_1^2\, p_{e_1}}\right]p'_w(x) \tag{8}$$

The Inner Shear-Disturbance Layer. This very thin inner deck contains the significant viscous and turbulent shear stress disturbances due to the interaction, plus the small upstream influence and an important contribution to the viscous displacement effect. In fact it lies well within the Law of the Wall region of the incoming turbulent boundary layer profile and also (for the high Reynolds numbers of interest here) below the sonic level of the profile as our resulting theory indeed confirms aposteriori. The original work of Lighthill [4] and others treated the problem by further neglecting the turbulent stresses altogether and considering only the laminar sublayer effect; while this greatly simplifies the problem and yields an elegant analytical solution, the results can be significantly in error at high Reynolds numbers and cannot explain (and indeed conflicts with) the ultimate asymptotic behavior pertaining to the $Re_\ell \to \infty$ limit. The present theory remedies this by extending Lighthill's approach to include the entire Law of the Wall region turbulent stress-effects; the resulting general shear-disturbance sublayer theory provides a non-asymptotic treatment which encompasses the complete range of Reynolds numbers. It is noted in this connection that our consideration of the entire Law of the Wall combined with the use of the effective inviscid wall concept to treat the inner deck displacement effect eliminates the need for the "blending layer" [6] that is otherwise required to match the disturbance field in the laminar sublayer region with the middle inviscid deck; except for higher order derivative aspects of asymptotic matching, our inner solution effectively includes this blending function since it imposes a boundary condition of vanishing total shear disturbance at the outer edge of the deck. In addition, our retention of the explicit disturbance pressure gradient term for the inner deck not only provides the correct physics at practical Reynolds numbers but also correctly models the situation near separation ($\tau_w \to 0$) where this term becomes of dominant importance.

To facilitate a tractible thoery, we retain only the main physical effects by introducing the following simplifying assumptions. (a) The incoming boundary layer is free from any post-transitional memory or low Reynolds number effects and its Law of the Wall region is characterized by a constant total (laminar plus turbulent eddy) shear stress and a Van Driest-Cebeci type of damped eddy viscosity model [21]. This model is known to be a good one for a wide range of upstream non-separating boundary layer flow histories. (b) For the weak incident shock strengths of present interest, the sublayer disturbance flow is assumed to be a small perturbation upon the incoming boundary layer; in the resulting linearized disturbance equations, however, all the physically-important effects of streamwise pressure gradient, streamwise and vertical acceleration, and both laminar and turbulent disturbance stresses are retained. Although the resulting theory necessarily becomes inaccurate near separation, it provides a valuable physical insight to the interactive physics close to the wall for non-separating flow and a firm basis for subsequent improvement. Moreover, it can be shown that the form of the particular set of linear equations used here is in fact unaltered by non-linear effects; hence even the quantitative accuracy of the present theory is expected to be good until rather close to separation. (c) For adiabatic flows at low-to-moderate external Mach numbers, the undisturbed and perturbation flow Mach numbers are both quite small within the shear disturbance sublayer; consequently the treatment of compressibility effects therein can be greatly simplified without any significant error. Thus, the influence of the density perturbations on the sublayer disturbance flow may be neglected altogether, while the corresponding modest compressibility effect on the Law of the Wall portion of the undisturbed profile is quite adequately treated by the Eckert reference temperature method [22] wherein incompressible relations are used based on wall recovery temperature properties. Excluding hypersonic flow this is equivalent in accuracy to (but easier than) the use of Van Driest's compressible Law of the Wall profile [23]. (d) The turbulent fluctuations and the small interactive disturbances are assumed uncorrelated in both the lower and middle decks. (e) The thinness of the inner deck allows the boundary layer-type approximation of neglecting its lateral pressure gradient; the wall pressure distribution $p'_w(x)$ is taken equal to the overlaying pressure perturbation field along the bottom of the middle deck.

Under these assumptions, the disturbance field is governed by the following continuity and momentum equations:

$$\frac{\partial u'}{\partial x} + \frac{\partial v'}{\partial y} = 0 \tag{9}$$

$$U_o \frac{\partial u'}{\partial x} + v' \frac{dU_o}{dy} + (\rho_{w_o}^{-1}) \frac{\partial p'_w}{\partial x} = \frac{\partial}{\partial y}\left(\nu_{w_o} \frac{\partial u'}{\partial y} + \varepsilon_{T_o} \frac{\partial u'}{\partial y} + \varepsilon_T \frac{dU_o}{dy}\right) \tag{10}$$

$$\partial p'/\partial y = 0;\ p' = p'(x) \approx p'_w(x) \tag{11}$$

where ρ_{w_o} and ν_{w_o} are evaluated at the adiabatic wall recovery temperature and where it should be noted that the kinematic eddy viscosity perturbation ε_T is being taken into account. The corresponding undisturbed turbulent boundary layer Law of the Wall profile $U_o(y)$ is governed by

$$\tau_o(y) = \text{const.} = \tau_{w_o} = [\mu_{w_o} + \rho_{w_o}\ \varepsilon_{T_0}(y)]\ \frac{dU_o}{dy} \tag{12}$$

where according to the Van Driest-Cebeci eddy viscosity model with $y+ = (y\sqrt{\tau_{w_o}/\rho_{w_o}})/\nu_{w_o}$

$$\varepsilon_T = [.4ly\ (1-e^{-y+/A})]^2\ \frac{\partial u}{\partial y} \tag{13A}$$

which yields for non-separating flow disturbances that

$$\varepsilon_{T_o} = [.4ly(1-e^{-y+/A})]^2\ \frac{dU_o}{dy} \tag{13B}$$

$$\varepsilon_T' \approx \left(\frac{\partial u'/\partial y}{dU_o/dy}\right)\varepsilon_{T_o} \tag{13C}$$

Here, A is the so-called Van Driest damping "constant"; we used the commonly-accepted value A = 26 although it is understood that a larger value may improve the experimental agreement in regions of shock-boundary layer interaction [24]. Substituting (13c) into (10) we thus have the disturbance momentum equation

$$U_o\ \frac{\partial u'}{\partial x} + v'\ \frac{dU_o}{dy}\ (\rho_{w_o}^{-1})\ \frac{\partial p'_w}{\partial x} = \frac{\partial}{\partial y}\ [(\nu_{w_o} + 2 \cdot \varepsilon_{T_o})\ \frac{\partial u'}{\partial y}] \tag{14}$$

from which we seen that inclusion of the eddy viscosity perturbation has exactly doubled the turbulent shear stress disturbance term.

We seek to solve Eqs. (9) and (14) within the incoming flow $U_o(0) = u'(x,o) = v'(x,o) = 0$ plus an initial condition $u'(-\infty,y) = 0$ requiring that all interactive disturbances vanish far upstream of the impinging shock. Furthermore, at some distance δ_{SL} sufficiently far from the wall, u' must pass over to the inviscid solution u'_{inv} along the bottom of the middle deck, this latter being governed by Eq. (9) plus

$$U_o \frac{\partial u_{inv}}{\partial x} + v'_{inv} \frac{dU_o}{dy} + (\rho_{w_o})^{-1} \frac{dp_w}{dx} \simeq 0 \tag{15}$$

with δ_{SL} defined as the height where the total shear disturbance (proportional to $\partial u'/\partial y$) of the inner solution vanishes to a desired accuracy.

Following Lighthill [4], it proves convenient to convert the foregoing problem into one involving the normal disturbance velocity field $v'(x,y)$. Differentiating Eq. (14) w.r.t. x, substituting Eq. (9) so as to eliminate u' and then differentiating the result w.r.t. y so as to eliminate p'_w by virtue of Eq. (11), one thus obtains the following fouth-order equation for v':

$$\frac{\partial}{\partial x}\left(U_o \frac{\partial^2 v'}{\partial y^2} - \frac{d^2 U_o}{dy^2} v'\right) = \frac{\partial^2}{\partial y^2} \left[(\nu_{o_w} + 2\varepsilon_{T_o}) \frac{\partial^2 v'}{\partial y^2}\right] \tag{16}$$

This differential equation contains a three-fold influence of the turbulent flow: the profile $U_o(y)$, its curvature d^2U_o/dy^2 (non-zero outside the laminar sublayer) and a new eddy disturbance stress term $2\varepsilon_{T_o}$. Equation (16) is to be solved together with (11) and (13b) subject to the wall boundary conditions $v'(x,o) = \partial v'/\partial y(x,o) = 0$. A third condition involving v' is obtained by satisfying the x-momentum equation (14) at the wall; when this is differentiated w.r.t. x and Eq. (9) used together with the fact that $\varepsilon_T \to 0$ at $y = 0$ we obtain the non-homogeneous condition

$$\frac{\partial^3 v'}{\partial y^3}(x,o) = -(2\mu_{o_w}^{-1}) \frac{d^2 p'_w}{dx^2} \tag{17}$$

The fourth boundary condition is the v' equivalent of the outer inviscid matching requirement (15), which yields $v'(x,\delta_{SL}) = v'_{inv}(x,\delta_{SL})$ with the inviscid solution governed by

$$\frac{\partial}{\partial x}\left(U_o \frac{\partial^2 v'_{inv}}{\partial y^2} - v'_{inv} \frac{d^2 U_o}{dy^2}\right) = 0 \tag{18}$$

with δ_{SL} pertaining to $\partial^2 v'/\partial y^2 \simeq 0$ (i.e., vanishing total disturbance shear) somewhere within the Law of the Wall region [12]. Once the $v'(x,y)$ field is obtained, the attendant streamwise velocity and disturbance shear stress fields may be found from

$$u' = \int_{-\infty}^{x} \frac{\partial u'}{\partial x}\, dx = - \int_{-\infty}^{x} \frac{\partial v'}{\partial y}\, dx \tag{19}$$

$$\frac{\tau'}{\mu_{w_o}} = \left(1+2\,\frac{\varepsilon_{T_o}}{\nu_{o_w}}\right)\frac{\partial u'}{\partial y} = -\left(1+2\,\frac{\varepsilon_{T_o}}{\nu_{o_w}}\right)\cdot \int_{-\infty}^{x} \frac{\partial^2 v'}{\partial y^2}\, dx \tag{20}$$

An important and useful feature of this approach is the definition of an "effective inviscid wall" position (or displacement thickness) that emerges from the asymptotic behavior of v' far from the wall [4]; see Figure 2. As schematically

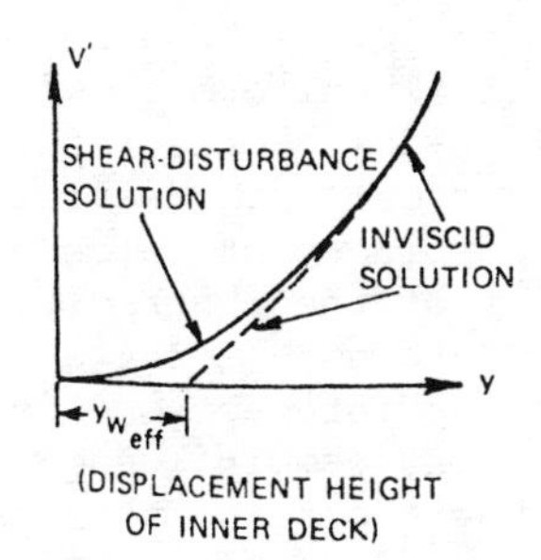

Fig. 2. Effective Wall Concept of Inner Deck

illustrated in Figure 3, this is defined by the value y_{weff}

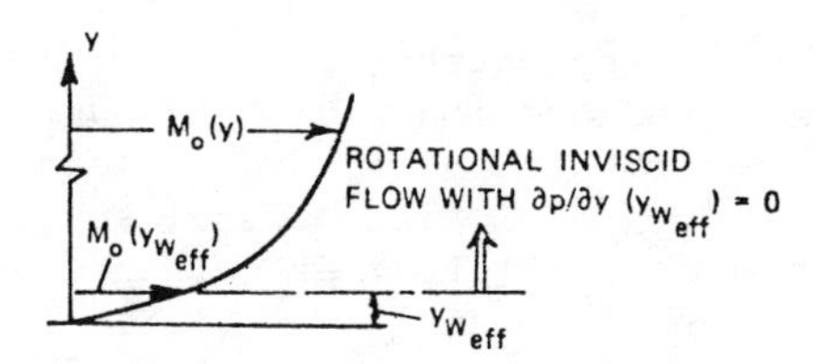

Fig. 3. Middle-Inner Deck Matching

where the "back projection" of the v_{inv} solution vanishes, this projection being the (generally non-linear) solution curve obtained by inwardly-integrating (18) and (12) in the negative y direction starting at δ_{SL}. Physically, y_{weff} thus represents the total mass defect height due to the shear stress perturbation field and hence the effective wall position seen by the overlying inviscid middle deck disturbance flow. As indicated in Figure 3, this concept serves to couple the inner and middle deck solutions in a direct physically-obvious way be providing the non-singular inner equivalent slip-flow boundary conditions $\partial p'/\partial y\ (y_{eff}) = v'_{inv}(y_{weff}) = 0$ at $U_o(y_{weff}) > 0$ for the middle deck solution of Eq. (6). In practice, this approach has proven quite useful for treating a variety of interaction problems.

2.3 Approximate solution by operational methods

An analytical solution is further achieved by assuming small linearized disturbances ahead of, behind and below the nonlinear shock jump plus an approximate treatment [31] of the detailed shock structure within the boundary layer, which gives reasonably accurate predictions for all the properties of engineering interest when $M_1 > 1.05$ (as far as the overall interaction properties are concerned, this non-linear shock jump provision plus the various non-uniform viscous flow effects within the boundary layer reduce the lower Mach number limit otherwise pertaining to the linearized supersonic theory in purely inviscid potential uniform flow). As described in detail in References [2] and [3], the resulting equations can be solved by Fourier transform methods yielding the interactive pressure rise and displacement thickness growth inputs to the above extended theory of the inner-disturbance sublayer. The resulting solution contains all the essential physics of the mixed transonic viscous interaction field for non-separating flows including the upstream influence, the lateral pressure gradient near the shock and the onset of incipient separation; numerous detailed comparisons with experiment [25] have shown that it gives a good account of all the important engineering features of the interaction over a wide range of Mach-Reynolds number conditions.

The matching of the outer two decks with the inner shear-disturbance deck in connection with the Fourier inversion process yields the determination of the upstream influence distance, $\ell_{up} = \varkappa^{-1}_{min}$; typical numerical results for it showing the important typical parametric effects of Reynolds number and shape factor can be found in Ref. [3]. The solution further yields the inner deck displacement thickness

$$\frac{y_{eff_w}}{\delta_o} \simeq .677\left[\frac{c_{f_o}}{2} Re^2_{\delta_o} (T_{e_o}/T_w)^{1+2\omega}\right]^{-1/3} (\varkappa_{min}\,\delta_o)^{-1/3}\, H(T)\, P^{1/3} \tag{21}$$

and the interactive skin friction relationship

$$\frac{\tau'_w(x)}{\tau_{w_o}} \simeq -\left(\frac{\varkappa_{min}}{\lambda}\right)^{2/3} \cdot S(T) \sqrt{\frac{\beta}{c_{f_o}} C_{p_w}}\; P^{-2/3} \tag{22}$$

where

$$P \equiv 3\,\varkappa_{min} \int_{-\infty}^{x} p'^{3/2}_w \, dx \Big/ 2\, p'^{3/2}_w \tag{23A}$$

$$T \simeq (.41)^2 \left[\frac{C_{f_o}}{2} Re_{\delta_o} (T_{e_o}/T_w)^{1+2\omega}\right]^{1/3} (\varkappa_{min}\,\delta_o)^{-2/3} \quad (23B)$$

$$\lambda\delta_o = .744\,\beta^{3/4} (C_{f_o}/2)^{5/4}\, Re_{\delta_o} (T_{e_o}/T_w)^{\omega+1/2} \quad (23C)$$

where the functions H(T) and S(T) are given in Figure 4 and

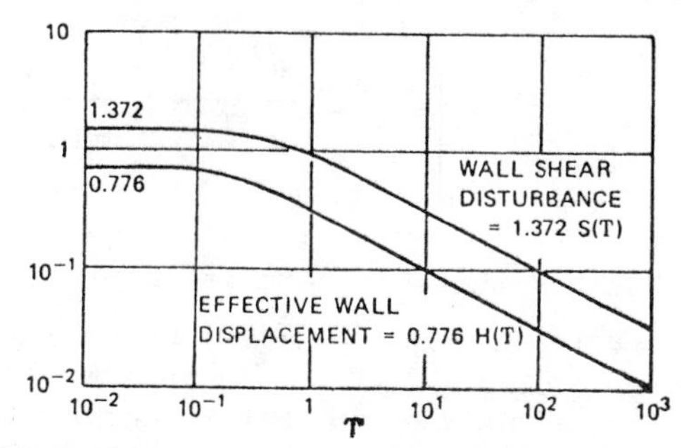

Fig. 4. Wall Turbulence Effect on Inner Deck Displacement Thickness and Disturbance Skin Friction

represent the wall turbulence effect on the interactive displacement effect and skin friction, respectively. Figure 4, in fact, is a central result of the general non-asymptotic theory, providing a unified account of the entire Reynolds number range in terms of the single new turbulent interaction parameter T (Eq. 23b): it ranges from the limiting behavior of negligible wall turbulence effect pertaining to lighthill's theory at $T \to 0$ (lower Reynolds numbers) to the opposite extreme of wall turbulence-dominated behavior at $T \gg 1$ pertaining to an asymptotic-type of theory at very large Reynolds numbers where the inner deck thickness and its disturbance field become vanishingly small. The relationship between these two heretofore-disparate theories has thus been explained and established: they belong at opposite extremes of a general non-asymptotic theory that describes the transition between them. Another important result emerging from Figure 4 is that the asymptotic trends occurring at very large Reynolds numbers cannot be extrapolated down to ordinary values; doing so can yield appreciable error in the inner deck properties of practical interaction problems. This would appear to explain the success of the Lighthill theory in correlating lower Reynolds number turbulent interactions in spite of the mathematical rigor of asymptotic theory: the former is simply closer to the actual physics and correctly predicts more significant interaction effects and scaling under the decidedly non-asymptotic conditions involved. By the same token, the extreme approximation involved in the $T = 0$ limit significantly breaks down at larger Re_ℓ's pertaining to $T \gg 1$, clearly warranting the use of the present theory to account for the increasing role of the wall turbulence effect on the interaction.

An important and useful final consequence of the foregoing analysis is that it yields an explicit analytical criterion for the onset of incipient separation due to an interactive pressure field; setting $\tau_w = \tau_{w_o} + \tau_w \simeq 0$, Eq. (22) predicts this to occur when

$$C_{p_w}\left[\frac{2\,C_{p_w}^{3/2}/\varkappa_{min}}{3\int_{-\infty}^{x}(C_{p_w})^{3/2}dx}\right]^{2/3} \geq \frac{\sqrt{C_{f_o}/\beta}}{S(T)}\left(\frac{\lambda}{\varkappa_{min}}\right)^{2/3} \qquad (24)$$

where it is re-emphasized that C_{p_w} here is the local <u>interactive</u> pressure distribution. Equation (24) bears a general resemblance to a Stratford-type [5] of incipient separation relation for turbulent flow, except that the present formula contains the integrated history effect along the interaction whereas Stratford's result involves purely local properties of C_p and dC_p/dx. It is understood, of course, that the present theory actually breaks down approaching such separation owing to the linearization assumptions and the Van Driest/Cebeci wall turbulence model used; nevertheless, Eq. (24) does give at least a roughly-correct indication of where this will occur and indeed does so without containing any adjustable empirical constants.

A computer program has been constructed to carry out the foregoing solution method; it involves the middle-deck disturbance pressure solution coupled to the inner deck by means of the effective wall shift (Eq. 21) combined with an upstream influence solution subroutine. The corresponding local total interactive displacement thickness growth and skin friction are obtained from Eqs. (8) and (22), respectively. If desired, the attendant boundary layer shape factor change along the interaction may also be calculated as $H = [\delta^*_o + \Delta\delta^*(x)]/\theta^*(x)$ with θ^* given by an x-wise integration of the overall momentum integral equation for the total local boundary layer since $p(x)$, θ^* and C_f are known. The incoming turbulent boundary layer is treated by the compressible version of a universal composite Law of the Wall - Law of the Wake model due to Walz [26] that not only has a convenient analytical form (see Appendix) but also provides a very general fundamental description of this boundary layer in terms of <u>three</u> arbitrary parameters; preshock Mach number, boundary layer displacement thickness Reynolds number, and the incompressible shape factor H_{i_1}. This enables us to account for the important influence of the upstream flow history (pressure gradient, suction, etc.) on the interaction.

2.4 Comparisons of the theory with experiment

Numerous comparisons of the present theory with experimental data from both wind tunnel and free flight experiments have been documented; a sample is presented in Figure 5 to illustrate the predicted behavior and good agreement for the

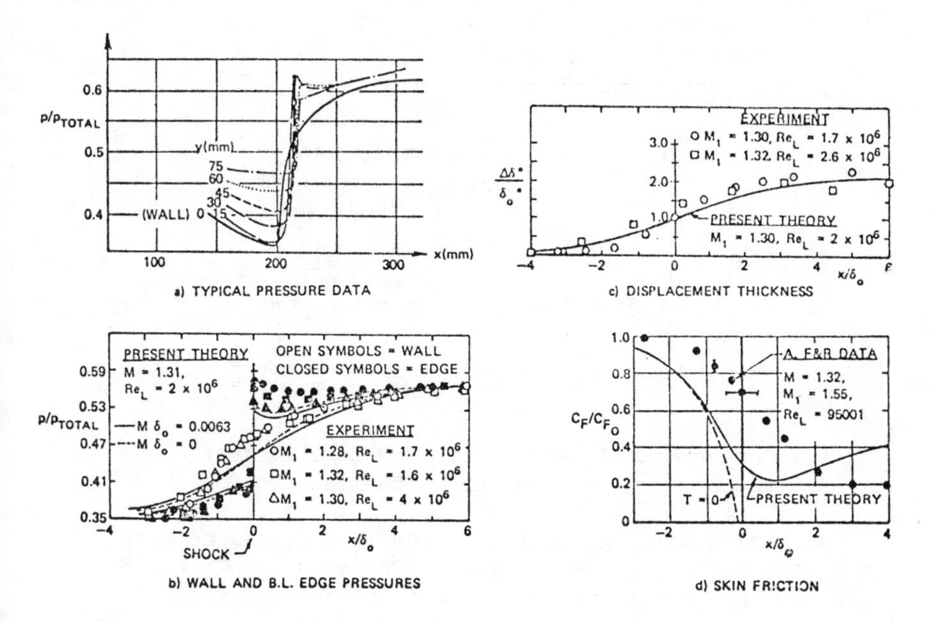

Fig. 5. Comparisons of Present Interaction Theory Results with Experimental Data of Ackeret, Feldman and Rott

interactive pressure, displacement thickness and skin friction distributions in a typical non-separating case. Note especially in Figure 5b the lateral pressure gradient that occurs in the vicinity of the shock. Regarding the skin friction comparison shown in Figure 5d, we note that the "experimental" values were actually inferred from measured velocity profiles along the interaction (hence Θ^* and δ^*) by means of the 2-D momentum integral equation; considering the well-known uncertainties involved in their experimental set up and this method, combined with the present theory's own limitations, the agreement is considered good as regards both the magnitude and shape of the C_f curve.

Figure 6 shows a comparison of the predicted local shape factor change along an interaction with some ONERA measurements [27]; it is seen that the theory nicely captures the characteristic local peaking of H near the shock foot as well as the overall behavior.

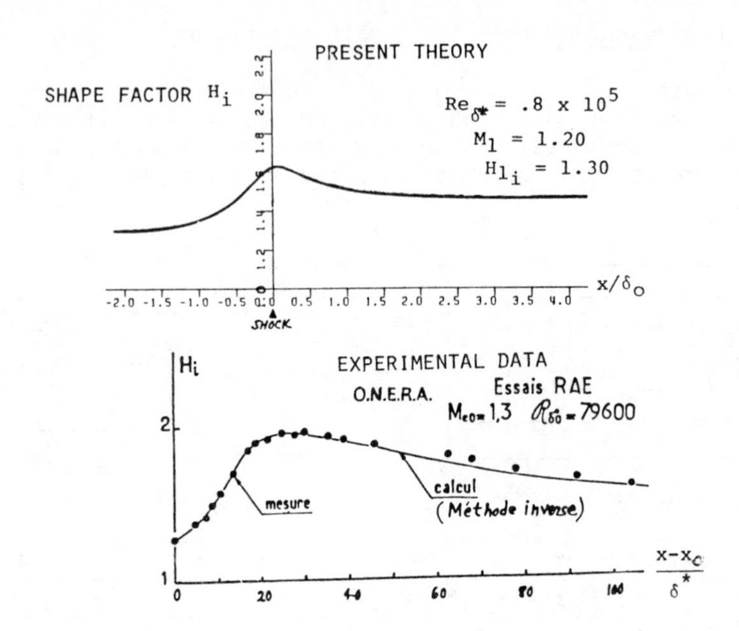

Fig. 6. Comparison of Theoretical and Experimental Shape Factor Distributions Along the Interaction

A particularly interesting comparison with some very recent DFVLR-AVA (Gö) experiments [28] is shown in Figure 7, where we compare with measurements along the non-separating interaction zones on two supercritical airfoils. The experimental C_f values were inferred from streamwise boundary layer profile surveys by means of the Ludwig-Tillman relationship

$$C_f = .246\,(T^*/Te)^{.796}\,Re_{\Theta^*}^{-.268}\,e^{-1.561\,H_i} \tag{25}$$

where H_i is given by

$$H_i \simeq (H - .273M_e^{\,2})/(1.0 + .1145M_e^{\,2}) \tag{26}$$

and $T^*/Te \simeq 1 + .14\,M_e^{\,2}$ for $\gamma = 1.40$ and $P_r \approx .7$ on an adiabatic wall. Theoretical predictions of the entire airfoil flow field were also made with a composite inviscid transonic - turbulent boundary layer - shock interaction numerical scheme in which the present theory was used as a local interactive module astride the inviscid shock location [28]. It is seen that the theory yields an excellent prediction of the local skin friction values upstream and slightly downstream of the shock (including the minimum value) and a good qualitative account of the overall shape of the streamwise distribution. Well downstream of the shock, the theory slightly overestimates the post-shock C_f recovery inferred by the Ludwig-Tillman relationship. Nevertheless, in view of the combined

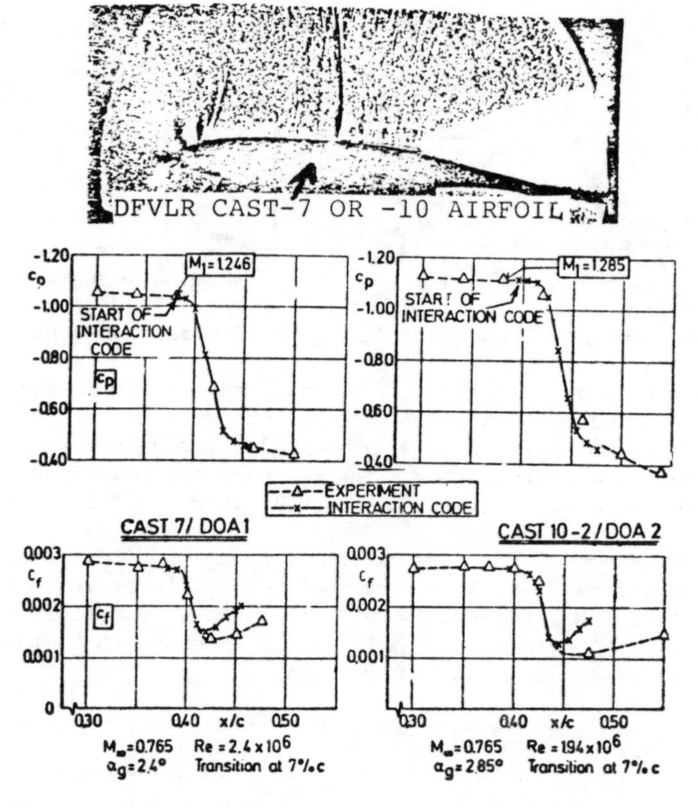

Fig. 7. Typical Comparison of SBLI Theory with Adiabatic Supercritical Airfoil Data

limitations of the experimental Ludwig-Tillman method and the present small disturbance theory, the overall agreement with the data is regarded as good.

3. APPLICATIONS

3.1 Parametric study of local interaction effects

The foregoing theory provides a fundamental analytical tool for examining the influence of the basic physical parameters on the local interactive properties of engineering interest. Such a study has been carried out and its results will be described in this section.

Influence of Mach Number (Shock Strength). The effects of increasing Mach number on the pressure, displacement thickness, skin friction and shape factor distributions are shown in Figs. 8 for fixed Reynolds number and shape factor values typical of a practical flight condition for an aerodynamic body. As expected, the theory predicts that the interactive pressure gradient, displacement thickening, skin friction drop

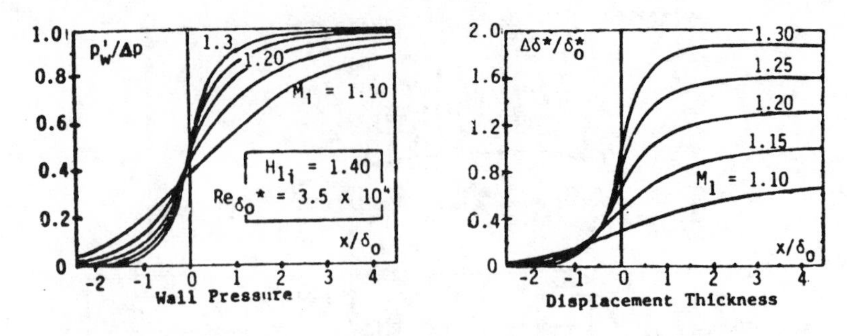

Fig. 8. Typical Properties Along the Interaction Zone

and local shape factor peak all increase with increasing pre-shock Mach number while the overall streamwise scale contracts. In particular, owing to the increased interactive pressure gradient with increasing M_1, we see the tendency of incipient separation to occur (i.e., $C_{f_{min}} \to 0$) as $M_1 > 1.30$ - see below.

Influence of incoming boundary layer profile shape. The predicted effects of the incoming boundary layer incompressible shape factor H_i, related to the adiabatic compressible value to a good approximation by Eq. 26, as is shown for a typical Reynolds number case in Fig. 9. In

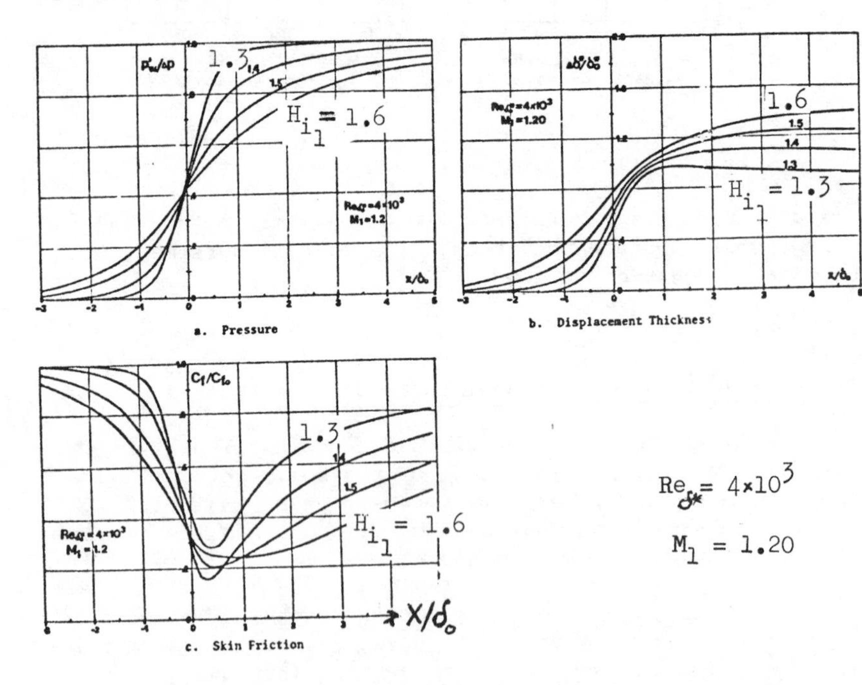

Fig. 9. Theoretical Predictions for the Effect of Shape Factor on Transonic Shock-Turbulent Boundary Layer Interaction

general this effect is seen to be quite significant: increasing H (which corresponds to reduced profile fullness typical of an adverse upstream pressure gradient history and/or blowing history) causes the interaction pressure field to spread out with a resulting large increase in ints upstream influence distance (Fig. 9a) and displacement thickness growth, especially downstream (Fig. 9b). This displacement thickness sensitivity to H_{1_i} is of significant practical importance since δ^* can have a significant back-effect on the inviscid flow and shock position on airfoils or in channel flows.

The shape factor influence on the interactive skin friction (Fig. 9c) is also significant but not monotone in the vicinity of the shock foot; whereas increasing H_{1_i} yields a continual reduction of the downstream C_f level, its effect on $C_{f_{min}}$ is much less and in fact reverses at larger H_{1_i} values > 1.5 (such that further H_{1_i} increase slightly increases $C_{f_{min}}$) owing to the influence of the reduced interactive pressure gradient that also occurs.

<u>Reynolds number (scale) effect.</u> The predicted Reynolds number effect on the interaction field along a typical M_1 = 1.20 shock-boundary layer interaction is shown in Fig. 10,

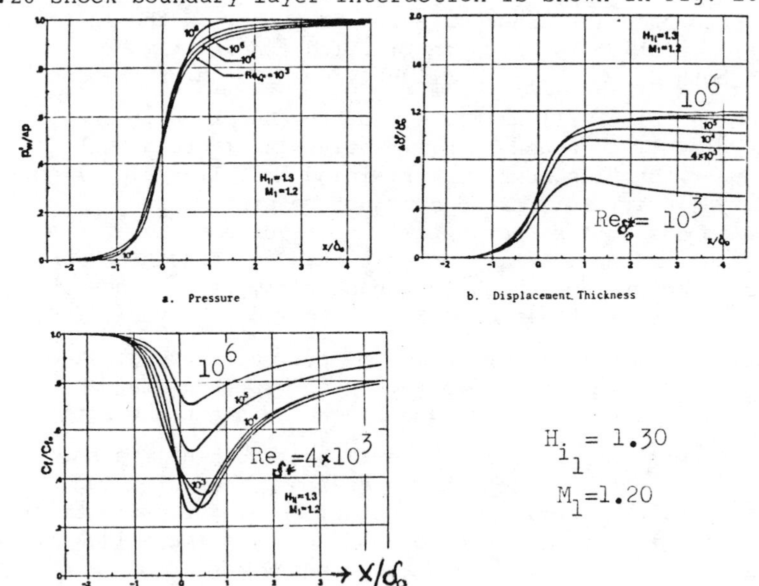

Fig. 10. Theoretical Predictions for Reynolds Number Effect on Transonic Shock-Turbulent Boundary Layer Interaction.

where it is noted that the abscissa of these curves has been normalized with respect to the undisturbed boundary layer thickness δ_o (which itself contains a Reynolds number effect). We observe from Fig. 10a that there is little influence of Reynolds number (other than that absorbed in δ_o) on the upstream pressure distribution; the upstream influence distance has the expected small value (~1.0 to 1.5 δ_o) that scales essentially with δ_o itself. In contrast, the downstream interaction region shows a Reynolds number effect involving a contraction of the non-dimensional streamwise scale with increasing $Re_{\delta}*$. These trends, including the tendency toward a very short range nearly step-like pressure rise at very high Reynolds numbers, are in agreement with experimental observation and the results of Navier-Stokes numerical simulation of turbulent boundary layer-shock wave interaction.

The corresponding displacement thickness predictions in Fig. 10b shows a similar behavior; only the downstream thickening exhibits a significant Reynolds number effect when plotted as $\Delta\delta^*/\delta_o^*$ vs. x/δ_o. Moreover, this effect evidently vanishes at higher Reynolds numbers for a given boundary layer shape factor, implying that $\Delta\delta^*$ becomes proportional to δ_o^*. The local shape factor distribution shown in Fig. 10c also exhibits this Re-independent limiting behavior at sufficiently layer $Re_{\delta}*$.

The predicted influence of Reynolds number on the interactive skin friction is shown in Fig. 10d, where it may be seen that both $C_{f_{min}}$ (located slightly downstream of the shock foot) and the value well downstream of the shock (which recovers quite slowly) decrease significantly with increasing $Re_{\delta}*$ at the larger Reynolds numbers pertaining to full-scale flight vehicles. However, this trend does not persist at low Reynolds numbers but in fact appears to reverse owing to the spreading out of the interaction and attendant reduction of the interaction pressure gradient which ultimately causes the $C_{f_{min}}$ to increase with reducing Re_{ℓ} at the lower Reynolds numbers. At sufficiently low Reynolds number and large shock strength, the present theory in fact predicts vanishing or slightly negative $C_{f_{min}}$ values and hence incipient separation with a very small separation bubble slightly behind the shock foot. Of course the present theory ceases to be valid for separated flow; nevertheless it does give a useful indication as to trends toward this situation and an approximate idea of when incipient separation occurs - see below for a more systematic examination of this behavior.

Viscous ramp modeling of the interaction. In certain engineering applications to global flow field analysis computer programs for wings or turbine blades, it has proven expedient to model the interaction as a simple local inviscid

δ^* - "bump" or "ramp". A serious deficiency of this approach is that it does not account for the dependence of the bump shape and size on Reynolds number, shock strength and boundary layer shape factor, while the additional interaction effects on the downstream boundary layer (such as C_f reduction) are ignored altogether. With the aforementioned parametric study results in hand, however, the present theory can provide (if desired) a much improved "viscous ramp" representation of the interaction whose key physical features have the correct dependence on M_1, Re_{δ^*} and H_{i_1}.

Results for these viscous wedge properties (see Fig. 11) are presented in Refs. 25 and 29 where the upstream and downstream influence distances, the slope and overall height of

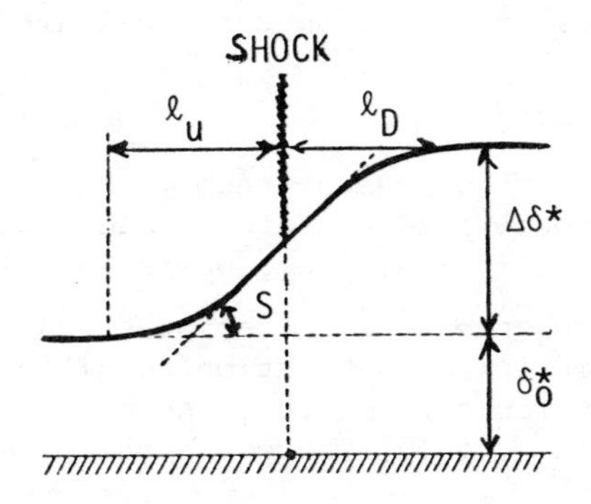

Fig. 11. Viscous Wedge Model

the δ^* - bump are given along with the downstream C_f/C_{f_o} values that may be needed along with δ^* to re-initialize a subsequent turbulent boundary layer calculation downstream. These results show that the viscous wedge slopes are in rough agreement with the maximum attached shock deflection value abserved empirically [30] although here of course there is a dependence on Re_δ and H_{i_1} as well as Mach number. All these results display a significant dependence on the incoming boundary layer shape factor that would appear to be an important consideration in practical applications.

3.2 Wall curvature and shock obliquity effects

The influence of local wall curvature is of interest since interactions often occur on curved surfaces (especially in cascade/turbine-blade applications) and because a singularity is associated with a normal shock in purely inviscid supercritical curved wall flow [31]. The present theory has therefore been extended to enable treatment of the viscous SBLI effects for such flows (see Ref. 32 for details); the results [32] show that the inviscid singularity disappears (the external inviscid disturbance flow now being quite regular) when the interactive boundary layer displacement effect is taken into account. Moreover, the small amounts of curvature typically found in practice moderately spread out and thicken the SBLI zone while slightly delaying the outset of shock-foot separation.

Attendant to the aforementioned study was an examination of the shock obliquity effect: it was found [30,32,33] that the interactive viscous displacement effect slightly alters the inviscid normal shock orientation so that it becomes slightly oblique (with an attendant lowered effective shock-strength) at the edge of the boundary layer. Additional studies have shown that this effective strength corresponds with very good approximation to maximum deflection-oblique shock, and that incorporation of this effect gives good agreement with detailed SBLI data on supercritical airfoils [28,30].

3.3 Incipient separation

As mentioned above the present theory, although it breaks down at separation, does yield a useful indication of incipient separation where $C_{f_{min}} \rightarrow 0$, i.e., where Eq. 24 is satisfied. Since this occurence is of great practical interest, a parametric study of incipient separation conditions inherent in the present theory was carried out; the results for a normal shock are presented in Fig. 12, where the shock Mach number

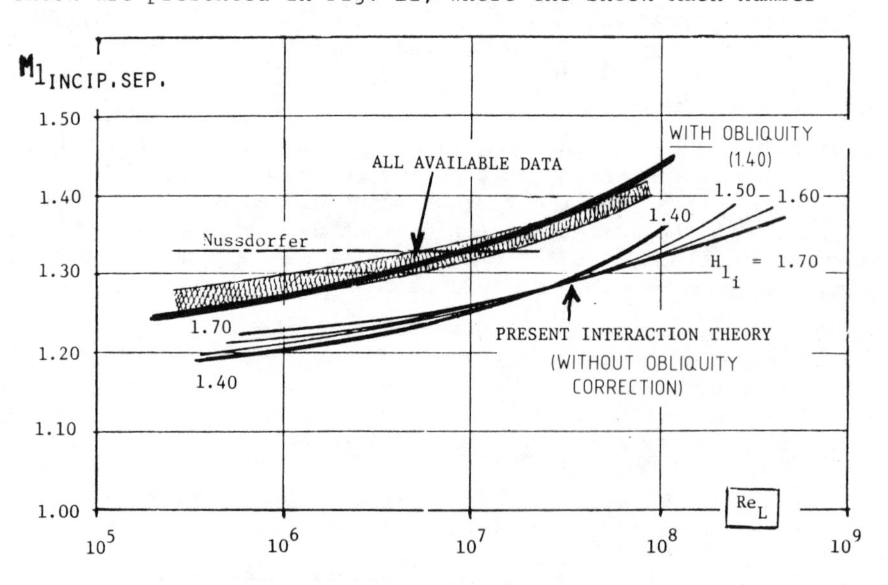

Fig. 12. Incipient Separation Shock Mach Number vs. Reynolds Number: Comparison of Theoretical Prediction with Experimental Data

$(M_1)_{incip.\ sep.}$ above which incipient separation occurs is plotted as a function of the Reynolds number with the shape as a parameter; also shown in the Figure is the approximate experimental boundary determined by a careful examination of a large number of transonic interaction tests, besides Nussdorfers [34] $M \sim 1.30$ criterion for turbulent flow. It is seen that the theoretical prediction of a gradual increase in the $(M_1)_{incip.sep.}$ value with Reynolds number is in agreement with the trend of the data; moreover, the theoretical prediction of only a small influence of shape factor on the incipient separation conditions is also bourne out by the lack of any consistent H-effect for the same Re discernable in the data. The absolute values of $M_{1_{incip.sep.}}$ predicted by the present interaction theory are seen to be consistently slightly lower than the average experimental value; this is attributable to the combined effects of the linearized inner deck theory (which overpredicts the pressure gradient effect on C_f and hence too small on incipient separation shock strength) and the assumption of a normal shock when in fact most of the experiments likely entail some shock obliquity (which also delays separation to somewhat higher shock-strengths). As shown above, wall curvature effects are expected to play a negligible role since they lie well within the data scatter.

3.4 Global flow field effects on supercritical airfoils

Parametric study of downstream effects of SBLI. In addition to the increased displacement thickness, the foregoing discussion shows that the skin friction level following the interaction is significantly reduced; combined with the attendant sensitivity to the profile shape, this suggests that the subsequent downstream boundary layer development may retain a memory of the interaction effects for a considerable distance (over and above a simple thickening), particularly as regards possible incipient separation in any adverse pressure gradients downstream. As indicated in Fig. 13, this downstream "interaction after effect" in the boundary layer influences

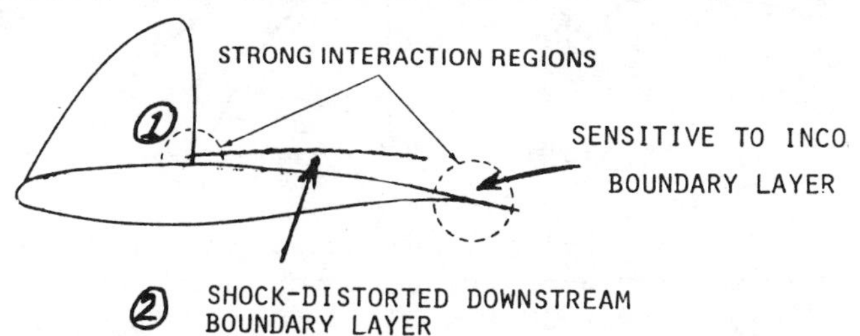

Fig. 13. The Global Viscous-Inviscid Interaction Problem for Supercritical Airfoils (Schematic)

the sensitive trailing edge region and thus may be important in the design and analysis of rear-loaded airfoils, especially at higher lift coefficients with increasingly-aft shock locations; it likewise may be important on three dimensional wing configurations where the shock interaction zones are well aft.

The aforementioned "after-effect" question was therefore subjected to detailed study using the two-layer turbulent boundary layer program of Moses [35] as a model of the downstream viscous flow; the program is coupled to the present interaction theory by initializing it with the post-interactive flow properties so as to account either fully (both C_f and δ^*), partially (δ^* only) or not at all for the changes across the interaction. Calculations were then made of the subsequent downstream turbulent boundary layer behavior (H, C_f, δ^*, δ^*) in various constant post-shock adverse pressure gradients typical of airfoils for different assumed local interactive shock strengths and positions or Reynolds numbers. The results serve as a paradigm of the downstream sensitivity question in real flows.

A variety of cases were studied [36], typical results of which are presented in Fig. 14 where we show the predicted behavior of the boundary layer shape factor and skin friction

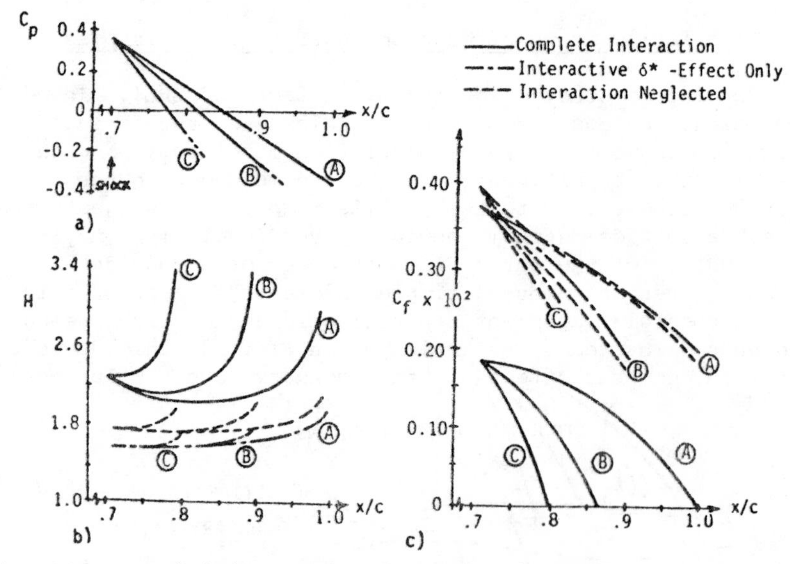

Fig. 14. Sensitivity study of Interaction Effects on Downstream turbulent boundary-layer behavior: a) Post-shock Pressure Gradients; b) Shape Factor; c) Skin Friction

in three increasingly-strong adverse pressure gradients downstream of an interaction occurring at a typically rearward position; the consequences of either fully, partially or negligently-treating the boundary layer changes across the interaction are indicated. Generally, it is seen that the downstream behavior of the boundary layer is indeed sensitive to detailed modeling of the interactive effects and that this sensitivity increases with the strength of the downstream adverse pressure gradient. The adverse pressure gradient magnifies the subsequent influence of the skin friction (as well as the δ^* rise) across the interaction so that downstream separation tends to occur earlier than would be predicted by either neglecting or treating only the δ^* effect of the upstream interaction. As shown in Fig. 15, these predictions are supported by a comparison with boundary layer measurements downstream of a non-separating shock interaction zone on a

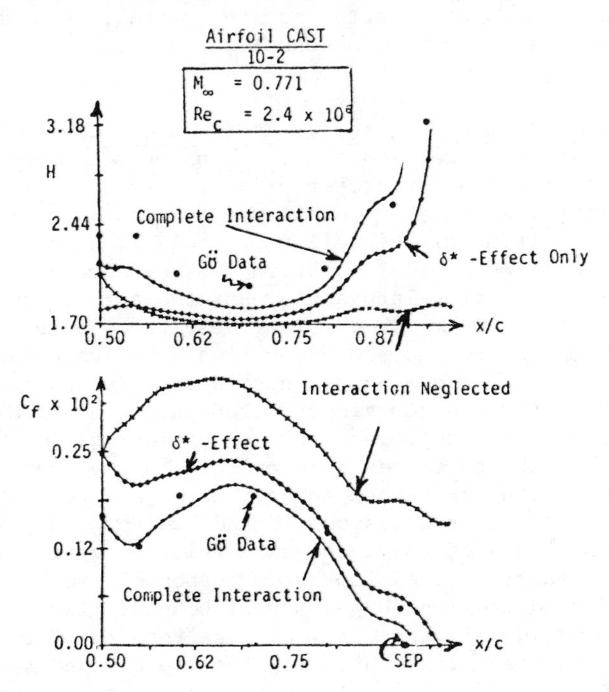

Fig. 15. Comparison of Postinteraction Turbulent Boundary-Layer Predictions with Experiment on a Supercritical Airfoil

supercritical airfoil; both the skin friction and shape factor data are poorly predicted when the interaction is neglected but are reasonably well predicted when the complete interaction effects are taken into account.

Examination of results for a wide variety of cases leads to the further conclusion that such interactive after-effects extend at least 20-30% chord distances downstream on a typical airfoil or wing, and increases (as expected) with either larger shock strength or decreasing Reynolds number. If the trailing edge region lies within this range of the shock, it is thus seen that a simple thickening effect alone is not sufficient to account for the interaction and may result in an inaccurate prediction of the rearward boundary layer shape factor, skin friction and incipient separation properties including their scaling. This is of practical importance for two major reasons: (1) in regions of sustained adverse pressure gradient that often follow the short-scale interaction zone, the shape of the velocity profile and streamwise shear stress distribution (as well as thickness) are of considerable importance to the aerodynamic design of an airfoil or wing, (2) the altered boundary layer properties (especially possible incipient separation) near the trailing edge and into the wake can further exert a powerful effect on the overall aerodynamic via their influence of the Kutta condition [37] and on possible buffet onset.

Applications of the SBLI theory as an interactive module. Nandanan et al. [28] have carried out an even more detailed study of interactions on actual supercritical airfoils including experimental comparisons. They developed a global computational method for transonic airfoil flow analysis which incorporates the present analytical solution for near normal shock/boundary-layer interaction into a state-of-the-art viscous inviscid computation code. Theoretical results obtained with this method were compared to representative data from boundary-layer and surface pressure measurements on three transonic airfoils in the DFVLR-AVA (Gottingen) Transonic Wind Tunnel; an example of these comparisons is shown in Fig. 16. The agreement between theory and experiment in both the boundary-layer displacement thickness and the surface pressure distributions was, for all test cases considered, quite good. The associated predictions of the local skin friction variation through the interaction zone also agree reasonably well with the values inferred from the experimental boundary-layer profiles via the Ludwig-Tillman relation. The results of this investigation indicated that threating the shock/boundary-layer interaction by conventional boundary-layer theory generally leads to a slight underprediction of the displacement thickness immediately downstream of the shock and, due to the amplifying effect of the sustained rear adverse pressure gradients, to an appreciable underestimation of the displacement thickness at the trailing edge. The latter is also clearly reflected in the pressure distribution and aerodynamic coefficients compared. Considering these results, one may conclude that it is generally necessary to include a physically correct treatment of shock wave/boundary-layer interaction in the analysis of transonic airfoil flow.

In a more recent study [38], the above SBLI solution *and* trailing edge interaction effects were included for non-separating turbulent boundary layer flows on supercritical airfoils, within the GRUMFOIL code developed by Melnik and coworkers. For non-separating airfoil operating conditions involving shock locations around mid-chord (such as a low

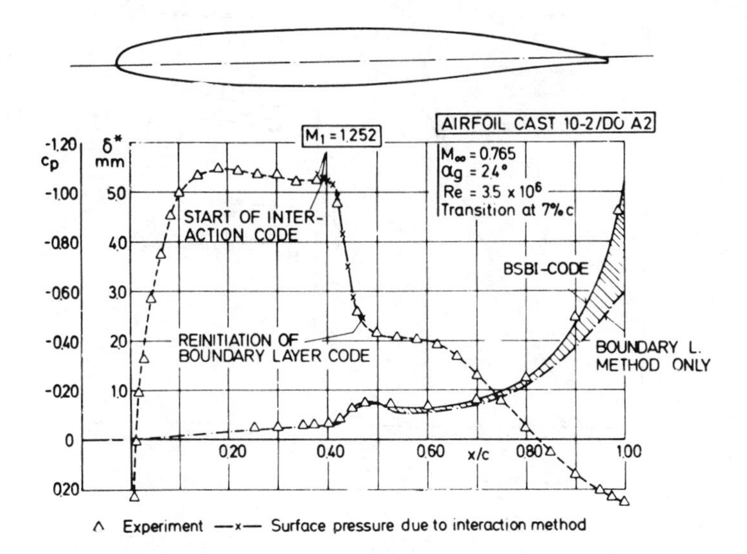

Fig. 16. Comparison of δ* distributions obtained by experiment, boundary layer code, and BSBI-code; airfoil CAST 10-2/DOA2.

angle of attack and moderate Mach number cruise flight condition), it was found that inclusion of proper SBLI-region modeling has little influence on the global aerodynamic properties although it does result in noticeable errors in the predicted displacement thickness and skin friction distributions for distances downstream of the shock. However, for airfoils that operate at higher angle of attack and/or Mach number (such as approaching incipient stall) where a more rearward (70% chord or more) shock locations occur, the details of the SBLI zone exert significant global effects on lift (10-15%), drag and pitching moment as well as very noticeable local flow field effects, see Fig. 17. These results are in agreement with the qualitative implications of several earlier studies. As a broad rule of thumb, when the shock is more than about 60% chord aft, the proper details of the SBLI zone should be included for accurate aerodynamic predictions as well as local boundary layer flow results downstream.

3.5 Multiple interaction effects on transonic bodies of revolution

Another important application of interaction theory has been to the analysis of the flow field around axi-symmetric bodies at supercritical transonic speeds. An interesting aspect of this problem for the boat-tailed bodies found in practice is the occurrence of two interaction regions in sequence, as illustrated in Fig. 18. The problem has been treated by a composite inviscid-boundary layer-shock interaction flow field analysis utilizing the present non-asymptotic

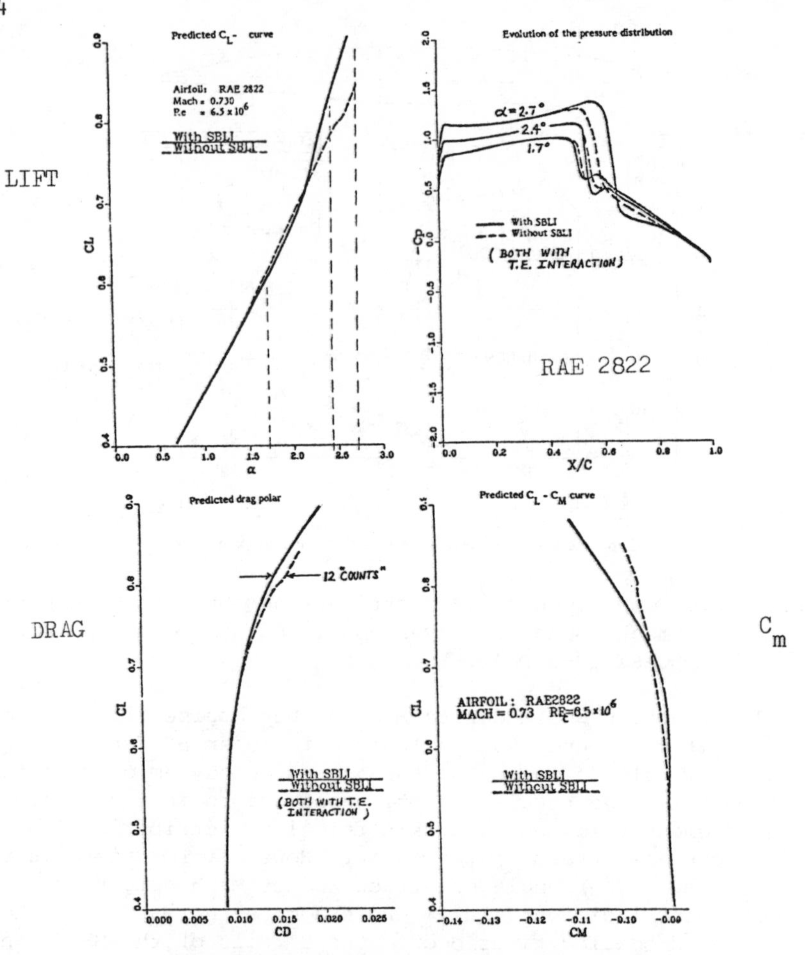

Fig. 17. SBLI Effects on Supercritical Airfoil Lift, Drag and Pitching Moment

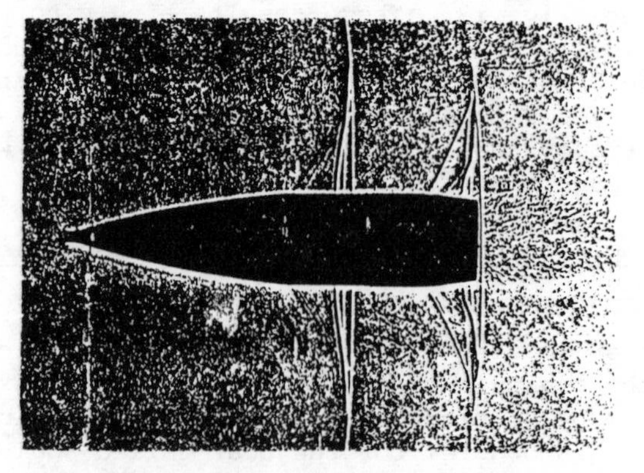

SECANT OGIVE-CYL.-BOATTAIL

$M_\infty = .908$, $\alpha = 0$

Fig. 18. Typical Double-Shock Pattern on a Transonic Projectile.

interaction theory.

Briefly, the approach consists of a global numerical inviscid flow region calculation based on a transonic small disturbance method coupled to a compressible turbulent boundary layer code [39] (which includes the influence of body spin, if present); the local transonic shock-turbulent boundary layer interaction effects are treated by interposing astride the inviscid shock location the axi-symmetric body extension [40,41] of the above non-asymptotic theory, the required inputs being the inviscid shock location (about which the interactive solution is "centered"), the corresponding streamwise component of the inviscid flow Mach number and the streamwise thickness Reynolds number plus the shape factor from the turbulent boundary layer code. The interactive solution is thus inserted as a local "module" at each shock location to produce a general combined inviscid-boundary layer-interaction solution code that is fully operational. In addition to the resultant rapid displacement thickness growth, the interactive distortion of the boundary layer profile shape and skin friction are also thereby accounted for. Furthermore, the influence of these changes on the subsequent turbulent boundary layer development downstream is included by appropriate post-interaction reinitialization using a composite Law of the Wall - Law of Wake profile model.

Typical results for a transonic boat-tailed body of revolution at a flight member of $M_\infty = .940$ and zero angle of attack are illustrated in Figs. 19(b)-(d), where the predicted pressure, displacement and skin friction distributions are shown along with some supporting experimental data obtained by Danberg [38]. Even though no separation occurs, the significant local interaction regions and their after-effect are clearly evident in these distributions. Another interesting set of results is presented in Fig. 20 for a slightly lower flight Mach case (.908), where the predictions of the composite theory are not only compared favorably with the data due to Danberg but also to predictions obtained by Nietubicz [42] using a Navier-Stokes numerical code. The ability of the composite approach to resolve the physical behavior in the smaller scale interaction zones is clearly evident.

4. CONCLUDING REMARKS

This paper has shown that it is now possible to incorporate as an interactive module within a global flow field analysis the correctly-modelled (and scaled) local shock-boundary layer interaction affects for the non-separating case. The non-asymptotic triple-deck interaction theory involved covers a wide range of practical Reynolds numbers and turbulent boundary layer profile shape factors including the effect of wall curvature; moreover, it gives an approxi-

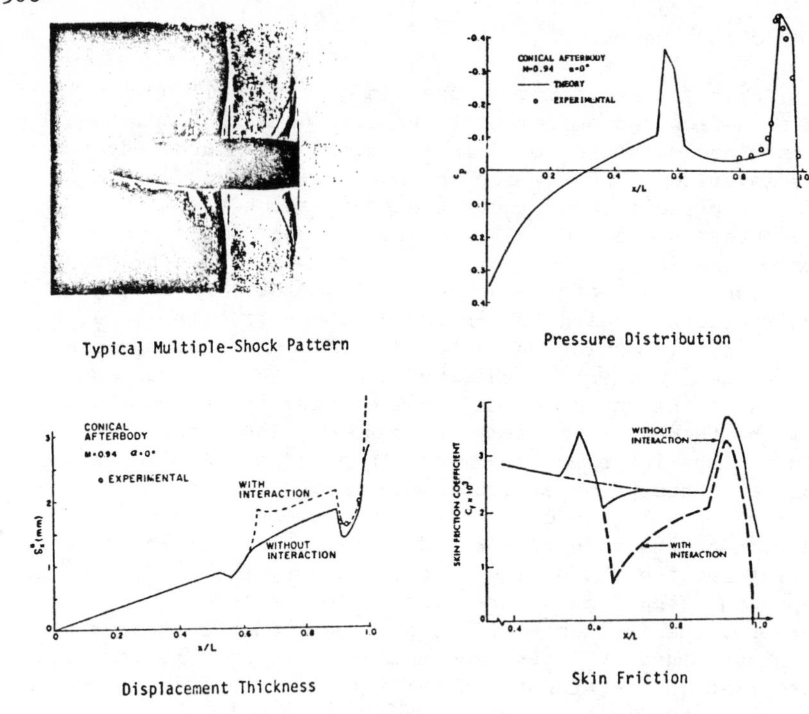

Fig. 19. Composite Inviscid-Boundary Layer-Interaction Theory Predictions for a Boattailed Body of Revolution at $M_{\infty} = .940$.

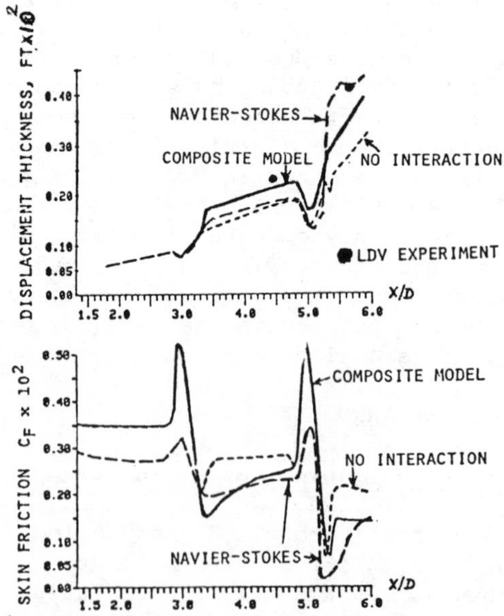

Fig. 20. Comparison of Composite Interaction Solution with Navier-Stokes Code Results and Experiment for a Transonic Body of Revolution ($M_{\infty} = .91$).

mate indication of when incipient separation occurs. It was further shown that such theory is generally desireable when accurate predictions are desired in the important trailing edge region of rear-loaded supercritical air-foils (or approaching the base of axi-symmetric projectile bodies), because the detailed changes across an upstream interaction can significantly alter the subsequent turbulent boundary layer behavior for appreciable distances downstream.

Three areas of future application warranting further study appear to be of practical interest. (1) Extension of the present interaction theory to the unsteady case (examining first the validity of the quasi-steady approximation) in order to study unsteady air loads due to flutter at transonic speeds. (2) Adaptation to three dimensional flow fields of finite-span wings, at least outside wing/fuselage juncture or tip - influence regions. (3) More detailed study of the effects of shock-boundary layer interactions on transonic internal flows within engine inlets and ducts and turbomachinery blade passages and cascades. The influence of these interactions on the resulting losses and downstream effects, especially with incipient separation, is important to understand and predict in practice.

APPENDIX A

COMPOSITE LAW OF THE WALL-LAW OF THE WAKE TURBULENT VELOCITY PROFILE RELATIONSHIPS

Because of its convenient analytical form, accurate blended representation of the combined Law of the Wall-Law of the Wake behavior and generality, we have adopted Walz's model for the incoming turbulent boundary layer upstream of the interaction. For the low Mach number adiabatic wall conditions appropriate to transonic applications, it may be satisfactorily corrected for compressibility effects by the Eckert Reference Temperature method (which under these conditions is, in fact, comparable in accuracy to, but far simpler to implement than, the Van Driest compressibility transformation approach).

Let π be Coles (incompressible) Wake Function, $\eta \equiv y/\delta$ and denote for convenience $R \equiv .41\ Re_{\delta_o^*}/[(1 + \pi)(Tw/Te)^{1+\omega}]$ where $T_w/T_e \simeq 1 + .18\ Me_1^2$ and $\omega \simeq .76$ for a perfect gas; then the compressible form of Walz's composite profile may be written

$$\frac{U_o}{Uo_e} = 1 + \frac{1}{.41}\sqrt{\frac{Cfo}{2}\left(\frac{Tw}{Te}\right)}\left[\left(\frac{R}{1+R}\right)\eta^2(1-\eta) + 2\pi\,\eta^2(3-2\eta) - 2\pi\right.$$

$$\left. + \ln\left(\frac{1+R\eta}{1+R}\right) - (2.15 + 1.235\ R\eta)e^{-.3R\eta}\right] \qquad \text{(A-1)}$$

subject to the following condition linking π to Cf_o and $Re\delta_o^*$ that derives from the $u \rightarrow ue$ matching condition at the boundary layer edge:

$$2\pi + 2.15 + \ln(1+R) = \frac{.41}{\sqrt{\frac{Cfo}{2}\left(\frac{Tw}{Te}\right)}} \qquad \text{(A-2)}$$

Eqs. (A-1) and (A-2) have the following desireable properties: (a) for η > .10 or so, Uo is dominated by a Law of the Wake behavior which correctly satisfies both the outer limit conditions $Uo/Ue \rightarrow 1$ and $du_o/dy \rightarrow 0$ as $\eta \rightarrow 1$; (b) on the other hand, for very small η values, Uo assumes a Law of the Wall-type behavior consisting of a logarithmic term that is exponentially damped out extremely close to the wall into a linear laminar sublayer profile $U/Ue \simeq R$ as $\eta \rightarrow 0$; (c) Eq. (A-1) may be differentiated W.R.T. η to yield an analytical expression for du_o/dy, which proves advantageous in solving the middle and inner deck interaction problems (see text) where dM_o/dy must be known and vanish at the boundary layer edge.

The use of the incompressible form of (A-1) in the defining integral relations for δ_i^* and θ_i^* yields the following relationship that links the wake parameter to the resulting incompressible shape factor $Hi_1 = (\delta_i^*/\theta_i^*)_1$:

$$\frac{H_{i1} - 1}{H_{i1}} = \frac{2}{.41}\sqrt{\frac{Tw}{Te}\frac{Cf_o}{2}}\left(\frac{1+1.59\pi + .75\pi^2}{1+\pi}\right) \qquad \text{(A-3)}$$

Eqs. (A-2) and (A-3) together with the defining relation for R enable a general and convenient parameterization of the profile (and hence the interaction that depends on it) in terms of the shock strength Me_1 the local displacement thickness Reynolds number $Re\delta_o^*$, and one additional physical parameter upstream of the shock. This parameter is taken to be the shape factor Hi_1 because it reflects the upstream history of the incoming boundary layer including possible pressure gradient and surface mass transfer effects. With this prescribed, the aforementioned three equations may be solved simultaneously for the attendant skin friction value Cf_o, the value of R and, if desired, the π value appropriate to these flow conditions.

REFERENCES

1. Ackeret, J., Feldman, F., and Rott, N., "Investigations of Compression Shocks and Boundary Layers in Gases Moving at High Speed," NACA TM-1113, January 1947.
2. Inger, G. R., and Mason, W. H., "Analytical Theory of Transonic Normal Shock-Boundary Layer Interaction," AIAA Paper 75-831, June 1975 (abbreviated version published in AIAA Journal 14, pp. 1266-72, September 1976).

3. Inger, G. R., "Upstream Influence and Skin Friction in Non-Separating Shock Turbulent Boundary Layer Interactions," AIAA Paper No. 80-1411, Snowmass, Colorado, July 1980. See also "Nonasymptotic Theory of Unseparated Turbulent Boundary Layer Interaction," in Numerical and Physical Aspects of Aerodynamic Flows (T. Cebeci, Ed.), Springer-Verlag, N.Y. 1983.
4. Lighthill, M. J., "On Boundary Layers and Upstream Influence; II. Supersonic Flow Without Separation," Proc. Royal Soc. A 217, 1953, pp. 578-507.
5. Stratford, B. S., "The Prediction of Separation of the Turbulent Boundary Layer," Jour. Fluid Mech. 5, pp. 1-16, 1959.
6. Melnik, R. E., and Grossman, B., "Analysis of the Interaction of a Weak Normal Shock Wave with a Turbulent Boundary Layer," AIAA Paper 74-598, June 1974.
7. Adamson, T. C., and Feo, A., "Interaction Between a Shock Wave and a Turbulent Layer in Transonic Flow," SIAM Journal Appl. Math 29, July 1975, pp. 121-144.
8. Gadd, G. E., "Interactions Between Wholly Laminar or Wholly Turbulent Boundary Layers and Shock Wave Strong Enough to Cause Separation," Jour. of the Aeronaut. Sci. 25, November 1958, pp. 729.
9. Honda, M., "A Theoretical Investigation of the Interaction Between Shock Waves and Boundary Layers," Jour. Aero/Space Sci. 25, November 1958, pp. 667-677.
10. Hankey, W., and Shang, J., "Numerical Solution of the Navier-Stokes Equations for Supersonic
11. Green, J. E., "Interactions Between Shock Waves and Boundary Layers," in Progress of Aero. Sci., Vol. 11, Pergamon, NY, 1965, pp. 319.
12. Rose, W. C., Murphy, J. D., and Watson, E. E., "Interaction of an Oblique Shock Wave with a Turbulent Boundary Layer," AIAA Jour. 7, December 1969, pp. 2211-2221.
13. Hankey, W. L., and Holden, M. S., "Two-Dimensional Shock Wave-Boundary Layer Interactions in High Speed Flow," AGARDograph 203, June 1974.
14. Burggraf, O. R., "Asymptotic Theory of Separation and Attachment of a Laminar Boundary Layer on a Compression Ramp," in AGARD CP-168 - Flow Separation, 1975.
15. Werle, M., and Bertke, S. D., "Application of an Interacting Boundary Layer Model to the Supersonic Turbulent Separation Problem," University of Cincinnati Report AFL 76-4-21, August 1976.
16. Le Balleur, J. C., Peyret, R., and Vivand, H., "Numerical Studies in High Reynolds Number Aerodynamics," in Computers and Fluids, Vol. 8, Pergamon, 1980, pp. 1-30.
17. Ragab, S. A., and Nayfeh, A. H., "A Second Order Asymptotic Solution for Laminar Separation of Supersonic Flows Past Compression Ramps," AIAA 78-1132, 1978.
18. Murman, E. M., and Cole, J. D., "Calculation of Plane Steady Transonic Flow," AIAA Jour. 9, January 1971, pp. 114-121.

19. Yajnik, A., "Asymptotic Theory of Turbulent Wall Boundary Layer Flows," JFM 42, 1970, pp. 411-427.
20. Rose, W. C., and Childs, M. E., "Reynolds Shear Stress Measurements in a Compressible Boundary Layer within a Shock Wave-Induced Adverse Pressure Gradient," JFM 65, 1, 1974, pp. 177-188.
21. Cebeci, T., and Bradshaw, P., "Momentum Transfer in Boundary Layers, McGraw-Hill/ Hemisphere," Wash., D.C., 1977, p. 365.
22. Burggraf, O. R., "The Compressibility Transformation and the Turbulent Boundary Layer Equation," Jour. of the Aerospace Sci. 29, 1962, pp. 434-439.
23. Van Driest, E. R., "Turbulent Boundary Layers in Compressible Fluid," Jour. of the Aeronaut. Sci. 18, March 1951, pp. 145-160.
24. Jobe, C. E., and Hankey, W. L., "Turbulent Boundary Layer Calculations in Adverse Pressure Gradient Flows," AIAA Paper 80-0136, Pasadena, January 1980.
25. Inger, G. R., "Application of a Shock-Boundary Layer Interaction Theory to Transonic Airfoil Analysis," AGARD CP-291, Colorado Springs, September 1980.
26. Walz, A., "Boundary Layers of Flow and Temperature," M.I.T. Press, Cambridge, Mass., 1969, pp. 113.
27. Sirieix, M., Delery, J., and Stanewsky, E., "High Reynolds Number Boundary-Layer Shock-Wave Interaction in Transonic Flows," Lecture Notes in Physics, Vol. 148, Springer-Verlag, Berlin, November 1981.
28. Nandanan, M., Stanewsky, E., and Inger, G. R., "A Computational Procedure for Transonic Airfoil Flow Including a Special Solution for Shock-Boundary Layer Interaction," AIAA Journal, Vol. 19, Dec. 1981, pp. 1540-46.
29. Inger, G. R., "Application of a Shock-Turbulent Boundary-Layer Interaction Theory in Transonic Flowfield Analysis" in Transonic Aerodynamics, Vol. 81 of Progress in Astronautics and Aeronautics, AIAA, 1981.
30. Stanewsky, E., Nanandan, N., and Inger, G. R., Proc. AGARD Symposium on "Computation of Viscous-Inviscid Interactions," AGARD CP-291, Colorado Springs, September 1980.
31. Oswaitisch, K. and J. Zierep, "Das Problem des Senkrechten Stosses an Einer Gerkrumpten Wand," ZAMM 40, 1960, p. 143.
32. Inger, G. R. and Sobieczky, H., "Normal Shock Interaction with a Turbulent Boundary Layer on a Curved Wall: VPI&SU Report Aero-088, Blacksburg, Va., Oct. 1978; see also AIAA Paper 81-1244, Palo Alto, Calif., June 1981 and Jour. of Aircraft 20, June 1983, pp. 571-74.
33. Inger, G. R. and Sobieczky, H., "Shock Obliquity Effect on Transonic Shock-Boundary Layer Interaction," Zeitschrift fur Angewante Math and Mechanik, Vol. 58T, 1978, pp. 333-335.

34. Nussdorfer, T. J., "Some Observations of Shock-Induced Turbulent Separation on Supersonic Diffusers," NACA RM E51L26, May 1956.

35. Moses, H. L., "A Strip-Integral Method for PRedicting the Behavior of Turbulent Boundary Layers," Proceedings of the Computation of Turbulent Boundary Layers - 1968 AFOSR-IFP Stanford Conference, Vol. 1, edited by S. Kline et al., Stanford U. Press, Calif., 1968.

36. Inger, G. R. and Cantrell, J. C., "Application of Shock-Turbulent Boundary Layer Interaction Theory to Transonic Aerodynamics," Proceedings of the 1979 USAF-Fed. Republic of Germany DEA Meeting, April 1979; see also J. C. Cantrell, M. S. Thesis, VPI&SU, Blacksburg, Va., June 1979.

37. Melnik, R. E., Chow, R. and Mead, H. R., "Theory of Viscous Transonic Flow Over Airfoils at High Reynolds Number," AIAA Paper 77-680, Albuquerque, N. Mex., June 1977.

38. Lekoudis, S. G., G. R. Inger and M. Khan, "Computation of the Viscous Transonic Flow Around Airfoils with Trailing Edge Effects and Proper Treatment of the Shock/Boundary Layer Interaction," (to be published in Jour. of Aircraft), AIAA Paper 82-0989, June 1982.

39. Reklis, R. P., J. E. Danberg and G. R. Inger, "Boundary Layer Flows on Transonic Projectiles," AIAA Paper 79-1551, Williamsburg, July 1979.

40. Inger, G. R., "Transonic Shock-Turbulent Boundary Layer Interactions on Spinning Axisymmetric Bodies at Zero Angle of Attack," BRL Report ARBRL-TR-02558, Aberdeen, Md., Jan. 1983.

41. Inger, G. R. and L. D. Kayser, "A Computational Study of the Aerodynamic Flow Field Around Supercritical Transonic Projectiles," AIAA Paper 83-2076, Aug. 1983.

42. Nietubicz, C., G. R. Inger and J. E. Danberg, "A Theoretical and Experimental Investigation of a Transonic Projectile Flow Field", AIAA Paper 82-0101, Jan. 1982 (to be published in AIAA Jour.).

A MULTI-DOMAIN APPROACH FOR THE COMPUTATION OF VISCOUS TRANSONIC FLOWS BY UNSTEADY TYPE METHODS.

L. Cambier, W. Ghazzi, J.P. Veuillot and H. Viviand

Office National d'Etudes et de Recherches Aérospatiales (ONERA)
92320 CHATILLON (FRANCE).

1 - INTRODUCTION -

Difficulties met with in the computation of fluid flows are usually linked to the existence of very different length scales (not to mention time scales, considering only steady flow problems) associated to different types of flow regions : inviscid flow region, thin viscous layer, shock wave, and also neighbourhood of a geometrical singularity (such as a corner).

A natural way, which is now attracting more and more attention, to deal with these difficulties is the multi-domain or zonal approach (for example, refs. [1] to [8]). In this paper we present a multi-domain technique for computing compressible flows by unsteady type methods. The matching of sub-domains is based on the use of selected compatibility relations associated with the hyperbolic system solved, and therefore it assumes that dissipative effects are negligible at the inner boundary between two sub-domains. This method is applied to the numerical simulation of a shock wave-turbulent boundary layer interaction in a plane transonic channel. Preliminary results obtained by this method for the same problem have been presented in [8]. Here we give a more detailed description of the inviscid matching technique, as well as of the dichotomy technique used to refine the mesh in the viscous layer, and we present new results obtained with a finer mesh.

2 - INVISCID MATCHING TECHNIQUE BASED ON COMPATIBILITY RELATIONS -

2.1 - General principle of boundary treatment in hyperbolic problems based on compatibility relations -

Consider a first order partial differential system written in conservation or divergence form :

$$\frac{\partial \underline{U}}{\partial t} + \frac{\partial \underline{F}_j}{\partial x_j} = 0 \tag{1}$$

The M components of the vector $\underline{U}$ are independent variables describing completely the physical phenomenon under consideration ; t represents the time and x_j the space coordinates (j varies from 1 to N, where N represents the dimension of the considered space R^N). In gas dynamics, various systems such as (1) may be considered. The basic system is the system of the Euler equations which express the conservation laws of mass, momentum and energy, taking $\underline{U} = (\rho, \rho\vec{V}, \rho E)^T$, where ρ is the density, $\vec{V}$ the velocity vector and E the specific total energy. Simplified systems derived from the Euler equations by using known exact properties of the solution may also be considered. These properties may be exact for unsteady solutions, or only for steady solutions. In that last case, we deal with pseudo-unsteady type methods in which the time t appears as an iterative variable. The reader may find examples and applications of pseudo-unsteady methods in the references [9] to [12] .

System (1) can be written under the following quasi-linear form :

$$\underline{\underline{A}}_0 \frac{\partial \underline{f}}{\partial t} + \underline{\underline{A}}_j \frac{\partial \underline{f}}{\partial x_j} = 0 \tag{2}$$

$\underline{\underline{A}}_0$ and $\underline{\underline{A}}_j$ are M x M matrices defined by $\underline{\underline{A}}_0 = \frac{\partial \underline{U}}{\partial \underline{f}}$, $\underline{\underline{A}}_j = \frac{\partial \underline{F}_j}{\partial \underline{f}}$. There is a one to one relation between $\underline{f}$ and $\underline{U}$. Note that the two systems (1) and (2) are equivalent for continuous solutions. But weak solutions may be defined only with system (1).

The matching technique we present here, is valid only for hyperbolic systems. We first recall the major properties of these systems (e.g. see [13] and [14]). System (2) is said to be t-hyperbolic if :

- i) for all unit vector $\vec{\eta} = (\eta_j)$ of R^N, the characteristic equation for λ ,

$$\det\left\{ \lambda \underline{\underline{A}}_0 - \eta_j \underline{\underline{A}}_j \right\} = 0 \tag{3}$$

admits M real roots, separate or not separate, eigenvalues of the matrix $\underline{\underline{A}}_0^{-1} (\eta_j \underline{\underline{A}}_j)$

- ii) the M left eigenvectors $\underline{\alpha}$ such that :

$$(4) \qquad \underline{\alpha} \cdot (\lambda \underline{\underline{A}}_0 - \eta_j \underline{\underline{A}}_j) = 0$$

are linearly independent.

Then, for every eigenvector $\underline{\alpha}$, there is a linear combination of system (2) defined by :

$$(5) \qquad \underline{\alpha} \cdot \left(\underline{\underline{A}}_0 \frac{\partial \underline{f}}{\partial t} + \underline{\underline{A}}_j \frac{\partial \underline{f}}{\partial x_j} \right) = 0$$

where only N partial derivatives appear instead of (N + 1) in the initial system (2). This combination, called compatibility relation, may be written under the following form :

$$(6) \qquad (\underline{\alpha} \cdot \underline{\underline{A}}_0) \cdot \left(\frac{\partial \underline{f}}{\partial t} + \lambda \frac{\partial \underline{f}}{\partial \eta} \right) + \underline{\tau}^{(k)} \frac{\partial \underline{f}}{\partial \xi^{(k)}} = 0 \qquad (k = 1 \text{ to } N-1)$$

$\partial/\partial\eta = \eta_j \, \partial/\partial x_j$ and $\partial/\partial\xi^{(k)} = \xi_j^{(k)} \partial/\partial x_j$ represent derivations along the directions $\vec{\eta}$, $\vec{\xi}^{(k)}$ of R^N, the (N - 1) vectors $\vec{\xi}^{(k)}$ forming a basis of the (N - 1) dimensional manifold normal to $\vec{\eta}$.

If the partial derivatives of $\underline{f}$ along the directions $\vec{\xi}^{(k)}$ are known at time t_0 , Eq. (6)may then be considered as a transport equation along the characteristic line of slope ($1/\lambda$) in the plane ($\vec{\eta}$, t) and enables us to determine the solution $\underline{f}$ at time $t_1 > t_0$. This interpretation is the basis of a general technique of boundary condition treatment [10], and the matching technique described in § 2.2 is only an extension of this general technique.

Let P be a point of the possibly moving boundary Σ of the computational domain $\mathcal{D}$; $\vec{\nu}$ denotes the unit outward normal to Σ at P and $\vec{w}$ represents the normal velocity of P : $\vec{w} = w\vec{\nu}$. At P , the hyperbolic system to be solved may be replaced by an equivalent system made up of the compatibility relations of the form (6) taking as vector $\vec{\eta}$ the outward normal $\vec{\nu}$. The directions $\vec{\xi}^{(k)}$ normal to $\vec{\nu}$ belong therefore to the hyperplane tangent at P to the boundary Σ ; then the partial derivatives of $\underline{f}$ along these directions $\vec{\xi}^{(k)}$ are known when $\underline{f}$ is known in the domain $\mathcal{D}$. The position of w with respect to the M eigenvalues $\lambda(\vec{\nu})$ arranged in increasing order, say :

$$(7) \qquad \lambda_1 \leq \ldots \leq \lambda_m < w \leq \lambda_{m+1} \leq \ldots \leq \lambda_M$$

shows that, among the M compatibility relations (6), m relations transport information from the outside of the computational domain to the inside, and $M-m$ relations correspond to a transport in the opposite direction. The m first compatibility relations cannot be used and must be replaced by m boundary conditions, the nature of which depends on the physical problem to be solved. Hence, we can state the following rule :

At a boundary point, the values of the variables are derived from a system consisting of $(M-m)$ compatibility relations associated with the eigenvalues such that :

(8) $$\lambda(\vec{\nu}) \geqslant w$$

and of m boundary conditions.

Note that, in the border-line case $\lambda = W$, the corresponding compatibility relation transports information along the boundary Σ and is used for the computation of the solution.

In practice, the discretization of these compatibility relations is easily achieved in the computational mesh starting from the linear combination of the conservative system (1) equivalent to (6) :

(9) $$\underline{\alpha} \cdot \left(\frac{\partial \underline{U}}{\partial t} + \frac{\partial \underline{F}_j}{\partial x_j} \right) = 0$$

After discretizing, this combination may be written in the following concise form :

(10) $$\underline{\alpha}^{n+\frac{1}{2}} \cdot \left(\frac{\underline{U}^{n+1} - \underline{U}^n}{\Delta t} + \left\{ \frac{\partial \underline{F}_j}{\partial x_j} \right\}^{n+\frac{1}{2}} \right) = 0$$

where the superscript n refers to the time level $t^n = n\,\Delta t$ and where $\{\partial \underline{F}_j / \partial x_j\}^{n+\frac{1}{2}}$ denotes a numerical approximation of the space derivatives. The simplest way to obtain this approximation consists in discretizing system (1) at the boundary point (the lack of mesh points outside the domain requiring a suitable adjustment of the numerical scheme). This discretized system gives a value noted $\underline{U}^*$:

(11) $$\frac{\underline{U}^* - \underline{U}^n}{\Delta t} + \delta_j \underline{F}_j^{n+\frac{1}{2}} = 0$$

and the approximation $\{\partial \underline{F}_j / \partial x_j\}^{n+\frac{1}{2}}$ is defined by the discretized term $\delta_j \underline{F}_j^{n+\frac{1}{2}}$. It must be pointed out that the variables $\underline{U}^*$ are not physically meaningful since they do not satisfy the boundary conditions. By eliminating the term $\delta_j \underline{F}_j^{n+\frac{1}{2}}$ between (10) and (11), we obtain the very simple relation :

(12) $$\underline{\alpha}^n \cdot \left(\underline{U}^{n+1} - \underline{U}^* \right) = 0$$

This relation can be transformed in terms of the variables $\underline{f}$:

(13) $$\underline{\omega}^n \cdot \left(\underline{f}(\underline{U}^{n+1}) - \underline{f}(\underline{U}^*) \right) = 0 \qquad (\underline{\omega} = \underline{\alpha} \cdot \underline{A}_0)$$

the vector $\underline{\omega}$ being much simpler to calculate than $\underline{\alpha}$ for a particular choice of $\underline{f}$ (see § 2.4).

Note that the relation (12) or (13) has been linearized by replacing $\underline{\alpha}^{n+\frac{1}{2}}$ by $\underline{\alpha}^n$ or $\underline{\omega}^{n+\frac{1}{2}}$ by $\underline{\omega}^n$.

The $(M-m)$ compatibility relations retained at a boundary point are thus discretized in the form (13) and the value $\underline{U}^{n+1}$ is determined as the solution of the system composed of these $(M-m)$ relations and of m boundary conditions.

2.2 - Matching of two adjacent sub-domains -

We purpose now to apply the general principle stated in § 2.1 to the treatment of an inner boundary Σ between two sub-domains $\mathcal{D}^{(1)}$ and $\mathcal{D}^{(2)}$ of the computational domain $\mathcal{D}$. Note that the solutions $\underline{U}^{(1)}$ and $\underline{U}^{(2)}$ respectively defined in $\mathcal{D}^{(1)}$ and $\mathcal{D}^{(2)}$ may be either continuous or discontinuous along the interface Σ. The normal velocity of a point P of Σ is denoted by : $\vec{W} = w^{(1)}\vec{\nu}^{(1)} = w^{(2)}\vec{\nu}^{(2)}$ $(w^{(1)} = -w^{(2)}, \vec{\nu}^{(1)} = -\vec{\nu}^{(2)})$

System (1) is solved in the two sub-domains. Let $\lambda_k^{(1)}$ and $\lambda_k^{(2)}$ be the eigenvalues, solutions of the characteristic equation (3), defined in the two sub-domains $\mathcal{D}^{(1)}$ and $\mathcal{D}^{(2)}$ ($\lambda_k^{(1)} = \lambda_k(\underline{U}^{(1)}, \vec{\nu}^{(1)})$, $\lambda_k^{(2)} = \lambda_k(\underline{U}^{(2)}, \vec{\nu}^{(2)})$). Among the M eigenvalues $\lambda_k^{(1)}$ (resp. $\lambda_k^{(2)}$), p (resp. q) verify the inequality (8), that is to say :

$$(14) \qquad \begin{cases} \lambda_1^{(1)} \leq \dots \leq \lambda_{M-p}^{(1)} < w^{(1)} \leq \lambda_{M-p+1}^{(1)} \dots \leq \lambda_M^{(1)} \\ \lambda_1^{(2)} \leq \dots \leq \lambda_{M-q}^{(2)} < w^{(2)} \leq \lambda_{M-q+1}^{(2)} \dots \leq \lambda_M^{(2)} \end{cases}$$

This being so, p (resp. q) compatibility relations must be retained in the sub-domain $\mathcal{D}^{(1)}$ (resp. $\mathcal{D}^{(2)}$) to calculate the M components of $\underline{U}^{(1)}$ (resp. $\underline{U}^{(2)}$) at P. Hence, on the whole, ($p+q$) relations must be used to determine $2M$ unknowns, to which we possibly have to add the normal velocity of the point P ($w^{(1)}$ or $w^{(2)}$). For the general problem of the matching of two sub-domains $\mathcal{D}^{(1)}$ and $\mathcal{D}^{(2)}$ to be well posed, we must add to the ($p+q$) compatibility relations, ℓ additional relations, ℓ being given by :

$$(15) \qquad \ell = 2M + \varepsilon - p - q$$

($\varepsilon = 1$ when the speed $w^{(1)} = -w^{(2)}$ is unknown, $\varepsilon = 0$ in the opposite case). Henceforth, these additional relations will be called matching relations.

2.3 - Various types of interfaces -

. Artificial interface -

In that case, the solution $\underline{U}$ is continuous across the interface Σ and the matching relations simply express the continuity of the M unknowns U_i at the point P of Σ, hence M relations : $U_i^{(1)} = U_i^{(2)}$, $i = 1, \dots, M$. An obvious consequence of the continuity of $\underline{U}$ is that the eigenvalues defined in $\mathcal{D}^{(1)}$ and $\mathcal{D}^{(2)}$ and the corresponding eigenvectors, $\alpha^{(1)}$ and $\alpha^{(2)}$ are connected by the following relations :

$$(16.a) \qquad \lambda_k^{(1)} = -\lambda_{M-k+1}^{(2)}$$

$$(16.b) \qquad \underline{\alpha}_k^{(1)} = \underline{\alpha}_{M-k+1}^{(2)} \qquad k = 1, \dots, M$$

We suppose first that $w^{(1)}$ is not equal to an eigenvalue $\lambda_k^{(1)}$. Relation (16a) implies that : $q = M - p$ and relation (16b) shows that the p compatibility relations to be used in $\mathcal{D}^{(1)}$ and the $q = M - p$ compatibility relations to be used in $\mathcal{D}^{(2)}$ are linearly independent ; these M independent compatibility relations enable us to determine the solution at P, the speed $w^{(1)}$ being arbitrarily given.

When the normal velocity of P is equal to an r-order eigenvalue ($r \geq 1$), r compatibility relations corresponding to this eigenvalue must be used in each sub-domain and q is equal to $M - p + r$. It seems that there are too many compatibility relations. Yet, the r compatibility relations corresponding to the r-order eigenvalue are identical in the two sub-domains since they only include derivatives along directions belonging to the hyperplane of $R^N \times [0,T]$ tangent at P to the trajectory of Σ, and these derivatives are identical by continuity. Finally, there are only M independent compatibility relations.

Thus in all cases, the problem of the continuous matching of two sub-domains is easily solved.

. Discontinuity -

We assume now that Σ is a discontinuity surface associated with a weak solution of system (1). We show how the results of § 2.2 allow to develop a numerical technique for fitting the discontinuity Σ. The matching relations between $\mathcal{D}^{(1)}$ and $\mathcal{D}^{(2)}$ are the jump relations associated with system (1). These M scalar relations, non linear with respect to the variables U_i are written in the form :

$$(17) \qquad -w\left(\underline{U}^{(2)} - \underline{U}^{(1)}\right) + \nu_j\left(\underline{F}_j^{(2)} - \underline{F}_j^{(1)}\right) = 0$$

with, for example, $w = w^{(1)}$ and $\vec{\nu} = \vec{\nu}^{(1)}$.

If the position of $w^{(1)}$ (or $w^{(2)}$) compared with the M eigenvalues $\lambda_k^{(1)}$ (or $\lambda_k^{(2)}$) is arbitrary, the problem of the matching of $\mathcal{D}^{(1)}$ and $\mathcal{D}^{(2)}$ generally is overdetermined or underdetermined. Indeed, contrary to the continuous case, the position of $w^{(1)}$ compared with the eigenvalues $\lambda_k^{(1)}$ defined in $\mathcal{D}^{(1)}$ does not a priori imply any particular position of $w^{(2)}$ compared with the eigenvalues $\lambda_k^{(2)}$ defined in $\mathcal{D}^{(2)}$. On the assumption that the M jump relations are independent and non-degenerate, two sub-domains $\mathcal{D}^{(1)}$ and $\mathcal{D}^{(2)}$ can be matched only if the total number of compatibility relations retained in the two sub-domains is equal to : $M+1$.

This propitious situation occurs when the discontinuity Σ is a k-shock, that is to say, when the eigenvalues $\lambda_k^{(1)}$

and $\lambda_k^{(2)}$ comply with the "entropy condition" of Lax [15] : there exists an integer k such that

$$\text{(18)} \qquad \begin{cases} \lambda_{k-1}^{(1)} < w^{(1)} < \lambda_k^{(1)} \\ \lambda_{M-k}^{(2)} < w^{(2)} < \lambda_{M-k+1}^{(2)} \end{cases}$$

Then, ($M-k+1$) compatibility relations must be used in $\mathcal{D}^{(1)}$ and k in $\mathcal{D}^{(2)}$, hence a total of ($M+1$) relations. These ($M+1$) compatibility relations joined with the M jump relations (17) enable us to calculate the ($2M+1$) unknowns at the point P of Σ , namely : the M components of the vector $\underline{U}^{(1)}$ in the sub-domain $\mathcal{D}^{(1)}$, the M components of the vector $\underline{U}^{(2)}$ in the sub-domain $\mathcal{D}^{(2)}$, and the normal velocity w of the discontinuity Σ .

If Σ is a k-contact discontinuity, that is to say, if an integer k exists for which the eigenvalues $\lambda^{(1)}$ and $\lambda^{(2)}$ satisfy the conditions :

$$\text{(19)} \qquad \lambda_k^{(1)} = w^{(1)} = - w^{(2)} = - \lambda_{M-k+1}^{(2)}$$

the conclusion is less obvious. Indeed, the total number of compatibility relations to be used in the sub-domains $\mathcal{D}^{(1)}$ and $\mathcal{D}^{(2)}$ is equal to $M+r$, where r denotes the order of the eigenvalue λ_k (or μ_{M-k+1}). The velocity $w^{(1)}$ (or $w^{(2)}$) being derived from the relation (19), the problem of the matching of two sub-domains $\mathcal{D}^{(1)}$ and $\mathcal{D}^{(2)}$ is then determined only if r jump relations (among the M relations (17)) are degenerate, that is to say, are automatically verified and disappear.

2.4 - Application to the Euler equations -

Let us now apply the general treatment of external or internal boundaries in the case when system (1)is composed of the 2-D Euler equations. We consider only the cases of boundary treatments encountered in the numerical application described in § 4.

a) External boundaries :

In the case of the Euler equations, the four eigenvalues associated with the unit normal $\vec{\nu}$ to a boundary Σ have the following expressions :

$$\text{(20)} \qquad \lambda_1 = V_\nu - a \ , \ \lambda_2 = \lambda_3 = V_\nu \ , \ \lambda_4 = V_\nu + a$$

where V_ν is the $\vec{\nu}$-component of the velocity ($V_\nu = \vec{V}.\vec{\nu}$) and a is the sound speed.

When the elements of $\underline{f}$ (see § 2.1) are the density ρ , the two components (V_ν , V_ξ) of the velocity and the static pressure p , the compatibility relation (6) may be written in the following concise form :

(21.a) $$-\rho a\left[\frac{\partial V_\nu}{\partial t}+(V_\nu-a)\frac{\partial V_\nu}{\partial \nu}\right]+\left[\frac{\partial p}{\partial t}+(V_\nu-a)\frac{\partial p}{\partial \nu}\right]=RHS_1$$

(21.b) $$\frac{\partial V_\xi}{\partial t}+V_\nu\frac{\partial V_\xi}{\partial \nu}=RHS_2$$

(21.c) $$-a^2\left[\frac{\partial \rho}{\partial t}+V_\nu\frac{\partial \rho}{\partial \nu}\right]+\left[\frac{\partial p}{\partial t}+V_\nu\frac{\partial p}{\partial \nu}\right]=RHS_3$$

(21.d) $$+\rho a\left[\frac{\partial V_\nu}{\partial t}+(V_\nu+a)\frac{\partial V_\nu}{\partial \nu}\right]+\left[\frac{\partial p}{\partial t}+(V_\nu+a)\frac{\partial p}{\partial \nu}\right]=RHS_4$$

The right hand sides contain only derivatives along the direction $\vec{\xi}$ which is normal to $\vec{\nu}$, hence tangent to the boundary Σ . Let us notice that :

- only derivatives of V_ξ appear in the left-hand side of the second relation (21.b) ;

- the third relation (21.c) expresses the conservation of entropy along a particle path ;

- only derivatives of V_ν and p appear in the left-hand sides of both relations (21.a) and (21.d).

Concerning the number m of boundary conditions to prescribe on a possibly moving boundary, the rule stated in § 2.1 leads to a simple discussion, where only the value of the relative normal Mach number $M'_\nu=(V_\nu-w)/a$ matters,

$M'_\nu<-1$	(supersonic upstream boundary)	: m = 4
$-1\leq M'_\nu<0$	(sonic or subsonic upstream boundary)	: m = 3
$0\leq M'_\nu<1$	(subsonic downstream boundary, or wall)	: m = 1
$1\leq M'_\nu$	(sonic or supersonic downstream boundary)	: m = 0

In the first case, all the variables must be prescribed at the boundary point ; in the fourth case, no boundary condition must be prescribed, and the variables are directly derived from the numerical scheme. Let us detail the second and the third cases.

. <u>Subsonic upstream boundary</u> :

In this case, we must use the compatibility relation corresponding to the eigenvalue $\lambda_4=V_\nu+a$, which yields :

(22) $$p^{n+1}+(\rho a)^n\,V_\nu^{n+1}=p^*+(\rho a)^n\,V_\nu^*$$

where the subscript * denotes values directly provided by the numerical scheme (Eq. (1)), and three boundary conditions must be imposed. A possible choice of these boundary conditions consists in prescribing the angle θ of the velocity vector with $\vec{\nu}$, the stagnation enthalpy H and the entropy S. These boundary conditions allow to consider the compatibility relation (22) as a non-linear equation for the magnitude V of the velocity vector :

$$p(V^{n+1}) + (\rho a)^n V^{n+1} \cos\theta = p^* + (\rho a)^n V_\nu^* \quad (23)$$

This equation can be solved by Newton's method, and knowing V^{n+1}, all the variables may be computed from the boundary conditions.

. Subsonic downstream boundary (and wall) :

In this case, we must use the three compatibility relations corresponding to the eigenvalues $\lambda_2 = \lambda_3 = V_\nu$, $\lambda_4 = V_\nu + a$ which yields :

$$\begin{cases} p^{n+1} - (a^2)^n \rho^{n+1} = p^* - (a^2)^n \rho^* \\ V_\xi^{n+1} = V_\xi^* \\ p^{n+1} + (\rho a)^n V_\nu^{n+1} = p^* + (\rho a)^n V_\nu^* \end{cases} \quad (24)$$

and one boundary condition must be imposed. For a subsonic downstream boundary, the generally used boundary condition is the static pressure value setting. Knowing p^{n+1}, system (24) immediately yields ρ^{n+1}, V_ξ^{n+1} and V_ν^{n+1}. For a wall, the normal speed W of which is known, the slip condition gives : $V_\nu^{n+1} = W$, and p^{n+1}, ρ^{n+1} and V_ξ^{n+1} are easily derived from (24).

b) Internal boundaries :

We examine now the case when Σ is an interface dividing two sub-domains $\mathcal{D}^{(1)}$ and $\mathcal{D}^{(2)}$. For the Euler equations, the number of compatibility relations to retain in a sub-domain $\mathcal{D}^{(i)}$ depends only on the value of the relative Mach number along the direction $\vec{\nu}^{(i)}$, namely $M'_{\nu(i)} = (V_{\nu(i)} - W^{(i)})/a$.

. Artificial interface (flow properties are continuous) :

Suppose for instance that the fluid goes through Σ from $\mathcal{D}^{(1)}$ to $\mathcal{D}^{(2)}$ with a subsonic normal velocity : $0 < M'_{\nu(1)} < 1$ (hence, $-1 < M'_{\nu(2)} < 0$). On this assumption, the three compatibility relations corresponding to the eigenvalues $\lambda_2^{(1)} = V_{\nu(1)}$, $\lambda_3^{(1)} = V_{\nu(1)}$, $\lambda_4^{(1)} = V_{\nu(1)} + a$ must be retained in $\mathcal{D}^{(1)}$ and the compatibility relation corresponding to the eigenvalue $\lambda_4^{(2)} = V_{\nu(2)} + a$ must be retained in $\mathcal{D}^{(2)}$. In order to simplify the notations, it is convenient to consider only one normal vector $\vec{\nu}$ to Σ ; for instance : $\vec{\nu} = \vec{\nu}^{(1)} = -\vec{\nu}^{(2)}$. Taking into account the matching relations which express the continuity across Σ, the discretized compatibility relations form the following system :

(25)
$$
\begin{cases}
V_\xi^{n+1} = [V_\xi^*]^{(1)} \\
p^{n+1} - (a^2)^n \rho^{n+1} = [p^* - (a^2)^n \rho^*]^{(1)} \\
p^{n+1} + (\rho a)^n V_\nu^{n+1} = [p^* + (\rho a)^n V_\nu^*]^{(1)} \\
p^{n+1} - (\rho a)^n V_\nu^{n+1} = [p^* - (\rho a)^n V_\nu^*]^{(2)}
\end{cases}
$$

Due to the continuity, it is not necessary to specify the sub-domain superscript in the left-hand sides. In return this superscript must be specified in the right-hand sides in order to know in which sub-domain the space derivatives are discretized. Solving system (25) is always possible without any difficulty.

. Shock :

Suppose for instance that the fluid goes through a shock Σ, from $\mathcal{D}^{(1)}$ to $\mathcal{D}^{(2)}$. For the Euler equations, the entropy condition of Lax (18) may be expressed :

(26) $$M'^{(1)}_{\nu_1} > 1 \quad \text{and} \quad -1 < M'_{\nu_2^{(2)}} < 0$$

Therefore, in $\mathcal{D}^{(1)}$, we must retain the four compatibility relations, which implies that the values $\underline{U}^{(1)*}$ given by the numerical scheme are kept, without any change ($\underline{U}^{(1)n+1} = \underline{U}^{(1)*}$), and in $\mathcal{D}^{(2)}$ we must retain the compatibility relation corresponding to the eigenvalue $\lambda_4^{(2)} = V_{\nu^{(2)}}^{(2)} + a^{(2)}$. This relation together with the four Rankine-Hugoniot jump relations form the following non-linear system in the unknowns $\underline{U}^{(2)n+1}$ and W^{n+1}

(27.a) $$p^{(2)n+1} - (\rho a)^{(2)n} V_\nu^{(2)n+1} = p^{(2)*} - (\rho a)^{(2)n} V_\nu^{(2)*}$$

(27.b) $$\rho^{(2)n+1} (V_\nu^{(2)n+1} - W^{n+1}) = \rho^{(1)n+1} (V_\nu^{(1)n+1} - W^{n+1}) \quad (= Q)$$

(27.c) $$p^{(2)n+1} + Q\, V_\nu^{(2)n+1} = p^{(1)n+1} + Q\, V_\nu^{(1)n+1}$$

(27.d) $$V_\xi^{(2)n+1} = V_\xi^{(1)n+1}$$

(27.e) $$\frac{\gamma}{\gamma-1} \frac{p^{(2)n+1}}{\rho^{(2)n+1}} + \frac{1}{2} (V_\nu^{(2)n+1} - W^{n+1})^2 = \frac{\gamma}{\gamma-1} \frac{p^{(1)n+1}}{\rho^{(1)n+1}} + \frac{1}{2} (V_\nu^{(1)n+1} - W^{n+1})^2$$

As in the previous paragraph, the normal vector $\vec{\nu}$ has been chosen equal to $\vec{\nu}^{(1)}\ (= -\vec{\nu}^{(2)})$, and W denotes the velocity of Σ along $\vec{\nu}$.

In the code, system (27) is solved as follows : equation (27.a) and equation (27.c) linearized by setting $Q = \rho^{(1)n+1}(V_\nu^{(1)n+1} - W^n)$ yield $p^{(2)n+1}$ and $V_\nu^{(2)n+1}$; then, $\rho^{(2)n+1}$ and W^{n+1} are derived from equations (27.b) and (27.e) ; at last, equation (27.d) directly gives $V_\xi^{(2)n+1}$.

3 - ZONAL GRID REFINEMENT TECHNIQUE -

3.1 - Mesh refinement by dichotomy technique -

For large Reynolds number flows, an accurate description of a viscous layer requires that the transversal mesh dimension be refined by a factor as high as several hundreds when going from the inviscid flow region to the laminar sublayer at the wall. Usually such a refinement is obtained by means of an analytical stretching transformation. Here we describe a zonal refinement technique, which has been developed at ONERA since several years [16], [17], and which combines the stretching transformation with a dichotomy technique.

In two dimensions, basic curvilinear coordinates ξ , η , are defined, analytically or numerically, such that the line $\eta = 0$ represents the wall and that there exists a curvilinear strip $0 \leq \eta \leq \eta_e(\xi)$ which contains the viscous layer. The first family of mesh lines is made of lines $\xi = Cst$ which go across the viscous layer. Here we are concerned with the determination of the mesh lines of the second family which we choose to be of the form $\eta/\eta_e(\xi) = Cst$; so we introduce a transversal variable Y by the relation :

(28) $$\eta/\eta_e(\xi) = g(Y), \quad \text{with } g(0) = 0, g(1) = 1$$

where $g(Y)$ is a stretching function. A first set of mesh lines $Y = Cst$, to be called the basic mesh, is then determined by the condition of uniform step in Y, ΔY :

(29) $$Y = Y_j = (j-1)\,\Delta Y \ , \quad j = 1, 2, \ldots J+1$$

with : $$J\,\Delta Y = 1$$

A classical form for the stretching function $g(Y)$ is the exponential :

(30) $$g(Y) = (e^{\beta Y} - 1)/(e^{\beta} - 1) \ , \quad \beta > 0$$

It has the property that the local mesh size ratio in terms of η :

(31) $$r_j = (\eta_{j+1} - \eta_j)/(\eta_j - \eta_{j-1})$$

where $\eta_j = \eta_e\, g(Y_j)$, is a constant :

(32) $$r_j = r = e^{\beta \Delta Y}$$

The global mesh size ratio for the basic mesh, in terms of η, is :

$$R_B = \frac{\eta_{J+1} - \eta_J}{\eta_2 - \eta_1} = e^{\beta(1-\Delta Y)} = r^{J-1} \tag{33}$$

In the application presented in § 4.5, the mesh used in the viscous sub-domain corresponds to $\xi = x$ and $\eta = y - y_0(x)$ where x, y are cartesian coordinates, so that r_j and R_B give respectively the local and global mesh refinement ratios in the physical plane. In the general case, these physical refinement ratios have more complex expressions involving the transformation formulae giving x, y as functions of ξ, η, and in particular r_j is not constant.

In usual methods, the refinement required in the viscous layer is obtained only from a stretching function such as (30). In practice, one must be careful that the local refinement ratio r be not too large compared to 1, otherwise truncation errors could become numerically important depending on how rapidly the flow properties vary. Therefore, if R_B is chosen, the condition $r \leq r_M$ leads to $J-1 \geq \mathrm{Log}\, R_B / \mathrm{Log}\, r_M$ which imposes a minimum number of mesh lines.

Another way to refine the mesh is to make a discontinuous change in mesh size in Y. In the dichotomy technique used here, the mesh size in Y is divided by 2, and this operation is done several times starting from the basic mesh, in such a way that the mesh becomes constituted of a series of zones each of them having a constant mesh size in Y. These zones will be referred to by an integer q, $q = 1$ to Q, increasing when going from the wall to the outer flow (fig. 1). Thus zone 1 is adjacent to the wall, and zone Q is part of the basic mesh. Let $\Delta Y^{(q)}$ be the mesh size in Y in zone q ; we have the relation :

$$\Delta Y^{(q)} = 2^{q-Q}\, \Delta Y \tag{34}$$

The mesh lines belonging to zone q will be numbered with a local index ℓ varying from 1 to $L^{(q)}$ in the direction of increasing Y (i.e. increasing η) (fig. 1).

With this dichotomy technique, the global mesh refinement ratio is now found to be :

$$R_G = \frac{\eta^{(Q)}_{L^{(Q)}} - \eta^{(Q)}_{L^{(Q)}-1}}{\eta^{(1)}_2 - \eta^{(1)}_1} = e^{\beta}\, \frac{1 - 1/r}{r^{2^{1-Q}} - 1} \tag{35}$$

where $\eta^{(q)}_\ell$ is the value of η for the mesh line ℓ of zone q. It can be shown that for the same value of β and the same global mesh refinement ratio, the dichotomy technique allows a saving of mesh lines (that is to say, of effective mesh lines on which the scheme is actually applied). An additional interesting feature of the dichotomy arises from the fact that the local mesh refinement ratio depends on the zone q and becomes closer to 1 as q decreases ; this ratio, $r^{(q)}$, is given by formula (32) but with $\Delta Y^{(q)}$ instead of ΔY, hence:

$$r^{(q)} = r^{2^{q-Q}} \quad , \quad \text{and} \quad r^{(q)} = \sqrt{r^{(q+1)}} \tag{36}$$

This effect is very favourable for the accuracy since the condition that $r^{(q)}$ be not too large compared to 1 is more stringent near the wall, where flow properties vary rapidly, than in the outer part of the viscous layer.

Discretization of the equations at a point where the mesh size is discontinuous is hazardous for the accuracy unless special complicated schemes are used ; in fact we are confronted again with the same problem as above concerning the limitation on r . This problem now disappears if we can avoid writing discretized equations at such a point ; this is possible if we make two adjacent zones overlap so as to transfer information between them. This matching technique is explained on fig. 1.

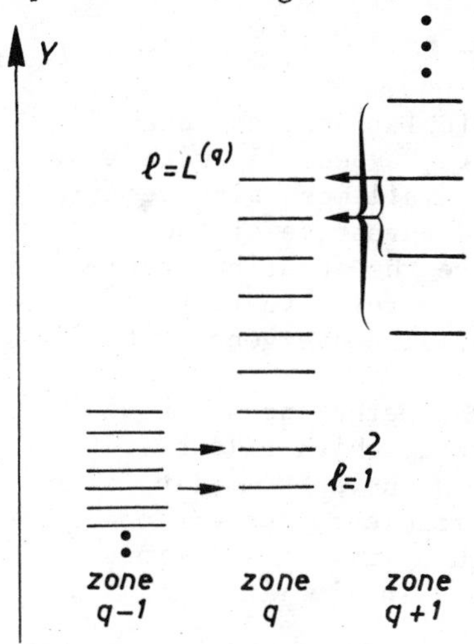

Fig. 1 — Dichotomy technique and matching of zones.

The extent of the overlap region is fixed by the numerical domain of dependence of the scheme. In the case of the MacCormack scheme applied to the Navier-Stokes equations (§ 4), this domain is centered at the point considered and extends over 2 cells in each direction ; therefore the scheme can be used in zone q only on the lines $\ell = 3$ to $\ell = L^{(q)}-2$. The values $\underline{U}_\ell^{(q)}$ of the unknowns $\underline{U}$ on the lines $\ell = 1, 2$, $L^{(q)}-1$ and $L^{(q)}$ are then determined from the knowledge of $\underline{U}$ in the two adjacent zones ($q-1$) and ($q+1$) through the relations :

$$(37)\quad \begin{cases} (a) \quad \underline{U}_1^{(q)} = \underline{U}_{L^{(q-1)}-4}^{(q-1)} \quad , \quad \underline{U}_2^{(q)} = \underline{U}_{L^{(q-1)}-2}^{(q-1)} \\[2ex] (b) \begin{cases} \underline{U}_{L^{(q)}}^{(q)} = \underline{U}_3^{(q+1)} \\[2ex] \underline{U}_{L^{(q)}-1}^{(q)} = \dfrac{1}{16}\left\{ -\underline{U}_1^{(q+1)} + 9\left[\underline{U}_2^{(q+1)} + \underline{U}_3^{(q+1)}\right] - \underline{U}_4^{(q+1)} \right\} \end{cases} \end{cases}$$

Figure 2 shows the zonal structure in the physical plane of the mesh used for the application presented in § 4.5 ; the figure is in two parts on different scales. The parameters for this mesh are :

$$Q = 6 \quad , \quad J \text{ (basic mesh)} = 19 \quad , \quad \beta = 2.474$$

$$L^{(1)} = 7 \quad , \quad L^{(2)} = L^{(3)} = L^{(4)} = L^{(5)} = 6 \quad , \quad L^{(6)} = 19$$

we deduce r = 1.139, R_B = 10.4 and R_G = 355.5. To reach the same global refinement ratio without dichotomy, with the same β , requires $J+1$ = 47 mesh lines, instead of 30 effective mesh lines in the present mesh. Moreover, without dichotomy, the local mesh refinement ratio would be constant and equal to 1.139 even near the wall, whereas in the present mesh this local ratio becomes very close to 1 in the zones 5 to 1 ($r^{(5)}$ = 1.067, $r^{(1)}$= 1.004). If the same global refinement ratio (R_G = 355.5) were to be obtained without dichotomy and with a local refinement ratio limited to r = 1.1, it would be necessary to have 64 mesh lines.

3.2 - Zonal time-marching procedure -

In high Reynolds number flow calculations, the stability condition can be very restrictive, especially with an explicit scheme, because of the very small mesh size required near the wall. It is possible to take advantage of the zonal structure of the mesh to advance the solution in time in each zone with the maximum time-step relative to this zone, thus greatly improving the overall convergence rate.

For steady flow calculations, the method need not be time-accurate, nor even time-consistent, which allows to apply the matching relations (37) with uneven time levels in the two sides of each relation. Nevertheless, convergence difficulties may arise from the matching if the solutions in the different zones do not progress in time in some orderly way.

We describe the zonal time-marching procedure which has been set up and which is valid for an arbitrary number of zones. It differs somewhat from the procedure used in [17], but it is always based on the choice of a time step $\Delta t^{(q)}$ for zone q proportional to $\Delta y^{(q)}$, hence from (34) :

(38) $$\Delta t^{(q)} = 2^{q-Q}\, \Delta t$$

where Δt is the time step used in zone Q . This choice corresponds approximately to the optimum value in each zone if the stability is governed by an inviscid CFL-type criterion. As a consequence, the number of iterations (time steps) to be made in zone q for one iteration in zone Q is $N^{(q)} = 2^{Q-q}$. How the iterations progress in the

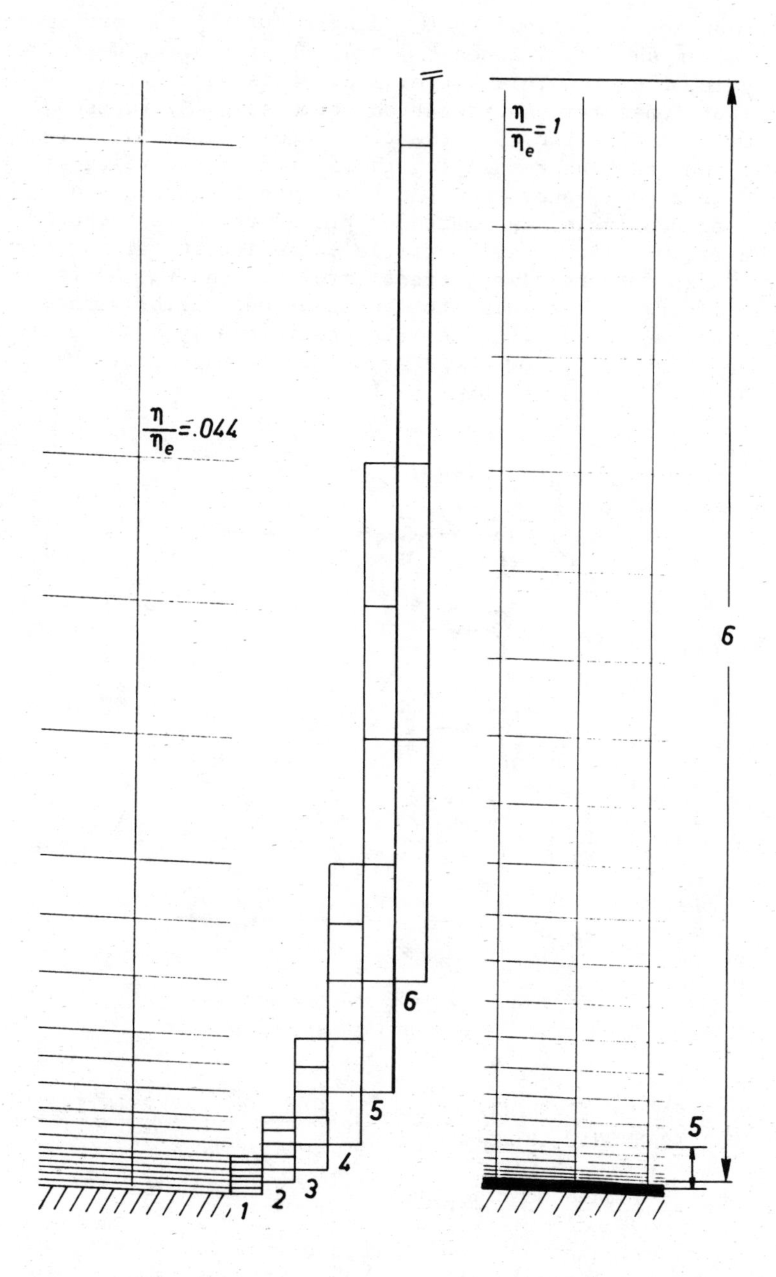

Fig. 2 – Mesh lines $Y = Cst$ in the physical plane and zonal structure ($Q = 6$).

different zones is shown by the diagram of fig. 3 corresponding to a mesh with 4 zones (Q = 4). This diagram describes one cycle, i.e. the full set of iterations made in the different zones for one iteration (time step Δt) in zone Q. The small circles indicate the time levels in the different zones, and the line connecting these circles, starting at $t = t^n$ in zone 1, shows how the time levels vary in the course of one cycle. The matching procedure is represented by the arrows placed near a circle : the latest values calculated in zone q are always transferred to zone ($q-1$) (arrow ←), if $q \neq 1$, according to the relations (37.b) with q replaced by ($q-1$), and they are conditionally transferred to zone ($q+1$) (arrow →) according to the relations (37.a) with q replaced by ($q+1$).

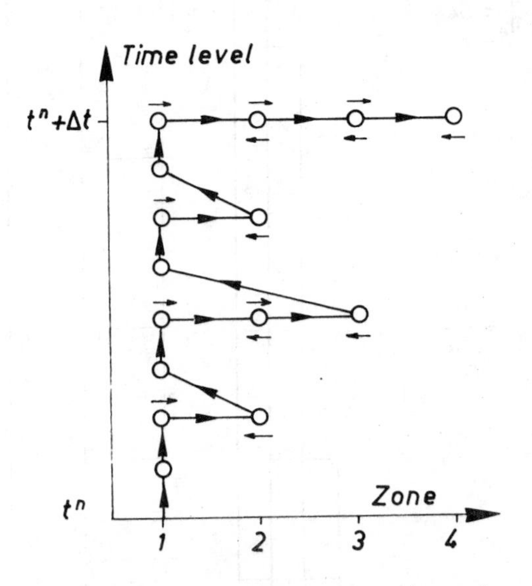

Fig. 3 – Zonal time-marching procedure and matching. Iterations for 1 cycle (Q = 4).

4 - APPLICATION TO SHOCK-TURBULENT BOUNDARY-LAYER INTERACTION IN A TRANSONIC CHANNEL -

4.1 - Multi-domain description of flow field -

We describe now an application of the sub-domain approach to the numerical simulation of a shock-turbulent boundary layer interaction in a transonic channel based on the time-averaged Navier-Stokes equations. This problem has already been treated by mono-domain methods, the shock being captured in the whole domain (e.g see [18] to [20]).

The sub-domain approach has three advantages : first, to limit the region where the Navier-Stokes equations are solved ; second, to simplify mesh problems and third, to fit the major part of the shock. A saving of computational time results from the first point. A saving of mesh points (hence of computational time) and a better accuracy result from the last two points.

The channel under consideration is plane and symmetrical. The computational domain (fig. 5) is bounded by the wall, the symmetry axis, an upstream section and a downstream section (located far enough from the interaction region) ; it is divided into three sub-domains as follows. A sub-domain $\mathcal{D}_v$ where viscous effects are important is included between the wall and a longitudinal interface Σ_1 . The remaining part $\mathcal{D}_p$ of the computational domain, where the inviscid approximation is justified, is divided into two sub-domains $\mathcal{D}_p^{(1)}$ and $\mathcal{D}_p^{(2)}$ separated by the shock-wave Σ_2 . So, the shock-wave is fitted in $\mathcal{D}_p$ and captured in $\mathcal{D}_v$.

The Euler equations are solved in the two sub-domains $\mathcal{D}_p^{(1)}$ and $\mathcal{D}_p^{(2)}$, and the time-averaged Navier-Stokes equations with an eddy viscosity type model are solved in the sub-domain $\mathcal{D}_v$. The basic variables $\underline{U}$ in $\mathcal{D}_v$ are the time-averaged values of the density, ρ , of the momentum, $\rho\vec{V}$ ($\vec{V}$ is therefore the mass-averaged mean velocity), of the internal energy density, ρe (e is therefore the mass-averaged mean specific internal energy).

4.2 - Boundary conditions -

On the wall we impose a no-slip condition ($\vec{V} = 0$) and a zero heat flux ($\vec{\nu} . \nabla e = 0$).

On the symmetry axis,a boundary of $\mathcal{D}_p$ we impose a slip condition ($\vec{V} . \vec{\nu} = 0$) which is equivalent in perfect fluid to the symmetry condition.

On the upstream boundary where the flow is supersonic except in a very small region near the wall, all flow properties are prescribed.

At last, on the downstream boundary where the flow is subsonic, we set the pressure value. The shock position essentially depends on this value.

4.3 - Turbulence model -

The numerical results presented in § 4.5 have been computed with the algebraic mixing length type model studied by Michel et al. [21] in the framework of Prandtl's equations, and valid for non separated boundary layers. For the Navier-Stokes equations, this model has been modified by replacing the normal derivative of the velocity by the opposite of the vorticity : $\zeta = \frac{\partial v}{\partial x} - \frac{\partial u}{\partial y}$. Then the eddy viscosity is given by :

$$\mu_T = \rho\, \ell^2 F^2(\Lambda) |\zeta| \tag{39}$$

The ratio of the mixing length ℓ to the boundary layer thickness δ is given by the classical law :

$$\frac{\ell}{\delta} = 0.085\ \mathrm{th}\left(\frac{K}{0.085}\frac{\eta}{\delta}\right) \tag{40}$$

where η is the distance normal to the wall and K is the Karman constant (K = 0.41).

The function F is a damping factor and depends on the quantity $\tau = (\mu + \mu_T)\zeta$, through the variable $\Lambda = \rho \ell^2 |\tau| / \mu^2$ (μ is the molecular viscosity) :

$$F(\Lambda) = 1 - \exp\left(-\sqrt{\Lambda}/26K\right) \tag{41}$$

The relation (39) giving μ_T is therefore implicit since F depends on μ_T . Actually μ_T/μ is found to be a function of the variable $z = |\zeta| \rho \ell^2/\mu$ which can be tabulated once for all.

4.4 - Numerical method -

a) Finite difference scheme :

The Navier-Stokes equations in $\mathcal{D}_V$ as well as the Euler equations in $\mathcal{D}_p^{(1)}$ and $\mathcal{D}_p^{(2)}$ are solved by a time marching method with an explicit predictor-corrector scheme. The equations are discretized in conservative form using MacCormack's finite difference scheme [22] directly in the physical plane on an arbitrary curvilinear mesh. This discretization method has been presented in [10], [17] and [23].

The numerical treatment of the upstream and downstream boundaries, of the symmetry axis and of the two internal boundaries is based on the use of compatibility relations as described in § 2.4. The detail of the computation at the wall can be found in [23].

b) Time varying mesh :

Contrary to some authors (see, for instance, [24]), who fit discontinuities in a fixed mesh, we consider discontinuities as moving mesh lines. In the application presented here, the shock is a downstream boundary mesh line of $\mathcal{D}_p^{(1)}$ and an upstream boundary mesh line of $\mathcal{D}_p^{(2)}$. Consequently, the computation must be carried out in a moving mesh. Note that the mesh is moving in $\mathcal{D}_v$, as well as in $\mathcal{D}_p$, for reasons discussed in the next paragraph.

The numerical scheme is defined in a fixed mesh. It could be applied in a moving mesh, like in [10], by replacing $\partial \underline{U}/\partial t$ in system (1) by $\partial \underline{U}/\partial \tau - \vec{W} . \nabla \underline{U}$ where $\partial/\partial\tau$ represents the time derivative for a given moving mesh point of velocity $\vec{W}$, and by integrating the modified system with respect to the variable τ . We use here another technique. $\underline{U}^{n+1}$ is computed in a fixed mesh (defined at time t^n), and is projected upon a new mesh defined at time t^{n+1}. If M and M' are the corresponding mesh points respectively in the first mesh (t^n-mesh) and in the second mesh (t^{n+1}-mesh), then the value at M' of a quantity ϕ known in the t^n-mesh (the time being frozen at t^{n+1}) is obtained from a first order Taylor's expansion :

$$\left\{ \begin{array}{l} \overrightarrow{MM'} = \vec{W} \Delta t \\ \phi(M') = \phi(M) + \overrightarrow{MM'} . \nabla \phi \end{array} \right. \tag{42}$$

The displacement MM' is always by far less than the size of the mesh cells.

c) Computation at the triple point :

The triple point is the intersection of the interface Σ_1 (or cut) and the fitted part Σ_2 of the shock wave. This point is a node, A , of the $\mathcal{D}_p^{(1)}$ mesh and a node, A' , of the $\mathcal{D}_p^{(2)}$ mesh. In $\mathcal{D}_v$, mesh lines i = const. (which are vertical lines) must be contracted in the shock region in order for the shock capturing process to be accurate enough, and this mesh contraction must follow the shock motion. Moreover special attention must be paid to the change from shock fitting in $\mathcal{D}_p$ to shock capturing in $\mathcal{D}_v$. For these reasons, the triple point will be attached to a given mesh line of $\mathcal{D}_v$ The sketch of figure 4 shows the details of the mesh in the neighbourhood of the triple point (the interfaces between the sub-domains are split into two lines in order to separate the nodes). It is clear that this triple point requires a suitable treatment which is developed just below.

The numerical scheme is not applied at the nodes A and A' in $\mathcal{D}_p$ but the values of $\underline{U}^*$ at these nodes (values which appear in the compatibility relation (13)) are derived from three upstream nodes for A , and from three downstream nodes for A' by means of parabolic extrapolations along the interface Σ_1 . The displacement of the triple point is not calculated from the shock velocity obtained by the standard shock treatment. In fact, in order to connect the fitted shock in $\mathcal{D}_p$ to the captured shock in $\mathcal{D}_v$, the triple point is defined, at each time step, as the intersection of the interface Σ_1 and the segment BS (fig. 4). The point S is the sonic point of the mesh line (j = const.) just below Σ_1 . For the mesh points of $\mathcal{D}_V$ located on Σ_1 between E and F (fig. 4), the unknown $\underline{U}$ is derived from a linear interpolation first between C and A , then between A' and D , and setting : $\underline{U}(A'') = (\underline{U}(A) + \underline{U}(A'))/2$. The shock thickness on Σ_1 , in $\mathcal{D}_V$, is thus restricted to two cells. At the $\mathcal{D}_V$ nodes E and F as well as at the $\mathcal{D}_P$ nodes C and D the standard artificial interface treatment is used.

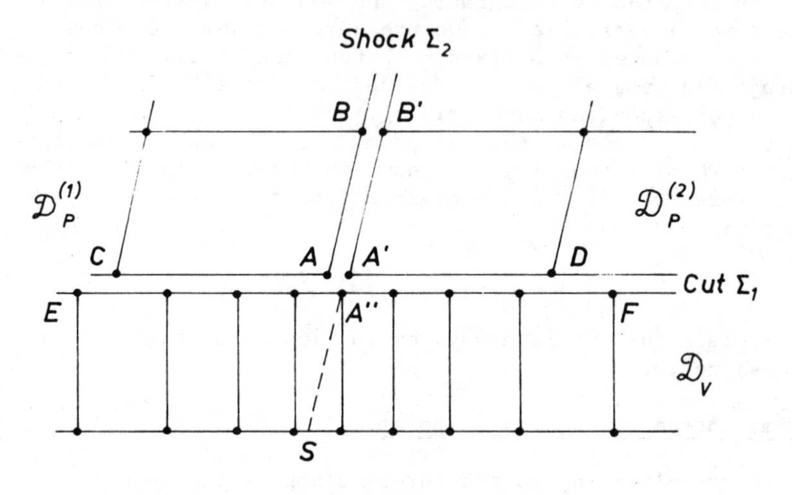

Fig. 4 — Sketch of the mesh near the triple point.

4.5 - <u>Numerical results</u> -

The results presented in this paragraph correspond to the conditions of an experimental study, conducted in the ONERA S8 transonic wind tunnel, of the shock-turbulent boundary layer interaction phenomenon [25]. The stagnation conditions are the following : p_{i_0} = 95 kPa, T_{i_0} = 300 K. The Reynolds number Re calculated from the height of the symmetric channel ($H = 10^{-1}$ m), the stagnation sound speed and the kinematic viscosity at stagnation condition, is equal to $2.078.10^6$. The turbulent Prandtl number is 0.9 ; the Prandtl number is 0.725 and the ratio of specific heats is 1.4. Lastly, the Mach number at the start of the interaction is about equal to 1.3.

In [8] the same problem was calculated with somewhat different upstream conditions and downstream pressure, and using a coarser mesh. The new calculation presented here is found to be in better agreement with the experiment.

The computational domain, which extends from x/H = 1.03 to x/H = 2.38, and the meshes (at convergence) are shown on fig. 5. The meshes are made up of (26 x 15) points in $\mathcal{D}_p^{(1)}$, (30 x 15) points in $\mathcal{D}_p^{(2)}$ and (65 x 35) points in $\mathcal{D}_V$. The zonal structure of the mesh in $\mathcal{D}_V$ is shown on fig. 2.

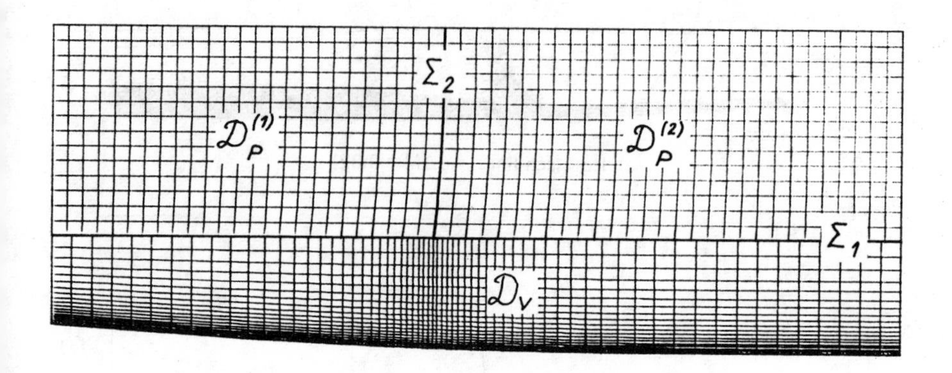

Fig. 5 — Computational domain and sub-domain meshes.

On the upstream boundary, flow conditions have been provided by a preliminary computation carried out in the upstream part of the nozzle ($-0.27 \leq x/H \leq 1.13$), in a shockless choked configuration. It has been verified that the results of the two computations in the overlapping region ($1.03 \leq x/H \leq 1.13$) are practically identical.

The initial conditions correspond to an inviscid one-dimensional flow with a shock located at its experimental position. On the downstream boundary, the pressure is imposed equal to a value ($p_2 = 0.655\, p_{i_0}$) which is different from the value ($p_2' = 0.672\, p_{i_0}$) of the pressure in this 1D inviscid approximation, and which is determined by taking into account the displacement effect (from the measured velocity profiles). The imposed downstream pressure is found to be different from the experimental value ($p_2'' = 0.636\, p_{i_0}$) because of side effects resulting from the presence of lateral boundary layers in the experiment.

The main features of the computed flow field are displayed on figs. 6a, b, c, which represent the contour lines respectively for the Mach number, for the static pressure and for the density. A rapid thickening of the boundary layer, due to the interaction with the shock wave can be noticed on fig. 6a. An enlargement of the Mach number field in the interaction region is shown on fig. 7. This enlargement points out the smooth transition between the fitted shock and the captured shock. A qualitative comparison between calculation and experiment is provided by figs 6c and 6d ;

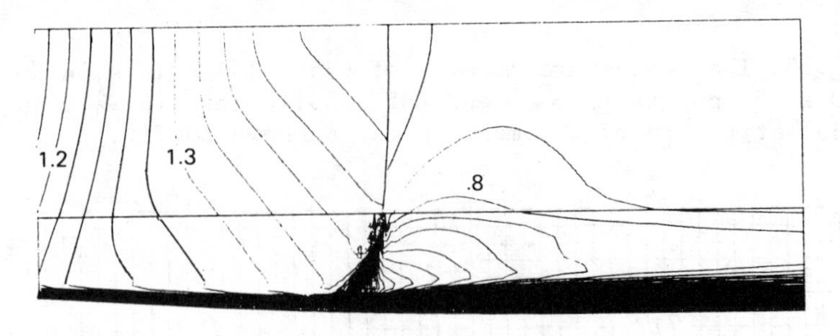

a) Iso-Mach lines $(\Delta M = 0.02)$

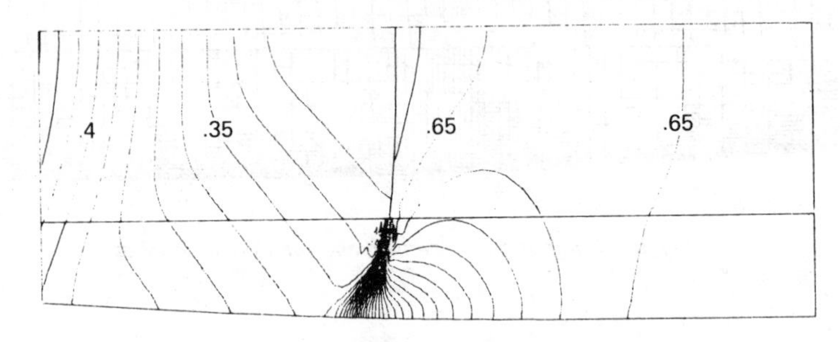

b) Isobaric lines $(\Delta p = 0.01\ p_{io})$

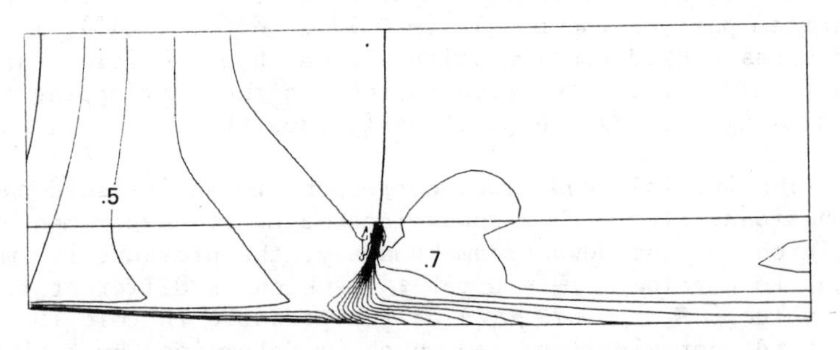

c) Iso-density lines $(\Delta \rho = 0.02\ \rho_{io})$

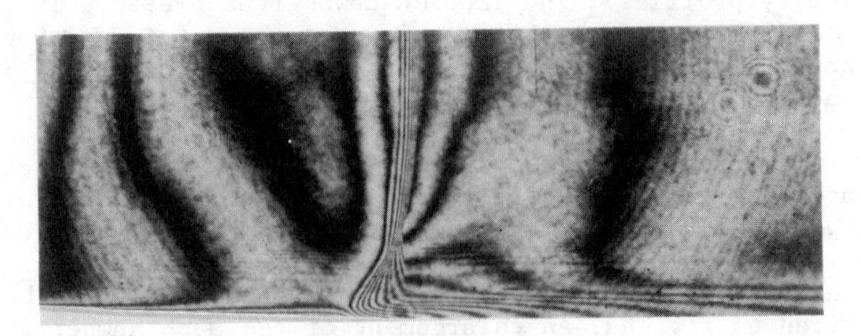

d) Interferogram

Fig. 6 – Flow field

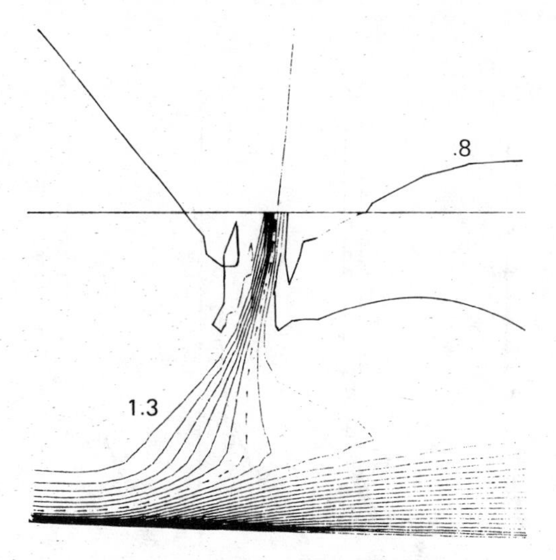

Fig. 7 – Iso-Mach lines (ΔM = 0.05) in the interaction region. (––– Sonic line).

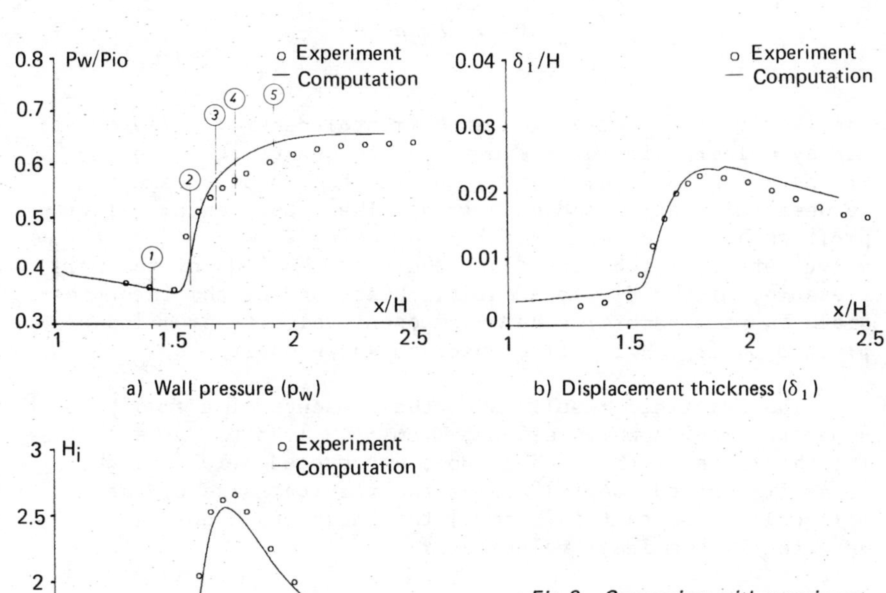

Fig. 8 – Comparison with experiment.

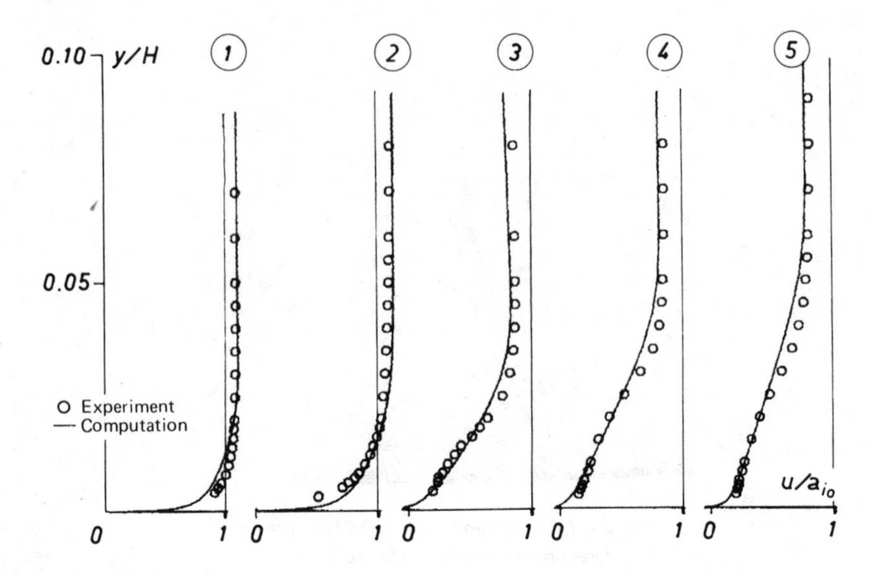

Fig. 9 — Velocity profiles.

the latter representing the experimental density field obtained by holographic interferometry. The quantitative comparisons shown in figs. 8 and 9 confirm the good agreement between calculation and experiment. The experimental velocity profiles have been measured [25] by using a two-color laser velocimeter. Fig. 8 represents the distributions of the wall pressure, of the displacement thickness and of the incompressible shape parameter. The x-component velocity profiles are plotted on fig. 9 for five abscissa shown on fig. 8a.

The numerical results show the presence of a small separated region which extends from $x/H = 1.58$ to $x/H = 1.84$. The thickness of this bubble does not exceed $8.10^{-4} H$, which is equivalent to about 13 times the vertical size of the wall cell. Note that this small thickness could not be detected by the laser velocimeter.

Acknowledgements.

This work was supported by DRET and STPA.

5 - REFERENCES -

1. BRUNE, G.W., RUBBERT, P.E. and FORESTER, C.K. - The Analysis of Flow Fields with Separation by Numerical Matching, AGARD-CP-168, 1975.

2. DINH, Q.V., GLOWINSKI, R., MANTEL, B., PERIAUX, J. and PERRIER, P. - Sub-Domain Solutions of Non-Linear Problems in Fluid Dynamics on Parallel Processors, Computing Methods in Applied Sciences and Engineering V, Ed. Glowinski, R. and Lions, J.L., North Holland, 1982.

3. MORCHOISNE, Y. - Résolution des équations de Navier-Stokes par une méthode spectrale de sous-domaines, Numerical Methods in Engineering, Ed. Lascaux, P., Pluralis, Paris, 1983.

4. LABBE, O. - Couplage fluide visqueux-fluide parfait par une méthode d'éléments finis, ONERA Note Technique, 1983.

5. ROSE, W.C. and SEGINER, A. - Calculation of Transonic Flow over Supercritical Airfoil Sections, J. Aircraft, vol. 15, n° 8, pp. 514-519, 1978.

6. LE, T.H. - Transonic Potential Flow Calculation about Complex Bodies by a Technique of Overlapping Sub-Domains, Lecture Notes in Physics, vol. 170, Springer-Verlag, 1982.

7. LE BALLEUR, J.C. - Strong matching method for computing transonic viscous flow including wakes and separations. Lifting airfoils, La Rech. Aérosp. n° 1981-3, English Edition, pp. 21-45, 1981.

8. CAMBIER L., GHAZZI, W., VEUILLOT, J.P. and VIVIAND, H. - Une approche par domaines pour le calcul d'écoulements compressibles, Computing Methods in Applied Sciences and Engineering V, Ed. Glowinski, R. and Lions, J.L., North Holland, 1982.

9. VIVIAND, H. - Pseudo-Unsteady Methods for Transonic Flow Computation, Lecture Notes in Physics, vol. 141, Springer-Verlag, 1981.

10. VIVIAND, H. and VEUILLOT, J.P. - Méthodes pseudo-instationnaires pour le calcul d'écoulements transsoniques, ONERA Publication n° 1978-4, 1978 (English translation : ESA TT 561).

11. VEUILLOT, J.P. and VIVIAND, H. - A Pseudo-Unsteady Method for the Computation of Transonic Potential Flows, AIAA Paper n° 78-1150, and AIAA J., vol.17, n° 7, 1979.

12. ENSELME, M., BROCHET, J. and BOISSEAU, J.P. - Low-Cost Three-Dimensional Flow Computations Using a Minisystem, AIAA Paper n° 81-1013 and AIAA J., Vol. 20, n° 11, 1982.

13. COURANT, R. and HILBERT, D. - Methods of Mathematical Physics, Vol. II, Interscience Publishers, 1966.

14. JEFFREY, A. - Quasi-linear Hyperbolic System and Waves, Pitman Publishing, 1976.

15. LAX, P.D. - Hyperbolic Systems of Conservation Laws II, Comm. on Pure and Appl. Math., Vol X, pp. 537-566, 1957.

16. VIVIAND, H. and GHAZZI, W. - Numerical Solution of the Compressible Navier-Stokes Equations at High Reynolds Numbers with Applications to the Blunt Body Problem, Lecture Notes in Physics, vol. 59, Springer-Verlag, 1976.

17. LE BALLEUR, J.C., PEYRET, R. and VIVIAND, H. - Numerical Studies in High Reynolds Number Aerodynamics, Computers and Fluids, Vol. 8, pp. 1-30, 1980.

18. MATEER, G.G., BROSH, A. and VIEGAS, J.R. - A Normal Shock-Wave Turbulent Boundary-Layer Interaction at Transonic Speeds, AIAA Paper n° 76-161, 1976.

19. SHEA, J.R. - A Numerical Study of Transonic Normal Shock-Turbulent Boundary Layer Interactions, AIAA Paper n° 78-1170, 1978.

20. VIEGAS, J.R. and HORSTMAN, C.C. - Comparison of Multi-equation Turbulence Models for General Shock-Boundary-Layer Interaction Flows, AIAA J., Vol. 17, n° 8, 1979.

21. MICHEL, R., QUEMARD, C. and DURANT, R. - Application d'un schéma de longueur de mélange à l'étude des couches limites turbulentes d'équilibre, Note Technique ONERA n° 154, 1969.

22. MACCORMACK, R.W. - The Effect of Viscosity in Hyper Velocity Impact Cratering, AIAA Paper n° 69-354, 1969.

23. HOLLANDERS, H. and VIVIAND, H. - The Numerical Treatment of Compressible High Reynolds Number Flows, Computational Fluid Dynamics, Vol. 2, Ed. Kollmann, W., Hemisphere Publ. Corp., 1980.

24. ZANNETTI, L., COLASURDO, G., FORNASIER, L. and PANDOLFI, M. - A Physically Consistent Time-Dependent Method for the Solution of the Euler Equations in Transonic Flow, Notes on Numerical Fluid Mechanics, Vol. 3, Ed. Rizzi, A. and Viviand, H., Vieweg, 1981.

25. DELERY, J. - Investigation of Strong Shock-Turbulent Boundary-Layer Interaction in 2-D Transonic Flows with Emphasis on Turbulence Phenomena, AIAA Paper n° 81-1245 and AIAA J., Vol. 21, n° 2, 1983.

Section 5:

SPECIAL METHODS

NUMERICAL SIMULATION OF SELF-EXCITED OSCILLATIONS IN FLUID FLOWS

Wilbur L. Hankey
Joseph S. Shang

Flight Dynamics Laboratory, Wright-Patterson Air Force Base, Ohio

1.0 INTRODUCTION

A self-excited oscillation is one in which the force that sustains the motion is created by the motion itself; when the motion ceases the alternating force disappears[1]. (In a forced vibration, the alternating force exists independently of the motion and persists even when the motion is stopped). Self-excited oscillations are encountered in mechanical and aero-mechanical systems as well as in other fields. Some examples are nose wheel shimmy, machine chatter, chalk screech, galloping transmission lines, Karman vortex trails, wing flutter and inlet buzz.

Den Hartog[1] analyzes self-excited oscillations of mechanical systems with particular attention given to the damping term. Consider a spring-mass system with viscous damping for which the motion may be described as follows:

$$m\ddot{x} + c\dot{x} + kx = 0 \tag{1.1}$$

The solution to this equation for constant coefficients is

$$x = x_o e^{\frac{-ct}{2m}} \cos(\omega t + \Theta) \tag{1.2}$$

The natural frequency of the system is

$$\omega = \left[\frac{k}{m}\left(1 - \frac{c^2}{4km}\right)\right]^{\frac{1}{2}} \tag{1.3}$$

Given an initial disturbance, the motion will grow or decay depending upon the sign of the damping term (c). Negative damping (c < 0) is necessary to produce a self-excited oscillation. For this linear system, the disturbance will be amplified and grow without bound. In nature, however, non-linear effects occur and both negative and positive damping exist during portions of the oscillation so that a "limit-cycle" can result. A balance is reached between energy production and dissipation so that the net work is zero during one cycle. This steady state periodic solution is the self-excited oscillation that we observe in nature.

The analysis for fluid flows is analogous to that for a mechanical system[2]. For simple parallel flows a small perturbation analysis produces a linear system of equations. Under certain flow conditions negative damping occurs and the flow field becomes unstable. This linear analysis will be reviewed in the next section.

2.0 LINEARIZED ANALYSIS OF UNSTEADY FLUID FLOWS

Consider a two-dimensional parallel flow of an incompressible, inviscid (but rotational) gas. The governing equations are as follows:

$$\nabla \cdot \underline{V} = 0 \tag{2.1}$$

$$\rho \frac{D\underline{V}}{Dt} = - \nabla p \tag{2.2}$$

The mean flow is assumed to be parallel and rotational with small perturbations of the following form assumed:

$$\begin{aligned} u &= \bar{u}(y) + u'(x,y,t) \\ v &= v'(x,y,t) \\ p &= p_\infty + p'(x,y,t) \end{aligned} \tag{2.3}$$

Inserting these relationships into the governing equations and retaining only first order terms produces a linear system of equations.

$$u'_x + v'_y = 0 \tag{2.4}$$

$$\zeta'_t + \bar{u}\zeta'_x = \bar{u}_{yy}\, v' \tag{2.5}$$

where

$$\zeta' = v'_x - u'_y = \text{vorticity} \tag{2.6}$$

Disturbances of the following form are assumed

$$v' = \phi(y)\, e^{i\alpha(x-ct)} \tag{2.7}$$

where $\quad c = c_r + i\, c_i \quad$ (2.8)

c_r = propagation velocity

c_i = amplification factor

α = wave number

Substitution of this relationship into the governing equation produces the Rayleigh equation[3] (which is a degenerate Orr-Sommerfeld equation appropriate for large Reynolds numbers)

$$\phi'' - (\alpha^2 + \frac{\bar{u}''}{\bar{u}-c})\ \phi = 0 \tag{2.9}$$

with boundary conditions requiring that disturbances vanish at the wall and at the undisturbed outer edge.

$$\phi(0) = 0 \qquad \text{and} \qquad \phi(\infty) = 0$$

For prescribed values of $\bar{u}$ this is an eigenvalue problem in which $c(\alpha)$ can be obtained subject to the boundary condition constraint. The resulting solution takes on the following form:

$$v' = \phi e^{\alpha c_i t}\ e^{i\alpha(x-c_r t)} \tag{2.10}$$

For positive values of c_i an instability occurs which is equivalent to a negative damping case. Rayleigh[3] first investigated this type of flow and proved that velocity profiles with inflection points are unstable. In order to further explore this fact, a class of separated flows was analyzed. The stability of Stewartson's Lower Branch solutions of the Falkner-Skan equation was investigated (Fig 2.1). The Rayleigh equation was solved for several different values of the pressure gradient parameter, β, for the entire range of separated flows from incipient to a free shear layer[4]. Figure 2.2 presents the values of the amplification factor for the unstable frequency range. (Note $f_A = \bar{\alpha}\ c_r/2\pi\delta$). For reference purposes, these amplification factors are nearly two orders of magnitude greater than the more familiar Tollmien-Schlichting waves. The propagation speed (c_r) for the disturbances was generally between 0.4 and 0.9 of $\bar{u}_e$ (Fig 2.3). Therefore, one can deduce from these results that flow instabilities do exist (positive c_i) but over a very limited frequency range for similar separated laminar boundary layers. By analogy, the frequency for which maximum c_i occurs can be viewed as the <u>natural frequency</u> of the shear layer. This corresponds to the most probable Strouhal Number likely to occur for periodic disturbances and is always numerically less than unity. In Ref 5, compressibility effects of a free shear layer were investigated and the instability was found to diminish as Mach number increased (Fig 2.4). Although only one class of flows with inflection points has been examined, one is tempted to generalize these findings for all separated flows. One can speculate that (1) separated flows become more unstable in progressing from incipient to fully separated; (2) separated flows possess a relatively low natural frequency for which they are most likely to be self-excited and are stable on either side of that frequency; (3) the instability diminishes as Mach number increases. Based upon these hypotheses, one can embark upon an analysis of self-excited flow problems.

3.0 LINEAR OSCILLATOR MODEL

Separated flows were shown to possess a natural frequency for which small disturbances are highly amplified over a limited frequency range. However, for a self-sustained oscillation to persist a continuous string of disturbances is required to excite the shear layer. In this section the mechanism necessary to attain this result will be discussed.

It is informative at this point to compare a fluid dynamic oscillator with an electronic oscillator[6]. To create an electronic oscillator the circuit must contain an amplifier with a positive feedback loop (Fig 3.1). A self-excited fluid dynamic oscillator therefore must also contain these two components. In the previous section the shear layer was shown to play the role of the amplifier. The feedback loop is postulated to be a subsonic path in which pressure waves (acoustical signals) are returned to the shear layer origin and selectively reamplified.

The condition for oscillator resonance can be ascertained by examining the transfer functions for the two components.

If A in Figure 3.1 is the transfer function (a complex number) of the amplifier and B is the transfer function of the feedback loop then the overall gain is as follows:

$$\text{Gain} = \frac{A}{1-AB} \qquad (3.1)$$

The existence of a frequency for which the return ratio, AB, equals unity is a sufficient condition for infinite gain and is hence the criterion for a sustained oscillation.

To summarize, three ingredients are necessary to produce an oscillator, i.e. (1) amplifier with, (2) positive feedback at (3) a return ratio of unity. The implication of these characteristics as related to a fluid dynamic self-excited oscillation now will be addressed.

The features may be analyzed best by examining a simple wave diagram (Fig 3.2). A forward traveling wave of propagation speed, c_1, is shown moving from left to right.

$$p_1 = \hat{p}_1 e^{i\alpha_1(x-c_1t)} \quad (3.2)$$

A rearward traveling wave is also shown propagating at speed, c_2, moving from right to left.

$$p_2 = \hat{p}_2 e^{i\alpha_2(L-x-c_2t)} \quad (3.3)$$

where α is the respective complex wave number and x=0 represents an upstream reflective surface and x=L a downstream reflective surface.

The transfer function for the amplifier, A, is the ratio of the pressure values at the two end points for the forward traveling wave.

$$A = \frac{p_1(L)}{p_1(0)} = \frac{\hat{p}_1 e^{i\alpha_1(L-c_1t)}}{\hat{p}_1 e^{-i\alpha_1 c_1 t}} = e^{i\alpha_1 L} \quad (3.4)$$

The amplifer frequency is

$$f_A = \frac{\omega_1}{2\pi} = \frac{\alpha_1 c_1}{2\pi} \quad (3.5)$$

The transfer function for the feedback loop, B, is similarly:

$$B = \frac{p_2(0)}{p_2(L)} = \frac{\hat{p}_2 e^{i\alpha_2(L-c_2t)}}{\hat{p}_2 e^{-i\alpha_2 c_2 t}} = e^{i\alpha_2 L} \quad (3.6)$$

The feedback frequency is

$$f_B = \frac{\omega_2}{2\pi} = \frac{\alpha_2 c_2}{2\pi} \quad (3.7)$$

For resonance to occur the frequencies must match

$$f_A = f_B \quad (3.8)$$

or $\alpha_1 c_1 = \alpha_2 c_2 = \omega$

Also the return ratio must be unity.

$$AB = 1 = e^{i2\pi m} \tag{3.9}$$

Using the derived values for the transfer functions produces a phase requirement at resonance.

$$AB = e^{i\alpha_1 L} e^{i\alpha_2 L} = e^{i2\pi m} \tag{3.10}$$

or $$\alpha_1 + \alpha_2 = \frac{2\pi m}{L} \tag{3.11}$$

Since the wave number is complex $\alpha = \alpha_r + i\alpha_i$, two conditions are produced.

Real Part:

$$\alpha_{r1} + \alpha_{r2} = \frac{2\pi m}{L} \tag{3.12}$$

Imaginary Part:

$$\alpha_{i1} + \alpha_{i2} = 0 \tag{3.13}$$

Also since $\alpha_1 c_1 = \alpha_2 c_2 = \omega$, the equation for the real part produces an expression for the resonant frequency.

$$\omega = \frac{2\pi m}{L(c_1^{-1} + c_2^{-1})} = 2\pi\ f_B \tag{3.14}$$

Since $c_1 = c_r$ propagation velocity for the amplifier
and $c_2 = a_o$ sound speed for the feedback loop

$$f_B = \frac{m\ u_e}{L(M_o + k^{-1})} \quad \text{Rossiter's Equation[7]} \tag{3.15}$$

where $$M_o = \frac{u_e}{a_o}$$

$$k^{-1} = \frac{c_r}{u_e}$$

u_e = reference flow velocity

This relationship has been called Rossiter's equation and is used by experimentalists to predict cavity resonance.

The equation for the imaginary part can be rewritten in the following form by using the interrelationship between spatial and temporal disturbances[8], i.e. $\alpha_i c_r = -\alpha_r c_i$.

$$\omega\left(\frac{c_{i_1}}{c_r^2} + \frac{c_{i_2}}{a_o^2}\right) = 0 \tag{3.16}$$

This merely indicates that net damping is zero during one cycle.

From this analysis of a linear oscillator the following resonant conditions are derived:

(1). The "natural frequency" of the amplifier, f_A, must equal the feedback frequency of the acoustical waves, f_B, at the resonance state.

(2) The frequency at resonance can be predicted by the Rossiter equation.

$$f_B = \frac{m\ Ue}{L(M_o + k^{-1})} \tag{3.17}$$

Examination of the linear equations for a self-excited oscillation identifies the essential components and predicts the resonant frequency. However, it provides no capability to predict the amplitude of the disturbance or produce detailed flow field features for large disturbances. To proceed further it becomes necessary to examine the non-linear characteristics, i.e. Navier-Stokes equations. This will be accomplished in the next section.

4.0 PROCEDURE FOR NUMERICALLY SOLVING THE NAVIER-STOKES EQUATIONS

4.1 Governing Equations

The two-dimensional or axisymmetric Navier-Stokes equations for a compressible perfect gas are listed below. This system or equations will be used to analyze several cases involving self-excited oscillations discussed later in this chapter.

$$U_t + E_x + y^{-k}\,(y^k F)_y = k\ y^{-k}\ H \tag{4.1}$$

where

$$U = \begin{vmatrix} \rho \\ \rho u \\ \rho v \\ \rho e \end{vmatrix} \quad ; \qquad E = \begin{vmatrix} \rho u \\ \rho u^2 - \sigma_{11} \\ \rho uv - \tau \\ \rho ue - u\sigma_{11} - v\tau - kT_x \end{vmatrix} \quad ;$$

$$F = \begin{vmatrix} \rho v \\ \rho uv-\tau \\ \rho v^2-\sigma_{22} \\ \rho ve-v\sigma_{22}-u\tau-kT_y \end{vmatrix} \quad ; \quad H = \begin{vmatrix} 0 \\ 0 \\ -\sigma_{\Theta\Theta} \\ 0 \end{vmatrix} \qquad (4.2)$$

and

$$\tau=\mu(u_y+v_x) \qquad \sigma_{11}= -p-2/3\mu\nabla\cdot\underline{V}+2\mu u_x$$

$$e=cvT+\frac{u^2+v^2}{2} \qquad \sigma_{22}= -p-2/3\mu\nabla\cdot\underline{V}+2\mu v_y$$

$$p=\rho RT \qquad \sigma_{\Theta\Theta}= -p-2/3\mu\nabla\cdot\underline{V}+2\mu\frac{v}{y}$$

$$\nabla\cdot\underline{V}=u_x+y^{-k}(y^k v)_y$$

$k = 0$ for 2 dimensional flow

$k = 1$ for axisymmetric flow

This system of equations contains four dependent variables (u,v,T,p) where μ, k, cv and R are prescribed for the gas. The three independent variables (x,y,t) are expressed in a Cartesian framework. A cartesian system is unsatisfactory for most problems and therefore a general coordinate transformation must be employed.

4.2 Coordinate Transformation

The most useful coordinate system should simplify the boundary conditions and reduce the magnitude of the higher order derivatives in order to minimize truncation errors arising from the finite difference procedure. A surface oriented coordinate system with arbitrary grid stretching provides this desired capability. Note that it is not necessary to transform the velocity components to accomplish our goal. At this point one might fear that the transformation could greatly complicate the governing equations. It will be shown that fortunately this is not the case.

Select a set of transformed variables

$$\xi = \xi(x,y) \qquad (4.3)$$
$$\eta = \eta(x,y)$$

The chain rule of differentiation is used to accomplish the transformation.

$$\frac{\partial}{\partial x} = \xi_x \frac{\partial}{\partial \xi} + \eta_x \frac{\partial}{\partial \eta}$$

$$\frac{\partial}{\partial y} = \xi_y \frac{\partial}{\partial \xi} + \eta_y \frac{\partial}{\partial \eta} \qquad (4.4)$$

The governing equations become as follows:

$$U_t+[\xi_x E_\xi+\xi_y\ y^{-k}(y^k F)_\xi]+[\eta_x E_\eta+\eta_y\ y^{-k}(y^k F)_\eta]=ky^{-k}H \qquad (4.5)$$

Divide this equation by Jy^{-k} where

$J = \xi_x\eta_y - \xi_y\eta_x$ and recall the inverse transformation

$$x_\xi = J^{-1}\ \eta_y \qquad y_\xi = -J^{-1}\ \eta_x$$

$$x_\eta = -J^{-1}\ \xi_y \qquad y_\eta = J^{-1}\ \xi_x \qquad (4.6)$$

Therefore the transformed equations can be regrouped in the following manner.

$$\hat{U}_t + \hat{E}_\xi + \hat{F}_\eta = kJ^{-1}H \qquad (4.7)$$

$$\hat{U} = J^{-1}\ y^k\ U$$

where $\hat{E} = y^k(y_\eta E - x_\eta\ F)$

$$\hat{F} = -y_\xi\ E + x_\xi\ F$$

The above analysis shows that the governing equations are not greatly complicated when transformed to a surface oriented system. The procedure does greatly simplify the description of the boundary conditions in that they occur along lines of either constant ξ or η.

4.3 Boundary Conditions

Four types of boundary conditions are required for the cases to be computed in this chapter, i.e. (a) wall, (b) inflow, (c) outflow and (d) symmetry surfaces. These shall now be addressed.

(a) Wall (η=0)

On an impermeable wall a no-slip condition for the velocity is required.

$$u(\eta=0)=0$$
$$v(\eta=0)=0$$

The wall temperature is also specified

$$T(\eta=0)=Tw$$

The pressure on the wall does not require a boundary condition but must be determined from the flow field equations. The finite difference algorithm does require specification of a pressure relation at the wall and therefore a "compatibility condition" is used which is obtained from a degenerate normal momentum equation, i.e.

$$\frac{\partial p}{\partial \eta}(\eta=0) \cong 0 \quad \text{to order } Re^{-1}$$

(b) Inflow ($\xi=0$)

At the inflow surface all flow variables are prescribed at the freestream value

$$\begin{aligned} u(\xi=0) &= u_\infty \\ v(\xi=0) &= 0 \\ T(\xi=0) &= T_\infty \\ p(\xi=0) &= p_\infty \end{aligned}$$

(c) Outflow ($\zeta=L$)

At a downstream boundary in which outflow occurs a simple wave equation is used to minimize reflections.

$$U_t + c_r U_s = 0$$

where s is aligned with the mean streamline.

c_r = propagation velocity of disturbances.

(d) Symmetry ($y=0$)

For axi-symmetric flows the axis requires a symmetry condition as follows:

$$\begin{aligned} v(y=0) &= 0 \\ u_y(y=0) &= 0 \\ T_y(y=0) &= 0 \\ p_y(y=0) &= 0 \end{aligned}$$

This concludes the description of the principle boundary conditions for the problems to be investigated. These four types are not implied to be a complete set to be used to investigate all flows but are representative of the type used in many present day calculations. Research is in progress in this area to improve the description of boundary conditions; especially for subsonic flows.

4.4 Finite Difference Algorithm

The system of equations in the chain rule conservation law form (equation 4.7) is solved by a two-step predictor and corrector scheme originated by MacCormack[9]. The numerical algorithm is an explicit, conditionally stable procedure of second order accuracy in spatial and temporal variables. The CFL condition on allowable time increment for generalized coordinates can be given as:

$$\Delta t_{CFL} = 1/\{u_\xi/\Delta\xi + u_\eta/\Delta\eta + c[(\frac{\xi_x}{\Delta\xi} + \frac{\eta_x}{\Delta\eta})^2 + (\frac{\xi_y}{\Delta\xi} + \frac{\eta_y}{\Delta\eta})^2]^{\frac{1}{2}}\} \quad (4.8)$$

where the contravariant velocity components are defined by

$$u_\xi = \xi_x u + \xi_y v \quad (4.9)$$

$$u_\eta = \eta_x u + \eta_y v \quad (4.10)$$

Since the stability analysis does not contain viscous terms the highest and most consistent CFL number used is 0.85.

For self-excited oscillations, the required numerical resolution in time is comparable to the allowable time step from the stability consideration of the finite difference approximation. Hence, the use of explicit method for unsteady flows does not result in a severe penalty that one would encounter in steady flow calculations.

For the most problems investigated, the unsplit version of MacCormack's algorithm is adopted to reduce the number of accessions of main memory, thereby minimizing the data movement to and from main memory of a vector processor (CRAY-1). For a two-dimensional or axisymmetric factored scheme, a field point requires three accessions of data in order to advance one time increment in either the predictor or corrector sweep:

$$u^{n+1} = L_\eta(\Delta t/2)\, L_\xi(\Delta t)\, L_\eta(\Delta t/2) u^n \quad (4.11)$$

However, the unsplit algorithm require only one accession of main memory to acquire the same end result:

$$u^{n+1} = [L_\xi(\Delta t) + L_\eta(\Delta t)] u^n \quad (4.12)$$

The algorithm adopted requires a combination of a forward and backward differencing procedure for the predictor and corrector sweeps. This requirement is easily achieved by using a different indexing procedure for the predictor and corrector operation. However, the source-like term, $\sigma_{\theta\theta}/y$, in the radial momentum equation is evaluated by a central differencing procedure. Again, for the axisymmetric problem the term contained in the gradient operator

$\frac{1}{y}\frac{\partial}{\partial y}(yv)$ is further expanded resulting in the expressions of $(\frac{\partial v}{\partial y} + \frac{v}{y})$. This additional operation alleviates the numerical difficulty encountered when an indeterminate form emerges on the axis of symmetry.

4.5 Numerical Procedure

Most of the computer codes used in the related studies were developed for the vector processor (CRAY-1 or CYBER 201). Careful attention has been paid to optimize the data flow so that the memory loading between the central memory unit and the vector registors is minimal[10]. For the two-dimensional and axisymmetric code, the coordinate transformation derivatives are computed once and stored for repeated applications. Therefore, the data processing rate equals 1.97×10^{-5} seconds per mesh point per iteration with a suitable vector length. The data processing rate is conventionally defined as the central processing time required per grid point per number of iterations. For the explicit numerical scheme, the data processing rates either using the CRAY-1 or the CYBER 201 are nearly identical. However, the vector processor outperforms the CYBER 750 computer by a factor of 29. The cost reduction for identical computations between the vector processor and the scalar processor is 1:7.3 in favor of the vector processor.

A wide range of mesh point systems was used to compute the self-sustained oscillatory flows. Usually two or more mesh spacings were implemented for a problem to numerically experiment with different boundary conditions and placement of the far field boundary. The finest mesh point distribution was designed to resolve the oscillatory frequency up to about 300 KHZ for the laminar stability study. In order to attain this goal, the streamwise mesh spacing was assigned the value of a quarter of the boundary layer thickness. (Δx=0.14 cm) at the Reynolds number of 3.6 million. In the transverse direction usually thirty to sixty mesh points were used. The transverse node distribution was clustered closely near the solid surface with the finest mesh spacing adjacent to the surface. Thus the boundary layer is generally defined within twenty mesh points. For laminar flows, the finest mesh step commonly is given a value inversely proportional to the square root value of the characteristic Reynolds number for turbulent flows. The finest grid spacing is controlled by the value of the law of the wall variable $y^+ \leq 10$, which defines the outer edge of the laminar sublayer.

Since the investigated flow fields frequently contain a strong bow shock and traveling shock waves, the fourth-order pressure damping by MacCormack[11] was adopted with modification. The net result is an artificial viscosity-like term of the form.

$$-\alpha_d \Delta t \Delta\xi^3 [|u_\xi| + (\xi_x^2+\xi_y^2)^{\frac{1}{2}} c] \frac{1}{p} |\frac{\partial^2 p}{\partial \xi^2}| \quad (4.13)$$

$$-\alpha_d \Delta t \Delta\eta^3 [|u_\eta| + (\eta_x^2+\eta_y^2)^{\frac{1}{2}} c] \frac{1}{p} |\frac{\partial^2 p}{\partial \eta^2}| \quad (4.14)$$

The result is identical to MacCormack's original expression if the Cartesian coordinate system were used. For the most of the results contained in the present effort, the damping constant α_d, is limited to the range of values from two to three. These damping terms are only of significant magnitude in regions of strong pressure gradient where the truncation error is already degrading the computations.

4.6 Turbulence Model

For the turbulent flow simulations, the closure of the system of equations is achieved by introducing the Cebeci-Smith turbulence[12] model and by assigning a turbulent Prandtl number of 0.9. In a numerical analysis of fluctuating fluid motion, the mesh distribution and the numerical algorithm dictate the range of the resolvable frequency range. The wave length of this resolvable fluctuating motion is directly proportional to the coarsest grid point spacing. Unfortunately, the currently available computer severely limits the capability to achieve the direct simulation of turbulence. Therefore the use of a turbulence model in conjunction with highly oscillatory fluid motion can be interpreted as a device to model the extremely small scale motion. Meanwhile the large scale oscillation will be simulated directed. This area requires extensive research to reveal the fundamental knowledge necessary to properly represent turbulence.

The turbulence model utilizes a two layer formulation. The inner layer represents the laminar sublayer region and the law of the wall domain. The outer layer depicts the law of the wake or the momentum defect law.

The inner layer

$$\varepsilon_i = \rho k^2 y^2 D^2 |\omega| \quad (4.15)$$

where k is the Von Karman constant[13] (0.4) and D is the Van Driest damping factor[14] for the description of the laminar sublayer region.

$$D = 1-\exp[-(\frac{\rho_w |\omega_w|}{\mu_w})^{\frac{1}{2}} y/26] \quad (4.16)$$

In the outer region, Clauser's defect law[15] is given as

$$\varepsilon_o = 0.0168\ \rho u_\infty\ \delta_i^* \tag{4.17}$$

where δ_i^* is the kinematic displacement thickness which is the basic length scale of the outer structure.

For each individual investigation reported here, several minor modifications may be utilized for a particular purpose. However, in general, only the secondary features are included such as the inclusion of the Klebanoff's intermittency correction and/or a slight alteration of the proportionality constants.

5.0 NUMERICAL RESULTS

In the previous section the numerical procedure was described for solving the time-dependent Navier-Stokes equations. In this section the results of several large scale computations will be explored. The configurations investigated include a cylinder, airfoil, open cavity, spike-tipped body, inlet and cone.

5.1 CYLINDER

The periodic shedding of large scale eddies from a cylinder immersed in a flowing stream is probably the most commonly recognized self-sustained oscillation in fluid mechanics. A stability analyses of a series of potential vortices representing this flow was accomplished by Von Karman[16] in 1911. This wake analysis of the "Karman Vortex Street" unfortunately contains little information about the true physics of the phenomenon and hence further investigation is required. Von Karman was limited at that time since the only tool available was linear potential theory. Today, however, the computer provides us with the ability to numerically integrate the Navier-Stokes equations and investigate problems of this type.

In Ref 17 the flow behind a cylinder at a Mach number of 0.6 and a Reynolds number of $1.7\text{x}10^5$ was computed and compared with an experiment for similar conditions[18]. By use of the techniques described in Section 4 the time dependent flow over a cylinder was determined by numerically integrating the Navier-Stokes equations. No turbulence model was used for this case since the cylinder was in the subcritical (laminar) regime at this Reynolds number. Therefore, all large scale "turbulent" eddies in the wake were computed based upon first principles.

The flow was impulsively started. All points in the

field were initially at the free stream state and suddenly the nonslip boundary condition on the cylinder walls applied. Symmetric vortices developed behind the cylinder which became asymmetric after one cycle period and developed into periodic asymmetric vortex shedding after about three periods. The vorticity contours and velocity vector field are shown for this state in Figures 5.1 and 5.2 respectively. The wall pressure history for the 90° and 270° polar angle location covering over twenty cycles of oscillation is shown in Fig 5.3.

The computed Strouhal number, fd/u, is 0.21; in general agreement with experiment. Comparison with experiment of the wake centerline survey is shown for both the mean and fluctuating velocity components (Figs 5.4 and 5.5). The comparison is favorable considering that only large scale, low frequency eddies are simulated with a grid of 122x60 used in this case. This investigation shows that it is possible to numerically generate the production of large scale turbulent eddies and generally duplicate the experiment without accurately simulating the dissipation of the fine scale structure. Numerically computed Reynolds stresses for the wake are presented in Fig 5.6. Although no experimental data were taken for comparison the computed results possess the correct trend.

The mean-velocity wake profile is shown in Figure 5.7. Note that this profile possesses two inflection points. A linear stability analysis of this profile[19] shows two unstable modes to exist due to these two inflection points (see Fig 5.8). The first mode produces an asymmetric oscillation while the second mode (lower amplification factor) produces a symmetric one. Since the asymmetric mode has the greater amplitude, this accounts for the observed asymmetric serpentine wake pattern behind the cylinder (Fig 5.1). The Strouhal number, $f_A d/u$, for which the amplification factor is maximum in Fig 5.8 is 0.2, which further confirms the hypothesis. The symmetric mode still exists, however, although lower in amplitude and higher in frequency. This tends to explain the modulation by a higher frequency of the wave form in Figure 5.3.

5.2 Airfoil at Angle of Attack

An airfoil at an attitude near stall will develop separated flow over the upper surface (Fig 5.9). The velocity profile possesses an inflection point and hence selectively amplifies disturbances of frequency, f_A (see Section 2.0). Numerical solution[20] of the Navier-Stokes equations for laminar flow over an NACA 0012 airfoil at a Reynolds number of 10^5 produced a regular pattern of eddies which shed from the upper surface (Fig 5.10). A time dependent sequence of streamline scenes permitted the tracking of the eddies from which both the propagation speed and growth rate of the eddies were ascertained. The propagation speed, determined from the slope of the wave diagram for a series of traveling eddies (Fig 5.11) was $c_r=0.4\ u_\infty$ with a corresponding wave length of 0.28L.

The eddy height (h) is amplified according to the following relationship:

$$h = h_o e^{\alpha c_i t}$$

From a plot of log h versus time (Fig 5.12) the amplification factor was found to be $c_i = 0.07$. Both the values of c_r and c_i are within the range of the unstable waves in separated boundary layers as predicted from linear stability theory. (See Figures 2.2 and 2.3).

Navier-Stokes solutions of the airfoil confirms the fact that self-excited oscillations can be numerically produced in flow fields with separated boundary layers. The frequency, propagation speed and amplification factor of the instability is in general agreement with linear theory.

5.3 Open Cavity

The second important component of a self-excited oscillation is the feedback loop. One configuration which is useful for exploring this feature is the open cavity. Transonic flow over a cavity has been investigated experimentally and found to produce severe pressure oscillations under certain conditions[21,22]. Such a flow obviously has an inflection point as shown in the velocity vector plot from a numerical computation (Fig 5.13). The cavity also has a large subsonic region in which the propagation of acoustic waves can be observed.

In Ref 23 the flow over an open cavity at a Mach number of 1.5, a Reynolds number of $2.6\text{x}10^7$ and a length to depth ratio of 2.25 was computed by solving the Reynolds averaged Navier-Stokes equations. A self excited oscillation was produced at an intensity and frequency similar to the wind tunnel data for the same conditions. The computed density contours for the flowfield at one instant of time are presented in Fig 5.14. A movie depicting a sequence of similar contours clearly showed traveling waves moving upstream in the cavity, thus proving the phenomenon of feedback. A wave diagram of the individual waves in the shear layer (Fig 5.15) indicates a propagation speed of $c_r = 0.51\ u_\infty$ and a wave length of 0.26L. These values are in good agreement with linear theory (Section

2). Linear theory predicts a natural frequency of this shear flow of the following value:

$$f_A = \frac{\alpha_{opt} c_r}{2\pi\delta} = 200 \text{ HZ}$$

Figure 5.16 shows the amplification curve (c_i vs f) for this flow with the maximum c_i occuring at 200 HZ. The feedback frequency (Eqn 3.17) is

$$f_B = \frac{mu_e}{L(M_o + k^{-1})} = 115 \text{ m HZ}.$$

where m equals the harmonic number. Figure 5.16 also shows the frequency for these harmonics up to m=8. Also shown in this plot is the experimental spectral analysis[24] confirming the existence of the feedback harmonics in the region of high shear layer amplification.

5.4 Buzz of Spikes

The numerical solutions of the time dependent Navier-Stokes equations for the cylinder, airfoil and open cavity have now confirmed the linear oscillation model for a self-excited oscillation, i.e. (a) the shear layer with an inflection point is the fluid dynamic amplifier; (b) feedback is achieved by acoustic waves returning to the origin; (c) resonance occurs at discrete integer values of the fundamental frequency when the return signal is "in phase" with the original disturbance. The existence of all three features is required to produce a self-excited oscillation. The removal of any one feature should eliminate the oscillation. (In most practical flight problems the oscillation is undesirable and must be avoided). The elimination characteristics of the oscillation shall now be discussed.

Obviously, the avoidance of separation will eliminate the amplifier. Many computations of unseparated flows using time dependent Navier-Stokes equations have converged to steady state solutions. This, of course, is necessary but not sufficient proof of the concept. A more useful proof would be to isolate the amplifier frequency (f_A) at a value far below the feedback frequency (f_B). i.e.

$$f_A < f_B$$

This would prevent a resonance condition from occurring. Fig 5.17 shows a typical case of resonance and also a case of "no resonance" in which all feedback frequencies are above the shear layer unstable region. For the later case, even though separation and feedback exist no resonance is possible.

A configuration to demonstrate this phenomenon is a blunt

body with a spike tip operating at supersonic speed. Spike-tipped bodies are noted for producing violent buzz under a restricted range of spike lengths[25]. Figure 5.18 shows the experimental pressure intensity for different spike lengths at a Mach number of three. Buzz exists, but only for spike lengths above 20 mm for this configuration. Oscillations are not encountered at shorter lengths. Separated flow will always occur in the concave region between the spike and face of the blunt-nose body, and hence amplification is always present. However, resonance will not occur if

$$f_A < f_B$$

or

$$\frac{\alpha_{opt} c_r}{2\pi\delta} < \frac{m u_e}{L(M_o + k^{-1})}$$

This produces a condition for a limiting spike length to avoid resonance (m=1)

$$\frac{L}{\delta} = \frac{2\pi}{\alpha_{opt}(1+kM_o)} \simeq 3$$

Two numerical calculations were conducted[26] for spike lengths of 13 mm and 39 mm. The shorter spike length had an $\frac{L}{\delta}$ of 1.5 and resulted in a stable flow. However, the longer spike length (39mm) had an $\frac{L}{\delta} = 9$ (Fig 5.19) and produced a self excited oscillation comparable to the experiment. The shock pattern is shown in Fig 5.20 and the spike pressure history comparison in Fig 5.21. The spectral analysis for both computation and experiment is depicted in Fig 5.22. Outstanding agreement is observed between the computation and experiment for frequency, amplitude and wave form, showing the ability of the numerics to simulate self-excited oscillations.

Additional information on feedback was obtained in this case by tracking the shock front and the vortex core (Fig 5.23). The vortex front is observed to move upstream at a rate approximately equal to the speed of sound.

5.5 Inlet Buzz

A supersonic inlet operating at subcritical flow conditions possesses the necessary features for buzz, i.e. a large region of intermittent separated flow and a downstream interface to reflect acoustical signals. When an inlet with a supersonic diffuser is throttled back to subcritical flow conditions, the normal shock is expelled from the diffuser causing separation on the centerbody. This separated shear layer is unstable and the principal cause of the oscillation. Standing waves occur in the duct. The upstream end of the inlet behaves as an open end (pressure node) while the downstream end behaves as a closed end (pressure antinode). Antisymmetric modes occur with all harmonics being odd (m=odd). Two very significant results can be obtained from this standing wave analysis. First, the frequencies should be commensurable in which harmonics occur at exact integer values of the fundamental frequency. Secondly, antisymmetric mode shapes occur in the duct. One can also anticipate frequency modes to jump discretely to the next integer eigenvalue as flow conditions are changed by different throttle settings. This indeed is the experimental finding[27].

One calculation[28] using the complete Navier-Stokes equations for inlet buzz has been accomplished to date and compared with experimental data[29]. The external compression axisymmetric inlet and diffuser configuration is shown in Fig 5.24. Flow conditions correspond to a Mach two free stream with a Reynolds number based upon 6 cm diameter of $Re_D = 2.36 \times 10^6$. Because the turbulence model was found to artificially damp the occurrence of instabilities it was deleted from the program. The justification for the omission is that the numerical code is capable of resolving a finite number of low frequency components up to the shortest wave length ($2\Delta x$). Current turbulence models over-predict the appropriate eddy viscosity. When the turbulence model is omitted, the turbulent transport process is resolvable while the turbulent dissipative process is not. This approach was used to compute through three buzz cycles. The instability developed immediately as a consequence of the non-equilibrium state of the initial conditions. A sequence of Mach contours covering the third buzz cycle is shown in Figure 5.25. During buzz the bow shock was forced to the tip of the centerbody as a result of the interaction with a reflected compression wave. In the expulsion phase a region of reverse flow extended between the base of the bow shock and the cowl lip. As the shock reached the centerbody tip, the shear layer ruptured and flow was spilled. The bow shock remained in this position for a time corresponding to the propagation and reflection of an expansion wave from the downstream choked throat. The inlet then ingested mass and the shock retreated to the cowl lip with the flow reattaching to the centerbody. A comparison with experiment of pressure histories for several different diffuser locations is shown in Fig 5.26.

The wave propagation involved in the oscillatory cycle is best shown by a sequence of instantaneous mass flux plots (Fig 5.27). An expansion wave can be followed as it propagates downstream and reflects back upstream. This is followed by a compression wave, covering the second half of the first buzz cycle. The ingestion and expulsion of mass during the oscillation is near the same magnitude as the supercritically captured mass flux under steady operation.

5.6 Boundary Layer Transition

As alluded to in previous sections, a portion of the turbulence spectrum can be resolved in computing self-excited oscillations. One then wonders if boundary layer transition can be simulated numerically as a self-excited oscillation on todays' computers. The onset of transition has been computed for a hypersonic boundary layer[30]. For highly compressible flow, the generalized inflection point, $(\rho u')'=0$, replaces the low speed Rayleigh condition of $u''=0$. At hypersonic speed, linear theory shows that the second mode instability dominates (which is fortunately two dimensional). Examination of this case by use of the linear stability theory results of Mack[31] showed the numerical computation to be feasible. Numerical solution of the time dependent Navier-Stokes equations was accomplished (with steady boundary conditions) using step sizes sufficiently small to resolve the unstable waves predicted from linear theory. The configuration was a 7° cone at a Mach number of 6.2 and a Reynolds number of 10^6 based on a one meter length (Fig 5.28). Temperature fluctuations with a regular periodic behavior were obtained (Fig 5.29). The amplitude of these self-excited waves varied across the boundary layer with a maximum occurring near the edge. This is in good agreement with experiment for both the temperature and velocity fluctuations (Fig 5.30).

A comparison of the computed spectral analysis (Fig 5.31) with experiment shows agreement in the frequency at peak amplitude however,the experiment has a broader band. Further research is required to resolve the disparity, however, the results are encouraging for the first phase in the prediction of turbulence.

5.7 Edgetones

Edgetones have been of interest to the scientist for many years. Edgetones (musical notes) are created by inserting an orifice plate (or similar reflecting surface) into a small jet (Fig 5.33). This self-excited phenomenon can be explained by use of the linear oscillator model.

Jets possess two inflection points in the velocity profile which serve as amplifiers. The frequency of the instability is

$$f_A = \frac{\alpha c_r}{2\pi\delta} \tag{3.5}$$

The shear layer thickness, δ, for a laminar jet is the following:

$$\frac{\delta}{L} = \frac{c}{\sqrt{Re_L}}$$

The Strouhal number is therefore a function of Reynolds number.

$$\frac{f_A d}{u} = \frac{\alpha k}{2\pi c} \left(\frac{d}{L}\right)^{\frac{1}{2}} (Re_d)^{\frac{1}{2}}$$

where $\alpha = 1 \pm 0.5$ [Ref 19]

The shaded portion of Fig 5.32 depicts a band of frequencies for which disturbances of a jet are amplified as a function of Reynolds number.

The feedback frequency is the following:

$$f_B = \frac{m\, u_e}{L(M_o + k^{-1})}$$

The Strouhal number for feedback is

$$\frac{f_B d}{u} = \left(\frac{d/L}{M_o + k^{-1}}\right) m$$

Hence only discrete frequencies occur during the feedback cycle. Resonance occurs when $f_A = f_B$. These frequencies are also shown in Fig 5.32.

Observe that instead of a continuous band of frequencies occuring, only one or two frequencies are possible at each Reynolds number. The later case in which double-valued frequencies occur leads to a hysteresis in that the prevailing frequency depends upon the initial condition (or history).

This phenomenon is demonstrated by the data of Ref 32. Fig 5.33 of Ref 32 shows a limited region of Reynolds number in which a jump between the 3rd and 4th harmonic exists. A hysteresis is observed in the overlap region.

This simple explanation of the edge tone phenomenon provides further confirmation of the linear oscillator model for describing self-excited oscillations in fluid mechanics.

This technique also has been used to explain other self-excited oscillations such as autorotion[33], oil whip[34], whistler nozzles[35] and rotating stall[36].

6.0 SUMMARY

Self-excited oscillations in fluid flows have been analyzed. The concept of a fluid amplifier within a separated shear layer was presented. Signals are selectively amplified over a limited frequency range and returned through a feed-back loop as acoustic pressure waves. Resonance occurs when the return signal is "in phase" with the original disturbance wave. Under these circumstances no external forcing function is required and a self-excited oscillation can occur with steady boundary conditions. Linear theory is useful in predicting the frequency of the instability and providing a qualitative description of the phenomenon. Quantitative results are possible through the numerical solution of the time-dependent Navier-Stokes equations.

REFERENCES

1. DEN HARTOG, J.P. - Mechanical Vibrations, McGraw-Hill Book Co., Inc, 1947.

2. HANKEY, W.L. and SHANG, J.S. - Analysis of Self-Excited Oscillations in Fluid Flows. AIAA-80-1346, Snowmass, CO, 14-16 July, 1980.

3. RAYLEIGH, LORD - On the Stability of Instability of Certain Fluid Motion. Scientific Papers, Vol. 1, pp. 474-484, Cambridge University Press, 1880.

4. VERMA, G., HANKEY, W.L. and SCHERR, S. - Stability Analysis of the Lower Branch Solutions of the Falkner-Skan Equations. AFFDL-TR-79-3116, July 1979.

5. ROSCOE, D. and HANKEY, W.L. - Stability of Compressible Shear Layers. AFWAL-TR-80-3016, April 1980.

6. GLASFORD, G.M. - Linear Analysis of Electronic Circuits, Addison-Wesley, 1965.

7. ROSSITER, J.E. - Wind-Tunnel Experiment on the Flow Over Rectangular Cavities at Subsonic and Transonic Speeds. R & M No. 3438, British ARC, Oct 1964.

8. LIN, C.C. - The Theory of Hydrodynamic Stability. Cambridge University Press, 1955.

9. MACCORMACK, R.W. - Numerical Solutions of the Interactions of a Shock Wave with Laminar Boundary Layer. Lecture notes

in physics, Vol. 59, Springer-Verlag, 1976.

10. SHANG, J.S., BUNING, P.G., HANKEY, W.L. and WIRTH, M.C. - Performance of a Vectorized Three-Dimensional Navier-Stokes Code on the CRAY-1 Computer. AIAA J., Vol. 18, No. 9, pp. 1073-1079, Sept 1980.

11. MACCORMACK, R.W. and BALDWIN, B.S. - A Numerical Method for Solving the Navier-Stokes Equations with Applications to Shock-Boundary Layer Interactions. AIAA Paper 75-1, Jan 1975.

12. CEBECI, T., SMITH, A.M.O. and MOSINSKIS, G. - Calculations of Compressible Adiabatic Turbulent Boundary Layers. AIAA Journal, Vol. 8, pp. 974-982.

13. SCHLICHTING, H. - Boundary Layer Theory, Pergamon Press, New York, 1955.

14. VAN DRIEST, F.R. - On Turbulent Flow Near A Wall. Journal of the Aeronautical Sciences, Vol. 23, No. 11, pp 1007-1011, 1956.

15. COLES, D. - The Law of the Wake in the Turbulent Boundary Layer. Journal of Fluid Mechanics, Vol. 1, Pt. 2, pp 191-226, 1956.

16. VON KARMAN, T. and nachr. d. wiss. ges. gottinger, - Mathe. Phys. Klasse, 509 (1911).

17. SHANG, J.S. - Oscillatory Compressible Flow Around a Cylinder. AIAA Paper No. 82-0098, 1982.

18. OWEN, F.K. and JOHNSON, D.A. - Measurements of Unsteady Vortex Flow Field. AIAA Journal, Vol. 18, pp 1173-1179,

19. BETCHOV, R. and CRIMINALE, W. - Stability of Parallel Flows, New York Academic Press, pp 37-40, 1967.

20. HODGE, J.K. and COOPER, W.H. - Solution of Navier-Stokes Equations for Airfoils at Angle of Attack Near Stall. AFFDL-TM-77-94-FXM, Nov 1977.

21. HELLER, H., HOLMES, G. and COVERT, E. - Flow-Induced Pressure Oscillations in Shallow Cavities. AFFDL-TR-70-104, Dec 1970.

22. HELLER, H. and BLISS, D. - Aerodynamically Induced Pressure Oscillations in Cavities: Physical Mechanisms and Suppression Concepts. AFFDL-TR-74-133, WPAFB, Ohio, 1974.

23. HANKEY, W. and SHANG, J. - The Numerical Solution of Pressure Oscillations in an Open Cavity. AIAA Paper, 79-0136, 17th Aerospace Sciences Meeting, New Orleans, Jan 1979.

24. MAURER, O. - Investigation and Reduction of Weapons Bay Pressure Oscillations Expected in the B-1 Aircraft. AFFDL-TM-41-101, WPAFB, Ohio, 1974.

25. HARNEY, D.J. - Oscillating Shocks on Spiked Nose Tips at Mach 3. AFFDL-TM-79-9-FX, 1979.

26. SHANG, J., SMITH, R. and HANKEY, W. - Flow Oscillations at Spike-Tipped Bodies. AIAA Paper 80-0062.

27. HANKEY, W. - Analysis of Inlet Buzz. AFWAL-TM-FIMM-1980.

28. NEWSOME, R.W. - Numerical Simulation of Near-Critical and Unsteady Subcritical Inlet Flow Fields. AIAA Paper No. 83-0175, Jan 1983.

29. NAGASHIM, T., OBKATA, T. and ASANUMA, T. - Experiment of Supersonic Air Intake Buzz. Report No. 481, University of Tokyo, 1972.

30. HANKEY, W.L. and SHANG, J.S. - Natural Transition - A Self-Excited Oscillation. AIAA Paper No.

31. MACK, L.M. - The Stability of the Compressible Laminar Boundary Layer According to a Direct Numerical Solution. AGARDograph 87, Pt. 1, pp 329-362, May 1965.

32. CHARAUD, R.D. and POWELL - Some Experiments Concerning the Hole and Ring Tone, A.J., Acoust. Soc. AM. 37 902-11, 1965.

33. GALLAWAY, C., GRAHAM, J. and HANKEY, W. - Flight Analysis of a Free-Falling Autorotating Plate. AFWAL-TM-83-

34. HANKEY, W.L. - Self-Excited Oscillations in Journal Bearings. AFWAL-TR-82-

35. HANKEY, W.L. - Analysis of the Whistler Nozzle. AFWAL-TM-82-168-FIMM, April 1982.

36. HANKEY, W.L. - Rotating Stall Analyzed as a Self-Excited Oscillation. AFWAL-TM-82-

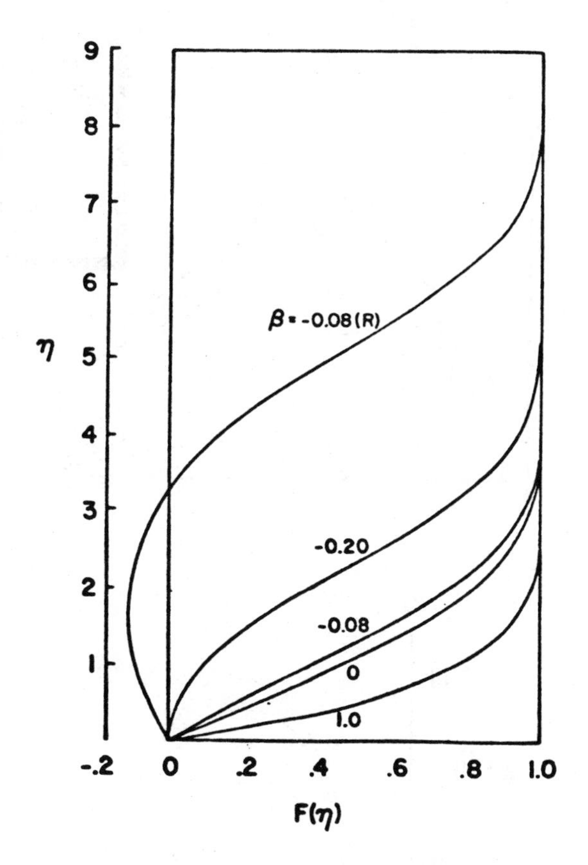

Fig 2.1 Similar Velocity Profiles for Falkner-Skan Flow

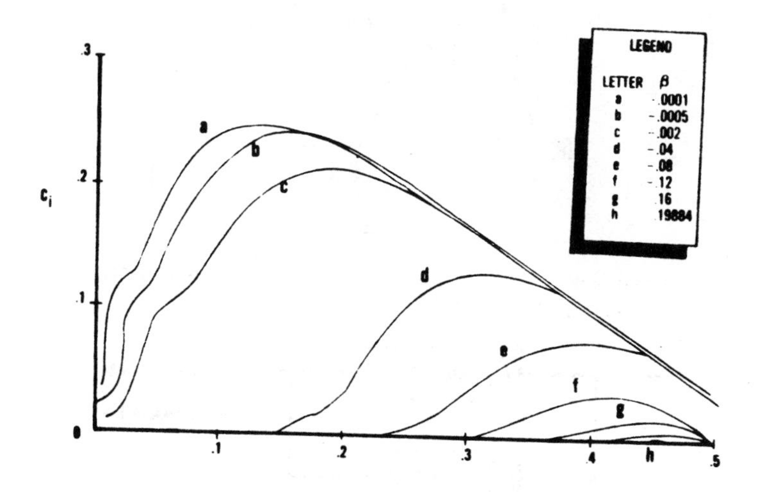

Fig 2.2 Amplification Factor vs. Wave Number for Various β (Ref 4)

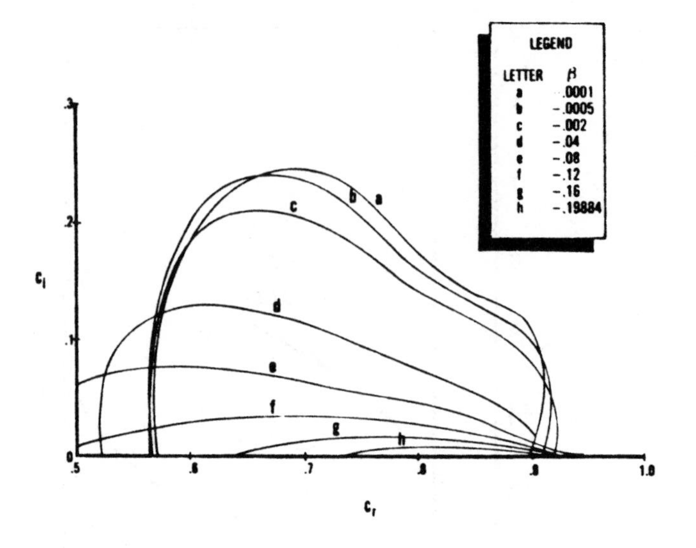

Fig 2.3 c_i vs. c_r for Various β (Ref 4)

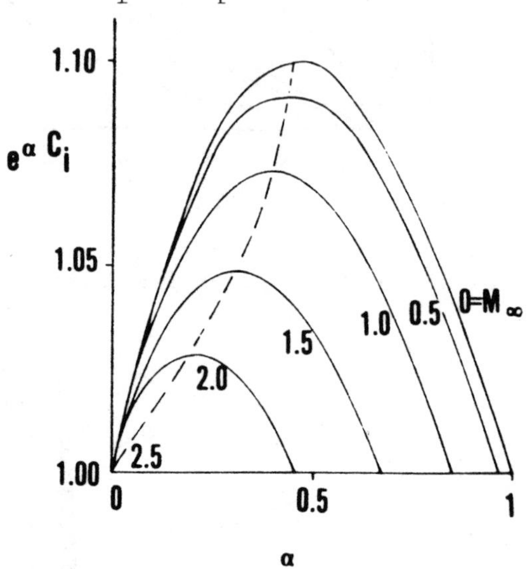

Fig 2.4 Influence of Mach Number on the Amplification Factor (Ref 5)

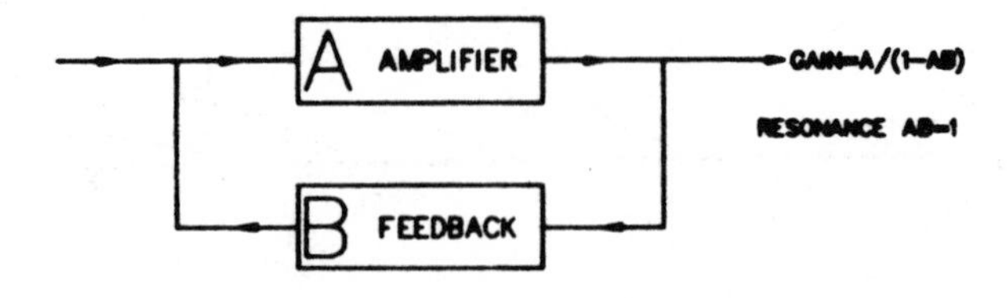

Fig 3.1 Diagram of Oscillator with Feedback

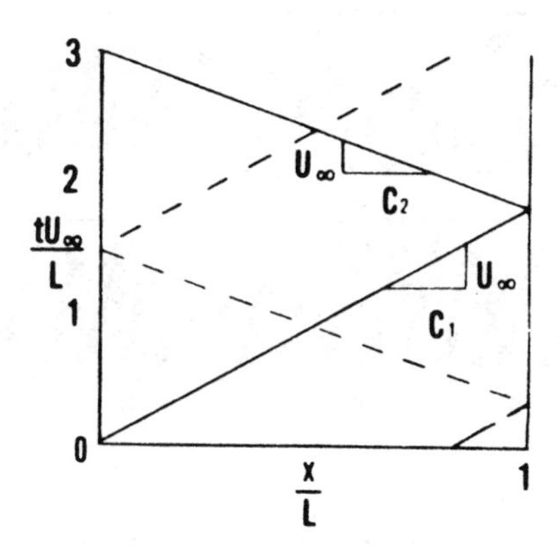

Fig 3.2 Wave Diagram for Pressure Oscillation

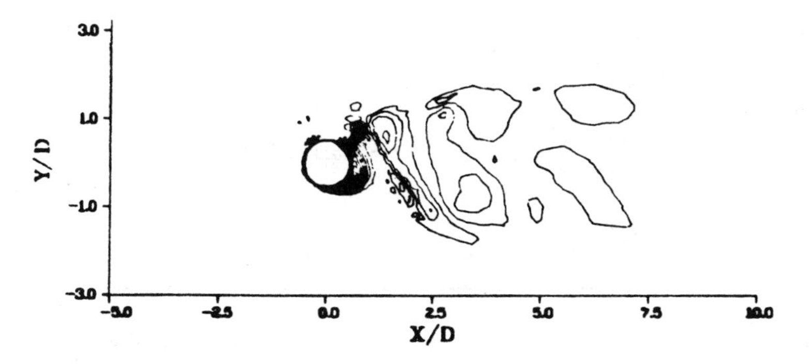

Fig 5.1 Temporal Development of Vorticity Field

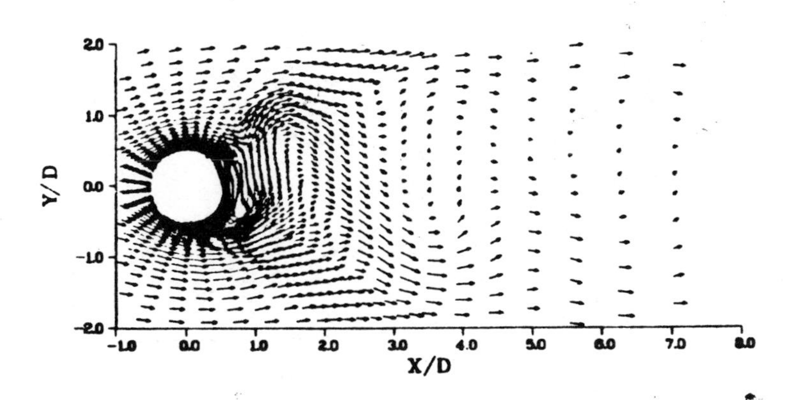

Fig 5.2 Temporal Development of Velocity Field

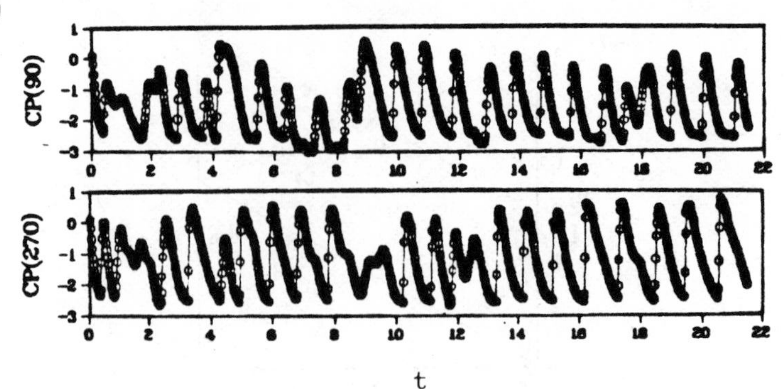

Fig 5.3 Temporal Evolution of Coefficients

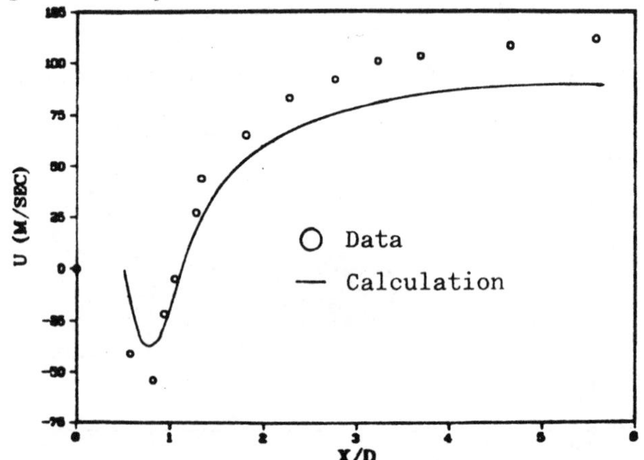

Fig 5.4 Comparison of Wake Centerline Velocity (Data Ref; F.K. Owen and D.A. Johnson, "Measurements of Unsteady Vortex Flow Field," AIAA J. Vol. 18, 1981, pp 1173-1179)

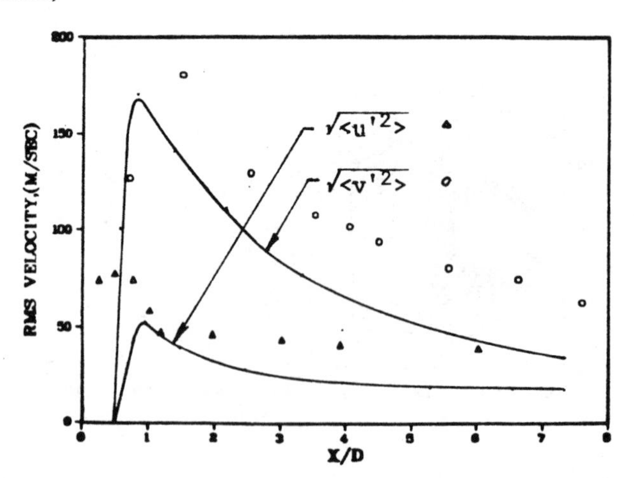

Fig 5.5 Comparison of Turbulent Intensity (Data Ref; F.K. Owen and D.A. Johnson, "Measurements of Unsteady Vortex Flow Field," AIAA J. Vol. 18, 1981, pp 1173-1179)

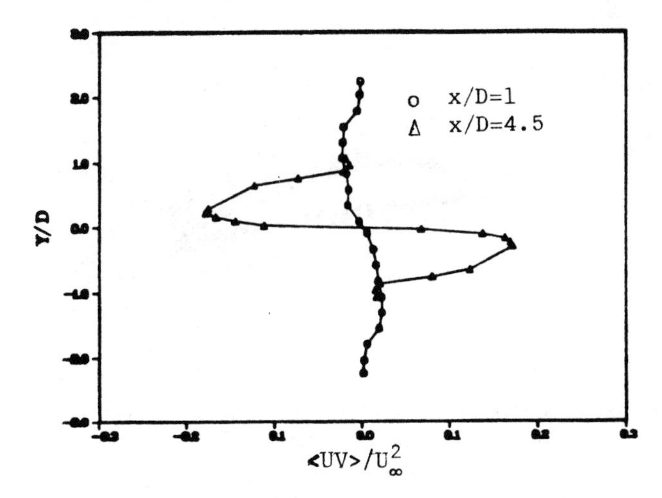

Fig 5.6 Computed Cross Correlation of u and v

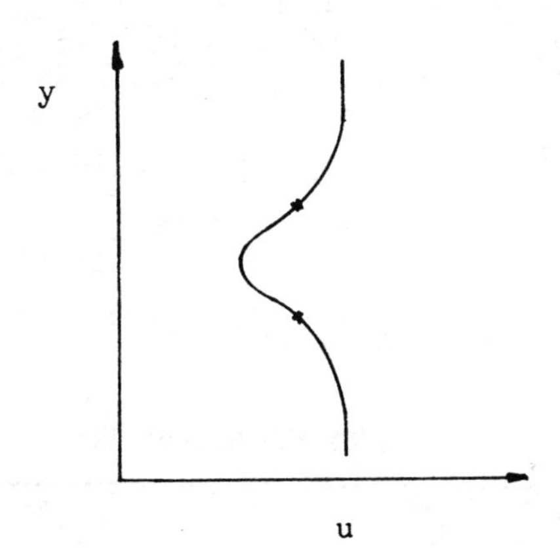

Fig 5.7 Wake Profile Showing Inflection Points

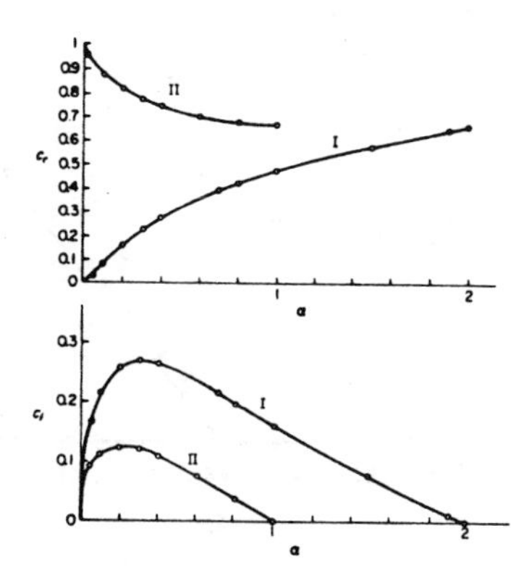

Fig 5.8 Amplification Factor Showing Antisymmetric (I) and Symmetric (II) Modes (Ref 19)

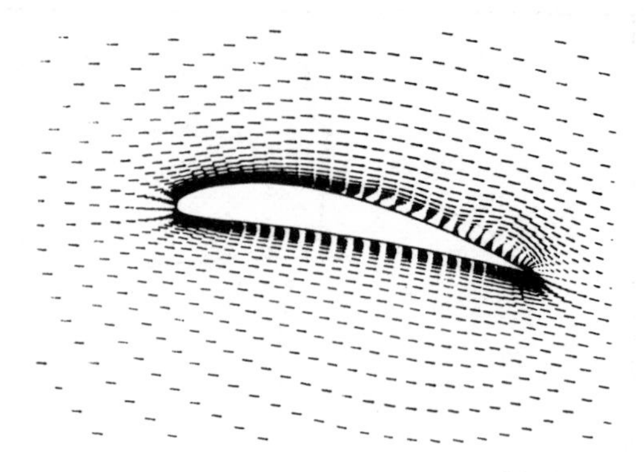

Fig 5.9 Velocity Field Over NACA 6412 Airfoil
Depicting Inflection Points

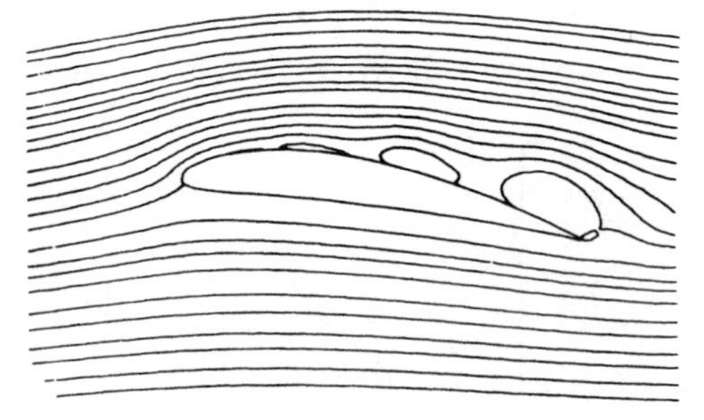

Fig 5.10 Instantaneous Streamline Distribution
Over NACA 6412 Airfoil

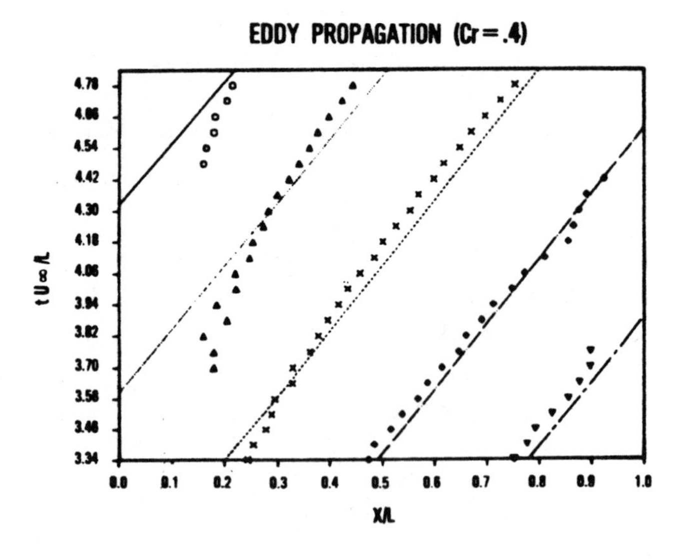

Fig 5.11 Wave Diagram of Shedding Eddies for Airfoil

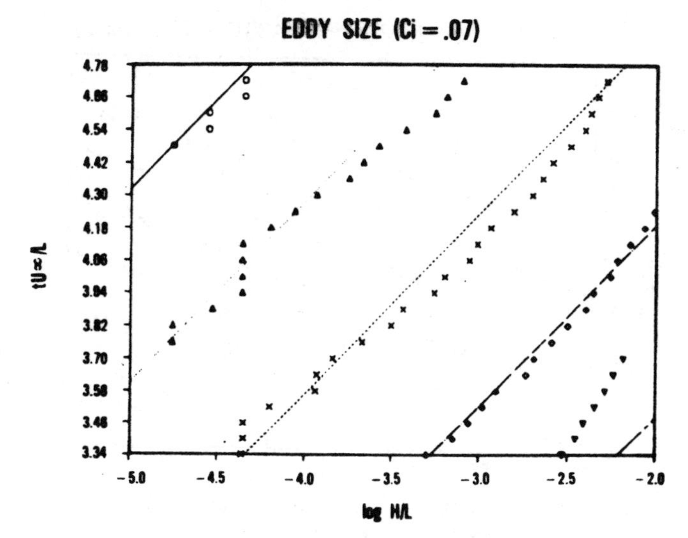

Fig 5.12 Growth Rate of Shedding Eddies for Airfoil

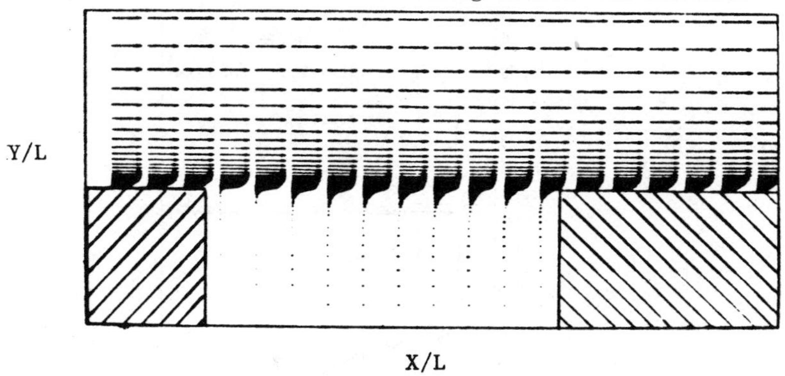

Fig 5.13 Velocity Distribution for Flow Over an Open Cavity

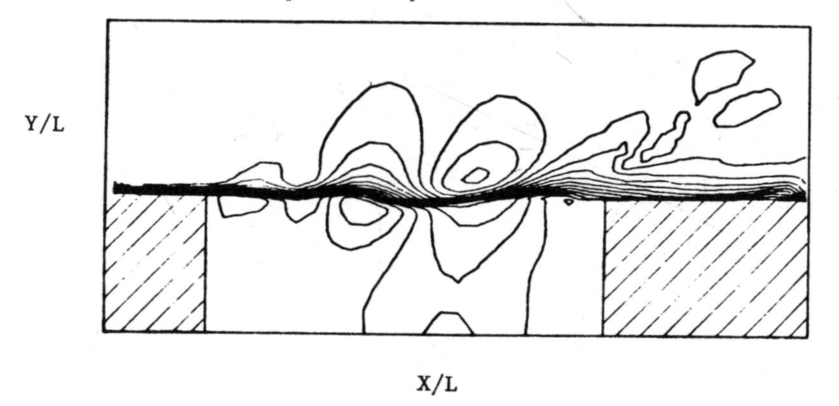

Fig 5.14 Density Contours for Open Cavity

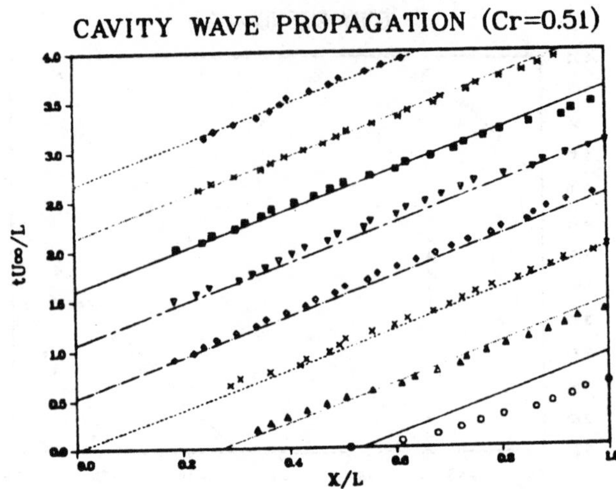

Fig 5.15 Wave Diagram for the Numerical Solution of the Open Cavity

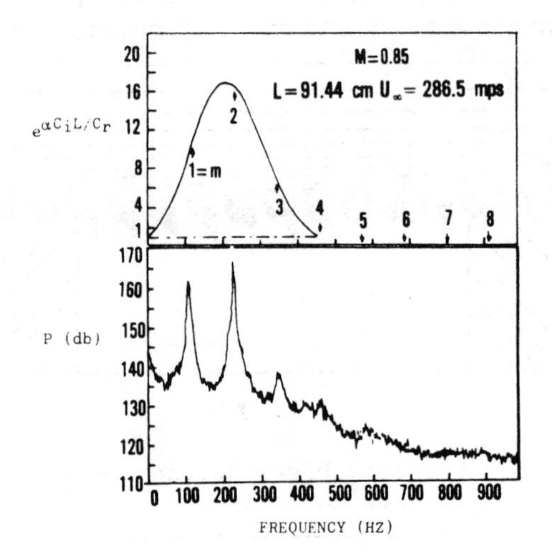

Fig 5.16 Comparison of Predicted and Experimental Spectral Analysis for Open Cavity

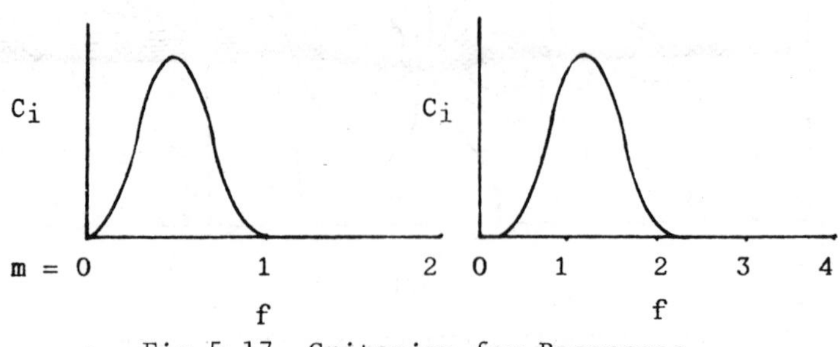

Fig 5.17 Criterion for Resonance

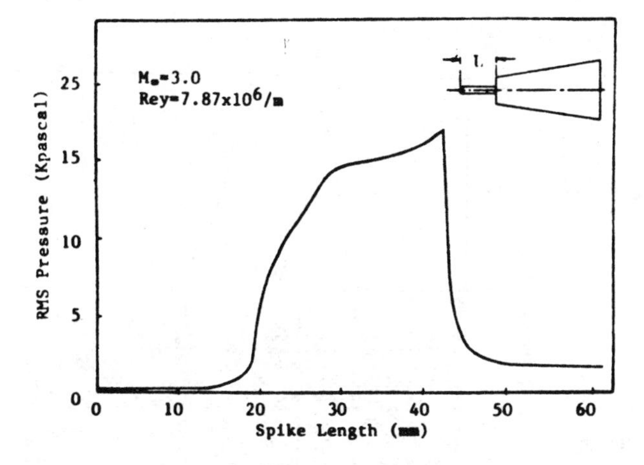

Fig 5.18 RMS Pressure Level for Spike Buzz

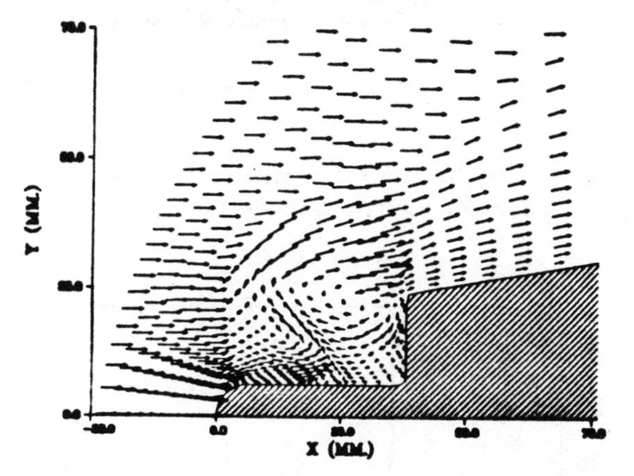

Fig 5.19 Instantaneous Velocity Field Over Spike-Tipped Body

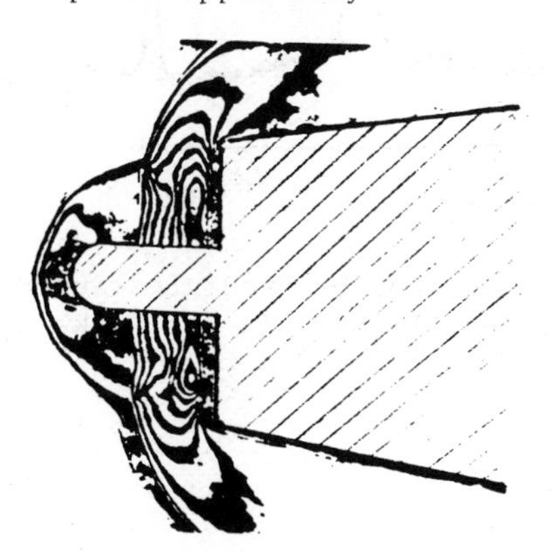

Fig 5.20 Schlieren of Spike Buzz

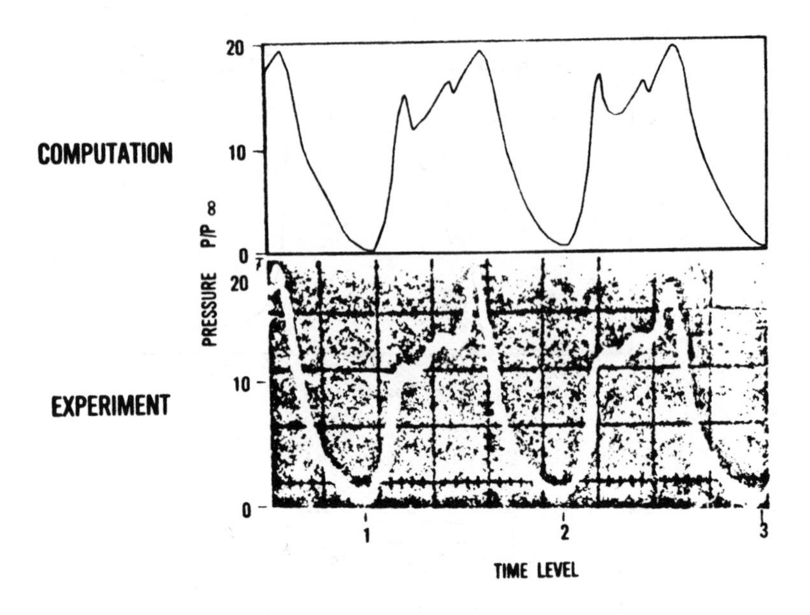

Fig 5.21 Comparison of Predicted and Experimental Wave Forms for Spike-Buzz

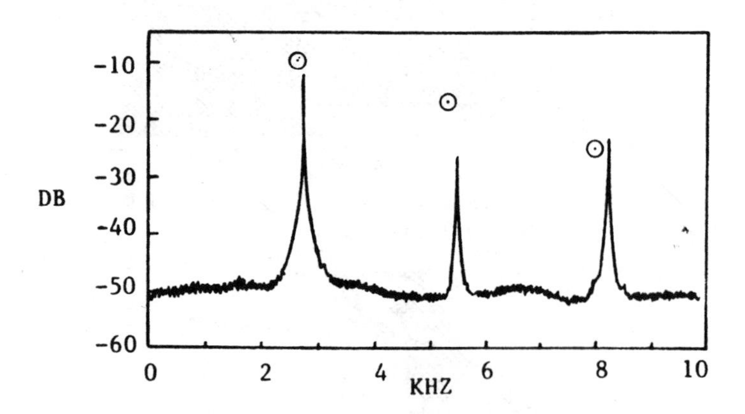

Fig 5.22 Comparison of Predicted and Experimental Spectral Analysis for Spike-Buzz

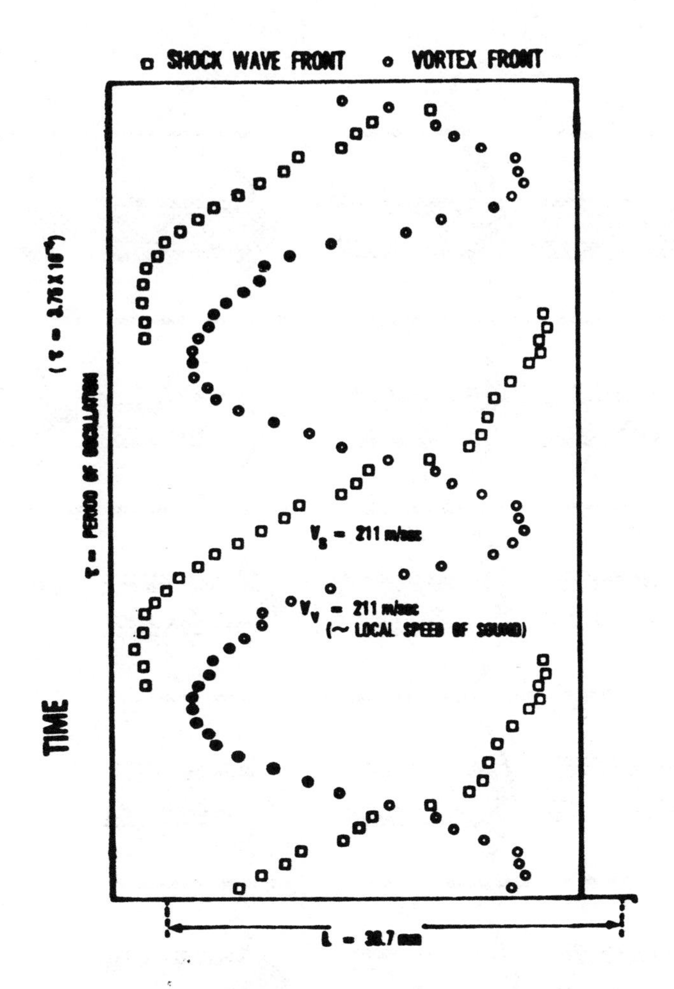

Fig 5.23 Wave Diagram of Spike-Buzz for Numerical Computation

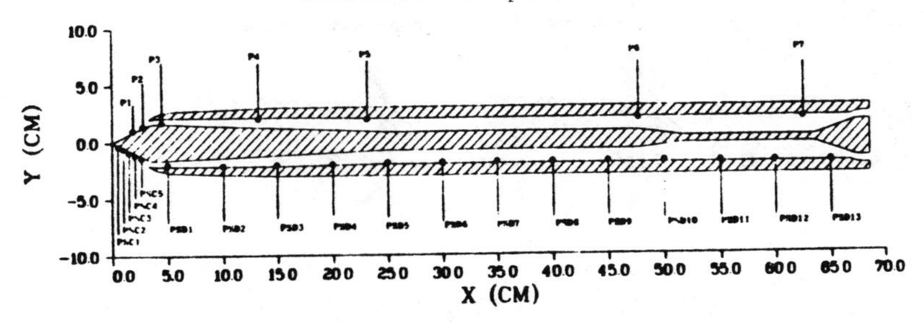

Fig 5.24 Computed Inlet Geometry with Static Pressure Probe Locations (TR=1.42, AR=1.16, L/D=15.88)

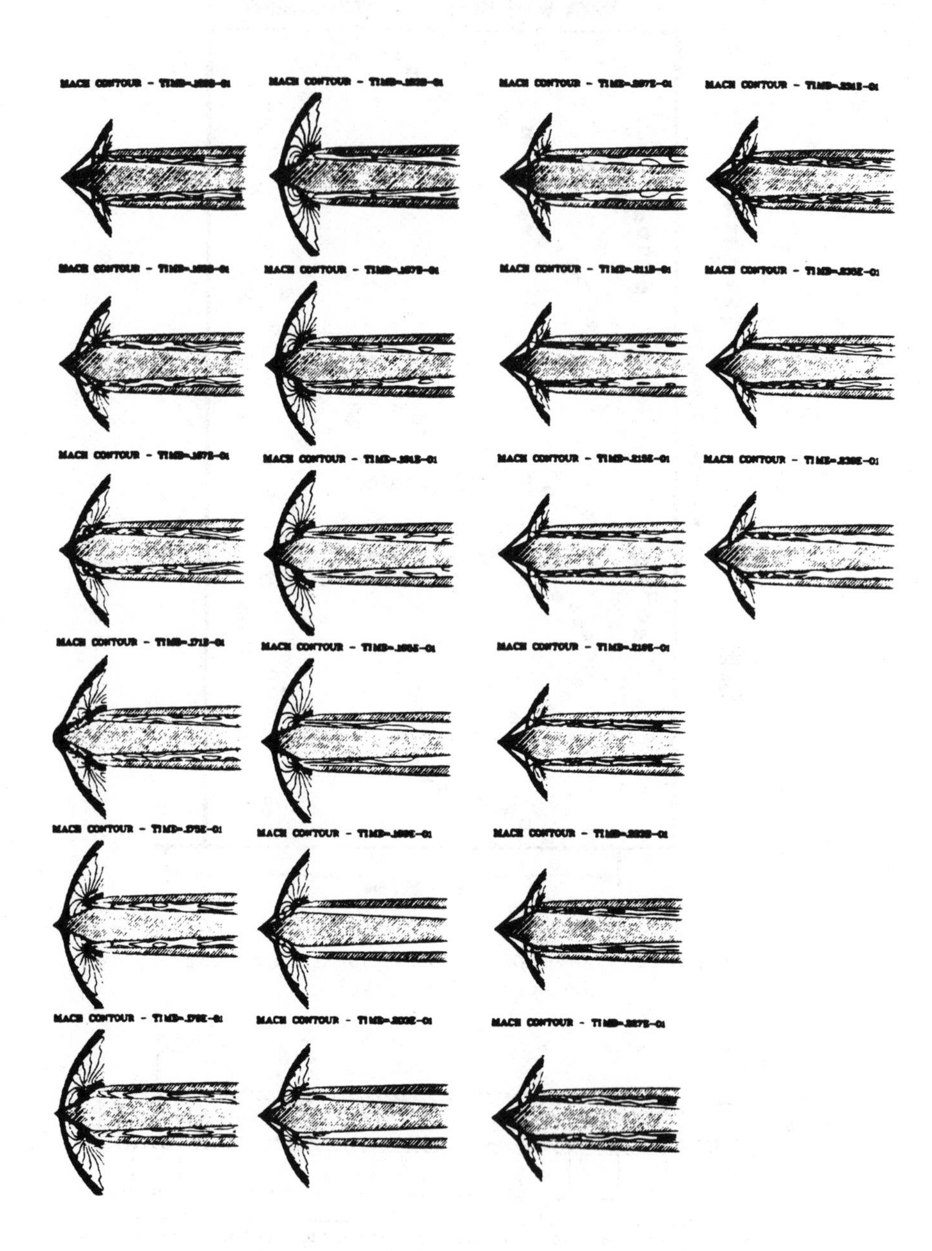

Fig 5.25 Forebody Flow Field, Third Buzz Cycle

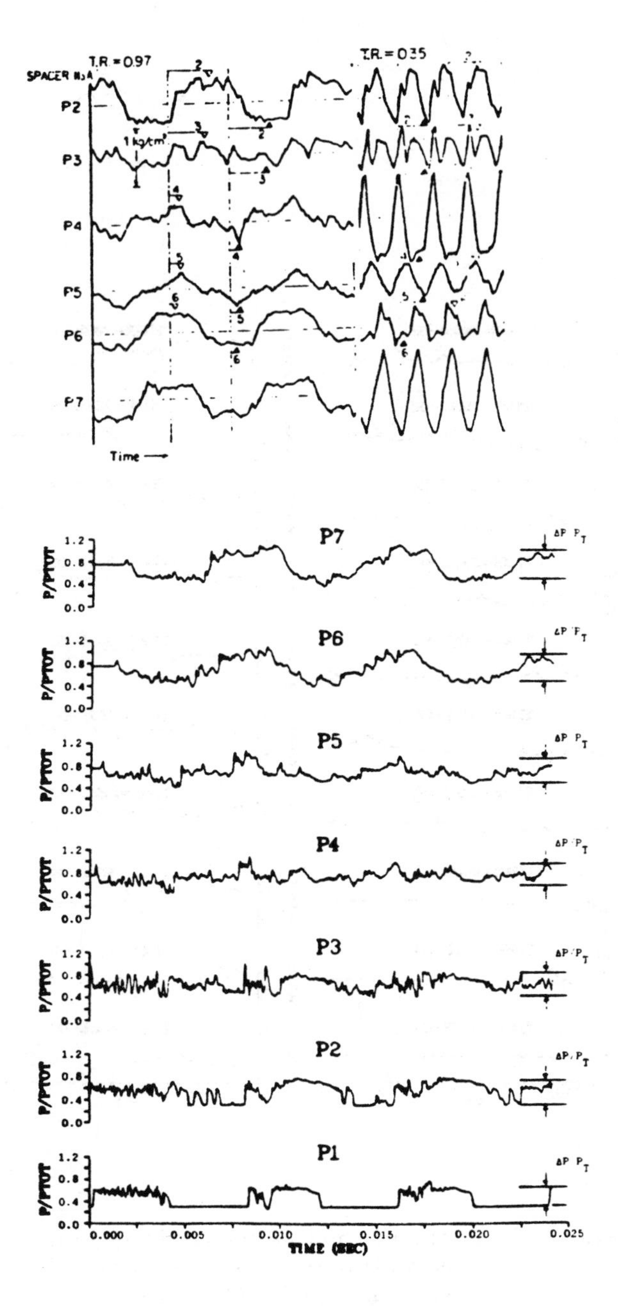

Fig 5.26 Computed Pressure Fluctuations for Experimental Probe Locations, Fig. 5 (TR=0.0)

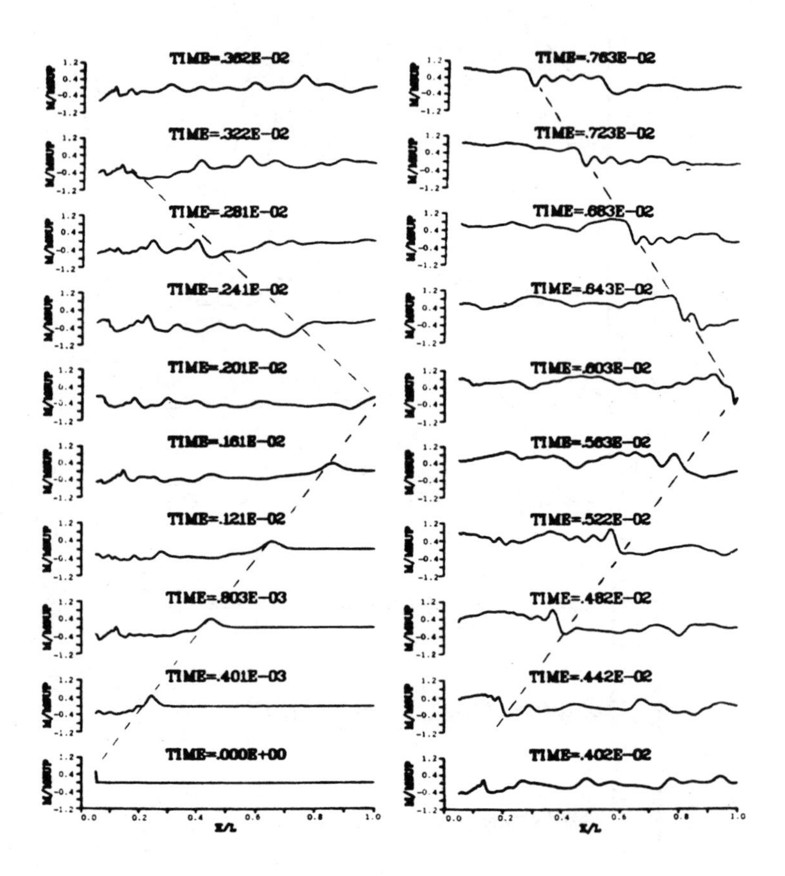

Fig 5.27 Ratio of Instantaneous Mass Flux to Supercritically Captured Mass Flux First Buzz Cycle (TR=0.0)

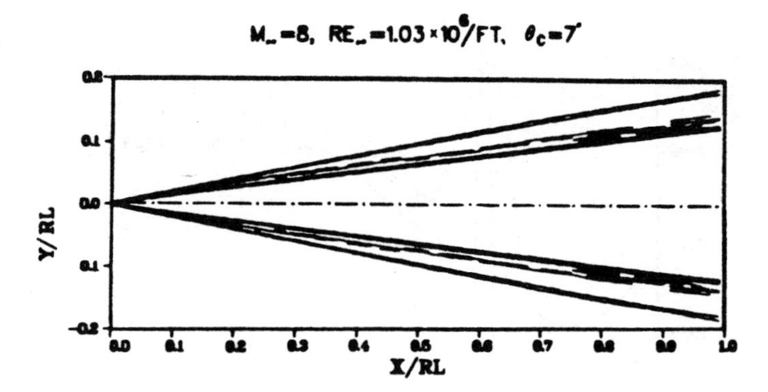

Fig 5.28 Mach Number Contour

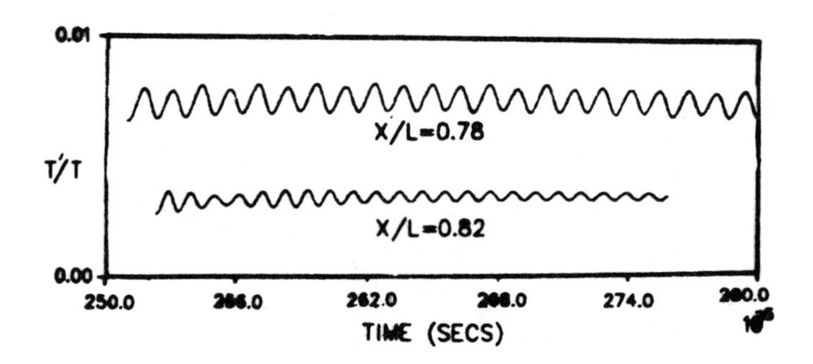

Fig 5.29 Typical Histogram of Temperature Oscillation

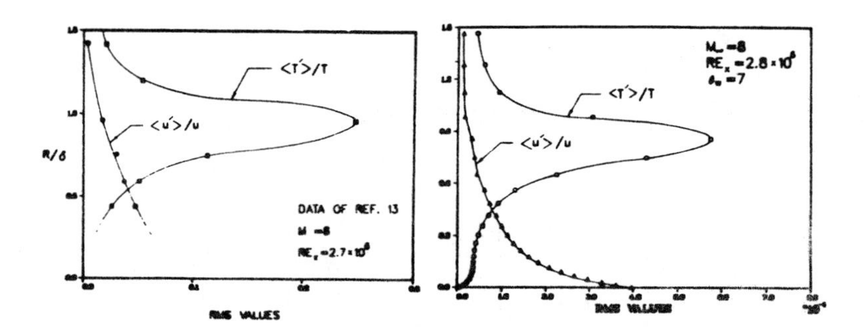

Fig 5.30 Computed RMS Profiles of Temperature and Velocity (Data Ref: A. Demetriades, "Hydrodynamics Stability and Transition to Turbulence in the Hypersonic Boundary Layer Over a Sharp Cone," AFOSR-TR-75-1435, July, 1975)

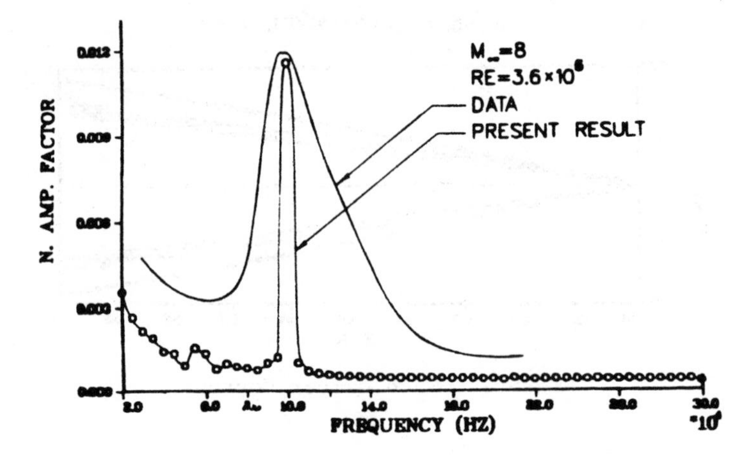

Fig 5.31 Spectral Analysis

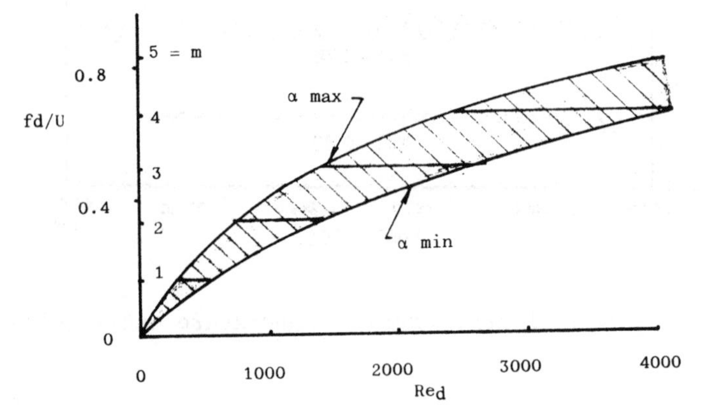

Fig 5.32 Theoretical Values of Edgetone Frequency

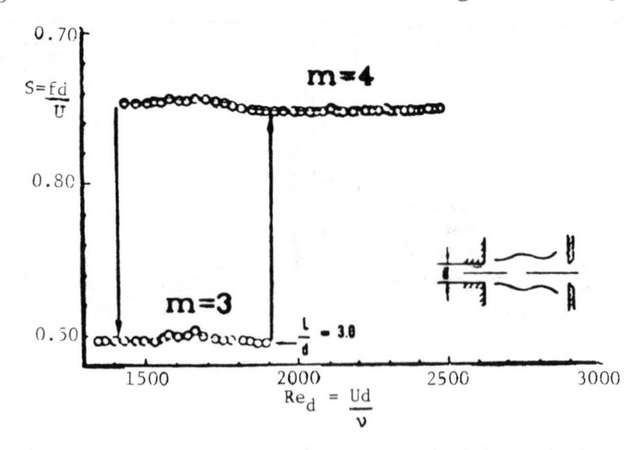

Fig 5.33 Jumps in Strouhal Number for Edge Tones (Ref 32)

TURBULENT FLOW AND HEAT TRANSFER IN COUPLED SOLID/FLUID SYSTEMS

C. TAYLOR and K. MORGAN

DEPARTMENT OF CIVIL ENGINEERING, UNIVERSITY COLLEGE OF SWANSEA, U.K.

1. INTRODUCTION

The advent of numerical techniques has proved invaluable when attempting to simulate, via the Navier-Stokes equations, both laminar and turbulent flow. Most flow situations in practice are turbulent exhibiting characteristic fluctuations whose random nature invalidates the use of the governing equations in their traditionally accepted form. It has been estimated [1] that even with present day computational facilities the analysis of relatively simple turbulent flows via a direct solution of the Navier-Stokes equations would involve, measured in hundreds of years, considerable time on a computer of almost limitless capacity. Probabilistic approaches can be used to quantify the variables but, for engineering purposes, a time averaged definition of these variables is far more convenient. Manipulation of the governing equations leads to further equations which express the mean flow variables relative to mean values of cross correlations of deviations from these means. This leads to a system of equations which contain more unknowns than equations. Additional, required, relationships are provided by various turbulence models so that closure of the equations can be effected.

The main objective of this current chapter is to utilise well established turbulence models to analyse turbulent flow with heat transfer. A two equation model of turbulence is used and, for each problem analysed, the transfer of energy is by forced convection.

2. MATHEMATICAL CONCEPTS

2.1 Hydrodynamic Principles

Limiting the physical problem to the steady state turbulent flow of an incompressible, viscous, Newtonian fluid, a normally accepted form of the governing, Navier-Stokes equations in two dimensions is,

$$u_j \frac{\partial u_i}{\partial x_j} = -\frac{1}{\rho}\frac{\partial p}{\partial x_i} + F_i + \frac{\partial}{\partial x_j}\left[\nu_e\left(\frac{\partial u_i}{\partial x_j} + \frac{\partial u_j}{\partial x_i}\right)\right] \qquad (1)$$

$$i,j = 1,2$$

in which F_i is a body force, over and above the local pressure, the velocity u_i and pressure p are time averaged mean values and ν_e a local value of the effective kinematic viscosity,

$$\nu_e = \nu + \nu_t \qquad (2)$$

in which ν_t is the turbulent kinematic viscosity.

When defining the turbulent viscosity, an analogy is drawn between laminar and turbulent flow. For laminar conditions a local shear stress, τ, is related to the gradient in velocity by,

$$\tau = \rho\, \nu \frac{\partial u_i}{\partial x_j}$$

in which ρ is the density of the fluid. For turbulent flow the same concept is employed and the shear stress is now defined by,

$$\tau = \rho\, \nu_e \frac{\partial u_i}{\partial x_j}$$

A full exposé of this concept is readily available in other texts [2] and the above definition is accepted.

Equation (1) together with the equation of continuity,

$$\frac{\partial u_i}{\partial x_i} = 0 \qquad (3)$$

form the basis of a significant proportion of the solution techniques currently being utilised to solve problems of turbulent flow.

A major variation between advocated solution techniques is the definition of the magnitude of the physical property ν_e. The method of approach used in the present chapter is the so-called two equation model of turbulence. Two further transport type equations are used to define the spatial variation in two turbulence related parameters which are subsequently used to evaluate the effective kinematic viscosity. This is defined utilising the Prandtl [3] and Kolmogorov [4] relationship,

$$\nu_t = \frac{\mu_t}{\rho} = C_\mu \; k^{\frac{1}{2}} \ell \tag{4}$$

where C_μ is a constant, k is the turbulence kinetic energy and ℓ a characteristic length referred to as the 'mixing length' [2]. Denoting the instantaneous fluctuations in the mean velocity by u_i', the turbulence kinematic energy is defined,

$$k = \frac{1}{2} \sqrt{(u_i')^2} \tag{5}$$

Closure of the mathematical model represented by (1), (3), (4) and (5) can now be effected by defining the spatial distribution of k and ℓ within and on the boundary of the flow domain. This can be achieved by,

i) defining both k and ℓ via algebraic expressions

or

ii) defining k using a further differential equation, similar in form to (1), and representing ℓ by an algebraic equation - the one equation model of turbulence

or

iii) defining both k and ℓ using transport type differential equations, again similar in form to (1), - the two equation model of turbulence.

The third option is used in which the spatial variation in the turbulence kinetic energy is assumed to comply with,

$$\underset{\text{(CONVECTION)}}{u_j \frac{\partial k}{\partial x_j}} = \underset{\text{(DIFFUSION)}}{\frac{\partial}{\partial x_j}\left[\left(\frac{\nu_t}{\sigma_k} + \nu\right)\frac{\partial k}{\partial x_j}\right]} + \underset{\text{(PRODUCTION)}}{\nu_t \frac{\partial u_i}{\partial x_j}\left[\frac{\partial u_i}{\partial x_j} + \frac{\partial u_j}{\partial x_i}\right]} - \underset{\text{(DISSIPATION)}}{C_D \frac{k^{3/2}}{\ell}} \tag{6}$$

as advocated by Wolfshtein [5]. Here, $\frac{\nu_t}{\sigma_k}$ represents a type of turbulent diffusion coefficient for turbulence kinetic energy and both σ_k and C_D are also assumed to be constant.

The term σ_k is, in fact, a form of turbulent Schmidt number, S_c, where

$$S_c \equiv \frac{\nu_t}{D_1} = \sigma_k \tag{7}$$

and D_1 is a diffusion coefficient, in a conventionally accepted form, related to the diffusion of turbulence kinetic energy.

An equation representing the dissipation of turbulence kinetic energy, ε, is usually written [6],

$$\underset{\text{(CONVECTION)}}{u_i \frac{\partial \varepsilon}{\partial x_j}} = \underset{\text{(DIFFUSION)}}{\frac{\partial}{\partial x_j}\left[\left(\frac{\nu_t}{\sigma_\varepsilon} + \nu\right)\frac{\partial \varepsilon}{\partial x_j}\right]} + \underset{\text{(PRODUCTION)}}{C_1 k \frac{\partial u_i}{\partial x_j}\left(\frac{\partial u_i}{\partial x_j} + \frac{\partial u_j}{\partial x_i}\right)} - \underset{\text{(DISSIPATION)}}{C_2 \frac{\varepsilon^2}{k}} \tag{8}$$

in which the rate of dissipation of the turbulence kinetic energy is,

$$\varepsilon = \frac{k^{3/2}}{\ell}$$

Again, $\frac{\nu_t}{\sigma_\varepsilon}$ can be interpreted as a turbulent diffusion coeff-ient for ε and σ_ε a form of Schmidt number. The values of C_1, C_2 and σ_ε are assumed to be constant.

The two 'transport' equations (6) and (8) are by no means unique in form and a number of variations have been suggested. Kolmogorov [4] advocated a frequency type parameter, in preference to the rate of dissipation of turbulence kinetic energy. This was subsequently used, the k-w model [7], where w is related to the vorticity of large scale eddies within the flow field. The Wilcox-Traci model [8] is also based on such a definition. A major factor in the retention, by most researchers, of the ε equation is the ease with which this parameter can be defined close to solid boundaries. This will be demonstrated by example. A more recent model, [9], is a further variation on the theme involving frequency and is, in a finite element context, reported to improve both solution convergence and stability. Since no upwinding [10], is introduced into the finite element model, and therefore no artificial damping, turbulence models which aid convergence and stability are particularly advantageous. For present purposes, however, the conventional k-ε model is used and a simultaneous solution of the relevant equations is obtained using simple relaxation techniques.

2.1.1 Boundary Conditions

On part of the boundary, not associated with a solid surface, the following boundary conditions apply,

$$\left.\begin{array}{l} u_i = \hat{u}_i \\ k = \hat{k} \\ \varepsilon = \hat{\varepsilon} \end{array}\right\} \text{ on } \Gamma_1 \tag{9}$$

and on Γ_2, a constant is usually placed on either first order gradients,

$$\left.\begin{array}{l} \dfrac{\partial u}{\partial x_n} = \hat{q}_u \\ \dfrac{\partial k}{\partial x_n} = \hat{k}' \\ \dfrac{\partial \varepsilon}{\partial x_n} = \hat{\varepsilon}' \end{array}\right\} \text{ on } \Gamma_2 \tag{10}$$

where the circumflex in (9) and (10) denotes specified magnitudes and suffix 'n' implies a direction normal to the boundary surface.

Another form of boundary condition, now used quite extensively on Γ_2, is the specification of a traction type boundary condition,

$$\left.\begin{array}{l} T_n = \hat{T}_n \\ T_s = \hat{T}_s \end{array}\right\} \text{ on } \Gamma_2 \tag{11}$$

where suffix 's' denotes a direction tangential to the boundary. For the momentum equations these take the form,

$$\begin{aligned} \hat{T}_n &= -p + 2\mu_e \frac{\partial u}{\partial x_n} \\ \hat{T}_s &= \mu_e \left(\frac{\partial u_n}{\partial x_s} + \frac{\partial u_\varepsilon}{\partial x_n} \right) \end{aligned} \tag{12}$$

Similar 'traction type' boundary conditions can be defined for both the k and ε equations.

On surfaces of symmetry the tangential traction and normal velocities can be assumed to be zero.

Although the specification of the velocity on a boundary is self-explanatory some special consideration must be given to the variation of velocity in a near wall region. Large velocity gradients exist in such regions particularly when zero slip is imposed. In addition, as the solid boundaries are approached, the regime reverts to laminar flow. In such circumstances the 'law of the wall' technique has been proved to be extremely useful. Its utilisation enables the fluid zone to be curtailed some small distance away from the wall and the governing equations solved in this region only, Figure 1.

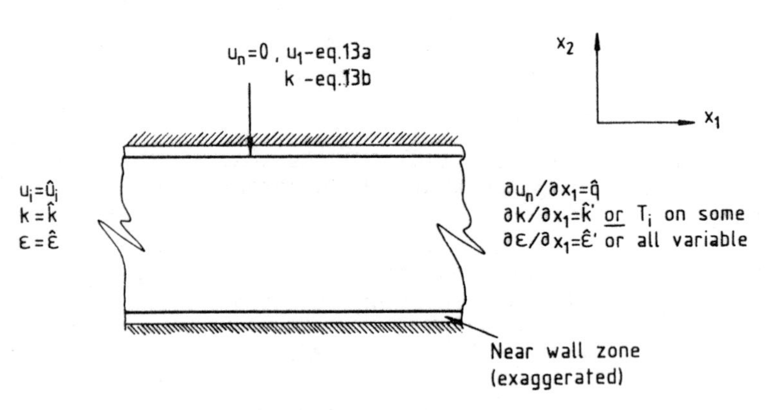

FIGURE 1 DEFINITION OF "NEAR WALL" CONCEPT AND "TYPE" BOUNDARY CONDITIONS

The variables are matched in the near wall zone with algebraic expressions. One form of these equations is, [11],

Velocity

$$\left.\begin{aligned} u_s^+ &= x_n^+ && 0 \leqq x_n^+ \leq 5 && \text{(a)} \\ u_s^+ &= (-3.05+5.0 \log_n x_n^+)\,\frac{\tau_w}{|\tau_w|} && 5 \leqq x_n^+ \leq 30 && \text{(b)} \\ u_s^+ &= (-5.5+2.5 \log \; x_n^+)\,\frac{\tau_w}{|\tau_w|} && x_n^+ \geq 30 && \text{(c)} \end{aligned}\right\} \quad (13)$$

where,
$$u_s^+ = \frac{u_s}{\sqrt{\frac{\tau_w}{\rho}}}$$

$$x_n^+ = \left(\frac{x_n}{\mu}\right)\sqrt{\left|\frac{\tau_w}{\rho}\right|}$$

and the value used for k^+ is given by,

$$k^+ = (C_\mu\ C_D)^{-0.5} \qquad (13d)$$

where C_μ and C_D are constants and the shear stress at and to within a small distance from the wall is given by,

$$\tau_w = \mu \frac{\partial u_s}{\partial x_n} \qquad (14)$$

Once the shear stress is known, therefore, values of the velocity in the near wall zone can be calculated.

2.2 The Energy Equation

When dealing with heat transfer problems, part of the body force in the momentum equations, F_i, could be dependent on temperature. This is attributable to the dependence of the fluid density on the local temperature relative to a datum temperature and hence density. Such 'buoyancy' contributions are oriented in the local direction of the acceleration field and are written in the form,

$$F_i = a_i\ \beta\ \Delta\theta \qquad (15)$$

where a_i is the local acceleration in the ith coordinate direction, β the coefficient of thermal expansion of the fluid and $\Delta\theta$ the temperature rise relative to a datum value. For problems of free convection such terms can constitute the major forcing function whereas in forced convection these can usually be ignored.

Invoking the 'time averaging' concept previously adopted a form of the energy equation can be written,

$$u_i \frac{\partial\theta}{\partial x_i} = \frac{\partial}{\partial x_i}\left[D_h \frac{\partial\theta}{\partial x_i}\right] + g \qquad (16)$$

in which D_h denotes a thermal diffusivity coefficient and g a heat source per unit volume.

In a purely conducting medium, D_h can be defined in terms of the fluid density, ρ, the specific heat at constant pressure, c, and the thermal conductivity of the material, k^0,

$$D_h = \frac{k^0}{\rho c} \tag{17}$$

For heat transfer in turbulent flow of a fluid then D_h can be interpreted as,

$$D_h = \left(\frac{\nu}{\sigma} + \frac{\nu_t}{\sigma_t}\right) \tag{18}$$

where σ and σ_t is the laminar and turbulent Prandtl number, respectively.

2.2.1 Boundary Conditions

The imposition of boundary conditions associated with the energy equation pose similar problems to those when evaluating the distribution of velocity. In this instance the turbulent Prandtl number changes markedly in the near wall region, $0 < x_n^+ < 30$. Although a value of 0.9 is widely accepted for the fully turbulent region in confined flows, its value in the near wall region is still the subject of some speculation.

Cebeci [12] concluded that the turbulent Prandtl number increases as the wall is approached and attains a constant value in the viscous sub-layer. Na and Habib [13] suggested a value of 1.43; Wassel and Cotton [14] 1.32; Sleitcher [15] 1.4 and Sherwood et al. [16] a value of $Pr^{-\frac{1}{2}}$.

In all cases the basic concept that the most likely trend in the value of the turbulent Prandtl number is towards a constant value within the sublayer has a theoretical basis. Meranay [17] and Orlando et al. [18] both concluded that a constant value would conform to theoretical predictions. Indeed the latter suggested a value of 1.4 in this zone. The accepted trend is, therefore, a sharp rise in the value of the number as the sublayer is approached.

The reverse of the above trend has been advocated by Antonia [19] who suggested a Taylor series expansion of the correlations $\overline{u_i' u_j'}$, $\overline{u_j \theta'}$, u_i and θ in terms of x_n^+. Here the prime denotes, as in conventional turbulence theory, instantaneous fluctuations from the mean and the overbar time averaged values. The theory leads to an expression of the form,

$$Pr_{(t)} \simeq 0.61 \frac{(1 - 0.045\, x_n^+)}{(1 - 0.05\, x_n^+)} \tag{19}$$

in which the turbulent Prandtl number $Pr_{(t)}$ is asymptotic to 0.61 as x_n^+ approaches zero. The value of (19) increases sharply near the edge of the sublayer and becomes essentially constant, at 0.9, when $x_n^+ > 40$.

Having decided upon suitable values of $Pr_{(t)}$ the energy equation can now be solved and is usually subject to boundary conditions of the form,

$$\theta = \hat{\theta} \text{ on } \Gamma_3$$
$$\frac{\partial \theta}{\partial n} = \hat{q}_\theta \text{ on } \Gamma_4 \tag{20}$$

A simple 'wall function' approach can again be used to depict the variation of θ in near wall regions. The variation is assumed small parallel to the wall such that,

$$\frac{\partial \theta}{\partial x_n} >> \frac{\partial \theta}{\partial x_s} \tag{21}$$

This implies that,

$$\frac{\partial}{\partial x_n}\left(\frac{\nu_t}{\sigma_t}\frac{\partial \theta}{\partial x_n}\right) = 0 \tag{22}$$

or,

$$\frac{\partial \theta}{\partial x_n}\ \frac{\nu_t}{\sigma_t} = \lambda = \frac{C_\mu}{\sigma_t}\ \rho\ k^{\frac{1}{2}}\ \frac{\partial \theta}{\partial x_n} \tag{23}$$

in which λ is some constant. Defining,

$$\theta^+ = \frac{\theta - \theta_w}{\lambda}\sqrt{\frac{\tau_w}{\rho}} \tag{24}$$

where θ_w is the temperature at the wall. Following standard procedure [20] and utilising equations (23) and (24) an expression for θ^+ can be derived in the form,

$$\theta^+ = \sigma_t\left[2.5\ \ell_n\left(\frac{x_n}{\nu}\sqrt{\frac{\tau_w}{\rho}}\right) + 5.5 + P\left(\frac{\sigma}{\sigma_t}\right)\right] \tag{25}$$

where $P\left(\frac{\sigma}{\sigma_t}\right)$ is the Spalding-Jayatalika function,

$$P\left(\frac{\sigma}{\sigma_t}\right) = 9.24\left[\left(\frac{\sigma}{\sigma_t}\right)^{0.75} - 1\right]\left[1 + 0.28\ \exp\left(-0.007\ \frac{\sigma}{\sigma_t}\right)\right] \tag{26}$$

Within the laminar sub-layer (16) reduces to,

$$\frac{\partial}{\partial x_n}\left(\frac{\nu}{\sigma}\frac{\partial \theta}{\partial x_n}\right) = 0 \qquad (27)$$

which, on integration, becomes

$$\theta = \left.\frac{\partial \theta}{\partial x_n}\right|_w x_n + \theta_w \qquad (28)$$

where $\left.\frac{\partial \theta}{\partial x_n}\right|_w$ denotes the temperature gradient at the wall.

In the intermediate zone, $5 \leq x_n^+ \leq 30$, (25) can be re-written,

$$\theta^+ = \sigma_t \left[5 \,\ell_n \left(\frac{x_n}{\nu}\sqrt{\frac{\tau_w}{\rho}}\right) - 3.05 + P\left(\frac{\sigma}{\sigma_t}\right)\right] \qquad (29)$$

For situations where heat exchange occurs, for instance between a solid and fluid, an overall heat balance and interfacial flux balance could be imposed as additional conditions for closure. Such refinements will be discussed, as necessary, when dealing with specific examples.

3. SOLUTION PROCEDURE

The spatial domain, either solid or fluid, is discretised utilising conventional eight noded isoparametric finite elements [21]. Special elements could be used as the final near wall zone element in the fluid to allow a better approximation of the shear stress. Imposing a C' continuity at the limit of the fluid zone enables an explicit evaluation of the first order gradient of the fluid velocity and therefore a direct evaluation of the shear stress [9]. Another innovation is the utilisation of 'logarithmic' elements [22] which results in a mapping of the variable concerned right up to a solid boundary. Both the above innovations have proved useful but not essential and, in the present chapter, conventional elements are employed.

A straightforward solution procedure was adopted and consisted of the following basic steps,

i) Set all variables to zero and assume laminar properties.

ii) Estimate near wall boundary conditions and impose known boundary conditions.

iii) Solve for u_i, k, ε, p and θ.

iv) Evaluate μ_t knowing k and ℓ.

v) use the wall functions to evaluate the wall shear stress, τ_w, and boundary conditions on u, θ, k and ε.

vi) Check on convergence of all variables.

A simple relaxation technique can be used, relaxation factor of 0.5. The rate of convergence of the solution technique could be increased by separating the ε equation and conducting some two to three iterative steps on the remaining equations before introducing the variations in ε. There are a number of variations on this theme and more efficient algorithms could undoubtedly be evolved.

4. APPLICATIONS

4.1 Flow with Separation

One of the most difficult problems to analyse, when using a discrete form of the governing equations, is when flow separation and possible points of re-attachment are evident in the flow regime. In the case of a sudden expansion, Figure 2, a singularity exists at both the abrupt expansion and point of re-attachment. Again, the use of special elements, to ensure compatibility of shear stress, could be used at such locations.

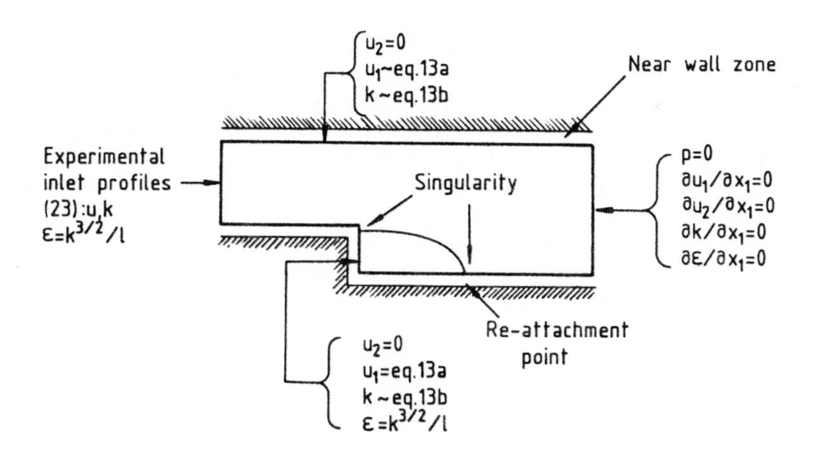

FIGURE 2 BOUNDARY CONDITIONS FOR STEP

Alternatively, a smoothing process could be adopted at the expansion such that the gradient of the stream function is maintained at zero at the point adjacent to the abrupt expansion, Figure 2, [20]. Whichever technique is employed the solution procedure is necessarily iterative and particular attention has to be paid to the accuracy of the wall shear stress.

The boundary conditions imposed are defined on Figure 2 and the only values that need further comment are the magnitudes of k and ε. The values of k are those obtained experimentally [23]. Preliminary calculations can be undertaken to find values of ε to impose on the upstream boundary. These are taken to be equal to the fully developed values obtained when analysing flow, of equal magnitude, through a straight channel of the same width as that upstream of the expansion. This does not, as expected, affect the downstream, fully developed, values of turbulence kinetic energy. However, better correlation is obtained at intermediate sections. The values given to the various coefficients are,

$$C_\mu = 0.22, \; \sigma_k = 1.0, \; C_1 = 1.45, \; C_2 = 0.18,$$

$$\sigma_\varepsilon = 1.3 \text{ and } C_D = 0.092$$

as used previously [24].

Although the length of re-attachment was found to be smaller than measured values, approximately 4.5 x step height as opposed to approximately 6 x step height, the velocity distribution, except in the vicinity of the point of re-attachment, were found to be quite close to experimental values and are adequately reported elsewhere [20]. A closer inspection of the recirculation zone, Figure 3, with regard to the evaluation of effective viscosity, reveals that the assumptions inherent in the 'law of the wall' model, particularly uni-dimensionality, are inadequate and a more sophisticated model is desirable in this area. The introduction of two dimensional 'logarithmic' elements could well lead to a better correspondence with measured values.

Within the separation zone the rate of heat transfer varies considerably. This is particularly noticeable near the point of re-attachment where the heat transfer to the walls could increase as a direct result of the reduction in the thickness of the laminar sub-layer, due to increased diffusion. The resulting gradients in temperature could be significant, which, in a design context, is of particular importance.

In accordance with the values given by Seban, a heat flux is imposed on the upper surface of the expansion and those relating to fully developed conditions are imposed upstream, obtained in a manner similar to that used when evaluating

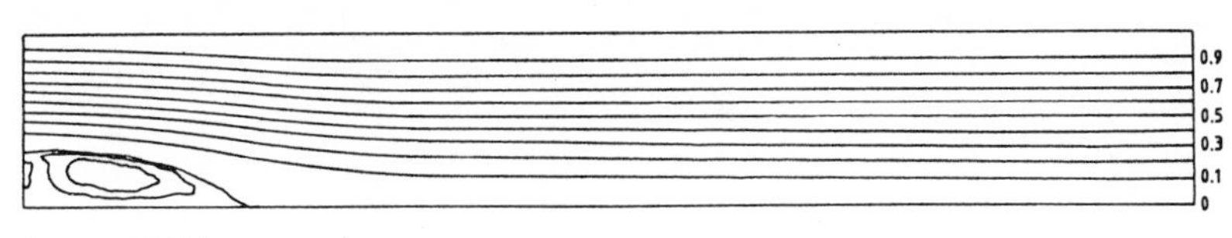

Re = 100000

(a) Stream function

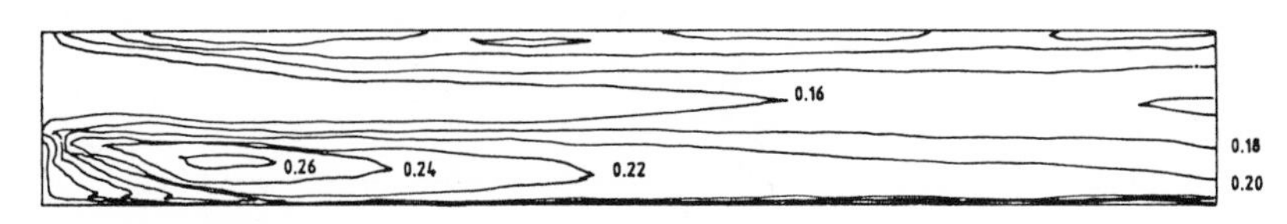

Re = 100000

(b) Turbulent kinetic energy

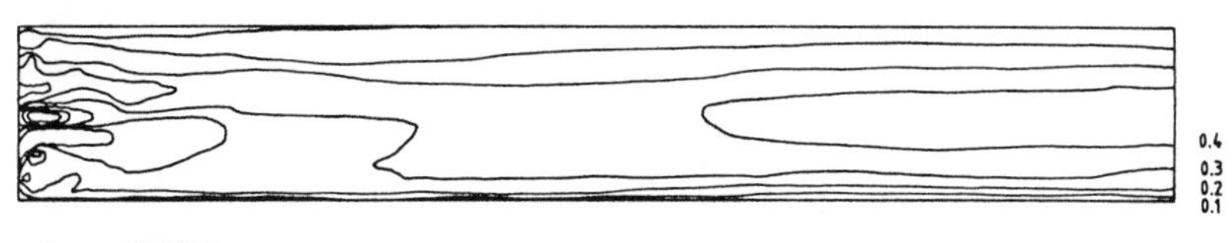

Re = 100000

(c) Length scale

FIGURE 3 STEP-PLOT OF STREAM FUNCTION TURBULENT KINETIC ENERGY AND LENGTH SCALE

upstream values of k, and the gradient is assumed to be zero on the downstream boundary. At near wall points, guessed initial values of the temperature are imposed and are updated, after each iteration, using the appropriate wall functions.

Assuming that the velocity distribution is known, the temperature distribution is determined and compared with previously published numerical and experimental results.

The resulting distribution of temperature is as shown on Figure 4 with the corresponding longitudinal variation in the Nusselt number, Figure 5. For present purposes the variation

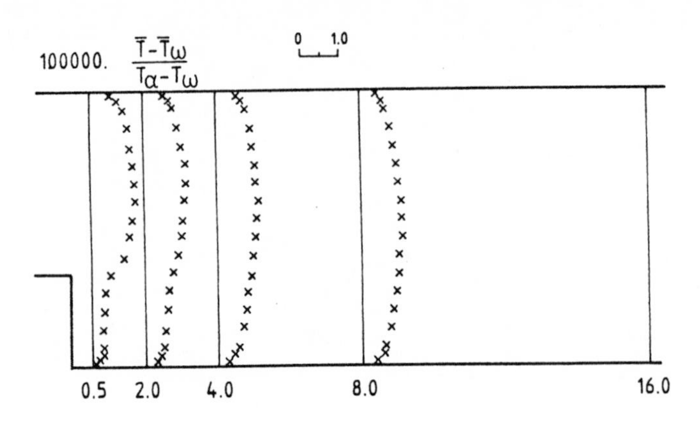

FIGURE 4 TEMPERATURE DISTRIBUTION DOWNSTREAM OF STEP

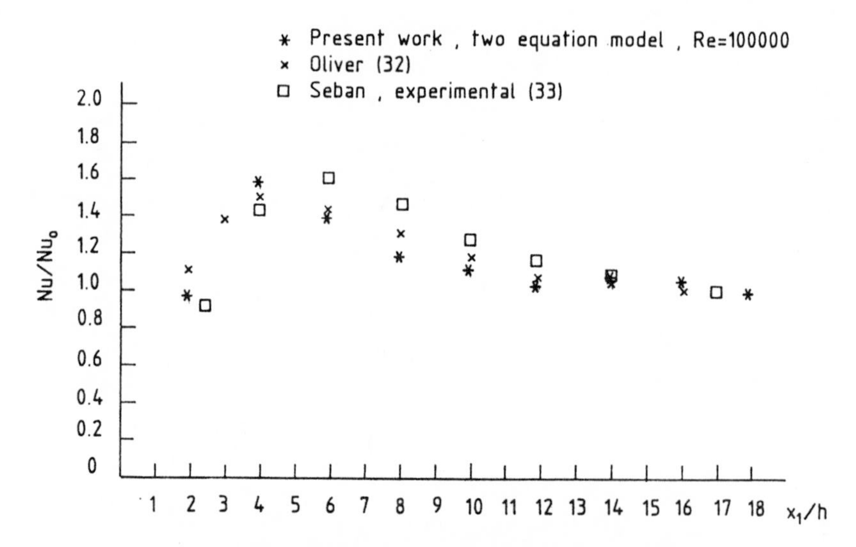

FIGURE 5 STREAMWISE VARIATION OF NUSSELT NUMBER DOWNSTREAM OF STEP

in Nusselt number utilises a sectionally averaged temperature, θ_{AV},

$$\theta_{AV} = \frac{1}{A}\int_{\Omega} \theta \, dA \tag{30}$$

and

$$Nu = h \frac{\partial\theta}{\partial x_n} (\theta_{ST}-\theta_{AV}) \tag{31}$$

in which A is local cross sectional area and θ_{ST} is the local temperature of the fluid at step level. A turbulent Prandtl number of 0.9 was used in the fully developed region and 1.4 in the laminar sub-layer. In the intermediate zone, $5 < x_n^+ \leq 30$, a linear interpolation on σ_t was used.

It is apparent from Figure 5, that the maximum heat transfer occurs between four and five step lengths downstream of the step. The location of the maximum transfer is, as expected, coincident with the predicted location of the reattachment point.

4.2 Cooling of Rod Bundles

An accurate prediction of the flow characteristics of the inter-rod coolant on the overall heat transfer are essential features of rod bundle design. The more recent numerically based investigations of these problems are found in references [25] to [26].

Slager, [26], using a finite element approach, the one equation model of turbulence, and Meyder's [27] mixing length concept obtained velocity distributions which compared favourably with experimental results [28]. Flow and associated heat transfer were also analysed where both one and two equation models were utilised.

Since the secondary flows, in the plane of Figure 6, are usually small, < 5% of the flow parallel to the rod longitudinal axes, the time averaged velocities u_1 and u_2, in the transverse plane, can be ignored. Fully developed flow conditions are assumed and buoyancy terms are negligible.

A sub-channel approach is adopted where, due to symmetry, only the shaded portion of Figure 6 is subjected to hydrodynamic analysis. This inherently assumes that all sub-channels are identical which, at the physical extremities of the rod bundle, is not strictly valid. However, such areas can be considered in isolation and linked to the analyses associated with the sub-channels. Using this concept, boundary conditions can be defined in the following manner,

Surfaces AB, BC and CD $\quad \dfrac{\partial u_3}{\partial n} = 0, \ \dfrac{\partial k}{\partial n} = 0$

and on AD $\quad u_3 = k = 0$

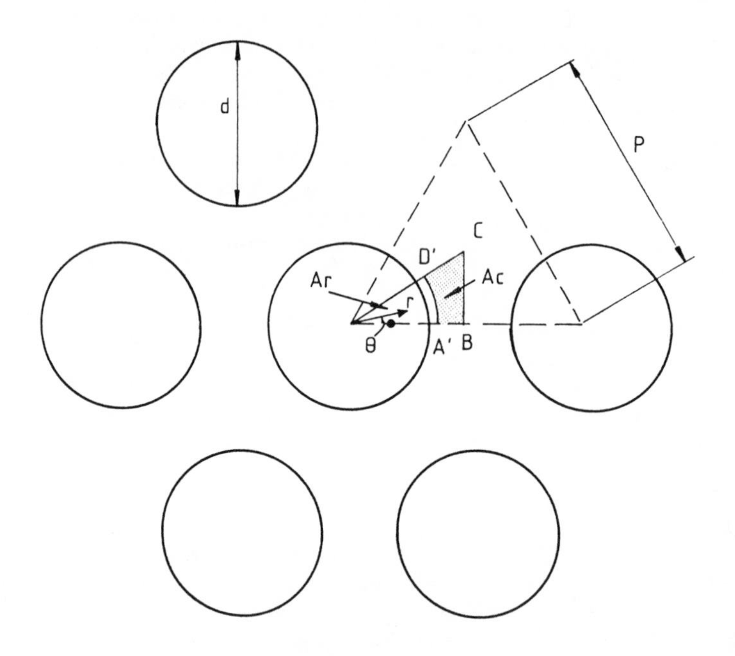

FIGURE 6 ROD BUNDLE GEOMETRY

Again, wall functions are used and, since there are no abrupt geometric discontinuities confining the fluid regime, the solution of the governing equations does not present undue difficulties. In this case the near wall boundary is placed within the fully turbulent region and the wall functions are,

$$u_s = u_3 = \frac{1}{K}\left(\frac{\tau_w}{\rho}\right)^{\frac{1}{2}} \ell_n \, [E\ r^+] \tag{32}$$

$$k = \frac{(C_\mu\ \tau_w)}{(K^2\ \rho)} \tag{33}$$

where

$$r^+ = \left(\frac{r\ \rho}{\mu}\right)\left(\frac{\tau_w}{\rho}\right)^{\frac{1}{2}} \tag{34}$$

in which r represents the radial distance from the solid, rod, surface and τ_w is defined by,

$$\tau_w = \mu \left.\frac{\partial u_s}{\partial r}\right|_{wall} = (\mu + \mu_t) \left.\frac{\partial u_s}{\partial r}\right|_{\text{near wall boundary}}$$

In the present example heat transfer is effected by conduction within the solid rod and by conduction-convection within the coolant. Assuming that the rods are homogeneous the energy equation reduces to,

$$\frac{\partial}{\partial x_1}\left(k_r \frac{\partial \theta_r}{\partial x_1}\right) + \frac{\partial}{\partial x_2}\left(k_r \frac{\partial \theta_r}{\partial x_2}\right) + q = 0 \qquad (35)$$

where the suffix r appertains to the rod.

For the coolant an additional, axial only, conductive term is introduced,

$$\frac{\partial}{\partial x_1}\left[\left(\frac{\nu}{\sigma} + \frac{\nu_t}{\sigma_t}\right)\frac{\partial \theta_c}{\partial x_1}\right] + \frac{\partial}{\partial x_2}\left[\left(\frac{\nu}{\sigma} + \frac{\nu_t}{\sigma_t}\right)\frac{\partial \theta_c}{\partial x_2}\right] = u_3 \frac{\partial \theta_c}{\partial x_3} \qquad (36)$$

in which suffix c refers to the coolant.

Imposing an overall energy balance between the rods and coolant, equations (35) and (36) can be combined to yield,

$$\frac{\partial}{\partial x_1}\left[\left(\frac{\nu}{\sigma} + \frac{\nu_t}{\sigma_t}\right)\frac{\partial \theta_c}{\partial x_1}\right] + \frac{\partial}{\partial x_2}\left[\left(\frac{\nu}{\sigma} + \frac{\nu_t}{\sigma_t}\right)\frac{\partial \theta_c}{\partial x_2}\right] = \frac{u_3}{u_3^*}\frac{A_r}{A_c} g^* \frac{\nu}{k_c \sigma} \qquad (37)$$

in which k_c is the effective thermal diffusivity of the fluid, u_3^* the average velocity of the fluid,

$$u_3^* = \frac{1}{A_c}\int_{A_c} u_3 \, dx_1 \, dx_2 \qquad (38)$$

and g^* the average heat generation per unit volume

$$g^* = \frac{1}{A_r}\int_{A_r} g \, dx_1 \, dx_2 \qquad (39)$$

where suffices are again as defined previously.

The set of equations (37) to (39) represent the equations which can be used to evaluate the distribution of temperature within the solid/fluid system provided g, k_c, μ_t and the velocity distribution can be specified.

Although the Prandtl-Kolmogorov relationship can again be employed to define local values of the turbulent dynamic viscosity a more relevant model, for the present problem, is that suggested by Buleev et al. [29]. A relationship of the form,

$$\frac{\mu_t}{\mu} = 0.2\ f_o(\eta)\ f_1(\eta)\ \gamma^* \tag{40}$$

and

$$\frac{\mu_t}{\mu\sigma_t} = 0.2\ f_o(\eta)\ f_1(\alpha\eta)\ \gamma^* \tag{41}$$

are used. These result in,

$$\sigma_t = \frac{f_1(\eta)}{f_1(\alpha\eta)} \tag{42}$$

where

$$\alpha = 0.8 + \frac{0.2}{\sigma^{0.67}} \qquad \sigma < 1$$

$$\alpha = 1 \qquad \sigma > 1$$

and $\gamma^* = \dfrac{\rho L_1^{\,2}}{\mu}\ \dfrac{\partial u_3}{\partial n}$

The functions $f_o(\eta)$ and $f_1(\eta)$ are defined by,

$$f_o(\eta) = \exp(-\eta)$$

and

$$f_1(\eta) = \frac{[1-\exp(-\eta)]}{\eta}$$

Here $\eta = \dfrac{65}{\gamma^*}$

The additional length scale, L_1, is given by, refer Figure 7,

$$L_1 = PQ\ (1.0 - 0.5\ \frac{PQ}{QS}) \tag{43}$$

At the solid/fluid interface a flux balance can be checked, subsequent to the evaluation of the temperature distribution, to ensure solution accuracy in the interfacial zone

$$\frac{\partial\theta_c}{\partial n} = \frac{k_r}{k_c}\frac{\partial\theta_r}{\partial n} \qquad (44)$$

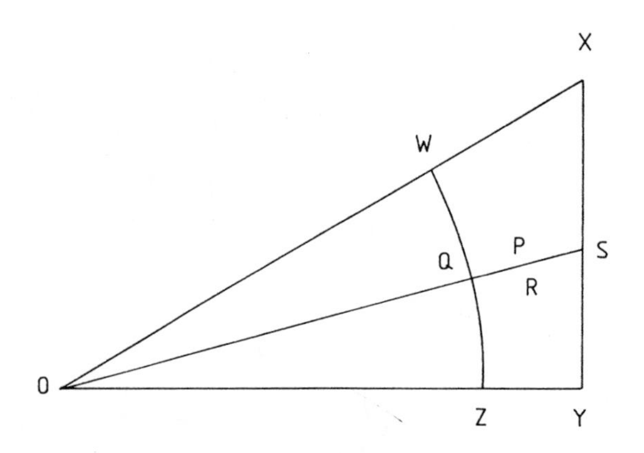

FIGURE 7 DEFINITION OF BULEEV'S LENGTH SCALE

The finite element mesh, shown on Figure 8, has sufficient local resolution to ensure an accuracy of variable evaluation within required tolerances for the following conditions,

$$C_\mu = 0.18,\quad C_D = 0.30,\quad \sigma_k = 1.0,\quad K = 0.4186,$$

$$E = 9.8,\quad C_1 = 1.45,\quad C_2 = 0.18,\quad \sigma_\varepsilon = 1.0,$$

$$\frac{g^*}{g} = 1.0,\quad \frac{D}{d} = 1.217,\quad Re = 270{,}000.$$

The Reynolds number, Re, is defined,

$$Re = \frac{\rho\, u_B\, m}{\mu}$$

where u_B is the mean velocity of flow and m is the hydraulic diameter of the sub-channel. The eddy conductivity $\frac{\mu_t}{\sigma_t}$ is evaluated from equation (40) - (42).

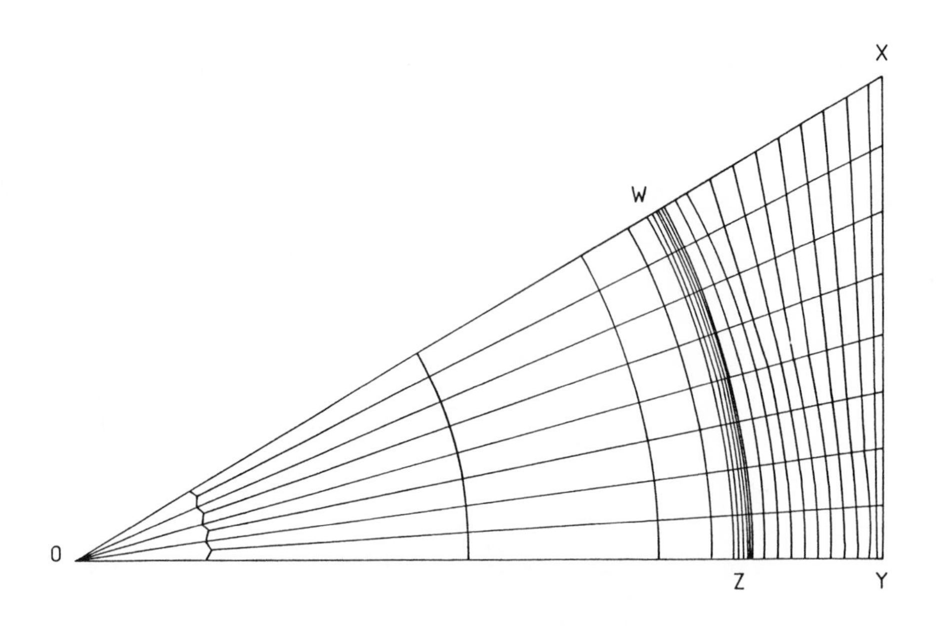

FIGURE 8 FINITE ELEMENT MESH FOR TEMPERATURE DISTRIBUTION

The value of wall shear stress, usually a good indicator of accuracy, obtained using the two equation model is presented on Figure 9. These compare favourably with experiment [30]. It has been reported [20] that the results obtained using a one equation model were less accurate.

A comparison of calculated length scales and those measured [24] or evaluated from empirical relationships are presented on Figure 10. Although these follow the trend advocated by semi-empirical relationships they deviate considerably from the measured values, particularly as the outer extremity of the sub-channel is approached. The effect corresponding to varying the turbulent Prandtl number is clearly illustrated on Figures 11 and 12 which illustrate the additional cooling as σ_t increases.

It is clearly evident, from the present chapter and that written by A.J. Baker, that finite element modelling of turbulent flow problems is now well established. However, there are areas open to further investigation utilising the conventionally accepted procedures, namely the one and two equation models. These are themselves open to criticism and some research is necessary into the basic concepts of turbulence modelling. Accepting that such models are reasonably effective there are innumerable areas, in a numerical modelling

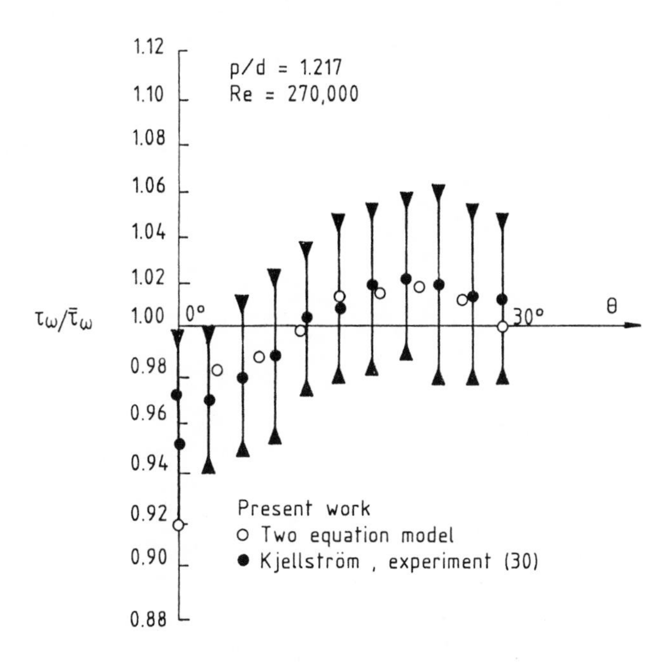

FIGURE 9 VARIATION OF WALL SHEAR STRESS

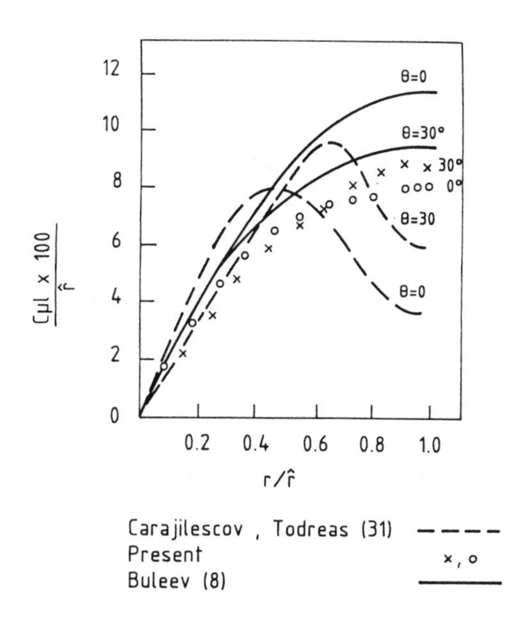

FIGURE 10 COMPARISON OF TWO EQUATION MODEL LENGTH SCALES

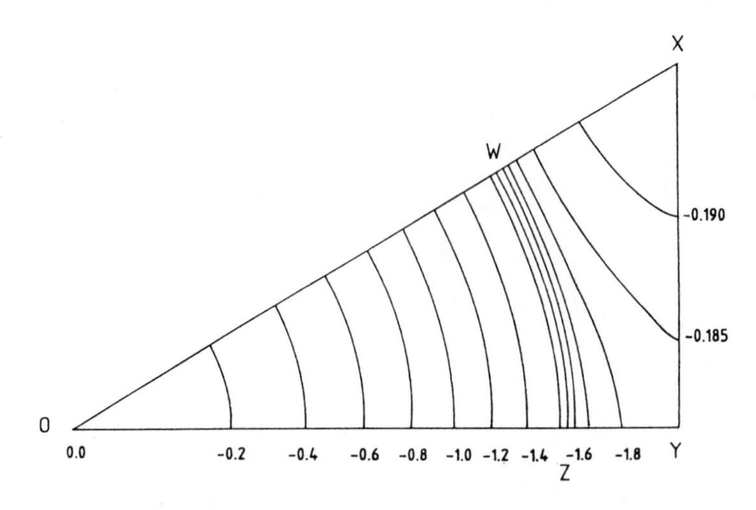

FIGURE 11 DIMENSIONLESS TEMPERATURE CONTOURS
BULEEV'S MODEL FOR EDDY CONDUCTIVITY
$k_r/k^1=1.5$

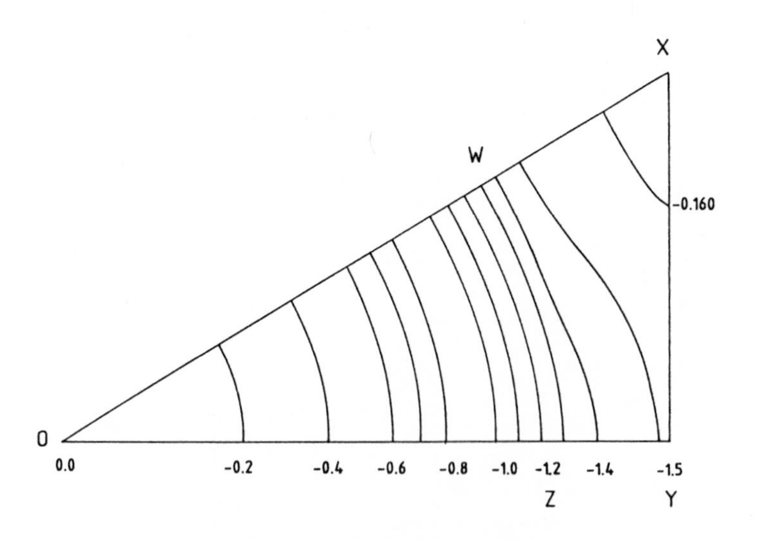

FIGURE 12 DIMENSIONLESS TEMPERATURE CONTOURS
BULEEV'S MODEL FOR EDDY CONDUCTIVITY
$k_r/k^1=1.9$

context, which need further scrutiny. Not least among these are

(i) the effect of releasing boundary constraints and updating - particularly Neumann and traction boundaries

(ii) the introduction of sophisticated elements to emulate more accurately the variation of the variables in near wall regions

(iii) introducing more efficient solution algorithms both with respect to c.p.u. time and core requirements

and

(iv) exploration of new turbulence modelling procedures which obviate solution instabilities

In each of the above areas work is progressing steadily and new concepts will undoubtedly occur quite rapidly over the next decade.

REFERENCES

1. CEBECI, T. and SMITH, A.M.O., Analysis of Turbulent Boundary Layers, Academic Press, 1974.

2. SCHLICHTING, H., Boundary Layer Theory, translated by Kestin, J, McGraw-Hill, 1960.

3. PRANDTL, L., 'Uber ein neues Formelsystem fur die ausgebildete Turbulenz', Nachr. Akad. der Wissenshafft, Gottingen, 1945.

4. KOLMOGOROV, A.H., 'Equations of turbulent motion of an incompressible fluid', Izv. Akad. Nauk. SSSR. Ser. Phys., VI (1-2), pp. 56-58, 1942.

5. WOLFSHTEIN, M., 'Convection Processes in Turbulent Impinging Jets', Ph.D. Thesis, University of London, 1967.

6. HARLOW, F.H. and NAKAYAMA, P.I., 'Transport of Turbulence Energy Decay', Phys. of Fluids, 10(11), pp. 2323-2332, 1967.

7. SPALDING, D.B., 'The k-w model of turbulence', Imperial College, Mechanical Engineering Department, Report TM/TN/A/16, London, 1972.

8. WILCOX, D.C. and TRACI, R.M., 'A Complete Model of Turbulence', A.I.A.A., Paper 76-361, AIAA 9th Fluid and Plasma Dynamics Conference, San Diego, California, 1976.

9. SMITH, R.M., 'A Practical Method of Two-equation Turbulence Modelling Using Finite Elements', C.E.G.B. Report No. TPRD/B/0182/N82, 1983.

10. SMITH, R.M., 'On the Finite Element Calculation of Turbulent Flow Using the k-ε model', C.E.G.B. Report No. TPRD/B/0161/N82.

11. DAVIS, J.T., Turbulence Phenomena, Academic Press, New York, 1972.

12. CEBECI, T., 'A Model for Eddy Conductivity and Turbulent Prandtl number', J. Heat Transfer, 95, pp. 227-234, 1973.

13. NA, T.Y. and HABIB, I.S., 'Heat Transfer in Turbulent Pipe Flow Based on a New Mixing Length Model', Appl. Scient. Res. 28, pp. 302-314, 1973.

14. WASSEL, A.T. and COTTON, I., 'Calculation of Turbulent Boundary Layers over Flat Plates with Different Phenomenological Theories of Turbulence and Variable Turbulent Prandtl Number', Int. J. Heat and Mass Transfer, 16, pp. 1547-1563, 1973.

15. SLEITCHER, Jr., C.A., 'Experimental Velocity and Temperature Profiles for Air in Turbulent Pipe Flow', J. Heat Transfer 80, pp. 693-704, 1958.

16. SHERWOOD, T.K., SMITH, K.A. and FOWLES, P.E., 'The Velocity and Eddy Viscosity Distribution in the Wall Region of Turbulent Pipe Flow', Chem. Engng., Sci. 23, pp. 1225-1234, 1968.

17. MERONEY, R.N., 'Turbulent Sub-layer Temperature Distribution including Wall Injection and Dissipation', Int. J. Heat and Mass Transfer 11, pp. 1406-1408, 1968.

18. ORLANDO, A.F., MOFFAT, R.J. and KAYS, W.M., 'Turbulent Transport of Heat and Momentum in a Boundary Subject to Deceleration, Suction and Variable Wall Temperature', Report No. HMT-17. Thermosciences Div., Stanford Univ., 1974.

19. ANTONIA, R.A., 'Behaviour of the Turbulent Prandtl Number Near the Wall', Int. J. Heat and Mass Transfer, 23, pp. 906-908, 1980.

20. THOMAS, C.E., Analysis of Confined Turbulent Flows, Ph.D. Thesis, Univ. of Wales, 1982.

21. TAYLOR, C., HUGHES, T.G., 'Finite Element Programming of the Navier-Stokes Equations', Pineridge Press, U.K. 1981.

22. TAYLOR, C., HUGHES, T.G. and MORGAN, K., 'A Numerical Analysis of Turbulent Flow in Pipes', J. Comp. and Fluids, 5, pp. 191-204, 1977.

23. DENHAM, M.K., BRIARD, P. and PATRICK, M.A., 'A Directionally Sensitive Laser Anemometer for Velocity Measurements in Highly Turbulent Flow', J. of Phys. E.: Scientific Instruments, 8, pp. 681-683, 1975.

24. ATKINS, D.J., 'Numerical Studies of Separated Flows, Ph.D. Thesis, Univ. of Exeter, 1974.

25. TAYLOR, C., THOMAS, C.E. and MORGAN, K., 'Turbulent Heat Transfer via the F.E.M. : Heat Transfer in Rod Bundles', Int. J. Num. Meth. Fluids (to appear), 1983.

26. SLAGER, W., 'Finite Element Solution of Axial Turbulent Flow in a Bare Rod Bundle Using a One-Equation Turbulence Model', Nucl. Sc. Engng., 82, pp. 243-259, 1982.

27. MEYDER, R., 'Turbulent Velocity and Temperature Distribution in the Central Channel of Rod Bundles', Nucl. Eng. Design, 35, pp. 181-189, 1975.

28. REHME, K., 'Turbulent Strömung in einen Wandkanal eines Stabbundels', KFK-2617, Kernforschungszentrum Karlsruhe, 1978.

29. BULEEV, N.I. and MIRONOVICH, R. YO., 'Heat Transfer in Turbulent Flow in a Triangular Array of Rods', Phys. and Power Inst., Translated from Teplofizika Vysokobh Temperatur, 10, 5, pp. 1031-1038, 1972.

30. KJELLSTROM, B., 'Studies of Turbulent Flow Parallel to a Rod Bundle of Triangular Array', AE-487, A.B. Atomenergi, Sweden, 1974.

31. CARAJKESCOV, P. and TODREAS, N.E., 'Experimental and Analytical Study of Axial Turbulent Flows in an Interior Sub-Channel of a Bare Rod Bundle', Symp. Turbulent Shear Flows, 1, Univ. Park, Penn., 1977.

32. OLIVER, A.J., 'A Finite Difference Solution for Turbulent Flow and Heat Transfer over a Backward Facing Step in an Annular Duct', Num. Meth. in Laminar and Turbulent Flows, Ed. Taylor, C., Morgan, K. and Brebbia, C.A. Pentech Press, pp. 467-478, 1978.

33. SEBAN, R.A., 'Heat Transfer to the Turbulent Separated Flow of Air Downstream of a Step in the Surface of a Plate', J. Heat Transfer, 86(2), pp. 259-264, 1964.

EFFECT OF THE SOLID PARTICLE SIZE IN TWO PHASE FLOW AROUND A PLANE CYLINDER

B.V.R. Vittal and W. Tabakoff

Department of Aerospace Engineering and Applied Mechanics
University of Cincinnati
Cincinnati, Ohio 45221 U.S.A.

1. INTRODUCTION

In many industrial and military applications, the erosive action of particulate flows results in performance deterioration. Many solid rocket propellants contain fine metallic particles [1], which may have concentrations as high as 40% in the combusted gas mass flow and affect the fluid flow properties considerably. For vehicles flying through dusty atmosphere or rain, it is important to know the impact velocity with which small particles hit the vehicle surface. The inability of solid particles to follow fast flow changes is applied in helicopter engine particle separators, and industrial cyclones.

Many gas turbine engines often operate in contaminated environment. Aircraft engines, during the flight, may encounter sand, dust, salt or water droplets, depending on their mission. Solid particles are also ingested by gas turbine engines in ground vehicles, auxiliary power units and military tanks. Industrial gas turbines burning synthetic fuels or pulverized coal are exposed to particulate flows.

Considerable research work has been done in multiphase flow fields for inviscid carrier fluid. These studies have concentrated on the calculation of particle trajectories in various flow systems. References [2] and [3] are some good examples. Particle trajectories are calculated by integrating the equations of motion of the particle, as it proceeds through the flow field. There are a number of published reports available on the prediction of erosion of materials due to particle impacts. However, there is very little published literature concerned with the effect of the presence of the particles on the carrier fluid properties.

Several numerical models for one dimensional gas-particle flows in which the mutual effects of both the particles and the gas are taken into account have been reported. Tabakoff et al. [3] have studied the effect of solid particles on the turbomachinery cascade performance using a simplified inviscid model. However, for viscous fluids, complications arise due to the nonlinearity of the governing equations of motion [4]. The problem of the flow induced by the impulsive motion of an infinite flat plate parallel to its own plane, has been extended to two-phase flows by several authors. Liu [5], and Marble [6] studied the flow induced in a dusty gas; the interphase force in their analyses being Stokes drag force. Hamed and Tabakoff [4] were the first to include in their analysis, the antisymmetric stresses. Immich [7] extended the application by means of a volume averaging method.

None of the above studies on particle-fluid interaction (two-way coupling) has taken into account the rebounding character of the solid particles when they impact on the solid boundaries. Therefore, as a first step in a program of study of particulated flow systems, we have started with a simple case of flow over a two dimensional cylinder.

The study of plane uniform viscous flow around a cylinder is a fundamental problem because, all difficulties which arise in it are amplified for obstacles of other shapes. Its field of application is very large, being basic to the calculation of more complex flows such as flow over airfoils. Most of the work done so far on two phase flow (solid particle-gas) over cylinder is based on the inviscid approach [8, 9]. The path of the particles changes appreciably when viscosity of the continuous phase is taken into account. All of the above studies are based on the assumption that the particles do not affect the fluid flow field. The present analysis of viscous fluid-particle flow over a two dimensional cylinder takes into account the effect of the presence of the particles everywhere, including those rebounded off the cylinder surface.

2. GOVERNING EQUATIONS

In case of solid particle-gas flow, where the volume fraction of the particles is low, the Lagrangian formulation of the equations of motion is most appropriate for the particulate phase, while the Eulerian approach is convenient for the continuous phase. Further, if Eulerian approach is to be used for the particle phase, it becomes extremely difficult to define the boundary conditions at the solid boundaries, where the particles deflect after impact.

2.1 Eulerian Formulation for the Fluid

For two dimensional incompressible flow, if the volume fraction of the solid particles is low, the governing equations in Cartesian coordinates are [6],

$$\frac{\partial u'}{\partial x'} + \frac{\partial v'}{\partial y'} = 0 \tag{1}$$

$$\rho'(\frac{\partial u'}{\partial t'} + u' \frac{\partial u'}{\partial x'} + v' \frac{\partial u'}{\partial y'}) = - R_x' - \frac{\partial p'}{\partial x'} + \mu(\frac{\partial^2 u'}{\partial x'^2} + \frac{\partial^2 u'}{\partial y'^2}) \tag{2}$$

$$\rho'(\frac{\partial v'}{\partial t'} + u' \frac{\partial v'}{\partial x'} + v' \frac{\partial v'}{\partial y'}) = - R_y' - \frac{\partial p'}{\partial y'} + \mu(\frac{\partial^2 v'}{\partial x'^2} + \frac{\partial^2 v'}{\partial y'^2}) \tag{3}$$

where R_x' and R_y' are interphase force terms depicting the presence of the solid particles. Eliminating the pressure terms from the above equations and introducing vorticity, $\omega' = \frac{\partial v'}{\partial x'} - \frac{\partial u'}{\partial y'}$, one obtains

$$\rho'(\frac{\partial \omega'}{\partial t'} + u' \frac{\partial \omega'}{\partial x'} + v' \frac{\partial \omega'}{\partial y'}) = - \frac{\partial R_y'}{\partial x'} + \frac{\partial R_x'}{\partial y'} + \mu(\frac{\partial^2 \omega'}{\partial x'^2} + \frac{\partial^2 \omega'}{\partial y'^2}) \tag{4}$$

Substituting for u' and v' in terms of stream functions ψ and nondimensionalizing the various quantities with respect to free stream velocity U_∞ and radius r of the cylinder, one obtains the following equation

$$\frac{\partial \omega}{\partial t} + \frac{\partial \psi}{\partial y}\frac{\partial \omega}{\partial x} - \frac{\partial \psi}{\partial x}\frac{\partial \omega}{\partial y} = - \frac{\partial R_y}{\partial x} + \frac{\partial R_x}{\partial y} + \frac{1}{Re_r}(\frac{\partial^2 \omega}{\partial x^2} + \frac{\partial^2 \omega}{\partial y^2}) \tag{5}$$

For general use of these equations, it is convenient to express them in a coordinate system in which ξ and η are two independent spatial variables. Equation (5) becomes,

$$J \frac{\partial \omega}{\partial t} + \frac{\partial \psi}{\partial \eta}\frac{\partial \omega}{\partial \xi} - \frac{\partial \psi}{\partial \xi}\frac{\partial \omega}{\partial \eta}$$

$$= y_\xi \frac{\partial R_y}{\partial \eta} - y_\eta \frac{\partial R_y}{\partial \xi} + x_\xi \frac{\partial R_x}{\partial \eta} - x_\eta \frac{\partial R_x}{\partial \xi}$$

$$+ \frac{1}{Re_r}[\frac{\partial}{\partial \xi}(\frac{\alpha}{J}\frac{\partial \omega}{\partial \xi} - \frac{\beta}{J}\frac{\partial \omega}{\partial \eta}) + \frac{\partial}{\partial \eta}(\frac{\gamma}{J}\frac{\partial \omega}{\partial \eta} - \frac{\beta}{J}\frac{\partial \omega}{\partial \xi})] \tag{6}$$

where

$$\alpha = x_\eta^2 + y_\eta^2 \quad , \quad \beta = x_\xi x_\eta + y_\xi y_\eta \quad ,$$

$$\gamma = x_\xi^2 + y_\xi^2 \quad , \quad J = x_\xi y_\eta + x_\eta y_\xi \quad . \tag{7}$$

The relationship between vorticity, ω, and the stream function, ψ, in the general coordinate system is given by,

$$-J\omega = \frac{\partial}{\partial\xi}\left(\frac{\alpha}{J}\frac{\partial\psi}{\partial\xi} - \frac{\beta}{J}\frac{\partial\psi}{\partial\eta}\right) + \frac{\partial}{\partial\eta}\left(\frac{\gamma}{J}\frac{\partial\psi}{\partial\eta} - \frac{\beta}{J}\frac{\partial\psi}{\partial\xi}\right) \qquad (8)$$

2.2 Lagrangian Formulation for the Particles

The trajectory of a particle in a moving fluid is governed by the vector balance of its rate of change of momentum and the external forces acting upon it. There are many external forces which may act on the particles. For large ratios of particle material density to gas density, the force acting on a spherical particle due to the flow pressure gradient can be neglected when compared to the drag force on it. If the volume fraction of the particles is low, particle-particle interaction is neglected.

An excellent review of the forces acting on the particles and their relative importance in the fluid flow calculations is given in reference [10]. The drag of the particle, which is the predominant force, is usually calculated based on empirical relations for perfectly spherical particles. The drag on a nonspherical particle depends upon its shape and its orientation with respect to the direction of relative motion. Sometimes an overall shape factor is made use of for nonspherical particles. The particle drag is also affected by bounded walls. Although, some experimental studies are available to show the effect of proximity of walls on the drag coefficient of the particles, there is no correlation available. Various studies on the effect of accelerated motion of particles in the fluid suggest, that this effect decreases with decreasing fluid densities. The motion of small spheres in air should show only a small effect due to acceleration, compared to the drag force. When a spherical particle moves relative to a shear flow, a lift force in the direction normal to the main flow acts on the particle. Staffman [11] evaluated this force component analytically. However, this is valid for very low particle Reynolds numbers. No correlations are available for the torque acting on a single particle rotating relative to the flow field. It is extremely difficult to determine experimentally the rotational velocity of a solid particle in the flow field. Other forces such as molecular impact and Van der Waal's forces are negligible. However, in the present study, only drag force is considered as the force of integration between the phases. The equations of motion of a single particle in Cartesian coordinates are given by:

$$m_p' \frac{du_p'}{dt'} = \frac{\pi}{8} C_d \, \rho' \, (D_p')^2 \, (u_p' - u')^2 \quad , \qquad (9)$$

$$m'_p \frac{dv'_p}{dt'} = \frac{\pi}{8} C_d \rho' (D'_p)^2 (v'_p - v')^2 \tag{10}$$

2.3 Impact and Rebound Phenomenon of Particles

When a solid particle moves in a stream of gas, it does not in general follow the streamlines taken by the gas due to its higher inertia. They collide with the solid boundaries. After hitting, the particles experience a loss in their momentum relative to the wall and change the direction of their motion. The value of the particle velocity and direction of its motion as it rebounds from the surface after collision must be known in order that the solution of the particle equations of motion be continued beyond the points of collision. Particle rebound characteristics have been studied experimentally by Tabakoff et al. [12]. The change in particle momentum due to impact is found to be mainly a function of the particle velocity and incidence angle β_p. The following empirical relations for the rebound to impact restitution ratios, are used in trajectory calculations.

$$V'_{p_{2_T}} / V'_{p_{1_T}} = 1.0 - 2.12\,\beta'_p + 3.0775\,\beta'^2_p - 1.1\,\beta'^3_p$$

$$V'_{p_{2_N}} / V'_{p_{1_N}} = 1.0 - 0.4159\,\beta'_p - 0.4994\,\beta'^2_p - 0.292\,\beta'^3_p \tag{11}$$

where V'_{p_N} and V'_{p_T} represent the particle velocity components normal and tangent to the solid surface, and the subscripts 1 and 2 refer to the conditions before and after impact, respectively. In the above equations β'_p is the angle between the impact velocity and the tangent to the surface in radians.

3. NUMERICAL PROCEDURE

The stream function and vorticity transport form of the equations of motion of fluid including the interphase force terms, are solved using a factored ADI scheme developed by Davis et al. [13, 14]. Symmetry is assumed along the stagnation streamline; therefore the solution is presented only for the upper half plane. The flow in the upper half is bounded by the half cylinder of one radius; the outer flow boundary is at a distance of 49 times the radius. This region is transformed to an ξ,η solution plane in such a way that ξ is a coordinate tangent to the body surface and η is a coordinate normal to the body surface. The equations are solved employing a (65 x 65) grid for the solution plane. Symmetry boundary conditions of $\psi = 0$ and $\omega = 0$ are used along symmetry boundaries. For the outer flow, the vorticity boundary condition, $\omega = 0$, is used. The stretching in the physical plane is accomplished by using appropriate stretching parameters.

The interphase force terms in the fluid flow equations are evaluated separately using particle trajectory calculations as employed by Crowe [15]. Particles are introduced upstream of the cylinder at various heights. Uniform entry of the particles with the flow is approximated by 250 discrete entry locations, each location is assumed to carry a fraction of the total particle mass. First the flow field is established by solution without solid particles. Using this flow field, the particle flow path and its properties along the path are calculated. Referring to Fig. 1, the force per unit volume, R_p', at the grid point (i,j) is given by:

$$R_p' = m_p' \, a_p' \tag{12}$$

where

$$m_p' = MF_p \, \tau / V_{i,j} \quad ,$$

MF_p is the mass fraction of the particles carried by the particular trajectory, and τ is its residence time in volume Vol(i,j).

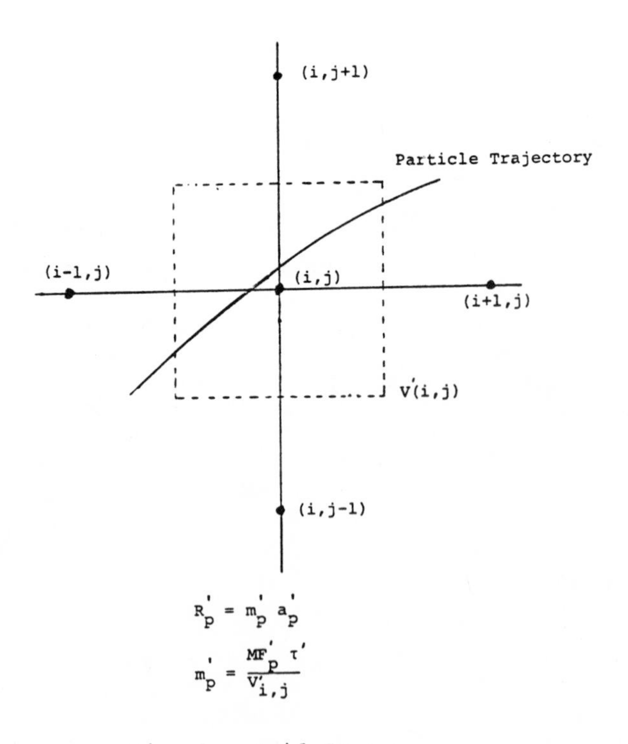

Fig. 1. Grid System.

The total mass fraction of the particles is defined as the ratio of the particle mass flow to that of the gas flow. Equation (12) is integrated along the particle trajectory for each time step and for each grid point. The total force R'_p at the grid point (i,j) is the sum of all the forces due to all the particle trajectories passing through the mesh control volume.

$$R'_{p(i,j)} = \sum_{k=1}^{N} (m'_p \alpha'_p)_k \quad . \tag{13}$$

4. RESULTS AND DISCUSSION

Numerical solutions are obtained for two dimensional incompressible viscous particulate flow over a cylinder. Air at atmospheric pressure and temperature is considered for the continuous phase and quartz solid particles are used for the particulate phase. The flow calculations are made for a cylinder diameter of 3.175 mm. At low cylinder Reynolds numbers, the coefficient of drag of the cylinder changes appreciably for slight changes in flow conditions. So any change in the flow properties due to the presence of the particles would be reflected on the fluid streamline pattern and the coefficient of drag of the cylinder. The following operating conditions are used:

Free stream pressure = 10,366 kg/m^2
Free stream temperature = 288°K
Air density = 0.01273 kg/m^3
Cylinder diameter = 3.175 mm
Particle material = quartz
Particle material density = 2444 kg/m^3
Kinematic viscosity of air = 1.46×10^{-5} m^2/sec
Cylinder Reynolds number = $U_\infty D/\nu$; where D is the diameter of the cylinder.

4.1 Comparison of Particle Trajectories in Viscous and Inviscid Fluid

The particle trajectory is affected by fluid flow path and its properties. Viscosity, being one of the properties of the fluid, would obviously affect the particle trajectory, especially near the solid boundaries. Figure 2 shows the particle trajectories for different particle sizes. Besides the obvious difference in particle trajectories among particles of different sizes, there is considerable difference in the particle trajectories between viscous and inviscid solutions. It can be observed that smaller particles are deviated away from the cylinder and do not impact on the cylinder surface. Larger particles, because of their inertia, tend to keep their initial momentum, impact on the surface of

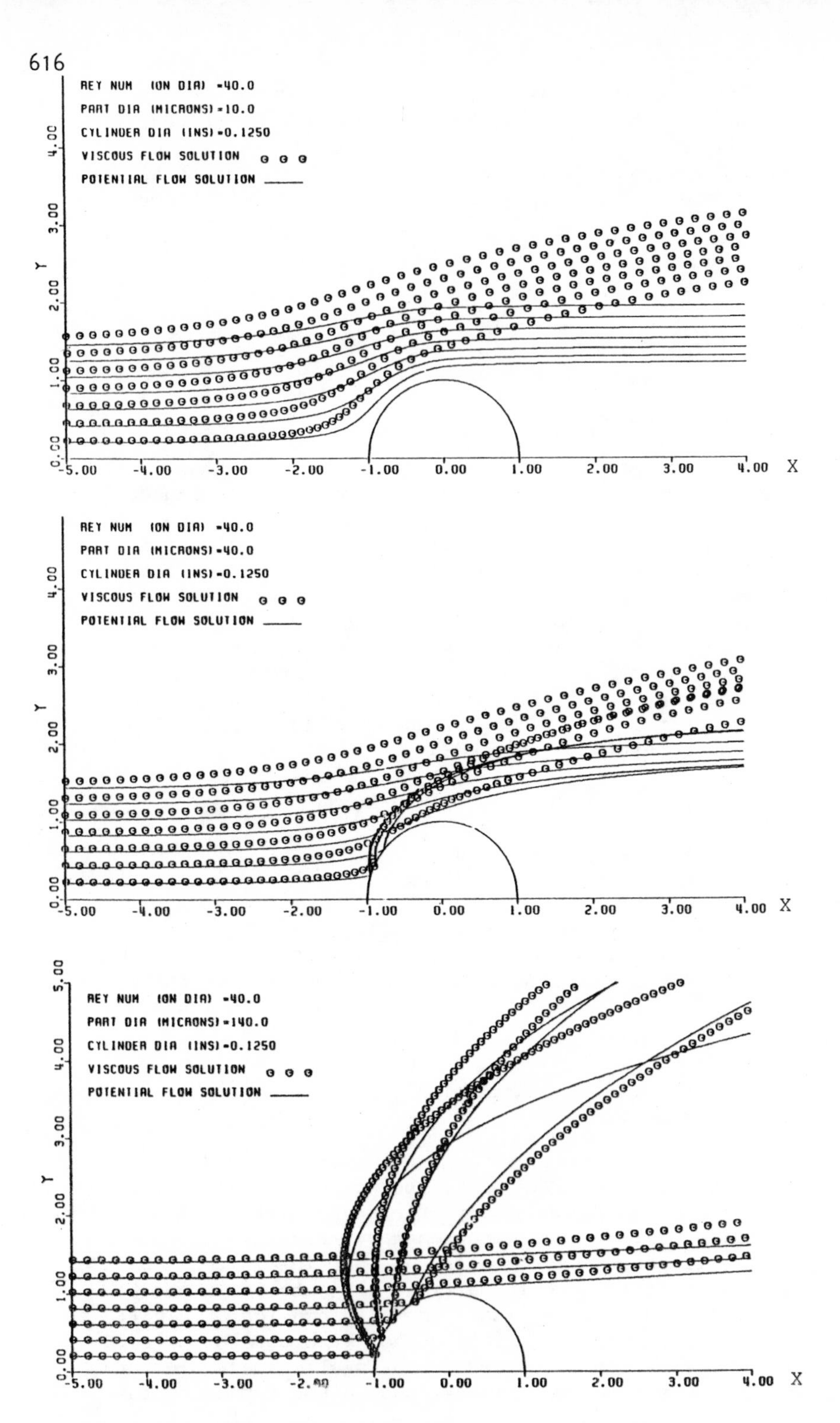

Fig. 2. Particle Flow Path Over Cylinder.

the cylinder and get deflected. In the case of viscous flow solutions, the boundary layer creates a displacement thickness altering the apparent size of the cylinder, thus changing the particle path when viscosity is taken into account. Consequently, the smaller particles are deviated more when compared with inviscid flow solution.

In Fig. 3, the collection efficiency or the strike efficiency is given for various particle sizes, for both viscous and inviscid flow cases. This efficiency, a number frequently quoted in industrial literature, is the ratio of the number of particles impacting with the object, to the number which would impact if they followed straight line trajectories without deflection by the gas. Clearly, the effect of viscosity is to reduce the strike efficiency for a given particle size. Most of the particles less than 20 microns in diameter do not strike the cylinder altogether.

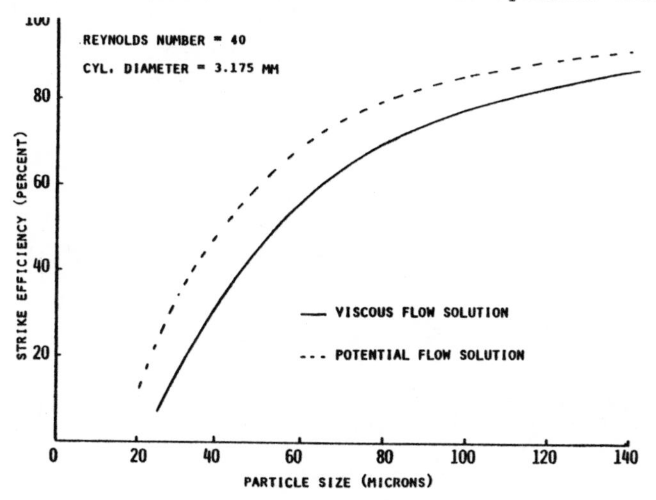

Fig. 3. Effect of Particle Size on Strike Efficiency.

4.2 Effect of Particles on Fluid Flow Properties

The presence of particles in the fluid could affect flow properties such as streamline pattern, vorticity, coefficient of drag, separation angle, recirculation eddy and so on. The overall effect of the particles would be readily reflected on the coefficient of drag of the cylinder. Figure 4 shows the percentage change in coefficient of drag of the cylinder plotted against particle size for different particle mass fractions. It is interesting to observe that initially, as the particle size increases, the coefficient of drag increases. With the increase in size of the particle, there is a bigger difference in relative velocities between the particle and the air because of higher inertia, thus resulting in higher interphase forces. This changes the fluid velocities, resulting

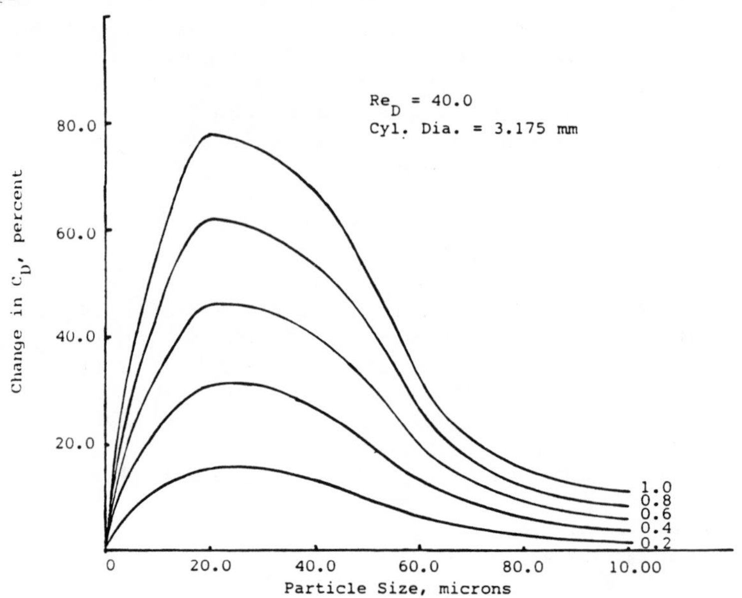

Fig. 4. Effect of Particle Size on Cylinder, C_D.

in increased coefficient of drag of the cylinder. However, with the increase in particle size, for the same mass fraction of the particles, there is less number of particles present in the flow. This decreases the interphase forces, thereby affecting the coefficient of drag of the cylinder. Thus, the effect of particle is a compromise between the above two factors. This can be clearly observed in Fig. 4. As the particle size increases, initially, there is an increase in coefficient of drag of the cylinder. The C_D increases and reaches a maximum for 20 micron particles and then starts decreasing with increase in particle size, as the number of particles decreases for the same mass fraction of the particles. As explained earlier, the particle mass fraction is defined as the ratio of the particle mass flow to that of the gas flow. For the same mass fraction, there are less number of larger diameter particles when compared to smaller diameter particles, and the total surface area of the particles exposed is lesser.

Figure 5 shows the increase in coefficient of drag of the cylinder plotted against particle mass fraction for different particle sizes. The increase in coefficient of drag is a direct linear function of particle mass fraction. The coefficient of drag is a very important parameter, as it indicates the magnitude of losses in the flow system. Any increase in drag of the bodies, e.g., airfoils, increases the inefficiency of the flow systems which employ them. This is especially very important in gas turbine applications, where a decrease

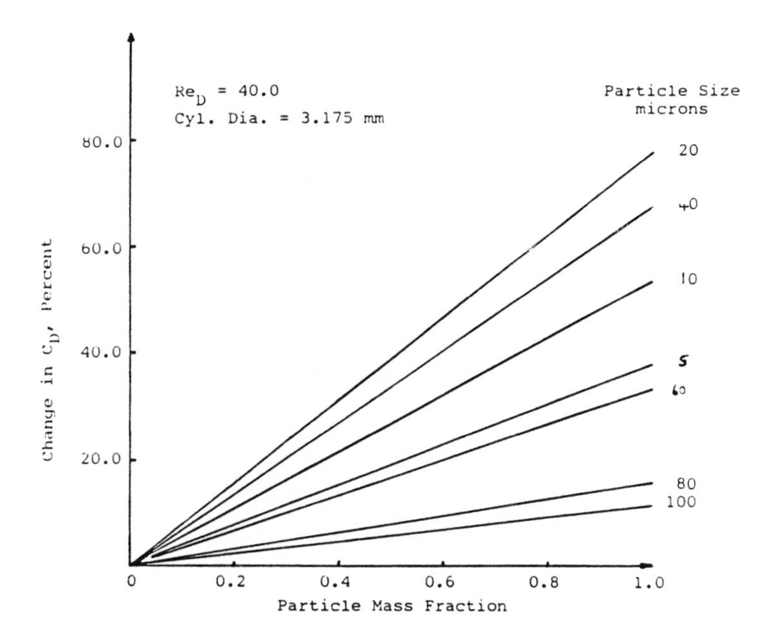

Fig. 5. Effect of Particle Mass Fraction on Cylinder C_D.

in component efficiency, would drastically affect the specific fuel consumption of the gas turbine engine.

The presence of the particles is felt on all the fluid flow parameters. Figures 6 and 7 show the effect of particles on the pressure distribution over the cylinder surface. The pressure coefficient, C_p, is presented at various angular positions on the cylinder.

The effect of particles on fluid flow can be easily seen in the way it changes the fluid streamline pattern. Figure 8 shows the effect of particle concentration on fluid streamline pattern for 10 micron diameter particles. It can be observed that there is a shift in the streamlines with the presence of the particles. There is a change in the recirculation pattern of the fluid downstream of the cylinder. With increase in particle mass fraction, the size of the recirculation zone increases. It may be recalled that smaller diameter particles in the 10 micron range do not impact the cylinder. As the diameter of the particle increases, larger number of particles impact on the cylinder front face. This is reflected on the fluid streamlines and recirculation zone, as shown in Fig. 9. With larger diameter of particles, there is a reduction in the length of the recirculation eddy with increase in particle concentration. The larger diameter particles which impact and rebound off the surface of the cylinder, oppose the motion of the fluid for a short distance from the cylinder, which reduces the velocity of the oncoming

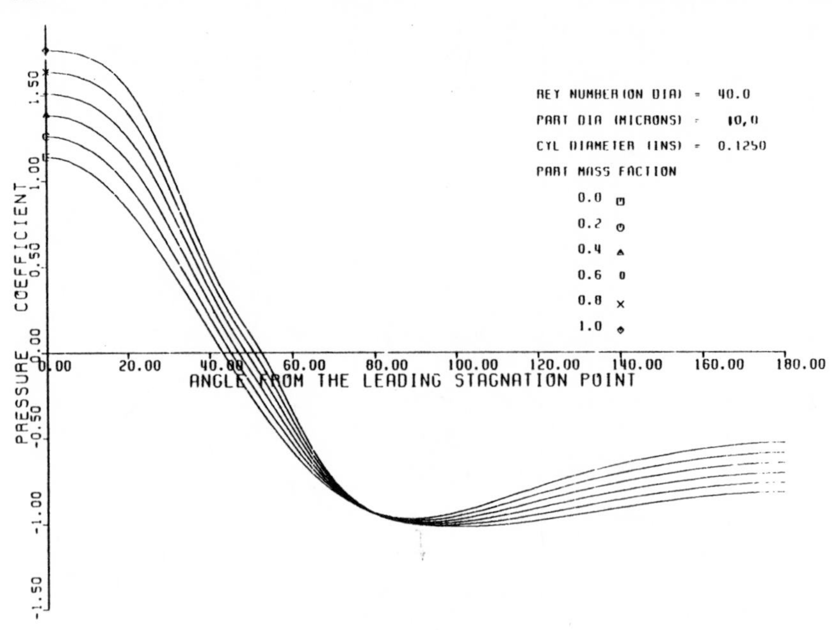

Fig. 6. Pressure Coefficient Variation with Particle Mass Faction.

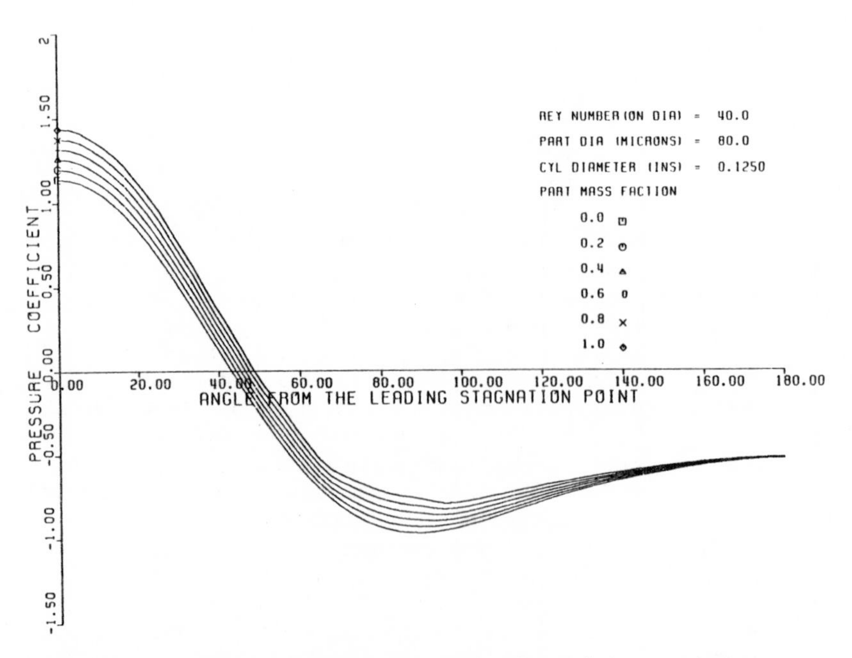

Fig. 7. Pressure Coefficient Variation with Particle Mass Faction.

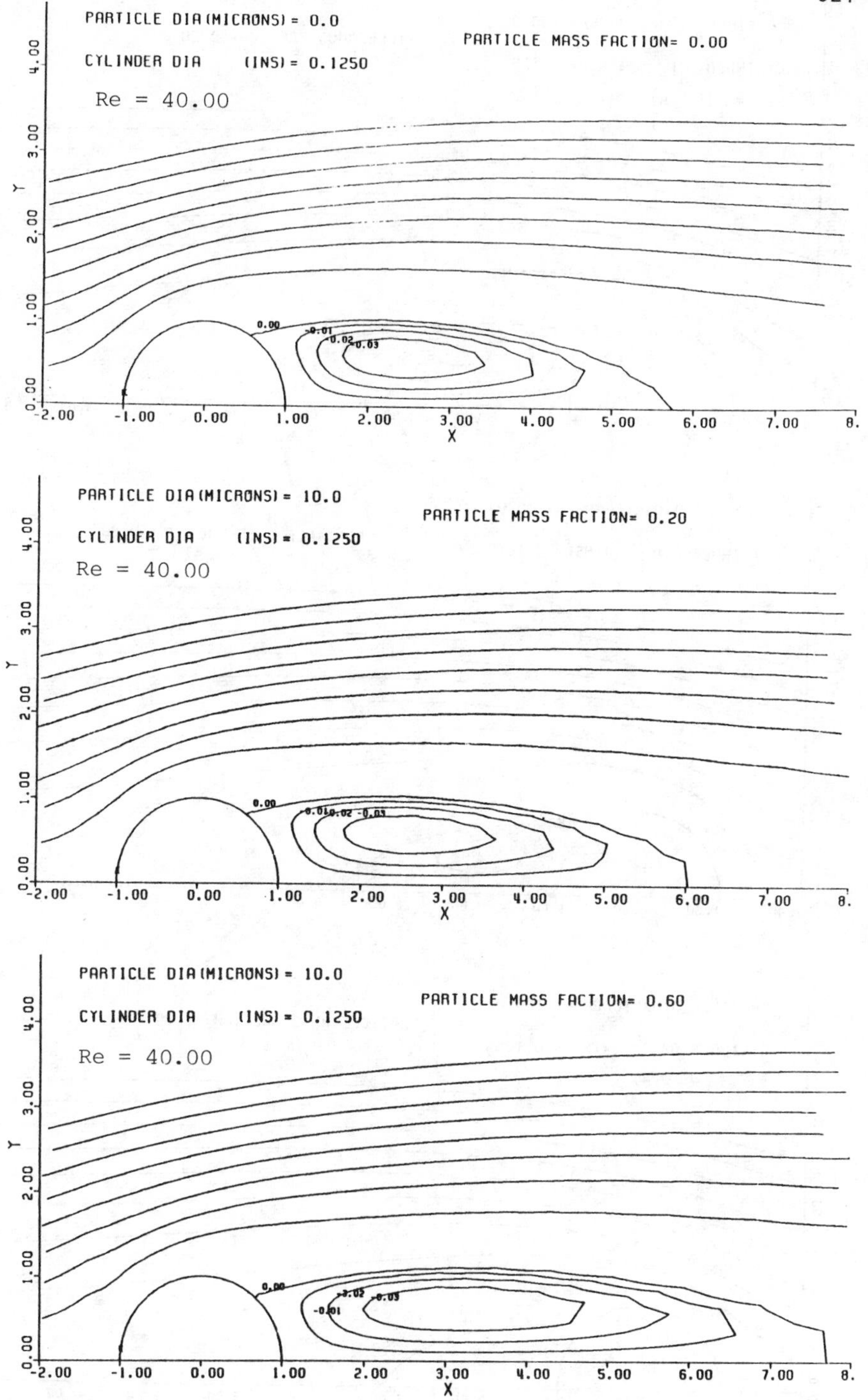

Fig. 8. Streamlines for Different Mass Fractions.

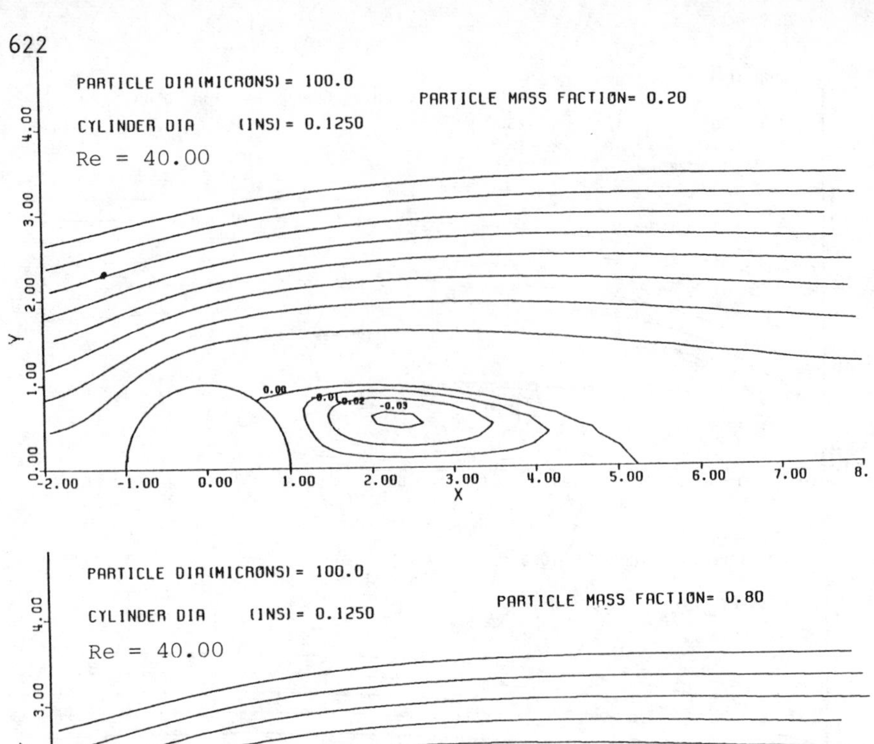

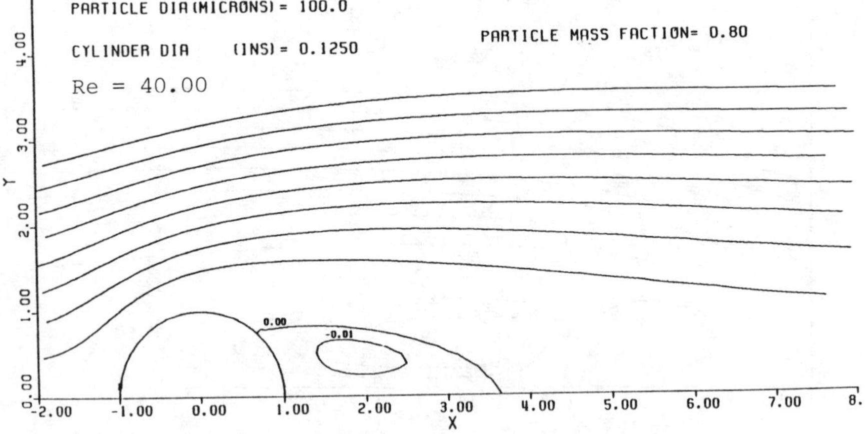

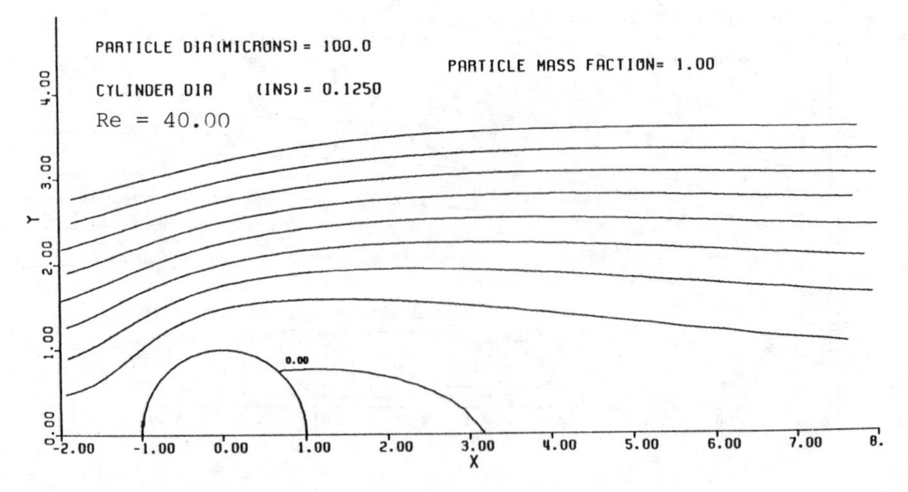

Fig. 9. Streamlines for Different Mass Fractions.

fluid. The fluid gives away its momentum to drive the particles, the effect of which is as if the cylinder is being run at lower free stream velocity and thus at lower Reynolds number.

One of the important flow properties which is indicative of the flow situation near the solid boundaries, is the surface vorticity. Figures 10 and 11 show the effect of particle concentration on the variation of surface vorticity for 80 and 100 micron particles, respectively. The values of dimensionless vorticity are plotted at various angular positions on the cylinder. It can be observed that as the particle concentration increases, the peak value of vorticity changes. Also, there is a change in the pattern of vorticity distribution over the cylinder surface. The particles have an effect on the flow separation angle as well. The change in flow separation angle with particle concentration is shown in Fig. 12. Once again, it can be observed that the particles have a substantial effect on the separation angle. All these changes in the flow properties are reflected on the coefficient of drag of the cylinder.

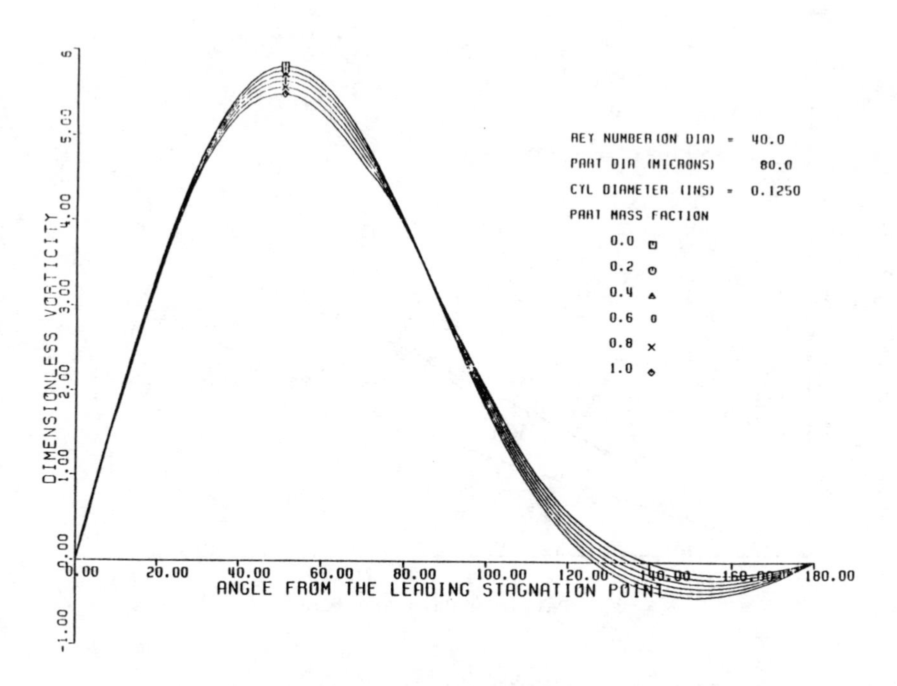

Fig. 10. Surface Vorticity Variation with Particle Mass Fraction.

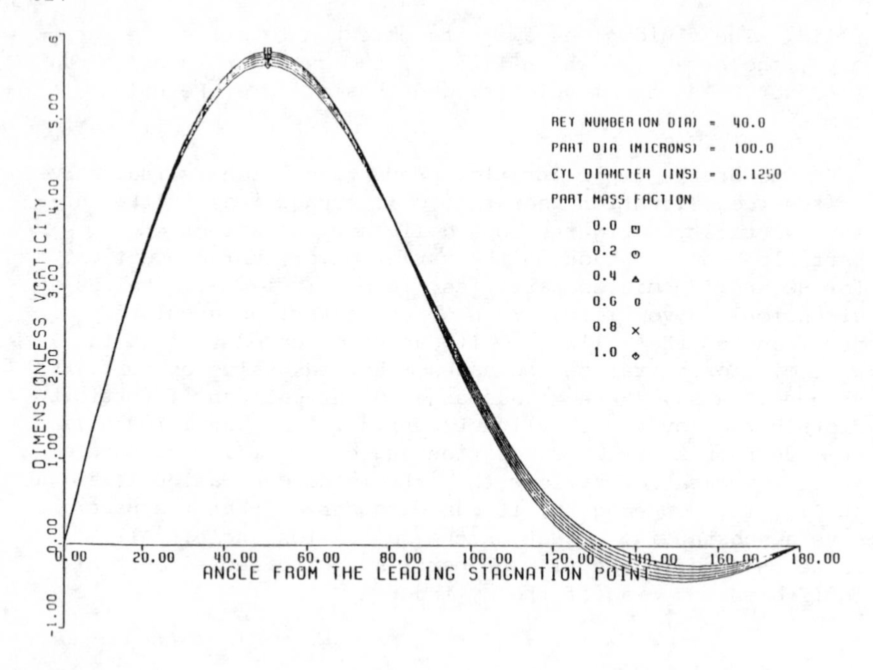

Fig. 11. Surface Vorticity Variation with Particle Mass Fraction.

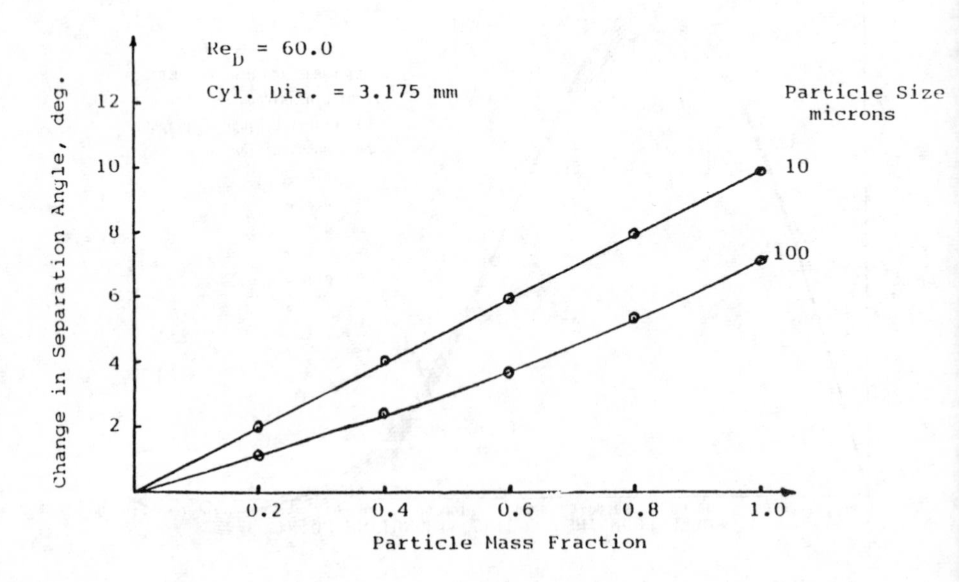

Fig. 12. Effect of Particle Mass Fraction on Separation Angle.

5. CONCLUSION

Comparisons between solid particle trajectories over a two dimensional cylinder, in viscous and inviscid fluids are made. The particle trajectories are shown to be different for different particle sizes. The strike or collection efficiency is found to be different for viscous and inviscid solutions. It has been shown that the presence of the particles have a substantial effect on the fluid flow properties. They change important flow parameters such as coefficient of drag, separation zone and recirculation pattern. Since the study of flow over a cylinder is a fundamental problem, it may find wide range of applications, especially in turbomachinery.

ACKNOWLEDGEMENT

This research was sponsored by the U.S. Army under Contract No. DAAG29-82-K-0029.

REFERENCES

1. RUDINGER, G., "Fundamentals and Applications of Gas-Particle Flow," AGARD-AG-222, 1976.

2. BEACHER, B. and TABAKOFF, W., "Improved Particle Trajectory Calculations Through Turbomachinery Affected by Coal Ash Particles," ASME Transactions, Journal of Engineering for Power, Vol. 104, January 1982, pp. 64-68.

3. TABAKOFF, W. and HUSSEIN, M.F., "Effect of Suspended Solid Particles on the Properties in Cascade Flow," AIAA Journal, Vol. 9, No. 8, August 1971, pp. 1514-1519.

4. HAMED, A. and TABAKOFF, W., "Solid Particle Demixing in a Suspension Flow of a Viscous Gas," Trans. of ASME, Journal of Fluids Engineering, March 1975, pp. 106-111.

5. LIU, J.T.C., "Flow Induced by the Impulsive Motion of an Infinite Flat Plate in a Dusty Gas," Astronautica Acta, Vol. 13, No. 4, July-August 1967, pp. 369-377.

6. MARBLE, F.E., "Dynamics of Dusty Gases," Annual Review of Fluid Mechanics, Vol. 2, 1970, pp. 369-377.

7. IMMICH, H., "Impulsive Motion of a Suspension: Effect of Antisymmetric Stresses and Particle Rotation," International Journal of Multiphase Flow, Vol. 6, 1980, pp. 441-471.

8. MORSI, S.A. and ALEXANDER, A.J., "An Investigation of Particle Trajectories in Two Phase Flow System," Journal of Fluid Mechanics, Vol. 55, 1972, p. 193.

9. VAL HEALY, J., "Perturbed Two Phase Cylindrical Type Flows," The Physics of Fluids, Vol. 13, March 1970, p. 551.

10. HAMED, A., "Analysis of Nonequilibrium Flow of Viscous Fluid With Suspended Solid Particles," Ph.D. Dissertation, University of Cincinnati, 1972.

11. STAFFMAN, P.G., "The Lift of a Small Sphere in a Slow Shear Flow," Journal of Fluid Mechanics, Vol. 31, 1968, p. 624.

12. TABAKOFF, W. and HAMED, A., "Aerodynamic Effects on Erosion in Turbomachinery," JSME and ASME Paper No. 70, Joint Gas Turbine Conference, Tokyo, May 1977.

13. DAVIS, R.T., "Numerical Solution of the Navier-Stokes Equations for Symmetric Laminar Incompressible Flow Past a Parabola," Journal of Fluid Mechanics, Vol. 51, 1972, p. 417.

14. HILL, J.A., DAVIS, R.T. and SLATER, G.L., "Development of a Factored ADI Scheme for Solving the Navier-Stokes Equations in Stream Function-Vorticity Variables," University of Cincinnati, Report No. AFL 79-12-49, 1979.

15. CROWE, C.T. and STOCK, D.E., "A Computer Solution for Two Dimensional Fluid-Particle Flows," International Journal for Numerical Methods in Engineering, Vol. 10, 1976, p. 185.

NOMENCLATURE

A	surface area of the particle
a	acceleration
c_d	coefficient of drag of the particle
D	diameter of the cylinder
D_p	diameter of the particle
MF_p	mass fraction of the particles carried by a particular trajectory (Eq. 12)
m_p	mass of the particle
r	radius of the cylinder
Re_d	Reynolds number of the cylinder based on cylinder diameter
Re_r	Reynolds number of the cylinder based on cylinder radius
R_x, R_y	interphase force components
t	time
U_∞	free stream velocity
u, v	velocity components
V	volume of the element
x, y	independent Cartesian spatial coordinates
β	angle between impacting particle velocity and the surface
μ	coefficient of viscosity
ξ, η	independent spatial coordinates in transformed plane
ρ	density
τ	residence time of particle (Eq. 12)
ψ	stream function
ω	vorticity

Subscripts and Superscripts

N	component normal to the surface
p	particle
T	component tangential to the surface
1,2	before and after impact, respectively
ξ, η	partial derivatives with regard to ξ and η, respectively
	dimensional quantities

THE ELLIPTIC SOLUTION OF 3-D INTERNAL VISCOUS FLOW USING THE STREAMLIKE FUNCTION

A. Hamed and S. Abdallah
Dept. of Aerospace Engineering and Applied Mechanics
University of Cincinnati
Cincinnati, Ohio U.S.A.

S. Abdallah
Applied Research Laboratory
Pennsylvania State University
State College, Pennsylvania
U.S.A.

1. INTRODUCTION

The prediction of the complex 3-D flow field in turbomachinery blade passages continues to be the subject of many viscous and inviscid flow studies. Recently developed 3-D inviscid methods [1-2] are capable of predicting 3-D flow characteristics such as secondary flows [3]. While the secondary flow is caused by vorticity which is produced by viscous forces, these inviscid methods cannot predict the viscous produced losses in the blade passage. Internal viscous flow solution methods have been developed using parabolized Navier-Stokes equations. While these methods are very fast, their application is limited due to their inabilities to simulate downstream blockage and strong curvature effects. Partially parabolized methods maintain the advantages of the parabolized methods in that the streamwise diffusion of mass, momentum and energy are still neglected but the elliptic influence is transmitted upstream through the pressure field. These well developed methods are not discussed here since the reader can refer to the extensive review of Davis and Rubin [4] and Rubin [5]. Our following discussion will be limited to the fully elliptic methods for the solution of the 3-D Navier-Stokes equations for internal flows.

Aziz and Hellums [6] developed the vector potential-vorticity formulation, and used it to solve the problem of laminar natural convection. This approach is an extension of the well known 2-D stream function vorticity formulation to 3-D problems. The equation of conservation of mass is identically satisfied through the definition of the vector potential. In this formulation, the resulting governing equations consist of three 3-D Poisson equations for the vector potential and three vorticity transport equations. Williams [7] used a time marching method for the solution of the laminar incompressible flow field due to thermal convection in a rotating annulus using primitive variables. In this case, the pressure field is computed from the solution of a 3-D Poisson equation with Neumann boundary conditions. Both these methods therefore require the solution of three parabolic transport equations. The first method requires in addition the solution of the 3-D Poisson equations with Dirichlet boundary conditions along two boundaries and zero gradient Neumann conditions along the third boundary, while the second has only one additional 3-D Poisson equation for the pressure, with Neumann boundary conditions over all the boundaries. Roache [8] discussed the relative merits of these two methods in terms of CPU time, computer storage requirements and program development time when iterative and direct methods are used in the solution of Poisson equation. Rosco obtained an elliptic solution for viscous flow in a straight pipe [9] and curved duct [10] from the governing equations in primitive variables using a new finite difference scheme. Two problems were encountered in the application of his method to through flow calculations; the tendency of the solution to flutter and a loss of mass between the duct inlet and exit in the computed results. More recently, Dodge [11] introduced a velocity split procedure in which the velocity is separated into viscous and potential components, and the pressure field is determined from the potential velocity component. The governing equation for the viscous velocity vector component is obtained from the momentum equation with the pressure gradient expressed in terms of the derivatives of the velocity potential vector component, while the governing equation for the velocity potential vector component is obtained from continuity. Beyond this formulation, Dodge's numerical solution procedure is partially parabolic since he neglected the streamwise diffusion of momentum in pursuing a marching solution for the viscous velocity components. The analysis itself, in terms of the type of the governing equations and their boundary conditions, is comparable to the velocity pressure formulation since the governing equation for the velocity potential is a 3-D Poisson equation. Dodge did not discuss the boundary conditions for the velocity potential. He only mentioned that it can be complex and that he used zero potential gradient normal to the wall in his numerical solution. Other formulations that can lead to elliptic solution, were described in references [5] and [8], however they will not be discussed here since they have not yet been applied to 3-D flow computations.

In summary, existing elliptic solvers for the Navier-Stokes equations require the solution of one or three 3-D Poisson equations in addition to the three momentum or the three vorticity transport equations. In the case of the velocity pressure formulation, the velocity vector is evaluated from the momentum equation and the continuity equation is satisfied indirectly in the pressure equation. The convergence of the iterative numerical procedure would be very slow if direct solvers are not used for the solution of the 3-D Poisson equation with Neumann boundary conditions for the static pressure. On the other hand, the continuity equation is satisfied identically in the vector potential-vorticity formulation, but it leads to three 3-D Poisson equations for the vector potential equations. Convergence of the iterative method does not present a problem in this case, since Dirichlet conditions are imposed on parts of the boundary, but computer storage requirements are greatly increased by two additional 3-D arrays.

The present work represents a new formulation for the 3-D Navier-Stokes equations that leads to a very economical elliptic solution. The formulation is based on the use of 2-D streamlike functions [12] to identically satisfy the continuity equation for 3-D rotational flows. In addition to the vorticity transport equation, the present formulation results in 2-D Poisson equations with Dirichlet boundary conditions for three streamlike functions. The presented method is very general, in that inviscid flow solutions can be obtained in the limit where $Re \to \infty$. In fact, numerical solutions have been obtained for inviscid rotational incompressible [13] and compressible [14] flows in curved ducts and it was demonstrated that the method predicts the secondary flow development due to inlet vorticity. The validity of the method for predicting secondary flow which are indirectly produced through the effects of viscosity have been demonstrated for inviscid flow [13, 14]. The following work represents an extension of this formulation to viscous flow problems. The analytical formulation is developed for general 3-D flows but the results of the computation are presented for the simple viscous flow problem of a uniform entrance flow in a straight duct. The final goal is to develop a method for predicting the three dimensional flow field, and the losses in internal flow fields with large surface curvature and significant downstream effects as in the case of turbine blade passages.

2. ANALYSIS

The governing equations are the vorticity transport equation and the equation of conservation of mass, which are given below in nondimensional vector form for incompressible flow [15]

$$\frac{\partial \bar{\Omega}}{\partial t} + (\bar{V} \cdot \nabla)\bar{\Omega} - (\bar{\Omega} \cdot \nabla)\bar{V} - \frac{1}{Re} \nabla^2 \bar{\Omega} = 0 \qquad (1)$$

and

$$\nabla \cdot \bar{V} = 0 \qquad (2)$$

where

$$\bar{\Omega} = \nabla \times \bar{V} \tag{3}$$

The Reynolds number ($Re = \frac{V^*D}{\nu}$), appears in the equations as a result of normalizing the velocities by V^*, the space dimensions by D and the vorticity by V^*/D, where ν is the kinematic viscosity.

Equations (1-3) are written in Cartesian coordinates as follows.

Vorticity Transport Equations

$$\frac{\partial \eta}{\partial t} = - \bar{V}\cdot\nabla\eta + \bar{\Omega}\cdot\nabla u + \frac{1}{Re}\nabla^2\eta \tag{4}$$

$$\frac{\partial \xi}{\partial t} = - \bar{V}\cdot\nabla\xi + \bar{\Omega}\cdot\nabla v + \frac{1}{Re}\nabla^2\xi \tag{5}$$

$$\frac{\partial \zeta}{\partial t} = - \bar{V}\cdot\nabla\zeta + \bar{\Omega}\cdot\nabla w + \frac{1}{Re}\nabla^2\zeta \tag{6}$$

where η,ξ,ζ are the components of the vorticity vector, $\bar{\Omega}$, in the x, y and z directions, respectively; u,v,w are the components of the velocity vector $\bar{V}$.

Equation of Conservation of Mass

$$\frac{\partial u}{\partial x} + \frac{\partial v}{\partial y} + \frac{\partial w}{\partial z} = 0 \tag{7}$$

Vorticity Velocity Relations

$$\eta = \frac{\partial w}{\partial y} - \frac{\partial v}{\partial z} \tag{8}$$

$$\xi = \frac{\partial u}{\partial z} - \frac{\partial w}{\partial x} \tag{9}$$

$$\zeta = \frac{\partial v}{\partial x} - \frac{\partial u}{\partial y} \tag{10}$$

The Streamlike Function Formulation

Three streamlike functions [12] $\chi_h(x,y)$, $\chi_v(y,z)$, $\chi_c(x,z)$ are defined to identically satisfy the equation of conservation of mass in the case of internal rotational flows [13, 14]. The velocity field is determined from the streamlike functions according to the following relations:

Streamlike Functions Velocity Relations

$$u_h = \frac{\partial \chi_h}{\partial y} - \int_{x=0}^{x} \frac{\partial w_v}{\partial z} dx \quad (11)$$

$$v_h = - \frac{\partial \chi_h}{\partial x} \quad (12)$$

$$w_v = - \frac{\partial \chi_v}{\partial y} - \int_{z=0}^{z} \frac{\partial u_h}{\partial x} dz \quad (13)$$

$$v_v = \frac{\partial \chi_v}{\partial z} \quad (14)$$

$$u_c = - \frac{\partial \chi_c}{\partial z} + \int_{x=0}^{x} \frac{\partial w_v}{\partial z} dx \quad (15)$$

$$w_c = \frac{\partial \chi_c}{\partial x} + \int_{z=0}^{z} \frac{\partial u_h}{\partial x} dz \quad (16)$$

where the subscripts h, v and c refer to solutions on the horizontal, vertical and cross-sectional planes respectively with

$$u = u_h + u_c; \quad v = v_h + v_v \quad \text{and} \quad w = w_v + w_c \quad (17)$$

Equations (11-17) satisfy the equation of conservation of mass (7) identically.

Streamlike Functions Equations

Substituting equations (11)-(17) into equations (8)-(10) one obtains

$$\frac{\partial^2 \chi_h}{\partial x^2} + \frac{\partial^2 \chi_h}{\partial y^2} = \zeta^* - \zeta + \frac{\partial}{\partial y} \int_{x=0}^{x} \frac{\partial w_v}{\partial z} dx \quad (18)$$

$$\frac{\partial^2 \chi_v}{\partial z^2} + \frac{\partial^2 \chi_v}{\partial y^2} = \eta^* - \eta - \frac{\partial}{\partial y} \int_{z=0}^{z} \frac{\partial u_h}{\partial x} dz \quad (19)$$

$$\frac{\partial^2 \chi_c}{\partial x^2} + \frac{\partial^2 \chi_c}{\partial z^2} = \xi^* - \xi + \frac{\partial}{\partial z} \int_{x=0}^{x} \frac{\partial w_v}{\partial z} dx - \frac{\partial}{\partial x} \int_{z=0}^{z} \frac{\partial u_h}{\partial x} dz \quad (20)$$

where

$$\eta^* = \frac{\partial w_c}{\partial y} - \frac{\partial v_h}{\partial z} \tag{21}$$

$$\xi^* = \frac{\partial u_h}{\partial z} - \frac{\partial w_v}{\partial x} \tag{22}$$

$$\zeta^* = \frac{\partial v_v}{\partial x} - \frac{\partial u_c}{\partial y} \tag{23}$$

The governing equations (4)-(6) and (18)-(20) are solved for the vorticity components η, ξ, ζ and the streamlike functions χ_h, χ_v and χ_c, respectively. The boundary conditions used for the solution of these equations are given for the viscous flow in a square duct. The inlet station is extended far upstream where uniform incoming flow is assumed (Fig. 1). This case simulates flow in cascades with zero turning where periodic conditions apply over the extended boundaries up to the duct entrance. Because of symmetry, only one quarter of the square duct is considered in the derivation of the boundary conditions. The coordinates x and z are measured from the duct centerline and y from the duct entrance as shown in Fig. 1.

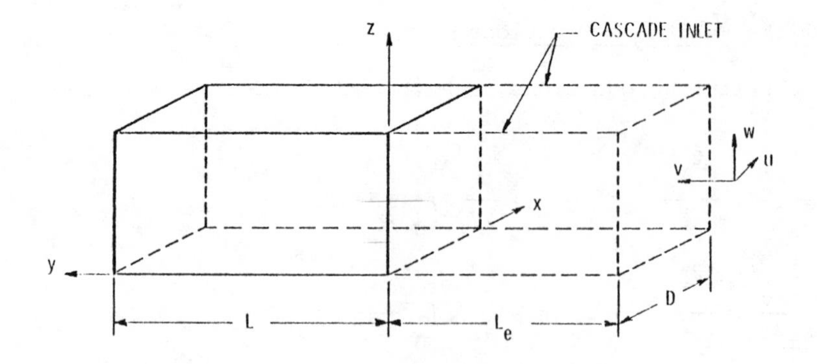

Fig. 1. A Schematic of the Duct with Cascade Entrance.

The Boundary Conditions for the Streamlike Functions

i. At the duct boundaries:

Referring to Fig. 2, the boundary conditions for χ_h, χ_v and χ_c are obtained from the substitution of the no flux condition into equations (11)-(17). The symmetry conditions are used to determine the rest of the boundary conditions at x = 0, z = 0. In addition the following condition is satisfied at the duct boundaries

$$\frac{\partial \chi_v}{\partial x} + \frac{\partial \chi_c}{\partial y} + \frac{\partial \chi_h}{\partial t} = 0$$

The above equation is used in the determination of the streamlike functions in the planes coinciding with the duct boundaries.

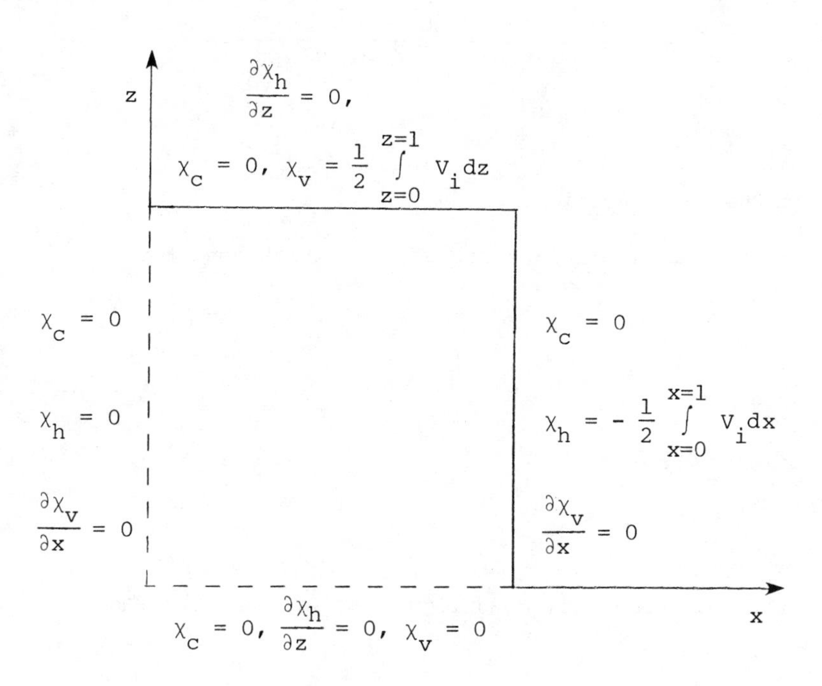

Figure 2

ii. Along the cascade entry:

Figure 3 shows the boundary conditions for the streamlike functions at the extension of the duct boundaries. The conditions at the planes of symmetry are unchanged.

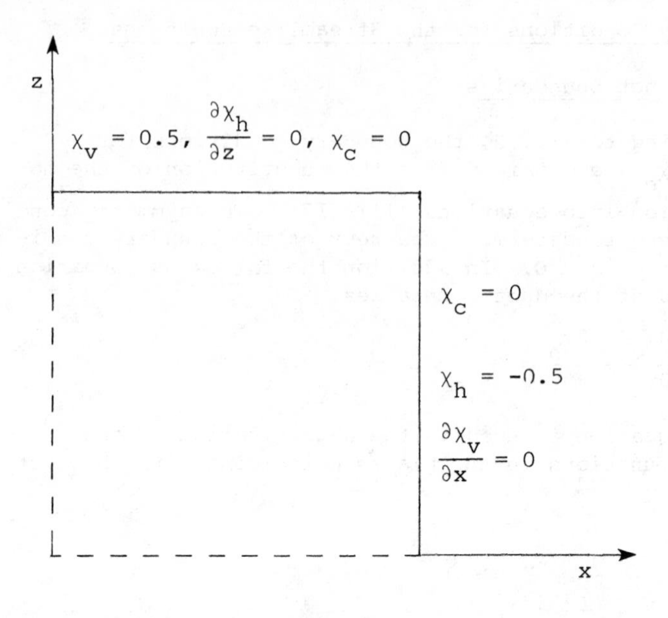

Figure 3

iii. <u>At the inlet station:</u>

The following relations satisfy the inlet condition $u = w = 0$

$$\frac{\partial \chi_h}{\partial y} = 0$$

$$\frac{\partial \chi_v}{\partial y} = 0$$

iv. <u>At exit:</u>

Fully developed flow conditions are assumed, leading to

$$\frac{\partial \chi_h}{\partial y} = 0$$

$$\frac{\partial \chi_v}{\partial y} = 0$$

The Boundary Conditions for the Vorticity Components

i. At the duct boundaries:

The no-slip condition is satisfied through the vorticity boundary conditions. The symmetry conditions are used along the vertical and horizontal central planes to determine the rest of the boundary conditions shown in Fig. 4.

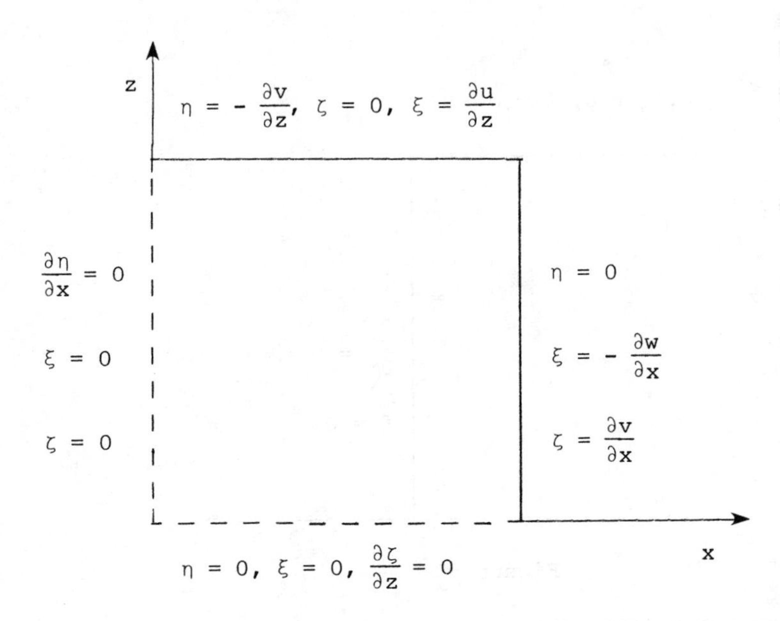

Figure 4

ii. Along the cascade entrance

Figure 5 shows the boundary conditions at the extension of the duct boundaries.

iii. At the inlet station:

$\xi = 0$

$\eta = 0$

$\zeta = 0$

iv. At exit:

Fully developed flow conditions are assumed, leading to:

$$\frac{\partial \eta}{\partial y} = 0$$

$$\frac{\partial \zeta}{\partial y} = 0$$

$$\xi = 0$$

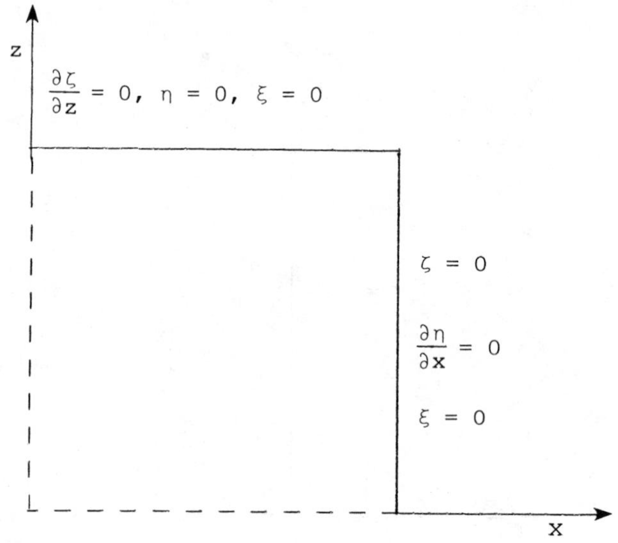

$$\frac{\partial \zeta}{\partial z} = 0, \ \eta = 0, \ \xi = 0$$

$$\zeta = 0$$

$$\frac{\partial \eta}{\partial x} = 0$$

$$\xi = 0$$

Figure 5

3. RESULTS AND DISCUSSION

The results of the numerical computations are presented in one quarter of a square duct. In this case, only two of the vorticity equations [eqs. (4) and (5)] and two of the streamlike function equations [eqs. (19) and (20)] are solved, since $\chi_h = \chi_v$ and $\zeta(x,y,z) = -\eta(z,y,x)$. Referring to Fig. 1, the solution was obtained for Re = 50 in a duct with L/DRe = 0.1 and Le/DRe = 0.01 using SOR and a (11 x 11 x 34) grid.

Figures 6a and 6b show the through flow velocity contours at the duct entrance and exit. The influence of the cascade entering on the elliptic solution is demonstrated in contours of Fig. 6a.

The contours for the secondary velocity component, w, are shown at y/DRe = 0.0 and 0.0075 in Figs. 7a and 7b. From these figures, one can see a large change in both the magnitude and the location of the maximum secondary velocities along the duct. The development of the through flow velocity profiles along the plane of symmetry, x = 0, is shown in

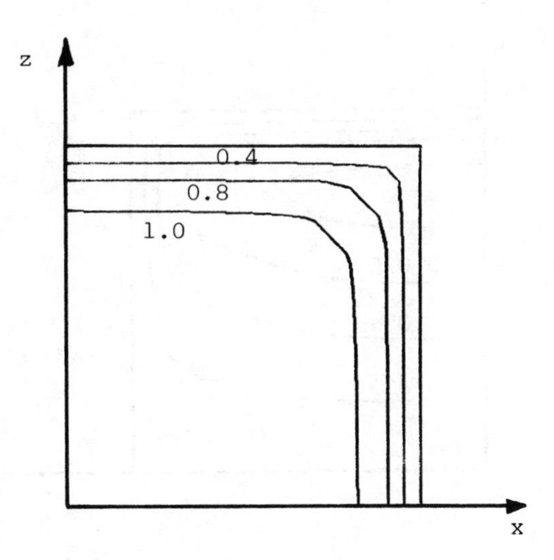

Fig. 6a. Through Flow Velocity Contours at the Duct Entrance.

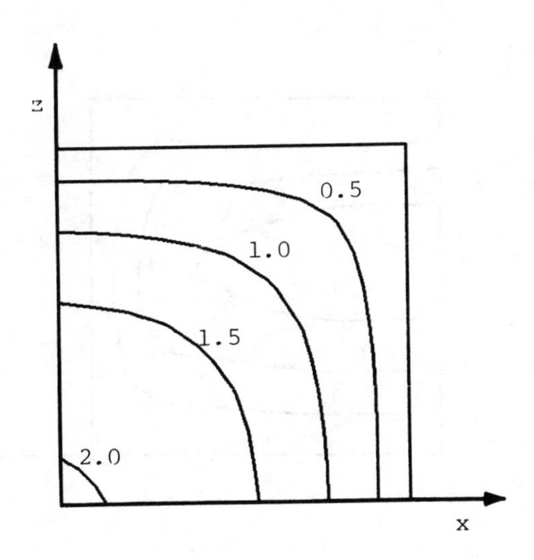

Fig. 6b. Through Flow Velocity Contours at the Duct Exit (y/DRe = 0.1).

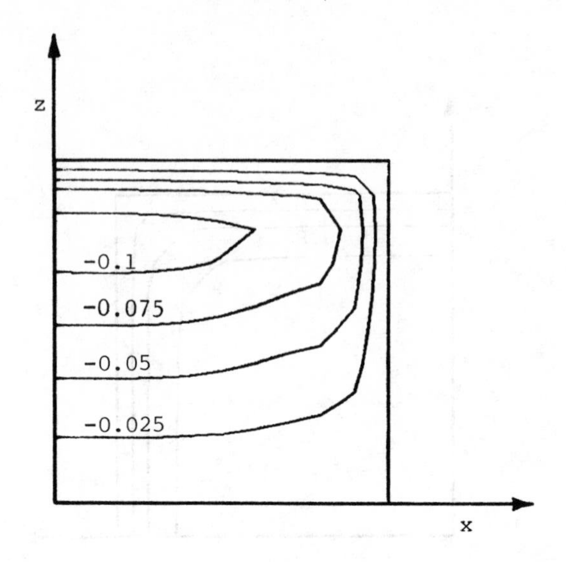

Fig. 7a. Secondary Velocity Contours at y/DRe = 0.0

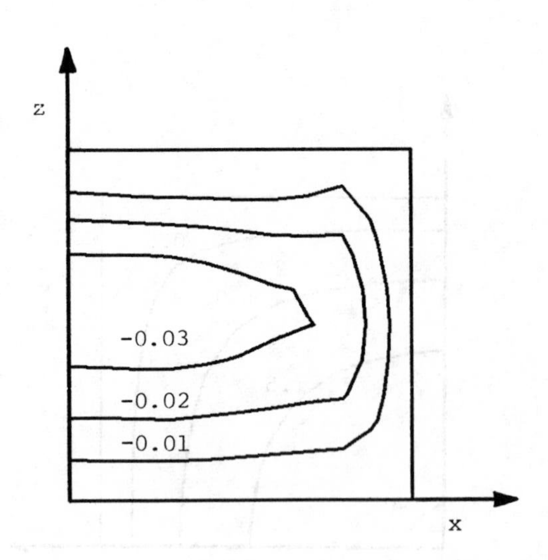

Fig. 7b. Secondary Velocity Contours at y/DRe = 0.0075

Fig. 8 for the computed results and the experimental measurements of reference [16]. One can see that the computed results are in good agreement with the experimental measurements, at y/DRe = 0.0075 and 0.02, but that the computed through flow velocities at y/DRe = 0.1 did not reach the experimentally measured fully developed profile.

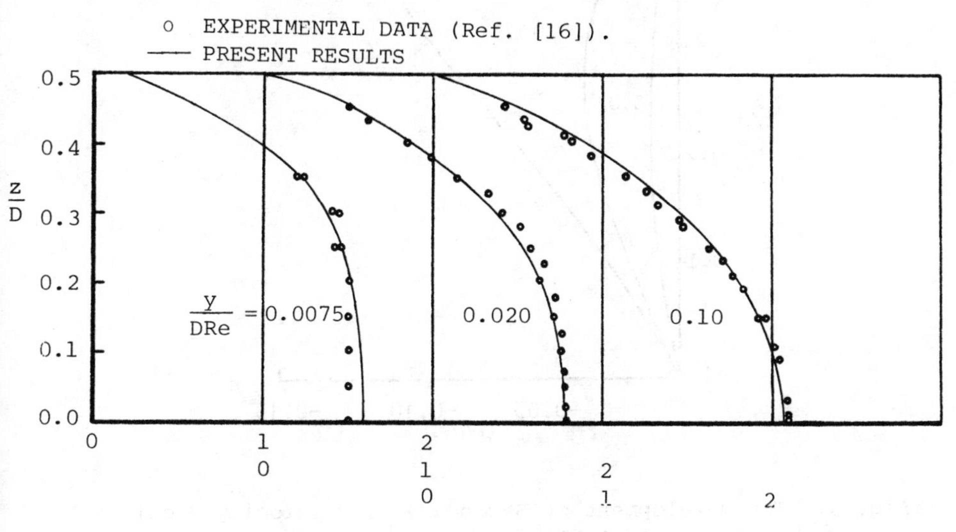

Fig. 8. Development of the Through Flow Velocity Profile at the Central Plane x = 0.

Figure 9 shows the development of the secondary velocity component, w, along the duct plane of symmetry, x = 0. No experimental measurements are available for comparison with the computed secondary velocities. Figure 9 shows that the maximum secondary flow is initially located near the solid boundaries, then moves towards the center of the duct and decreases as the flow proceeds towards fully developed conditions. The computed through flow velocity development along the duct centerline are compared with the experimental measurements of reference [16] in Fig. 10. One can see in this figure that the elliptic solution predicts an increase in the centerline through flow velocity in the cascade entry region preceding the actual duct entry. Figure 10 shows that the computations slightly underestimate the centerline velocity. Considering the coarse grid used in the numerical computations [(11 x 11 x 34) grid points in the duct], the agreement of the computed results with the experimental data as shown in Figures 8 and 10 is very satisfactory.

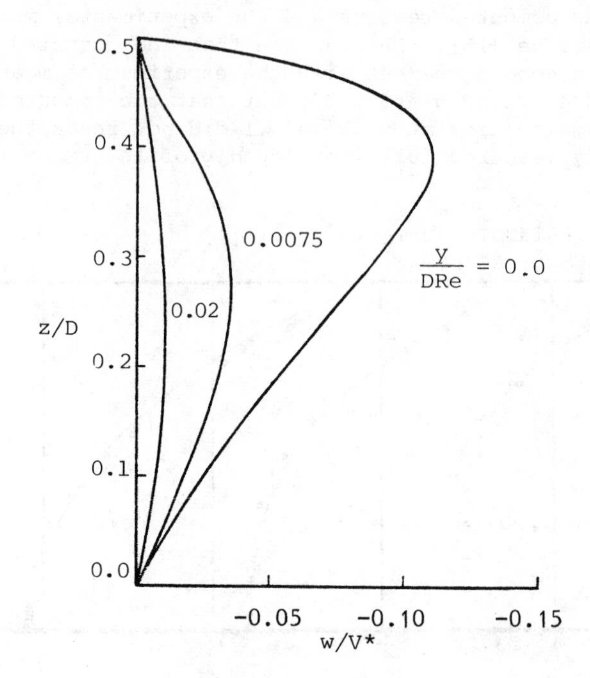

Fig. 9. The Development of Secondary Flow Velocity Profile at the Central Plane x = 0.

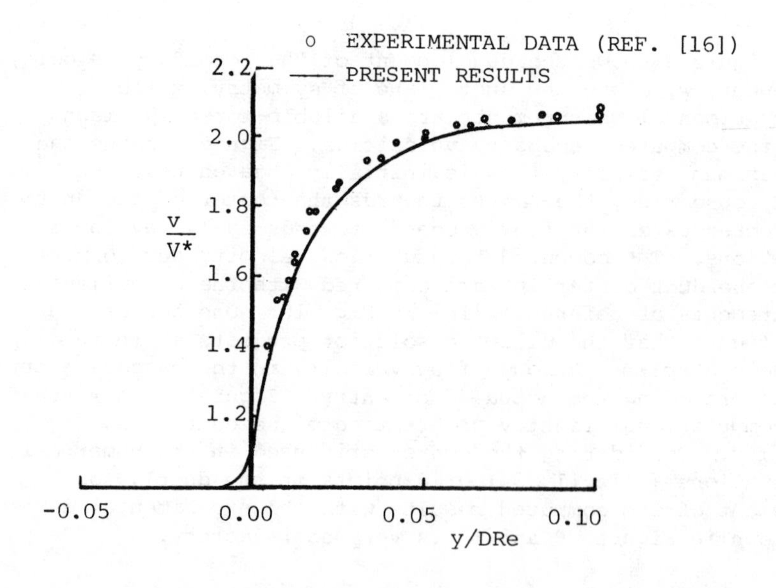

Fig. 10. Through Flow Velocity Development Along the Duct Centerline.

4. SUMMARY

This paper presents a fast efficient method for the 3-D elliptic solution of the Navier-Stokes equations. It is based on the streamlike-function vorticity formulation which leads to 2-D Poisson equations with Dirichlet boundary conditions for the three streamlike functions, in addition to the vorticity transport equations. The method is more economical than existing Navier-Stokes elliptic solvers, yet does not have the limitations of the parabolized and partially parabolized procedures. It offers a useful tool for the numerical solution of 3-D internal viscous flow fields where surface curvature and downstream effects are significant, as in turbine blade passages. The results of the computations are presented for the viscous flow in a constant area duct, as a corner stone for more general future applications.

ACKNOWLEDGEMENT

This work was supported by the Air Force Office of Scientific Research under Grant No. 80-0242.

REFERENCES

1. DENTON, J.D., "A Time Marching Method for Two and Three Dimensional Blade-to-Blade Flows," ARC R&M 3775, 1975.

2. PANDOLFI, M. and GOLASURDO, G., "Three-Dimensional Inviscid Compressible Rotational Flows - Numerical Results and Comparison with Analytical Solutions," ASME Flow in Primary Non-Rotating Passages in Turbomachines, ASME Symposium, pp. 141-149, 1979.

3. BARBER, T. and LANGSTON, L.S., "Three Dimensional Modelling of Cascade Flow," AIAA Paper 79-0047, 1979.

4. DAVIS, R.T and RUBIN, S.G., "Non-Navier-Stokes Viscous Flow Computations," Computers and Fluids, Vol. 8, 1982, pp. 101.

5. RUBIN, S.G., "Incompressible Navier-Stokes and Parabolized Navier-Stokes Solution Procedures on Computational Techniques," VKI Lecture Series on Computational Fluid Dynamics, 1982.

6. AZIZ, A. and HELLUMS, J.D., "Numerical Solution of the Three Dimensional Equations of Motion for Laminar Natural Convection," The Physics of Fluids, Vol. 10, No. 2, February 1967, pp. 314-324.

7. WILLIAMS, G.P., "Numerical Integration of the Three-Dimensional Navier-Stokes Equations for Incompressible Flow," Journal of Fluid Mechanics, Vol. 37, Part 4, 1969, pp. 727-250.

8. ROACHE, P.J., Computational Fluid Dynamics, Hermosa, Albuquerque, New Mexico, 1972.

9. ROSCOE, D.F., "The Numerical Solution of the Navier-Stokes Equations for a Three-Dimensional Laminar Flow in Curved Pipes Using Finite Difference Methods," Journal of Engineering Mathematics, Vol. 12, No. 4, October 1978, pp. 303-323.

10. ROSCOE, D.F., "The Solution of the Three-Dimensional Navier-Stokes Equations Using a New Finite Difference Approach," Intn. Journal of Numerical Methods in Engineering, Vol. 10, 1976, pp. 1299-1308.

11. DODGE, P.R., "Numerical Method for 2-D and 3-D Viscous Flows," AIAA Journal, Vol. 15, No. 7, 1977, pp. 961-965.

12. HAMED, A. and ABDALLAH, S., "Streamlike Function: A New Concept in Flow Problems Formulation," Journal of Aircraft, Vol. 16, No. 12, December 1979, pp. 801-802.

13. ABDALLAH, S. and HAMED, A., "The Elliptic Solution of the Secondary Flow Problem," ASME Paper No. 82-GT-242, 1982.

14. HAMED, A. and LIU, C., "Three Dimensional Rotational Compressible Flow Solution in Variable Area Channels," AIAA Paper No. 83-0259, 1983.

15. PEYRET, R. and TAYLOR, T., "Computational Methods for Fluid Flow," Springer Series in Computational Methods, Springer-Verlag, New York, Heidelberg, Berlin, 1983.

16. GOLDSTEIN, R.J. and KREID, D.K., "Measurement of Laminar Flow Development in a Square Duct Using a Laser-Doppler Flow Meter," Journal of Applied Mechanics, December 1967, pp. 813-818.

NOMENCLATURE

D	duct height
L	duct length
Re	Reynolds number
t	time
u,v,w	velocity components in x, y and z directions
$\bar{V}$	velocity vector
V^*	normalizing velocity at inlet
x,y,z	Cartesian coordinates
χ_h, χ_v, χ_c	streamlike functions
ν	kinematic viscosity
$\bar{\Omega}$	vorticity vector
ξ,η,ζ	vorticity components in x, y and z directions

Subscripts

c	refers to cross sectional plane
h	refers to horizontal plane
i	refers to inlet plane
v	refers to vertical plane